MODERN AND ANCIENT GEOSYNCLINAL SEDIMENTATION

*Proceedings of a Symposium Dedicated to Marshall Kay and
Held at Madison, Wisconsin, November 10–11, 1972
Sponsored by the Department of Geology and Geophysics of the
University of Wisconsin, Madison, and by the National Science Foundation*

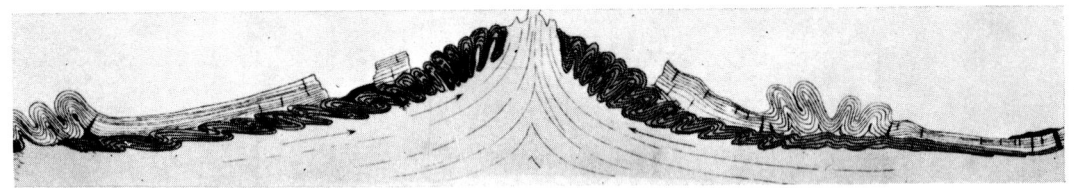

G. P. Scrope's Ideal Transverse Section of a Mountain Range (1825)

Edited by
R. H. Dott, Jr., *University of Wisconsin, Madison*
and
Robert H. Shaver, *Indiana University, Bloomington*

Copyright © 1974 by
SOCIETY OF ECONOMIC PALEONTOLOGISTS AND MINERALOGISTS
Special Publication No. 19

Tulsa, Oklahoma, U.S.A.   May 1974

A Publication of

The Society of Economic Paleontologists and Mineralogists

a division of

The American Association of Petroleum Geologists

Marshall Kay

PREFACE

In the two decades since the publication of Marshall Kay's *North American Geosynclines,* students of both modern and ancient sediments have compiled an immense body of knowledge relevant to the geosynclinal concept. The idea of a specialized research conference to integrate some of that knowledge was first conceived in about 1965, and the unprecedented reception of plate tectonics beginning in 1968 assured it. As an outgrowth of conversations with James Helwig and others, a decision was made to dedicate the conference to Marshall Kay in recognition of his many contributions to our knowledge of geosynclines. After more than three decades of vigorous research and teaching, Marshall was working at that time in the forefront of geologic implications of the new tectonics.

The Kay Conference was held in Madison, Wisconsin, November 1972, and was attended by 180 persons who represented 15 countries. This symposium volume contains the texts of papers presented at Madison. It is organized in a topical manner, and, in most areas of discussion, modern analogues and ancient examples together provide a comparative basis for evaluating sedimentary models for geosynclines. The grouping of papers under the subheadings shown in the Table of Contents inevitably is somewhat artificial, but it helps to give structure to the book.

The two introductory papers (by Dott and by Dietz) are largely historical, but they also highlight the present status of the geosynclinal concept. Many workers have argued that plate tectonics renders the geosynclinal concept completely obsolete. Must an entire ocean basin now be regarded as the geosyncline? Is the eugeosyncline merely scrapings from subducted oceanic plates, an island arc complex, or simply the collapsed continental rise of Atlantic-type continental margins? Is there any validity to the long-popular geotectonic cycle and its handmaiden, continental accretion? While final answers to these and other important problems are not yet possible, the papers contained in this volume go far toward a better framing of the critical tectonic questions to ask of the earth. Continents are mosaics of various old orogenic belts, and orogenic belts themselves can be thought of as collages of diverse tectonic elements (see Helwig's paper). Therefore, the sedimentologist's contribution to global tectonics is to infer the depositional processes and environments for the various sediments occurring within different tectonic elements so that the history and palinspastic restoration of each member of the collage can be reconstructed as fully as possible. To realize these goals, the sedimentologist must develop the best possible models for deciphering each of the different assemblages encountered in the collage.

The papers by Moore and Curray, Bird, Harms and Pray, and to some extent those by Hoffman, Dewey, and Burke and by Bernoulli and Jenkyns illustrate deposits formed on passive (nonorogenic), midplate or Atlantic-type continental margins. The distinction of such deposits coincidentally caught up in ancient orogenic belts from those strata deposited during active orogenesis on Andean-, western Pacific-, or Mediterranean-type plate margins is of fundamental importance. Hoffman, Dewey, and Burke focus attention upon a cratonic tectonic feature, the *aulacogen,* well known in the USSR but new elsewhere. Aulacogens are large filled grabens or rifts presumed to have genetic relations with the incipient break-up of continental plates. Examples may include, besides the Russian and Canadian ones, the Paleozoic depression of southern Oklahoma, the late Cenozoic Rhine Graben, and the late Precambrian Keweenawan Trough of the central United States.

Because submarine channels and fans are now recognized as of great importance in understanding the distribution of many geosynclinal sediments, especially of flysch type, extensive coverage of these features is included. Papers by Harms and Pray, Kulm and Fowler, Moore and Curray, Mutti, Nelson and Nilsen, Normark, Pícha, Scholl and Marlow, Stanley, and Whitaker bear upon these features. In their oral presentation, Moore and Curray presented evidence of contemporary subduction of the huge Bengal submarine fan beneath Burma, which underscores the prospect that *any* sediment assemblage can be caught up by orogenesis. Deep-sea sediments and ophiolites, which are recognized increasingly within ancient orogens, are treated by Fischer, Bernoulli and Jenkyns, Kanmera, and Churkin. Kanmera and also Blake and Jones describe ophiolitelike sequences that do not seem to have originated at ocean ridges. Arc-trench deposits, long considered trademarks of eugeosynclines, are discussed extensively both from modern and ancient viewpoints by Dickinson, Kulm and Fowler, Scholl and Marlow, and Stanley. Stanley shows that differences among Quaternary flysch deposits in various portions of the eastern Mediterranean compare closely with the differences among several early Cenozoic flysch formations of the Alps. Late orogenic and post-orogenic successor-basin deposits (largely molasse) are discussed by Van Houten and Eisbacher, and Crowell elucidates the varied conditions of sedimentation associated with transcurrent faults.

As a consequence of the new, extremely mobile view of the earth, much of the problem of interpreting ancient rocks of orogenic belts turns upon formidable palinspastic problems. The proper unravelling of different structurally juxtaposed tectonic elements, many of which originally had neither a close genetic nor geographic relation to one another, is a prerequisite to interpretation of the sediments associated with them. Yet this may be a formidable and subjective task. Crook and Okada illustrate how sandstone petrology can help identify source terranes found in different tectonic elements. Churkin weighs evidence for limited-versus-extreme mobilist views of the Cordillera. Hsü, Wood, and Blake and Jones discuss the diverse origins and implications of mélanges, which are being recognized widely, and Dewey illustrates the unique importance of ophiolite terranes in palinspastic analysis. Helwig emphasizes the important role of different basement types in the collage of tectonic elements comprising orogenic belts. Finally, Kay offers a retrospect and prospect of geosynclinal research.

A number of philosophic issues are either explicit or implicit in many of the contributions. Perhaps the most fundamental one concerns models. The conference took place during the height of a model-building era, and some participants worried lest there be more concern with models than with reality. We can only hope that it is the model, not the geology, that will be discarded when the two disagree; a valid model must have more relation to reality than an abstract painting. Of the several kinds of models (e.g., scale, mathematical, conceptual and actualistic), all are intended to provide simplifying analogies for explaining the complex realities of nature. Simply put, the basis for drawing the analogies is the cliché, "the present is a key to the past." But *is* the present a very good key to the past? Clearly it is not for such phenomena as evolution that have long time periods. Nor is it for continental shelves and rises, according to Moore and Curray, nor for trench sedimentation, according to Scholl and Marlow, because of the uniqueness of Pleistocene sea level fluctuations and erosional vigor. Any strictly uniformitarian modern-ancient analogy between such large features is fraught with pitfalls; merely renaming the comparison "actualism" because that term is judged more respectable than uniformitarianism is no help (see Dott paper). Harms and Pray note that no close modern analogue for the rocks they describe is known, and they make the important point that not all sedimentary models must be based upon modern deposits. Mutti also shows that the "past may be a key to the present," in that ancient rocks provide larger three-dimensional views of sediments than do the methods of marine geology.

Obviously, insights from both ancient and modern sediments are complementary as exemplified by a comparison of the papers on deep-sea fans. The modern provides knowledge of surficial dimensions, form, and processes, but the ancient must provide the details of internal make-up and history. The Nelson and Nilsen paper is the most complete example of truly actualistic reasoning in the symposium. It is a self-contained modern-ancient comparison of submarine fan deposits for which processes, three-dimensional morphology, textural features, and history are all compared.

Another general issue is that of causality or determinism. The hindcasting of ancient causal relationships is fraught with uncertainties. Selecting the correct modern analogue commonly is very difficult and may be colored with causal assumptions. As Kay notes, "frequency of association need not imply genetic relationship" among rocks and structural features. As important as deep-sea fan deposits are in ancient orogens, excellent modern examples occur also in structurally passive as well as in active regions. And Crowell argues convincingly that the deposits described by him from transcurrent fault systems are not causally restricted to orogens. The orogenic cycle and continental accretion are perhaps the most deterministic of all concepts associated with geosynclines. Indeed, they are so causally charged that complete abandonment of the tectonic cycle seems necessary, although Eisbacher, Crook, and Van Houten make a case for a limited tectonic evolutionary pattern.

The old contest over periodicity versus continuity of global tectonics has been reopened by plate theory, and some hints of this issue appear in the symposium. Undeniably the stratigraphic record contains some impressive discontinuities that are geologically synchronous on a worldwide scale. A few examples include late Paleozoic mountain building, mid-Cretaceous transgression, and a mid-Cenozoic tectonic reorganization. For the first time, plate tectonic theory provides plausible reasons why at least some degree of synchroneity of tectonic events in widely separated regions should be expected.

Is the geosyncline extinct, after all, or has the leopard merely changed its spots? Fantastic variability of sediments called "geosynclinal," as well as the complexity of jumbled tectonic elements in orogens, is inescapable. The difficulty of generalizing, together with the misleading deterministic overtones and semantical confusion associated with "geosyncline," may be sufficient argument for abandoning most existing terminology. It may be best to clean the slate and develop new language.

ACKNOWLEDGMENTS

The National Science Foundation generously supported the research conference with funds to underwrite the travel of 11 of the foreign participants and to defray many of the organizational costs for the meeting. Other incidental expenses of the meeting, as well as some editorial expenses incurred in preparing this volume, were financed by contributions from Esso Production Research Company, Shell Oil Corporation, Mobil Oil Corporation, Marathon Oil Company's Denver Research Center, and the Atlantic-Richfield Company. The Society of Economic Paleontologists and Mineralogists has enthusiastically supported the publication of the conference proceedings. It is impossible to acknowledge every individual who aided in organizing the Kay Conference and in expediting publication. Special thanks, however, are due James Helwig, who was especially helpful at every stage; he suggested the title for the conference. John V. Byrne, Michael Churkin, David L. Clark, W. R. Dickinson, C. L. Drake, A. G. Fischer, Hugh Gabrielse, D. S. Gorsline, J. W. Kerr, Bernhard Kummel, J. D. Lowell, D. B. McKenzie, Jr., R. G. Ray, D. J. Stanley, and R. G. Walker also were very helpful during the planning of the meeting. A. W. Bally, Creighton A. Burk, I. W. D. Dalziel, Zoltan de Cserna, Charles L. Drake, Stanislaus Dzulynski, Donn S. Gorsline, H. R. Gould, George deVries Klein, Earle F. McBride, J. R. Moore, H. G. Reading, John Rodgers, Adolph Seilacher, L. L. Sloss, R. G. Walker, and E. L. Winterer effectively chaired sessions at the conference. The Wisconsin Center provided very pleasing and efficient services, which made for an exceptionally smooth meeting. Appreciation is expressed to all who reviewed manuscripts (see acknowledgments accompanying each paper). Finally, special thanks go to my coeditor, Robert H. Shaver, who has done a most efficient final editing of the manuscripts, and to Menzi Behrnd for her indispensible secretarial help.

R. H. DOTT, JR.
Convener
June 1, 1973

CONTENTS

Preface... iii

INTRODUCTORY

The Geosynclinal Concept..R. H. Dott, Jr. 1
Collapsing Continental Rises: Actualistic Concept of Geosynclines—a Review...................
....Robert S. Dietz and John C. Holden 14

CONTINENTAL TERRACES AND RELATED CRATONIC ACCUMULATIONS

Midplate Continental Margin Geosynclines: Growth Processes and Quaternary Modifications.....
....David G. Moore and Joseph R. Curray 26
Evolution of the Western Appalachian Continental Margin.....................John M. Bird 36
Erosion and Deposition along the Mid-Permian Intracratonic Basin Margin, Guadalupe Mountains,
Texas...John C. Harms and Lloyd C. Pray 37
Aulacogens and Their Genetic Relation to Geosynclines, with a Proterozoic Example from Great
Slave Lake, Canada.........................Paul Hoffman, John F. Dewey, and Kevin Burke 38

SUBMARINE CANYON AND FAN DEPOSITS

Submarine Canyons and Fan Valleys: Factors Affecting Growth Patterns of Deep-Sea Fans.......
....William R. Normark 56
Depositional Trends of Modern and Ancient Deep-Sea Fans.....C. Hans Nelson and Tor H. Nilsen 69
Examples of Ancient Deep-Sea Fan Deposits from Circum-Mediterranean Geosynclines..........
....Emiliano Mutti 92
Ancient Submarine Canyons and Fan Valleys.........................J. H. McD. Whitaker 106
Ancient Submarine Canyons of the Carpathian Miogeosyncline..................František Picha 126

DEEP-SEA PELAGIC SEDIMENTS AND OPHIOLITE ASSEMBLAGES

The Odd Rocks of Mountain Belts...A. G. Fischer 128
Alpine, Mediterranean, and Central Atlantic Mesozoic Facies in Relation to the Early Evolution of
the Tethys...Daniel Bernoulli and Hugh C. Jenkyns 129
Paleozoic and Mesozoic Geosynclinal Volcanism in the Japanese Islands and Associated Chert Sedi-
mentation...Kametoshi Kanmera 161
Paleozoic Marginal Ocean Basin-Volcanic Arc Systems in the Cordilleran Foldbelt...............
....Michael Churkin, Jr. 174

DEPOSITS IN MAGMATIC ARC AND TRENCH SYSTEMS

Sedimentary Sequence in Modern Pacific Trenches and the Deformed Circum-Pacific Eugeosyncline
....D. W. Scholl and M. S. Marlow 193
Cenozoic Sedimentary Framework of the Gorda-Juan de Fuca Plate and Adjacent Continental
Margin—a Review...L. D. Kulm and G. A. Fowler 212
Sedimentation within and beside Ancient and Modern Magmatic Arcs.......William R. Dickinson 230
Modern Flysch Sedimentation in a Mediterranean Island Arc Setting........Daniel Jean Stanley 240

SUCCESSOR BASIN ASSEMBLAGES

Northern Alpine Molasse and Similar Cenozoic Sequences of Southern Europe....F. B. Van Houten 260
Evolution of Successor Basins in the Canadian Cordillera.........................G. H. Eisbacher 274
Sedimentation along the San Andreas Fault, California.........................John C. Crowell 292

PROBLEMS OF PALINSPASTIC RESTORATION

Lithogenesis and Geotectonics: the Significance of Compositional Variation in Flysch Arenites
(Graywackes)..Keith A. W. Crook 304
Migration of Ancient Arc-Trench Systems...Hakuyu Okada 311
Melanges and Their Distinction from Olistostromes...............................K. J. Hsü 321
Ophiolites, Melanges, Blueschists, and Ignimbrites: Early Caledonian Subduction in Wales?........
....Dennis S. Wood 334
Origin of Franciscan Melanges in Northern California.......M. C. Blake, Jr., and David L. Jones 345
Ophiolite Generation and Emplacement: a Key to Alpine Evolution..............John F. Dewey 358
Eugeosynclinal Basement and a Collage Concept of Orogenic Belts..................James Helwig 359

REFLECTIONS

Geosynclines, Flysch, and Melanges...Marshall Kay 377

INTRODUCTORY

THE GEOSYNCLINAL CONCEPT

R. H. DOTT, JR.
University of Wisconsin, Madison

ABSTRACT

Whether they formed under shallow or deep water, thick geosynclinal sediments were regarded until recently as essential precursors to mountains. Every orogenic belt presumably had evolved stage by stage from geosyncline to mountain, ultimately producing a peripheral accretion to some evergrowing continental craton. Conversely, by implication, thick sediment prisms along any present continental margin inevitably should lead to mountains. These long-standing deterministic generalizations were hardly justified, however, for modern orogenic belts are not consistently located at continental margins, nor do they all contain thick sediments. Moreover, it is impossible to designate any uniquely geosynclinal sediment type. Most geosynclinal sediments are results more than causes of orogenesis; an orthogeosyncline is simply a sediment-filled orogenic belt.

The early, strictly uniformitarian sea-floor spreading model for geosynclines was unacceptable because it regarded continental terrace sediment prisms (miogeoclines) formed on passive or nonorogenic continental margins as essential, evolutionary precursors to mountain building—a holdover from the venerable tectonic cycle. But most existing continental terraces are almost 200 million years old and still show practically no tectonic mobility. Genetically, these miogeoclines belong to a different genus than accumulations formed in active orogenic zones. Miogeoclines form on passive trailing edges of diverging continents, whereas orthogeosynclines form near active leading edges of converging lithosphere plates. Plate tectonics shows how these two genetically distinct sediment prisms may become coincidently crushed together in orogenic belts. Rather than being simple concentric accretions of successive orogenic belts, continents are mosaics of very complexly truncated, overprinted, and even rifted tectonic elements containing haphazard relics of former plate margins.

The geosynclinal concept has suffered from an analogue syndrome. Dogmatic generalizations were applied to all cases from an incomplete list of supposed modern analogues, and tectonic environments were confused with sedimentary ones. But intensive marine research over the past two decades has provided many more well-documented possible analogues for the testing of truly actualistic models. Armed with these, as well as with new tectonic insights and new vocabulary, geosynclinal studies can advance from a long descriptive phase to a more genetic one.

INTRODUCTION

Actualism and analogy.—In the two decades since publication of Marshall Kay's (1951) *North American Geosynclines,* students of both modern and ancient sediments have compiled an immense body of knowledge relevant to the geosynclinal concept. Moreover, the new theory of plate tectonics has seemed overnight to require a complete reassessment of the geosyncline as well as of orogenesis. The purpose of this symposium volume is to evaluate by comparison of modern and ancient sediments a number of depositional models applicable to the great variety of strata seen in orogenic belts and heretofore called "geosynclinal."

Studies of ancient sedimentary rocks have relied heavily upon analogies with modern features of known origin for genetic interpretations. Large-scale analogues, such as deltas and deep-sea trenches, have long been invoked in trying to understand geosynclines. Such comparisons represent familiar applications of the uniformitarian doctrine. Of course, what is involved is simply inductive reasoning from present, more or less ideal dynamic features to allow inferences about some now-static and imperfect fossilized phenomenon. Being based upon simple morphologic comparisons, such inferences represent a *strict uniformitarianism* (of Dott and Batten, 1971, or *substantive uniformitarianism* of Gould, 1965). But if we are concerned with the full genesis of ancient sediments, then we must make more abstract inferences about processes as well as about morphology. These inferences involve more sophisticated analogies, but with only the single methodological assumption of the invariance of natural laws through geologic time. This restricted assumption was termed *actualisme* by Constant Prevost in 1825, seven years before *uniformitarianism* was coined. It allows the induction of ancient causes through knowledge gained from modern processes without also requiring a one-to-one identity of form, dimensions, rates, and the like between ancient and modern phenomena. The distinction urged here is not trivial because allegedly actualistic models are being invoked carelessly in a wide variety of contexts; a truly actualistic hypothesis or model is something other than merely a one-to-one, modern-ancient analogy (see Hooykaas, 1963). Long ago actualism was given a more precise meaning than was uniformitarianism. As the single most important philosophic doctrine for the study of earth history, that meaning should be strictly maintained.

Geology is greatly dependent upon comparative reasoning as is illustrated throughout this volume. But "... it will always be a hypothetical investigation. It cannot be in any way made a series of facts—its production must be *probable* analogies" (Davy, 1840, p. 185; italics mine). The history of geology is rife with abuses of analogical reasoning. As the phenomena being compared differ more and more in scale, details of similarity, or age, analogy becomes increasingly tenuous. Because analogical reasoning has no internal criterion of truth, "it is (either) the bane or the blessing of science" (Hooykaas, 1963, p. 157). Too often we either have chosen in ignorance the wrong modern analogue (the apple-and-orange fallacy) or have seized zealously upon only one of several possible analogues. Moreover, once a comparison is drawn, we have tended to dogmatize and fall victim to the "tyranny of the pigeonhole" (John Rodgers, 1970 oral presidential address, Geological Society of America). A classic example of such tyranny is the widespread use of Kay's 1951 Ordovician cross section through New England as *the* model for miogeosynclinal and eugeosynclinal relationships for all times and places, whereas it was intended simply as one illustrative case.

THE CLASSIC CONCEPT

Early American and European views.—Early literature on geosynclines reminds one of the Indian fable of "The Blind Men and the Elephant," for everyone was partly right, yet partly wrong about his example. Most writers, that is, those from 1859 to 1900, acknowledged a great thickness of sediments within orogenic belts, but agreement ceased there (fig. 1). Early writers on the Appalachians (e.g., James Hall and J. D. Dana) regarded all the geosynclinal sediments as shallow-marine deposits. Meanwhile, European writers such as Bertrand (1897), Haug (1900), and Suess (1909), considering the Alpine orogen, regarded the sediments most typical of geosynclines as relatively deep marine (bathyal) deposits (fig. 1). Haug, especially, thought bathymetry and biofacies more important characteristics than thickness—to him a geosyncline began as an elongated trough receiving pelagic sediments. Soon after 1900, most Europeans tended to assume that geosynclinal accumulation was initiated by the ophiolite sequence, consisting of oceanic ultramafic and mafic igneous rocks succeeded by deep-marine siliceous and argillaceous sediments (Pantanelli, 1883; Steinmann, 1905). According to Hsü (1972), Suess even drew one-to-one analogies between such deep-marine sediments and mafic rocks on mid-Atlantic islands. Analogies also were drawn between deep-sea sediment samples from the *Challenger* expedition and Alpine strata by T. Fuchs, Emil Haug, G. Steinmann, and A. Heim (Bernoulli and Jenkyns, this volume). Ultramafic and mafic rocks also were recognized in the Appalachian belt, but Americans tended to interpret them as intrusions and extrusions within the orogen, that is, as emplaced during deformation. Were ophiolites extruded from the mantle up through sialic crust, or were they an oceanic basement beneath the sediments? To find deep-sea rocks now in mountains on continents seemed to be a threat to an American assumption of continental permanency, but, by repeatedly drilling through abyssal sediments into mafic basement, the recent Deep Sea Drilling Program has vindicated the European view.

Based upon the Appalachian example, Americans concluded that geosynclines and their apparent descendants, mountains, formed at the margins between continents and ocean basins. Hall (1859) challenged lateral compression due to earth shrinkage as the cause of deformation of strata, which was first elucidated by Elie de Beaumont in 1825. He believed, instead, that sediment loading was the sole cause of both the tenfold greater subsidence of the Appalachian region than of the continental interior and the subsequent deformation of the strata. From Hall (1859):

> "The line of greatest depression would be along the line of greatest accumulation (p. 70) ... (thus) the course of the original transporting current (p. 73). By this process of subsidence ... the diminished width of surface above, caused by this curving below, will produce wrinkles and folding of the strata (p. 70)."

J. D. Dana, however, postulated downbending of a *geosynclinal* complemented by an upwarped *geanticlinal* resulting from shrinkage of a supposedly cooling earth having a liquid or plastic interior (fig. 1). Unequal contraction had first given rise to continents and ocean basins; continuing contraction caused bending due "to lateral pressure from the contraction of that crust" (Dana, 1873, pt. V, p. 170). What could be more logical than for buckling to occur primarily at the juncture between two fundamentally different types of crust?

> "... The position of mountains on the borders of continents ... is due to the fact that the oceanic

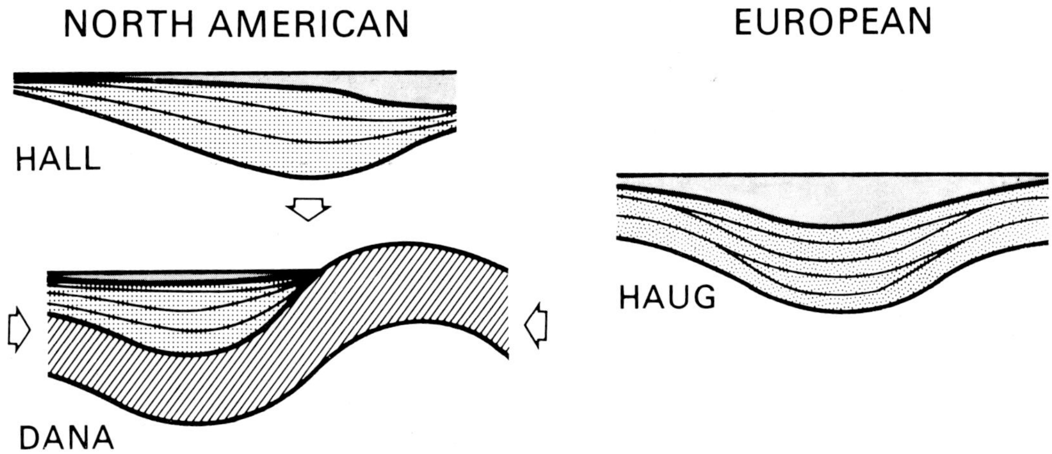

NORTH AMERICAN

HALL

DANA

EUROPEAN

HAUG

FIG. 1.—Early conceptions of the geosyncline: American shallow versus European deeper marine views; James Hall's subsidence by sedimentation (upper left) versus J. D. Dana's subsidence due to crustal buckling with complementary upwarping of a geanticline (lower left). Emil Haug's view of a symmetrical, intercratonic trough is idealized at right.

areas were much the largest, and were the areas of greatest subsidence under continued general contraction of the globe . . . the oceanic crust had the advantage through its lower position of leverage, or more strictly speaking, of obliquely upward thrust against the borders of the continents." (Dana, 1873, p. 170–171.)

Meanwhile, again largely on the basis of the Alpine example, Europeans (e.g. Haug, 1900) inferred that geosynclines could form between two continents (fig. 1). Moreover, because of its apparent intercontinental position and the presence of continental (Hercynian) basement rocks in its core, it appeared that the Alpine geosyncline had been initiated by the foundering of continental crust (see Bernoulli and Jenkyns, this volume). Thus were planted the seeds of a long controversy about the ensialic-versus-ensimatic locus of geosynclines (Wells, 1949).

The tectonic cycle.—By the turn of the century the notion of distinct stages of sedimentation in all orogens had developed from the recognition in Europe of the vertical succession of chert and shale through flysch to molasse, the general significance of which was first advocated by Bertrand (1897) and Haug (1900). Daly (1912), Kossmat (1921), and Kraus (1927) broadened the concept to include sequential igneous and structural events. By mid-20th century, both in Europe and North America, the stages of the sedimentary-tectonic cycle had become *preorogenic* or *geosynclinal* (graywacke), *orogenic* (arkose), and *cratonic* (orthoquartzite) (e.g., Krynine, 1948). There are many variations, including the incorporation of metamorphism and ore deposition by many authors (see Knopf, 1948; Aubouin, 1965; Coney, 1970; and Hsü, 1972).

Even before plate tectonics was formulated, incongruities revealed by detailed features of orogenic belts seemed glaring enough for one to question the validity of the geotectonic or geosynclinal cycle (e.g., Dott, 1964a; 1964b). With the advent of plate theory, Coney (1970) argued emphatically for its complete abandonment. The tectonic cycle implied that all orogens developed in just the same manner. There is, however, little consistency of stage of occurrence within the idealized cycle, either of sediment or of igneous types from belt to belt or even within a single belt. For example, flysch and molasse are diachronous, occurring at different times in different loci and grading laterally as well as vertically into one another; in some belts flysch may not be present at all. Such sediment differences are, after all, a function of local depositional environment, which is only indirectly related to stage of tectonic development (fig. 2). Another great shortcoming of the tectonic cycle is its implication of regularity of timing of tectonic events and the seductive implication of a simple evolutionary continuity from a tranquil continental terrace to an unstable orthogeosyncline through time.

In fact, there have been profound genetic discontinuities between overprinted tectonic patterns, and the developmental history of different orogenic belts is similar in only the most general way. To insist upon fitting each belt into one simplistic, man-constructed cycle obscures the element of uniqueness of such belts and thus impedes understanding more than it helps. The most serious defect of the tectonic cycle, how-

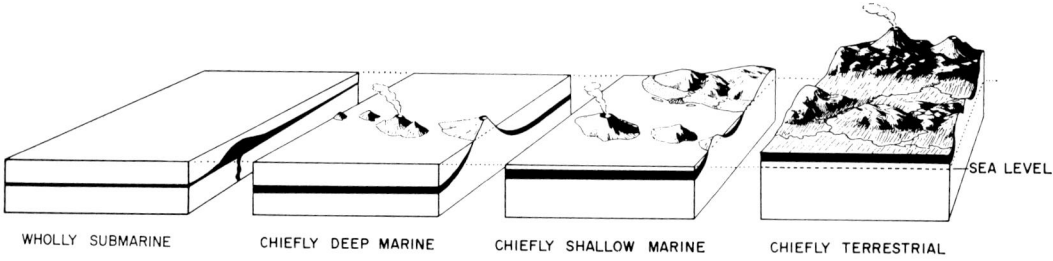

FIG. 2.—Diagrammatic portrayal of great diversity of sedimentary environments represented by geosynclinal sediments and volcanic rocks found in orogenic belts. No implication of a temporal sequence (cycle) is intended. (After Dott and Batten, 1971, p. 251).

ever, was its strong implication that thick sediments caused orogenesis. Although there were a few doubters, such as Eduard Suess and H. H. Hess, a geosyncline was generally considered to be a prerequisite for orogenesis, and, conversely, the absence of thick sediments should preclude significant mountain building. "The greater the accumulation, the higher will be the mountain chain," said Hall (1859, p. 83).

Continental accretion.—A logical handmaiden of the tectonic cycle was the important concept of enlargement of continents through the development of successive, more or less concentric orogenic belts. In America this idea became firmly established through the writings especially of Dana (1873) and in Europe, of Suess (1909), Bertrand (1887), Haug (1900), and Stille (1936, 1941).

"That each epoch of mountain-making ended in annexing the region upturned, thickened and solidified, to the stiffer part of the continental crust, and that consequently the geosynclinal that was afterward in progress occupied a parallel region more or less outside the former. . . ." (Dana, 1873, p. 171).

Accretion generally dictated that juvenile continental crust was generated by regional metamorphism, volcanism, and granitic plutonism accompanying orogenesis in former geosynclinal belts. Moreover, it became a common uniformitarian assumption that continental crust has increased in volume and area at a more or less linear rate through time, a view allegedly supported by geochronology (e.g., Wilson, 1949; Hurley and others, 1962; Engel, 1963). On the other hand, lead-isotope evolution (Patterson, 1964), as well as the increasing documentation of sharply discordant, isotopic-date provinces and of successive metamorphic overprintings or orogenic reworkings (Gastil, 1960; Dott, 1964b), suggested anything but constant-rate accumulation of continental crust and anything but temporal and spatial regularity of orogenesis. Although limited concentric growth has occurred (cf. Okada, this volume), one was confronted with a very complex picture of continents as mosaics laced by orogenic belts of widely varying ages and patterns. Recognition today that plate spreading can extend from oceans into continents casts doubt upon a long-popular implication of irreversibility of accretion.

Continental accretion implied that most orogenic belts must be at least partially ensimatic and marginal to continents initially if new continental crust was to be generated therein. Early studies suggested that, in North America, at least, Phanerozoic orogens supported most of the implications of accretion. The theory became less popular abroad, however, because of the obviously sialic basement found in many belts, the clear truncations of Caledonian structures by Hercynian ones, the overprinting in turn of Alpine upon Hercynian trends, the present intercratonic position of the Urals and Himalayas, and the location of several modern arc-trench systems far from any continent in a wholly oceanic environment. It seemed to many workers that orogenic belts could be, after all, either ensialic or ensimatic and that they had not formed in any simple concentric fashion. Hess (1960) argued that volcanic arcs were oceanic and not precursors to Alpine-type mountains; the two represent different responses to the same tectonic regime in different geologic environments.

An additional objection to accretion exists if new granitic crust forms solely through granitization of sediments, an idea popular around 1940. The sediments themselves were formed largely by erosion of older continental crust and of volcanic rocks, which were commonly assumed to represent crustal melting. But without additions of juvenile material from outside the continental realm, obviously no increase of crustal volume would be possible. Recent proposals that andesitic eruptions in magmatic arcs probably originate by fractional melting well below the continent provides an appealing answer to this dilemma (e.g., Kuno, 1959; Dickinson and Hatherton, 1967; Dickinson, 1970). If granitic plutons as well as andesites do originate from such melting, a mechanism for accretion of new continental rocks by igneous differentiation is provided after all.

Sedimentation and paleogeography.—European workers have envisioned the geosyncline as a deep sea-floor trough or furrow receiving only fine pelagic sediments at first. Subsequently the sand of the flysch was introduced from terrigenous sources, which were interpreted as embryonic mountain ridges called *cordilleras* that were raised as elongate islands within the geosynclinal belt (Argand, 1916). Flysch was deposited between the cordilleras, but, with full uplift of most of the geosyncline, coarser, largely nonmarine molasse was deposited at the margins of the rising mountains (fig. 2).

Meanwhile in America, the prominence of sandstones and conglomerates in middle and late Paleozoic rocks of the Appalachian Mountains led to the view of geosynclinal sediments all being shallow-marine and nonmarine deposits. Eastward coarsening of these strata and the presence in adjacent western New England of crystalline rocks, which were assumed until the 1930's to be entirely pre-Paleozoic, gave rise to the concept of a persistent eroding borderland of ancient rocks lying next to the subsiding geosyncline. James Hall had already recognized vaguely an eastern land source for Ordovician and later clastic sediments, but Dana (1873, p. 171) was more specific in stating that:

". . . on the ocean side of the progressing geosynclinal referred to, there has been generally, as the first effect of the thrust against the continental border, a progressing geanticlinal, which usually disappeared in the later history of the region. . . ."

In 1882 Chamberlin showed such a borderland on an early Paleozoic paleogeographic map for North America (apparently the first of its kind). Williams (1897) coined the name "Appalachia" for the land, and Schuchert (1910) then generalized *borderland* (or geanticline) sources for all American geosynclines. Deltaic and fluvial deposits were recognized widely among the later Paleozoic strata of the Appalachians, and uniformitation analogies were drawn with the Mississippi River delta system (e.g., Barrell, 1913-1914; Storm, 1945), but the origin of Appalachian Ordovician and Devonian flyschlike deposits remained a mystery.

In 1937 Kay recognized *tectonic lands raised within* the Appalachian belt by episodic uplift of earlier geosynclinal rocks to provide sources of younger deposits laid in still-subsiding portions of the belt (fig. 3). These were much like the European cordilleras. Kay's interpretation was given strong impetus through the discovery by Harvard geologists of Paleozoic marine fossils in the alleged pre-Paleozoic crystalline complex of New England, that is, in the heart of old Appalachia (see Kay, this volume).

At the same time, Hans Stille noted the rather consistent arrangement of belts of different rock types within most ancient orogens. His subdivisions focussed upon two parallel zones, one having volcanic and plutonic rocks (including ophiolites)—the eugeosynclinal belt—and one lacking such rocks—the miogeosynclinal belt. Although Stille was European, his principal examples were chosen from the North American Cordillera (Stille, 1936; 1941). Meanwhile, Kay had compiled for teaching purposes the distributions of volcanic rocks and sedimentary facies for various rock systems in North America and found the same patterns to hold for long spans of history in every orogenic belt that he studied. The prominence of volcanic rocks in the eugeosynclinal zones and the physical continuity of several such belts with modern volcanic arcs then led to a general paleogeographic analogy between the two features (Hess, 1939; Kay, 1947; Eardley, 1947). Was this not the first actualistic geosynclinal model? Early in this century, Haug had compared geosynclines to the newly discovered deep troughs of Indonesia, but it was the Dutch studies of the 1930's that gave the greatest impetus to the volcanic arc-trench analogue (Umbgrove, 1933; Vening Meinesz, 1940; Van Bemmelen, 1949). Direct comparisons of Indonesia with the Alpine geosyncline were especially popular, for the trench seemed fully compatible with the long-standing European assumption of deep-marine geosynclinal sedimentation. An unfortunate tendency to equate trenches and eugeosynclines on a one-to-one basis developed and still persists (e.g., Hsü, 1972, p. 34-35). The analogy intended by Kay and others was more general and would encompass both the volcanic lands (the arc) *and* the trench as a possible modern counterpart for an ancient eugeosyncline (see Scholl and Marlow, also Dickinson, this volume). It must be stressed, however, that both "eugeosyncline" and "miogeosyncline" were descriptive terms referring to tangible subdivisions of ancient orogens rather than to any modern morphologic features, sedimentary facies, or hypothetical paleogeographic restoration. So viewed, geosynclines may be *elucidated*

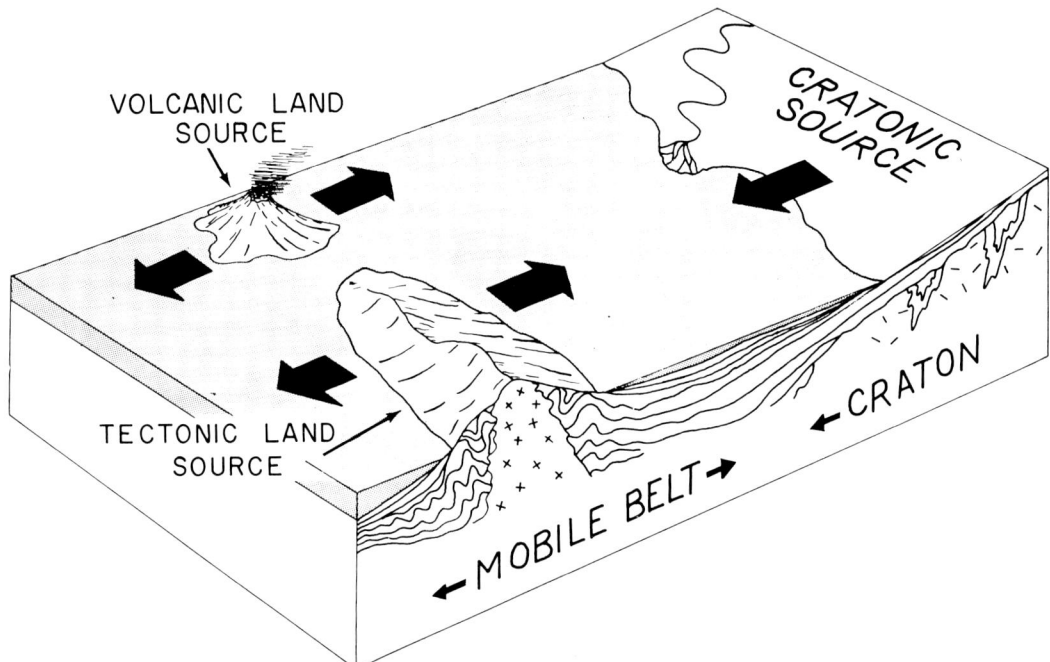

Fig. 3.—Different major types of source lands for clastic geosynclinal sediments; purely diagrammatic in terms of spatial dimensions and number of source types present at any given time. Note cannibalism of slightly older geosynclinal sediments so characteristic of tectonic lands. (After Dott and Batten, 1971, p. 248).

by reference to modern features, but they can not be *defined* by them (see also Aubouin, 1965, p. 3).

In Britain, E. B. Bailey (1930), with his distinction of graded from cross-stratified sandstone suites, was among the first to perceive that much coarse clastic sediment might be deposited in deep-marine environments by some kind of gravity processes. However, O. T. Jones (1938), in spite of his pioneering work with slump structures, maintained that no great depth necessarily prevailed during deposition of graywackes. Yet, in California Natland (1933) had noted that conglomerates as well as sandstones were interstratified with Cenozoic abyssal shales, and in 1950 Kuenen and Migliorini finally resolved the long-standing European-American bathymetric dilemma with their diagnosis of turbidity current transport. Through actualistic reasoning, ancient, coarse submarine fan deposits are being recognized widely in geosynclines (e.g., Sullwold, 1960; Dzulynski and others, 1959; Walker, 1966; and Kulm and Fowler, Mutti, Nelson and Nilsen, and Normark, all this volume). Now we know that a great deal of gravel as well as sand has reached deep-sea fans via submarine canyons, although the exact mechanism of transport of the coarsest materials is just beginning to be understood (Stauffer, 1967; Aalto and Dott, 1970; Middleton, 1970; Hampton, 1972). Very coarse slide breccias (olistostromes and some wildflysch) also are recognized increasingly, but some are difficult to distinguish from tectonic melanges (see Hsü, Blake and Jones, and Wood, all this volume). Widespread application of paleocurrent analysis pioneered by F. J. Pettijohn and his students in the 1950's also has shed much light upon paleobathymetry and paleogeography in general over the past two decades. The importance of diverse source terranes for geosynclinal sediments—cratonic, volcanic, and tectonic lands (fig. 3)—has been confirmed beyond question through such investigations coupled with sandstone petrography (see Crook, this volume).

By at least 15 years ago, it had become clear that most geosynclines at different times and in different places have had practically every conceivable type of sediment deposited within them (table 1). Schwab (1971) noted that evaporites are the only sediments almost totally lacking. Carbonate rocks and mature sandstones are certainly more abundant in miogeosynclines, and graywacke and chert in eugeosynclines, but none are mutually exclusive. Nonetheless, by 1960 unfortunate dogmatic generalizations were entrenched. For example, "graywacke," "turbi-

TABLE 1.—PERCENTAGE PROPORTIONS OF ROCK TYPES IN SOME REPRESENTATIVE NORTH AMERICAN ORTHOGEOSYNCLINES

Rock type	Eugeosynclines Including volcanics Dott[2]	Eugeosynclines Including volcanics Schwab[3]	Eugeosynclines Without volcanics Dott[2]	Eugeosynclines Without volcanics Schwab[3]	Miogeosynclines Schwab[3]
Shale	40	26	50	33	26
Volcanic	20	21	—	—	—
Graywacke[1]	10	30	13	38	2
Carbonate	10	5	13	6	44
Quartz arenite	9	4	11	5	12
Chert	9	7	11	9	1
Conglomerate	2	?	2	?	?
Arkose and subgraywacke	?	7	?	9	15

[1] Includes tuffaceous sandstones.
[2] Point counts of relative areas on geologic maps from eight regions (done for writer by L. J. Suttner, Univ. Wisconsin Research Assistant, 1965).
[3] Percentage of total thickness in two stratigraphic sequences from Appalachian region and two from Cordilleran region (Schwab, 1971).

dite," and "flysch" all became loosely equated with "eugeosyncline" even in the absence of associated volcanic rocks. In fact, many examples of clastic wedges within miogeosynclinal (or, strictly, exogeosynclinal) belts exist. Such wedges are comprised either of flyschlike graywacke deposited by turbidity currents (e.g., Ordovician of the Appalachians, upper Paleozoic of the Ouachita-Marathon region, and some classic flysch of the Northern Alps) or of molasselike deposits (e.g., Old Red Sandstone of Britain, Catskill and upper Paleozoic strata of the Appalachians, and the Cretaceous of the Rocky Mountains). Conversely, mature quartz arenites occur in volcanic eugeosynclinal belts. In the North American Cordillera, pure quartzites occur within Ordovician black slates, where they comprise as much as 25 percent of the sequence (Ketner, 1966; Churkin, this volume), and within Lower Jurassic carbonates (Stanley, and others, 1971). In the middle Precambrian orogen of the Great Lakes region, quartzites and quartz wackes are widespread (Dott, 1964b; 1972).

Environments of deposition, and thus also paleogeography, were as variable as the sediments themselves (fig. 2). Many workers have assumed unconsciously that most geosynclinal sediments must be marine; no rational reason exists, however, for excluding thick nonmarine ones. Great volumes of nonmarine sediments, most notably the famous molasse (Van Houten and Eisbacher, both this volume), are contained within every classic geosyncline. And it is the rule that simultaneously along a single belt nonmarine, shallow- and deep-marine environments coexisted. Moreover, sediments formed by the same processes may occur in diverse sedimentary or tectonic environments (Crowell, Harms and Pray, Normark, all this volume). For example, submarine fans built by turbidity currents can occur in cratonic basins, continental rises, arc-rear basins, and trenches (see Reading, 1972).

Stille and Kay's various subclasses of geosynclines were based upon tectonic setting and upon presence or absence of certain igneous rocks, rather than upon sedimentary rock type or inferred environment of deposition. "Geosynclinal facies are characterized by thickness rather than kind" (Kay, 1951, p. 1).

THE TECTONIC REVOLUTION

The causality myth.—As noted above, causality very early became a part of the geosynclinal concept, especially in North America. It was an article of faith that "thick sediments must invariably lead to mountains—in fact are a prerequisite—and mountain building leads to enlargement of continents." In reality, however, ancient orogenic belts and modern magmatic arcs, which clearly are genetically related species, by no means seem to be consistently marginal to continents. Considering the diversity, their role in crustal evolution seemed unclear to some workers a decade ago (e.g., Hess, 1960; Dott, 1964b). All that did seem definite was that the tectonically mobile belts of the earth, regardless of age or location, represent first and foremost a tectonic environment. The sediments of geosynclines—strictly, the *orthogeosynclines* of Stille and Kay—must be quite secondary! This crucial point was appreciated by Suess, Kay, Hess, and others, but not by many other workers. Modern belts such as the Marianas arc-trench system are impoverished of sediments and may always be so, yet surely they are experiencing a tectonic regime similar to, say Japan's or to that of the northeastern Mediterranean. This line of reasoning led to the suggestion that orthogeosynclines are best thought of as sediment-filled mobile belts (Dott, 1964a, 1964b; Dott and Batten, 1971); "sediments are but innocent bystanders" (Hsü, 1972. p. 33). Conversely, then, because structure is the key to orogens and controls sedimentation to a great extent, the Gulf Coast type of so-called '"geosyncline" must belong to a genus wholly different from the orthogeosyncline. The Gulf Coast type owes its origin to a tranquil progradation of a continental terrace and apparently has nothing causal to do with orogenesis. Ironically, this *paraliageosyncline* of Kay (1951) (synonymous with the *miogeocline-eugeocline* couplet of Dietz and Holden, 1966) fits closely James Hall's original, wrongly conceived model for sedimentary accumulation on the postulated site of future mountains.

The miogeocline paradox.—Counter to the above reasoning, Drake and others (1959) already had suggested a possible analogy between young shelf-rise (continental terrace) sediment prisms and ancient orthogeosynclines. Dietz (1963) then incorporated this view and that of Hess (1962) about sea-floor spreading into an inclusive model for geosynclines and mountain building (see also Dietz, this volume). Whereas to me a great degree of tectonic mobility had seemed to be a prerequisite for all true orthogeosynclines, Dietz argued that such belts began first as passive continental shelf-rise prisms of the Atlantic or Gulf Coast type. The key point is that his miogeocline was first conceived as an essential stage in the evolution of orthogeosynclines. But why was there so little tectonic mobility evidenced in modern miogeoclines? And where was there any compelling evidence that a continental rise (Dietz' eugeocline) on an Atlantic-type margin could be a youthful eugeosyncline? Submarine volcanism on such rises was a guess only. The entire model, although allegedly actualistic, was in fact strictly uniformitarian, for it involved a one-to-one analogy between a modern topographic entity, the continental terrace, and complexly deformed ancient rocks, which in fact share only limited lithologic similarities (Stanley, 1970; Moore and Curray, this volume). As outlined in the introduction, a truly actualistic model would imply a more complete analogy of tectonic and sedimentary processes in addition to primarily morphologic comparisons (e.g., Stanley, this volume). Moreover, the uniqueness of the Quaternary System makes modern shelf-rise sediments doubtful analogues for ancient ones (Moore and Curray, this volume).

Salvation through plate tectonics?—The paradox between the sea-floor-spreading orogenic model of Dietz and the concepts of orthogeosynclines derived from studies of ancient rocks seems to have been resolved dramatically with the appearance of papers on the geologic implications of plate tectonic theory (e.g., LePichon, 1968; Isacks and others, 1968; Dewey and Bird, 1970a; Dickinson, 1970; North, 1971). These papers came close on the heels of Dietz' (1963) and Rodger's (1968) inference that strata of the Appalachian region ranging in age from earliest Cambrian through Early Ordovician represent a stable Atlantic-type continental-shelf sequence caught up by orogenesis, which began abruptly in medial Ordovician time (see Bird, this volume). The new global tectonic theory showed how a stable continental margin could become involved coincidentally at any time in orogenesis at an active convergent plate margin. Apparently, many ancient miogeosynclines were, in part, paraliageosynclines that had suffered exactly such a fate (fig. 4). The important Drake-Dietz-Rodgers analogy between youthful Atlantic-type continental terraces and some of the (early) deposits now seen within orogenic belts could be accepted so long as a genetic discontinuity separated such so-called "miogeoclinal" or nonorogenic rocks from subsequent, truly orthogeosynclinal or orogenic sequences. The crucial point revealed by plate theory is that the two do not represent a single, continuous evolutionary succession, but a coincidental association, an overprinting. The discontinuity typically is marked by an abrupt reversal of source direction and maturity of major clastic sediment contributions, and it may involve considerable lateral translation or telescoping of quite different terranes (fig. 4, VI). The significance of such temporal and spatial discontinuities can not be overemphasized. Even if many ancient orogens have one, a continental terrace record (or miogeocline in some usage) is not a requirement for every mobile belt—witness the Marianas arc built in a wholly oceanic realm. Miogeoclines are coincidental to orogenesis, but as Mitchell and Reading (1969) and Dickinson (1971) have stressed, one consequence of plate theory is that most passive continental margins are almost certain sooner or later to become tectonically crumpled by some active plate boundary.

The new global tectonics is so attractive because it seems, more than any older global theory, to generalize and simplify so many diverse phenomena. Moreover, some of those phenom-

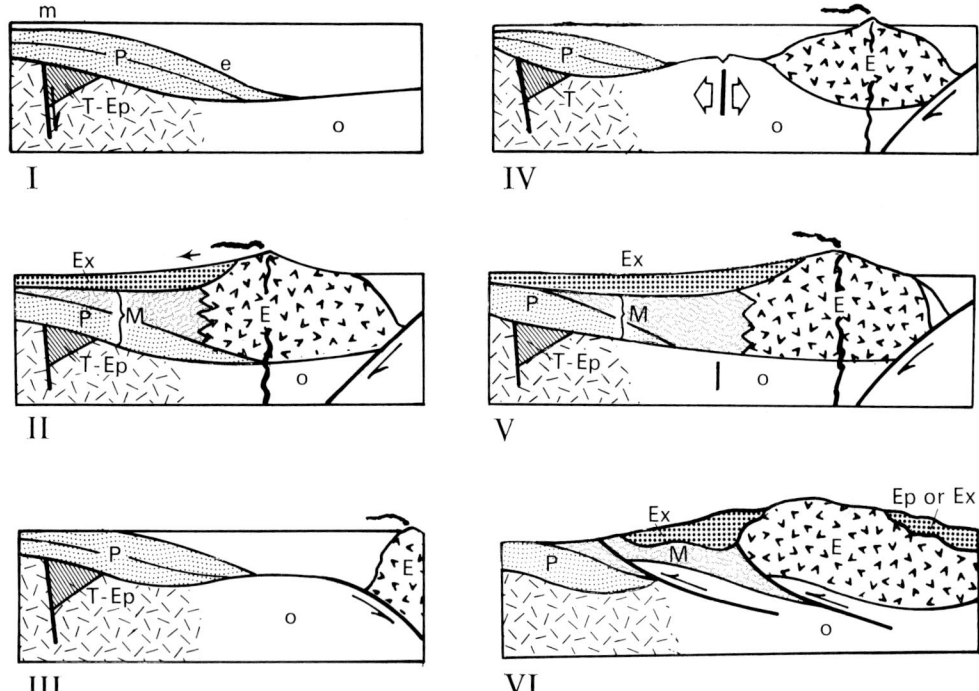

Fig. 4.—Possible genetic relations among some Kay-Stille geosynclinal species, especially showing ways that nonorogenic paraliageosynclines (P) [or miogeocline-eugeocline couplets (m and e)] can be coincidentally associated with orthogeosynclines. Not to scale: o, oceanic crust and mantle, including ophiolite suites; I, continental terrace on a passive margin following rifting [taphrogeosynclines (T) and certain epieugeosynclines (EP)]; II, subsequent development of an eugeosyncline-miogeosyncline-exogeosyncline complex (E-M-Ex) when the continental margin became orogenically active; III, a nonorogenic trailing-edge terrace destined to be involved in orogenesis by propagation toward it of an arc-trench system; note problem of distinguishing after the fact the polarity of this arc system from that of II and IV; IV, continental terrace facing a small-ocean basin having a local spreading site and an orogenic outer margin; V, marginal ocean basin after cessation of spreading therein followed by filling, which juxtaposed exogeosynclinal strata over continental terrace strata; VI, orogenic telescoping of rock suites and basements as might accompany the closing of an ocean basin (as in III, IV, V) or the termination of a continent-margin orogenic belt (as in II). Dual exogeosynclines shown characterize especially intercratonic orogens; in such belts, thrusting also may be bilateral. This figure is not intended to advocate perpetuation of the terms.

ena seemed almost totally unrelated before, for example, rifting in one region contemporaneously with orogenesis far away. Significantly, the new theory accommodates the several species recognized by Stille and Kay within the broad genus "orthogeosyncline." Twenty-five years ago species distinctions were necessarily descriptive and were conceived in order to focus upon what appeared to be tectonically significant differences. Now, at last, the new tectonics provides a rational genetic explanation for each species (fig. 4; see also Dewey and Bird, 1970b).

Loose ends.—The new tectonics seems to constitute a scientific revolution and to have produced a sweeping new *paradigm* in the sense of Kuhn (1962). Plate theory and its many implications—some entirely new—are shared by a large scientific community, which adopted them with unprecedented rapidity, presumably because of a so-called "crisis" of rapidly accumulating anomalies that were less satisfactorily explained by older theories. Moreover, the new tectonics has almost overnight provided a host of new problems for investigation as well as new ways of looking at old problems. Plate tectonics indeed seems to represent a quantum jump rather than a cumulative or inexorable evolutionary progression in geologic thought. It has provided a new tradition—a "reconstruction of the field from new fundamentals" (Kuhn, p. 85). For example, the mobilist view of the earth that involves breathtaking translations of tectonic elements, for which we were partially prepared by Wegener, Carey, and others, has rapidly become widely accepted as commonplace.

Significant as it seems to be, however, I doubt that plate tectonics has changed the goals and methods of geology to anything like the degree that the revolutions discussed by Kuhn changed astronomy, physics, and chemistry. Moreover, as Kuhn himself would predict (p. 79), the new tectonic paradigm has not completely resolved all old problems. Critics cannot condemn it (or any theory) for that, although they might scold its advocates for a disturbing degree of dogmatism acquired astonishingly quickly. We should remind ourselves that plate theory is not a catechism, but only a model, which must be held up for some time yet to the critical mirror of reality. The new theory is remarkably similar qualitatively to Taylor's (1910) and Holmes' (1930) early speculations during this century on continental drift as a mountain-building mechanism, and it bears some resemblance to the ideas of crustal downbuckle or tectogene that were popular in the 1930's. Finally, we must not forget that, like Charles Darwin's theory of natural selection, plate tectonics will remain an incomplete theory so long as a driving mechanism remains unverified.

There are some questions to ask about the new orthodoxy. Why do some orogenic belts seem to lack any clear oceanic relics? Perhaps they really were ensialic—"not organs of continental growth, but wounds healed by orogenesis," to paraphrase Aubouin (1965, p. 220). Were marginal and interarc ocean basins of such universal significance in the past as is currently suggested? If so, why are not clear suture zones representing their former positions more widely recognizable? Is it simply because of the varied structural levels that are exposed? How can we tell which chert-basalt assemblages are actually ocean-ridge deposits (Kanmera, this volume)? Why do modern ocean trenches not show more obvious commonalities with parts of ancient eugeosynclinal belts? Many of the latter, for example, lack tectonic mélanges and blueschists, which are assumed by many to be identifying characteristics of the trench tectonic environment. Can subduction zones flip so easily and frequently as some authors suggest? Lastly, must an orthogeosyncline now be viewed as an entire half-ocean basin riding on a conveyor belt, so to speak, away from its parental spreading site so that most of its rocks simply are scraped off in a trench?

Regardless of the ultimate fate of plate tectonics, which history tells us must be at least somewhat ephemeral, a number of concepts developed as handmaidens to it seem to stand on their own merits. Most important to the venerable geosynclinal concept is to distinguish the nonorogenic, miogeoclinal (or paraliageosynclinal or Atlantic-type continental marginal) sequences and some aulacogen sequences (Hoffman and others, this volume) that are now seen in orogenic belts from the rock sequences that developed within the genetic context of active orogenesis (or Pacific- and Mediterranean-type convergent plate margins). Also of great significance is the growing recognition of the enormous importance of submarine channels (Whitaker, also Picha; this volume) and fans to sedimentation within developing orogenic belts as well as on passive continental margins. But how, in ancient rocks, can we discriminate fans formed on stable margins from those formed either in more active marginal sea areas or in trenches? We see ever-increasing evidence of mass gravity failures both in the ancient and recent records, but their distinction from tectonic mélanges commonly presents serious problems. We have a number of bathymetric indicators (e.g., foraminiferal ratios, trace fossils, vesiculation in submarine flow rocks, and relative position of carbonate compensation), but still-more-accurate criteria would be welcome.

We need to distinguish among ancient turbidites formed on passive continental rises, arc-rear basins, arc fronts and trenches, and abyssal plains (see Crook, this volume for a partial key). Paleogeographic criteria for polarity of arc-trench systems to supplement the geochemical gradients for igneous rocks as proposed by Dickinson (1970) are needed as well, for in ancient rocks chemical alteration and metamorphism commonly have rendered that approach impossible. Difficulties of polarity interpretation are underscored, for example, by Stanley's (this volume) documentation of flysch derived from the stable craton of North Africa being sandier than flysch derived from the Hellenic-arc (opposite) side of the eastern Mediterranean. We need to distinguish among sea-floor, arc, and rift volcanics, all of which may be found together in the same belt. Ignimbrites, for example, are most common in arcs sited upon continental crust. Finer distinctions of volcanic products, including their trace-element chemistry, may be the most important means of unravelling the mysteries of the bewildering array of rock sequences found today jumbled together in orogenic belts (see Helwig, this volume).

CONCLUSION

"When I use a word, it means just what I choose it to mean—neither more nor less." (Humpty Dumpty in Lewis Carroll's *Through the Looking Glass*).

Classification and terminology are essential to all disciplines, especially if the phenomena under study are themselves both as complex and poorly understood genetically as was the geosyncline twenty years ago. Without some ordering of material, the human cannot cope. Whether nature really exists "out there" or only in our minds, classification is an indispensible part of thought. The earlier phase of study of orogenic belts necessarily was a descriptive or taxonomic one in which classifications of geosynclinal species served to focus upon important space-time differences. "The concern about the terms has resulted in more penetrating analyses of the history of the rocks; the very endeavor to classify has been rewarding" (Kay, 1967, p. 315). With the advent of a powerful, unifying explanation of orogenic belts, however, we move into a new, more genetic phase of study in which taxonomy should be far less important.

What is likely to be the fate of the geosyncline in the next 20 years? Some would argue that the entire concept is dead or dying and that the term itself should be put quietly to rest in the archives of science history. Others, however, have sought to rationalize every detail of older geosynclinal concepts with the new global tectonics (e.g., Wang, 1972). While most of the various descriptive species designated over the years can now be seen to have some genetic significance within the context of plate tectonics, there is little utility—and much potential danger—in perpetuating all of them. Long before the new global tectonics was formulated, one was confronted with such a multitude of contradictory and overlapping usages of terms that much of the utility of a scientific shorthand was lost. The most obvious anomaly is that noted by Aubouin (1965), whereby "geosyncline" is loosely applied to individual local depositional basins or furrows within an orogen as well as to the sedimentary and volcanic fill of the entire orogen.

If much of the old terminology is retained, then communication problems are likely to be aggravated with the advent of new hypotheses. Toulmin (1953) argued that major new breakthroughs in physical science involve the discovery chiefly of new modes of representation, thus new generalized models. The best model, which is "something more than a simple metaphor" (Toulmin, p. 39), should inspire new ways of drawing inferences and thus of formulating and testing hypotheses. It follows that a revolution in scientific theory need not involve discovery of any new evidence, but it must involve the discovery of a new model, or, in the approximately equivalent, but more colorful language of Kuhn, a new paradigm. It must provide a new conceptual and methodological framework. As a new general model is discovered, the bases for classification will change, which results in an important language shift (Toulmin, 1953, p. 13). In short, changing concepts require changing language! Terminology conceived in the context of, and useful to, an earlier general model or concept is almost certain to be constraining in the context of a new one (see also Coney, 1970). A language shift could, of course, take place in either of two ways: first, by a redefinition of all old terms, or, secondly, by the construction of an entirely new terminology. Different workers will have their own preferences, but the clutter of different shades of meaning attached during the past two decades to geosynclinal and tectonic nomenclature should give one considerable pause before he opts for the former course. So many of the premises associated with the older taxonomy are incompatible with the new idiom that, if perpetuated, such terminology would certainly act as a mental straitjacket. Conversely, exploitation of the new idiom's truly revolutionary implications could be greatly accelerated by a clean break with the old language.

ACKNOWLEDGMENTS

It is impossible to recall, much less to cite, all who have contributed materially over the years to the views expressed here. Most important, certainly, has been Marshall Kay's influence on my thinking. Previous historical reviews of the geosynclinal concept, such as those by Glaessner and Teichert (1947), Knopf (1948), Aubouin (1965), Coney (1970), and especially a recent one by Hsü (1972), were very helpful in composing this paper. The reader should consult them for a broader historical perspective. The manuscript was materially improved by criticism at various stages by I. W. D. Dalziel, James Helwig, W. R. Dickinson, J. F. Dewey, and H. G. Reading. Much of the substance of this paper was first presented orally as an invited paper in September 1971 at Edinburgh, Scotland, for a symposium on "Global Topics in Sedimentology."

REFERENCES

AALTO, K. R., AND DOTT, R. H., JR., 1970, Late Mesozoic conglomeratic flysch in southwestern Oregon, and the problem of transport of coarse gravel in deep water: in LAJOIE, J. (ed.), Flysch sedimentology in North America, Geol. Assoc. Canada Special Paper 7, p. 53–65.
ARGAND, E., 1916, Sur l'arc des Alpes occidentales: Eclogae Geol. Helvetiae, v. 14, p. 145–191.
AUBOUIN, JEAN, 1965, Geosynclines: Amsterdam, Elsevier Pub. Co., 335 p.
BAILEY, E. B., 1930, New light on sedimentation and tectonics: Geol. Mag., v. 67, p. 71–92.

Barrell, Joseph, 1913–1914, The Upper Devonian delta of the Appalachian Geosyncline: Am. Jour. Sci., v. 36, p. 429–472; v. 37, p. 87–109.
Bertrand, Marcel, 1887, La chaîne des Alpes et la formation du continent européen: Soc. géol. France Bull., v. 15, p. 423–447.
———, 1897, Structures des Alpes francaises et recurrence de certains facies sedimentaires: 6th Cong. Geol. Internat. (1894), Compte Rendu, Lausanne, p. 161–177.
Chamberlin, T. C., 1882, Geology of Wisconsin: Madison, Wisconsin, v. 4, 779 p.
Coney, P. J., 1970, The geotectonic cycle and the new global tectonics: Geol. Soc. America Bull., v. 81, p. 739–748.
Daly, R. A., 1912, Geology of the North American Cordillera at the 49th parallel: Geol. Survey Canada Mem. 38, pt. 2, p. 547–857.
Dana, J. D., 1873, On some results of the earth's contraction from cooling, including a discussion of the origin of mountains, and the nature of the earth's interior: Am. Jour. Sci., ser. 3, v. 5, p. 423–443; v. 6, p. 6–14; 104–115; 161–172.
Davy, John (ed.), 1840, The collected works of Sir Humphrey Davy: Cornhill, London, Smith, Elder and Co., v. 8 (Johnson Reprint Edition of 1972; The sources of science No. 114), 365 p.
Dewey, J. F., and Bird, J. M., 1970a, Mountain belts and the new global tectonics: Jour. Geophys. Research, v. 75, p. 2625–2647.
———, and ———, 1970b, Plate tectonics and geosynclines: Tectonophysics, v. 10, p. 625–638.
Dickinson, W. R., 1970, Relation of andesites, granites, and derivative sandstones to arc-trench tectonics: Rev. Geophysics and Space Physics, v. 8, p. 813–860.
———, 1971, Plate tectonic models of geosynclines: Earth and Planetary Sci. Letters, v. 10, p. 165–174.
———, and Hatherton, T., 1967, Andesitic volcanism and seismicity around the Pacific: Science, v. 157, p. 801–803.
Dietz, R. S., 1963, Collapsing continental rises: an actualistic concept of geosynclines and mountain building: Jour. Geology, v. 71, p. 314–333.
———, and Holden, J. C., 1966, Miogeoclines (miogeosynclines) in space and time: *ibid.*, v. 74, p. 566–583.
Drake, C. L., Ewing, M., and Sutton, G. H., 1959, Continental margin and geosynclines: the east coast of North America, north of Cape Hatteras, *in* Ahrens, L. H., and others (eds.), Physics and chemistry of the earth: London, Pergamon Press, v. 3, p. 110–198.
Dott, R. H., Jr., 1964a, Mobile belts and sedimentation (abs.): Geol. Soc. America Special Paper 76, p. 49–50.
———, 1964b, Mobile belts, sedimentation and orogenesis: New York Acad. Sci., Trans. ser. 2, v. 27, p. 135–143.
———, 1972, Implications of Precambrian quartz-rich sediments of the Lake Superior region for crustal evolution: 24th Internat. Geol. Cong. (Montreal), Proc., abs. v., p. 8–9.
———, and Batten, R. L., 1971, Evolution of the earth: New York, McGraw-Hill Book Co., 649 p.
Dzulynski, S., Ksiazkiewicz, M., and Kuenen, Ph. H., 1959, Turbidites in flysch of the Polish Carpathian Mountains: Geol. Soc. America Bull., v. 70, p. 1089–1118.
Eardley, A. J., 1947, Paleozoic Cordilleran Geosyncline and related orogeny: Jour. Geology, v. 55, p. 309–342.
Engel, A. E. J., 1963, Geologic evolution of North America: Science, v. 140, p. 143–152.
Gastil, R. G., 1960, Continents and mobile belts in the light of mineral dating: 21st Internat. Geol. Cong. Rept., pt. 9, p. 162–169.
Glaessner, M. F., and Teichert, Curt, 1947, Geosynclines: a fundamental concept in geology: Am. Jour. Sci., v. 245, p. 465–482; 571–591.
Gould, S. J., 1965, Is uniformitarianism necessary?: *ibid.*, v. 263, p. 223–228.
Hall, James, 1859, Description and figures of the organic remains of the lower Helderberg Group and the Oriskany Sandstone: New York Geol. Survey, Natural History of New York, pt. 6, Paleontology, v. 3, 532 p.
Hampton, M. A., 1972, The role of subaqueous debris flow in generating turbidity currents: Jour. Sed. Petrology, v. 42, p. 775–793.
Haug, Emil, 1900, Les géosynclinaux et les aires continentales: Soc. géol. France Bull., v. 28, p. 617–711.
Hess, H. H., 1939, Island arcs, gravity anomalies and serpentine intrusions, a contribution to the ophiolite problem: 17th Internat. Geol. Cong. Rept., v. 2, p. 263–282.
———, 1960, Caribbean research project, progress report: Geol. Soc. America Bull., v. 71, p. 235–240.
———, 1962, History of ocean basins, *in* Engel, A. E. J., and others (eds.), Petrologic studies: a volume in honor of A. F. Buddington: Geol. Soc. America, p. 599–620.
Holmes, Arthur, 1930, Radioactivity and earth movements: Geol. Soc. Glasgow Trans. (1928–29), v. 18, p. 559–606.
Hooykas, R., 1963, Natural law and divine miracle: the principle of uniformity in geology, biology and theology: Leiden, Brill, 237 p.
Hsü, K. J., 1972, The concept of the geosyncline, yesterday and today: Leicester Lit. and Philos. Soc. Trans., v. 66, p. 26–48.
Hurley, P. M., Hughes, H., Faure, G., Fairbairn, H. W., and Pinson, W. H., 1962, Radiogenic strontium-87 model of continent formation: Jour. Geophys. Research, v. 67, p. 5315–5334.
Isacks, B., Oliver, J., and Sykes, L. R., 1968, Seismology and the new global tectonics: Jour. Geophys. Research, v. 73, p. 5855–5899.
Jones, O. T., 1938, On the evolution of a geosyncline: Geol. Soc. London Quart. Jour., v. 94, p. lx–cx.
Kay, Marshall, 1937, Stratigraphy of the Trenton Group: Geol. Soc. America Bull., v. 48, p. 233–302.
———, 1947, Geosynclinal nomenclature and the craton: Am. Assoc. Petroleum Geologists Bull., v. 31, p. 1289–1291.
———, 1951, North American geosynclines: Geol. Soc. America Mem. 48, 143 p.
———, 1967, On geosynclinal nomenclature: Geol. Mag., v. 104, p. 311–316.
Ketner, K. B., 1966, Comparison of Ordovician eugeosynclinal and miogeosynclinal quartzites of the Cordilleran Geosyncline: U.S. Geol. Survey Prof. Paper 550C, p. 54–60.

Knopf, Adolph, 1948, The geosynclinal theory: Geol. Soc. America Bull., v. 57, p. 649–670.
Kossmat, F., 1921, Die mediterranen Kettengebirge in ihrer Berichtigung zum Gleichgewichtszustande der Erdrinde: Sachsische Akad. Wiss., Abh., Math.-Phys. Kl., v. 38, p. 46–68.
Kraus, E., 1927, Der orogene Zyklus und seine Studien: Centralbl. für Minéralogie, Abt. B, p. 216–233.
Krynine, P. D., 1948, The megascopic study and field classification of sedimentary rocks: Jour. Geology, v. 56, p. 130–165.
Kuenen, Ph. H., and Migliorini, C. I., 1950, Turbidity currents as a cause of graded bedding: *ibid.*, v. 58, p. 91–127.
Kuhn, T. S., 1962, The structure of scientific revolutions: Chicago, Univ. Chicago Press, 2d ed. (1970), 210 p.
Kuno, H., 1959, Origin of Cenozozoic petrographic provinces of Japan and surrounding areas: Bull. volcanol., v. 20, p. 37–67.
LePichon, Xavier, 1968, Sea-floor spreading and continental drift: Jour. Geophys. Research, v. 73, p. 3661–3697.
Middleton, G. V., 1970, Experimental studies relating to problems of flysch sedimentation, *in* Lajoie, J. (ed.), Flysch sedimentology in North America: Geol. Assoc. Canada Special Paper 7, p. 253–272.
Mitchell, A. H., and Reading, H. G., 1969, Continental margins, geosynclines, and ocean floor spreading: Jour. Geology, v. 77, p. 629–646.
Natland, M. L., 1933, Depth and temperature distribution of some Recent and fossil Foraminifera in the southern California region: Scripps Inst. Oceanography Bull., Tech. Ser., v. 3, p. 225–230.
North, F. K., 1971, Alpine serpentinites, oceanic ridges, and continental drift: Geol. Mag., v. 108, p. 81–192.
Pantanelli, D., 1883, I diaspri della Toscana e loro fossili: Reale Accad. Nazl. Lincei, v. 7, p. 13–14 (not seen; *fide* Aubouin, 1965).
Patterson, Clair, 1964, Characteristics of lead isotope evolution on a continental scale in the earth, *in* Craig, H., and others (eds.), Isotopic and cosmic chemistry; Amsterdam, North Holland Publishing Co., p. 244–268.
Reading, H. G., 1972, Global tectonics and the genesis of flysch successions: 24th Internat. Geol. Cong. Proc., Sec. 6, p. 59–66.
Rodgers, John, 1968, The eastern edge of the North American continent during the Cambrian and Early Ordovician, *in* Zen, E., and others (eds.), Studies of Appalachian geology: northern and maritime: New York, Wiley-Inter-Science, p. 141–149.
Schuchert, Charles, 1910, Paleogeography of North America: Geol. Soc. America Bull., v. 20, p. 427–606.
Schwab, F. L., 1971, Geosynclinal compositions and the new global tectonics: Jour. Sed. Petrology, v. 41, p. 928–938.
Stanley, D. J., 1970, Flyschoid sedimentation on the outer Atlantic margin off northeast North America, *in* Lajoie, J. (ed.), Flysch sedimentology in North America: Geol. Assoc. Canada Special Paper 7, p. 179–210.
Stanley, K. O., Jordan, W. M., and Dott, R. H., Jr., 1971, New hypothesis of Early Jurassic paleogeography and sediment dispersal for western United States: Am. Assoc. Petroleum Geol. Bull., v. 55, p. 10–19.
Stauffer, P. H., 1967, Grain-flow deposits and their implications, Santa Ynez Mountains, California: Jour. Sed. Petrology, v. 37, p. 487–508.
Steinmann, G., 1905, Die geologische Bedeutung der Tiefseeabsätze und der ophiolithischen Massengesteine: Naturf. Gesell. Freiburg Ber., v. 16, p. 44–65.
Stille, Hans, 1936, Wege und Ergebnisse der geologisch-tectonischen Forschung: Gesell. Wiss. Förh., 25 Jahre Kaiser Wilhelm, Bd. 2, p. 84–85 (not seen; *fide* Kay, 1951).
———, 1941, Einfuhrung in den Bau Amerikas: Berlin, Borntraeger, 717 p. (not seen; *fide* Kay, 1951).
Storm, L. W., 1945, Resumé of facts and opinions on sedimentation in the Gulf Coast region of Texas and Louisiana: Am. Assoc. Petroleum Geol. Bull., v. 29, p. 1304–1335.
Suess, Eduard, 1909, Das Antlitz der Erde: Leipzig, Freytag, v. 3, pt. 2, 789 p.
Sullwold, H. H., Jr., 1960, Tarzana fan, deep submarine fan of late Miocene age, Los Angeles County, California: Am. Assoc. Petroleum Geol. Bull., v. 44, p. 433–457.
Taylor, F. B., 1910, Bearing of the Tertiary mountain belt on the origin of the earth's plan: Geol. Soc. America Bull., v. 21, p. 179–226.
Toulmin, Stephen, 1953, The philosophy of science: New York, Harper Torchbook, paper ed. (TB 513), 176 p.
Umbgrove, J. H. F., 1933, Verschillende typen van Tertiare geosynclinalen in den Indischen Archipel.: Leidsche Geol. Meded., v. 6, p. 33–43.
Van Bemmelen, R. W., 1949, The geology of Indonesia: The Hague, Nijhoff, 997 p.
Vening Meinesz, F. A., 1940, The earth's crust deformation in the East Indies: Proc. K. Nederlandsch Akad. Wetensch. Proc., ser. B, v. 43, p. 278–306.
Walker, R. G., 1966, Deep channels in turbidite-bearing formations: Am. Assoc. Petroleum Geol. Bull., v. 50, p. 1899–1917.
Wang, C. S., 1972, Geosynclines and the new global tectonics: Geol. Soc. America Bull., v. 83, p. 2105–2110.
Wells, F. G., 1949, Ensimatic and ensialic geosynclines (abs.): *ibid.*, v. 60, p. 1927.
Williams, H. S., 1897, On the southern Devonian formations: Am. Jour. Sci., v. 3, p. 393–403.
Wilson, J. T., 1949, The origin of continents and Precambrian history: Royal Soc. Canada Trans., v. 43, ser. 3, p. 157–182.

COLLAPSING CONTINENTAL RISES: ACTUALISTIC CONCEPT OF GEOSYNCLINES—A REVIEW

ROBERT S. DIETZ AND JOHN C. HOLDEN

NOAA, Atlantic Oceanographic & Meteorological Laboratories, Miami, Florida

ABSTRACT

In the 1950's, geosynclinal theory was dominated by the tectogene concept and Marshall Kay's synthesis. These and earlier concepts were derived from field study of tectonized geosynclines on land. In 1959 C. Drake, Maurice Ewing, and G. Sutton, applying the data of marine geophysics, recognized that sedimentary prisms now being laid down along the eastern margin of the United States may represent nascent miogeosynclines and eugeosynclines. They assumed that there is a close parallel with Kay's model and included in their interpretation a shelf-edge basement high that supposedly is equivalent to the tectonic borderland and, also, a toe of sialic crust underlying the continental rise that supposedly makes the rise ensialic. The eugeosyncline then would be elevated eventually to continental level largely by sialization of oceanic crust and without horizontal translation of the prism.

Between 1963 and 1967, we have developed what may be called an actualistic concept of geosynclines that is based upon sea-floor spreading and collapsing continental rises. This, too, was based upon Kay's model, except that gross surgery was applied. The seaward half of the miogeosyncline was deleted, as though it never existed and making it a wedge that thickened out, so to speak, like the modern terrace wedge. Also omitted was the tectonic borderland; instead, a continental slope was inserted between the miogeocline and eugeocline. (For simplicity and since none of these sedimentary prisms are really synclinal in form, we prefer the terms miogeocline and eugeocline.) In this model, the miogeoclinal sediments were deposited ensialically on a downflexing continental margin and the eugeoclinal sediments ensimatically on oceanic crust. There seemed to be insufficient reason to equate the shelf-edge basement high with a tectonic borderland or to insert a sialic toe beneath the continental rise. Tectonization was envisioned as the result of underthrusting of the continental margin (subduction), which collapsed the continental rise, magmatized it, and inserted allochthonous crust and mantle rock within the eugeocline.

Our model is explicitly concerned with the mio-eugeoclinal couplet of the Atlantic type, such as would form marginal to a rift ocean on the trailing edge of a drifting continent. With the rapid development of plate tectonics and especially with the recognition of opening and closing ocean basins, much sophistication has recently been added to geosynclinal theory by J. Dewey, J. Bird, A. Mitchell, H. Reading, W. R. Dickinson, and many others.

INTRODUCTION

The purpose of this paper is to review briefly the development of actualistic geosynclinal theory[1] in North America between the years 1951 to 1967 as was suggested by the symposium convener. These dates are not arbitrary, as 1951 is a landmark year in which Marshall Kay, to whom this conference is appropriately dedicated, published his classic monograph, *North American Geosynclines* (1951). A geologic revolution occurred commencing in about 1967 with the wholesale acceptance of what is now called plate tectonics. This, in turn, has remarkably affected geosynclinal theory and resulted in a clear acceptance of actualism based upon interaction of subduction zones with continental margins. The extent of this revolution probably can be assessed by reading the other papers in this volume. Although lacking prior opportunity to do this ourselves, we are certain that essentially all authors will embrace plate tectonics and actualism. A conference on geosynclines convened in, say, 1966 would have resulted in quite a different set of papers in which at least some of the authors would insist that there are no modern-day equivalents of Paleozoic and Precambrian geosynclines.

Space limitations prevent any review of pre-Kay concepts except to pay homage to Stille's (e.g., 1936, 1941) views on orthogeosynclines as being marginal to cratons and comprised of a mio-eugeosynclinal couplet and, also, to Haug's (1900) correct opinion that eugeosynclinal sediments were laid down in the bathyal zone, although by this he referred to water a few hundred meters deep and not to the abyssal ocean floor. Prior to 1950 it was unpopular to suppose that sediments were carried beyond what was called wave base and over the edge of the continental slope. We should also recall that one of Schuchert's (1925) "certain facts" of geology was that, not only did the Paleozoic borderland of Appalachia lie off the eastern United States, but the continent was surrounded by seven other

---

[1] *Editor's note.* The authors prefer "actualistic" to "uniformitarian" whenever modern sea-floor phenomena are used as analogues for interpreting ancient rocks (R. S. Dietz, written communication, May 10, 1973). See p. 1–2 for a discussion of the history and meaning of actualism.

borderlands as well. We can also recall the vogue for tectogenes both as great downfolds within the craton and as features related to trenches. For these last two concepts, let it be sufficient to say, *requiescat in pacem*. An excellent summary of geosynclinal theory prior to 1950 has been provided by Glaessner and Teichert (1947).

KAY'S CONCEPT OF GEOSYNCLINES

The need for brevity prevents us from reviewing Kay's (1951) fundamental work except to reproduce his plate 9, our figure 1, which quickly became the textbook example of a mio-eugeosynclinal couplet even though it was not Kay's intent that this be so (personal communication). It should be recalled that at that time

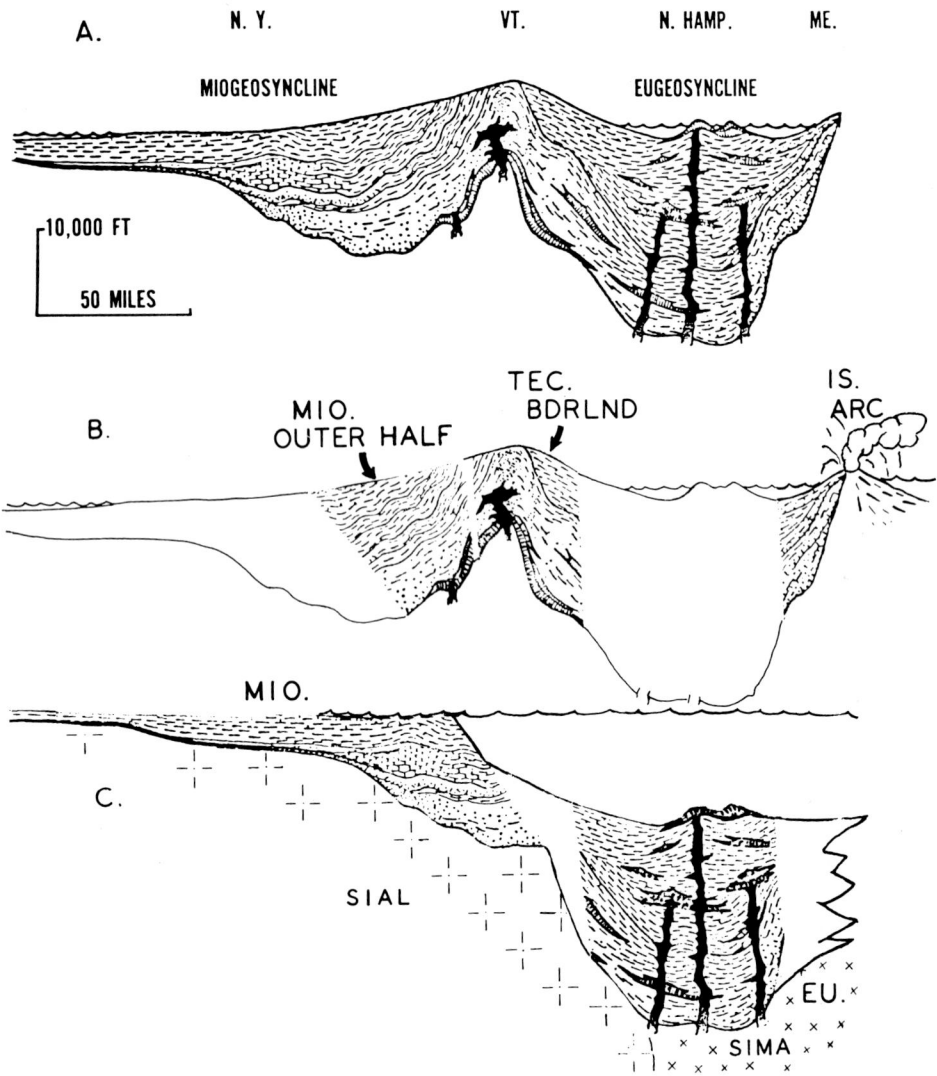

Fig. 1.—Kay's geosynclinal couplet. A drawing to show that, if three out of five elements are deleted from Kay's (1951) classical example of an ensialic mio-eugeosynclinal couplet, model is transformed into an ensialic-ensimatic actualistic geosynclinal couplet. Outer half of miogeosyncline, tectonic borderland, and island arc are eliminated; a continental slope is inserted, beyond which eugeocline is inserted. A, Mio-eugeosynclinal couplet along eastern North America palinspastically reconstructed as of mid-Ordovician when orogenesis began, according to Kay (1951); B, deleted elements, cut out of diagram A by scissors; C, new paste-up of mio-eugeoclinal couplet with new ensimatic eugeocline being downdropped along continental slope according to actualistic concept of geosynclines (Dietz, 1963a), by which pre-Middle Ordovician sedimentary prisms shown may be equated with sedimentary prisms along modern continental edge of eastern North America (adapted from Dietz and Sproll, 1968).

almost nothing was known about the ocean floor —almost nothing about its realms of sedimentation, nor was it even known that the ocean crust differed from continental crust inasmuch as the seismic refraction studies of Ewing and colleagues were just commencing. In fact, many then-extant misconceptions can now be dismissed with a smile. Considering the state of the art, we may conclude that Kay's synthesis showed remarkable insight. He, for example, recognized the marginal position of most orthogeosynclines and the possible construction of the continents by the accretion of geosynclinal foldbelts. On the other hand, the continental terrace sedimentary accumulations of the modern east coast and Gulf Coast were regarded, not as miogeosynclines, but as a new type of geosyncline, the paraliageosyncline, having uncertain affinities with miogeosynclines. Kay was naturally unaware of the enormous sedimentary prisms that lay at the base of modern continental slopes. While not subscribing to the then-usual belief that all geosynclines were necessarily ensialic, Kay made few inferences about the nature of the crust underlying geosynclines. The nature of this crust, although basic to a complete understanding of geosynclines, is nowhere shown on his figures. All of Kay's eugeosynclines were palinspastically reconstructed to lie at sea level with no indication of the position of that most important of all topographic boundaries, the continental slope. The role of turbidity currents in transporting sediments to the deep sea, so ably championed by Ph. H. Kuenen, was then only just beginning to be recognized. In figure 1 we also show how we suppose Kay's concept, with some modification, can be adapted to our actualistic concept of geosynclines.

MODERN ATLANTIC CONTINENTAL MARGIN
AND GEOSYNCLINES

The paper on continental margins and geosynclines by Drake and others (1959) was another important landmark in North American geosynclinal theory. The authors attempted to equate Kay's concept and his palinspastic reconstruction of the Paleozoic Appalachian orthogeosyncline (an inner miogeosyncline and an outer eugeosyncline) with the modern mid-Mesozoic to Recent shelf and continental rise prisms (fig. 2). Thus, it was an actualistic approach using the guiding precept that the present is the key to the past. Kay's example called for a geanticlinal barrier or tectonic borderland separating the miogeosyncline from the eugeosyncline. Drake and others, in turn, identified by seismic methods shelf-edge basement highs along the eastern United States and cited examples from other parts of the world that they supposed to be geanticlinal barriers in the sense of Kay. We doubt that on the basis of the seismic data available today one can argue any longer that such shelf-edge subsurface highs are typical of most shelves. In any event, as Burk (1968) has emphasized, buried shelf-edge highs may have many origins, some of which are nontectonic (fig. 3).

Drake and others (1959) then equated the continental rise prism with Kay's Appalachian eugeosyncline. They looked for evidences of volcanism within the prism, the hallmark of a eugeosyncline according to Stille, with only doubtful success. It was not recognized that an island-arc stage might appear later or that oceanic crust volcanics might be allochthonous, having been intercalated by sea-floor spreading. Nevertheless, their recognition of the continental rise as a nascent eugeosynclinal foldbelt was certainly a great step forward in actualism.

An interesting aspect of the continental rises of Drake and others (1959) with which we later disagreed (Dietz, 1963a) is their sialic underlining, termed transitional crust in the text but shown as sial rather than sima in the interpreted seismic profiles. We suppose that this view, which appears to be optional insofar as seismic velocities are concerned, was in accord with the tenor of the times, which held that both miogeosynclinal and eugeosynclinal prisms were laid down on sialic foundations. Also, the view fitted well with the idea that sialization of the upper mantle must eventually follow in order that the crustal thickness beneath continental rise prisms be increased to the continental thickness of about 35 km. From these beginnings, the concept of a sialic toe extending from the continental slope into the deep sea became commonplace a decade ago, especially among scientists of the Lamont Geological Observatory, and is still widely held. Heezen's and others' (1959) stylized section across the Atlantic Ocean strongly emphasizes both the shelf-edge anticline and the continental rise sialic toe (fig. 4). It is noteworthy that Dewey and Bird (1970, 1971) have adapted this sialic underliner to their synthesis of geosynclines and mountain building. In their view, however, the toe is not newly formed sial; rather, it is a remnant of continental rifting in which the pullapart was not a clean break but involved a taffylike necking and thinning. Earlier Hsü (1965) relied heavily upon this sialic toe as evidence of crustal thinning and for the eventual disappearance of Appalachian-type sialic masses.

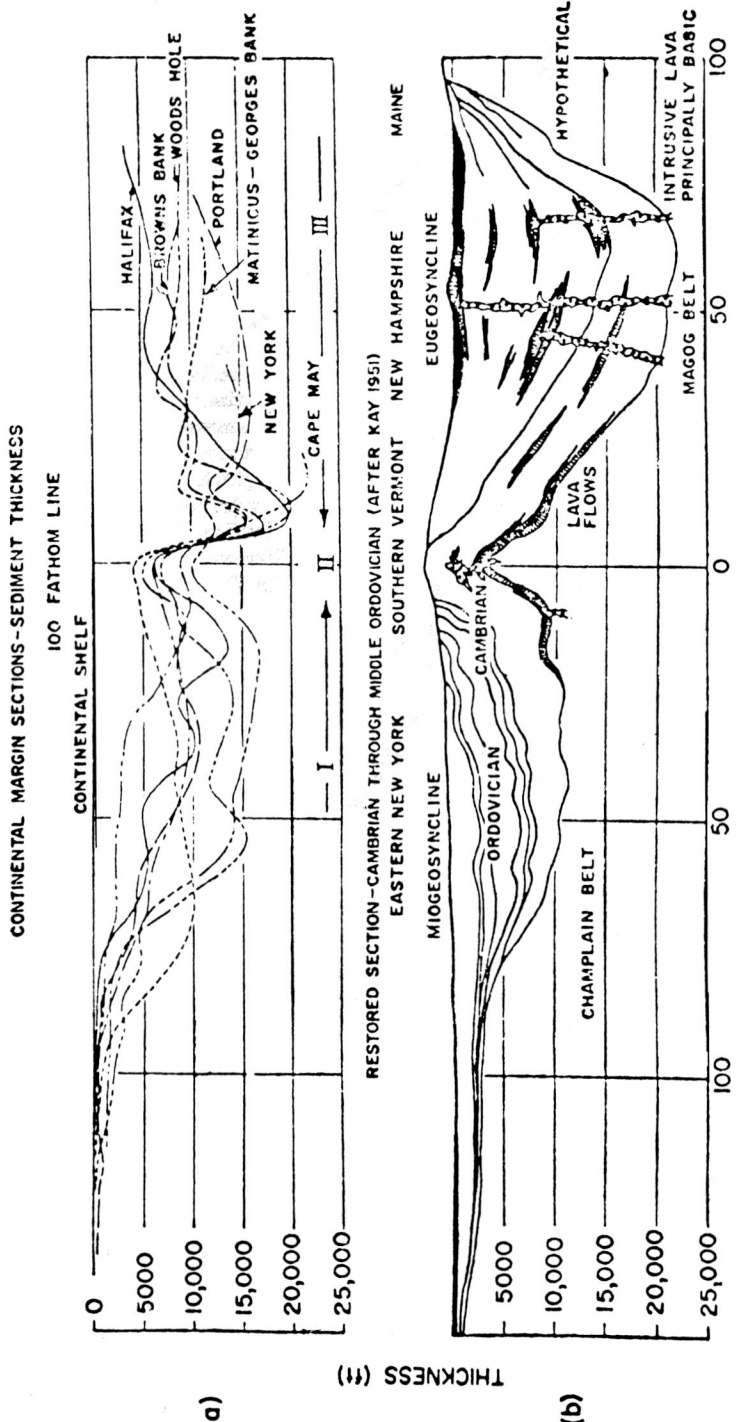

FIG. 2.—Paleozoic and modern geosynclines. This concept of Drake and others (1959) attempted to equate modern isopachous sediment sections off east coast of North America as revealed by seismic refraction with that of restored Middle Ordovician geosynclinal section of Appalachian Geosyncline of Kay (1951) (after Drake and others, 1959).

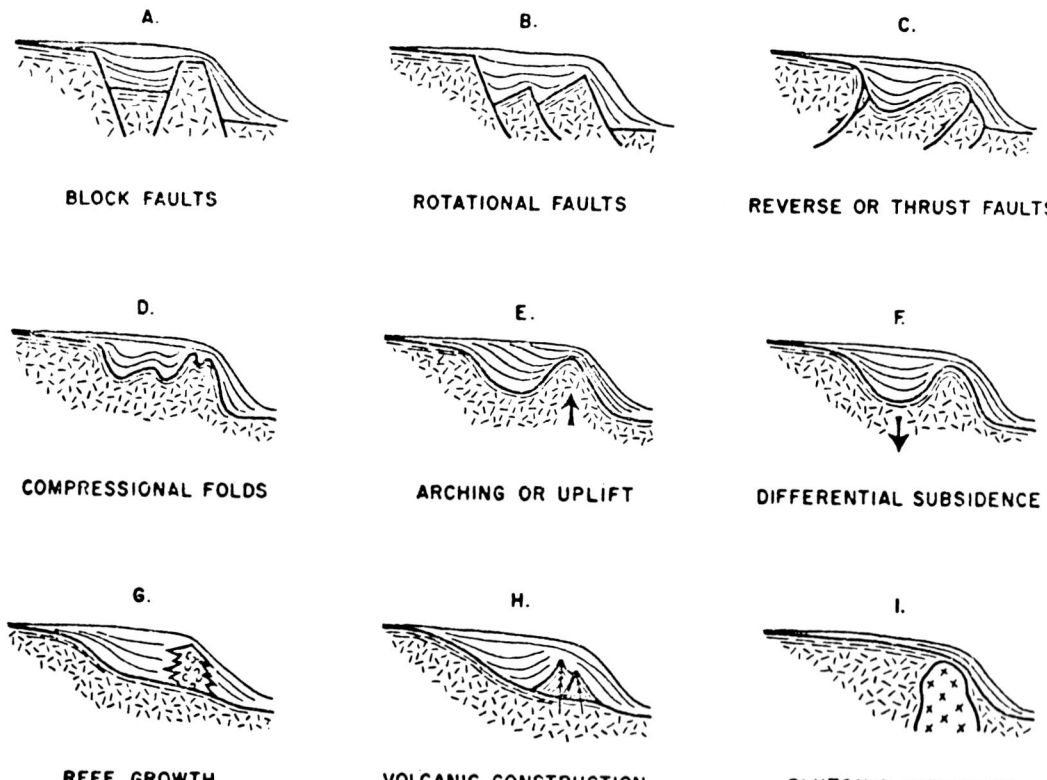

Fig. 3.—Shelf-margin ridges. Buried outer ridges discovered by seismic methods along some continental margins may have many origins, some of which are shown in this diagram from Burk (1968). Unless they deform the miogeoclinal wedge, indicating that they were tectonically active during its deposition, these highs cannot be considered as tectonic borderlands in sense of Kay (1951). This pertains to those highs off the eastern United States where an undeformed Jurassic-to-Recent wedge overlaps shelf-edge buried ridge, which must, therefore, be at least as old as Jurassic. One possible explanation of the east coast ridge is that it is the outer flank of a graben associated with initial continental rifting. More likely explanation is that ridge is an Early Cretaceous shelf-ridge reef, an extension of that known from the Gulf Coast, west margin of Florida, and Bahama platform. Continental drift reconstructions reveal that eastern seaboard was then nearer equator than now (Dietz and Holden, 1970).

MIOGEOSYNCLINES AND EUGEOCLINES

Several years ago we (Dietz and Holden, 1966) proposed the term *miogeocline* as a substitute word for *miogeosyncline*. This was partly in the interest of simplicity but, more importantly, to emphasize that miogeosynclines are seaward-thickening prisms of shallow-water sediments laid down mostly above surf base. Thus, miogeosynclines comprise only half of, or one limb of, a syncline. We pointed out that all folded and cratonized miogeosynclines seem to have this aspect. Further, it seemed unlikely to us that the outer limb of the miogeosyncline was lost by uplift, thrusting, and subsequent erosion, as some writers argued, in such manner that the outer limb would be lost from the geologic record. With our model it was possible to equate directly ancient miogeosynclinal prisms with such modern shelf-terrace prisms as that capping the coastal plain and continental shelf of the eastern United States (fig. 5).

The term miogeocline now has become widely accepted, as has the companion word *eugeocline* as well. We believe that this is proper, especially for those uses in which the plate-tectonic concept of geosynclines is implied. Certainly, if we equate eugeoclines with the continental rise prism, the sedimentary body is wedge shaped as well, but in this event the thickening is toward the continental slope. Only when the miogeocline and eugeocline are placed together as a couplet is the resulting sedimentary body synclinal in form (fig. 1).

ACTUALISTIC CONCEPT OF GEOSYNCLINES AND MOUNTAIN BUILDING

In 1963 we (Dietz, 1963a) took issue with previous thoughts on the geosynclinal cycles as

COLLAPSING CONTINENTAL RISES

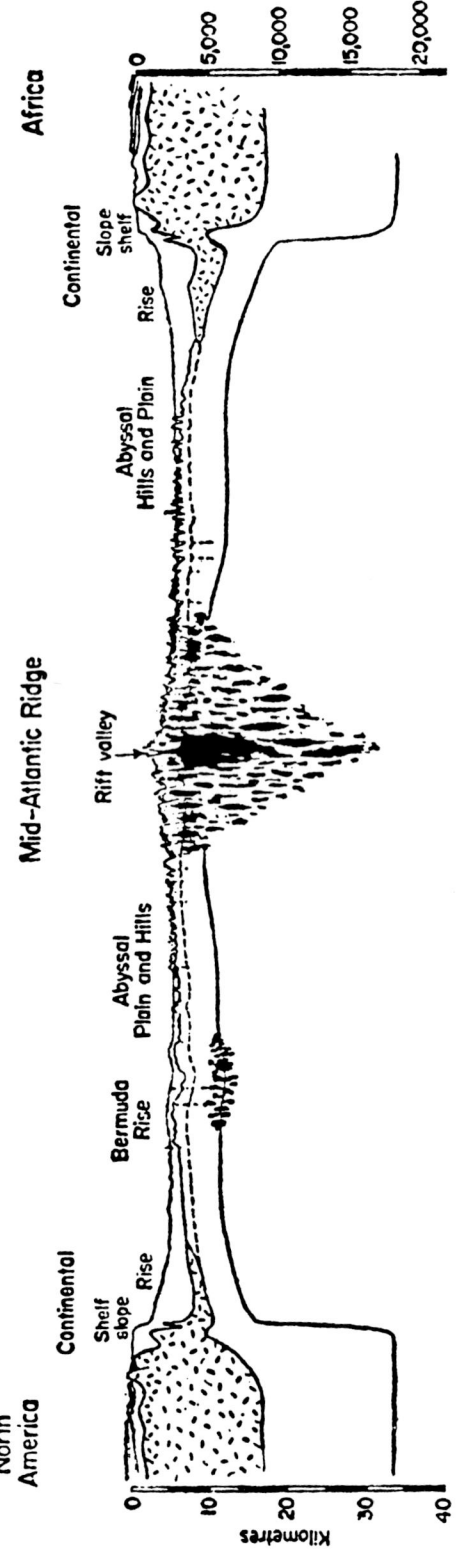

FIG. 4.—Generalized crustal section across North Atlantic according to Heezen and others (1959), reproduced here to show general emphasis formerly placed by Lamont Geological Observatory scientists upon a shelf-edge ridge and upon continental rise being underlain by a shelf-edge sialic toe. These features are not inferred to be present in similar section across the central Atlantic Ocean from Brazil to Africa by Leyden (1972).

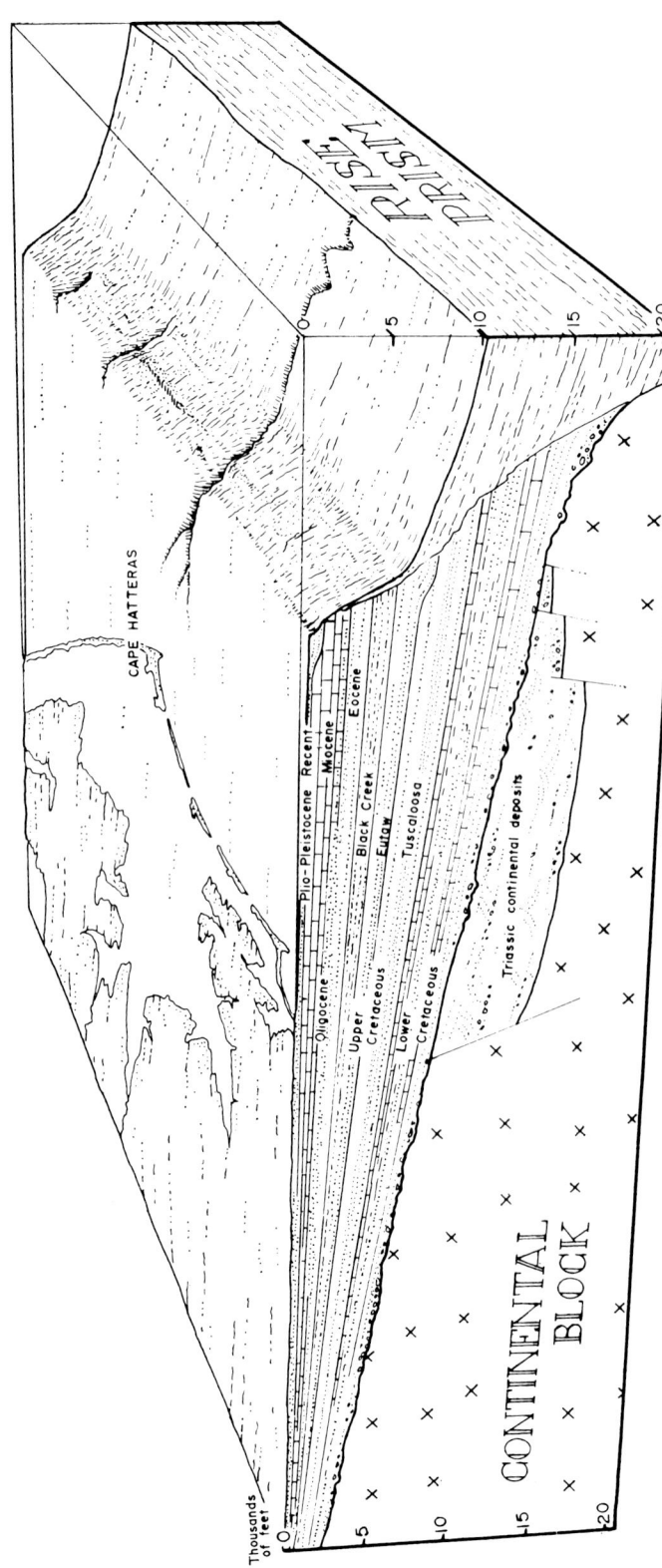

FIG. 5.—Modern miogeocline. Simplified block diagram of modern continental terrace wedge or miogeocline off eastern United States in vicinity of Cape Hatteras (from Dietz and Holden, 1966). Cretaceous and younger strata dip monoclinally seaward and thicken at continental slope. Wedge is laid down on a Lower Cretaceous peneplain, and basement rocks presumably are composed of intrusives and metasediments of crystalline Appalachian foldbelt. Triassic and Jurassic grabens presumably are also present, affecting basement associated with rifting of North Africa from North America about 180 my bp. Graben shown here is simply inferred for diagrammatic purposes. Stratigraphic data for the section has been adapted from Spangler (1950) and Heezen and others (1959).

applied to the eastern margin of North America. Utilizing Hess' (1962) concept, sea-floor spreading, translation of the ocean floor, and underthrusting (subduction) at continental margins was proposed as the driving mechanism. This was termed "an actualistic concept of geosynclines and mountain building," and continental rise prisms were regarded as nascent eugeoclines. The proposed model is best explained in a series of drawings (figs. 6–7), but a few words of explanation are needed as well.

Sedimentation phase.—Along the trailing edge of a drifting continent like North America, a large prism of terrigenous sediments accumulates at the base of a continental slope and builds a continental rise. These sediments would be of the poured-in type, carried down submarine canyons and deposited as turbidites, or, in the broad sense, would be equivalents of flysch. As it grows, this prism, a eugeocline, slowly subsides isostatically, causing downwarping of the adjacent sialic continental margin. Prograding paralic deposits build up a monoclinal wedge of shallow-water miogeoclinal deposits on this marginal flexure. Because sedimentation causes the subsidence, this entire geosynclinal development is gravitationally induced (Dietz, 1963a). We erred in 1963 in not recognizing the important role of lithospheric cooling (Sclater and Francheteau, 1970), which now appears to be an important cause of continental margin subsidence. From the simple dike-injection concept of sea-floor spreading, it is not immediately clear why the sialic craton should subside rather than only the ocean floor being created at a level of a few kilometers lower than the sial. Nevertheless, it appears, from the history of high Africa and from the arching of the sialic flanks of the Red Sea, that continental uplift and crustal thinning by erosion precedes continental rifting and the appearance of a midocean ridge. Mantle plumes and associated so-called hot spots may be involved. However, we should not entirely set aside regional isostatic downbowing owing to sediment loading as a companion cause. A recent analysis by Walcott (1972) found this effect to be both real and important.

Orogenic phase—

"Orogeny is ushered in by the simatic ocean floor moving toward and underthrusting the buoyant continent—presumably by the mechanism of sea-floor spreading. The eugeosyncline is compressed, folded, thrust, magmatized and metamorphosed. Ultrabasics, from the old sea floor upon which the sediments were deposited, are caught up in the folding. The miogeosyncline is also affected, but to a milder extent, as it is resting on tectonically passive sial— only the sima is active. . . . The eugeosyncline is accreted to the continent and becomes an intrinsic part of it—its outer margin forms a new continental slope" (Dietz, 1963a).

Late and postorogenic phase—

"The sea floor continues to underthrust, so now a trench forms at the continental margin. The sea floor, including its sedimentary layers and any detritus poured into the trench, is mostly carried beneath the continental raft and granitized. . . . Granite batholiths invade the continental margin, adding buoyancy and causing diapiric and general uplift of alpine mountains. Eventually the sea-floor thrusting ceases and the mountains are eroded. With erosion, further isostatic uplift occurs, but eventually a congealed and stable craton results. The stage is set once more for the sedimentation of a new continental rise and eventual development of new marginal orthogeosynclines" (Dietz, 1963a).

It is interesting to recall that the following five views, to which many of the arguments on the actualistic geosynclinal concept were addressed, were regarded as unorthodox at the time: (1) that deep-sea sediments, even as tectonized metasediments, are ever found on continents (permanency of continents and ocean basins); (2) that large amounts of terrigenous sediments reach the deep ocean floor and that the continental rise prism is a giant, isostatically downbowed turbidite prism; (3) that fragments of the ocean crust and upper mantle appear in eugeoclines so that it was not really necessary to drill a Mohole to obtain such samples—we have been unable to discover any friend of the defunct Mohole Project for several years now; (4) that eugeoclines are generally ensimatic and only the miogeoclines are ensialic; and (5) that miogeosynclinal sediments, prior to exogeosynclinal deposition, were derived from the landward side and that they thicken *away* from their source. Many recent papers on geosynclinal theory that embrace sea-floor spreading and plate tectonics seem now to use these views as explicit or implicit premises. Some of our own thoughts on geosynclines and their associated realms of sedimentation have been developed in several papers (e.g., Dietz, 1963b, 1963c, 1964, 1966; Dietz and Holden, 1966, 1967).

CONCLUDING REMARKS

In this brief review, no attempt has been made to bring the subject up to date inasmuch as the geosynclinal concept is now in a state of ferment occasioned by the advent of plate tectonics. Excellent treatments have recently been

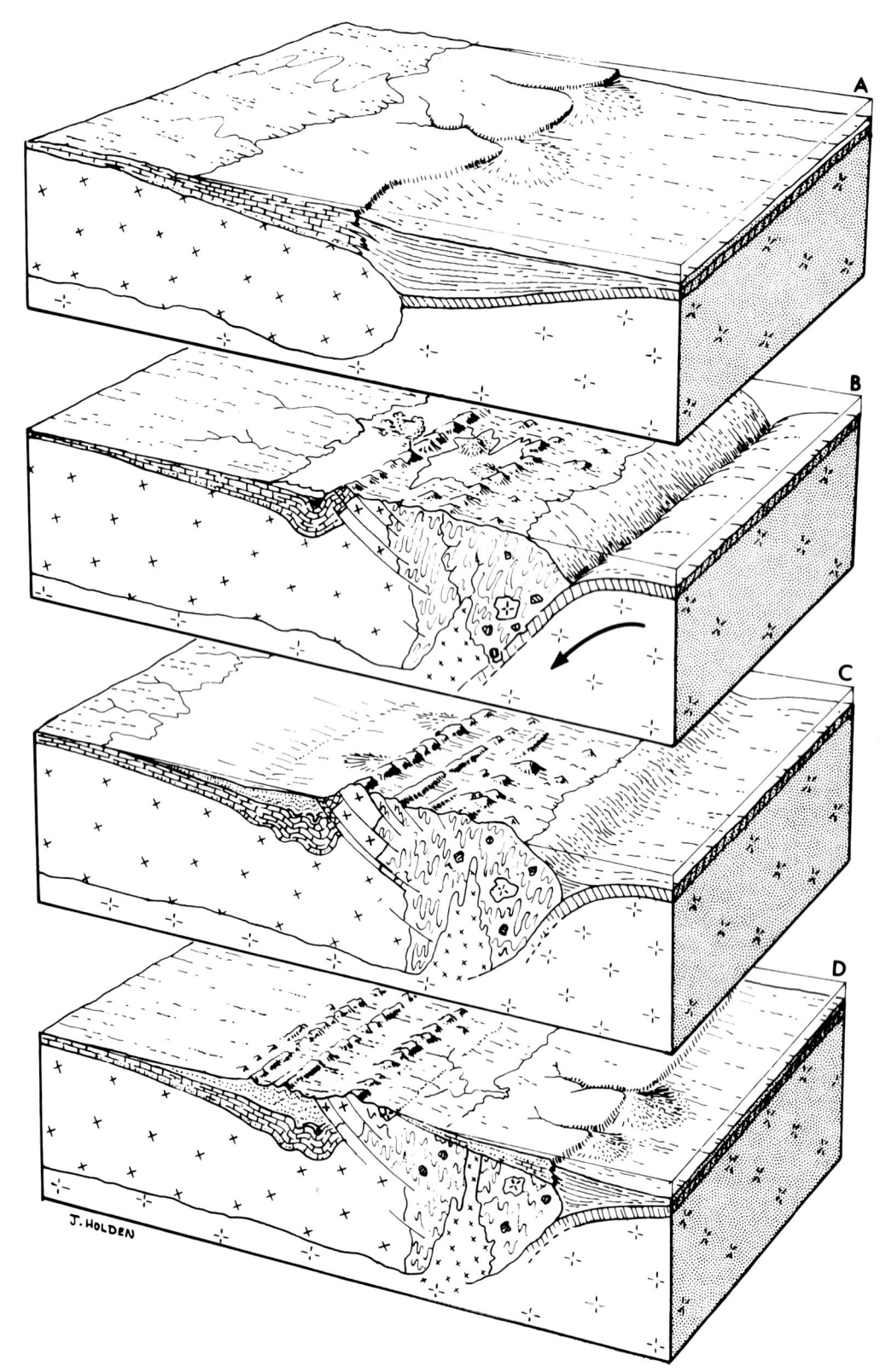

provided, for example, by Mitchell and Reading (1969), Dickinson (1970) and by Dewey and Bird (1970). This review also has been explicitly limited to the mio-eugeoclinal couplet such as would form along the trailing edge of a drifting continent. In such a regime, the margin is stable and is of the Atlantic type except that the subsidence is caused by regional isostatic compensation owing to the load of the continental rise prism. Added to this is lithospheric subsidence associated with cooling as newly formed ocean crust moves away from its place of origin at the midocean rift. Of course, it takes the creation of a new subduction zone to convert an opening ocean into a closing ocean. Such an event is also required to collapse a continental rise, a nascent eugeocline, into a folded eugeocline as a mountainous foldbelt or orogen.

Let it be sufficient to say, many and varied scenarios are possible along a trench ocean of the Pacific type having margins that are being elevated and frequently subjected to transcurrent faulting, especially where the strike-slip component of subduction is taken up, not in the trench axis, but within the arc-trench gap. For example, if the Franciscan mélange is composed of collapsed trench turbidites and large admixtures of skimmed-off oceanic crust and if the Great Valley Sequence was laid down in an arc-trench gap, these circumstances suggest a facies quite different from that resulting from the initial collapse of a mature continental rise of the Atlantic type.

The concept of actualistic geosynclines wherein sediments are deposited on the ocean floor and then accreted to continental margins by plate tectonics satisfactorily explains how sedimentary prisms are transformed into folded mountains. The close relationship between eugeoclines and foldbelts is not one of cause and effect, but one of jeopardy of position—sediments laid down on the ocean floor are returned to the continental margin by sea-floor spreading. An active continental margin is the locus of interaction between continents and subduction zones.

Considering the complex history of geosynclines, one may ask, should we not scrap this term, as it only leads to confusion? This, of course, is not our decision to make, as terms will thrive or fall depending upon their usefulness in geologic communication. We suppose that the term, together with its plethora of Greek prefixes, will remain as a basic concept in geology but also that a new classification will develop. Most significantly, such a classification will be based upon the world today and upon the fate of sedimentary deposits laid down at continental margins within the framework of plate tectonics. Geologists will no longer simply attempt palinspastically to erect nascent geosynclines from the *corpus delicti,* ancient, particularly Paleozoic foldbelts—at least not as type examples. Rather, we will compare these tectonized deposits to modern sedimentary prisms. This is the essence of actualism.

REFERENCES

BURK, C. A., 1968, Buried ridges with continental margins; New York Acad. Sci. Trans., ser. 2, v. 30, p. 397–409.
DEWEY, J., AND BIRD, J., 1970), Mountain belts and the new global tectonics: Jour. Geophys. Research, v. 75, p. 2625–2647.
————, AND ————, 1971, Origin and emplacement of ophiolite suite: Appalachian ophiolites in Newfoundland: *Ibid.,* v. 76, p. 3179–3205.
DICKINSON, W. R., 1970, Plate tectonic models of geosynclines: Earth and Planetary Sci. Letters, v. 10, p. 165–174.
DIETZ, R. S., 1963a, Collapsing continental rises: an actualistic concept of geosynclines and mountain building: Jour. Geology, v. 71, p. 314–333.

FIG. 6.—Collapsing continental rises. Miogeoclinal sediments deposited as marginal wedge on craton rather than in intracratonic trough (i.e., behind a shelf-edge ridge). A, Initially (late Precambrian to Middle Ordovician), miogeoclinal sediments are deposited on a marginally flexing continental edge similar to that now developing on the continental shelf off eastern United States; B, Orogeny collapses continental rise in Late Ordovician, terminating true miogeoclinal deposition and initiating flysch deposition (e.g., Martinsburg), within marginal cratonic basins or exogeosynclines; C, continued orogeny and further subsidence of miogeocline wedge and molasse deposition; D, Appearance as of today but with shoreline showing maximum incursion of prograding sea over coastal plain; not shown is probable collision of Africa with North America and formation of subduction zone at first in open Atlantic with an associated island followed by migration of this arc to continental margin. (From Dietz and Holden, 1966.)

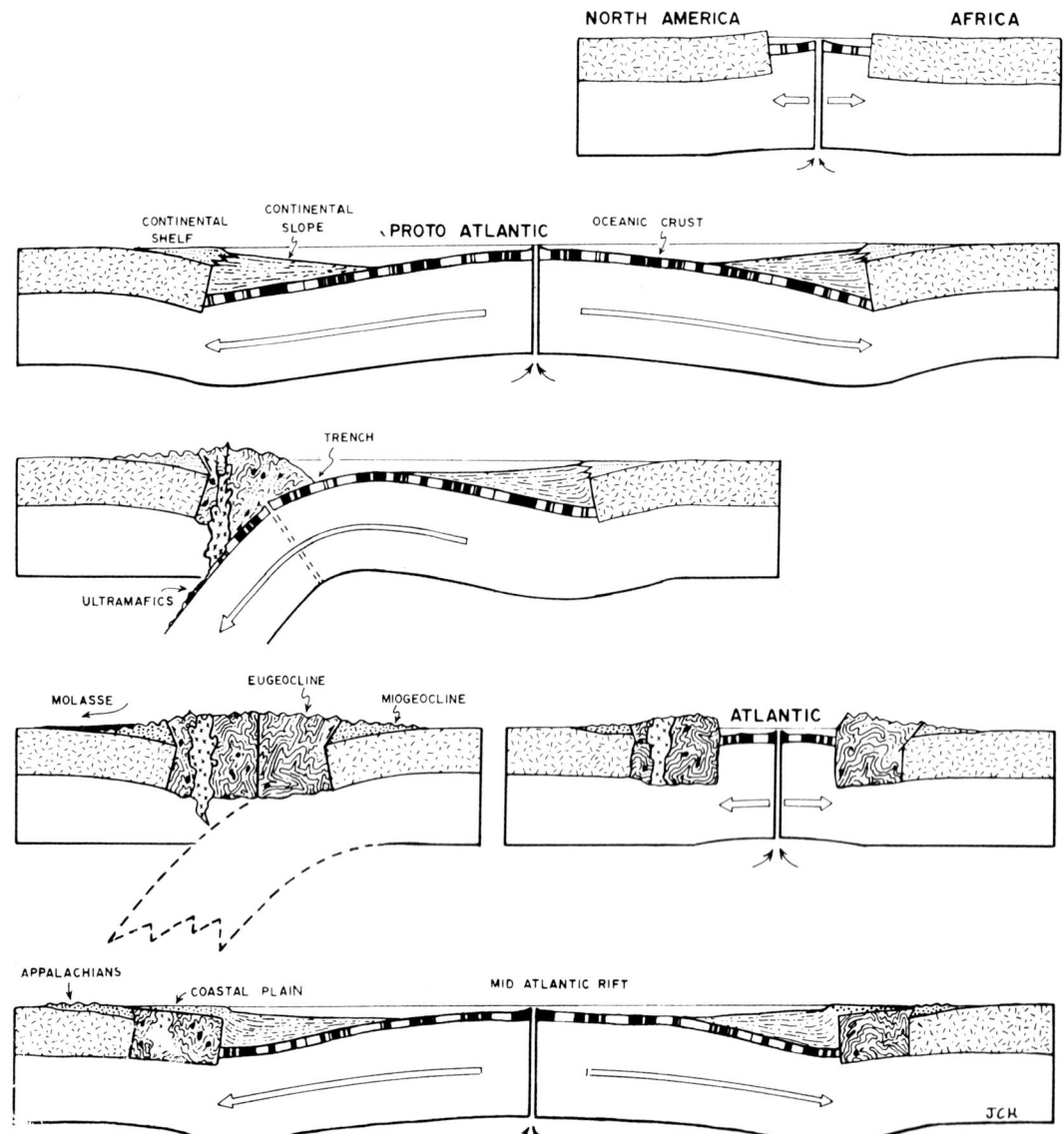

Fig. 7.—Closing and reopening of Atlantic Ocean. Opening, closing, and reopening of Atlantic Ocean with accompanying formation of marginal foldbelts is depicted (modified from Dietz, 1972). In late Precambrian, North America and Africa were split apart by spreading rift, which inserts a new ocean basin, the proto-Atlantic. By process of sea-floor spreading, ancestral Atlantic Ocean opens. New oceanic crust is created as lithospheric plates move apart. Normal and reversed magnetic anomalies diagrammatically shown within oceanic crust. On margins of each continent, sediments produce orthogeosynclinal couplet: miogeocline on continental shelf and eugeocline of adjacent ocean floor. Ancestral (Paleozoic) Atlantic next begins to close. Lithosphere breaks, forming a new plate boundary, and a subduction zone is produced as lithosphere descends into earth's mantle and is resorbed. Consequent subduction collapses eugeocline, creating Appalachian foldbelt. Eugeocline is intruded with ascending magmas that create granodioritic plutons and andesitic volcanoes. Proto-Atlantic then fully closed. Opposing cratons, each carrying a geosynclinal couplet, are sutured together, leaving only a transform fault. Suture is locus of squeezed-up pods of ultramafic mantle rock. Sediments eroded from mountainous foldbelt create deltas and fluvial deposits, that is, molasse. About 180 million years ago, present Atlantic reopened near old suture line. Today, central North Atlantic is opening at a rate of nearly 3 cm/yr, and new marginal miogeocline-eugeocline couplets are being deposited.

———, 1963b, Wave base, marine profile of equilibrium, and wave-built terraces: Geol. Soc. America Bull., v. 74, p. 971–990.
———, 1963c, Alpine serpentines as oceanic rind fragments: *ibid.*, p. 947–953.
———, 1964, Origin of continental slopes: Am. Scientist, v. 52, p. 50–69.
———, 1966, Passive continents, spreading ocean floors, and continental rises: Am. Jour. Sci., v. 246, p. 177–193.
———, 1972, Geosynclines, mountains and continent building: Sci. American, v. 226, p. 30–38.
———, AND HOLDEN, J. C., 1966, Miogeoclines (miogeosynclines) in space and time: Jour. Geology, v. 74, pt. 1, p. 566–583.
———, AND ———, 1967, Deep sea sediments *in* but not *on* continents: Am. Assoc. Petroleum Geologists Bull., v. 50, p. 351–362.
———, AND ———, 1970, The breakup and dispersion of continents: Permian to present: Jour. Geophys. Research, v. 75, p. 4939–4956.
———, AND SPROLL, W., 1968, Miogeoclines (miogeosynclines) in space and time: a reply: Jour. Geology, v. 76, 1, p. 113–116.
DRAKE, C., EWING, M., AND SUTTON, G., 1959, Continental margins and geosynclines: the east coast of North America, north of Cape Hatteras, *in* Physics and chemistry of earth: London, Pergamon Press, v. 3, p. 110–198.
GLAESSNER, M., AND TEICHERT, C., 1947, Geosynclines: a fundamental concept in geology: Am. Jour. Sci., v. 245, p. 465–482, 571–591.
HAUG, E., 1900, Les geosynclineaux et les Aires Continentales; Soc. géol. France, v. 26, p. 617–711.
HESS, H. H., 1962, History of ocean basins, *in* ENGEL, A., AND OTHERS (eds.), Petrologic studies: Boulder, Colorado, Geol. Soc. America, A. F. Buddington vol., p. 599–620.
HEEZEN, B. C., THARP, M., AND EWING, M., 1959, The floors of the oceans: I. North Atlantic: *ibid.*, Special Paper 65, 122 p.
HSÜ, K. J., 1965, Isostasy, crustal thinning, mantle changes, and the disappearance of ancient land masses: Am. Jour. Sci., v. 263, p. 97–109.
KAY, M., 1951, North American geosynclines: Geol. Soc. America Mem. 48, 143 p.
LEYDEN, R., SHERIDAN, R., AND EWING, M., 1972, Seismic refraction section across the equatorial Atlantic: Eos, v. 53, p. 171–173.
MITCHELL, A., AND READING, H., 1969, Continental margins, geosynclines and ocean floor spreading: Jour. Geology, v. 77, p. 629–646.
SCHUCHERT, C., 1925, Sites and nature of North American geosynclines: Geol. Soc. America Bull., v. 34, p. 151–230.
SCLATER, J., AND FRANCHETEAU, J., 1970, The implications of terrestrial heat flow observations on current tectonic and geochemical models of the crust and upper mantle: Royal Astron. Soc. Geophysics Jour., no. 20, p. 509–542.
SPANGLER, W. B., 1950, Subsurface geology of Atlantic coastal plain; Am. Assoc. Petroleum Geologists Bull., v. 34, p. 2054–2060.
STILLE, H., 1936, Wege und Ergebnisse der geologisch-tectonischen Forschung: (Kaiser-Wilhelm Gessel. Förh. Wiss. Festschr., v. 2.
———, 1941, Einführung in den Bau Amerikas: Berlin, Borntraeger.
WALCOTT, R., 1972, Gravity, flexure, and the growth of sedimentary basins at a continental edge: Geol. Soc. America Bull., v. 83, p. 1845–1848.

CONTINENTAL TERRACES AND RELATED CRATONIC ACCUMULATIONS

MIDPLATE CONTINENTAL MARGIN GEOSYNCLINES: GROWTH PROCESSES AND QUATERNARY MODIFICATIONS

DAVID G. MOORE AND JOSEPH R. CURRAY

Naval Undersea Center and Scripps Institution of Oceanography, San Diego and La Jolla, California

ABSTRACT

Multiple rapid sea level changes of the past two million years have resulted in major changes in the processes and structures that prevailed for tens of millions of years of pre-Quaternary time during the building of massive midplate continental margins. These changes have hindered interpretations of pre-Quaternary, or what may be called normal, growth processes and structures to the extent that fallacious concepts of the genetic processes have evolved. We submit that an idealized non-Quaternary midplate continental margin would have a shore zone of prograding beaches but would be devoid of estuaries. Barriers and lagoons would exist primarily in association with delta complexes. Muddy shelf facies would prevail, as no mechanism exists for moving sands out of the shore zone to the shelf in significant quantities. Slopes would be largely smooth and prograding by slow pelagic, hemipelagic, and low-density turbid layer deposition. Erosion by turbidity currents and slumping would be uncommon and largely restricted to the vicinity of a few large long-lived submarine canyons. Axiomatically, the rise would be poorly developed and appear lobate in the form of discrete deep-sea fans off the canyon mouths. Terrigenous turbidites would be mainly confined to these fans.

INTRODUCTION

Some of the best known midplate continental margins were formed during Mesozoic time with the rifting apart of super continents. They have evolved with the addition of new crust at trailing plate edges, which formed the seaward limbs of the new continental margins. For tens of millions of years following their early evolution and until late Tertiary time, most parts of these margins subsided more or less continually as the lithosphere aged (Menard, 1969). With subsidence, interior drainage of the bisected continental masses was diverted toward the youthful continental margins and brought sediment that in turn assisted subsidence by isostatic loading.

Some of these midplate continental margins have accumulated sufficient volumes of sediment that we consider them to be geosynclinal. By the analogy drawn by Drake and others (1959) and Dietz and Holden (1966), the section underlying the continental terrace (coastal plain, shelf, and slope) might be considered to be a modern miogeosyncline (miogeocline), and the section underlying the continental rise might be considered to be one possible type of modern eugeosyncline (eugeocline). Other possible analogues exist, especially for the modern eugeosyncline, but we prefer not to enter the controversy. It is more important to describe and understand modern geosynclinal accumulations of sediment than to engage in semantic debates on terminology based on poorly understood ancient sediment accumulations. Our usage will be to consider midplate continental margin accumulations of sediment to be geosynclines when they contain very large volumes of sediment. These midplate margin accumulations, which form under nonorogenic conditions, may later become incorporated in orogenic belts.

Sea level changes brought about by epeirogenetic movements, or even by eustatic changes, and induced by ocean-basin volume variations were very slow during pre-Quaternary time, and sedimentological equilibrium between environment and deposit must have prevailed. Quaternary glaciation and rapid eustatic sea level changes drastically disrupted the prevailing balance of processes and resultant deposition. Long-established depositional environments were subject to violent erosion, and, conversely, very rapid accumulation of sediments, which were deposited from turbidity currents, occurred in regions long accustomed to slow pelagic rain. These profound alterations of established processes and environments have hindered interpretation of important pre-Quaternary, or what we prefer to call normal, genetic processes and structure and have in some instances led to fallacious concepts.

The purpose of this paper, then, is to examine several aspects of long-term midplate continental margin growth that we believe have been dramatically modified or even completely changed by Quaternary processes and to suggest models of pre-Quaternary growth. Most of the Quaternary modifications we discuss have been recognized by ourselves and other authors previously, but we feel that the nature of pre-Quaternary, or normal, processes and structure as they existed prior to modification have been insufficiently stressed.

Our discussion will be primarily concerned with regions of dominantly terrigenous sedimentation. We recognize that pre-Quaternary biogenous shelf sands may have been normal in some regions.

ACKNOWLEDGMENT

This contribution from the Scripps Institution of Oceanography was supported by budgets of the Naval Undersea Center and the Office of Naval Research.

QUATERNARY CHANGES IN SHELF, SLOPE, AND RISE ENVIRONMENTS

Some important aspects of modern midplate continental margins that are vestiges of Quaternary modifications include the following:

Continental Shelf and Shore Zone:
1. Prevalence of estuaries, barrier-lagoon systems, and young deltas in the shore zone
2. Uniform depth of shelves
3. Widespread exposure of terrigenous sands on the shelf

Shelf-Slope Transition:
1. Uniform depth
2. Sharpness of shelf break

Continental Slope:
1. Ubiquity of gullies and canyons that commonly have heads at the shelf edge and distal ends at the base of the slope
2. Topography and structure indicating mass movements of large and small scale
3. Common exposures of Tertiary and older strata

Continental Rise:
1. Topographic continuity
2. Prevalence of structure produced by turbidity current channels and slumps

Periods of glaciation in the Quaternary have resulted in and been accompanied by an indeterminate number of sea level fluctuations (fig. 1). The magnitude and chronology of the changes are controversial, as only relative sea level changes can be observed, and distinction between eustatic and tectonic causes is a matter of interpretation. For present purposes, two factors emerge as being much more important than an absolute chronology of sea level changes. First is the fact that there have been numerous large-amplitude and geologically rapid alterations of sea level during the Quaternary period. The second factor is that most published curves for late Quaternary sea level show a pronounced decrease in the rate of rise at about 7,000 years bp.

Shelf and shore zone.—Estuaries, barrier islands, lagoons, and deltas in our modern shore zone environments began to form only within the past 7,000 years during the time of the slower sea level rise. Shore-zone environments continue today to evolve rapidly, and many estuaries and lagoons have already been filled and have become part of the coastal plain. Others are filling rapidly, and projections into the future suggest that a high percentage of these features in the world may be filled within the next few hundred to few thousands of years. Thus, if we consider the normal, slow epeirogenic subsidence of an idealized midplate continental margin, it appears probable that topographic irregularities would be quickly smothered by sedimentation. Estuaries probably could not exist, and barrier and lagoon complexes could exist perhaps only in specialized environments associated with large deltas. Rather than estuaries and barrier-lagoon complexes, we would expect that under normal, slowly subsiding conditions, the shore zone would be largely sandy beaches and dunes and that beach ridges might offer a principal means of progradation of the shore zone sand bodies.

The morphology of present shelves was most certainly strongly modified, if not largely shaped, by erosion and deposition associated with the repeated transgressions and regressions of the Pleistocene. Worldwide statistics on shelf depths and shapes (Shepard, 1963) indicate that, although there is a relatively wide range in depths to the flattest part of the shelf and to the outer shelf edge, variations around the mean are rather small, and in general there is a remarkable uniformity of the depth and profile of shelves. The worldwide average of the flattest part of the shelves is about 65 m, and that of the outer edge of the shelf is about 130 m. Uniformity of depths and profile, as well as the abrupt break in the slope which divides the continental shelf from the continental slope, exists today because of control by the multiple fluctuations of sea level. Pre-Quaternary shelves probably did exist, but they must not have been as uniform in depth and profile, nor did they display as distinct shelf breaks as they do today.

The widespread occurrence of exposures of terrigenous sand on contemporary continental shelves is most certainly the result of Quaternary modification. The very rapid transgression of the sea ending about 7,000 years ago resulted

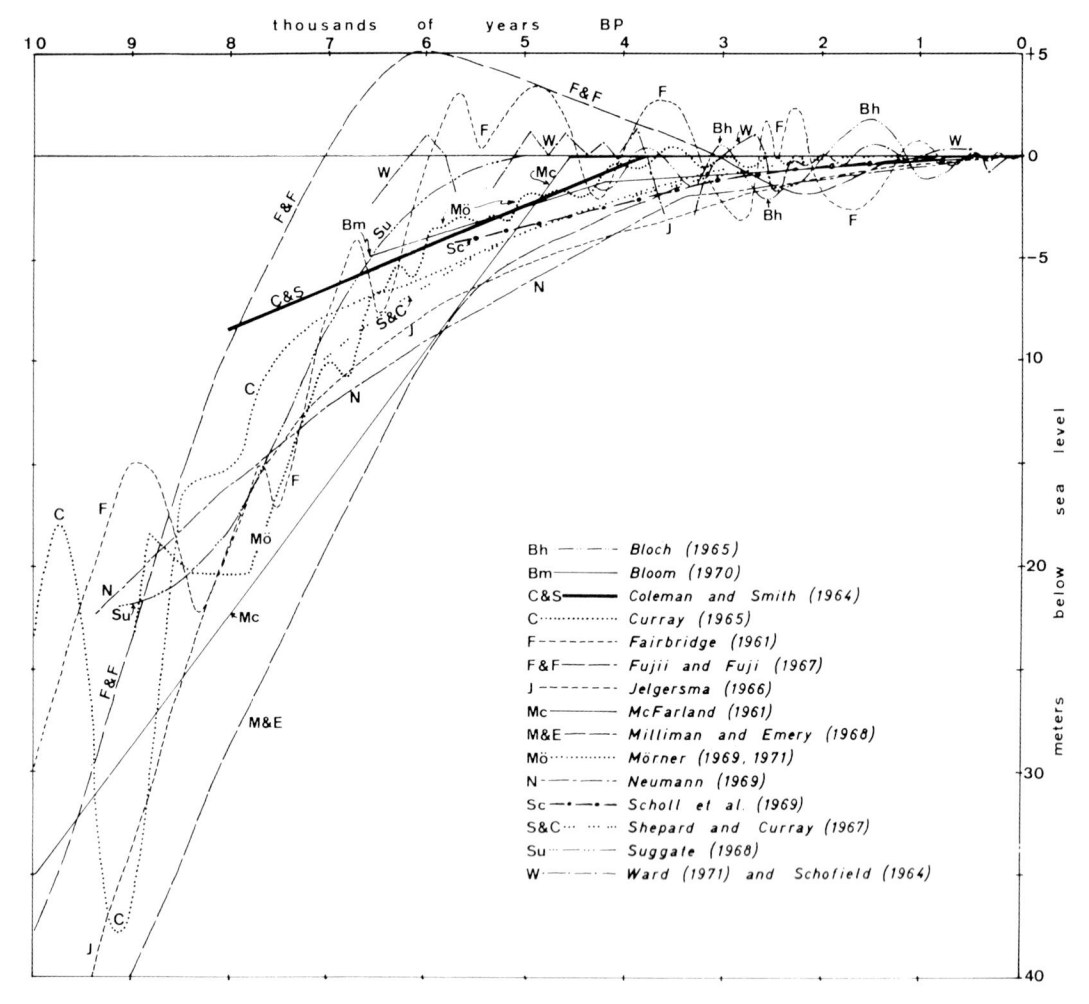

Fig. 1.—Composite late Quaternary sea level curves for last 100,000 years. Redrawn from published sources and reproduced from Curray (in press).

in the wave-activated shore zone plowing its way across most of the shelf surface, leaving shore-zone sands in its wake. Off most slowly subsiding or essentially steady state midplate continental margins, there has been insufficient time for the rivers to supply fine-grained sediments to cover the sands of the last transgression. With exposure on the shelf bottoms, the sands commonly are being reworked or modified to take on characteristics of the present environment of deeper water (Swift and others, 1971), or some sands are even being moved about and across the shelves by strong tidal currents (Belderson and others, 1971) to thus enter a state of pseudo-equilibrium with their present environments (Curray, in press). The fact remains that the sands were not demonstrably carried to their present shelf environment from the rivers which ultimately must supply the sands. In a sense, then, these studies of the alteration of relict sandy shelf sediments and the movement of highly altered shelf sands by strong tidal currents are adding a confusing background to what we believe to be one of the most important problems in sedimentary geology. That is, can shore-zone sands be moved through the surf-zone barrier? Is there a mechanism which will allow sands to move from the shore zone to accumulate as shelf sands? Review of the literature and studies of many modern environments lead us to conclude that there is no known mechanism for transporting significant quantities of sand-sized sediments from the shore zone to accumulate as sheet sands on the continental shelf. River-borne sands are transported both as bed load and in suspension. As a river enters the sea, the bed load is dropped immediately, and, because of the decrease in velocity, suspended sands are the first of the suspended load to settle and be deposited. In either case, sands are deposited very close to the river mouth, generally well within the influence of wave action. We believe that, under normal conditions on a midplate continental margin gradually subsiding without influence of rapid Quaternary-type sea level changes, there would be no sand exposed on the central or outer shelf. There would instead be a sandy shore zone and offshore muds, the thickness of which would be a function of the regional supply of sediment and subsidence rate.

Continental slope.—It has become almost trite to remind geologists that during periods of glacially lowered sea level the shore line lay very close to, or at the edge of, the continental shelves. Many sources supplied sediment directly to the continental slopes, and large volumes of sediment were subsequently transferred to the base-of-slope environments by slumping and erosional turbidity currents. Nonetheless, we believe that there is not yet a general realization of the profound effects on continental slopes of these Quaternary catastrophes. Inspection of detailed bathymetric maps of midplate continental margins (for example, fig. 2) reveals that the present continental slope does indeed bear the scars of the remarkably erosional Quaternary environment. This very striking erosional aspect of most sections of the continental slopes in modern midplate continental margins has led some geologists to hypothesize that continental slopes in general are realms of erosion and translation of sediment and have always been such. This very widely held view is not, however, borne out by the facts as revealed by reflection profiling on the continental margins of western Europe (Curray and others, 1966), Africa (Emery, 1972), eastern United States (Moore and Curray, 1963; Uchupi and Emery, 1967), and parts of South America (Butler, 1970); nor as revealed by deep-sea drilling on the slopes off New York (Hollister and others, 1972), above the Blake Plateau (Bunce and others, 1965), and on many other slopes (unpublished data). Instead, pre-Quaternary slopes were largely depositional.

The internal structure of continental margins varies widely, of course, depending on initial structure, on the regional rates of subsidence, and on the balance between subsidence and rate of supply and deposition of sediments. But in examining numerous profiles, we are impressed with the recurrence of several features common to many of these slopes. First, the slopes are indeed cut by many canyons and gullies of various sizes. But what are these erosional channels cut into? Generally, they are cut into seaward dipping slope deposits, not into horizontally bedded outcropping shallow-water sediments (fig. 3).

Secondly, if we look at the deepest reflectors in these records of continental slopes, we find that the intensity of channeling and disturbance decreases rapidly and that generally there are relatively undisturbed slope deposits lying beneath all but the most deeply incised valleys (fig. 4). Slope sediments are now commonly exposed by erosion, which results from turbidity currents or slumping. Where these newly exposed sections have been sampled by coring, dredging, and most recently by deep-sea drilling, rocks of pre-Quaternary, Cenozoic, and Mesozoic age have generally been recovered. The importance of these findings can hardly be overlooked if we are interested in determining dominant processes of normal or non-Quaternary midplate continental margins. Clearly, before Tertiary slope deposits can be eroded by Quaternary processes, the continental terrace must first be prograded by slope deposition throughout the many millions of years of Tertiary time. Undeniably, there must have been submarine canyons and turbidity-current action on midplate margins in pre-Quaternary time, but we submit that these were generally restricted to a few large active canyons that were able to extend themselves across whatever shelf may have existed and that commonly would lie off the mouths of major rivers. Along most of their lengths the slopes must have been the sites of slow pelagic, hemipelagic or quiet, low-density turbid-layer deposition. The presence of erosional remnants of younger Tertiary and exposed older Tertiary deposits, furthermore, implies that there was a long-term balance between the rate of subsidence and the rate of deposition on shelves and slopes of midplate continental margins.

The question of the importance of slumping as a major factor in shaping normal or pre-Quaternary continental slopes is a more difficult one than that of turbidity currents and channel erosion. The interpretation of what constitutes a slump or slump structure, as recorded in bathymetry or on seismic reflection profiles, is much more subjective than is that of turbidity-current channels, and, as we shall elaborate later, we feel that in many examples structure attributed to slumping has been misinterpreted. Certainly the very slow uniform deposition, which must have prevailed on most parts of the continental slopes prior to lowered Quaternary sea levels, would not be conducive to generating in-

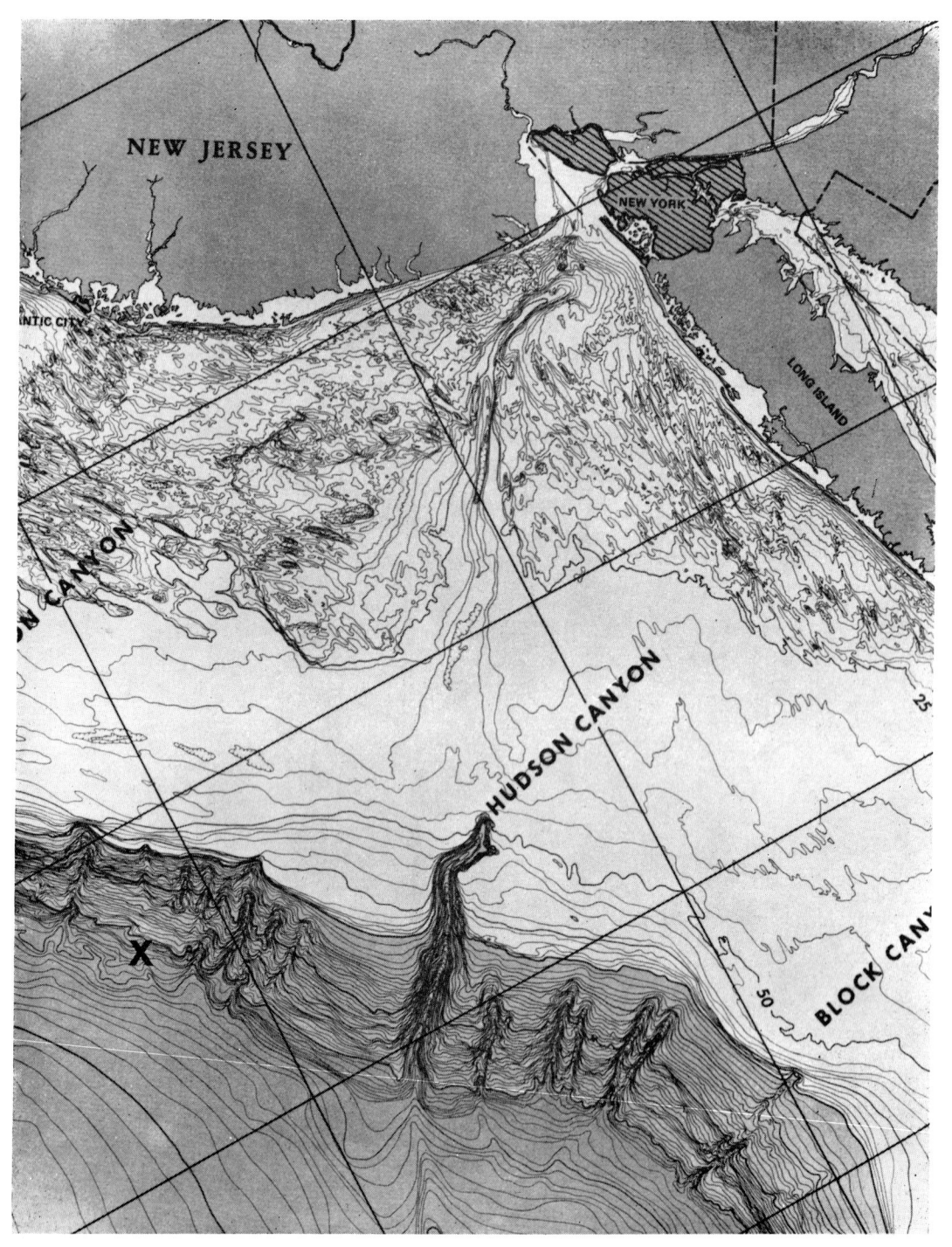

Fig. 2.—Continental shelf and slope off New York. Note many small canyons to north and south of Hudson Canyon. Small canyons head at edge of shelf and have distal ends at base of slope. X marks locality on slope south of Hudson Canyon of JOIDES site 108. Reproduced from Bathymetric map 1, Eastern continental margin, U.S.A. (Am. Assoc. Petroleum Geologists, 1970).

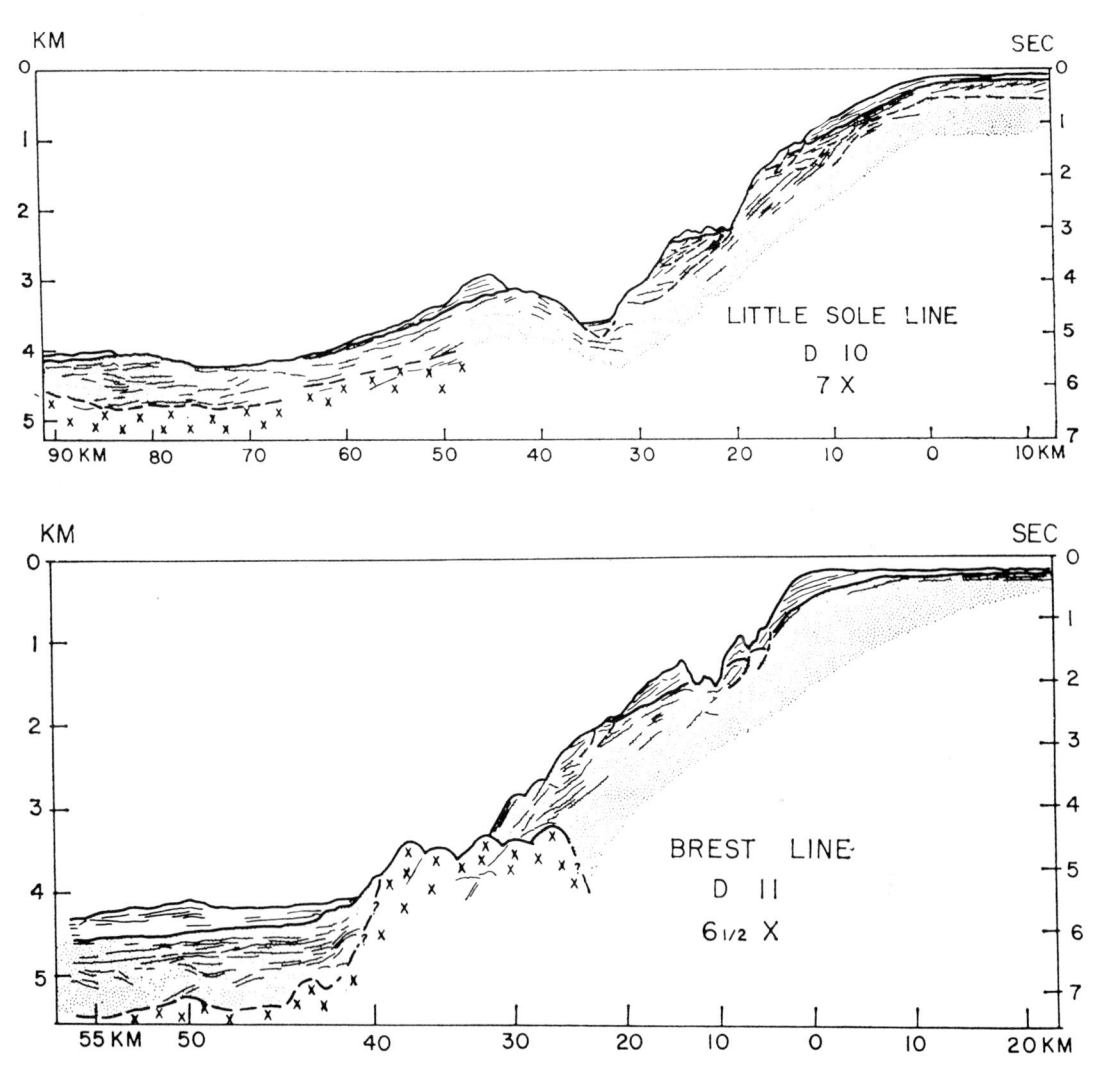

FIG. 3.—European continental margin. Line drawing of reflection-profiler traverses. Vertical scales are two-way travel time (seconds) and water depth (kilometers); horizontal scale varies with ship's speed; vertical exaggeration is an average value. Shaded stratigraphic sequence is interpreted as Tertiary sedimentary rock, and X pattern indicates probable basement rock. Little Sole Line: Note axis of a submarine canyon at 25 to 45 km and slumps on the shoaler flank of canyon. Brest Line: Note outcrop of possible basement rock near foot of slope and erosional remnants of younger slope sediments cut by submarine canyons and modified by slumping. Reproduced from Curray and others, 1966.

stability on the relatively gentle average continental slopes. On the contrary, studies in soil mechanics have shown that deposits that are laid down at a slow uniform rate and that are relatively homogenous tend to be quite stable even on relatively steep slopes (Moore, 1961). There is no question, however, that the undercutting of slopes by channeling and the imposition of sudden loads by rapid deposition have resulted in an abnormally common incidence of slumping during the Quaternary.

Continental rise.—If we are correct in our assumption that the multitude of canyons and gullies that crease the slopes of midplate continental margins are vestigial from low sea level stands of the Quaternary, then it is axiomatic that most of the large volumes of sediment in base-of-slope environments of Quaternary to modern vintage must be of turbidity-current origin. Certainly, there can be little doubt that in primary or major flow events large volumes of sediment are deposited when overflow and levee-topping occurs at the mouths of large submarine canyons in these same environments.

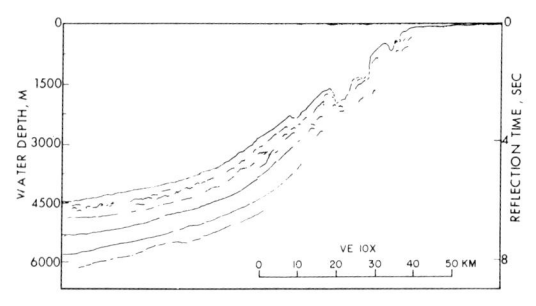

FIG. 4.—Continental margin of west Africa south from Cape Palmas, eastern Liberia. Turbidity-current erosion and small channeling have cut surface of slope and have marked upper strata of slope and base-of-slope environments. Note that older, probably Tertiary strata are unmarked and smooth. Section traced from line 31 of U.S. Geol. Survey GD-72-006, 1972.

The turbidity currents that originate in the upper parts of canyons or gullies to flow through the channel system may be considered as either primary or secondary events. We believe that the shape of individual fans and the slope of coalesced fans that formed base-of-slope continental rises are fundamentally a function of primary turbidity-current events, that is, of those currents which overflow and spill out of the channel as the current emerges from the canyon at the base of the slope. Natural levees are formed in this manner, and generally these are largest on the upper parts of the fans or rises (fig. 5a). The overflow from primary turbidity currents spreads out as sheet flow across the fan to flow through or partially fill preexisting topography. These sheet flows are probably largely deposited on the rise or fan. Secondary events would be classed as those that do not overtop channels and that are largely confined to the channels throughout their flow. Some of the residual flow of primary events, and possibly also parts of the secondary events may, indeed, bypass the entire rise or fan to deposit sands and fine-grained turbidites on the abyssal plains. Recent studies of the very large deep-sea fan in the Bay of Bengal have shown that even the largest of these channels may meander, may be braided, and may migrate gradually or break abruptly through levees and shift to entirely new courses (Curray and Moore, 1971; Moore, Emmel, and Curray, in preparation). This kind of channel migration can create very complex structures of natural levee deposits and filled channels piled one on top of the other (fig. 5). Such complex structures may be easily misinterpreted in seismic reflection records, and we believe that it is possible that many of the deformed sections (fig. 6), which have been interpreted as slumped sediment in fans and continental rises, may, in fact, be complex structures created by migrating channels.

The relative importance of turbidity-current versus geostrophic-current transport is also controversial. Proponents of the importance of geostrophic currents in transporting and depositing continental rise sediments (Heezen and others, 1966) believe that virtually all sediments on the continental rise are emplaced as the result of deposition out of contour currents. As we have noted previously, our studies of fans do not support this view, and Stanley and others (1971) have suggested that contour currents may not be as important as previously believed. They find that discrete fans, at least off the major canyon systems, can in fact be seen in survey lines that are run parallel to slope contours. There appears to be little doubt, however, that geostrophic currents move significant amounts of sediment, which is carried to the rise by turbidity currents, laterally along the contour to the south off eastern United States. These geostrophic currents are intensified on the western sides of the oceans, but may also occur elsewhere. Velocities as high as 20 cm per second are reported off the east coast of the United States over the continental rise. The effectiveness of these currents in smoothing topography and modifying final sites of deposition of sediments on the rise is shown by the fact that minor channels or distributaries are generally not recorded off the numerous small gullies and canyons of the east coast slope and rise. The fact that individual fans with distributaries having natural levees have been recorded off some of the larger canyons (Stanley and others, 1971; Emery and others, 1970) is strong evidence that the major body of continental rise deposition is of turbidity-current origin.

If we are correct in our conclusion that pre-Quaternary, or normal, continental slopes were largely stable depositional features and devoid of gullies and canyons, except possibly for a few major canyons off large rivers, then it follows that pre-Quaternary continental rises must not have had the abundant sediment supply, and hence the resulting physiographic continuity, displayed by modern examples. Our model, in fact, would suggest that a less well developed continental rise, having more pronounced lobate bulges or discrete deep-sea fans off the few major canyons, might have predominated in pre-Quaternary time.

SUMMARY

Sedimentological, bathymetric, and geophysical studies of midplate continental margins as

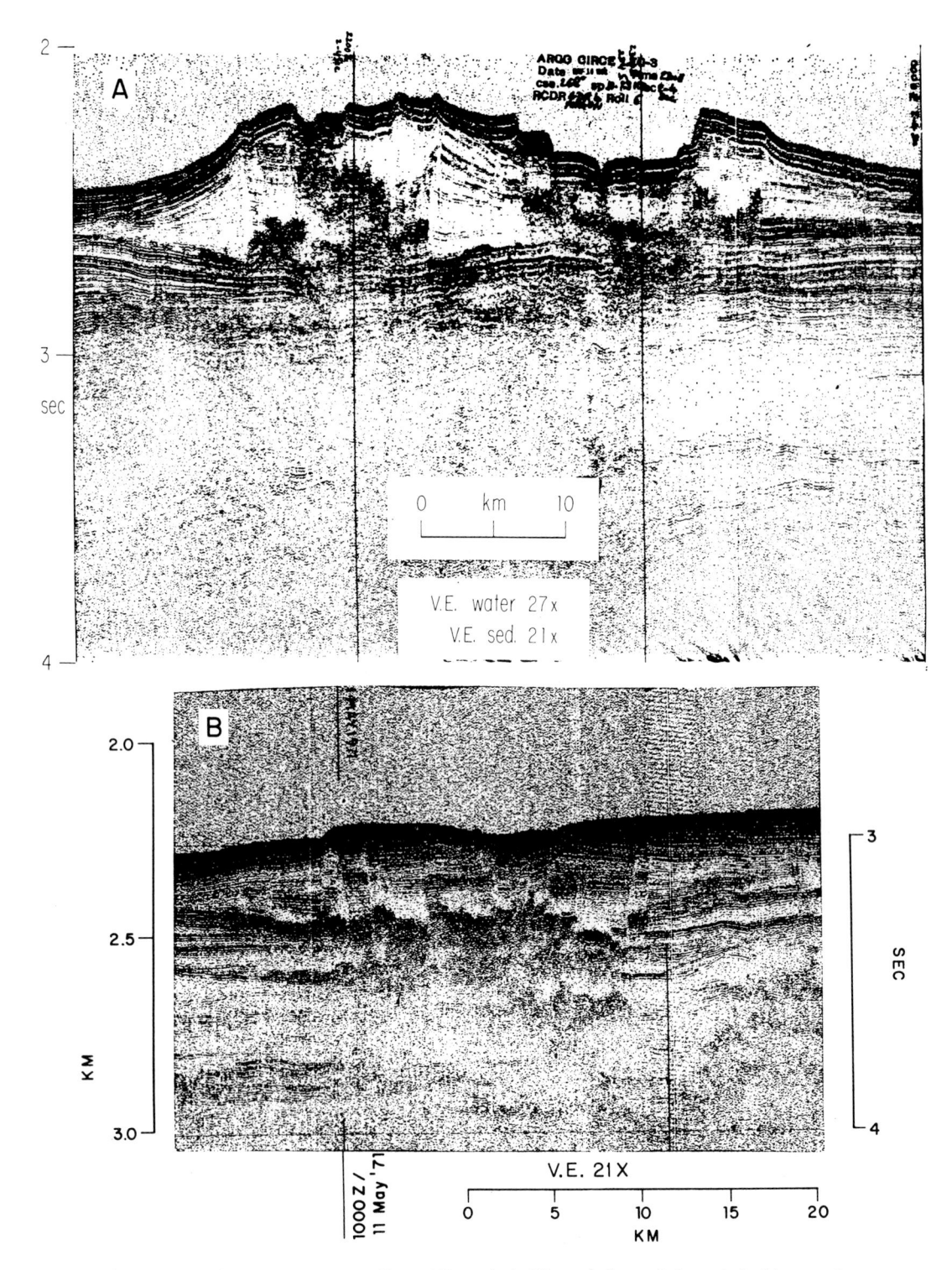

FIG. 5.—Profiles of upper Bengal Fan, Bay of Bengal. A, View of elevated channels looking north upstream; natural levee deposits and channel cut-and-fill structure form complex structure resembling slumping or faults; nearly filled leveed channel on left is now abandoned, and large active channel on right is built on its eastern flank. B, Complex internal structure resulting from filled and buried channels and natural levees; note irregular dissected surface about 0.3 sec below sea floor, abrupt right-hand-side termination at the channel wall about 10 km from right side of section, and uneven nature of fill above these features; all these are erosional and depositional turbidity-current features unrelated to either slumping or tectonics.

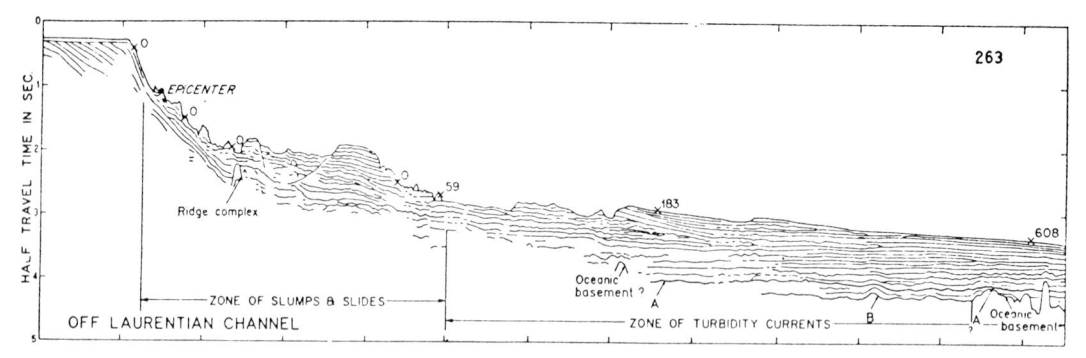

Fig. 6.—Slope off Laurentian Channel interpreted by Emery and others (1970) as zone of slumps and slides; interpreted here as erosional remnant and channel cut-and-fill structure. Section from Emery and others, 1970.

they exist today lead to the conclusion that many aspects of modern margins are vestigial from Quaternary periods of eustatic low sea levels. Shelf and shore-zone features common today are relict from transgressions and regressions of the surf zone across the shelves, this being in accord with glacially controlled rapid sea level changes. The commonly erosional nature of midplate continental slopes stems from turbidity currents generated by rapid deposition on upper slopes at low sea level stands. The deposition of the erosion-derived sediments in base-of-slope environments is similarly of glacially controlled origin.

All these continental margin modifications of the past few million years tend to mask growth processes, which predominated for tens of million years during pre-Quaternary time. Although the Quaternary Period certainly was not a unique part of geological time, it was very unusual. Transgressions and regressions have always occurred but usually at rates that were significantly slower, except perhaps during the other, rare glacial periods. Extensive sheet sands, for example, exist in older sediments, but we must interpret them in terms of modern process, and not necessarily in terms of Holocene examples that are inherited from the Pleistocene.

An idealized non-Quaternary midplate continental margin would have a shore zone of prograding beaches, mainly devoid of estuaries. Barriers and lagoons would exist primarily in association with delta complexes. Muddy shelf facies would prevail, as no mechanism exists for moving sands out of the shore zone to the shelf in significant quantities. Slopes would be largely morphologically smooth and prograding by slow pelagic, hemipelagic, and low-density turbid-layer deposition. Erosion by turbidity currents and slumping would be uncommon and largely restricted to the vicinity of a few large, long-lived submarine canyons. Axiomatically, the rise would be poorly developed and lobate in form off these canyons where deep-sea fans would evolve. Terrigenous clastic turbidities would be mainly confined to these fans.

REFERENCES

BELDERSON, R. H., KENYON, N. H., AND STRIDE, A. H., 1971, Holocene sediments on the continental shelf west of the British Isles, in DELANY, F. M. (ed.), The geology of the east Atlantic continental margin: Inst. of Geol. Sci. Rept. 74/14, p. 157–170.

BLOCH, M. R., 1965, A hypothesis for the change of ocean levels depending on the albedo of the polar ice caps: Paleogeography, Paleoclimatology, and Paleoecology, v. 1, p. 127–142.

BLOOM, A. L., 1970, Paludal stratigraphy of Truk, Ponape, and Kusaie, eastern Caroline Islands: Geol. Soc. America Bull., v. 81, p. 1895–1904.

BUNCE, E. T., AND OTHERS, 1965, Ocean drilling on the continental margin: Science, v. 150, p. 709–716.

BUTLER, L. W., 1970, Shallow structure of the continental margin, southern Brazil and Uruguay: Geol. Soc. America Bull., v. 81, p. 1079–1096.

COLEMAN, J. M., AND SMITH, W. G., 1964, Late Recent rise of sea level: ibid., v. 75, p. 833–840.

CURRAY, J. R., 1965, Late Quaternary history, continental shelves of the United States, in WRIGHT, H. E., JR., AND FREY, D. G. (eds.), The Quaternary of the United States: Princeton, New Jersey, Princeton Univ. Press, p. 723–735.

———, in press, Marine sediments, geosynclines and orogeny, in JUDSON, S. AND FISCHER, A. G. (eds.), Petroleum and global tectonics: Princeton, New Jersey, Princeton Univ. Press.

———, AND MOORE, D. G., 1971, Growth of the Bengal deep-sea fan and denudation in the Himalayas: Geol. Soc. America Bull., v. 82, p. 563–572.

———, AND OTHERS, 1966, Continental margin of western Europe: slope progradation and erosion: Science, v. 154, p. 265–266.

DIETZ, R. S., AND HOLDEN, J. C., 1966, Miogeosynclines (miogeoclines) in space and time: Jour. Geology, v. 74, p. 566–583.
DRAKE, C. L., EWING, M., AND SUTTON, G. H., 1959, Continental margins and geosynclines: the east coast of North America north of Cape Hatteras, *in* Physics and chemistry of the earth: London, Pergamon Press, v. 3, p. 110–198.
EMERY, K. O., 1972, Eastern Atlantic continental margin: some results of the 1972 cruise of the *R. V. Atlantis II:* Science, v. 178, p. 298–301.
———, AND OTHERS, 1970, Continental rise off eastern North America: Am. Assoc. Petroleum Geologists Bull., v. 54, p. 44–108.
FAIRBRIDGE, R. W., 1961, Eustatic changes in sea level, *in* Physics and chemistry of the earth: London, Pergamon Press, v. 4, p. 99–185.
FUJII, S., AND FUJI, N., 1967, Postglacial sea level in the Japanese Islands: Osaka City Univ. Jour. Geosci., v. 10, p. 43–51.
HEEZEN, B. C., HOLLISTER, C. D., AND RUDDIMAN, W. F., 1966, Shaping of the continental rise by deep geostrophic contour currents: Science, v. 152, p. 502–508.
HOLLISTER, C. D., AND OTHERS, 1972, Site 108—continental slope: Initial reports of the Deep Sea Drilling Project, v. 11, p. 357–364, Washington.
JELGERSMA, S., 1966, Sea-level changes during the last 10,000 years, *in* Proc. Internat. Symposium on World climate from 8000 to 0 BC, Imp. Coll., London, April 18–19, 1966: Royal Meteorol. Soc. London, p. 54–71.
MCFARLAN, E., JR., 1961, Radiocarbon dating of late Quaternary deposits, south Louisiana: Geol. Soc. America Bull., v. 72, p. 129–158.
MENARD, H. W., 1969, Elevation and subsidence of oceanic crust: Earth and Planetary Sci. Letters, v. 6, p. 275–284.
MILLIMAN, J. D., AND EMERY, K. O., 1968, Sea levels during the past 35,000 years: Science, v. 162, p. 1121–1123.
MOORE, D. G., 1961, Submarine slumps: Jour. Sed. Petrology, v. 31, p. 343–357.
———, AND CURRAY, J. R., 1963, Sedimentary framework of continental terrace off Norfolk, Virginia, and Newport, Rhode Island: Am. Assoc. Petrol. Geologists Bull., v. 47, p. 2051–2054.
———, EMMEL, F., AND CURRAY, J. R., in preparation, Morphology and sedimentary processes of the Bengal deep-sea fan.
MÖRNER, N. A., 1969, Climatic and eustatic changes during the last 15,000 years: Geol. en Mijnb., v. 48, p. 389–399.
———, 1971, The Holocene eustatic sea level problem: *ibid.,* v. 50, p. 699–702.
NEUMANN, A. C., 1969, Quaternary sea-level data from Bermuda: 8th Internat. Quaternary Assoc. Cong., Paris, Abs., p. 228–229.
SCHOFIELD, J. C., 1964, Post-glacial sea levels and isostatic uplift: New Zealand Jour. Geology and Geophysics, v. 7, p. 359–370.
SCHOLL, D. W., CRAIGHEAD, F. C., AND STUIVER, M., 1969, Florida submergence curve revised: its relation to coastal sedimentation rates: Science, v. 163, p. 562–564.
SHEPARD, F. P., 1963, Submarine geology: New York, Harper and Row, 567 p.
———, AND CURRAY, J. R., 1967, Carbon-14 determination of sea level changes in stable areas, *in* Progress in oceanography, v. 4, The Quaternary history of the ocean basins: Oxford, England, Pergamon Press, p. 283–291.
STANLEY, D. J., SHENG, H., AND PEDRAZA, C. P., 1971, Lower continental rise east of the Middle Atlantic States: predominant sediment dispersal perpendicular to isobaths: Geol. Soc. America Bull., v. 82, p. 1831–1840.
SUGGATE, R. P., 1968, Post-glacial sea-level rise in the Christchurch metropolitan area, New Zealand: Geol. en Mijnb., v. 47, p. 291–297.
SWIFT, D. J. P., STANLEY, D. J., AND CURRAY, J. R., 1971, Relict sediments on continental shelves: a reconsideration: Jour. Geology, v. 79, p. 322–346.
UCHUPI, E., AND EMERY, K. O., 1967, Structure of continental margin off Atlantic coast of United States: Am. Assoc. Petroleum Geologists Bull., v. 51, p. 223–234.
U.S. GEOL. SURVEY, 1972, Acoustic reflection profiles, Liberian continental margin: U.S. Geol. Survey G.D.-72–006, p. 28.
WARD, W. T., 1971, Postglacial changes in level of land and sea: Geol. en Mijnb., v. 50, p. 703–718.

EVOLUTION OF THE WESTERN APPALACHIAN CONTINENTAL MARGIN

JOHN M. BIRD
Cornell University, Ithaca, New York

ABSTRACT

Recently, the carbonate belt of the northwestern side of the northern Appalachian Orogen has been interpreted as an early Paleozoic shelf within a continental margin province that evolved synchronously with the opening and closing of an ocean, from late Precambrian through Devonian time.

This plate tectonic model incorporates results of various studies of the past century, commencing with the original Taconic controversy over the age of Taconic rocks. The last Taconic controversy revolved around the interpretation of Taconic sedimentary rocks that are surrounded by carbonate. Either they are overthrust, as originally suggested by Keith, or they are autochtonous lateral facies within a basin surrounded by the carbonates, as originaly proposed by Dale. The overthrust interpretation was subsequently modified by Ruedemann to become the Taconic klippen hypothesis that, during the 1960's, has been virtually proven by many workers, but principally by Zen. Kay, as a result of his classic work on the Trenton Group of New York, proposed a miogeosyncline-eugeosyncline (shelf-island arc) couple model for the pre-Middle Ordovician, western Appalachian belt, a concept which was to become the key for future refinement of our understanding of regional stratigraphic-structural relations of the various facies and volcanic assemblages. It provided the basis of the collapsing continental rise model of Dietz and of Dietz and Holden that followed Drake, Ewing, and Sutton's comparison of Kay's model with the present North American Atlantic continental margin. Rodgers then proposed that the southeastward termination of the carbonate shelf was represented by breccia facies that can be observed in the field, that the carbonates and breccias were of a Bahama bank and bank-edgelike environment, and that this termination was the termination of the Ordovician continental margin. More recently Bird and Dewey incorporated these models into the plate tectonics model of an evolving so-called Appalachian Atlantic Ocean, following Wilson's proposal of a driven-out Proto-Atlantic Ocean and Dewey and Kay's and Dewey's models of a continuous Appalachian-Caledonian orogenic belt.

Commencing in the Ordovician, the Appalachian portion of the enormous sheet of Cambro-Ordovician carbonates of central and eastern North America was involved in the extensive diachronous deformation that formed the Appalachian Orogen. Therefore, reconstruction of the continental margin to its condition before deformation involves the sorting out of bulk stratigraphic-tectonic units, the comparison of their relative chronologies, and the recognition of various sedimentary and structural environments. Using present-day lithosphere plate relations and the analogy of an Atlantic type of continental shelf-rise-abyss facies relationship, a plate tectonics model can be constructed from these various stratigraphic-tectonic elements of the carbonate belt and associated marginal, overlying, and allochthonous rocks that integrates otherwise seemingly diverse and unrelated aspects and provides an actualistic model for the evolution of the North American Appalachian continental margin.

The key to reconstructing the pre-Taconian or pre-Middle Ordovician relations is in the stratigraphic-structural assemblages of the Taconian thrust belt. By paleontological considerations of the various structural units or klippen one by one, the chronostratigraphic relations between the carbonate shelf assemblage and the sediments of the thrust sheets can be determined. Essentially, the earliest emplaced thrust sheets contain sediments whose age range matches that of the underlying shelf (autochthon), which was as pointed out a number of years ago by Zen. In addition, the facies of these sediments fit very well the model of shelf-continental rise assemblages that accumulated in a starved environment. The bulk of the initially emplaced klippen, the Giddings Brook Slice, is composed of these offshelf and synchronous sediments. The lowest known fossils in these sediments are in Lower Cambrian orthoquartzite, which matches petrographically and paleontologically the basal orthoquartzite of the carbonate shelf.

The bulk of the overthrust sediments, however, are pre-Lower Cambrian shales and clastics, which are thousands of feet thick and which in the lowest portion are apparently nonmarine graben assemblages containing extrusive basaltic rocks. All these rocks are at least slightly metamorphosed in subgreenschist facies. Additionally, similar facies occur in the autochthon below the Lower Cambrian basal unconformity of the carbonate platform, which also locally contains extrusive basalt and rhyolites. These relationships were discussed in detail by Bird and Dewey in 1970.

Essentially then, with the eastward limit of the carbonate belt being taken as a shelf edge, reconstruction of the overlying structural assemblages indicates a history of late Precambrian continent separation beginning with early horst and graben tectonics and sedimentation, followed by establishment of an Atlantic-type continental margin through to Late Cambrian and Early Ordovician time. Then, ocean closing by subduction along the continental margin converted the margin basement and sediment assemblage to an Andean-like system, diachronously through to Middle to Late Devonian continent-to-continent collision. The carbonate belt then lay along the north side of a Himalayan-like mountain system. Its present geographic position is a consequence of subsequent plate evolution commencing in the Late Triassic as indicated by the Newark Basin assemblage.

EROSION AND DEPOSITION ALONG THE MID-PERMIAN INTRACRATONIC BASIN MARGIN, GUADALUPE MOUNTAINS, TEXAS

JOHN C. HARMS
Marathon Oil Company Research Center, Littleton, Colorado
AND
LLOYD C. PRAY
University of Wisconsin, Madison

ABSTRACT

Sediments, processes, and the morphologic profile at intracratonic basin margins commonly are similar to those of the continental shelf, slope, and rise along open oceanic margins. However, the increased potential for sharp density stratification of intracratonic basin waters and for generation of density currents on surrounding epicontinental shelves can markedly influence depositional and erosional processes on the basin margin or floor and can create distinctive sedimentary features that help to differentiate intracratonic and ocean-margin environments of the geologic record. The mid-Permian outcrops of the Guadalupe Mountains provide excellent examples of both depositional and erosional features of an intracratonic basin margin where sharp density stratification and persistent density currents formed by temperature or salinity differences, rather than by suspended clay, were important sedimentologic factors.

The mid-Permian (Leonardian to early Guadalupian) northwestern margin of the Delaware Basin probably had a normal shelf-slope-rise profile, having several hundred meters of relief, and slopes of a few degrees or less along the basin-margin depositional slope. The exposed 1,000 m of mid-Permian basin-facies strata consist mostly of finely textured dark carbonate rocks, fine-grained sandstones, and siltstones. Carbonate sands and allochthonous carbonate conglomerates and megabreccias derived from the bank or bank margin are locally conspicuous (but minor) interbedded strata. The Leonardian rock units are the contemporaneous bank and basin facies, Victoria Peak dolomites and Bone Springs limestones, respectively. The early Guadalupian rock units are the Cutoff "Shale," composed mostly of basin-facies limestone, and the overlying Brushy Canyon Formation, composed mostly of detrital sandstone and siltstone. The Cutoff strata lie above and parallel to the basin-sloping unconformity that truncates Leonardian basin-margin deposits. Brushy Canyon strata unconformably onlap both Leonardian and Cutoff strata.

The abruptness and position of the Victorio Peak-to-Bone Springs facies change indicate the sharpness and persistence of a euxinic interface along the lower part of the Leonardian basin-margin slope. Currents were generally weak or absent near the interface, but erosion surfaces, some overlain by sheets and channel fills of bank-derived carbonate sands, indicate episodes of higher competence of bottom currents.

Intra- and interformational erosion features are more prevalent in Guadalupian strata. Two major erosional phases created unconformities at both the base of the Cutoff and the Brushy Canyon rock units. The unconformities at the basin margin slope 5° to 10° basinward. The lower one truncates about 250 m of Leonardian basin-margin strata, and its carving required appreciable retreat and steepening of the basin-margin depositional slope. The upper unconformity forms the onlap surface for more than 300 m of Brushy Canyon deposits. Several steep-sided, narrow channels as much as 40 m deep incise the sloping unconformity surfaces. Erosion concomitant with sedimentation of basin facies persisted throughout early Guadalupian deposition, and basin-trending channels are especially well displayed in the Brushy Canyon. Brushy Canyon intraformational channel dimensions are substantial, as depths may exceed 25 m, widths 1 km, and lengths many kilometers. Brushy Canyon channels are filled in part by beds of sandstone containing upper flow-regime features that conform to the flatter channel floors and that abut adjacent channel walls. Finely laminated siltstone beds mantle channel floors, walls, and interchannel areas and form the bulk of the Brushy Canyon deposits.

The erosive agents that cut both channels and unconformities left clean, smooth contacts but little evidence of their nature. We believe that density currents were the major erosive agent and that all erosion occurred in a relatively deep submarine environment. Evidence for submarine origin includes the basin-facies character of all deposits overlying erosional surfaces, the similarity of small and large scale erosion surfaces, the similarity of the Brushy Canyon erosional features to those of later Guadalupian deposits of established deep-basin origin, and the absence of recognized features of subaerial, vadose, or shallow-marine environments. If sea-level changes were involved, the sea may have been deeper rather than shallower during the carving of the major unconformities.

The erosional and depositional features of the mid-Permian basin margin are compatible with a basin having sharp density stratification and with frequent spilling of shelf-generated cold or saline density currents down the shelf margin. Denser, bottom-hugging currents carved the channels and probably the unconformities and deposited the coarser grained carbonate and detrital sands. Less dense currents moved partly down the slope and then spread far out over the basin as interflows, creating a rain of finer grained sediment on the deeper basin floor. Density currents may have been frequent, of long duration, and not limited to master channels, thus minimizing proximal-to-distal and fan apex-to-interfan contrasts. Episodic phenomena expectable on any marginal slope, such as debris flows that carried very coarse clasts several kilometers into the basin, or slumps, or perhaps deep wave action, contributed to the sedimentary features. The mid-Permian sedimentary prism of the intracratonic Delaware Basin provides some marked contrasts as well as similarities to sedimentary prisms fronting open ocean basins. In its overall features, it is significant that here is another example from the geologic record for which there appears to be no reasonably close modern analog.

AULACOGENS AND THEIR GENETIC RELATION TO GEOSYNCLINES, WITH A PROTEROZOIC EXAMPLE FROM GREAT SLAVE LAKE, CANADA

PAUL HOFFMAN, Geological Survey of Canada, Ottawa; JOHN F. DEWEY AND KEVIN BURKE, State University of New York, Albany

ABSTRACT

Aulacogens are long-lived deeply subsiding troughs, at times fault-bounded, that extend at high angles from geosynclines far into adjacent foreland platforms. They are normally located where the geosyncline makes a reentrant angle into the platform. Their fill is contemporaneous with, as thick as, and lithologically similar to the foreland sedimentary wedge of the geosyncline but in addition has periodically erupted alkalic basalt and fanglomerate. Although many aulacogens have suffered mild compressional deformation, tectonic movement within them is mainly vertical; large-scale horizontal translations are rare. Aulacogens are known throughout the Proterozoic and Phanerozoic, and incipient aulacogens occur at reentrants on modern continental margins.

The 1700-to-2200-million-year-old Athapuscow Aulacogen of Great Slave Lake began as a deeply subsiding transverse graben during the early miogeoclinal stage of the Coronation Geosyncline. During the orogenic stage of the geosyncline, the aulacogen became a broader downwarp that received abnormally thick exogeosynclinal sediments from the orogenic belt. The aulacogen was compressed mildly, prior to a final stage involving transcurrent faulting, one-sided uplift, and continental fanglomerate sedimentation. The aulacogen is distinguished from the foreland sedimentary wedge of the geosyncline by having paleocurrents parallel rather than transverse to its structural trend, by having high-angle faults rather than low-angle thrusts, by its alkalic basalt volcanism, and by the lack of metamorphism.

It is hypothesized that deep-mantle convective plumes produce three-armed radial rift systems (rrr triple junctions) in continents stationary with respect to the plumes. If only two of the arms spread to produce an ocean basin, the third remains as an abandoned rift extending into the continental interior from a reentrant on the new continental margin. For example, the Benue Trough, located in the Gulf of Guinea reentrant on the west coast of Africa, may be such an abandoned rift arm formed during the Cretaceous period at the time of initial rifting of Africa and South America. Inasmuch as new continental margins are predestined to become geosynclines, such abandoned rift arms are juvenile aulacogens. In this model, aulacogens and geosynclines have a common origin but differ in the extent of rifting.

INTRODUCTION

Aulacogens (literally, "born as furrows") are transverse linear troughs that extend from geosynclines far into the interiors of foreland platforms (figs. 1 and 2). They are generally located where geosyncline make reentrant angles at the margins of the platforms, and they gradually die out toward the interiors of the platforms. Aulacogens are usually long-lived, subsiding during almost the entire history of the related geosyncline, and are susceptible to subsequent reactivation. Most aulacogens begin as narrow fault-bounded grabens and later become broader downwarps that may themselves be ultimately broken by faults. In all examples, they profoundly influence sedimentation on the foreland, and some are characterized by evaporites or by intermittent basaltic and rhyolitic volcanism, which are absent from other parts of the foreland.

Aulacogens were first recognized as a unique and important class of tectonic structures by the brilliant Soviet geologist Nikolai S. Shatski. The type examples are the subsurface Pachelma and Dnieper-Donets Aulacogens in the southeast of the Russian Platform [Shatski, 1946a, 1946b, 1947, 1955, all published in German *in* Schatski (sic), 1961; Bondarchuk, 1956; Subbotin, 1958; Chirvinskaya, 1958, 1959; Shatski and Bogdanov, 1960; Nalivkin, 1963; Bogdanov, 1964; Gavrish, 1965; Hope, 1965; Chekunov, 1967; Dolenko, Varichev, and Galabuda, 1970; von Gaertner, 1969]. The presence of subsurface aulacogens in the eastern part of the Siberian Platform has also been suggested (Salop, 1967; Salop and Scheinmann, 1969), but these are less certain.

Outside the U.S.S.R., geologists have been slow to recognize aulacogens. Shatski (1946b) pointed out the most obvious North American example, the northwest-trending faulted and folded Paleozoic trough that extends from the Ouachita Geosyncline across the foreland platform of southern Oklahoma (figs. 1 and 2) and shows up so well on the Tectonic Map of North America (King, 1969). That this, the deepest Paleozoic basin of the North American Platform, is an aulacogen is dramatically evident from the work of Ham, Denison, and Merritt (1964); Ham and Wilson (1967); and Ham (1969). The trough began as a graben underlain by Precambrian granitic rocks and was filled by perhaps as much as 5000 m of coarse immature clastics and Early to Middle Cambrian basalt, spilite, and rhyolite volcanics and hypabyssal sills. From Late Cambrian to Late Ordovician time, the trough was a broad downwarp that received up to 3100 m of carbonates. The rate of subsidence of the downwarp decreased with passing time so that the 355 m of Siluro-Devonian sediments there are no more than elsewhere on the foreland platform. During late Paleozoic time, the downwarp was mildly compressed and broken by a braided system of faults, some of which have transcurrent as well as vertical displacements. This resulted in a complex pattern of paired uplifts and fault basins, for example, the Wichita Uplift and Anadarko Basin in the west and the Arbuckle Uplift and Ardmore Basin in the east. These basins received as much as 7000 m of Mississippian to Permian clastics. Here, then, is a trough of almost geosynclinal proportions, but one which, unlike geosynclines, is transverse to the platform margin and simply dies out in the platform interior.

The distinction between aulacogens and geosynclines becomes critical in the debate as to whether folded mountain belts arise through primarily vertical or horizontal tectonics. The concept of vast horizontal translations, involving the opening and closing of ocean basins, has had unparalleled success in rationalizing the geology of geosynclinal mountain belts (J. T. Wilson, 1966, 1968; Mitchell and Reading, 1969; Dewey and Bird, 1970; Dickinson, 1971). It is equally clear, however, that tectonic movements in aulacogens are primarily vertical, a fact that takes on some historical interest with the realization that much of the field experience of Vladimir Beloussov (1938-40), the principal modern proponent of primary vertical tectonics, was in the Greater Caucasus, itself perhaps another aulacogen (von Gaertner, 1969).

To help make the point that aulacogens really constitute a discrete *class* of structures, not just a diverse assortment of geodeviants, the recently discovered Athapuscow Aulacogen of the northwest Canadian Shield is described below. Following, we note two examples of what may be modern aulacogens extending inland from reentrants on the coast of Africa. Finally, a genetic model is proposed to explain the origin and development of aulacogens within the context of plate tectonics and to reconcile the contrasting tectonics of aulacogens and geosynclines.

Acknowledgment.—The original manuscript was reviewed by John Rodgers of Yale University, who pointed out the important papers by N. S. Shatski, W. E. Ham, and H. R. von Gaertner.

AULACOGENS OF THE NORTHWEST CANADIAN SHIELD

Regional Geologic Setting

The northwest corner of the Canadian Shield (figs. 3 and 4) is divided into three structural provinces (M. E. Wilson, 1939; J. T. Wilson, 1949; Stockwell, 1961, 1970; Price and Douglas, 1972). The Bear Province is the orogenic zone of the Coronation Geosyncline (Hoffman, Fraser, and McGlynn, 1970; Fraser, Hoffman, Irvine, and Mursky, 1972; Hoffman, 1973a), a precursor and look-alike of the Cordilleran Geosyncline, and the Slave Province is its foreland platform. The Churchill Province was also part of the foreland platform during the evolution of the Coronation Geosyncline, but shortly thereafter it was broadly uplifted and dextrally displaced relative to the Slave Province. Uplift reset, so to speak, the K-Ar but not the Rb-Sr radiometric clocks in the Churchill Province but seems not to be genetically related to the Coronation Geosyncline. Rather, uplift may have been related to collision with the Superior Province, 950 km southeast of the aulacogen, the Churchill Province thereby being analogous to the modern Himalayan Plateau in the Asian hinterland north of the collision suture with India (Dewey and Burke, 1973).

Bear Province.—The westernmost of three belts comprising the Bear Province is the Great Bear Batholith, consisting of coalesced epizonal granodiorite to granite plutons that intruded a comagmatic cover of silicic, welded ash-flow tuffs, alkalic basalt flows, hypabyssal porphyries, and locally derived sediments. The western margin of the batholith is covered beneath the Mackenzie Lowlands, and the eastern margin is the 400-km-long Wopmay River Fault of unknown displacement. East of the fault is the Hepburn metamorphic-plutonic belt, where intensely deformed, so-called eugeosynclinal pelites, turbidities, and pillow basalts, for which no basement is known, were intruded by mesozonal diapirs of porphyroblastic granodiorite. Metamorphism is of a low-pressure facies series (Miyashiro, 1961) and attained K-feldspar-sillimanite grade around the granodiorite diapirs. The eastern limit of regional cleavage and metamorphism is called the tectonite front, beyond which is the Epworth fold and thrust belt, where miogeoclinal orthoquartzite and dolomite overlain by exogeosynclinal flysch have been thrust eastward over granitic basement. The fold and thrust belt consists of broad, gently folded synclinoria, in which flysch is exposed, separated by relatively narrow, tightly compressed anticlinoria of miogeoclinal rocks in

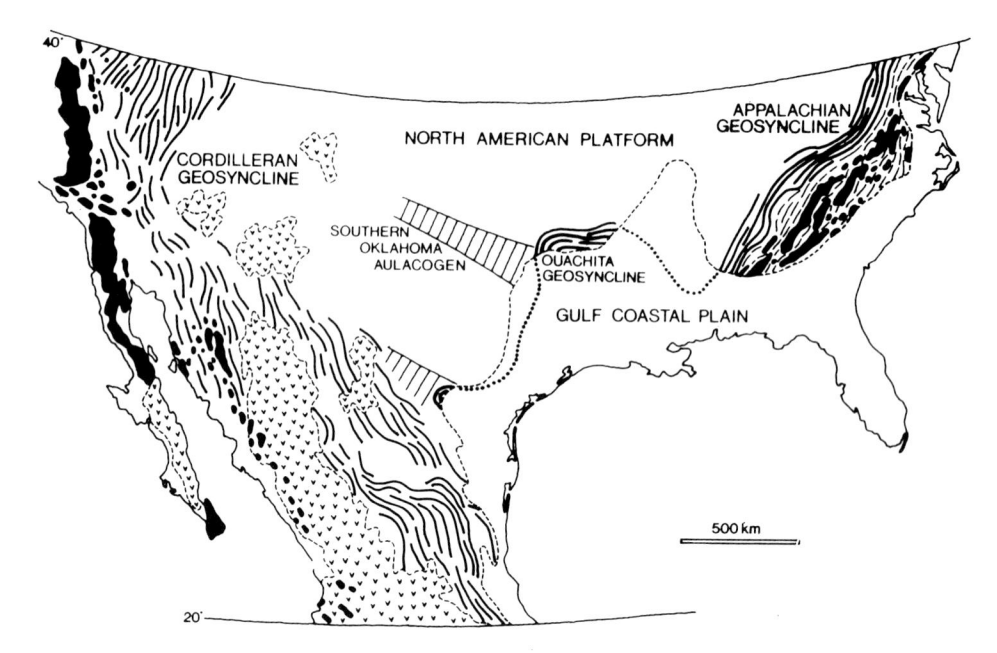

Fig. 1.—Map of part of North America showing major geologic structure and location of Southern Aulacogen (modified from King, 1969). Aulacogen is located where Ouachita extension of Appalachian orogen makes reentrant into North American Platform. See figure 2 for evolution of aulacogen.

which the thrust planes come to the surface. East of the thrust front are autochthonous sediments of the foreland platform. The full width (tectonically compressed) of the miogeocline is unmetamorphosed only in the northern part of the Bear Province. Southward, the miogeoclinal facies belt is truncated by plutons, which, in the extreme south, intruded granitic basement covered only by thin platform equivalents of the miogeocline.

Slave Province.—Most of the Slave Province consists of exhumed plutonic and metamorphic basement rocks. Autochthonous equivalents of the Coronation Geosyncline doubtless once covered the entire province but are now preserved only in a narrow strip east of the thrust front and in and around the two aulacogens. The Athapuscow Aulacogen is coextensive with the northeast-trending McDonald Fault system, and the Bathurst Aulacogen, which is less well known, is coextensive with the southeast-trending Bathurst Fault system. Supracrustal fill in the aulacogens is much thicker and more deformed than the adjacent platform cover, which is nearly flat lying except around the overturned margins of large, cold anticlinal basement uplifts east of the thrust front.

Churchill Province.—Like the Slave Province, the Churchill Province consists mainly of metamorphic and plutonic basement rocks but was later subjected to pervasive mylonitization, minor plutonism, and regional uplift that shed great volumes of fanglomerate into the Athapuscow Aulacogen and the intermontane Nonacho Basin. In the aulacogen, fanglomerate derived from the LaLoche River Uplift to the southeast lies unconformably on the southwesterly derived clastics of the Coronation exogeosyncline, indicating that the uplift postdates orogenesis in the Coronation Geosyncline. The uplift so denuded the southeastern margin of the aulacogen that its earlier history can be guessed at only by examining the boulders in the fanglomerate. They suggest that the Churchill Province, like the Slave Province, had been a stable platform. Whether the asymmetry of the aulacogen developed only in its final stage or from the beginning is the most important unanswered question regarding the evolution of the aulacogen.

Geochronology.—Radiometric age determinations in Canada have been summarized by Wanless (1970). Basement rocks of the Slave Province generally have ages of approximately 2500 Ma, the oldest K-Ar age being a 2555 Ma biotite from basement granodiorite near the northeast end of the Athapuscow Aulacogen. Rocks in the anticlinal basement uplifts east of the Epworth thrust belt have a broad range of K-Ar ages, from 2500 Ma to 1765 Ma, the

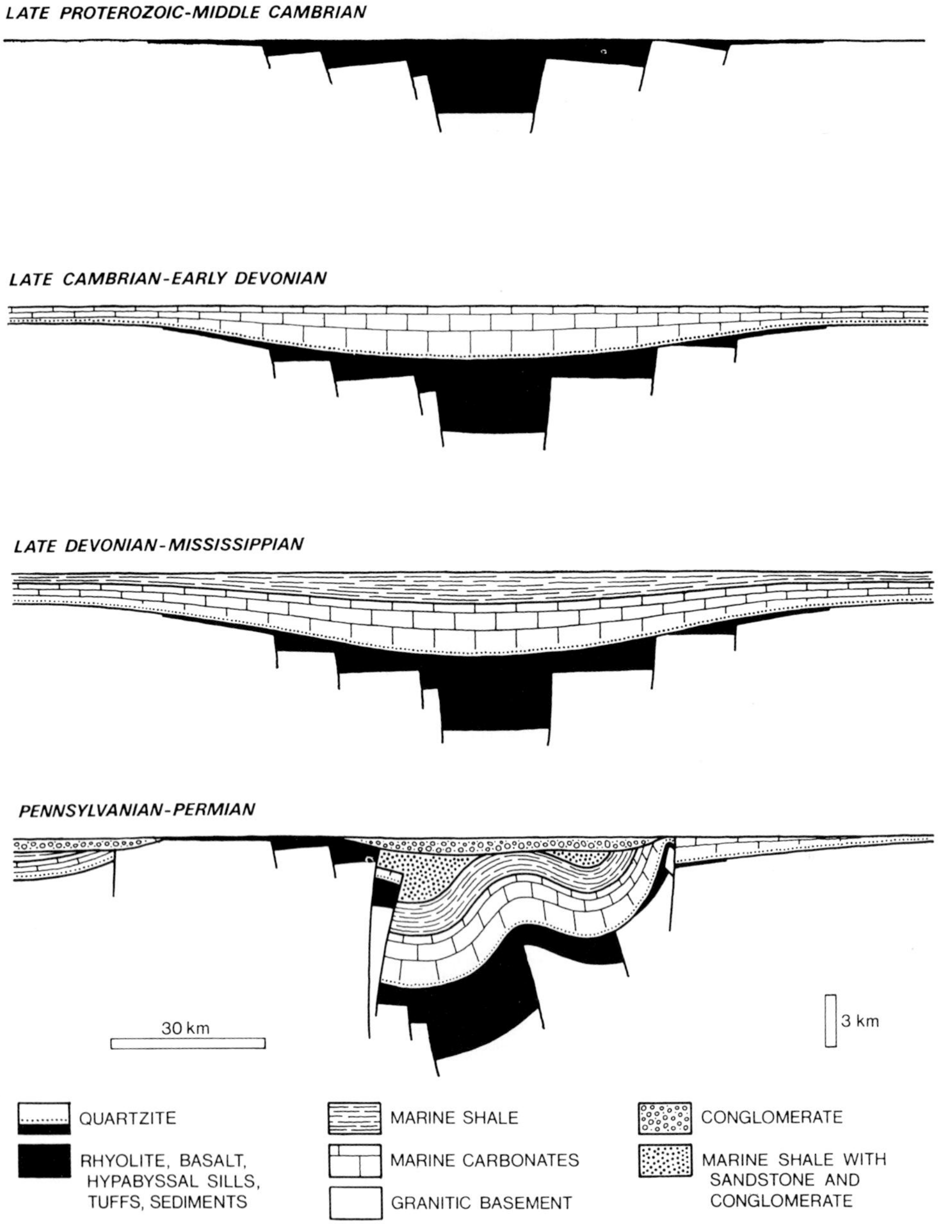

AULACOGENS AND THEIR RELATION TO GEOSYNCLINES

LATE PROTEROZOIC-MIDDLE CAMBRIAN

LATE CAMBRIAN-EARLY DEVONIAN

LATE DEVONIAN-MISSISSIPPIAN

PENNSYLVANIAN-PERMIAN

	QUARTZITE		MARINE SHALE		CONGLOMERATE
	RHYOLITE, BASALT, HYPABYSSAL SILLS, TUFFS, SEDIMENTS		MARINE CARBONATES		MARINE SHALE WITH SANDSTONE AND CONGLOMERATE
			GRANITIC BASEMENT		

FIG. 2.—Series of schematic transverse cross sections showing evolution of Southern Oklahoma Aulacogen (data from Ham, 1969). Aulacogen begins with graben stage of block faulting and volcanism, evolves into broad unfaulted downwarp, and ends with compressional stage of folding, faulting, and fanglomerate sedimentation. See figure 1 for relationship to Appalachian orogen and North American Platform.

youngest perhaps being close to the age of uplifting. Similarly, basement rocks of the La-Loche River Uplift in the Churchill Province have a broad range of ages, from 2575 Ma to 1650 Ma, the youngest perhaps indicating the age of mylonitization, uplift, and fanglomerate deposition in the Athapuscow Aulacogen. Basement rocks in the aulacogen are intruded by 2170 Ma dikes that do not cut the miogeoclinal facies equivalents there, thus providing a maximum age for deposition related to the Coronation Geosyncline.

Minimum ages for the Coronation Geosyncline include 1790 Ma for quartz diorite laccoliths in the Athapuscow Aulacogen, 1765 Ma for granodiorite in the Hepburn metamorphic-plu-

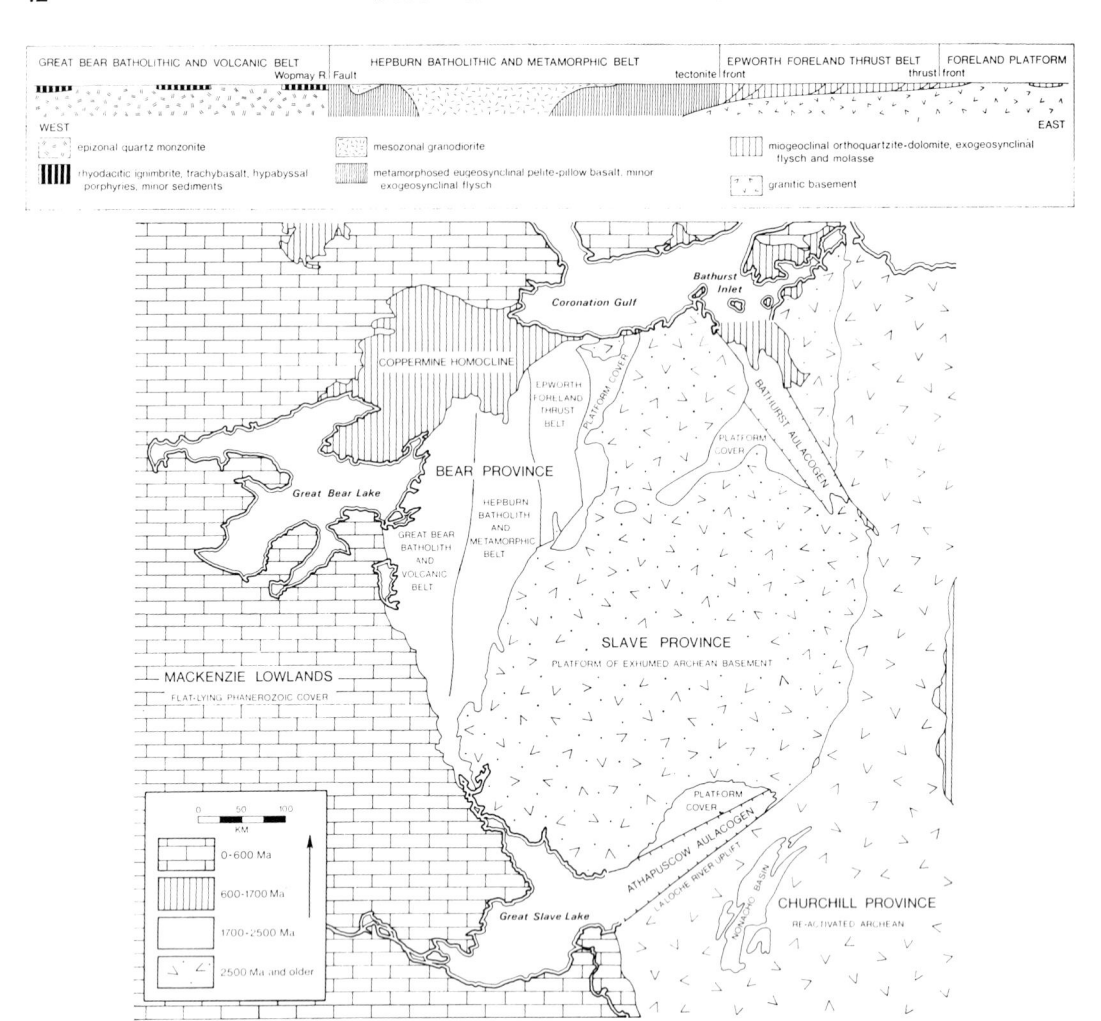

FIG. 3.—Tectonic map of northwestern part of Canadian Shield (modified from Stockwell, 1968) and structural cross section of type area of Coronation Geosyncline near north end of Bear Province (data from Hoffman, 1973a). See correlation chart in figure 4.

tonic belt, and 1740 Ma for the Great Bear Batholith. A minimum age for the postgeosynclinal fanglomerate in the aulacogen is provided by 1200 Ma diabase dikes. Geochronologic precision will be greatly improved with completion of Rb-Sr and U-Pb dating programs, but, for now, deposition in the geosyncline is best assumed to have occurred within the time interval of 2200 to 1700 Ma, which is termed Aphebian in the time-stratigraphic classification used by the Geological Survey of Canada (Stockwell, 1964).

Coronation Geosyncline

Five tectonic stages are recognized in the development of the foreland of the Coronation Geosyncline, involving eight phases of deposition (fig. 5) to which the evolution of the aulacogen can be well correlated (table 1). The first three depositional phases constituted the miogeoclinal sequence, interpreted as a westward facing, midplate, continental shelf sequence. The fourth phase was transitional and signaled foundering of the continental shelf. The following three phases constituted the exogeosynclinal sequence, a westerly derived, shoaling-upward clastic wedge beginning with flysch and ending with molasse. Compressional deformation of the foreland succession, probably in part coeval with exogeosynclinal sedimentation, was followed by transcurrent faulting and deposition of postgeosynclinal fanglomerate, now minimally preserved.

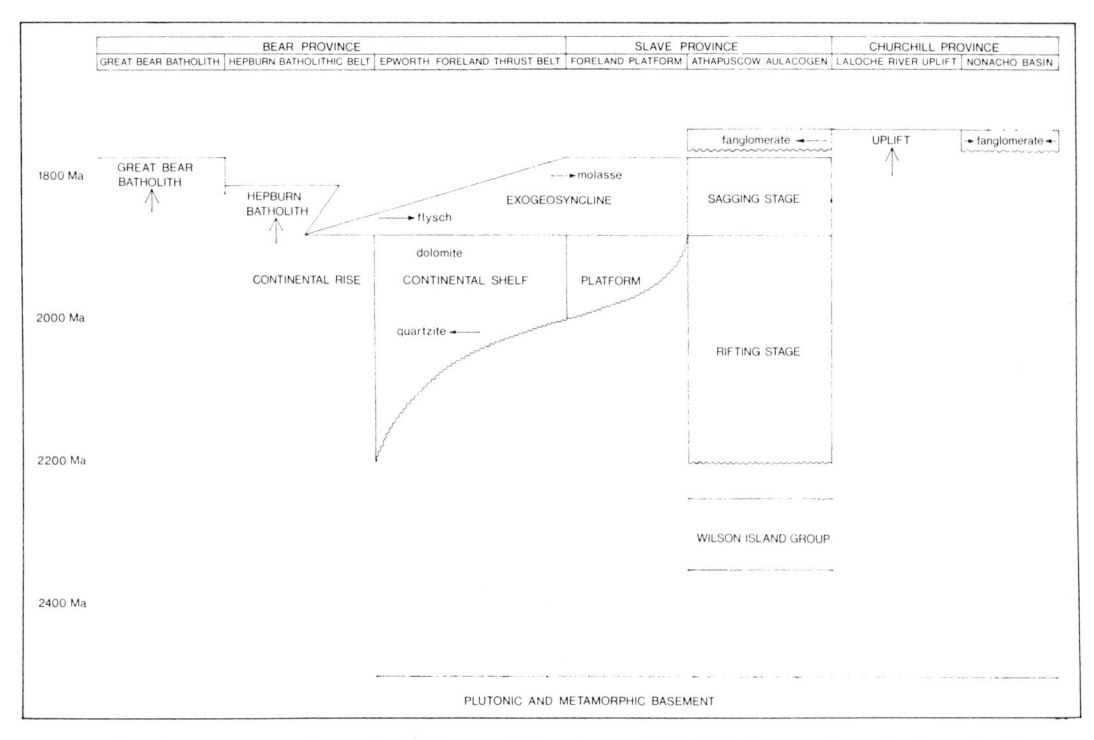

FIG. 4.—Correlation chart to be used with figure 3. Based on published K-Ar age determinations cited in text and on unpublished Rb-Sr isochrons.

Stage I—miogeoclinal stage

Prequartzite phase: East of the tectonite front, rocks of the prequartzite phase were deposited on granitic basement, but, to the west, no basement is known. Near the thrust front, there are but 50 m of varicolored laminated mudstone capped by tuffaceous and stromatolitic dolomite. The sequence thickens westward, and, near the tectonite front, 250 m of cherty stromatolitic dolomite with lenses of quartz sand overlie many hundreds of meters of pillow basalt and basalt breccia. Thick sequences of pillow basalt in the western part of the Hepburn metamorphic-plutonic belt, possibly correlative with the prequartzite phase, overlie great thicknesses of mudstone, arkose, and granite pebble-

TABLE 1.—CORRELATION OF STAGES OF DEVELOPMENT OF CORONATION GEOSYNCLINE AND ATHAPUSCOW AULACOGEN

Stage	Geosyncline	Aulacogen
Stage V *Fanglomerate phase*	*Postgeosyncline stage* Unnamed formation	*Postgeosyncline stage* Et-then Group
Stage IV	*Compressional stage*	*Compressional stage*
Stage III *Molasse phase* *Calc-flysch phase* *Flysch phase*	*Exogeosynclinal stage* Takiyuak Formation Cowles Lake Formation Recluse Formation	*Downwarping stage* Christie Bay Group Pethei Group Kahochella Group
Stage II *Preflysch phase*	*Transitional stage* Basal Recluse Formation	*Transitional stage* Seton Formation
Stage I *Dolomite phase* *Quartzite phase* *Prequartzite phase*	*Miogeoclinal stage* Rocknest Formation Odjick Formation Unnamed formation	*Graben stage* Duhamel Formation Hornby Channel Formation Union Island Formation

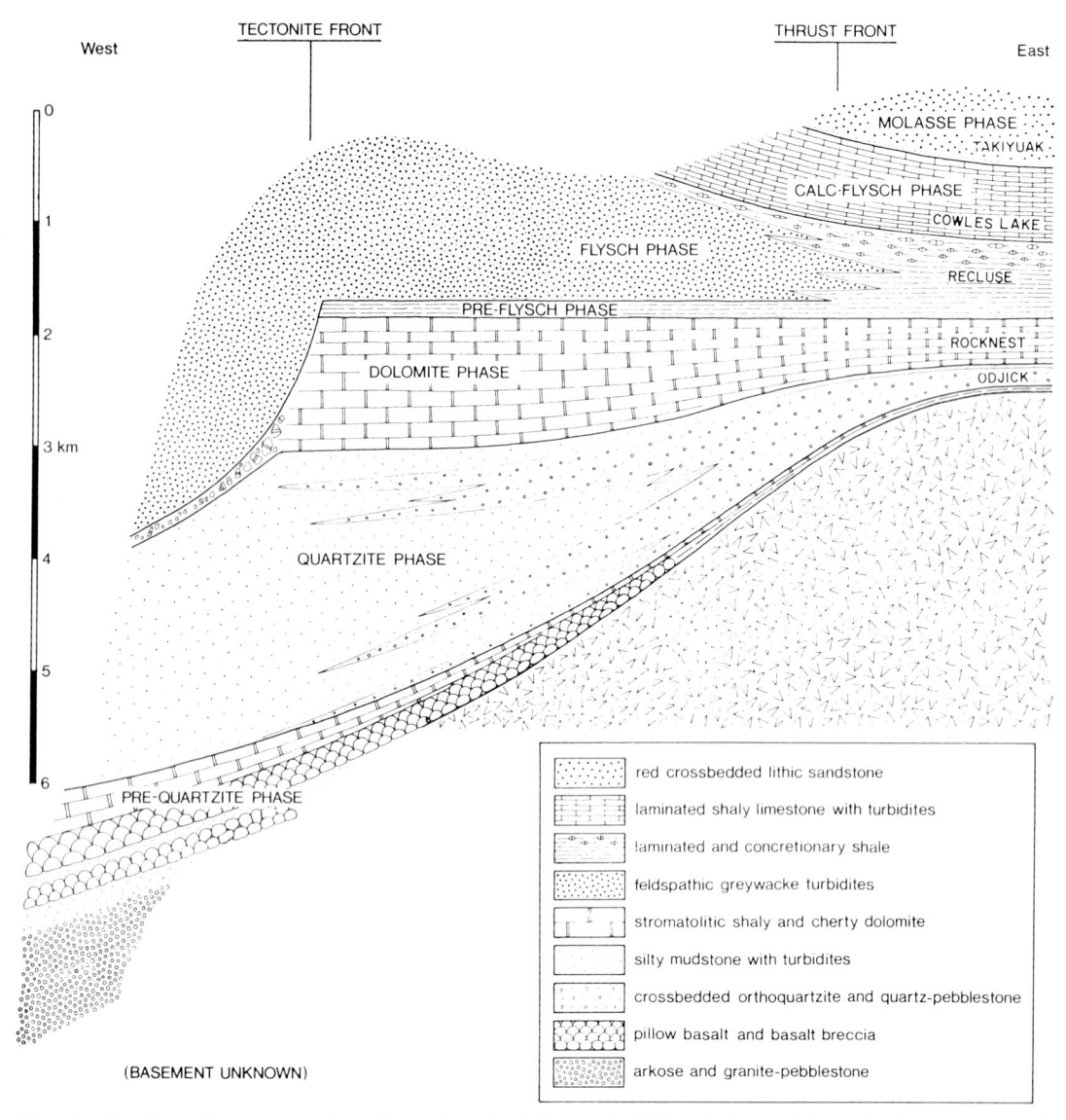

FIG. 5.—Stratigraphic cross section of north end of Epworth fold and thrust belt and adjacent parts of Coronation Geosyncline (data from Fraser and Tremblay, 1969; Hoffman, 1973a).

stone intruded by basic and silicic sills, all intensely deformed.

Quartzite phase: Varicolored crossbedded orthoquartzite, as little as 140 m thick east of the thrust front, becomes progressively more argillaceous westward and thickens to 750 m at the thrust front and perhaps to as much as 3000 m at the tectonite front. To the west are monotonous sequences of laminated silty mudstones and quartzite turbidites and slump breccias. Paleocurrents were from east to west, transverse to the regional trend of the shelf.

Dolomite phase: Thickening westward from 500 m east of the thrust front to 1200 m at the tectonite front is a magnificently exposed sequence of alternating cherty stromatolitic dolomite and laminated dolomitic shale. Most of the sequence consists of stacked, shoaling-upward cycles, each 2 to 15 m thick, resulting from progradation of tidal flats across shallow shelf lagoons (Hoffman, 1973b). The sharply delineated outer edge of the shelf is close to the tectonite front and is characterized by less apparent cyclicity and by large mounds of columnar stromatolites. Only a kilometer to the west is a starved foreshelf slope sequence, 110 m thick, of shale and dolomite breccia beds containing blocks as much as tens of meters in diameter.

Stage II—transitional stage

Preflysch phase: The top of the miogeocline is draped by a few tens of meters of black, pyritic, laminated shale with thin, easterly derived, quartz-siltstone turbidites at the base. This starved sequence was deposited as the underlying shelf foundered and the foreshelf basin filled with westerly derived flysch, soon to spill out onto the drowned shelf.

Stage III—exogeosynclinal stage

Flysch phase: More than 1400 m of coarse, feldspathic graywacke turbidites overlie the preflysch at the outer edge of the shelf and the foreshelf slope to the west. The graywacke tongues thin eastward and pinch out near the thrust front into an 800-m-thick sequence of dark-green to black shale containing calcareous concretions in its upper part.

Calc-flysch phase: East of the thrust front, the shaly distal equivalents of the flysch phase are overlain by approximately 500 m of finely laminated shaly limestone and scattered fine-grained graywacke turbidites and slump breccia beds.

Molasse phase: The calc-flysch is capped by a 10-m-thick olistostrome in which stromatolitic limestone blocks are dispersed in a red mudstone matrix. The olistostrome is overlain by nearly 125 m of red, mud-cracked silty mudstone followed by more than 500 m of westerly derived, red, cross-bedded lithic sandstone.

Stage IV—compressional stage

Compressional deformation of the foreland supracrustal wedge is genetically related to the rise and dilation of the granodiorite diapirs, the largest of which is 125 km long and 30 km wide, in the Hepburn metamorphic-plutonic belt. Deformation was probably diachronous, becoming younger eastward, and may have been in part coeval with exogeosynclinal sedimentation. The anticlinal basement uplifts east of the thrust front rose contemporaneously with the advancing thrust sheets of the Epworth belt. With the exception of the basement uplifts, the vergences of which are highly variable, all structures indicate eastward translation.

Stage V—postgeosynclinal stage

Fanglomerate phase: Friable conglomerate containing pebbles of molasse sandstone and older rocks, including basement rocks, occurs east of the thrust front in a single location close to one of the many northeast-trending, dextral transcurrent faults that displace the thrust traces and basement uplifts.

The depositional history and structural zonation of the foreland supracrustal wedge of the Coronation Geosyncline are fundamentally similar to those of the Paleozoic Ouachita Geosyncline (Flawn, Goldstein, King, and Weaver, 1961), to which W. E. Ham's Southern Oklahoma Aulacogen is related.

Athapuscow Aulacogen

Rocks of the Athapuscow Aulacogen, exposed along the folded and complexly faulted southeast half of the east arm of Great Slave Lake (fig. 6), thicken southwestward from 2200 m to at least 7000 m. To the north, contiguous sediments only 1400 m thick dip gently off the Slave Province platform (fig. 7).

Representatives of all eight phases of deposition in the geosyncline occur in the aulacogen (table 1). During the miogeoclinal stage of the geosyncline, the aulacogen was a narrow fault-bounded rift valley, or graben, apparently having elevated margins from which the first three phases of deposition are absent (fig. 8). During the transitional and exogeosynclinal stages, the margins sagged, and the aulacogen became a broad downwarp, ultimately suffering mild transverse compression. Finally, the aulacogen again became an active fault zone, then having dextral transcurrent movement in addition to vertical movement, and alluvial fanglomerates of the Et-then Group were shed into the aulacogen, mainly from the LaLoche River Uplift to the south.

Stage I—graben stage

Prequartzite phase: These rocks, up to 1000 m thick, are locally preserved in the southwestern half of the aulacogen and were tilted and faulted before deposition of the succeeding quartzite phase. Their stratigraphy varies locally but normally begins with nonstromatolitic dolomite, at the base of which is a discontinuous veneer of arkosic sand and breccia overlying the basement granite. The dolomite is overlain by black, pyritic, and carbonaceous slate, which is overlain in turn by a thick complex of pillow basalt, pillow breccia, and gabbro sills that are capped by cherty dolomite bearing lenses of quartz pebblestone.

Quartzite phase: Rocks of this phase are coarser grained and more feldspathic in the aulacogen than in the miogeocline, probably because of the active fault scarps bordering the aulacogen. Pebbly trough-cross-bedded subarkose thickens from less than 200 m at the northeast end of the aulacogen to more than 1600 m

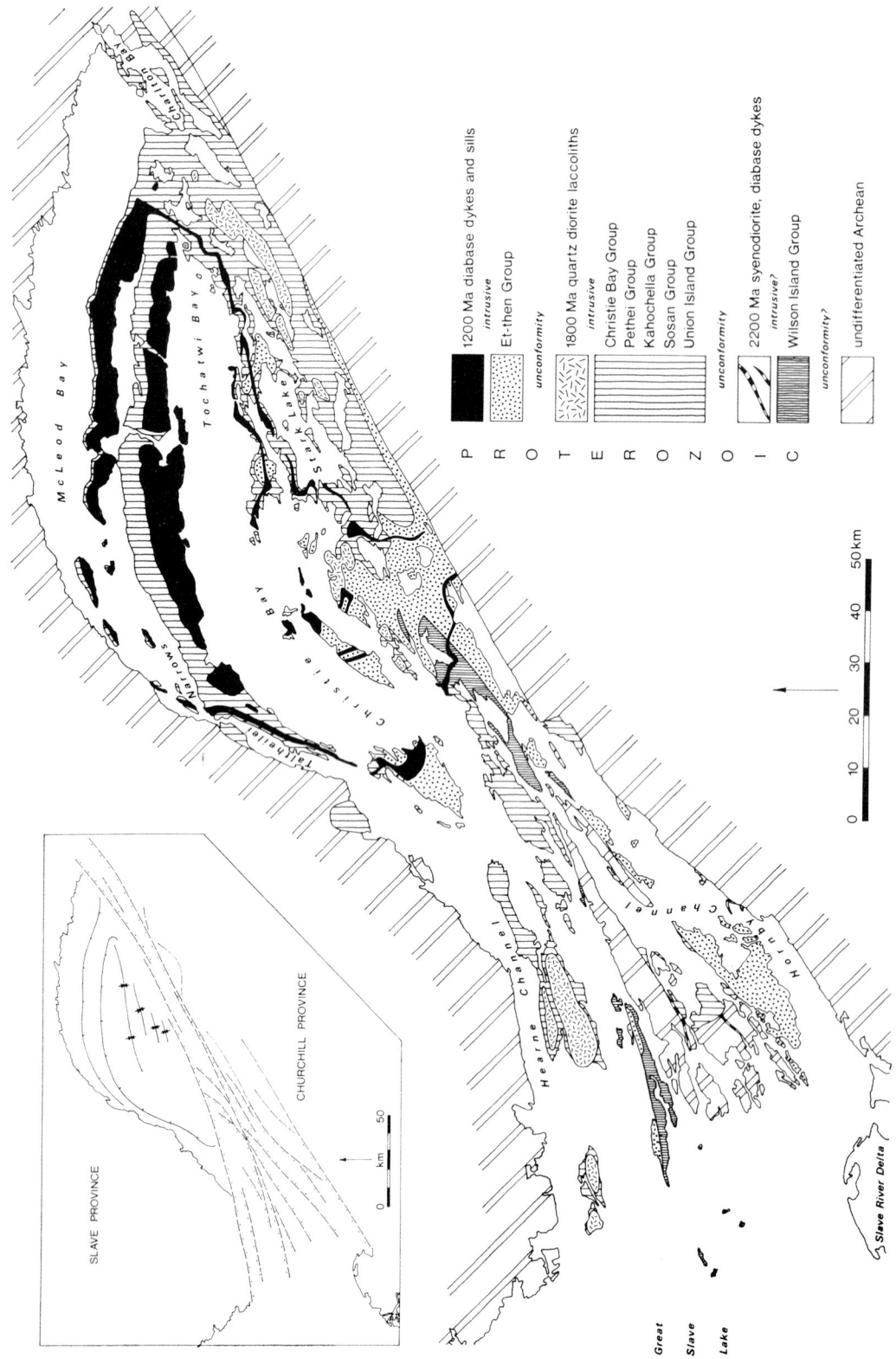

FIG. 6.—Geologic and tectonic (inset) maps of east arm of Great Slave Lake (modified from Stockwell, 1936).

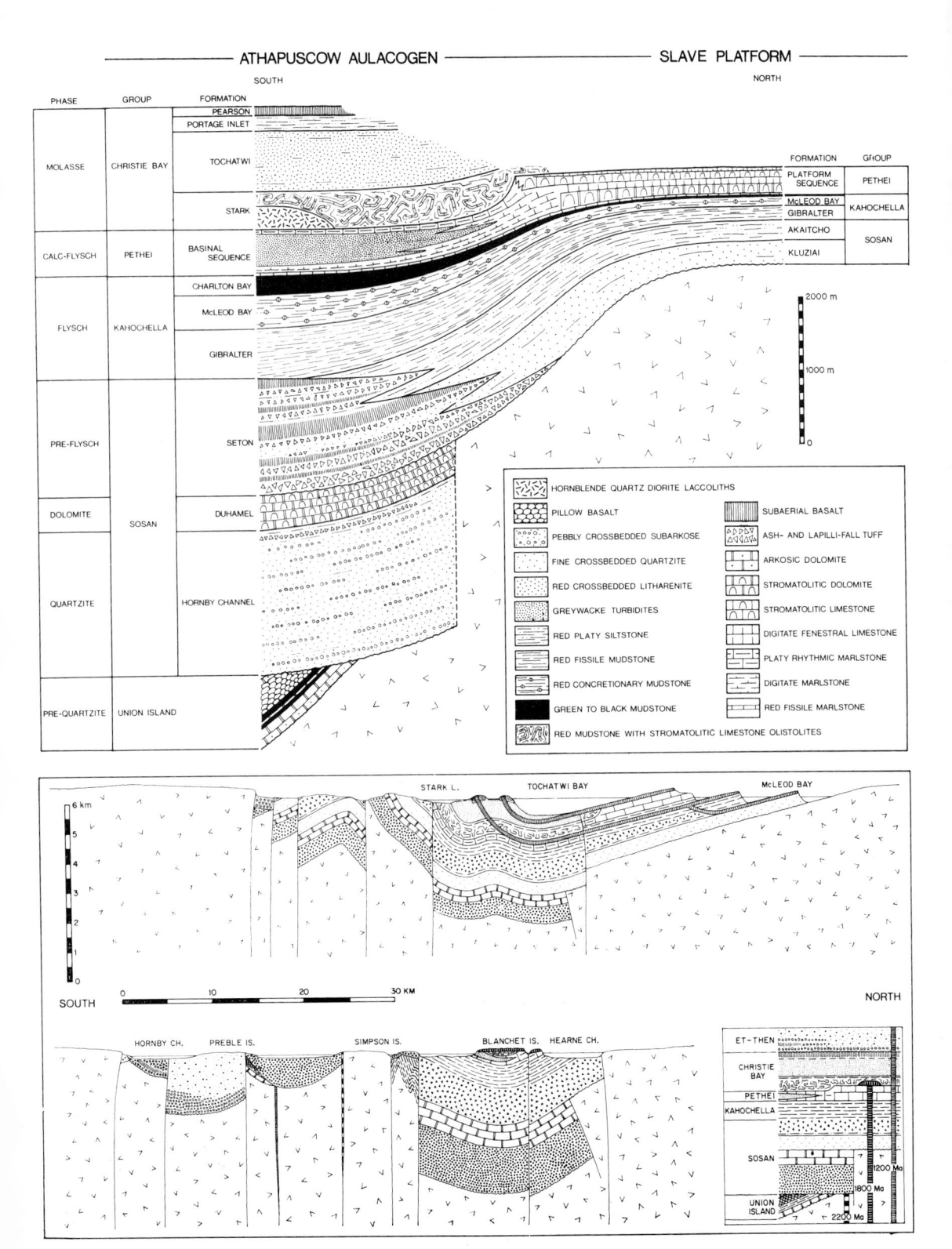

FIG. 7.—Stratigraphic cross section of Athapuscow Aulacogen and adjacent platform (data from Hoffman, 1968, 1969, 1973a). Structural cross sections are representative of northeast half (above) and southwest half (below) of Aulacogen.

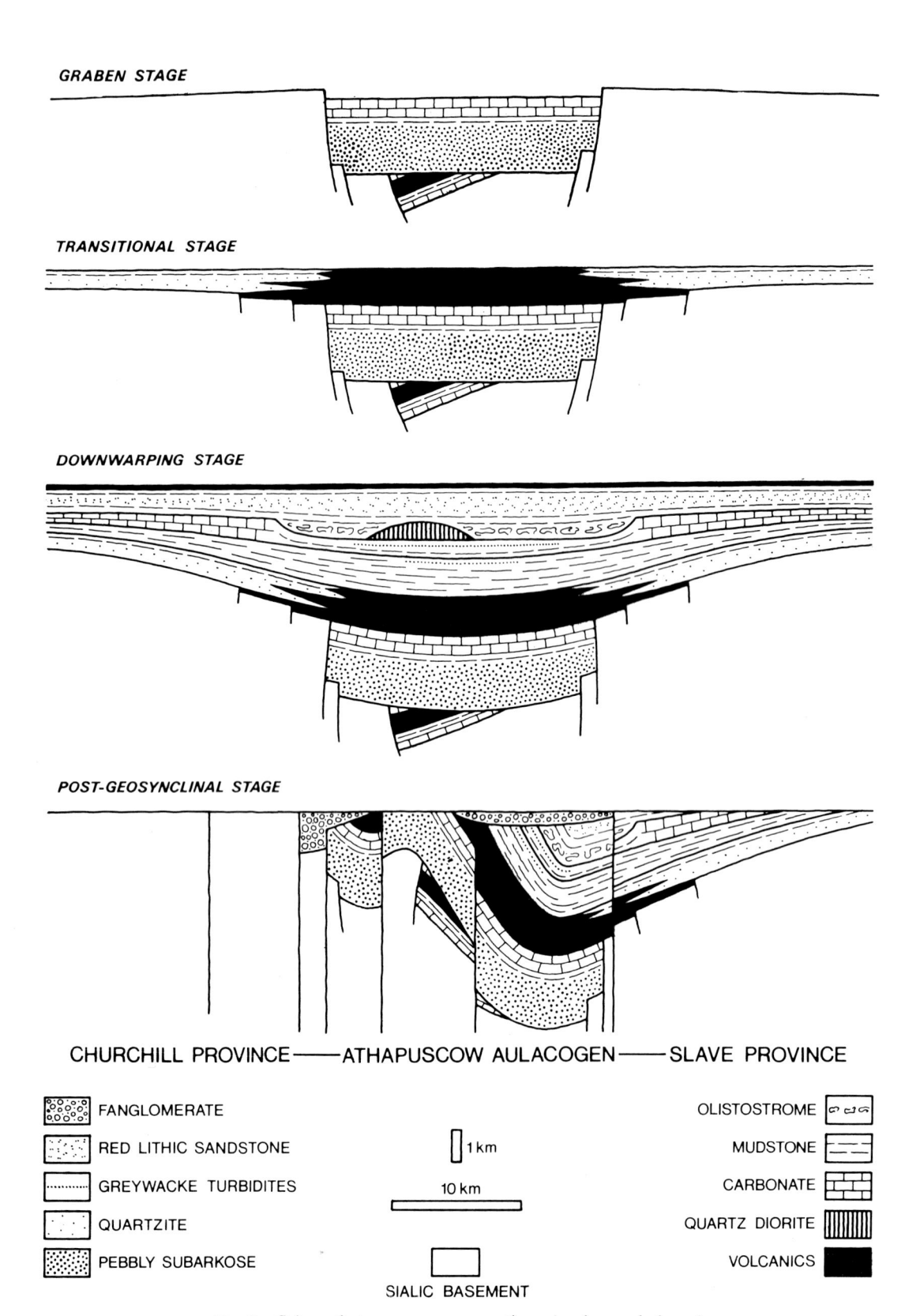

FIG. 8.—Schematic transverse cross sections showing evolution of Athapuscow Aulacogen. Compare with figure 2.

at the southwest end. These monotonous sequences were deposited by braided rivers that flowed southwestward down the axis of the rift valley. Glauconitic quartzite, together with flaser-bedded silty mudstone and basic ash-fall tuff, occurs at the top in the transition to the overlying dolomite.

Dolomite phase: The dolomite phase consists of the same cyclic alternation of laminated dolomitic shale and cherty stromatolitic dolomite as in the miogeocline. However, there are, in addition, beds of white cross-bedded orthoquartzite, probably derived from the fault scarps bordering the aulacogen.

Stage II—transitional stage

Preflysch phase: During this phase, markedly different from the starved black shale phase of the geosyncline, sediments for the first time accumulated on the platform adjacent to the aulacogen. On the platform and in the northeastern part of the aulacogen are perhaps 470 m of pink and gray, fine-grained, even-textured, cross-bedded quartzite overlain by 280 m of red, rippled, and mudcracked siltstone. Toward the southwest end of the aulacogen, the sequence thickens to 1400 m and contains increasing proportions of volcanic rocks, mainly spilitic and quartz keratophyric tuffs, spilitic and trachytic columnar flows, minor rhyolitic flows, kilometer-wide vent breccias, and hypabyssal albite granophyres and porphyries (Olade and Morton, 1972).

Stage III—downwarping stage

Flysch phase: Red and green shales, distal equivalents of the graywacke turbidites in the geosyncline, thin from 1450 m at the southwest end of the aulacogen to 350 m at the northeast end of and on the adjacent platform. The lower parts of the sequence, particularly on the platform, contain thin beds, interpreted as paleosols, of granular hematite ironstone associated with spherulitic and stromatolitic carbonate rock, gypsum casts, chert nodules, and intraformational conglomerate. The upper parts of the sequence are highly concretionary, bearing lenses of concretion-pebble conglomerate.

Calc-flysch phase: The contrast in facies between the platform and the aulacogen is dramatic (Hoffman, 1974). On the platform are 400 m of stromatolitic, loferitic (Fischer, 1964), oolitic, and intraclastic limestone and dolomite that was deposited in shallow water. At the southwest end of the aulacogen, correlative rocks consist of 500 m of relatively deep-water facies: rhythmically thin-bedded lithographic limestone having shale partings, calcareous shale bearing columnar calcareous growth structures that resemble stromatolites but that are poorly laminated and formed by *in situ* carbonate precipitation, and tongues of westerly derived graywacke deposited by turbidity currents that flowed along the axis of the aulacogen parallel to the edge of the carbonate platform. The platform edge is marked by a belt of stromatolite mounds, up to 80 m across and 20 m thick, separated by transverse tidal channels, and filled with cross-bedded intraclast grainstone. Beds of limestone breccia, some with displaced stromatolitic blocks, occur locally in the foreslope facies close to the edge of the platform. Toward the northeast end of the aulacogen, the graywacke turbidite tongues pinch out, and the basinal facies is reduced to 250 m in thickness.

Molasse phase: Sharply overlying the basinal facies of the calc-flysch phase is a gigantic olistostrome, in which angular blocks of stromatolitic dolomite and limestone, distinct from any rocks in the calc-flysch phase, are chaotically dispersed in a red mudstone matrix. The largest blocks, many recumbently folded or emplaced upside-down, reach 45 m in thickness and a kilometer in length. Quartz diorite to granodiorite laccoliths, up to 25 km in lenth, were intruded into the olistostrome, perhaps contemporaneously with its emplacement, along nearly the entire length of the aulacogen. The olistostrome is overlain by a coherent sequence, up to 450 m thick, of red mudstone with siltstone turbidites, ubiquitous hopper-shaped halite casts, and local pillow lava flows. This sequence shoals at the top, where rippled and mud-cracked red siltstone is overlain by up to 850 m of red cross-bedded lithic sandstone, which was deposited by rivers that flowed from the southwest along the axis of the aulacogen. Above the sandstone are 230 m of red mud-cracked siltstone containing halite and gypsum casts, and 180 m of columnar alkalic basalt flows.

Stage IV—compressional stage

A stage of mild compressional deformation separated the calc-flysch phase of deposition from the unconformably overlying fanglomerate. Locally in the axis of the aulacogen, cleavage is well developed and fold limbs are overturned, but, unlike the foreland of the geosyncline, there is no low-angle overthrusting and no consistent direction of translation.

Stage V—postgeosynclinal stage

Fanglomerate phase: Overlying the upturned edges of older rocks are as much as 4000 m of friable red and buff boulderstone and pebbly

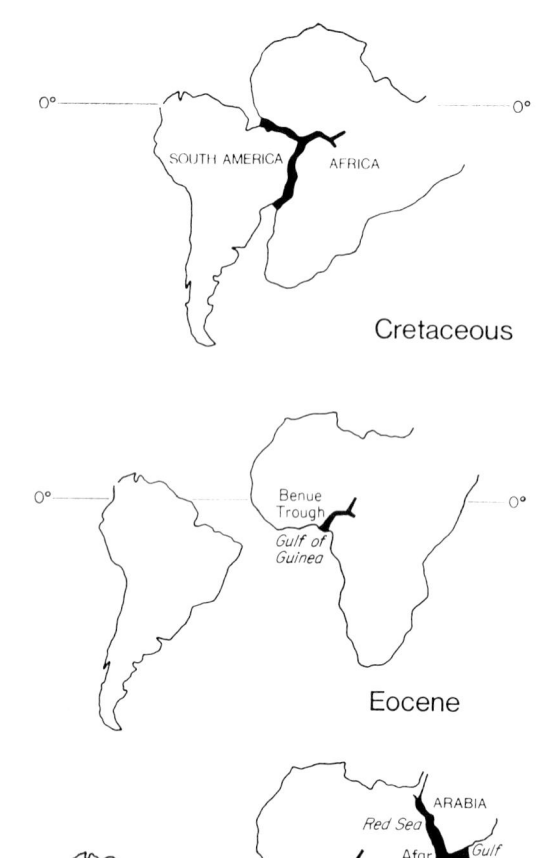

FIG. 9.—Maps showing origin of Benue Trough as rift arm abandoned during continental rift separation of Africa and South America in Cretaceous time and, similarly, Ethiopian Rift valley during separation of Africa and Arabia in last 25 million years (data from Smith, Briden, and Drewry, 1973). Plume-induced, three-armed rift systems are believed to have radiated during Cretaceous time from what is now mouth of Benue Trough and from Afar region at present. Rifting is believed to mark times when continent was stationary with respect to deep mantle plumes.

sandstone, locally enclosing alkalic basalt flows that were derived from the LaLoche River Uplift and from horst blocks within the aulacogen. The clasts record progressive erosion of platform cover from the uplift and unroofing of the basement. The tendency of the fanglomerates to occur in homoclines dipping parallel to the trend of the faults from which they are derived is typical of transcurrent fault basins (Crowell, this volume). Dextral transcurrent movement on the McDonald Fault system is indicated by mylonites in basement rocks of the LaLoche River Uplift (Reinhardt, 1969), but the age of mylonitization relative to deposition in the aulacogen is uncertain.

The evolution of the Athapuscow Aulacogen parallels that of the Paleozoic Southern Oklahoma Aulacogen. The quality of exposures is such that errors of interpretation must be the fault of the geologist, not the rocks.

MODERN AULACOGENS OF AFRICA

Two particularly interesting examples of transverse troughs located at reentrants on modern continental margins occur in Africa (fig. 9)—namely, the Benue Trough (Cratchley and Jones, 1965; Burke, Dessauvagie, and Whiteman, 1971; Murat, 1970; Grant, 1971; Nwachukwu, 1972; Burke and Whiteman, 1973) and the Afar Depression (Falcon, Gass, Girdler, and Laughton, 1970; Schilling, 1973). To the extent that continental margins are potential geosynclines, such transverse troughs are juvenile aulacogens. Significantly, both troughs developed during periods of continental breakup.

Benue Trough. — This northeast-trending trough extends into the continental interior from the Gulf of Guinea reentrant on the west coast of Africa. It began as an Early Cretaceous graben that developed contemporaneously with the initial rift separation of Africa and South America. However, rifting in the trough was short lived, producing oceanic crust only near its mouth, and by Late Cretaceous time the trough was subjected to compressional deformation. The southwest end of the trough continues to subside and is now the site of sedimentation from the Niger River and its prograding delta.

Afar Depression.—The Afar Depression is located where the main Ethiopian rift emerges at the coastal reentrant that was produced by rift separation, beginning in the Miocene, of Africa and Arabia along the Red Sea and Gulf of Aden. Although the Ethiopian Rift has been active almost as long as the Red Sea and Gulf of Aden have existed and has been a zone of almost continuous alkalic volcanism, oceanic crust has been produced only near its mouth, and there only during the Recent Epoch. Whether in the future the Afar-Ethiopian rift will be abandoned, as was the Benue Trough, or will lead to the continental separation of Somalia from Africa will determine whether the rift becomes an aulacogen or an ocean.

AULACOGENS AND PLATE TECTONICS

What is required is a genetic model that explains the origin and development of aulacogens within the context of plate tectonics. The model must provide a coherent rationale for: (1) the transverse orientation of aulacogens, (2) their location at reentrants of geosynclines or continental margins, (3) the contemporaneity of their initiation with periods of continental breakup, (4) their evolutionary trend from narrow grabens to broad downwarps, (5) their predominantly alkalic volcanicity, (6) the compression of their supracrustal fill, and (7) the peculiar combination of increased crustal thickness and positive Bouguer anomaly coincident with the Athapuscow and Southern Oklahoma Aulacogens (Tryggvason and Qualls, 1967; Barr, 1971; Goodacre, 1972, table 1).

Where local upwellings or plumes of hot material rise beneath a continent stationary with respect to the plumes, domal uplifts that ultimately break into three-armed crestal rift systems, dominated by alkalic volcanism, are to be expected (Cloos, 1939; Burke and Wilson, 1972; Burke and Whiteman, 1973). If only two of the rift arms are accommodated by the world plate system and lead to continental separation, creating a new ocean basin, the abandoned arm will remain as a transverse trough located at a reentrant on the new continental margin (fig. 10). Inasmuch as the poles of rotation of most plates are at high latitudes, east-west trending rift arms have perhaps a greater likelihood of being abandoned than north-south rift arms. So long as there is mantle upwelling beneath the trough, the trough will retain the form of the classic graben: elevated margins, normal boundary faults, and lack of transverse compression. However, cessation of upwelling, either by death of the plume or drifting of the continent above it, leads to heat dissipation, subcrustal contraction, and collapse of the graben (Hinze, 1972). At this stage, the domal uplift deflates, and the graben becomes centered over a broad downwarp. The lower crust of the central block, previously injected with basic magma during the stage of incipient rifting, may be converted to garnet granulite or eclogite and sink even deeper into the lighter subjacent mantle (Goodacre, 1972). Sinking of the central block results in inward sagging of the old graben walls. The end result is that the originally normal faults become high-angle reverse faults, the supracrustal fill suffers compressional deformation, and the dense crust of the central block is thickened, producing the geophysical anomalies observed. Thus, all seven requirements of the model are accounted for, and, to the extent that the model is correct, aulacogens support the contention that mantle plumes are important in initiating continental breakup.

A point worth making in closing is that aulacogens seem to be especially susceptible to reactivation and may control sedimentation even after long periods of dormancy. Examples of such behavior include the transcurrent faulting and resulting sedimentation of the fanglomerate phase in the Athapuscow Aulacogen and the much later localization of the mineralized Middle Devonian Presqu'ile reef trend by the McDonald Fault system (Bassett and Stout, 1967).

CONCLUSIONS

Aulacogens are long-lived deeply subsiding linear troughs that radiate from the interior of a continental platform and deepen outward toward the platform margin where they merge, at a high angle, into a geosyncline with which their fill is contemporaneous. They normally emerge where the geosyncline makes a reentrant angle into the platform. Aulacogens evolve from narrow grabens to broad downwarps, ultimately suffering transverse compressional deformation, and are susceptible to subsequent reactivation.

Aulacogens, being intracontinental, must be distinguished from geosynclines when one attempts to delineate ancient continental margins or continental collision sutures. Paleocurrents in aulacogens were mainly longitudinal rather than transverse as in many (but not all) geosynclines, their structures are dominated by high-angle faults rather than low-angle thrusts, their volcanism is mainly alkalic (commonly bimodal basalt-rhyolite associations) rather than tholeiitic and calc-alkaline, and they are generally unmetamorphosed. Unlike geosynclines, their tectonic evolution was dominated by vertical rather than horizontal crustal movements.

Aulacogens may originate as abandoned arms of three-armed radial rift systems located where mantle plumes rise beneath continents stationary with respect to the plumes. Geosynclines develop only on continental margins, produced by continental rift separation along the active arms. In this model, aulacogens and geosynclines may be somewhat intergradational, differing genetically in the extent of rifting. In geosynclines, rifting is sufficient to produce large areas of oceanic crust, later to be subducted. In aulacogens, rifting is insufficient to produce

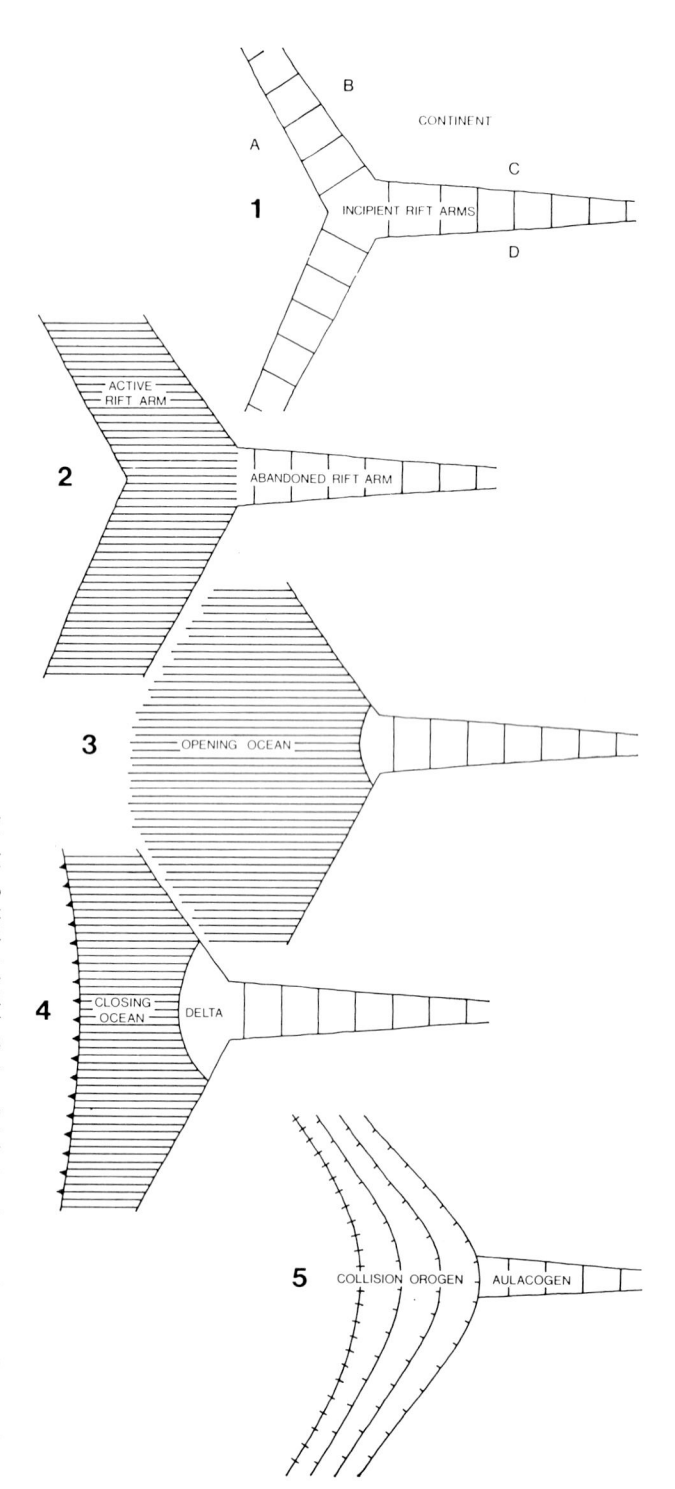

FIG. 10.—Genetic model of evolution of an aulacogen and related geosyncline. *1*, A three-armed radial rift system generated by crustal doming above mantle plume. *2*, Two rift arms spread to produce narrow rift ocean similar to Red Sea; rifting of third arm insufficient for continental separation. *3*, Spreading of two active rift arms produces large ocean basin. Downwarping and miogeoclinal sedimentation occur along new aseismic continental margin. Third arm is abandoned and remains as transverse trough located at reentrant on continental margin. Having lost support of plume from beneath, it evolves from incipient rift to broad downwarp. *4*, Ocean is closed by subduction along trench, producing adjacent volcanic arc. History of ocean closure may take many paths, only one of which is shown here. *5*, Closing of ocean ultimately results in continental collision and development of collision orogen similar to that of Himalayas. Abandoned rift arm is preserved as an aulacogen located where orogen makes reentrant into its foreland. Aulacogen is further loaded with exogeosynclinal sediments from advancing orogen, and its medial block founders, resulting in final stage of compressional deformation and faulting. Diagram not intended to imply precise temporal correlation of events in aulacogen and geosyncline.

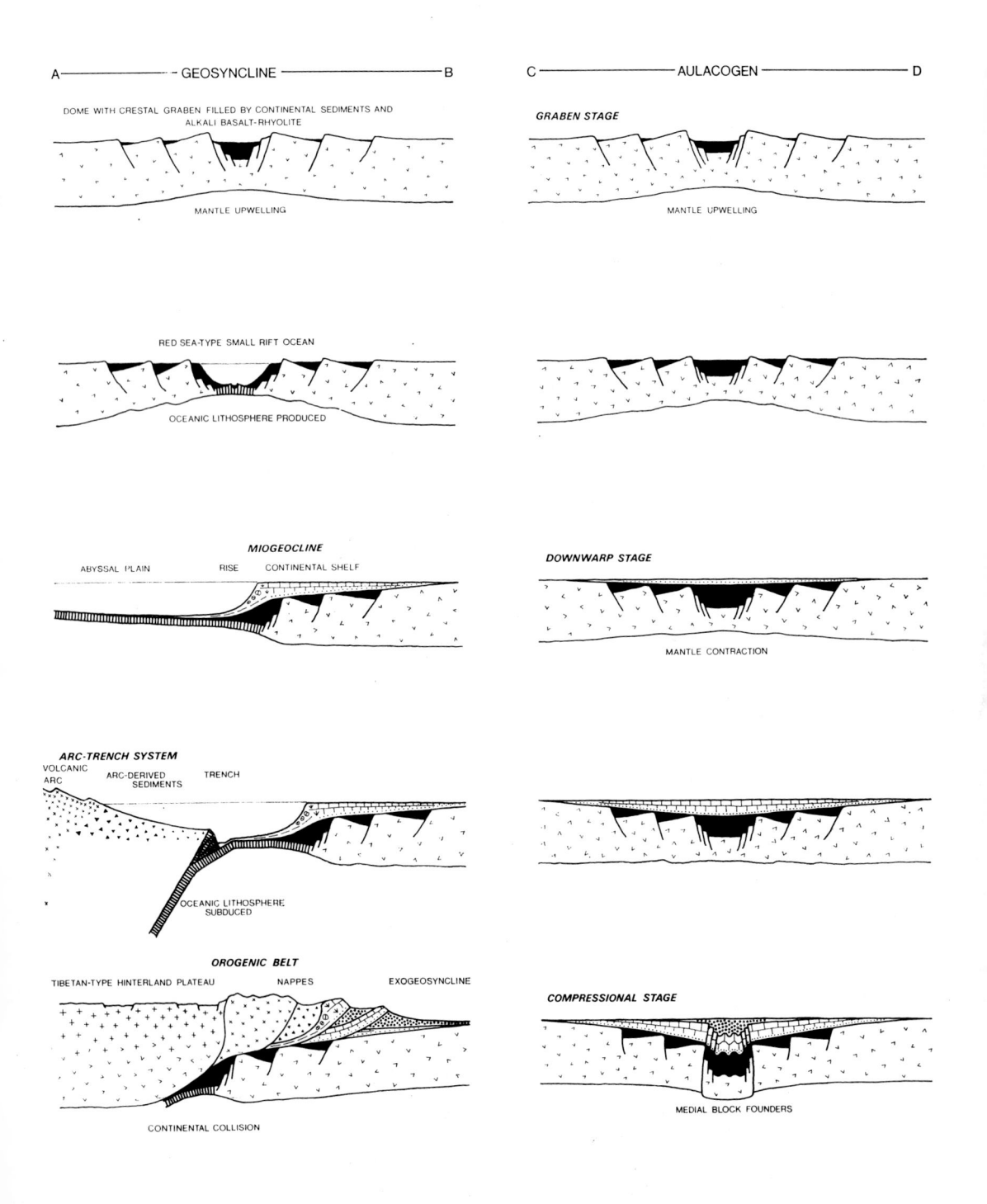

AULACOGENS AND THEIR RELATION TO GEOSYNCLINES

GEOSYNCLINE (A—B)

DOME WITH CRESTAL GRABEN FILLED BY CONTINENTAL SEDIMENTS AND ALKALI BASALT-RHYOLITE

MANTLE UPWELLING

RED SEA-TYPE SMALL RIFT OCEAN

OCEANIC LITHOSPHERE PRODUCED

MIOGEOCLINE

ABYSSAL PLAIN — RISE — CONTINENTAL SHELF

ARC-TRENCH SYSTEM

VOLCANIC ARC — ARC-DERIVED SEDIMENTS — TRENCH

OCEANIC LITHOSPHERE SUBDUCED

OROGENIC BELT

TIBETAN-TYPE HINTERLAND PLATEAU — NAPPES — EXOGEOSYNCLINE

CONTINENTAL COLLISION

AULACOGEN (C—D)

GRABEN STAGE

MANTLE UPWELLING

DOWNWARP STAGE

MANTLE CONTRACTION

COMPRESSIONAL STAGE

MEDIAL BLOCK FOUNDERS

much oceanic crust, and, therefore, there can be little or no subduction or arc-related magmatism. The controversy between proponents of primary vertical and horizontal tectonics may derive in part from the failure to distinguish aulacogens from geosynclines in the ancient record.

REFERENCES

BARR, K. G., 1971, Crustal refraction experiment: Yellowknife 1966; Jour. Geophys. Research, v. 76, p. 1929–1948.
BASSETT, H. G., AND STOUT, J. G., 1967, Devonian of western Canada, *in* Oswald, D. H. (ed.), International symposium on the Devonian System Calgary, Alberta Soc. Petroleum Geologists, v. 1, p. 717–752.
BELOUSSOV, V. V., 1939–40, The Greater Caucasus: parts I-III: TSNIIGRI Trans., v. 108, 121, 126.
BOGDANOV, A. A., 1964, Some general questions of the tectonics of ancient platforms (taking the East European Platform as an example) : Sov. Geologiya, no. 9.
BONDARCHUK, V. G., 1956, Tectonics of the Great Donets Basin and the origin of trough-like downwarpings of platforms: USSR, Akad. Nauk Geol. Zhur., v. 16.
BURKE, K., DESSAUVAGIE, T. F. J., AND WHITEMAN, A. J., 1971, Opening of the Gulf of Guinea and geological history of the Benue Depression and Niger delta: Nature and Phys. Sci., v. 233, p. 51–55.
———, AND WHITEMAN, A. J., in press (1973), Uplift, rifting and the break-up of Africa, *in* TARLING, D. H., AND RUNCORN, S. K. (eds.), Continental drift, seafloor spreading and plate tectonics: New York Academic Press.
———, AND WILSON, J. T., 1972, Is the African plate stationary?: Nature and Phys. Sci., v. 239, p. 387–390.
CHEKUNOV, A. V., 1967, Mechanism responsible for structure of the aulacogen type (taking the Dnieper-Donets Basin as an example) : Geotectonics, no. 3, p. 137–144.
CHIRVINSKAYA, M. V., 1958, Tectonic structure of the Dnieper-Donets depression and the Pripet Trough, *in* Tectonics of oil bearing Regions: Gostoptekhizdat, v. 2.
———, 1959, A concept of the tectonics of the Dnieper-Donets depression based on the results of geophysical research: Kiev, Izd. Ukr. NTO neft. i gaz. Prom.
CLOOS, H., 1939, Hebung-Spaltung-Vulcanismus: Geol. Rundsch., v. 30, p. 405–527.
CRATCHLEY, C. R., AND JONES, G. P., 1965, An interpretation of the geology and gravity anomalies of the Benue valley: Nigerian Overseas Geol. Survey Geophys. Paper 1, 26 p.
DEWEY, J. F., AND BIRD, J. M., 1970, Mountain belts and the new global tectonics: Jour. Geophys. Research, v. 75, p. 2625–2647.
———, AND BURKE, K., in preparation (1973), Tibetan, Variscan and Precambrian basement reactivation: products of continental collision.
DICKINSON, W. R., 1971, Plate tectonic models of geosynclines: Earth and Planetary Sci. Letters, v. 10, p. 165–174.
DOLENKO, G. N., VARICHEV, S. A., AND GALABUDA, N. I., 1970, Paleozoic stage of the Dnieper-Donets trough: Geotectonics, no. 1, p. 27–30.
FALCON, N. L., AND OTHERS, 1970, A discussion on the structure and evolution of the Red Sea and the nature of the Red Sea, Gulf of Aden and Ethiopia rift junction: Royal Soc. London Philos. Trans., ser. A, v. 267, p. 1–417.
FISCHER, A. G., 1964, The Lofer cyclothems of the Alpine Triassic, *in* MERRIAM, D. F. (ed.), Symposium on cyclic sedimentation: Kansas Geol. Survey, Bull. 169, p. 107–150.
FLAWN, P. T., AND OTHERS, 1961, The Ouachita System: Univ. Texas Pub. 6120, 401 p.
FRASER, J. A., AND OTHERS, 1972, The Bear Province, *in* PRICE, R. A., AND DOUGLAS, R. J. W. (eds.), Variations in tectonic styles in Canada: Geol. Assoc. Canada Special Paper 11, p. 453–503.
———, AND TREMBLAY, L. P., 1969, Correlation of Proterozoic strata in the northwestern Canadian Shield: Canadian Jour. Earth Sci., v. 6, p. 1–9.
GAERTNER, H. R. VON, 1969, Zur tektonischen und magmatischen Entwicklung der Kratone (Pyrenäen und Kaukasus als extreme Fälle der Aulacogene) : Geol. Jahrb. Beih., v. 80, p. 117–145.
GAVRISH, V. K., 1965, The role of abyssal faults in the formation of the structures of the Dnieper-Donets depression. USSR Akad. Nauk Geol. Zhur., v. 25.
GOODACRE, A. K., 1972, Generalized structure and composition of the deep crust and upper mantle in Canada: Jour. Geophys. Research, v. 77, p. 3146–3161.
GRANT, N. K., 1971, South Atlantic, Benue Trough and Gulf of Guinea Cretaceous triple junction: Geol. Soc. America Bull., v. 82, p. 2295–2298.
HAM, W. E., 1969, Regional geology of the Arbuckle Mountains, Oklahoma: Oklahoma Geol. Survey Guidebook 17, 52 p.
———, DENISON, R. E., AND MERRITT, C. A., 1964, Basement rocks and structural evolution of southern Oklahoma: *ibid.*, Bull. 95, 302 p.
———, AND WILSON, J. L., 1967, Paleozoic epeirogeny and orogeny in the central United States: Am. Jour. Sci., v. 265, p. 332–407.
HINZE, W. J., 1972, The origin of late Precambrian rifts: Geol. Soc. America Abs. with Programs, v. 4, p. 723.
HOFFMAN, P. F., 1968, Stratigraphy of the lower Proterozoic Great Slave Supergroup: Geol. Survey Canada Paper 68–42, 93 p.
———, 1969, Proterozoic paleocurrents and depositional history of the east arm fold belt, Great Slave Lake: Canadian Jour. Earth Sci., v. 6, p. 441–462.
———, 1973a, Evolution of an early Proterozoic continental margin: the Coronation Geosyncline and associated aulacogens of the northwestern Canadian Shield, *in* SUTTON, J., AND WINDLEY, B. F. (eds.), Evolution of the Precambrian crust: Royal Soc. London Philos. Trans., ser. A, v. 273, p. 547–581.

──, 1973b, Shoaling-upward shale-to-dolomite cycles in the Rocknest Formation, *in* GINSBURG, R. N., AND KLEIN, G. DEV. (eds.), Tidal deposits, a compilation of examples: Fisher Island, Miami, Florida, Comparative Sedimentology Lab., p. 1–6.

──, in press (1974), Shallow and deepwater stromatolites in a lower Proterozoic platform-to-basin facies charge, Great Slave Lake, Canada: Am. Assoc. Petroleum Geologists Bull., v. 58.

──, FRASER, J. A., AND MCGLYNN, J. C., 1970, The Coronation Geosyncline of Aphebian age, *in* BAER, A. J. (ed.), Basins and geosynclines of the Canadian Shield: Geol. Survey Canada Paper 70–40, p. 200–212.

HOPE, E. R., 1965, Translator's comments on paper by Nalivikin, V. D., 1963, Graben-like trenches in the east of the Russian Platform: Directorate Sci. Inform. Services, Defence Research Board Canada, Pub. T4OOR, p. i–ix.

KING, P. B., 1969, Tectonic map of North America: U.S. Geol. Survey.

MITCHELL, A. H., AND READING, H. G., 1969, Continental margins, geosynclines and ocean floor spreading: Jour. Geology, v. 77, p. 629–646.

MIYASHIRO, A., 1961, Evolution of metamorphic belts: Jour. Petrology, v. 2, p. 277–311.

MURAT, R. C., 1970, Stratigraphy and paleogeography of the Cretaceous and lower Tertiary in southern Nigeria, *in* DESSAUVAGIE, T. F. J., AND WHITEMAN, A. J. (eds.), African geology: Ibadan, Nigeria, Univ. Ibadan, p. 251–266.

NALIVKIN, V. D., 1963, Graben-like trenches in the east of the Russian Platform: Sov. Geologiya, no. 1, p. 40–52.

NWACHUKWU, S. O., 1972, The tectonic evolution of the southern portion of the Benue Trough, Nigeria: Geol. Mag., v. 109, p. 411–419.

OLADE, M. A. D., AND MORTON, R. D., 1972, Observations on the Proterozoic Seton Formation, east arm of Great Slave Lake: Canadian Jour. Earth Sci., v. 9, j. 1110–1123.

PRICE, R. A., AND DOUGLAS, R. J. W., 1972, Variations in tectonic styles in Canada: Geol. Assoc. Canada Special Paper 11, 688 p.

REINHARDT, E. W., 1969, Geology of the Precambrian rocks of Thubun Lakes map-area in relationship to the McDonald Fault system: Geol. Survey Canada paper 69–21.

SALOP, L. J., 1967, Geology of the Baikal Mountain region: Moscow, Nyedra, v. 1, 515 p., v. 2, 699 p.

──, AND SCHEINMANN, YU. M., 1969, Tectonic history and structures of platforms and shields: Tectonophysics, v. 7, p. 565–597.

SCHILLING, J.-G., 1973, A far mantle plume: rare earth evidence: Nature and Phys. Sci., v. 242, no. 114, p. 2–5.

SHATSKI, N. S., 1946a, Basic features of the structures and development of the East European Platform. Comparative tectonics of ancient platforms: SSSR, Akad. Nauk Izv., Geol. ser., no. 1, p. 5–62.

──, 1946b, The Great Donets Basin and the Wichita System. Comparative tectonics of ancient platforms: *ibid.*, no. 6, p. 57–90.

──, 1947, Structural correlations of platforms and geosynclinal folded regions: *ibid.*, no. 5, p. 37–56.

──, 1955, On the origin of the Pachelma Trough: Moskvskogo Obshchestva Lyubiteley Prirody Byull., Geol. sec., no. 5, p. 5–26.

──, 1961, Vergleichende Tektonik alter Tafeln: Berlin: Akademie-Verlag, 220 p.

──, AND BOGDNAOV, A. A., 1960, La carte tectonique de l'Europe au 2,500,000^e: SSSR Akad. Nauk Izv., Geol. ser., no. 4.

SMITH, A. G., BRIDEN, J. C., AND DREWRY, G. E., 1973, Phanerozoic world maps, *in* HUGHES, N. F. (ed.), Organisms and continents through time: London, Paleont. Assoc. Special Papers in Paleontology, no. 12, p. 1–42.

STOCKWELL, C. H., 1936, Eastern portion of Great Slave Lake: Geol. Survey Canada Maps 377A and 378A.

──, 1961, Structural provinces, orogenies, and time-classification or rocks of the Canadian Precambrian shield: *ibid.*, Paper 61–17.

──, 1964, Fourth report on structural provinces, orogenies and time-classification of rocks of the Canadian Precambrian shield: *ibid.*, 64–17.

──, 1968, Tectonic map of Canada: *ibid.*, Map 1251A.

──, 1970, Geology of the Canadian Shield. Introduction, *in* DOUGLAS, R. J. W. (ed.), Geology and economic minerals of Canada: *ibid.*, Econ. Geology Rept. 1, p. 44–54.

SUBBOTIN, S. I., 1958, Concerning the formation of crustal downwarps and the tectonics of the basement of the Dnieper-Donets trough: Geol. Zhur., v. 18, no. 6.

TRYGGVASON, E., AND QUALLS, B. R., 1967, Seismic refraction measurements of crustal structure in Oklahoma: Jour. Geophys. Research, v. 72, p. 3738–3740.

WANLESS, R. K., 1970, Isotopic age map of Canada: Geol. Survey Canada Map 1256A.

WILSON, J. T., 1949, Some major structures in the Canadian Shield: Canadian Inst. Mining and Metallurgy Trans., v. 52, p. 231–242.

──, 1966, Did the Atlantic close and then reopen?: Nature, v. 211, p. 676–681.

──, 1968, Static or mobile earth: the current scientific revolution, *in* Gondwanaland revisited: new evidence for continental drift: Am. Philos. Soc. Proc., v. 112, p. 309–320.

WILSON, M. E., 1939, The Canadian Shield, *in* Geologie der Erde, Geology of North America, v. 1, 232 p.

SUBMARINE CANYON AND FAN DEPOSITS

SUBMARINE CANYONS AND FAN VALLEYS: FACTORS AFFECTING GROWTH PATTERNS OF DEEP-SEA FANS

WILLIAM R. NORMARK
University of Minnesota, Minneapolis

ABSTRACT

Dispersal of sediment across a submarine fan is controlled by a distributary system of migrating fan valleys, which are generally contiguous with a feeding canyon that provides a point source for the sediments moving onto the sea floor. Under conditions of fan growth, one active, leveed fan valley on the upper fan leads to a suprafan, a depositional bulge with an irregular surface that is probably the site of most rapid aggradation on the fan. Below the suprafan, the fan surface is relatively smooth and apparently free of distributary channels. Fan-building processes are sensitive to changes in the rate of sediment supply, grain-size distribution within the sediments, and tectonic disturbances within fan and source areas, but they are relatively insensitive to the shape of the depositional basin or to scale factors (ultimate size of the deposit). Specific deep-sea fans included in this discussion are listed in order of decreasing radial dimensions: Bengal Fan (3000 km), Monterey Fan (300 km), La Jolla Fan (80 km), San Lucas Fan (60 km), Navy Fan (60 km), Coronado Fan (50 km), and a small fan in western Lake Superior (5 km).

Submarine canyons allow coarse sediment to bypass shelf environments. To remain active, a canyon must maintain its head in or near the surf zone to intercept the littoral drift; a rapid rise in sea level can cut off its major source of sediment. Ascension Canyon, one of several canyons leading to the Monterey Fan, is now relatively inactive as its head lies near the outer shelf. In the California Continental Borderland, Coronado Canyon likewise has been inactive during the present interglacial period, and reflection-profiling data suggest that it was probably inactive during earlier interglacial periods. The rapid sea-level fluctuations of greater than 100 m during the Pleistocene Epoch may be considered extreme, but similar effects may occur when submarine canyons experience large, apparent sea-level changes due to tectonic tilting. In either event, sudden changes affecting sediment supply to canyon heads cause a marked shift in the locus of turbidite deposition.

Depositional sites on a deep-sea fan also may change in response to fluctuating sea levels even when the canyon continues to receive sediment. If headward erosion within a canyon head keeps pace with transgression, incision of the fan valley on the upper and middle fan results; the La Jolla fan valley has cut across the entire fan area, and deposition is now confined to a suprafan farther down the San Diego Trough.

INTRODUCTION

Lithofacies and petrofacies variations in submarine fan deposits can be used to suggest the location and types of sediment sources available as well as the depositional sequence (e.g., Dickinson and Rich, 1972). However, it is often difficult to determine what external events, such as sea level fluctuations and tectonic disturbances, ultimately affect the composition and geometry of such deposits. This discussion focuses on the depositional effects of events occurring within the submarine canyon and fan-valley transport systems.

The first step is to suggest a steady-state model for fan deposition that defines the basic morphology, shallow structure, and surface sediment distribution of undisturbed deep-sea fans. This assumes that, under conditions of uniform (and known) rate of sediment supply and sediment-size distribution and no tectonic disturbance of the source, canyon, or fan systems, the shape of the fan can be predicted. I referred to such a steady-state product as the growth pattern of a deep-sea fan (Normark, 1970a). It must be emphasized that the growth patterns for different deep-sea fans may vary greatly, reflecting the differences in the sediment supplied; a fan built primarily from muddy turbidity currents should have predictable differences from those resulting from sand-depositing turbidity currents. It is suggested that this model for fan growth is reasonably independent of water depth and scale factors, that is, independent of the ultimate size of the deposit. This model, therefore, is based on studies from a wide range of deep-sea fans, including differing sizes, ages, water depths and lithologies, as it would be difficult to isolate the effects of any single parameter if only one fan setting were chosen.

Ideally, the next step is to determine how changing or cutting off the supply of sediment may alter the growth pattern of a deep-sea fan where eustatic fluctuations in sea level effected a change in the growth pattern by altering the sediment supply. Three different examples are discussed.

Tectonic disturbance of the deep-sea fan area itself also can drastically alter the growth pattern of the fan. It would be nice to study a fan that is disrupted only by tectonic movements within the area of deposition to isolate these effects, but, unfortunately, most modern fans show primary alteration due to fluctuating sea levels. Redistribution of sediment due to bottom currents or large-scale slumping of fan sediments or nearby slope sediments onto the fan are considered as modifying processes operating directly on the fan area, and they may be very important when considering turbidite deposition of geosynclinal proportions.

METHODS

Although this paper emphasizes geophysical studies of the structure and morphology of fans, the distribution and lithology of surface sediments are considered whenever sufficient bottom samples are available. Particular emphasis is placed on features 2 to 50 m in relief and from tens of meters to a few kilometers in length. These are on a scale smaller than that resolvable by most conventional marine geophysical techniques but larger than the rather limited observations provided by bottom photographs and core samples. I would emphasize that in water depths associated with continental rises, features less than 1 km in width cannot be defined by conventional sensing techniques. Thus, emphasis is placed on observations resulting from my own studies of deep-sea fans using the deeply towed instrument package (Spiess and Mudie, 1970) of the Marine Physical Laboratory of Scripps Institution of Oceanography. The geophysical and photographic sensors on the deep tow (table 1) usually are maintained between 10 and 100 meters above the sea floor during normal operating conditions. Spiess and Mudie (1970) provided a detailed description of the deep-tow systems, and the application of the instrument to deep-sea fan studies was covered by Normark (1970a) and Mudie and others (1970). Only selected sites on four fans have been studied with the deep tow, as slow towing speeds (2 to 3 km/hr) and the operational range of the acoustic transponders (15 to 20 km) limited surveys to an area 10 to 15 km on a side.

Conventional echo sounding at either 12 kHz or 3.5 kHz and low frequency air-gun reflection-profiling data were collected from the surface ship during deep-tow surveys whenever possible. Interpretation of these conventional records in light of the associated deep-tow data provides new insights for comparison with fan studies described in the literature and with other areas where I have only conventional data. The sediment samples collected within deep-tow study areas are used primarily for extension of interpretation of the 3.5-kHz reflection data in order to discuss the variation in the relative proportions of sands and muds within fan environments. Detailed analysis of core se-

TABLE 1.—GEOGRAPHIC SETTING, SURVEY EMPHASIS, AND TYPES OF DATA FOR SUBMARINE FANS

Name	Geographic setting	Radial dimension (km)	Range of water depths (m)	Survey emphasis	Types of data[1]
San Lucas	Tip of Baja California, Mexico	60	2500–3060	Suprafan, adjacent lower fan	a, c, d, e
La Jolla	Borderland[2]	80	470–1100	Upper and lower fan	a, d
Navy	"	60	1600–1900	Entire fan; deep-tow upper fan	a, b, c, d, f
Coronado	"	50	700–1400	Reconnaissance only	a, b, c
Monterey	Central California	300	3050–4600	Major fan valleys; deep-tow lower fan	a, c, d, f
Bengal	Bay of Bengal	3000	1500–5000	Reconnaissance only	a, b, c, e
Reserve	Lake Superior	5	150–250	Entire fan	b, f

[1] Types of data: a, echo sounding (surface)
    b, 3.5-kHz reflection profiling (surface)
    c, air-gun reflection profiling
    d, deep tow[3]
    e, bottom samples <50 cores
    f, bottom samples >50 cores

[2] California Continental Borderland
[3] Deep-tow data systems most useful in deep-sea fan studies include:
    a, narrow-beam echo-sounding with 2-m resolution
    b, side-looking sonar, which can resolve features as small as 5 m in relief and 10 to 20 m in width
    c, stereocamera system with 5- to 8-m field of view
    d, acoustic transponder positioning system with an accuracy of 20 m or better within a 100-km^2-survey area (Olsen, 1971; McGehee and Boegeman, 1966; Lowenstein, 1966, 1968)

quences on submarine fans are not discussed in this review (see Wilde, 1965; Shepard and others, 1969; Piper, 1970; Piper and Normark, 1971; and Nelson and Nilsen, this volume).

STUDY AREAS

To develop a comprehensive model for fan sedimentation, in which both sedimentological and tectonic factors could be evaluated, I felt it necessary to research a number of fans from different geomorphic and geologic settings and covering different scales. This effort is still continuing, but table 1 lists the seven subaqueous fans from which I have a sufficient amount of data to determine at least part of their growth histories. The extent of the fan areas surveyed and types of data collected are extremely variable (table 1). Each fan or fan area was chosen for a specific aspect of the problem.

The La Jolla Fan provided a reference or calibration area for the deep tow because of the earlier detailed studies by Shepard and Buffington, 1968; Shepard and others, 1969; Shepard and Dill, 1966; Shepard and Einsele, 1962; Piper, 1970. The San Lucas Fan was the primary example for the model reviewed here, although little sediment sampling was done there. The Navy and Reserve Fans were chosen for comprehensive studies of the entire fan areas, and about one hundred sediment cores were available for each. On the Monterey Fan, the deep tow was used to study fan-valley development on the upper fan and to look at a lower fan area well away from any presently active distributaries. My comments on the Bengal Fan stem from participation on a cruise under J. R. Curray and D. G. Moore during April and May 1971 and on earlier reflection records described by them (Curray and Moore, 1971). The Bengal Fan, by far the largest modern fan and of geosynclinal proportions itself, is one to three orders of magnitude larger than the others.

The Reserve Fan provides a unique opportunity to study turbidite sedimentation. The fan formed as a result of the discharge of tailings during the past 16 years from a taconite pellitizing operation along the north shore of Lake Superior at Silver Bay, Minnesota. The bulk of the tailings have formed a subaerial delta, but the morphology within the prodelta environment is typical of deep-sea fan deposits and is comparable to the Rhone delta of Lake Geneva (Houbolt and Jonker, 1968). I regard this particular deposit more as a large laboratory system than as a natural deposit because, unlike the other fans studied, we know the total amount of sediment supplied, the rate of supply, the grain-size distribution, and the shape of the lake floor before deposition of the tailings. In addition, the relative amount of shoaling in the area of fan deposition in Lake Superior (about 150 to 250 m) means that conventional 3.5-kHz seismic reflection records are more nearly comparable to deep-tow observations than similar work in the deep ocean. The results discussed here are considered tentative, as a much more comprehensive profiling and sampling program is now in progress.

MODEL GROWTH PATTERN

In my growth pattern for deep-sea fans (Normark, 1970a), I considered depositional processes resulting in a distinct morphological division of the fan surface (fig. 1): (1) the upper fan, which is characterized by leveed fan valleys; (2) the middle fan, where rapid deposition at the end of the leveed fan valleys builds a suprafan; and (3) the lower fan, which is apparently free of any major topographic relief and may correspond to a ponded turbidite sequence if the fan is building into a restricted basin.

Upper Fan

Leveed, or depositional, fan valleys are usually confined to the upper fan (Bates and others, 1959; Normark, 1970a). In many examples,

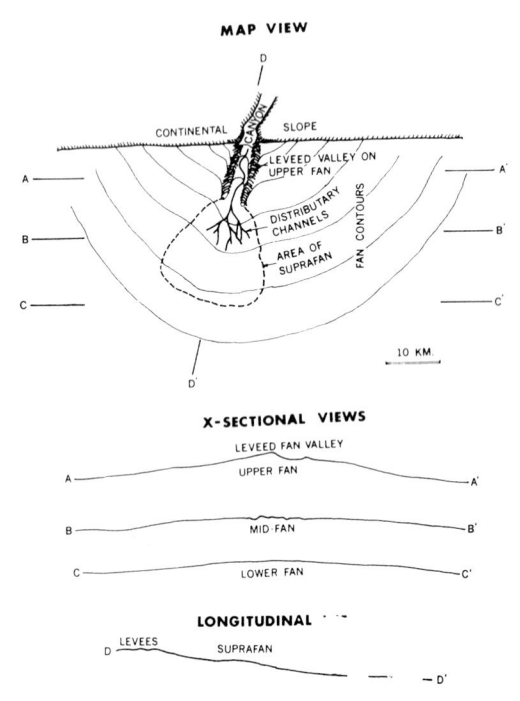

FIG. 1.—Model for submarine fan growth showing suprafan and schematic bathymetry for topographic divisions of fan (from Normark, 1970a).

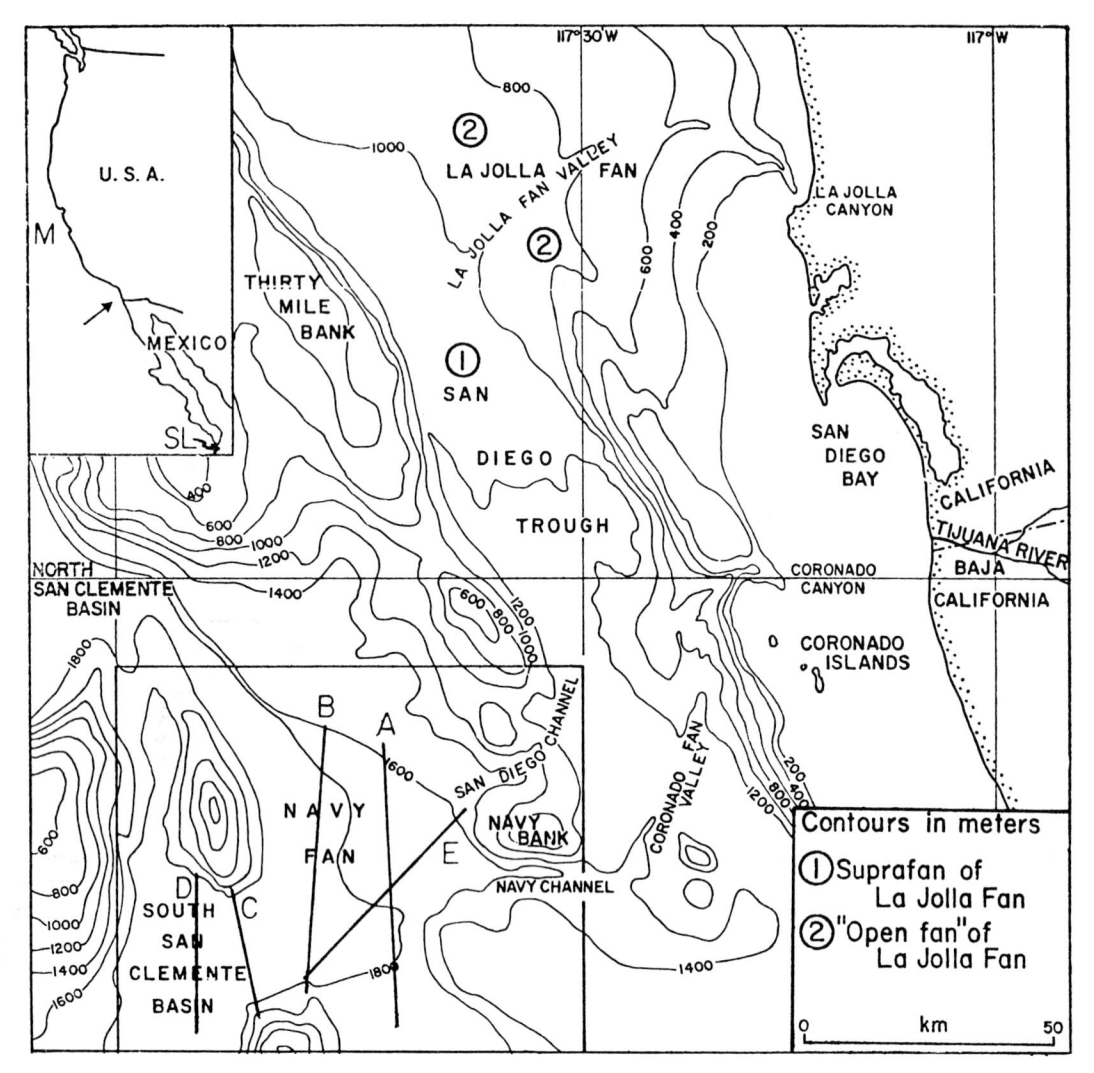

FIG. 2.—Bathymetric map for La Jolla, Navy, and Coronado Fans and adjacent features in the central California Continental Borderland. Inset map shows approximate location for Monterey (M) and San Lucas (SL) Fans. The 1800-m contour on Navy Fan outlines suprafan lobe; locations of profiles in figure 5 are also shown (modified from Normark and Piper, 1972).

there is only one active valley continuous with the submarine canyon, which is the ultimate source of the coarser sediment for the fan. The wide, flat valley floors characteristic of depositional fan valleys are built above the surface of the adjacent, open fan during periods of normal fan growth. In the northern hemisphere, the right-hand levee (looking downstream) is characteristically higher (Menard, 1955; Hamilton, 1967). A comparison of the upper portions of the Monterey Fan (off central California) and Navy Fan (fig. 2) led me (Normark, 1971) to suggest that fan-valley dimensions in accreting fans are proportional to the radial size of the fan: on the Navy Fan (radius, 50 km), the leveed valley is 13 km long and 5 to 8 km wide; on the Monterey Fan (radius 300 km), the original Ascension fan valley (Normark, 1970b) is about 80 km long and 25 to 30 km wide. Levee-to-valley-floor relief and the elevation of the valley floor above the surrounding fan are also greater in larger fans. However, no simple proportionality is found if one considers all of the fans listed in table 1.

Using low-frequency reflection profiles, Hamilton (1967) and Normark (1970b) suggested that depositional valleys are formed by rather uniform deposition over the whole area but that

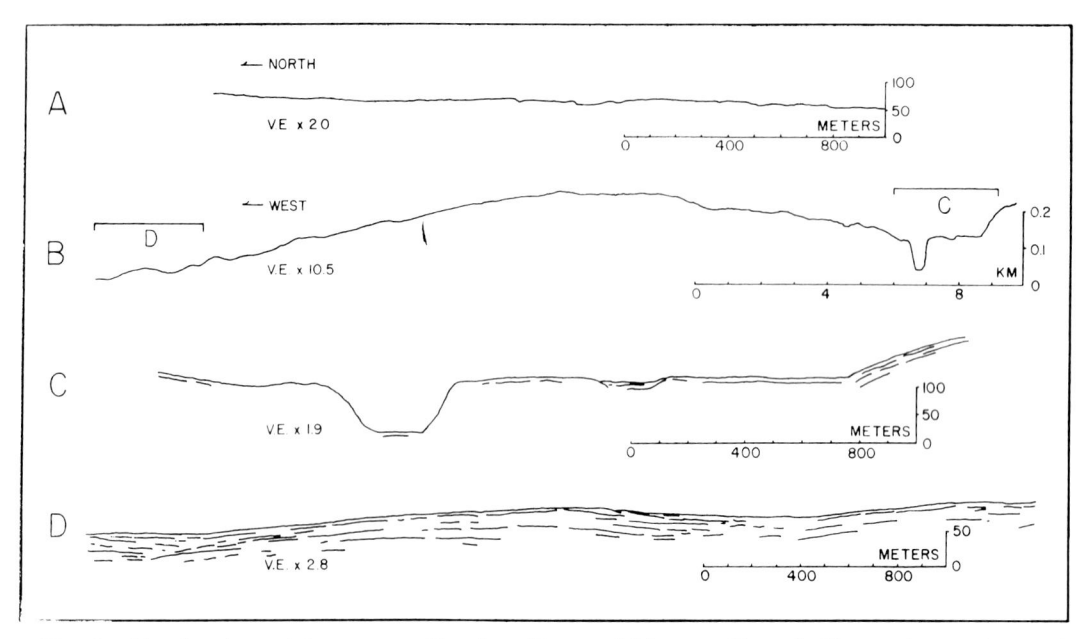

FIG. 3.—Line drawings of deep-tow profiles from Navy and Monterey Fans. A, Narrow-beam echo-sounding profile over floor of Navy fan valley; near line E, figure 2. B, Narrow-beam echo-sounding profile over Ascension fan valley, showing broad western levee with undulating topography. Narrow and deep (90 m) channel in floor of valley is result of headward erosion from its intersection with deeply incised Monterey fan valley. C, 3.5-kHz reflection profile over floor of Ascension fan valley; see line B for location. D, 3.5-kHz reflection profile over several undulations on back side of levee of Ascension fan valley; see line B for location.

rates of accumulation are slightly greater on the levee sites than elsewhere. Little structure is discernible within the sediments, and most internal reflectors follow the sea floor (Hamilton, 1967, fig. 10; Normark, 1970b, fig. 3, lines A and B).

However, when seen from the deep tow, the flat valley floors exhibit much relief of the order of 10 m. Reflection profiles (3.5 kHz) taken with the deep tow over the floor of the Navy fan valley (fig. 3A) show a general lack of subbottom reflectors, which is indicative of areas of sand deposition (Normark, 1970a; Normark and Piper, 1972); a pebbly sand was cored near the upper end of the valley. The Navy fan valley is crossed by shallow (4 to 8 m), irregular channels from 50 to 200 m wide. The channel walls are relatively steep and, even on the deep-tow profiles, appear as hyperbolic returns from near the upper edges of the channels. The channels are near the limit of resolution for the side-looking sonars, but irregular returns suggest that individual features may exceed 150 m in length. The existence of several such channels may result from meandering or braiding of the thalweg during aggradation of the valley floor.

The Ascension Canyon system (near M in fig. 2 inset) is no longer active, and its fan valley is blanketed by a 4- to 7-m layer of acoustically transparent material (fig. 3B and C). Below this layer, few subbottom reflectors were recorded with the 3.5-kHz system on the deep tow, indicating that sand was being deposited during the period of fan growth. Irregular channels on the floor of the valley similar in size to those in the Navy fan valley probably have the same origin. Slumping of sediments from the levees onto the floor of the valley may be common. The deep-tow profiles showed that the interbedded muds of the levee crests on the Ascension Fan valley dip toward the valley floor at about 10° (fig. 3C, east side); comparison with other depositional channels (Hamilton, 1967) suggests that slopes of this magnitude may be common.

The levee crests are smooth, slightly convex upward, and free from any topographic irregularities even when viewed from the deep tow. Limited acoustic penetration with the 3.5-kHz reflection system over the levee on the Ascension fan valley indicated uniform deposition and relatively continuous reflectors parallel to the sea floor. On the north levee of the Navy fan valley, the sea floor is smooth, but no acoustic penetration was achieved.

Particularly on the larger depositional valleys in the northeastern Pacific, the back side (away from the thalweg) of the higher, right-hand levee has undulating or terraced topography having spacing of as much as 1 km between high points and relief of as much as 40 m (fig. 3B; Hamilton, 1967; Normark, 1970b). The height of these irregularities (hummocks?) decreases as the relief of the levee over the surrounding fan decreases. Hamilton (1967) suggested that these hummocks are probably depositional features although erosional channelling or slumping of the sediments are also possible causes.

The deep-tow seismic profiles on the Monterey Fan over these features do not show sharp discontinuities indicative of channelling, and in cross section they resemble dunes (fig. 3D). Internal reflectors within them tend to overlap reflectors within the upslope hummock, suggesting that deposition leads to up-levee migration. The acoustic penetration achieved by the 3.5-kHz system suggests that, compared to the valley floor, this part of the levee is relatively free of sand deposition. Individual reflectors are several hundred meters in length and appear less continuous than those across the levee crest.

Deposition on the Reserve Fan in the last 16 years has already resulted in the formation of a prominent leveed valley on the upper fan, which extends for over 2 km from the abrupt break in slope at the delta toe (fig. 4). There is no major channel extending down the delta slope; and this fan valley, as well as several smaller, subparallel and discontinuous channels to the east, head against the surprisingly uniform deltaic slope. The right-hand (western, in this example) levee of the largest valley is both higher and wider than the eastern levee, and it has an irregular back-side slope. Its crest is as much as 15 m above the valley floor (fig. 4). Little internal structure was recorded from the levees, however. Unlike most deep-sea leveed valleys in the northern hemisphere, which appear to migrate to the left (Menard, 1955), this valley appears to have migrated westward, or to the right, into its large levee. The valley floor has not yet built above the surrounding Lake Superior floor. It appears that during the initial stages of sediment discharge, the turbidity currents eroded channel(s) into the predelta lake clays, and only sometime later, when the valley reached its present position, has aggradation begun to elevate the valley floor. The bulk of the sediments in the valley are silty sands decreasing laterally in grain size on either side; this contrasts with the coarse sand and pebbles of the delta slope.

Middle Fan

The distinguishing feature of the middle-fan region is the depositional bulge, which appears as a convex-upward segment on a radial profile (fig. 1). It is termed the suprafan (Normark, 1970a) and has a hummocky appearance on surface echo-sounding profiles, but, unlike the rolling topography on the back side of the levees, the relief on the suprafan is due to the presence of many isolated depressions, which are probably remnants of a channel-distributary system. Deep-tow profiles over the suprafan on the San Lucas Fan at the tip of the peninsula of Baja California, Mexico, show many features comparable to erosional channels; terraces are common on the sides of these depressions, and the outside wall is steeper where the feature is arcuate (Normark, 1970a, figs. 18, 19, and 20). It should be noted that the side-looking sonar data was most useful in confirming that the depressions on the San Lucas Fan were indeed not interconnected. Side-looking sonar records over probable relict-suprafan topography on the lower part of the La Jolla Fan show many small, irregular depressions closer in size to the valley-floor features previously described on other fans (Normark, 1971).

Coarse sediment (up to pebble grade on the San Lucas Fan) is found within these depressions, but interdepression areas are covered by interbedded muds and fine sands. The general lack of subbottom reflectors within areas of sandy sediments hinders attempts to relate deposition of these coarser sediments to the formation of the isolated depressions; the most likely explanations are channel-fill sands abandoned by channel migration or a surface layer of sand as a result of channel overflow from nearby active distributaries. Slump features also may be common within suprafan sediments where channel relief is locally great. Failure of the channel walls as a result of oversteepening could provide a local source for angular mudlumps, which would contrast with more rounded clasts found in an up-current direction that have been transported along the leveed fan-valley floor (Moore, 1965).

To build up the typical low cone shape of most submarine fans, the suprafan area must migrate across the midfan region. This can only occur, however, in response to events within the fan valley through which the sediments must pass. The Navy Fan may be unique in that the fan valley is confined between basement highs and the suprafan has probably occupied about the same position for most of the time (fig. 2; Normark and Piper, 1972). In this example, the suprafan surface still shows the characteristic

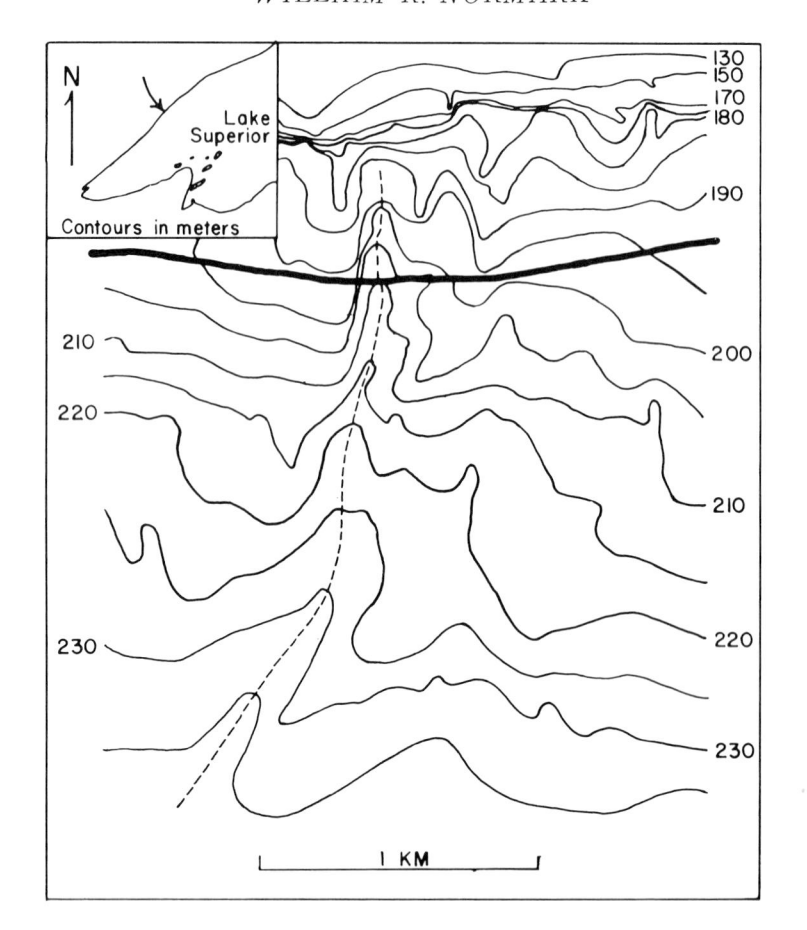

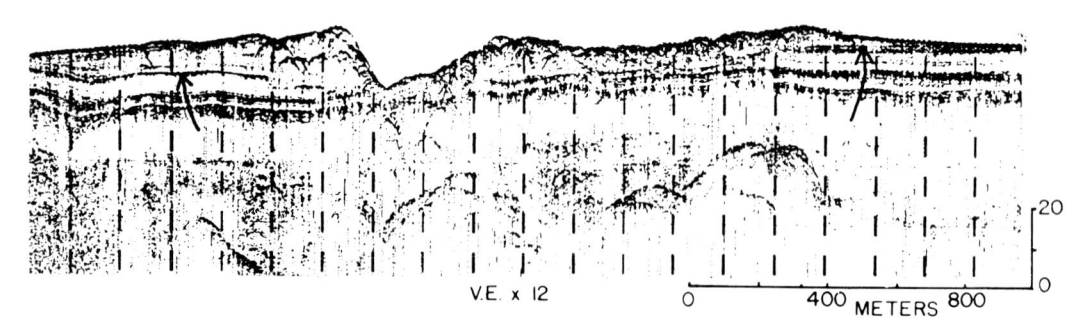

FIG. 4.—Upper: bathymetric map of upper part of Reserve Fan (only area surveyed in 1970) showing prominent leveed fan valley; heavy line near 200-m contour shows location of reflection profile. Lower: 3.5-kHz reflection profile across fan valley; view is to north (up-valley). Arrows locate prefan lake floor.

hummocky topography produced on conventional surface echo-sounding records over an area of numerous irregular, discontinuous channel segments. No continuous fan valley crosses the midfan, although one may not be detectable without deep-tow data. The suprafan area on the Navy Fan is marked by the lack of acoustic penetration into the sediment when using 3.5-kHz profiling (fig. 5). Extensive sediment sampling on the fan illustrated that sands are commonest on the suprafan and that sands and gravels occur within the floor of the fan valley above this area. Figure 5 clearly shows that the suprafan is building out over older, well-stratified sediments, which dip beneath the suprafan from all sides. Coring (Normark and Piper, 1972) and

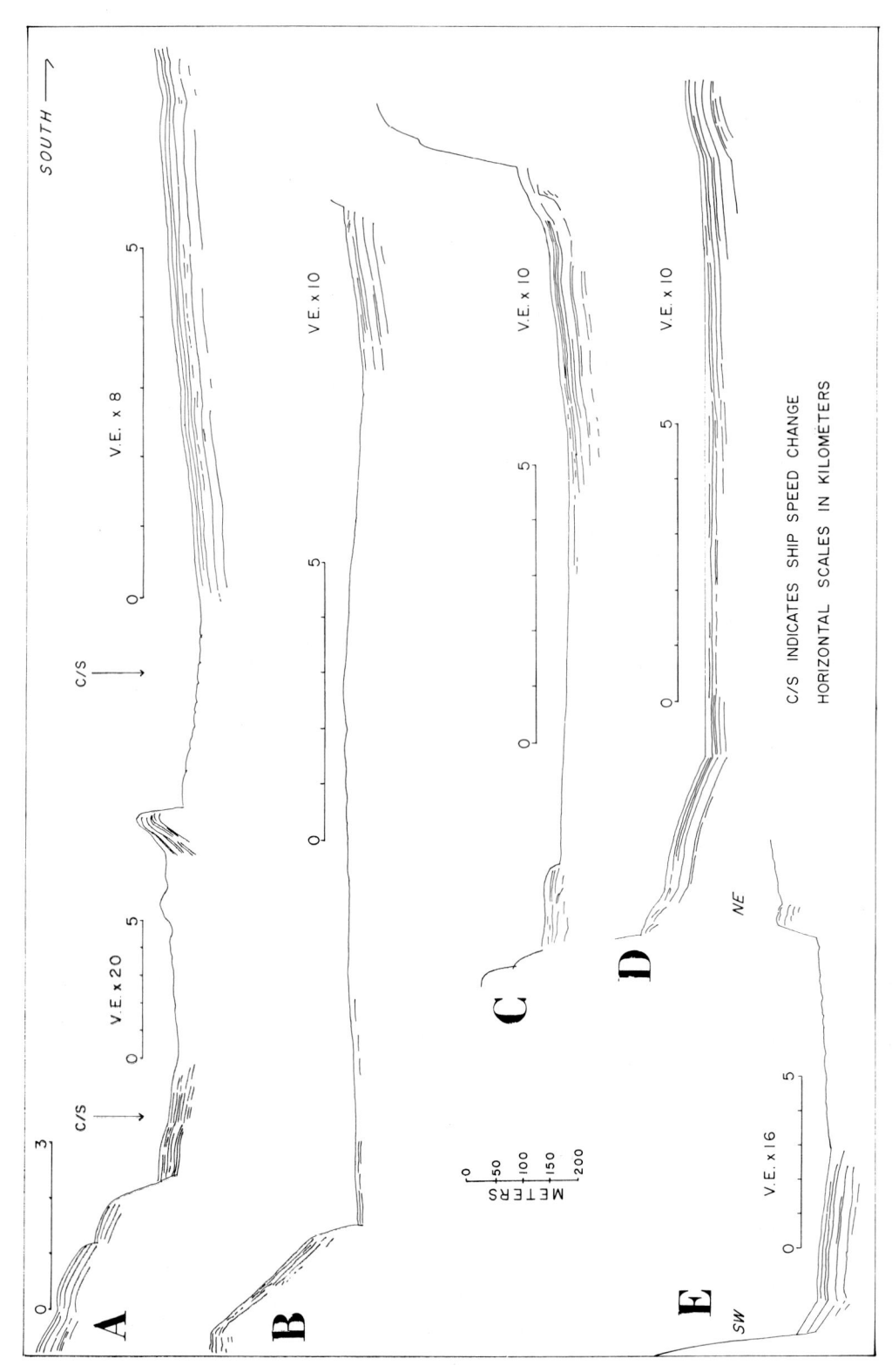

FIG. 5.—3.5-kHz reflection profiles from Navy Fan (from Normark and Piper, 1972); see figure 2 for profile location.

limited acoustic penetration over the toe of the fan indicated that mostly muds are being deposited.

The Reserve Fan also exhibits some suprafan morphology, which is somewhat surprising because the fan deposits are only a few meters thick around the end of the leveed fan valley. At least two separate suprafan areas are recognized, each with a short fan-valley segment leading to them. The main leveed valley has the largest suprafan at its termination. Coring on the Reserve Fan showed that the suprafan areas probably are the farthest extent of sand deposition in the lake. Deposition on the Reserve Fan supports Normark's (1970a) suggestion, based on the study of the sublacustrine fan of the Rhone delta in Lake Geneva by Houbolt and Jonker (1968), that deposition in the prodelta or lower delta slope environments is comparable to that of submarine fans.

Lower Fan

The lower fan, which is the largest in areal extent of the three fan divisions, is free of major topographic relief when viewed either from the surface or from the deep tow. Admittedly, comparatively little survey time has been expended on the lower fan environment because of its regularity. Four days of deep-tow work on the San Lucas lower fan showed the surface to be gently rolling and to have no local relief (insofar as resolution of the instruments permitted detection). However, a detailed survey (Olsen, 1971) on a portion of the Monterey lower fan (275 km southwest of the fan apex) delineated a group of subparallel depressions that are 2 to 6 m deep, 20 m wide, and up to 4 km in length. These linear depressions are mostly asymmetric in profile and trend normal to the fan slope, which is slight (4 m/km). Most of these depressions are so small that they were not resolved even with the deep-tow, narrow-beam echo sounder. Both the plane and cross-sectional shapes were determined by use of the side-looking sonars. It is possible that similar features may have been overlooked during deep-tow surveys on other fans; these records were enhanced for the Monterey Fan by towing the instrument very close to the sea floor while searching for submarine wreckage. The origin of these furrows may be related to sediment transport across the toe of the fan.

This growth pattern seems easily reconciled with other modern fans described in the literature, particularly the Astoria Fan (Nelson, 1968; Nelson and Byrne, 1968; and Nelson and others, 1970). On the Astoria Fan, the middle-fan segment and much of the lower fan (Nelson, 1968) appears to fit the suprafan surface morphology of Normark (1970a), and the lower fan area was not completely surveyed. The sediment distribution of the Astoria Fan is reviewed by Nelson and Nilsen (this volume).

Facies Relationships

Using the growth pattern outlined above, at least five depositional environments (table 2) may be distinguished under equilibrium growth

TABLE 2.—SEDIMENTARY FACIES FOR SUPRAFAN GROWTH PATTERN

Site	Sand/mud ratio	Sediments	Geometry of beds	Facies designations[1]
Floor of leveed fan valley	High	Predominantly sands up to coarse sands and mud-lumps; only likely environments for fluxoturbidites	Thick, up to 10 m; lateral dimensions up to several hundreds of meters; sand bodies lenticular	A1, A3, B1, F, ?A2, ?A4
Levee crest	Intermediate	Coarse materials absent	Regular and continuous; on order of 1 km in length; generally <1 m thick	E
Back side of levee	Low	Sands less common than on levee crests	Continuous; up to 1 km in length; irregular thicknesses related to up-levee migration of undulations	G (hemipelagics according to Mutti)
Suprafan	High	Predominantly sands up to pebble size; slump deposits common	Very irregular; channelling common; braided channel environment	B2, A4, ?A3, some C, D
Lower fan	Low	Little or no sand; mud beds thicker here than anywhere else; bioturbation	Very regular; continuous; up to kilometers in length	D, G

[1] Facies designation according to Walker and Mutti, 1973

conditions (Normark, 1971). The sedimentary facies suggested in table 2, therefore, apply to those deep-sea fans formed when appreciable amounts of sandy sediments are available. Even so, complete facies characterizations requires more information, particularly on grain-size distributions and sedimentary structures. Sheer numbers of samples from deep-sea fans does not provide the necessary details, as coarse-grained sediment is badly undersampled with coring devices currently in use, and prohibitively large samples would be required to recognize some types of indicative sedimentary structures. Thus, the reader should supplement these facies sketches with observations from ancient rocks (see Mutti, this volume; Mutti and Ricci Lucchi, 1972; Walker and Mutti, 1973).

GROWTH PATTERN MODIFICATIONS

The model for fan sedimentation does not imply that all accreting fans must exhibit the growth pattern explained above; important differences can result from changes in grain size and rate of sediment supply, tectonic movements, bottom currents, and sediment slumping.

Grain-Size Distribution Effects on the Growth Pattern

Although most deep-sea fans have a leveed fan valley near their apex, many apparently do not exhibit the suprafan deposit. In some fans, the suprafan has been destroyed, but other fans, such as the Bengal Fan, may never have developed one. The Bengal upper fan is characterized by an extensive system of large leveed fan valleys (Curray and Moore, 1971); because only one appears to be a continuation of the so-called "Swatch-of-No-Ground" submarine canyon, the others are probably a succession of abandoned distributaries. Although these leveed valleys are comparable in size to those on the Monterey Fan, they show little internal structure within the levees, which exhibit prominent terraces on the inner levee walls (Curray and Moore, 1971, fig. 3). Many of these valleys cross almost the entire fan, and none appears to terminate in a suprafan. Limited core data from the active distributary fan valley shows that almost no coarse sediment is being supplied to the Bengal Fan at present.

I would suggest that the growth pattern of the Bengal Fan differs from the suprafan model because of the relatively fine grain size of the sediments. All fans bearing a suprafan deposit have appreciable amounts of coarse sediment, which probably was transported as the bed load of turbidity currents. Rapid deposition and the formation of braided channels occurs where the currents exit from the confines of the leveed valley. On the other hand, if there is relatively little coarse sediment, overbank deposition of the suspended material enhances levee development. Relatively straight leveed channels will grow rapidly downfan, preventing the development of a suprafan lobe on the fan surface.

Changes in Sediment Supply

For fans that receive sediment as a result of the interception of littoral drift in the nearshore zone, a rapid change in sea level resulting in the migration of the littoral drift zone relative to the head of a canyon, can lead to a marked change in the rate of sediment supply. Eustatic changes in sea level during the Pleistocene Epoch have markedly altered the growth patterns of some modern deep-sea fans as shown in the three examples that follow.

Monterey Fan.—Three major submarine canyon groups have provided the bulk of the sediments for the Monterey Fan (Shepard and Dill, 1966; Normark, 1970b); during lowered sea level, the now inactive Ascension Canyon was a major conduit for sediments supplied to the fan. It produced the large, leveed fan valley, which is continuous with Ascension Canyon and is the most prominent depositional valley on the fan. However, no suprafan has been recognized for the Ascension fan valley. The Holocene transgression resulted in migration of the sediment source away from the canyon heads, and sedimentation within the fan valley ceased. On the other hand, the presently active Monterey Canyon maintained its head in the nearshore zone and is still feeding sediment to the fan (Shepard and Dill, 1966). In addition, its associated fan valley pirated the lower end of the Ascension system, and extensive downcutting was initiated (Normark, 1970b). This would have dissected any suprafan associated with the previous termination of the Ascension fan valley.

Coronado Fan.—The Coronado Canyon also heads near the outer shelf (Shepard and Dill, 1966), and it, too, was left inactive as a result of the Holocene transgression. The primary effect in this case was seen on the Navy Fan and not on the Coronado system. The Navy Fan receives its sediment from Navy Channel (fig. 2), a narrow notch that has been cut into the sill separating the San Clemente Basin from the San Diego Trough (Normark and Piper, 1972). Navy Channel was cut after the San Diego Trough was filled to sill depth and after sediments began to flow across the sill into the San Clemente Basin. This headward erosion continued northeastward until the Coronado fan valley was intercepted. Continued headward erosion appears to have incised most of the Coronado fan valley. At least during the last low stand of sea level, much of the sediment entering Coronado Canyon bypassed its associated fan and was deposited directly on the Navy Fan. Seismic reflection data from the Coronado upper fan valley suggests that deposition within the area was halted during at least one earlier interglacial period as well.

La Jolla Fan.—Farther north in the San Diego Trough, rapid headward erosion within La Jolla Canyon kept the canyon head in the shore zone during the rise in sea level. This led to extensive downcutting within the canyon—fan-valley system (Normark and Piper, 1969), and, as a result of this incision, the fan valley crosses the entire fan. Sediments now moving through La Jolla Canyon bypass the fan area. Deposition on the fan during the Holocene (Shepard and others, 1969) can account for only 10 percent of the sediment thought to have been provided during this period (Piper, 1970). Normark (1970a) suggested that a suprafan, if one existed for the La Jolla Fan, could be developed below the toe of the fan, which was the limit of his original study. Subsequent reflection profiling in the San Diego Trough has indicated an area of hummocky suprafan morphology and an overall convex-upward profile around the 1200-m contour (fig. 2); this is about 100 m deeper than the termination of the incised valley on the fan and about 15 km southeast down the San Diego Trough. Further study, preferably with the deep tow, would be necessary to substantiate this observation.

Tectonic Disturbance of the Fan Area

The California Continental Borderland is an active margin associated with many turbidite basins. One possible example of the effects of fault displacement in the fan area has been described by Haner (1971), who studied the Redondo Fan in the San Pedro Basin near Los Angeles. Unfortunately, this study did not distinguish sufficiently between products of normal growth and the results of tectonic displacements, which, in my judgment, resulted from the fact that Haner adopted the terminology commonly used for alluvial fans in arid regions (e.g., see Hooke, 1967) and assumed that similar features on subaerial and submarine fans have similar origins. As a result, Haner may have confused tectonic segmentation of fan surfaces (Bull, 1964; Hooke, 1967) with a natural growth profile for a deep-sea fan.

The simplified radial profiles and horizontal profile along the active fan valley on the Redondo Fan (Haner, 1971, figs. 8 and 9) both show a large, convex-upward bulge at the end of the fan valley on the upper fan. This bulge is identified as the middle and lower fan segments. Haner (1971, p. 2414) arbitrarily decided that the fan area ended where channels were no longer resolvable by surface echo-sounding techniques, so the lower fan environment (in my terminology) was never studied. However, the results of turbidite depositional processes in completely enclosed (ponded) basins are comparable to the broad expanses of most lower fan divisions, such expanses commonly greatly exceeding half the total fan area. For these reasons, I would prefer to correlate Haner's middle and lower fan segments with the normal suprafan topography described above. This conclusion is supported by the existence of a braiding and meandering channel system on this suprafan; Haner (1971) has called this the active fan area.

It was hoped that the Redondo Fan would demonstrate the effects of tectonic movements within the fan area on the growth pattern. Some of the older abandoned channels, which do not connect with the active channel system, cross one of the faults thought to be active in the fan area. Haner (1971) suggested that these may have been incised by headward erosion after fault movement. The incision of the main fan valley on the upper fan as suggested by Haner may be due either to headward erosion within the canyon in response to the Holocene transgression or to movement on the basin-slope fault. It does appear that fault movement, if any, during the growth of the Redondo Fan has had minimal effect on its growth pattern.

The Coronado upper fan valley is cut by a fault, which has been active during, and probably after, deposition. Internal structure within the levee is disrupted as are sediments within the adjacent areas of the San Diego Trough. Here also, however, the dominant factor controlling deposition on the fan was the capture and subsequent incision of the fan valley by Navy Channel, not tectonic displacements.

Other Modifying Effects

In areas of high sedimentation rates, such as on the Mississippi fan, large-scale slumping of unconsolidated sediments (Walker and Massingill, 1970) may modify or obscure normal depositional features. Walker and Massingill (1970) used 3.5-kHz reflection profiles to suggest that two large submarine slumps have covered much of the fan surface. These slumps are recognized primarily on the basis of lack of subbottom penetration or of internal reflectors within the slump area. The slumps appear to correspond with the major channel areas on the fan; profiles only from the edge of the slump area were presented, however, so no relationship between distributary-channel systems and slumping can be inferred (see also Shepard, 1955). Large slumps from adjacent continental slopes may be effective in diverting fan-valley distributaries on some fans (see Mutti, this volume).

The importance of normal bottom currents in redistributing sediment on a fan should not be overlooked. Arguments attempting to resolve the relative importance of normal bottom currents vs. turbidity current action in submarine fan environments are beyond the scope of this review. That contour currents over the continental rise of the western North Atlantic are extremely effective in reshaping the continental rise is well known (Heezen and others, 1966; Schneider and others, 1967; Heezen and Hollister, 1971). Little doubt exists that such contour currents could effectively remake the shape of a submarine fan and eliminate any vestige of a growth pattern.

CONCLUSIONS

Deposition within a wide range of submarine fan environments produces relatively uniform growth patterns that are independent of the size of the fan. The primary differences may be related to the grain-size distribution of the sediment; deposition of sands is frequently associated with a suprafan morphology, and the absence of coarse sediments leads to long leveed fan valleys. The modification of any growth pattern commonly reflects changes in the sediment supply, in some examples as a result of relative sea level changes around the canyon head. Little is known about the effects of tectonic displacements within the fan area during deposition.

ACKNOWLEDGMENTS

The writer is indebted to F. N. Spiess, J. R. Curray, and D. J. W. Piper, who provided support, guidance, and assistance during much of the shipboard work germane to this review, and to the Reserve Mining Company at Silver Bay, Minnesota, for providing copies of seismic reflection data from their 1970 survey of the tailings deposit. H. Reading, D. J. W. Piper, and R. H. Dott, Jr., provided many needed improvements in the manuscript.

REFERENCES

Bates, C. C., Mooney, A. R., and Bershad, S. F., 1959, Worldwide evidence of deltas off the mouths of submarine canyons (abs.), in Sears, M. (ed.), Preprints of the First International Oceanographic Congress: Am. Assoc. Adv. Sci. Pub. 67, p. 595–597.
Bull, W. B., 1964, Geomorphology of segmented alluvial fans in western Fresno County, California: U.S. Geol. Survey Prof. Paper 352-E, p. 79–129.
Curray, J. R., and Moore, D. G., 1971, Growth of the Bengal deep-sea fan and denudation in the Himalayas: Geol. Soc. America Bull., v. 82, p. 563–572.
Dickinson, W. R., and Rich, E. I., 1972, Petrologic intervals and petrofacies in the great valley sequence, Sacramento Valley, California: ibid., v. 83, p. 3007–3024.
Hamilton, E. L., 1967, Marine geology of abyssal plains in the Gulf of Alaska: Jour. Geophys. Research, v. 72, p. 4189–4213.
Haner, B. E., 1971, Morphology and sediments of Redondo submarine fan, southern California: Geol. Soc. America Bull., v. 82, p. 2413–2432.
Heezen, B. C., and Hollister, C. D., 1971, The face of the deep: Oxford Univ. Press, p. 335–421.
———, ———, and Ruddiman, W. F., 1966, Shaping of the continental rise by deep geostrophic contour currents: Science, v. 152, p. 502–508.
Hooke, R. L., 1967, Processes on arid region alluvial fans: Jour. Geology, v. 75, p. 438–460.
Houbolt, J. J. H. C., and Jonker, J. B. M., 1968, Recent sediments in the eastern part of the Lake of Geneva (Lac Leman): Geologie en Mijnb., v. 47, p. 131–148.
Lowenstein, C. D., 1966, Computations for transponder navigation: Natl. Marine Navigation Mtg., Proc., p. 305–311.
———, 1968, Position determination near the sea floor, in Marine sciences instrumentation 4: New York, Plenum Press, p. 319–324.
McGehee, M. S., and Boegeman, 1966, MPL acoustic transponder: Rev. Sci. Instrumentation, v. 37, p. 1450–1455.
Menard, H. W., 1955, Deep-sea channels, topography, and sedimentation: Am. Assoc. Petroleum Goelogists Bull., v. 39, p. 236–255.
Moore, D. G., 1965, The erosional channel wall in La Jolla sea-fan valley seen from bathyscaph Trieste II: Geol. Soc. America Bull., v. 76, p. 385–392.
Mudie, J. D., Normark, W. R., and Cray, E. J., Jr., 1970, Direct mapping of the sea floor using side-scanning sonar and transponder navigation: ibid., v. 81, p. 1547–1554.
Mutti, Emiliano, and Ricci Lucchi, Franco, 1972, Le torbiditi den 'Appennino settentrionale: introduzione all'analisi di facies: Soc. Geol. Italiana Mem., v. 11, p. 161–199.
Nelson, C. H., 1968, Marine geology of Astoria deep-sea fan (Ph.D. thesis): Corvallis, Oregon State Univ., 287 p.

———, AND BYRNE, J. V., 1968, Astoria fan: a model for deep-sea fan deposition (abs.): Geol. Soc. America Ann. Mtg., Mexico City.
———, CARLSON, P. R., BRYNE, J. V., AND ALPHA, T. R., 1970, Physiography of the Astoria canyon fan system: Marine Geology, v. 8, p. 259–291.
NORMARK, W. R., 1970a, Growth patterns of deep-sea fans, Am. Assoc. Petroleum Geologists Bull., v. 54, p. 2170–2195.
———, 1970b, Channel piracy on Monterey deep-sea fan: Deep-Sea Research, v. 17, p. 837–846.
———, 1971, Mini-topography of deep-sea fans: geometric considerations for facies interpretations in turbidites: Soc. Econ. Paleontologists and Mineralogists, Pacific Sec., Field Trip Guidebook, p. 22–36.
———, AND PIPER, D. J. W., 1969, Deep-sea fan-valleys, past and present: Geol. Soc. America Bull., v. 80. p. 1859–1866.
———, AND ———, 1972, Sediments and growth pattern of Navy deep-sea fan, San Clemente Basin, California borderland: Jour. Geology, v. 80, p. 198–223.
OLSEN, W. L., 1971, Use of deep tow to survey a submerged wreckage: Scripps Inst. Oceanography Ref. 71-14, 13 p.
PIPER, D. J. W., 1970, Transport and deposition of Holocene sediment on La Jolla deep sea fan, California: Marine Geology, v. 8, p. 211–227.
———, AND NORMARK, W. R., 1971, Re-examination of a Miocene deep-sea fan and fan-valley, southern California: Geol. Soc. America Bull., v. 82, p. 1823–1830.
SCHNEIDER, E. D., FOX, P. J., HOLLISTER, C. D., NEEDHAM, D., AND HEEZEN, B. C., 1967, Further evidence for contour currents in the western North Atlantic: Earth and Planetary Sci. Letters, v. 2, p. 351–359.
SHEPARD, F. P., 1955, Delta-front valleys bordering the Mississippi distributaries: Geol. Soc. America Bull., v. 66, p. 1489–1498.
———, AND BUFFINGTON, E. C., 1968, La Jolla submarine fan-valley: Marine Geology, v. 6, p. 107–143.
———, AND DILL, R. F., 1966, Submarine canyons and other sea valleys, Chicago, Rand McNally and Co., 381 p.
———, AND VAN RAD, U., 1969, Physiography and sedimentary processes of La Jolla submarine fan and fan-valley, California: Am. Assoc. Petroleum Geologists Bull., v. 53, p. 390–420.
———, AND EINSELE, G., 1962, Sedimentation in San Diego Trough and contributing submarine canyons: Sedimentology, v. 1, p. 81–133.
SPIESS, F. N., AND MUDIE, J. D., 1970, Small-scale topographic and magnetic features, *in* MAXWELL, A. E. (ed.), The seas, v. 4 pt. 1, New York, Interscience Pub,., p. 205–250.
WALKER, J. R., AND MASSINGILL, J. V., 1970, Slump features on the Mississippi fan, northeastern Gulf of Mexico: Geol. Soc. America Bull., v. 81, p. 3101–3108.
WALKER, R. G., AND MUTTI, EMILIANO, 1973, Facies and facies associations, Part IV, *in* BOUMA, A. H., AND MIDDLETON, G. V., Turbidites and deep-water sedimentation: Soc. Econ. Paleontologists and Mineralogists, Pacific Sec. short course, p. 119–158.
WILDE, P., 1965, Recent sediments of the Monterey deep-sea fan (Ph.D. thesis): Cambridge, Massachusetts, Harvard Univ., 153 p.

DEPOSITIONAL TRENDS OF MODERN AND ANCIENT DEEP-SEA FANS

C. HANS NELSON AND TOR H. NILSEN
U.S. Geological Survey, Menlo Park, California

ABSTRACT

Many flysch, turbidite, fluxoturbidite, and grain-flow sequences from ancient geosynclines probably have been deposited in deep-sea fans adjacent to continental margins. We have obtained stratigraphic and sedimentologic criteria for recognizing ancient fan deposits by comparing the Astoria Fan, a large open-ocean fan off the coast of northern Oregon, with the Eocene Butano Sandstone, an ancient continental borderland fan deposit of similar size in the Santa Cruz Mountains, California.

Deep-sea fan deposits consist of channel and interchannel facies. Both facies change significantly downfan and laterally across the fan as a result of decreasing current velocities during each turbidity current and as a result of the lateral migrations of channels through time. Geologic mapping reveals thick-bedded, coarser grained, and lens-shaped channel deposits intermixed with thin-bedded and finer grained interchannel deposits. Sand-shale ratios are high within channels and low within upper fan interchannel and distal fan areas. The coarsest grained and thickest bedded gravels and sands are deposited by channelized sediment gravity flows in the submarine canyons and upper fan valleys. These ungraded, poorly sorted, and massive channel sediments change by midfan to thinner bedded, finer grained, vertically graded, and better sorted turbidite sands that contain sedimentary structures in Bouma sequences. Turbidites in interchannel areas are formed by overbank spilling and consist of thin-bedded fine-grained sands and silts characterized by Bouma *cde* and *de* sequences.

The delineation of fan margins and paleogeography is aided by the lateral and downfan changes in thickness, texture, composition, and paleocurrent directions of fan sediments. High contents of terrigenous debris are present in the sand fractions of hemipelagic muds deposited near the continental margin, and this may help delimit the shoreward boundary of the fan; in contrast, gradation to high contents of pelagic material indicates the direction of the seaward edge of the fan. Radially oriented paleocurrent patterns define the fan apex but are typically complex because of lateral overflow out of and away from channels and because of meandering and lateral shifting of fan channels. The outlining of fan geometry and major channels also is assisted by the decrease of the maximum clast size and of thickness of turbidite beds both downfan and laterally from channels.

Variations in morphology, stratigraphy, sedimentary facies patterns, grain-size distribution, sediment composition, and sediment dispersal patterns help identify fans from different geosynclinal settings such as restricted borderland or marginal sea basins, open ocean continental rises, and deep-sea trenches.

INTRODUCTION
General

Descriptions of both modern deep-sea fans (Shepard and others, 1969; Nelson and others, 1970; Normark, 1970a; Piper, 1970a; Haner, 1971; Normark and Piper, 1969, 1972; Nelson, 1974; Nelson and Kulm, 1973) and inferred ancient deep-sea fan deposits (Sullwold, 1960; Walker, 1966, 1970; Jacka and others, 1968; Piper, 1970b; Stanley, 1969b; Hubert and others, 1970; Mutti and Ricci Lucchi, 1972) have become more common in the geologic literature. Modern fan systems have been shown to vary greatly in size and shape, but the character of the sediments and nature of the depositional processes on them seem comparable. Similarly, although many different tectonic and paleogeographic settings have been ascribed to ancient deep-sea fan deposits, the stratigraphic sequences, sedimentary structures, and character of the sedimentary rocks seem comparable. The common occurrence of modern deep-sea fans in presumed geosynclinal settings such as the intersection of continental margins with spreading oceanic plates (Dietz, 1963; Dewey and Horsfield, 1970), marginal-sea basins, and continental borderland basins makes a comparison of the modern and ancient depositional systems a worthwhile exercise for understanding geosynclinal sedimentation.

Some characteristics of deep-sea fan sedimentation are best examined in modern deposits, others in ancient deposits. In this paper we will compare the morphologic and sedimentary features of the modern Astoria Fan, located off the coast of Oregon and studied by Nelson (1968), with the Butano Sandstone, an inferred Eocene deep-sea fan deposit located in the central Coast Ranges of California and studied by Nilsen (1971). Six important characteristics of the deep-sea fans and their deposits will be compared: (1) morphology and physiography, (2) stratigraphy and sedimentary facies, (3) sedimentary structures, (4) sediment grain-size distributions, (5) sediment compositions, and (6) sediment-dispersal systems. Our objectives are to outline the depositional processes acting in deep-sea fan systems and to establish some criteria for the recognition of ancient deep-sea fan deposits and their geosynclinal settings.

Butano Sandstone

The Butano Sandstone is found in the Santa Cruz Mountains south of San Francisco, California (fig. 1). It was named by Branner and others (1909) for sandstones and conglomerates that crop out on Butano Ridge. Foraminiferal studies indicate that the Butano ranges in age from the Penutian (early Eocene) to the Narizian (late Eocene?) provincial foraminiferal stages of Mallory (1959), although most of it apparently accumulated during Narizian time (Brabb, 1960; Sullivan, 1962; Clark, 1966; Fairchild and others, 1969). The formation was deposited at lower bathyal to abyssal depths in a basin that had unrestricted access to the ocean (Cummings and others, 1962). It unconformably overlies the marine Locatelli Formation and is conformably overlain by the marine Two-bar Shale Member of the San Lorenzo Formation (Brabb, 1964). No complete section of the Butano is exposed, so its total thickness is not known; its minimum thickness is approximately 1,500 m, the maximum perhaps 3,000 m.

Sedimentological studies by Nilsen (1970, 1971) and Nilsen and Simoni (1973) indicate that the Butano Sandstone forms the southwestern part of an Eocene deep-sea fan. Evidence that it originated as a deep-sea fan includes deposition in deep marine waters, abundant grain-flow and turbidite deposits, radially oriented and fan-shaped paleocurrent patterns, strongly defined proximal to distal stratigraphic and sedimentologic relations, and prominent development of channel and interchannel facies.

The Butano was deposited in an elongate basin within the early Tertiary continental borderland formed by the Salinian block (fig. 1). The northeastern part of the ancient fan has been displaced by the San Andreas Fault and is represented by the Eocene Point of Rocks Sandstone of the Temblor Range, southern California Coast Ranges. The two segments of the fan are separated by about 305 km of right-lateral offset along the San Andreas Fault (Clarke and Nilsen, 1972). The reconstructed fan in the Butano-Point of Rocks Sandstone is approximately 120 to 160 km long and about 80 km wide.

Astoria Fan

The modern Astoria Fan is comparable in size, shape, and thickness to the reconstructed ancient Butano fan but has a different basin setting because of its location on the continental rise off the coast of Oregon (fig. 1). The wedge of fan sediments radiates asymmetrically southward from the mouth of Astoria Canyon, which heads off the Columbia River. From the canyon mouth, the fan extends about 100 km to its western boundary, Cascadia Channel. The morphology of the fan ends 160 km south of the canyon mouth, although the depositional basin extends southward for another 150 km to Blanco Trough (Nelson and others, 1970). The Astoria Fan is 284 m thick at its lower end, 1,000 m thick at the base of the continental slope, and more than 1,000 m thick at its upper end (von Huene, Kulm, and others, 1973). The Pleistocene sediments of the Astoria Fan rest unconformably on a thin sequence of Pliocene abyssal plain turbidites (Kulm, von Huene, and others, 1973; Kulm and Fowler, this volume).

CHARACTERISTICS OF THE BUTANO AND ASTORIA FAN SYSTEMS

Physiographic Setting and Morphology

The main sediment source for the Butano Sandstone probably was an island or peninsula to the south underlain by granitic rocks; a less important source area may have been located to the northwest (fig. 1). No record remains of possible connections such as submarine canyons between the fan deposits and the source areas. However, the distribution of thick sequences of coarse-grained conglomerates, which are interpreted to represent channel deposits, indicates that at least two major channels were present in the proximal part of the ancient Butano and Point of Rocks fan (fig. 2).

Astoria fan also contains two major channel systems, the Astoria and slope-base fan valleys (fig. 1). Both fan valleys connect to Astoria Canyon at the present time; however, in the past, Willipa Canyon may have alternated with Astoria Canyon in funneling sediment from the edge of the continental shelf to the continental rise. A variety of sediments have been transported to the canyon by the Columbia River, the third largest river of North America. Source areas included the Coast Range, Cascade Range, Columbia Plateau, and the Rocky Mountains.

Tectonic movements, paleogeographic features, and depositional processes have influenced the shapes of the Astoria Fan and of the Butano and Point of Rocks fan. The Astoria Fan fills and covers an apparent trench at the eastern edge of the subducting Gorda-Juan de Fuca plate (Silver, 1972; Kulm and Fowler, this volume). The fan thus constitutes a thick north-south oriented wedge of sediment along the margin of the continental slope, rather than a simple cone-shaped body. The shape of the reconstructed Butano and Point of Rocks fan is similarly elongate; paleogeographic reconstructions indicate that upland areas of the con-

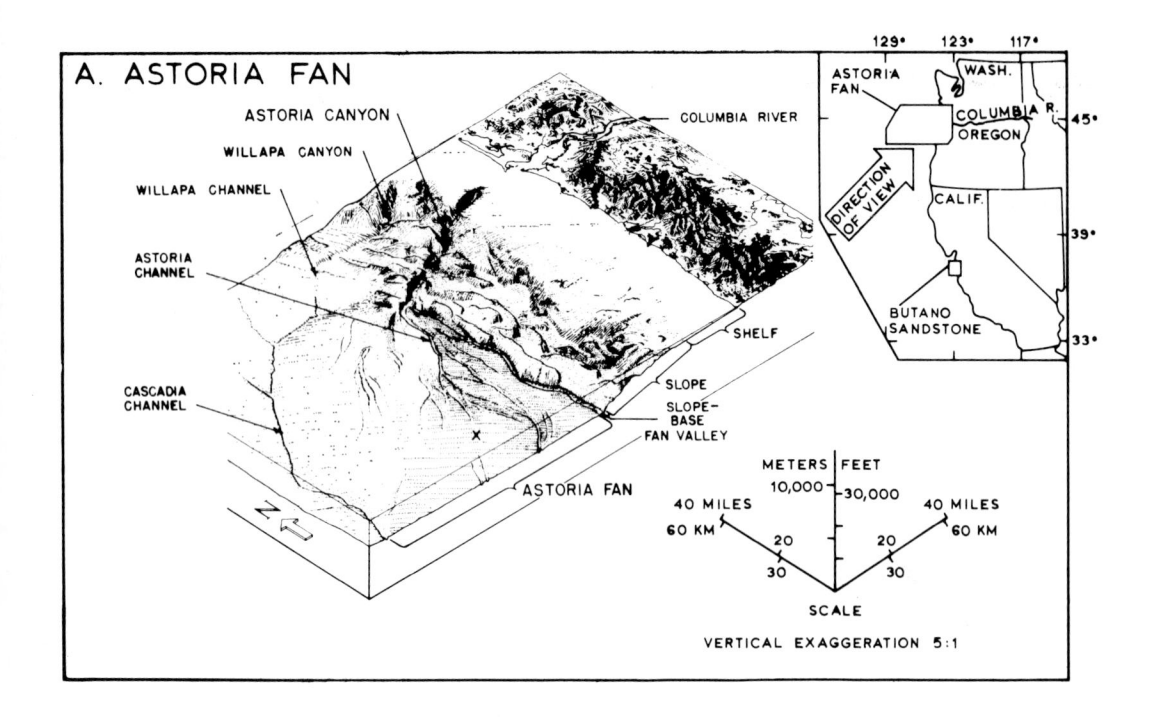

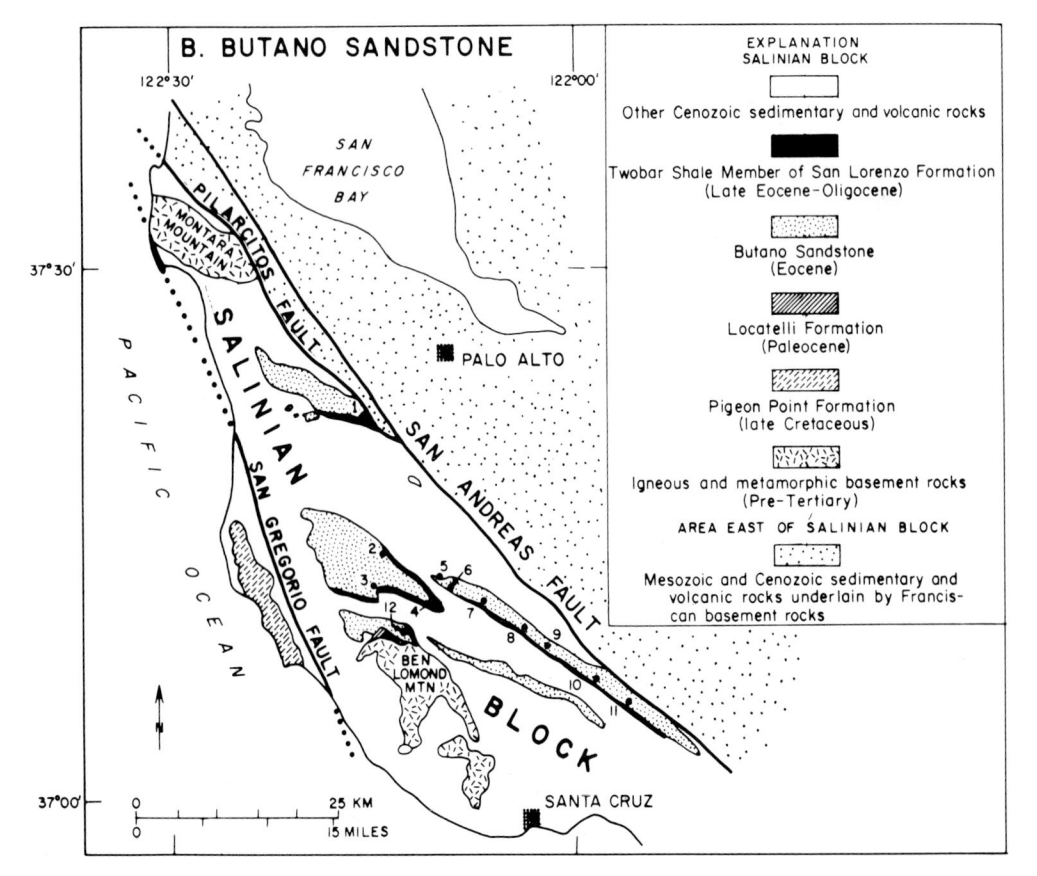

FIG. 1.—Location and setting of Astoria Fan and Butano Sandstone. X on Astoria Fan denotes location of Deep Sea Drilling Site 174 described by Kulm, von Huene, and others (1973).

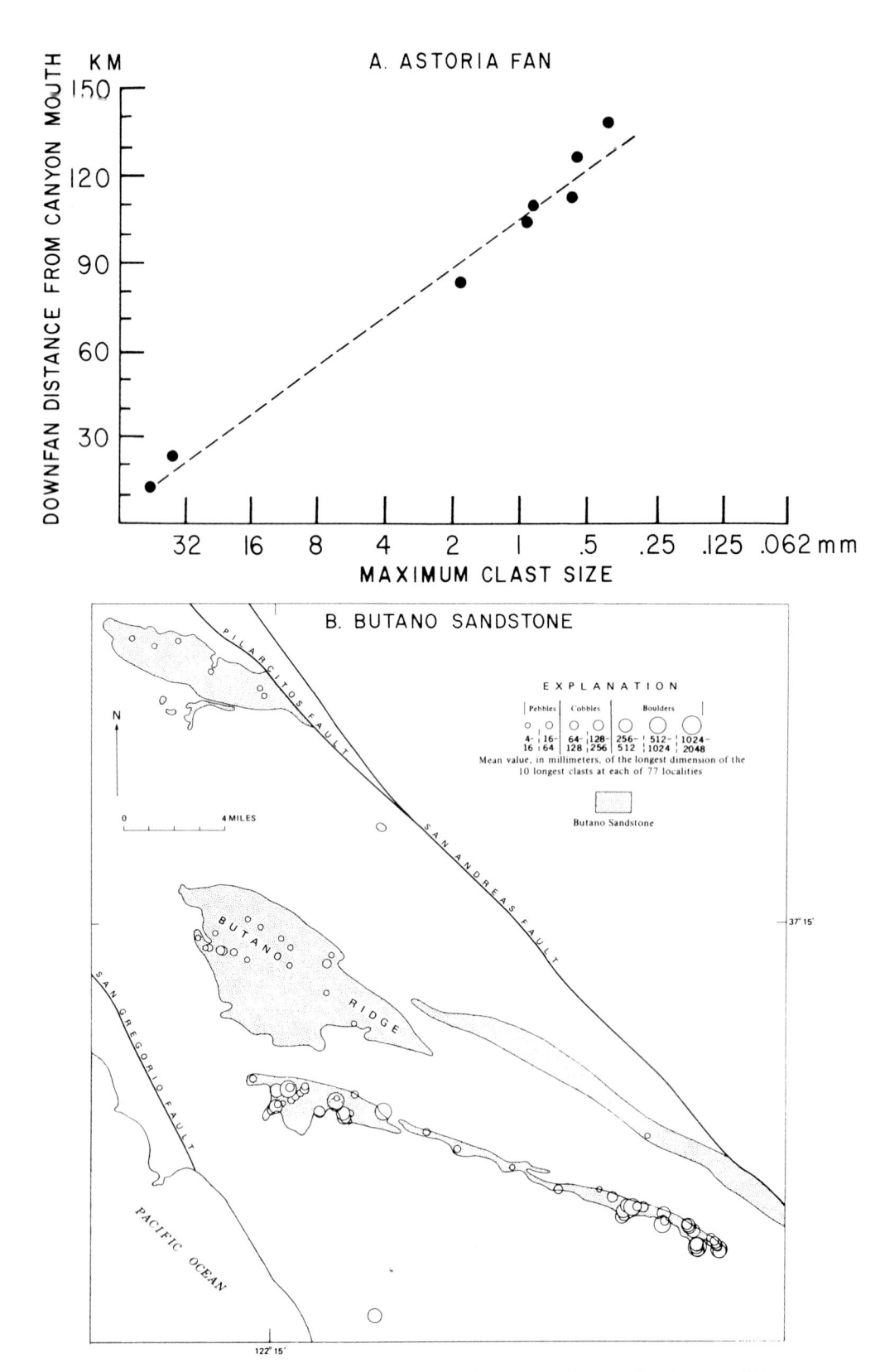

FIG. 2.—Distribution of largest clasts in Astoria Fan and Butano Sandstone turbidites.

tinental borderland to the west and a west-facing submarine slope to the east restricted the growth of the fan to the west and east (Nilsen, 1973; Nilsen and Clarke, 1973).

Cascadia Channel, located along the northern and western margins of the Astoria Fan, appears to have restricted the seaward and radial growth of the fan by limiting and capturing the flow of sediment westward from Astoria Canyon. In addition, the elongate growth of the Astoria Fan to the south was encouraged by the progressive leftward shift of the Astoria fan-valley systems through time (fig. 1; Nelson and others, 1970). When the fan-valley systems reached a position parallel to the base of the continental slope, all the sediments on the fan were transported southward. The youngest sedimentary rocks of the Butano and Point of Rocks Sandstones are located to the southwest, or on the left side of the ancient fan, suggesting a similar leftward growth pattern for this system. Menard (1955) has explained this type of growth pattern for deep-sea fans by postulating that the Coriolis effect causes sediment-laden currents to build higher channel levees on the right, thus encouraging deep-sea fan channels in the northern hemisphere to shift leftward with time.

Limited observations of channels within the Butano Sandstone suggest that the morphology of its fan valleys was similar to those of the Astoria Fan. A few large and deep channels characterize the proximal part of the fan; these are at least 1 to 3 km wide in the Butano and are 2 to 5 km wide on the upper part of the Astoria Fan. Prominent levees and levee deposits border these proximal channels in the Astoria Fan; similar deposits are found adjacent to channels in the proximal portions of the Butano Sandstone. The 200-m relief of the channels on the upper part of the Astoria Fan rapidly flattens to about 80 m in the middle part of the Astoria Fan. The main fan valleys generally break into numerous distributary channels in the middle fan, although some main channels continue throughout the length of the Astoria Fan, maintaining channel depths of 25 to 50 m and widths of 5 to 20 km.

Lateral gradation to smaller and shallower distributary channels continues to the distal fringes of both fan systems. It is difficult to establish the presence of small channels less than 200 m wide and 10 m deep on the lower part of the Astoria Fan because of poor resolution of sounding instruments. However, numerous small channels less than 10 m wide and 2 m deep are present in distal deposits of the ancient Butano and Point of Rocks fan.

In addition to the changes in the number and size of channels, the shapes of channels in both fans also change in the downfan direction. In the upper part of the Astoria Fan and locally in the proximal parts of the Butano Sandstone, channel walls are steep; in the Astoria Fan, channels have U-shaped profiles with flat floors. The walls gradually become less steep, and main channels assume a shallow saucer-shaped profile toward the lower fan.

Stratigraphy and Sedimentary Facies

Sources of data.—Forty piston cores provide a means of establishing Holocene and late Pleistocene stratigraphy for the Astoria Fan. The cores contain unique, correlative coarse-grained beds containing ash from the Mt. Mazama eruption 6,600 years ago (fig. 3; Nelson and others, 1968). Only limited information is available from beneath the upper several meters of the Astoria Fan, including data from site 174 of the Deep Sea Drilling Project (von Huene, Kulm, and others, 1971) and a few previously unpublished reflection profiles (Kulm and Fowler, this volume); this information suggests that the facies relationships seen in the cores persist at depth.

The stratigraphy of the Butano Sandstone is defined by 12 columnar sections measured from the upper contact with the Twobar Shale Member of the San Lorenzo Formation (fig. 3). The thickest section totals 855 m. Ten of the sections were measured in the middle fan, one each in the lower and upper fan. Unfortunately, no marker beds similar to the Mazama ash horizons of the Astoria Fan are available for correlation between the Butano sections (fig. 3).

Channel and interchannel facies.—Similar channel and interchannel facies that change from upper to lower parts of the fans are observed in both the Astoria Fan and the ancient Butano fan, even though there are correlation difficulties and stratigraphic data limit our comparison to the upper 6 m of Astoria Fan with the upper 855 m of the Butano Sandstone. It is generally impossible to correlate adjacent measured sections or cores bed by bed except for the tuffaceous layers on the Astoria Fan (fig. 3). However, the coarse-grained layers from the upper fan channels, middle to lower fan channels, and interchannel areas of the Astoria Fan have distinctive lithologic characteristics that are also present in apparent channel and interchannel facies of the Butano Sandstone.

In the Astoria Fan, the channels contain thicker layered gravels and sands whereas the interchannel areas include thinner layered fine-grained sands and silts (fig. 4). The lateral gra-

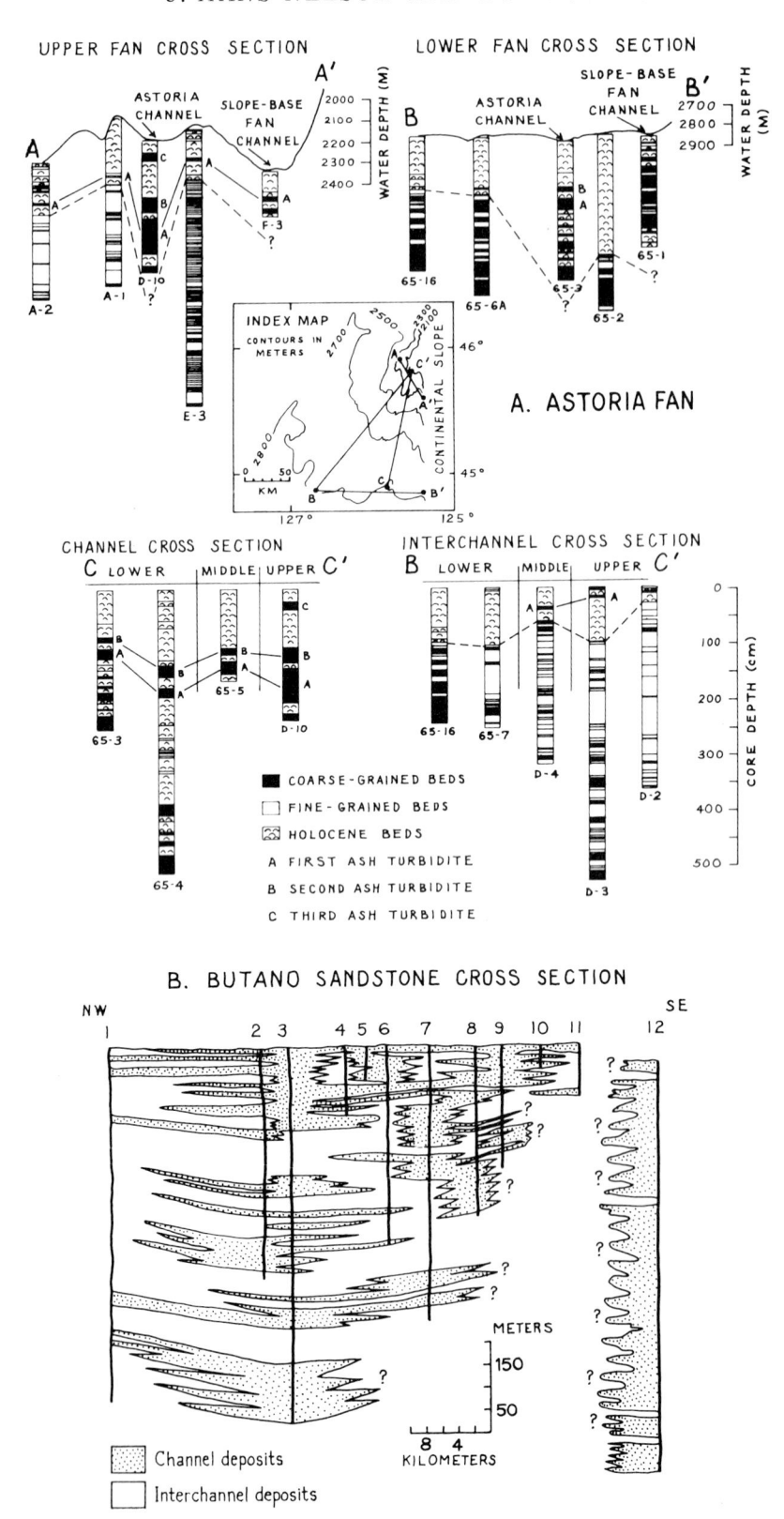

FIG. 3.—Stratigraphic cross sections of Astoria Fan (A-D) and Butano Sandstone (1-12). Location of 12 measured sections of Butano Sandstone shown in figure 1; location of core samples from Astoria Fan shown on index map for this figure.

BUTANO SANDSTONE
(Outcrops)

ASTORIA FAN
(Cores)

A
Upper fan channel deposits

B
Interchannel deposits

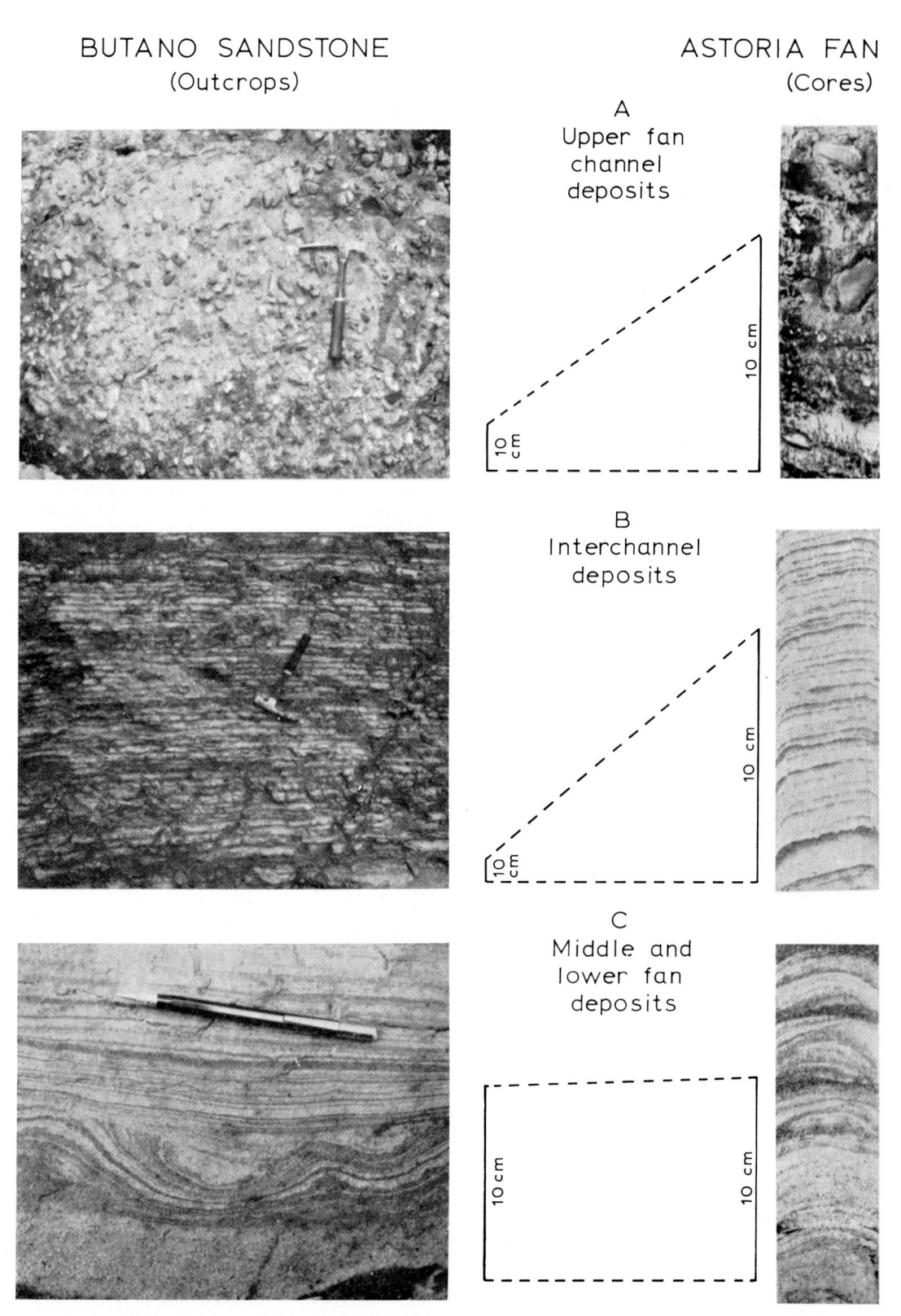

C
Middle and lower fan deposits

FIG. 4.—Comparison between sediment types and sedimentary structures of Astoria Fan and Butano Sandstone.

dation from channel to interchannel facies is most abrupt in the upper fan where thick irregular layers on the channel floor change into thin regularly layered levee deposits; these in turn grade laterally into more thinly layered and less regular interchannel deposits (figs. 3, A-A'; 4A,B).

The channel-to-interchannel facies changes are not as evident on the lower part of the Astoria Fan; there, prominent thinly layered levee deposits are not found, and thick sand and silt layers are widespread throughout channel and interchannel areas (fig. 3, B-B'). These thick layers found throughout the lower fan may have resulted from deposition in numerous shifting distributary channels or possibly from major overbank flows during the Pleistocene (Nelson and Kulm, 1973).

The channel deposits of the Butano Sandstone have been defined arbitrarily in the measured sections as those containing very thick-bedded conglomerates and medium- and coarse-grained sandstones. Interchannel deposits have been delineated by thick sequences of thinly interbedded fine-grained sandstones, siltstones, and shales. These two types of deposits are not mutually exclusive because some coarse-grained sequences are found in apparent interchannel facies and the reverse. The presumed channel and interchannel deposits alternate irregularly within the measured sections, resulting in an irregular distribution of channel deposits generally enclosed by interchannel deposits (fig. 3). This is expected because channels on modern fans are known to meander and migrate laterally, and new channels replace or capture old ones (Shepard, 1966; Normark, 1970b).

Proximal and distal facies.—The channel deposits of both the Astoria Fan and the Butano Sandstone change character from proximal to distal regions. The thickness and mean grain size of the tuffaceous sand and gravel layers in the upper part of Astoria Channel are twice that of the correlative sand layers in the channel in the midfan area (fig. 3—see layers A,B—and fig. 4). Downchannel from this point, however, the correlative beds are variably thicker or thinner from one location to the next but, in general, gradually thin from middle to distal parts of Astorial Channel. Mean grain size continues to decrease nearly linearly down the entire channel (Nelson and Kulm, 1973).

Thick conglomerate sequences are present in the two apparent channels located in the proximal part of the ancient Butano and Point of Rocks fan (figs. 2, 4) and beds as much as 20 m thick have been observed in these apparent channel deposits. In the distal channel deposits of the reconstructed ancient fan, only medium- to coarse-grained sandstone as much as 5 m thick is locally present. These data, plus the occurrence of a sand bed 8 m thick on the lower part of the Astoria Fan (Kulm, von Huene, and others, in press, 1973), suggest that relatively thick coarse-grained sandstones may be deposited in proximal, middle, and distal channels but that gravels appear to be restricted to proximal channels.

Sand-shale ratios of channels decline similarly downfan in both the modern and ancient systems even though thick coarse-grained beds are present in channel deposits of all fan areas. Proximal channel deposits of the Astoria Fan have sand-shale ratios[1] of about 4:1, and those of the Butano Sandstone have ratios greater than 5:1 (fig. 3, sec. 12). Downfan, ratios of 2:1 are maintained in the lower 75 km of Astoria Channel, and similar ratios are present in the comparable paleogeographic region of the ancient Butano and Point of Rocks fan.

The proximal to distal variations in sand-shale ratios from interchannel deposits of the Astoria and Butano and Point of Rocks fans are different, but the difference may result from insufficient data. Sand and silt layers are thicker and sand-shale ratios change from less than 1:9 to more than 3:1 from the proximal to distal parts of the Astoria Fan. In contrast, the ratios grade from 1:1 in the midfan region to 1:10 in the distal fringe of the Butano Sandstone, where deposits consist primarily of thinly interbedded sandstones and shales. The lack of distal gradation to finer and thinner sands and silts in the Astoria Fan may be explained by the prevention of distal growth on the western fan margin because of the presence of Cascadia Channel (fig. 5A). The distal facies may be present south of the fan margin, but no sampling has been done there.

Sedimentary Structures

The cores from the Astoria Fan provide sufficient data for a useful comparison of sedimentary structures in the ancient and modern fans even though they are only 6 cm wide and 600 cm long (table 1). However, the variety, lateral variation, and distribution of structures are more easily seen in the Butano Sandstone because of its extensive exposures. Except for the proximal channel deposits, both modern and ancient coarse-grained layers are typically graded and have sharp commonly erosional basal con-

---

[1] Astoria Fan calculations are based on the assumption that compaction of mud to shale is approximately 35 percent of the former thickness (Emery and Bray, 1962).

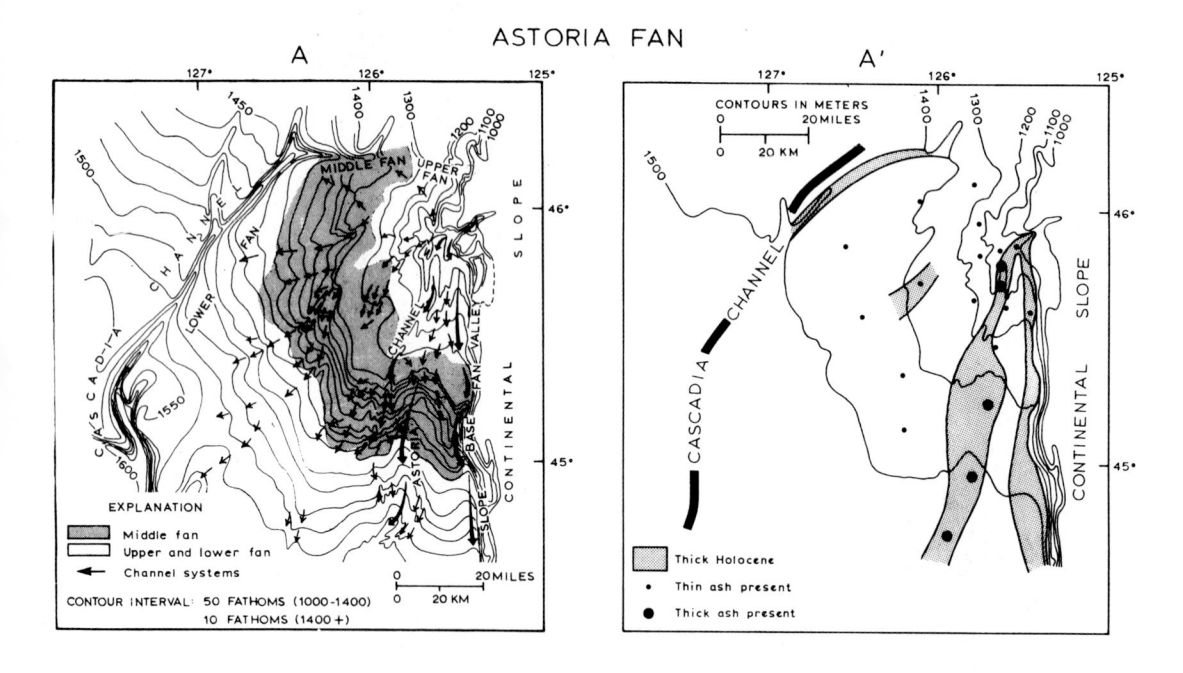

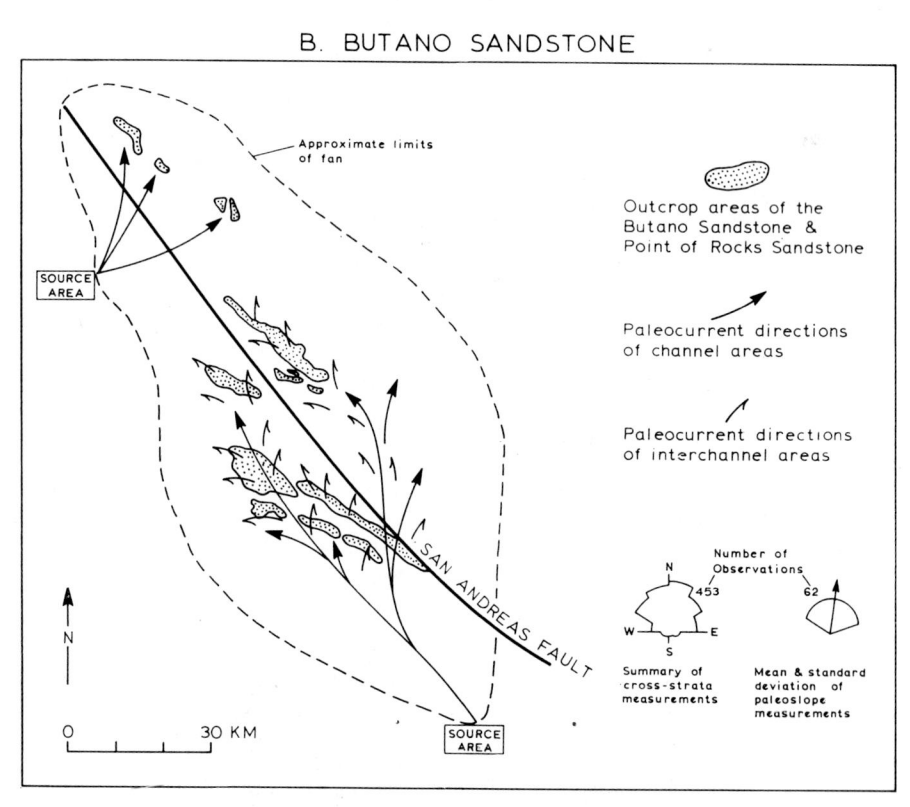

Fig. 5.—Sediment dispersal patterns of Astoria Fan and restored Butano and Point of Rocks fan. Data on Point of Rocks Sandstone from Clarke (1973).

TABLE 1.—COMPARISON OF TYPES AND DISTRIBUTION OF SEDIMENTARY STRUCTURES[1]
IN ASTORIA FAN SEDIMENTS (A) AND THE BUTANO SANDSTONE (B)

Sedimentary structure	Channel deposits			Interchannel deposits
	Upper fan	Middle fan	Lower fan	
Bouma sequences	A, B	A, B	A, B	A, B
a and ae	A, B	B		
abcde		A, B		
cde-de			A, B	A, B
Gravels, conglomerates	A, B	B		
Ungraded bedding	A, B	B		
Amalgamated sandstones	B	B		
Dish structures	B	B		
Mudstone ripups	A, B	A, B	B	
Flat stratification	A, B	A, B	A, B	A, B
Flute casts, groove casts, load casts	B	B	B	
Graded bedding		A, B	A, B	A, B
Cross stratification, current-ripple markings, convolute laminations		A, B	A, B	A, B
Flame structures		B	B	A, B
Bioturbation	A, B	A, B	A, B	A, B
Contorted stratification		B	B	B
Rotational slumps	A, B			B
Structures indicating sand-flow movements	B	B		
Sandstone dikes		B	B	

[1] Only the most common occurrence of the sedimentary structures and sequences is indicated in the chart; most of the structures may be found locally in other depositional environments. Evidence for Astoria Fan features mainly from photographs in Carlson and Nelson (1969).

tacts. Structures typical of Bouma sequences (Bouma, 1962) are the most abundant in these layers and also have distinctive vertical and lateral distribution patterns (fig. 4; table 1; Nelson and Kulm, 1973). Complete Bouma sequences may be found in channel deposits, whereas incomplete sequences are characteristic of interchannel deposits.

The proximal channel deposits of the two fans consist of Bouma *a* and *ae*[2] sequences. The Bouma *a* unit in the channels of upper Astoria Fan is generally massive but may contain some laminations and evidence of scour. It commonly contains mudstone ripup clasts. In the Butano and Point of Rocks fan, many types of conglomerate are present in the proximal channel deposits, but all have matrices that are composed primarily of sandstone. In contrast, the matrices of gravels in channel floors of the upper part of the Astoria Fan are commonly muddy, and gravels are irregularly layered (figs. 4A, 6A— see massive sand and gravel; Carlson and Nelson, 1969).

The conglomerates in the Butano range from well-bedded finely conglomeratic sandstones having well-defined clast orientations to massively or indistinctly bedded boulder conglomerates having chaotic clast orientations. The conglomerates grade upward into massively bedded sandstones of the Bouma *a* unit and flat-stratified sandstones of the Bouma *b* unit, or they may be overlain by thin Bouma *e* mudstones comparable to those separating *a* units in the proximal deposits of Astoria Channel (fig. 4A).

The Bouma *a* units from the Butano, that are composed wholly of sandstone, range up to 60 feet or more in thickness. They commonly are amalgamated or separated from overlying similar Bouma *a* sandstones only by thin discontinuous mudstones of the Bouma *e* unit. These sandstones are generally ungraded and massive but may contain diffuse parallel laminations, dish structures, irregular erosional surfaces, and scattered pebbles. Many are characterized by delayed grading, in which only the uppermost part of the bed grades abruptly upward into finer grained sediment. Mudstone and siltstone ripup clasts that are contorted and irregular in size and shape are commonly found in the *a* units of the Butano. Locally many feet long, they may be concentrated in the upper parts of the *a* units, randomly distributed, or segregated into irregular layers. The basal contacts of the Bouma *a* sequences in the Butano commonly are erosional and are either gently disconformable with or channeled into underlying sediments;

[2] We refer to thick, ungraded, and internally structureless sandstone beds separated from one another by thin mudstone or shale layers as Bouma *ae* sequences in this paper, even though the sandstones are not graded and probably were not deposited by turbidity currents.

large irregular flute- and groove-cast sole markings and load casts may be present at the contacts. Many basal contacts, however, are remarkably smooth, flat, and conformable.

A greater variety of sedimentary structures is present in the middle and distal channel deposits than in the proximal channel deposits of both fans. Complete Bouma *abcde* sequences are most often found in the middle channel deposits, although the individual sequences are generally not as thick or as coarse grained as the Bouma *ae* sequences of the proximal channel deposits. The *a* unit of the midfan deposits is generally thinner and graded. Smaller scale flute and groove casts, bounce marks, tool marks, prod marks, and other sole markings characteristic of turbidites are more common at the base of the *a* units in the Butano. The *b* unit also becomes thinner downfan and may be missing in the lower fan of both systems. Consequently, the Bouma *c* unit is dominant in the distal channel deposits. It exhibits current-ripple markings, small-scale cross strata, convolute laminations, flame structures and contorted bedding (Carlson and Nelson, 1969); in addition to these structures, ball-and-pillow structure has been noted in the Butano *c* units. The *d* unit in both fans is generally thin and consists of flat-laminated fine-grained sandstones, siltstones, and mudstones.

The interchannel deposits of the two fans consist primarily of Bouma *cde* sequences of fine sand and silt that rhythmically alternate with interbedded muds (fig. 4B). The fine sands of the *c* unit generally contain small-scale cross strata and are overlain by flat-laminated to unlaminated silts and muds of the *d* and *e* units. The small-scale cross strata in the Butano Sandstone were produced by migrating current ripples and are locally associated with or replaced by convolute laminae. Locally, entire sequences of beds in the Butano are contorted, probably as a result of downslope movement under the influence of gravity. Toward the distal part of the Butano fan, in interchannel deposits, Bouma *de* deposits become more common, and the *e* unit increases in thickness and is more widely distributed. Hemipelagic mudstones commonly comprise 50 percent or more of the interchannel deposits at the distal end (fig. 3B).

The Holocene hemipelagic deposits of the Astoria Fan as well as most interchannel deposits of the Butano Sandstone were typically extensively bioturbated, probably reflecting slow rates of sedimentation. In the Astoria Fan this resulted in a mottled structureless sediment containing scattered fecal pellets (Carlson and Nelson, 1969). Bioturbation of the interchannel deposits of the Butano Sandstone consists of burrows and borings of variable shape and size, most of which are oriented roughly parallel to the bedding planes. Reworking commonly extends downward from hemipelagic beds to disrupt the upper Bouma *de* layers in both fans.

Grain-Size Distributions

Similar grain-size distributions were determined for the following three sediment types of both fans: (1) thick to massive-bedded, ungraded sands and gravels from the *a* unit of Bouma *ae* sequences that are characteristic of proximal channel deposits; (2) sands from basal *ab* units of more complete Bouma sequences that are characteristic of middle and lower channel deposits; and (3) fine sands and silts from upper *cd* units of more complete Bouma sequences that are present everywhere except in proximal channel deposits (fig. 6; table 2). The deposits of incomplete Bouma *ae* sequences from the Astoria Fan are more variable in grain size than those from the Butano and consequently have been divided into two subgroups, those that are very poorly sorted and contain abundant silt- to clay-sized matrix, and those that are better sorted and contain only a small amount of silt- to clay-sized matrix. Only the better sorted variety was noted in outcrops of the Butano; however, because the grain-size distribution of the gravel and clay-sized fractions was not determined, the more poorly sorted variety may not have been detected if present.

Sands and gravels from Bouma *ae* sequences of both fans are characterized by positive skewness (skewed toward finer grain sizes) and poor sorting (fig. 6; table 2). The extremely poor sorting and high positive skewness of some Bouma *ae* sands of the Astoria Fan can be attributed to the very high content of silt- and clay-sized matrix (table 2).

Sediments from *ab* units of the more complete Bouma sequences on both fans are finer grained, better sorted, and less positively skewed than those from the Bouma *ae* sequences.

The grain-size distributions of the Bouma *cd* units from both fans are very similar, although the content of silt- and clay-sized matrix again is higher in the Astoria Fan deposits than in those from the Butano (table 2). Sediments from the Bouma *cd* units generally are finer grained and contain more silt- and clay-sized material than the coarser grained *ab* sediments.

The interbedded hemipelagic muds from the Astoria Fan have been analyzed but not those from the Butano Sandstone. They are charac-

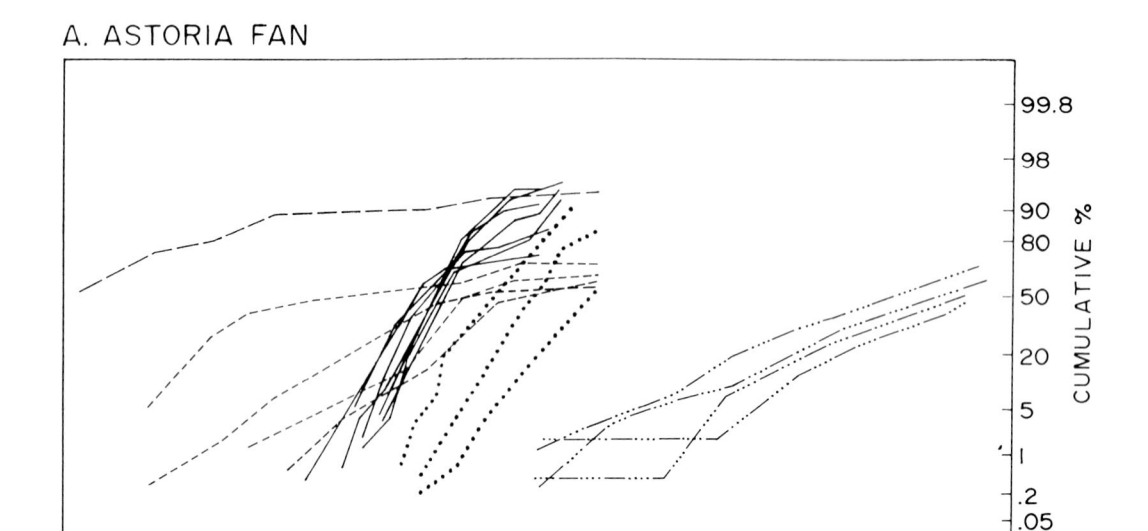

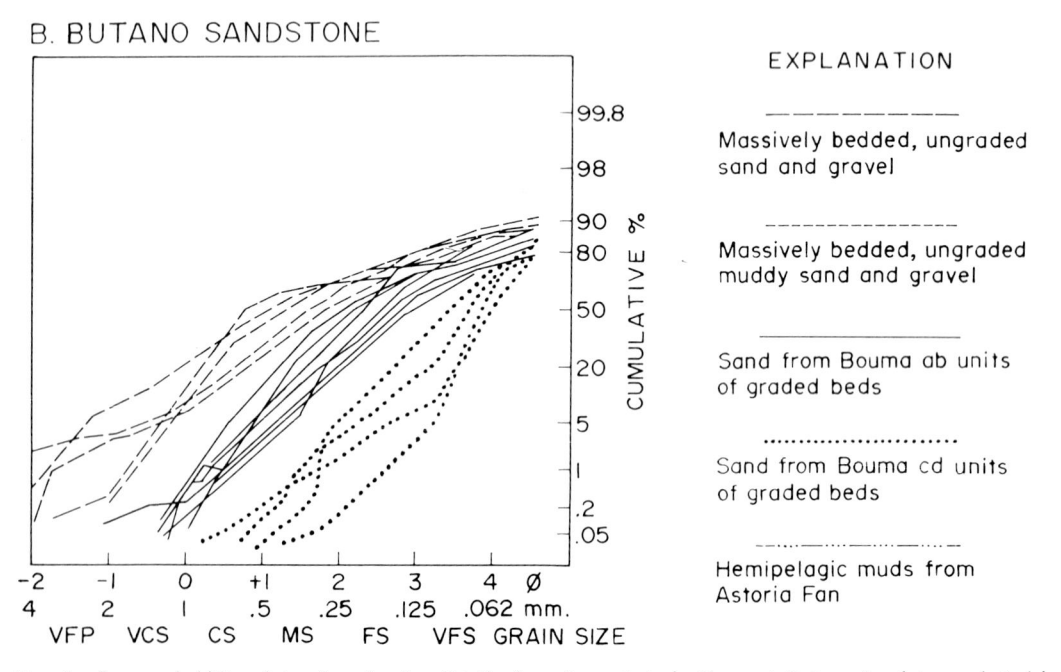

FIG. 6.—Log-probability plots of grain-size distributions from Astoria Fan and Butano Sandstone, plotted in straight line segments after method of Visher (1969).

terized by extremely fine-grain size and poor sorting, which also may be characteristic of hemipelagic mudstones of the Butano.

Because the Bouma *ae, ab,* and *cd* units typically occur in deposits from particular fan environments, the sorting, skewness, and other parameters (table 2) grade laterally over the whole fan. The hemipelagic muds also show lateral gradations over the Astoria Fan from silty clays at the base of the continental slope to fine clays at the western margin (Nelson and Kulm, 1973). As the channel deposits grade downfan from Bouma *ac* sequences to *abcde* sequences and laterally to *cde* sequences in interchannel deposits, the mean grain size decreases, the sorting increases, and the skewness becomes less positive.

Composition

The sediments of both fans have been derived from two separate sources: (1) the erosion of coarser grained clastic detritus from adjacent land areas and (2) the slow settling out in ocean waters of pelagic biogenic shell debris and suspended clay-sized terrigenous material. The different land-source areas of the two fans are indicated by the composition of the silt, sand, and gravel components of arkosic arenites in the Butano and lithic wackes or arenites in the Astoria Fan (fig. 7).

The finer grained hemipelagic sediments contain primarily *in situ* benthic fauna, whereas the coarse-grained sediments commonly contain reworked and displaced shallow-water biota such as shell fragments of megafauna and tests of benthic foraminifera. Consequently, biota of the hemipelagic muds best define the time-stratigraphic relations for fans. For example, the pelagic composition of the hemipelagic deposits of the Astoria Fan changed very abruptly from planktonic foraminifera to radiolarians during late Pleistocene and Holocene time, although no change in biota of the coeval coarse-grained sediments was detected.

Lateral changes in composition are present in both the hemipelagic and coarse-grained sediments of the Astoria Fan; similar changes probably are present in the Butano Sandstone although they have not been documented. The wackes of the upper part of the Astoria Fan typically change downfan to arenites as the amounts of matrix and lithic material become less (fig. 7). In addition, the content of platy and lower density constituents (i.e., mica, plant fragments, volcanic glass) decreases downchannel but progressively increases in interchannel fan deposits more distal from the channels (Nelson, 1974). The amount of terrigenous de-

TABLE 2.—SUMMARY OF GRAIN-SIZE DISTRIBUTION PARAMETERS[1] FOR ASTORIA AND BUTANO FAN DEPOSITS

Parameter	Massive-bedded ungraded sand and gravel			Bouma ab intervals		Bouma cd intervals		Holocene hemipelagic muds Astoria
	Astoria High matrix 3 samples	Astoria Low matrix 1 sample	Butano Low matrix 3 samples	Astoria 8 samples	Butano 11 samples	Astoria 3 samples	Butano 6 samples	4 samples
Median (mm)	0.07 to 0.27	4.03	0.29 to 0.59	0.15 to 0.19	0.13 to 0.23	0.042 to 0.085	0.07 to 0.12	.0033 to .0012
Mean (φ)	3.29 to 5.24	−1.84	1.53 to 1.89	2.48 to 3.01	2.34 to 3.15	3.55 to 5.12	3.25 to 4.01	8.0 to 9.56
Graphic standard deviation (sediment sorting)	3.03 to 4.08	2.47	1.30 to 1.70	0.52 to 1.26	1.06 to 1.31	0.72 to 1.94	0.48 to 0.87	1.48 to 2.91
Inclusive graphic skewness	+0.51 to +0.70	+0.35	+0.32 to +0.79	+0.62 to +0.15	+0.43 to −0.08	+0.56 to −0.04	+0.47 to −0.81	.11 to −.13
Graphic kurtosis	0.68 to 0.083	1.58	1.92 to 2.99	0.90 to 2.82	1.96 to 2.40	0.83 to 3.30	2.27 to 5.74	.90 to 1.23
Size (mm) of coarsest 1%	0.5 to 2.5	25.0	1.60 to 1.65	0.30 to 0.46	2.65 to 1.15	0.12 to 0.19	0.18 to 0.40	.14 to .062
Percent finer than 4φ	32.5 to 52.0	7	8 to 13	5 to 28	13 to 25	25 to 89	21 to 44	98.3 to 99.2
Velocity (cm/sec)[2] to transport coarsest 1%	125		25	30	50	25	30	

[1] The grain-size distribution characteristics of 350 samples from piston cores of the Astoria Fan were determined by (1) wet sieving the samples to separate the >62μ fractions from the silt- and clay-size fractions, (2) sieving the coarse fraction in ¼φ screen sizes or analysis in an Emery (1938) sedimentation tube as modified by Poole (1957), and (3) analyzing the fine fraction using the American Society for Testing Materials (ASTM) hydrometer method (1964). Twenty-one samples of the Butano Sandstone were sieved to ½φ intervals by T. R. Simoni and R. H. Wright of the U.S. Geological Survey. The samples were first disaggregated by means of water, hydrochloric acid, and mechanical devices. Because of both extensive calcite cementing, which formed concretions in which the detrital grains are infiltrated and broken, and the extensive weathering, which altered the feldspars to clay minerals, relatively few samples could be used for the sieving analyses. Statistical parameters for both data sets were derived utilizing the methods of Folk and Ward (1957).
[2] Velocities estimated by utilizing Sundborg's (1956) erosion-transport-deposition velocity fields.

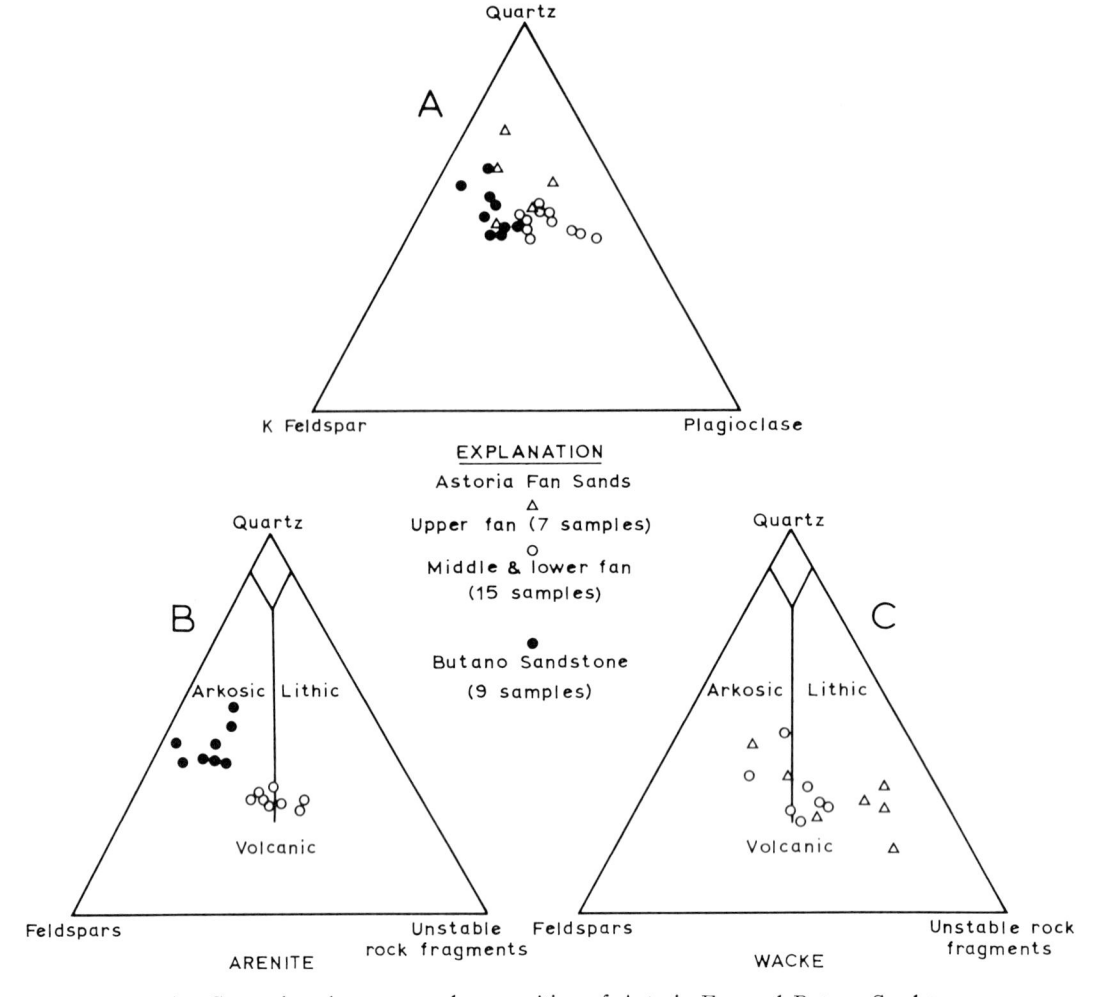

FIG. 7.—Comparison between sand composition of Astoria Fan and Butano Sandstone.

bris lessens and pelagic constitutents increase in hemipelagic deposits away from the continental slope (Nelson and Kulm, 1973).

Individual coarse-grained layers of both the Butano and Astoria fan systems display systematic vertical changes in composition. Detailed investigations on the Astoria Fan show that the most current-sensitive material, the platy and lower density debris, collects at the top of a coarse-grained layer. Consequently, when present, mica, plant fragments, volcanic glass shards, smaller foraminiferal species, and light minerals increase in abundance toward the top of a layer (Nelson, 1974). Plant debris appears to be more common in the upper parts of sandstone beds in the Butano, but information is lacking on the other platy constituents.

Because of the vertical and lateral gradations in composition, the lower parts of coarse-grained beds in the middle and lower fan contain less matrix and constituents of low density than the upper parts of the beds. Consequently, they are more likely to be arenites. In contrast, upper parts of coarse-grained beds in the middle and lower fan beds, in addition to all parts of upper fan interchannel beds, commonly are wackes because of their relatively high content of matrix and platy constituents. Lack of recognition of these compositional and textural trends may account for some of the past arguments about comparability of modern deep-sea sands versus ancient turbidites (Kuenen, 1964; Emery, 1965).

Sediment Dispersal Patterns

Directions of sediment transport on the Astoria Fan are inferred from the location, orientation, and distribution of fan channels (fig. 5).

Although sediment has not been directly observed in transport, the distribution of sedimentary facies (fig. 3) indicates that sediments are transported from the mouth of the Columbia River directly down Astoria Canyon to the head of the Astoria Fan and from there out onto the fan via the system of fan valleys. Many channels now exist on the fan surface, but development of present morphology (fig. 1) and distribution of the most recent tuffaceous sands (fig. 5A) suggest that only a few channels actively transport or fill with sediments at any particular time. The orientation of channel morphology and distribution of channel sediments, therefore, outline the pattern of sediment dispersal.

In ancient deep-sea fan deposits such as the Butano Sandstone, however, it is generally difficult to ascertain the location, orientation, and distribution of the fan channels because of incomplete exposures. Therefore, measurements of paleocurrent directions from both channel and interchannel deposits must be relied upon to determine the directions of sediment transport responsible for growth of the fan. An extensive paleocurrent study of the Butano, based on measurements of 565 paleocurrent indicators and 105 paleoslope indicators, provides a mappable pattern of sediment transport comparable to that inferred from the Astoria Fan (fig. 5; Nilsen, 1970; Nilsen and Simoni, 1973).

The dominant feature of both fan systems is the radial orientation of channels and apparent dispersal of sediment away from the fan apex. This pattern is clearly indicated by channel orientations on the Astoria Fan and by paleocurrent directions in the Butano Sandstone (figs. 1, 5). The dominant transport of sediments down the slope of the fans is indicated by the orientation of the surficial channels parallel to the slope of the Astoria Fan and by similar orientation of paleocurrents and paleoslopes in the Butano Sandstone (fig. 5). These data, give no indication that contour or bottom currents,[3] flowing at an angle to the fan slope, have had any effect on sedimentation in these two fans.

Directions of sediment transport from the interchannel deposits of the ancient Butano fan vary in azimuth orientation from about 270° to about 90° (fig. 5). Paleocurrent directions perpendicular to channel orientations suggest that interchannel deposition resulted from transport of sediments out of the channels toward the interchannel areas. Distribution of correlative thin beds of ash points to a similar type of dispersal pattern in interchannel areas of the Astoria Fan (fig. 5A).

DISCUSSION
Depositional Processes

The sedimentary processes responsible for deposition of the Astoria Fan and the Butano Sandstone are inferred to be similar. The fan shapes and morphology result primarily from channeled flow that has migrated laterally through time. Sediment dispersal patterns indicate that the coarsest debris is transported downfan along channels and that some of the associated finer grained sediment is transported out and away from the channels. Repeated alternations of thick coarse-grained sediment grading upward into thin interbeds of hemipelagic mud indicate that intermittent depositional events are mainly responsible for sedimentation on the fans. These events cause the episodic movement of coarse-grained sediments from shallower areas to the deep-sea fan.

The following five-point model for deep-sea fan depositional processes is based on our comparison of the modern Astoria Fan and the ancient Butano fan:

(1) The chaotic, massive-bedded, and very poorly sorted sands and gravels found at the canyon mouth and in upper channels of the As-

TABLE 3.—GENERAL MORPHOLOGIC CHARACTERISTICS OF MODERN AND ANCIENT DEEP-SEA FANS

Feature	Characteristics
Shape	Shape is more fanlike adjacent to straight open margins or if fan is too small to be disrupted by geographic elements; most fans are elongate or irregular because of growth constrictions by other geographic elements, multiple sources, dominant channel migration right or left, and(or) bottom-current patterns
Size	In modern fans ranges from 1,000 km ×2,500 km to 7×11 km; commonly about 150 km×300 km in open ocean basins and less than 100 km ×100 km in restricted basins
Thickness	In modern fans ranges from 0.3 km to 12 km; variable, probably within same limits for ancient fans
Channel and interchannel dispersal and paleocurrent patterns	Channel patterns generally oriented radially outward from fan apex; interchannel and levee patterns generally lateral away from channels; channel systems may shift to right or to left in southern or northern hemispheres respectively; all patterns may be modified by bottom and contour currents

---
[3] The term "bottom current" is used to indicate a traction current flowing because of the potential energy in a water mass, and not because of sediment load (Piper, 1970a).

toria Fan appear to have been deposited by debris flows[4] from the continental slope and canyon walls (fig. 3B).

(2) The massive-bedded, better sorted, and low-matrix Bouma *ae* sands and gravels found primarily in upper fan channel deposits probably result from fluidized sediment flows or grain flows;[5] these may be associated with sediment funneled directly into submarine canyons by longshore drift.

(3) The basal *ab* units of well-developed Bouma sequences appear to be deposited by turbidity currents[6] flowing downfan in channels. They represent bedload deposited under upper flow regime conditions (Harms and Fahnestock, 1965; Walker, 1965; Walton, 1967) and generally are restricted to middle and lower fan channels.

(4) The *cd* units of well-developed Bouma sequences appear to be deposited by turbidity currents flowing both in channels and outside of channels. They represent suspended load deposited under lower flow regime conditions (Harms and Fahnestock, 1965; Walker, 1965; Walton, 1967).

(5) Clay-sized materials containing a sand fraction of pelagic constituents are deposited by the continuous vertical settling of fine suspended debris from within the water column; hemipelagic sediment also may be contributed near the base of the continental slope by turbid layer transport[7] from the continental terrace.

From the data presented, the following history of an individual turbidity current can be postulated: The coarsest sediments of incipient turbidity currents are deposited in the most proximal channels as the massive *a* unit of incomplete Bouma *ae* sequences. The bedload of the turbidity current continues on in the channel and is deposited as Bouma *ab* units in the channels of the middle and lower fan. The finer grained debris of the turbidity current is enriched in low-density and platy material (mica, plant fragments, and volcanic glass shards); most of it is carried upward by the turbulence and lags behind the main channeled flow. Part of this cloud of fine debris spills out of the channel and flows laterally to deposit Bouma *cde* sequences in levee and interchannel areas. Some of this material remains behind in the channel to form Bouma *cde* units. Eventually, only the finest debris of the tail of the turbidity current is deposited in Bouma *cde* and *de* sequences on the distal fringes of the fan.

Resedimentation processes such as turbidity currents and grain flows interrupt the continuous rain of suspended debris accumulating as hemipelagic deposits. The suspended debris close to the source of sediments is enriched in terrigenous material, whereas toward the seaward edge of the fan it is enriched in pelagic materials. Movement of sediment by turbid-layer flow off the continental terrace onto the shoreward part of fans and down main channels is suggested by changes in clay-mineral composition along the axis of Astoria Channel and by greater thickness of the Holocene clays in the channel areas (fig. 5; Duncan and others, 1970; Nelson, 1974).

Comparison of Deposits and Depositional Processes

Both fans have similar facies changes and grain-size distribution variations in the downfan direction as well as from channel to interchannel areas (tables 3 and 4, fig. 3). Very definitive changes from thick-bedded conglomerates and sandstones to thin-bedded siltstones and mudstones are found in the downfan stratigraphy of the Butano Sandstone. A similar but less emphatic change may be present on the Astoria Fan, but Cascadia Channel on the west and the Gorda Rise and Blanco Gap on the south seem to have limited the development of the distal fringe facies. The youngest channel facies, in both fans, however, is generally parallel the basin slopes. This appears to be related to the tendency for a leftward shift of main channels through fan history (Menard, 1955; Nelson and Kulm, 1973).

The average grain size of deposits of the Astoria Fan and the ancient Butano fan is different, and it has varied from time to time within

---

[4] Debris flows are mixtures of granular solids and water that, in general, are of greater density and more sluggishly moving than turbidity currents (Hampton, 1970). For the most thorough discussion of this and other resedimentation processes described in footnotes 5 and 6, see Middleton and Hampton (1973).

[5] A grain flow is a mass flow of sand (Stauffer, 1967) in which the sediment is supported by direct grain-to-grain interactions (collisions or close approaches). Fluidized sediment flow is a similar flow in which the sediment is supported by the upward flow of fluid escaping from between the grains as the grains are settled out by gravity (Middleton and Hampton, 1973).

[6] We use the term turbidity currents in the context of Kuenen (*in* Sanders, 1965, p. 217) for "a current flowing in consequence of the load of sediment it is carrying and which gives it excess density" and in the context of Middleton (1969, p. GS-A-6) for "a density current in which the difference in density between the current and the ambient fluid (commonly sea water) is due to the presence of dispersed sediment" (see footnote 4 also).

[7] We use turbid layer transport in the context of Moore (1969, p. 83) as low velocity (<10 cm/sec) and low density turbidity currents of fine-grained debris flowing down the continental terrace and sometimes through canyons and fan valleys.

both fan systems. Deposits of the ancient Butano and Point of Rocks fan are generally coarser grained, thicker bedded, and contain larger amounts of Bouma a units than do deposits of the Astoria Fan. This is probably the result of deposition in a borderland basin, where only very narrow marine shelves are present between the source area and the deep-sea basin. Without an intervening shelf, sands and gravels from littoral drift are carried more directly from the source area down subsea slopes to deep water; the resulting fan developed at the base of the slope contains large amounts of thick-bedded coarse-grained deposits that are probably deposited primarily by grain flow or fluidized sediment-flow mechanisms rather than by turbidity currents. The narrow shelf probably also contributes to the formation of hemipelagic sediments that are rich in terrigenous debris and that are deposited mainly by turbid-layer transport.

The Astoria Fan, on the other hand, has been deposited at the base of the continental slope along the western margin of the North American continent. The continental shelf or terrace is wider than those typically surrounding borderland basins, so that less of the coarsest sediment reaches the fan. Because it has been fed primarily by a single major river, the Columbia, detritus available for deposition on the Astoria Fan is finer grained than littoral debris available for deposition in borderland fans. The Astoria Fan turbidites are thus typically finer grained, more thinly layered, and contain thinner and more complete Bouma sequences than do the Butano turbidites. This is particularly true for the turbidites of the Astoria Fan that were formed during Holocene time, when the continental shelf or terrace has been widest and glaciation in the source area has been at a minimum. Because of the narrower shelf during the Pleistocene, when sea level was lower, less coarse-grained sediment was trapped on the shelf as well as in estuaries, and more bypassed to the Astorian Fan.

The borderland setting of the Butano Sandstone and the lower sea levels of the Pleistocene for the Astoria Fan thus resulted in funneling of coarser grained, better sorted, and texturally more mature sand more directly from the shoreline to the deep sea floor. The higher sea level of the Holocene caused longer transport across a wider shelf and more fine-grained debris; consequently, more fine-grained matrix was incorporated during transport to the deep sea, and wackes became dominant in turbidites deposited on the Astoria Fan during the Holocene. At the same time, more debris flows resulted on the Astoria Fan apparently because of the finer grained sediments present on the continental slope and on the Astoria Canyon walls.

Both the Astoria and ancient Butano fan systems changed from growth regimes characterized by rapid sedimentation of coarse-grained sediment to nongrowth regimes characterized primarily by hemipelagic deposition. This occurred from the late Pleistocene to the Holocene in the Astoria Fan (Nelson, 1974) and during the late Eocene in the ancient Butano fan. The alternation from glacial to interglacial climates in the Pacific Northwest resulted in a reduced production of sediment and a rise of sea level that disrupted river transport of debris directly to canyon heads (Nelson and Kulm, 1973). Deposition of sand on the Butano and Point of Rocks fan ended abruptly with deposition of the Twobar Shale Member of the San Lorenzo Formation and of the Kreyenhagen Shale. (The Twobar and the Kreyenhagen overlie the Butano and Point of Rocks Sandstones respectively.) This rapid change may have resulted from tectonic activity, sea level fluctuation, climatic variation, or combinations of these effects.

Criteria for Recognizing Ancient Deep-Sea Fan Deposits

Most fans may be composed primarily of turbidites, but not all turbidites form deep-sea fans. The floor of some small local basins like Crater Lake, Oregon, may be covered by turbidites, but fan morphology is not developed (Nelson, 1967). Turbidite fills in some deep-sea trenches and other basins with extremely complex shapes also lack fan or cone shapes. Turbidite deposits of deep-sea channels are very linear and extend from hundreds to several thousand kilometers in length (Nelson and Kulm, 1973), whereas their abyssal plain terminations have irregular shape (Horn and others, 1971). On the other hand, fan shapes may not be associated with distal parts of extremely large fans (Curray and Moore, 1971) or coalesced fans. Nevertheless, deep-sea fans can be distinguished from other types of turbidite deposits on the combined basis of characteristic patterns of morphology, stratigraphy, sediment dispersal, sedimentary structures, texture of coarse-grained layers, and composition in hemipelagic muds as well as turbidites (tables 3, 4).

Distinguishing a radiating fan-valley system and(or) shape throughout an entire deposit is a fundamental key to recognizing deep-sea fans. If, as is common, the outcrop pattern does not permit this, the internal stratigraphy, facies, and lithology of deep-sea fan deposits can be identified in accord with the following six ob-

TABLE 4.—CHARACTERISTICS OF MODERN AND ANCIENT DEEP-SEA FANS AND CRITERIA FOR RECOGNITION OF ANCIENT DEEP-SEA FAN DEPOSITS[1]

Characteristic	Submarine canyon	Physiographic regions of fans			Depositional environments of fan regions		
		Upper fan	Middle fan	Lower fan	Fan fringe	Channel	Levee	Interchannel
A Physiography	Generally straight, V-shaped, and deep (hundreds of meters); commonly steep-walled with cliffs and overhangs	Few main channels (generally one or two) that are very deep (100+ m in large fans), steep-walled and wide (several kilometers); levees are well developed; most irregular channel topography of any fan region	Few main channels branching into tens of distributaries; (up to tens of meters deep); width variable (tens of meters to tens of kilometers); steep to gently sloping walls; channels may meander	Numerous shallow, narrow, braided, randomly shifting, distributary channels having gently sloping walls; limited development; limited to lack of levee development; size intermediate between middle and fan-fringe channels	Flat surface having a limited number of gentle channels of few meters width and depth in open basin fans or may lack channels in fans of restricted basins	U-shaped, leveed, flat-floored; typically depositional-erosional features on the fan surface	Sediment wedges up to tens of meters thick flanking channels in upper to middle fan areas; range from tens to hundreds of kilometers wide depending on fan size	Generally smooth, flat, gently sloping, and concave upward surfaces except in the middle fan where convex suprafan bulges may occur
B Stratigraphy and sedimentary facies	Thick (up to tens of meters) coarse-grained beds having highly variable and poorly developed bedding; fining upward to fewer and thinner coarse-grained beds in muds	Abrupt vertical and lateral distribution of channel and interchannel facies; channel bodies enclosed within interchannel facies, and isolated blocks of levee facies enclosed within channel facies	Like upper fan region evolving to distinctly bedded and structured channel facies having thin well-developed mud interbeds associated with channel interchannel facies	Dominant channel facies throughout section evolving to dominant interchannel facies containing fewer and thinner coarse-grained beds; channel beds as much as 5 to 10 m thick	Thin (5 to 20 cm) coarse-grained beds of even thickness over great lateral extent and a greater amount of mud beds	Like canyon but thick coarse-grained, lenticular beds dominate; mud interbeds cut out or incompletely developed except in fining upward sequence	Thin (<15 cm) well-bedded lenticular coarse-grained layers alternating with equal amounts of fine-grained beds in rhythmic sequence	Very thin and evenly bedded coarse-grained layers alternating with greater amounts of interbedded mudstone
C Sediment types	Dominantly sandstone conglomerate, or pebbly mud; silty mud important on canyon walls or in fining upward parts of canyon fill	Conglomerates and sandstones in channel beds flanked by fine sandstone and siltstone in levee and interchannel areas; silty mudstones limited in channel environments, but dominant in interchannel	Sandstone in channels; interchannel having sandstone and siltstone; limited silt and clay-sized mudstone	Medium- to fine-grained channel sandstone and interchannel sandstones and siltstones; clay-sized mudstone most common in interchannel areas	Clay-size mudstones dominant but numerous fine-grained sandstones and siltstones throughout area	Dominantly sandstones, conglomerates, or muddy conglomerates and limited mudstones	Fine-grained sandstone, siltstone, and mudstone in equal amounts	Sandstones, siltstones, and mudstones in variable amounts in different regions of the fan
D Sedimentary structures (see also table 1)	Bouma ae; typically lacking or poorly developed internal structure; may be highly bioturbated in muds; commonly very highly contorted; slump structures common; beds typically amalgamated in lower part of valley fill sequence	Channel is like canyon; Bouma ae predominant; some ab possible; coarsely laminated beds having dish structure, mudstone ripups and clasts, very highly contorted beds and slump structures common; channel sands and gravels injected into interchannel beds; interchannel like levee environment	Full Bouma abcde sequences common in channels; typically less complete Bouma sequences in interchannels; muds most bioturbated in interchannel	Bouma bcde sequences common in channels and Bouma cde in interchannels; base of beds flat and typically lacking sole marks; mud most bioturbated in interchannel	Bouma cde grading downfan to de; lamination dominant; cross lamination limited; muds most highly bioturbated of any fan environment	Bouma cde, typically poorly developed if present; beds displaying well-developed sole marks and channeled basal contacts typical	Some Bouma cde or de units convoluted and typically cross laminated, showing rippling and starved ripples characteristically at top of sandstone and siltstone beds	Similar to levee but generally thinner bedded and laminated in upper fan; in midlower fan, like channel deposits but thinner bedded and more base cutouts

TABLE 4.—Continued

Characteristic	Physiographic regions of fans					Depositional environments of fan regions		
	Submarine canyon	Upper fan	Middle fan	Lower fan	Fan fringe	Channel	Levee	Interchannel
E Grain-size distribution in coarse-grained beds	Very poorly sorted, very positively skewed	Very poorly sorted, very positively skewed; size grading vertically is generally slight or lacking	Poorly to moderately sorted, positive to neutral skewness, size graded vertically	Moderately sorted, slightly positive to negative skewness, size graded vertically	Positively skewed, poorly sorted, size graded vertically	Varies downchannel; see description from upper to middle fan areas	Moderately sorted, not highly skewed	Moderately to poorly sorted, not highly skewed
F Sandstone/shale ratio	Generally very high except in upper part of total channel fill	Very high in channel facies, very low in interchannel facies, and intermediate in levee facies	High in channel facies, low to intermediate in interchannel facies	High in channel, intermediate in interchannel facies	Low to very low throughout facies	High throughout but may become lower in upper part of channel fill	Intermediate and higher than interchannel and distal fringe areas	Low except in suprafan areas
G Conglomerate/sandstone ratio	Very high to 0	May be very high to 0 in channels	Generally low to 0 in channels	Generally 0 in channels		High to 0	Generally 0	Generally 0
H Maximum possible size	Boulders	Boulders in channels; sand in levee	Pebbles in channels; coarse sand in interchannel	Fine pebbles to coarse sand in channel; medium sand in interchannel	Fine sand in channel or interchannel	Boulders	Coarse sand	Medium sand
I Composition and textural maturity	Displaced shelf fauna characterizes coarse-grained beds; unique in situ species characterize hemipelagic mud; wackes dominate	See channel and levee; high % heavy minerals; high terrigenous content in sand fraction of hemipelagic mud; wackes typical	Composition of minerals and fauna graded vertically in each sand layer; wackes and arenites present in channel sands	Sand composition and maturity like middle fan; high pelagic content in sand fraction of hemipelagic beds	High content of mica and plant fragments and some displaced biota in coarse-grained beds but in situ fauna dominates; hemipelagic like lower fan, but pelagic oozes possible; sands typically wackes	Dominance of displaced biota and minerals from shelf occurs in coarse-grained beds; in situ pelagic and terrigenous debris forms hemipelagic mud	In coarse-grained beds high mica and plant fragments and fewer, smaller sized, and deeper environment displaced biota; hemipelagic like channel; wackes dominate	Coarse-grained beds like levee; hemipelagic muds like channel
J Depositional processes	Slumps, slides, grain flow, fluidized sediment flow, debris flow, and bottom currents dominant; turbidity currents limited; turbid layer flow dominant for hemipelagic deposition	Dominant downfan grain flow, fluidized sediment flow, and debris flow; some turbidity-current bed load in channels; suspension load typical in levees and interchannel areas; turbid layer flow most important in interchannel hemipelagic deposition	Turbidity-current bed load dominant in channels and suspension load throughout channel and interchannel areas; continuous particle by particle fall most important in interchannel hemipelagic deposition	Turbidity-current bed load and suspension load throughout channel and interchannel areas; hemipelagic like middle fan	Turbidity-current suspension load throughout entire area; continuous particle by particle hemipelagic deposition most important throughout area	Slumps, slides, grain flow, fluidized sediment flow, debris flow; turbidity currents and bottom currents; see canyon for hemipelagic deposition	Mainly turbidity-current suspension load; hemipelagic deposition mostly from continuous particle fall except near continental margin where turbid layer flow may dominate	Mainly turbidity-current suspension load; hemipelagic deposition like levee

[1] The characteristics have been synthesized from our research and the following studies: Haner (1971), Mutti and Ricci Lucci (1972), Normark (1970), Normark and Piper (1972), Piper (1970a, b), and Stanley and Unrug (1972).

servations: (1) lack of stratigraphic correlation between adjacent areas because of channel deposition and migration; (2) major abrupt lateral facies changes between channel, channel wall, levee, and interchannel deposits; (3) broad lenticularity of thick channel deposits and sequences of thick-bedded sands, channeling and channel walls being locally recognizable; (4) thickest bedding in the upper fan-channel deposits, the thinner bedding being found in channels downfan and laterally in interchannel areas; (5) turbidites containing Bouma sequences in which the completeness of the sequence, coarseness, thickness, sedimentary structures, and other parameters vary sharply from channel to interchannel deposits and from proximal to distal fan areas (table 1); and (6) decrease in grain size of hemipelagic deposits and turbidites in the downfan direction, the coarsest materials being deposited in the upper fan valleys and in submarine canyons.

The typical radial growth pattern of fans (Normark, 1970a) in addition to common tectonism in the region of fan development (Haner, 1971) results in the concept of stratigraphic superposition generally not being applicable to deep-sea fan deposits, just as it is not applicable to alluvial fan deposits (Bull, 1964). The youngest sediment of the fan may be deposited in the upper, middle, or lower fan region; this depends upon the energy of depositing currents, the tectonic framework of the fan, and whether or not the sediments bypass the upper and middle fan region to be deposited on the lower fan. Because channels tend to shift leftward in the northern hemisphere (Menard, 1955), younger fan deposits may be to the left of the older deposits of the fan, (and the reverse in the southern hemisphere) and yet may be at approximately the same topographic level.

Deep-Sea Fans and Geosynclines

Requirements for fan development include availability of a substantial source of sediment, a submarine canyon to conduct the sediment to deeper water, and a decrease in slope at the lower end of the canyon. The sediments are deposited in the area where the slope decreases, and a fan-shaped wedge of sediment forms that generally radiates outward from the lower end of the submarine canyon. Deep-sea fans already have been noted in a variety of modern continental margin settings, including continental borderlands, marginal seas adjacent to island arcs, mediterranean seas, continental rises, and deep-sea trenches. Inasmuch as these settings are the same as those inferred to have existed in ancient geosynclines, it should be possible to determine, in part, the nature of the continental margin from detailed studies of ancient deep-sea fan deposits.

Attempts should be made to define the local and regional paleogeographic setting of the fan and its relationship to the tectonic and sedimentologic history of a particular mobile belt or geosyncline. Although few such reconstructions have been accomplished, we can suggest some guidelines based on our comparative study. For example, in a tectonically stable basin, without major changes in climate or circulation of marine waters, relatively simple fan systems should develop. In geosynclines, where tectonic instability is generally the rule, a great variety of different types and styles of fan growth and development might be expected. Fans eventually may be wholly or partly destroyed by subduction in continental-rise and deep-sea trench settings.

There appear to be two categories of deep-sea fans with slightly differing characteristics and separate geosynclinal settings. The first type develops in more or less restricted basins such as continental borderlands, small marginal seas, and mediterranean seas. These fans are generally small in areal extent and relatively thin. The smallest may develop nearly perfect fan shapes, but larger ones in restricted basins may be sinuous, elongate, and irregularly shaped because of disruption by surrounding topographic features. The fans of restricted basins are generally characterized by (1) influx of sediments from multiple-source areas located at variable distances and directions from the basin floor, (2) well-developed proximal facies ("base of slope" facies of Stanley and Unrug, 1972) or "proximal exotic" facies of Walker and Mutti, 1973), (3) prominent depositional bulges in the middle fan region (suprafan of Normark, 1970a), and (4) formation of distinct distal facies because of ponding of distal flows in the restricted basin. The multiple sources and distal ponding result in abrupt lateral facies changes, distinct proximal-to-distal gradations in turbidite beds, and relatively coarse-grained sands, gravels, and interbedded hemipelaic muds.

The second category of fan develops in open ocean basins, typically on the continental rise. Because such basins are not restricted, larger and thicker fans develop that generally are regularly shaped but typically somewhat elongate. Generally, a large submarine canyon and a single source area control deposition, so that base-of-slope deposition along the continental margin is minor. The midfan or suprafan region commonly is not well developed. The distal parts of the fan may extend for many miles and grade imperceptibly into abyssal plains without a clearly defined fan edge. As a result, changes in proximal and distal facies in these fans are more gradual and take place over long distances. Also, hemipelagic muds and sands are generally finer grained because of the greater distances from source areas. Bottom currents generated by circulation in large bodies of water may be more effective in redistributing sediments.

Turbidite fills in elongate deep-sea trenches may be a third recognizable fanlike sequence derived from apparent geosynclinal settings (Piper and others, 1973). These deposits are characterized by a long, linear channel facies oriented parallel to the trench axis and by sediments of smaller fans built out perpendicularly to the continental-slope side of the trench. Because the sea floor may be spreading, so to speak, into the continent, the main channel facies tends to be pushed against and parallel to the base of the continental slope. Consequently, overbank-spill deposits of levee and interchannel facies are minimal on the slope side of the channel facies but are well developed on the seaward side of the channel facies. Such a linear yet asymmetric system of channel and interchannel turbidites may distinguish trench fill from the deep-sea fan deposits of restricted basins and open continental-rise settings.

ACKNOWLEDGMENTS

Work by Nilsen on the Butano Sandstone was funded by a U.S. Geological Survey Postdoctoral Research Associateship. Data collection by Nelson on the Astoria Fan was supported by the Office of Naval Research (contract Nonr 1286(10)) and by the Oceanography Department of Oregon State University. We thank Samuel H. Clarke, Jr., Department of Geology and Geophysics, University of California at Berkeley, for providing data on the Point of Rocks Sandstone, and Bradley Larsen and Tully Simoni of the U.S. Geological Survey for assistance in compiling data and preparing figures. Beneficial review comments were provided by Peter Barnes, H. Edward Clifton, R. H. Dott, Jr., and R. G. Walker. Extensive discussions on field trips with Franco Ricci Lucchi added to, strengthened, and corroborated our data for the summary table 4.

Publication has been authorized by the Director, U.S. Geological Survey.

REFERENCES

BOUMA, A. H., 1962, Sedimentology of some flysch deposits, a graphic approach to facies interpretation: Amsterdam, Elsevier Publishing Co., 168 p.
BRABB, E. E., 1960, Geology of the Big Basin area, Santa Cruz Mountains, California (Ph.D. thesis): Stanford, California, Stanford Univ., 191 p.
———, 1964, Subdivision of the San Lorenzo Formation (Eocene-Oligocene), west-central California: Am. Assoc. Petroleum Geologists Bull., v. 48, p. 670–679.
BRANNER, J. C., NEWSOME, J. F., AND ARNOLD, R., 1909, Description of the Santa Cruz Quadrangle, California: U.S. Geol. Survey Geol. Atlas, Folio 163, 12 p.
BULL, W. B., 1964, Geomorphology of segmented alluvial fans in western Fresno County, California: erosion and sedimentation in a semiarid environment: ibid., Prof. Paper 352-E, p. 89–128.
CARLSON, P. R., AND NELSON, C. H., 1969, Sediments and sedimentary structures of the Astoria submarine canyon-fan system, northeast Pacific: Jour. Sed. Petrology, v. 39, p. 1269–1282.
CLARK, J. C., 1966, Tertiary stratigraphy of the Felton-Santa Cruz area, Santa Cruz Mountains, California (Ph.D. thesis): Stanford, California, Stanford Univ., 179 p.
CLARKE, S. H., JR., 1973, The Eocene Point of Rocks Sandstone—provenance, mode of deposition and implications for the history of offset along the San Andreas Fault in central California (Ph.D. thesis): Berkeley, Univ. California, 302 p.
———, AND NILSEN, T. H., 1972, Postulated offsets of Eocene strata along the San Andreas Fault zone, central California: Geol. Soc. America Abs. with Programs, v. 4, p. 137–138.
CUMMINGS, J. C., TOURING, R. M., AND BRABB, E. E., 1962, Geology of the northern Santa Cruz Mountains, California, in BOWEN, O. E. (ed.), Geologic guide to the gas and oil fields of northern California: California Div. Mines and Geology Bull. 181, p. 179–220.
CURRAY, J. R., AND MOORE, D. G., 1971, Growth of the Bengal deep-sea fan and denudation in the Himalayas: Geol. Soc. America Bull., v. 82, p. 563–572.
DEWEY, J. F., AND HORSFIELD, B., 1970, Plate tectonics, orogeny and continental growth: Nature, v. 225, p. 521–525.
DIETZ, R. S., 1963, Collapsing continental rises: an actualistic concept of geosynclines and mountain building: Jour. Geology, v. 71, p. 314–333.
DUNCAN, J. R., KULM, L. D., AND GRIGGS, G. B., 1970, Clay mineral composition of late Pleistocene and Holocene sediments of Cascadia Basin, northeastern Pacific Ocean: ibid., v. 78, p. 213–221.
EMERY, K. O., 1938, Rapid method of mechanical analysis of sands: Jour. Sed. Petrology, v. 8, p. 105–111.
———, 1965, Turbidites—Precambrian to present, in YOSHIDA, Kozo (ed.), Studies on oceanography, dedicated to Professor Hidaka: Seattle, Univ. Washington Press, p. 486–493.
———, AND BRAY, E. E., 1962, Radiocarbon dating of California basins sediments: Am. Assoc. Petroleum Geologists Bull., v. 46, p. 1839–1856.

Fairchild, W. W., Wesendunk, P. R., and Weaver, D. W., 1969, Eocene and Oligocene foraminifera from the Santa Cruz Mountains, California: California Univ. Pub. Geol. Sci., v. 81, 93 p.

Folk, R. L., and Ward, W. C., 1957, Brazos River bar—a study in the significance of grain-size parameters: Jour. Sed. Petrology, v. 27, p. 3–26.

Hampton, M. A., 1970, Subaqueous debris flow and generation of turbidity currents (Ph.D. thesis): Stanford, California, Stanford Univ., 180 p.

Haner, B. E., 1971, Morphology and sediments of Redondo submarine fan, southern California: Geol. Soc. America Bull., v. 82, p. 2413–2432.

Harms, J. C., and Fahnestock, R. K., 1965, Stratification, bed forms, and flow phenomena (with an example from the Rio Grande), in Middleton, G. V. (ed.), Primary sedimentary structures and their hydrodynamic interpretation, p. 84–115: Soc. Econ. Paleontologists and Mineralogists Special Pub. 12, p. 84–115.

Horn, D. R., and others, 1971, Turbidites of the Hatteras and Sohm Abyssal Plains, western North Atlantic: Marine Geology, v. 11, p. 287–323.

Hubert, C., Lajoie, J., and Leonard, M. A., 1970, Deep-sea sediments in the lower Paleozoic Quebec Supergroup, in Lajoie, J. (ed.), Flysch sedimentology in North America: Geol. Assoc. Canada Special Paper 7, p. 103–125.

Jacka, A. D., and others, 1968, Permian deep-sea fans of the Delaware Mountain Group, Delaware Basin, in field trip guidebook for 1968 symposium, Guadalupian facies, Apache Mountain area, West Texas: Soc. Econ. Paleontologists and Mineralogists, Permian Basin Sec., Special Pub. 68, 11 p.

Kuenen, Ph. H., 1964, Deep-sea sands and ancient turbidites, in Bouma, A. H., and Brouwer, A. (ed.), Turbidites: Amsterdam, Elsevier Publishing Co., p. 3–33.

Kulm, L. D., Prince, R. A., and Snavely, P. D., 1973, Site survey of the northern Oregon continental margin and Astoria Fan, in Kulm, L. D., von Huene, R. E., and others, Initial reports of the Deep-Sea Drilling Project, vol. 18: Washington, D.C., U.S. Govt. Printing Office, p. 979–986.

———, von Huene, R. E., and others, 1973, Initial reports of the Deep Sea Drilling Project, v. 18: ibid., 1077.

Mallory, V. S., 1959, Lower Tertiary biostratigraphy of the California Coast Ranges: Tulsa, Oklahoma, Am. Assoc. Petroleum Geologists, 416 p.

Menard, H. W., 1955, Deep-sea channels, topography, and sedimentation: ibid., Bull., v. 39, p. 236–255.

Middleton, G. V., 1969, Turbidity currents, in Stanley, D. J. (ed.), The new concepts of continental margin sedimentation: Washington, D.C., Am. Geol. Inst., Short Course Lecture Notes, p. GM-A-1 to GM-A-20.

———, and Hampton, M. A., 1973, Part I: Sediment gravity flows—mechanics of flow and deposition, in Turbidites and deep-water sedimentation: Anaheim, California, Soc. Econ. Paleontologists and Mineralogists, Pacific Sec., Short Course Lecture Notes, p. 1–38.

Moore, D. G., 1969, Reflection profiling studies of the California continental borderland: Structure and Quaternary turbidite basins: Geol. Soc. America Special Paper 107, 142 p.

Mutti, E., and Ricci Lucchi, F. R., 1972, Le torbiditi dell'Appennino settentrionale: introduzione all'analisi di facies: Soc. Geol. Italiana Mem. 11, p. 161–199.

Nelson, C. H., 1967, Sediments of Crater Lake, Oregon: Geol. Soc. America Bull., v. 79, p. 833–848.

———, 1968, Marine geology of Astoria deep-sea fan (Ph.D. thesis): Corvallis, Oregon State Univ., 287 p.

———, 1974, Late Pleistocene and Holocene depositional trends, processes, and history of Astoria deep-sea fan: Marine Geology.

———, and Kulm, L. D., 1973, Part II: Submarine fans and deep-sea channels, in Turbidites and deep-water sedimentation: Anaheim California, Soc. Econ. Paleontologists and Mineralogists, Pacific Sec., Short Course Notes, p. 39–70.

———, and others, 1968, Mazama Ash in the northeastern Pacific: Science, v. 161, p. 47–49.

———, and ———, 1970, Development of the Astoria Canyon-Fan physiography and comparison with similar systems: Marine Geology, v. 8, p. 259–291.

Nilsen, T. H., 1970, Paleocurrent analysis of the flysch-like Butano Sandstone (Eocene), Santa Cruz Mountains, California: Geol. Soc. America Abs. with Programs, v. 2, p. 636.

———, 1971, Sedimentology of the Eocene Butano Sandstone, a continental borderland submarine fan deposit, Santa Cruz Mountains, California: ibid., v. 3, p. 660.

———, 1973, Facies relations in the Eocene Tejon Formation of the San Emigdio and western Tehachapi Mountains, California, in Sedimentary facies changes in Tertiary rocks—California Transverse and southern Coast Ranges: Anaheim, California, Soc. Econ. Paleontologists and Mineralogists, Pacific Sec. Field Trip Guidebook, Field Trip 2, p. 7–23.

———, and Clarke, S. H., Jr., 1973, Sedimentation and tectonics in the early Tertiary continental borderland of central California: Am. Assoc. Petroleum Geologists Bull., v. 57, p. 797.

———, and Simoni, T. R., Jr., 1973, Deep-sea fan paleocurrent patterns of the Eocene Butano Sandstone, Santa Cruz Mountains, California: U.S. Geol. Survey Jour. Research, v. 1, no. 4, p. 439–452.

Normark, W. R., 1970a, Growth patterns of deep-sea fans: Am. Assoc. Petroleum Geologists Bull., v. 54, p. 2170–2195.

———, 1970b, Channel piracy on Monterey deep-sea fan: Deep-Sea Research, v. 17, p. 837–846.

———, and Piper, D. J. W., 1969, Deep-sea fan-valleys, past and present: Geol. Soc. America Bull., v. 80, p. 1859–1866.

———, and ———, 1972, Sediments and growth pattern of Navy deep-sea fan, San Clemente Basin, California borderland: Jour. Geology, v. 80, p. 198–223.

Piper, D. J. W., 1970a, Transport and deposition of Holocene sediment on La Jolla deep-sea fan, California: Marine Geology, v. 8, p. 211–227.

———, 1970b, A Silurian deep-sea fan deposit in western Ireland and its bearing on the nature of turbidity currents: Jour. Geology, v. 78, p. 509–522.

———, von Huene, R. E., and Duncan, J. R., 1973, Sedimentation in a modern trench: Geology, v. 1, no. 1, p. 19–22.
Poole, D. M., 1957, Size analysis of sand by a sedimentation technique: Jour. Sed. Petrology, v. 27, p. 460–468.
Sanders, J. E., 1965, Primary sedimentary structures formed by turbidity currents and related resedimentation mechanisms, in Middleton, G. V. (ed.), Primary sedimentary structures and their hydrodynamic interpretation: Soc. Econ. Paleontologists and Mineralogists Special Pub. 12, p. 192–219.
Shepard, F. P., 1966, Meander in valley crossing a deep ocean fan: Science, v. 154, p. 385–386.
———, Dill, R. F., and von Rad, U., 1969, Physiography and sedimentary processes of La Jolla submarine fan and fan-valley, California: Am. Assoc. Petroleum Geologists Bull., v. 53, p. 390–420.
Silver, E. A., 1972, Pleistocene tectonic accretion of the continental slope off Washington: Marine Geology; v. 13, p. 239–249.
Stanley, D. J., 1969b, Submarine channel deposits and their fossil analogs ('fluxoturbidites'), in Stanley, D. J. (ed.), The new concepts of continental margin sedimentation: Washington, D.C., Am. Geol. Inst. Short Course Lecture Notes, p. DJS-9-1 to DJS-9-17.
———, and Unrug, R., 1972, Submarine channel deposits, fluxoturbidites, and other indicators of slope and base-of-slope environments in modern and ancient marine basins: Soc. Econ. Paleontologists and Mineralogists Special Pub. 16, p. 287–340.
Stauffer, P. H., 1967, Grain flow deposits and their implications, Santa Ynez Mountains, California: Jour. Sed. Petrology, v. 37, p. 481–508.
Sullivan, F. R., 1962, Foraminifera from the type section of the San Lorenzo Formation, Santa Cruz County, California: California Univ. Pub. Geol. Sci., v. 37, p. 233–352.
Sullwold, H. H., Jr., 1960, Tarzana Fan, deep submarine fan of late Miocene age, Los Angeles County, California: Am. Assoc. Petroleum Geologists Bull., v. 44, p. 433–457.
Sundborg, Åke, 1956, The river Klarälven— a study of fluvial processes: Geog. Annalav, Stockholm, v. 38, p. 125–316.
Visher, G. S., 1969, Grain size distributions and depositional processes: Jour. Sed. Petrology, v. 39, p. 1074–1106.
von Huene, R., Kulm, L. D., and others, 1971, Deep Sea Drilling Project leg 18: Geotimes, v. 16, p. 12–15.
Walker, R. G., 1965, The origin and significance of the internal sedimentary structures of turbidites: Yorkshire Geol. Soc. Proc., v. 35, p. 1–32.
———, 1966, Shale grit and Grindslow shales; transition from turbidite to shallow-water sediments in the Upper Carboniferous of northern England: Jour. Sed. Petrology, v. 36, p. 90–114.
———, 1970, Review of the geometry and facies organization of turbidites and turbidite-bearing basins, in Lajoie, J. (ed.), Flysch sedimentology in North America: Geol. Assoc. Canada Special Paper, 7, p. 219–251.
Walker, R. G., and Mutti, E., 1973, Part IV—Turbidite facies and facies associations, in Turbidites and deep-water sedimentation: Anaheim, California, Soc. Econ. Paleontologists Mineralogists, Pacific Sec. Short Course Lecture Notes, p. 119–157.
Walton, E. K., 1967, The sequence of internal structures in turbidites: Scottish Jour. Geology, v. 3, p. 305–317.

EXAMPLES OF ANCIENT DEEP-SEA FAN DEPOSITS FROM CIRCUM-MEDITERRANEAN GEOSYNCLINES

EMILIANO MUTTI
University of Turin, Italy

ABSTRACT

Examination of a number of ancient turbidite basins in the Alpine geosynclinal chains of the circum-Mediterranean region supports the assumption that many of the sediments therein were deposited in deep-sea fan environments. Sand—body geometry and vertical sequence analysis provide criteria for detecting associations of inner, middle, and outer fan facies in these turbidite sequences.

Examples of these three main facies associations are reported from selected Tertiary geosynclinal turbidites occurring in the northern Apennines (Ranzano Sandstone and Bobbio Formation) and on the island of Rhodes, Aegean Sea (Messanagros Sandstone). The proposed depositional model based on these studies is not unlike models of certain deltaic systems and emphasizes progradational, aggradational, and recessional events of turbidite sedimentation in complexes of ancient deep-sea fans.

Middle fan deposits commonly show thinning and (or) fining upward cycles developed within channel-fill sequences. Such cycles are readily comparable to those of abandoned fluvial channels that are also filled with similar sequences in delta-plain environments. In both cases, channel and interchannel areas indicate prevalent vertical accretion. Active deltaic outbuilding is expressed typically by stream-mouth bars whose progradational character is shown by the occurrence of thickening and(or) coarsening upward cycles. Detailed inspection of several northern Apennine turbidite formations has shown that sandstone bodies, closely exhibiting such a progradational pattern, are extremely widespread. These turbidite sediments are here interpreted as outer fan deposits, and are thought to be responsible for deep-sea fan growth in most ancient basins.

INTRODUCTION

As Sullwold (1961) realized, turbidite sediments are mainly deep-water deposits and should be interpreted paleoenvironmentally by means of concepts and terms that are compatible with criteria derived from geological studies of modern marine depositional analogs. Marine geologists have shown that turbidite deposition of terrigenous clastics derived from nearby exposed land masses is a function of morphological factors common to both intracontinental and continental margin basins. These factors include the presence of a platform area supplied by a source of terrigenous clastics, a slope area where canyons or other types of submarine valleys can first trap these clastics and later funnel them downslope, and a basin plain where the clastics attain final equilibrium. Typical features found within this depositional framework are deep-sea fans of various sizes that are constructed at the base of the slope, where the bulk of the coarse-grained clastics accumulate due to the abrupt change in depositional gradient.

The area and volume of recent deep-sea fans suggest that most ancient turbidites must somehow be associated with similar depositional environments in ancient basins. In the circum-Mediterranean region, geosynclinal turbidites ("flysch" of some usage) are particularly widespread in the chains of the Alpine orogenic cycle, reflecting various types of structural control and evolution. These sediments are actually found as fill of orthogeosynclinal (eugeosynclinal and miogeosynclinal), postorogenic, and intracontinental troughs, and, at least for the examples examined by this writer, they are related almost invariably to deep-sea fan systems. Recognition and paleoenvironmental subdivision of ancient deep-sea fan sediments are, therefore, basic tools for better understanding of the sedimentary evolution of the continental margins and adjacent ocean basins of such ancient mobile belts as orthogeosynclines.

Turbidite-facies associations indicative of a deep-sea fan environment have been reported and discussed by several authors (e.g., Dzulynski and others, 1959; Basset and Walton, 1960; Unrug, 1963; Walker, 1966; Walker and Sutton, 1967; Jacka and others, 1968; Kelling and Woollands, 1969; Mutti, 1969; Normark and Piper, 1969; Ricci Lucchi, 1969; Stanley, 1969; Piper, 1970a; Piper and Normark, 1971; Mutti and Ghibaudo, 1972; Mutti and Ricci Lucchi, 1972; Stanley and Unrug, 1972). Substantial agreement exists among these workers with respect to formulation of a general depositional model for these sediments. This model includes an inner fan characterized by deep submarine valleys (generally filled by coarse-grained sediments) within mudstone sequences; a middle fan associated with numerous, generally shallow channels contained within thin-bedded and current-laminated fine-grained sandstones and siltstones—in most instances these channels also are filled with thick-bedded coarse-grained deposits; and an outer fan, where coarse-grained sediments and associated channeling become less evident and are replaced by increasingly fine-grained well-bedded deposits that are indicative of basin-plain deposition.

There is a remarkable similarity between this pattern and a similar depositional pattern demonstrated by certain modern deep-sea fans. This is hardly accidental, and one can assume that this overall model can be applied with a reasonable degree of confidence to studies of most analogous ancient sediments. On the basis of this model, three turbidite sequences, here considered as products of inner, middle, and outer fan deposition respectively, are described briefly below. The three examples are: (1) inner fan, the Ranzano Sandstone (upper Eocene and Oligocene) in the northern Apennines; (2) inner and middle fan, the Messanagros Sandstone (Oligocene and lower Miocene) on the island of Rhodes in the Aegean Sea; and (3) outer fan, the Bobbio Formation (Miocene) in the northern Apennines.

The middle and outer parts of the tripartite model described above for ancient deep-sea fans can be significantly refined by using criteria derived from vertical sequence analysis of sandstone bodies. The resulting integrated model is presented here in the concluding section.

EXAMPLES OF ANCIENT DEEP-SEA FAN DEPOSITS

Inner fan deposits.—The facies association thought to represent an inner fan environment is characterized by large lenticular sand bodies enclosed by mudstone sequences that are virtually devoid of sandstone interbeds. Sand-body geometry indicates filling of submarine valleys ranging in width from hundreds of meters to a few kilometers and in depth from tens to hundreds of meters. These large-scale erosional features may be cut into older rocks forming part of the local substrata (e.g., van Hoorn, 1969; Mutti, 1969). More frequently, however, they are found within mudstone sequences that are only slightly older or even partly contemporaneous with the cutting and subsequent filling of the valleys themselves (e.g., Walker, 1966; Stanley, 1967; Mutti, 1969). Absence of sandstone interbeds within the mudstone sequences, even in the immediate vicinity of the valleys, indicates that the valleys were generally of sufficient depth and extent to receive the entire volume of sand brought downslope by currents. Thus, overbanking did not take place. These characteristic features can be related to depositional events that are confined either to lowermost reaches of submarine canyons or to valleys incised into the apex of a deep-sea fan. As far as the reconstruction of ancient basins is concerned, the two alternatives are of little consequence.

Very few examples of the depositional features of inner fan deposits are discussed in the literature. In the circum-Mediterranean region, published examples are confined to outcrops of the Annot Sandstone near Contes in the Maritime Alps (Stanley, 1967) and of the lower part of the Messanagros Sandstone on Rhodes (Mutti, 1969). (The last-named example is discussed in the next section.)

The Ranzano Sandstone (northern Apennines) is a noteworthy example of inner fan turbidites. This formation and the underlying and partly lateral equivalent, the Montepiano Marl, constitute a bathyal sedimentary sequence deposited in the Ligurian eugeosynclinal basins of northwestern Italy during late Eocene and Oligocene time. In the western part of the Ligurian realm, the sediments were deposited unconformably above already folded, abyssal plain turbidites and shales, as well as above oceanfloor ophiolites; to the east, equivalent sediments were deposited virtually in conformity with Paleogene basin-plain turbidites (Elter and others, 1966; Sestini, 1970; Boccaletti and others, 1971).

One of the best exposures of Ranzano deposits may be found in the Pessola Valley near Parma (fig. 1). Here, the sandstone fills a submarine valley about four 4 km wide to an apparent depth of 250 m (Ghibaudo and Mutti, 1973). Only the lower part of the sandstone is considered here because outcrops of upper sandstone are too small for reliable paleoenvironmental interpretations. As shown on the geological map of figure 1, the lower sandstone fills a submarine depression cut into the local Montepiano sequence. This depression is oriented northeast-southwest as inferred from sand-body geometry and paleocurrent trends. This large-scale erosional feature was cut into the Montepiano sequence through a series of smaller channels, which are progressively younger from west to east (fig. 2A). This erosional pattern gives way to a number of sandstone interfingers within the adjacent marl sequence in the east side of the submarine valley. Each sandstone interfinger fills an erosional depression and is conformably overlain by marl, the latter being eroded higher in the section and farther eastward by the successive channels. To the west, the individual channel fills can be traced into mostly nonchannelized deposits (fig. 2B). As a result, the lower sandstone displays channelized facies only along its basal erosional contact with the Montepiano sequence. Along this contact, the sandstone contains abundant

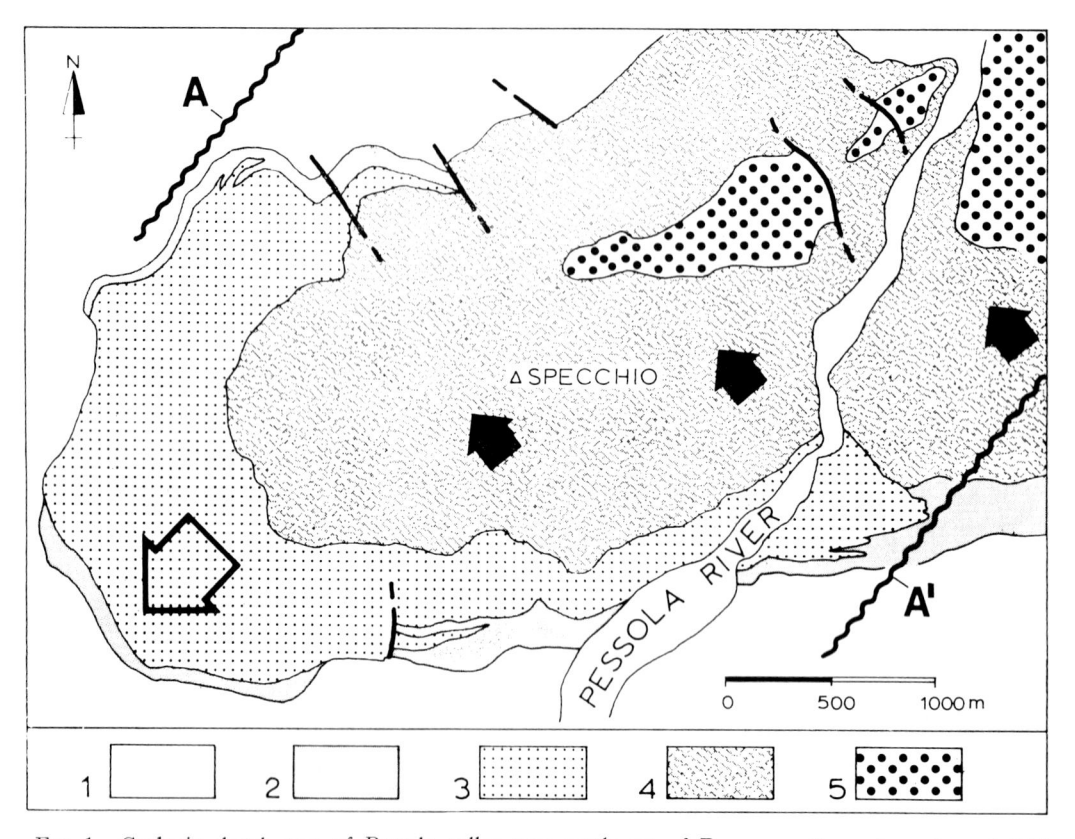

Fig. 1.—Geologic sketch map of Pessola valley area southwest of Parma, northern Apennines. *1*, Older sediments; *2*, Montepiano Marl; *3*, Ranzano Sandstone (lower sandstone part); *4*, Ranzano Sandstone (chaotic level); *5*, Ranzano Sandstone (upper sandstone part); *A-A'*, inferred maximum width of the submarine valley discussed in text. Black arrows indicate direction of movement of lateral slide, which halted sedimentation within the submarine valley. Large arrow indicates direction of sediment transport within submarine valley.

somewhat rounded Montepiano Marl blocks that slid down from the channel edges and were subsequently carried short distances along the valley floor.

Two main types of turbidite facies can be recognized within the valley-filling lower sandstone. The first type is developed in the channelized portions of the sandstone and immediately west of them. It consists of groups of thick conglomeratic sandstone beds. These are irregularly shaped due to channeling and amalgamation and are virtually devoid of mudstone partings. Internal structures include grading and thick crudely developed parallel laminae. The second facies type is characterized by thin beds of very poorly sorted fine- to coarse-grained sandstone. Lenticularity and considerable lateral discontinuity are conspicuous features of these

Fig. 2.—Sedimentary features of Ranzano Sandstone in Pessola area. *A*, Interfingers of coarse-grained Ranzano sandstone in Montepiano Marl on eastern side of the submarine fan valley. Note absence of thin sandstone interbeds (overbank deposits) near right end of flower interfinger. Each interfinger is a channel-fill body bounded by an erosional surface at the base and a depositional surface at the top. *B*, Erosional contact between Ranzano sandstones and underlying Montepiano deposits. Note large-scale, low-angle truncation at base of sandstones. Above, in valley-fill sediments, sandstones can be seen to grade to left (westward) into thinner bedded deposits. The latter, at point indicated by arrow, display clear depositional dips toward the right, that is, toward the deepest channelized portions of the valley-fill. Bedding surfaces above these sediments tend to conform to bedding surfaces of Montepiano sequence. See text for discussion.

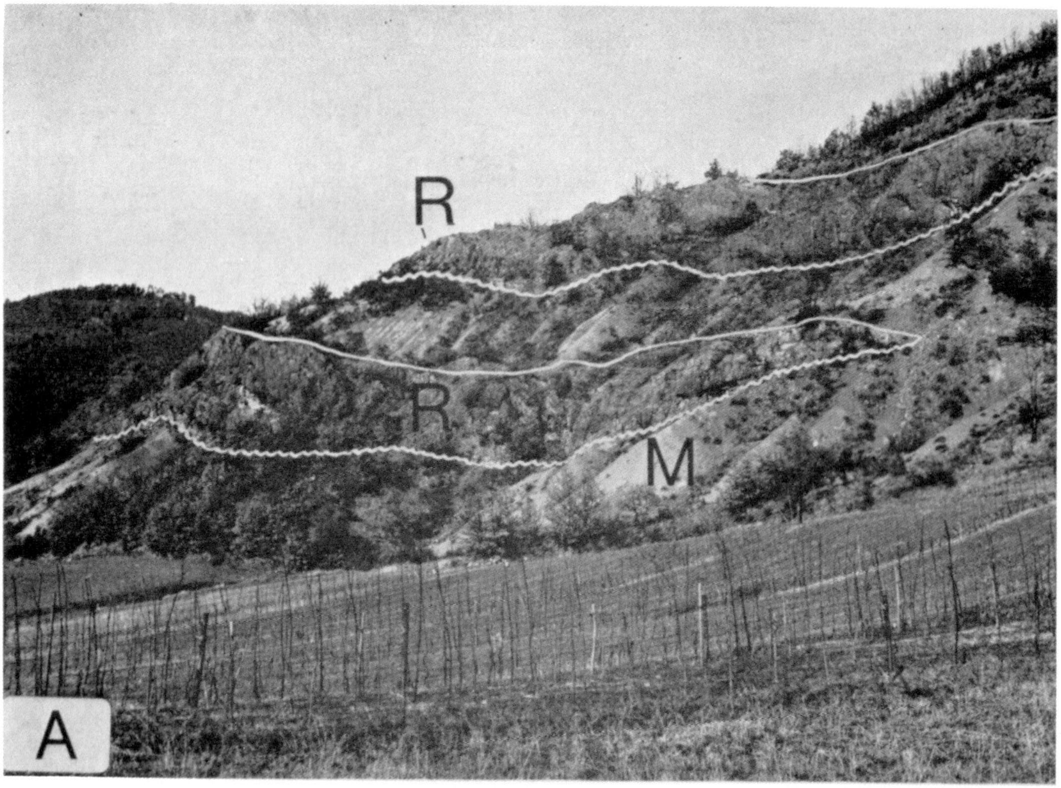
R
R
M
A

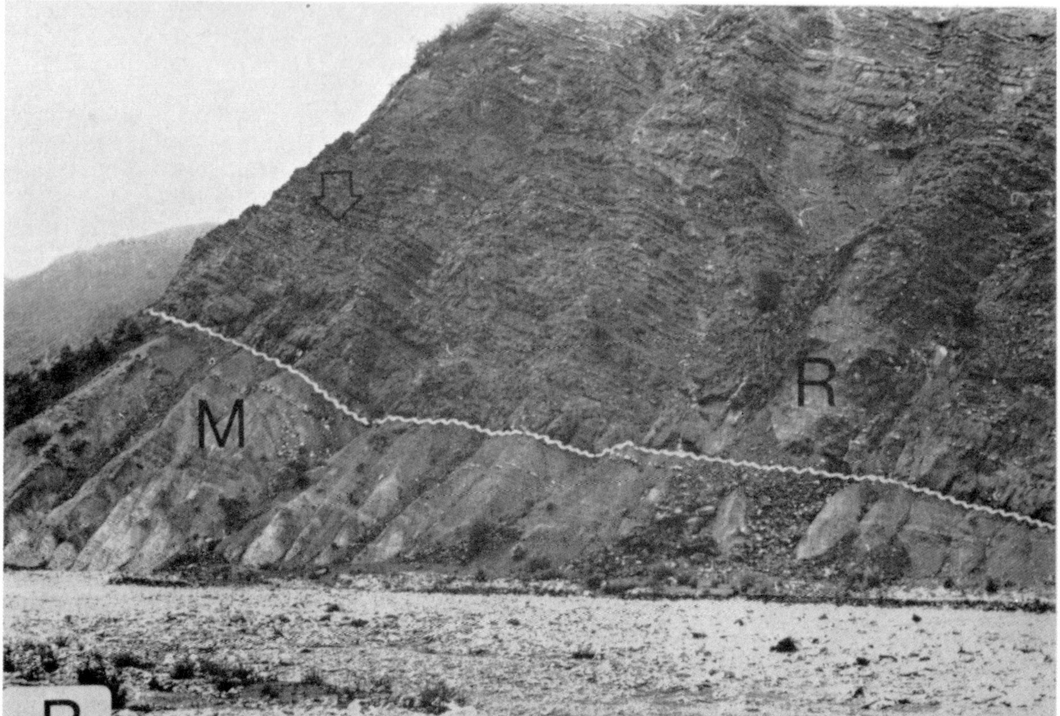
M
R
B

beds. Mudstone partings are common though sandstone prevails. The sandstone beds are more or less distinctly graded and commonly have sharp tops; well-developed current laminae are mostly absent. The thin-bedded facies comprises the bulk of the Ranzano sequence west of the channelized portions of the valley fill.

Stratigraphic relationships show that the two facies types are laterally equivalent, and their transition occurs primarily in a direction normal to the main paleocurrent trend. Coarse-grained and thick-bedded sandstones are confined to the channelized portions of the submarine valley, whereas the finer grained and thin-bedded sediments are found only on the west sides of the deepest parts of the channels. This asymmetric pattern of facies distribution, as well as the fact that the bedding surfaces of the thin-bedded sandstones often show a depositional dip toward the deepest part of the channels (fig. 2B), would seem to delineate a sedimentary pattern in which the lower sandstone filled up the local submarine valley through lateral accretion following channel migration from west to east in a meandering system not unlike the point-bar model.

Laterally to each channel-fill sandstone, the Montepiano sequence contains virtually no sandstone interbeds related to overbanking (fig. 2A). Either of two explanations is possible. First, the gradient of the open fan surface may have been so steep that occasional channel overflows proceeded downslope both laterally and parallel to the valley with such competence that no sandstone deposition took place. Second, transport and deposition of coarse-grained sediment within the channelized portions of the valley may have occurred primarily under bed-load conditions, thus preventing the formation of dilute suspensions that could have spread laterally away from the valley and deposited sand on the open fan surface.

Sedimentation within the valley was halted by a huge lateral slide from the southeast, probably as a result of local tectonic unstability coupled with submarine erosion in the valley. The slide includes large blocks ("olistoliths" of some European usage) of older eugeosynclinal sediments, of the Montepiano Marl, and of contorted and disrupted sandstone slabs that were deposited higher on the slope.

Middle fan deposits.—This facies association consists of coarse-grained thick-bedded channel-fill sandstones enclosed within finer grained thin-bedded deposits. Differentially developed mudstone sequences, devoid of sandstone interbeds, may be present. Turbidite sediments of this type, associated with conglomerates and chaotic deposits, are well known in the literature (Walker, 1966; Jacka and others, 1968; Mutti, 1969; Ricci Lucchi, 1969; Kelling and Woollands, 1969; Piper, 1970a; Piper and Normark, 1971; Mutti and Ricci Lucchi, 1972; Stanley and Unrug, 1972).

More or less contemporary deposition of both the coarse- and fine-grained material is clearly demonstrated by the vertical and lateral stratigraphic relationships. Paleocurrent directions indicate dispersion away from the channels followed by the deposition of finer sediments in interchannel areas. In recent sediments, this relationship has been clearly demonstrated by Haner (1971, p. 2431) for the Redondo deep-sea fan. In both recent and ancient examples, middle fan channels vary considerably in size. However, their dimensions, particularly depth, are always less than the depth of upper fan valleys (see comparative data of Haner, 1971).

Only two circum-Mediterranean associations of middle fan facies have been described so far: (1) the Marnoso-arenacea Formation, a thick Miocene sequence of the northern Apennines (Ricci Lucchi, 1969), and (2) the Messanagros Sandstone of Rhodes (Mutti, 1969). In the first example, the width of the channelized sand bodies (1.5 to 2 km) and the large-scale truncation of the underlying sediments suggest that these features are true upper fan valleys that have dissected the outer parts of the fan during a progradational phase (Mutti and Ricci Lucchi, 1972).

The Messanagros Sandstone offers a clearer and more pertinent example of simultaneous channel and interchannel deposition. The sandstone crops out in the central and southern parts of the Aegean island of Rhodes (fig. 3). It constitutes the uppermost portion of a thick sequence of deposits, known as the Vati Group, that accumulated during Oligocene and early Miocene time. This group is bounded stratigraphically at the top and base by unconformities and represents a postorogenic sequence deposited over folded geosynclinal sediments. The Vati Group is overlain by Pliocene fluvial and lacustrine deposits, the classic Levantinian facies of the eastern Mediterranean.

Figure 3 illustrates the stratigraphic relationships of the Vati Group and reconstructs the paleogeography of the Messanagros Sandstone. The Vati sequence begins in its lower part with the fluviodeltaic Koriati Conglomerate and continues upward with slope deposits of the Agios Minas Marl and with turbidites of the Messanagros Sandstone. Detailed field mapping, precise time correlations of ash layers, and an isopachous map of the underlying, partly time-equiva-

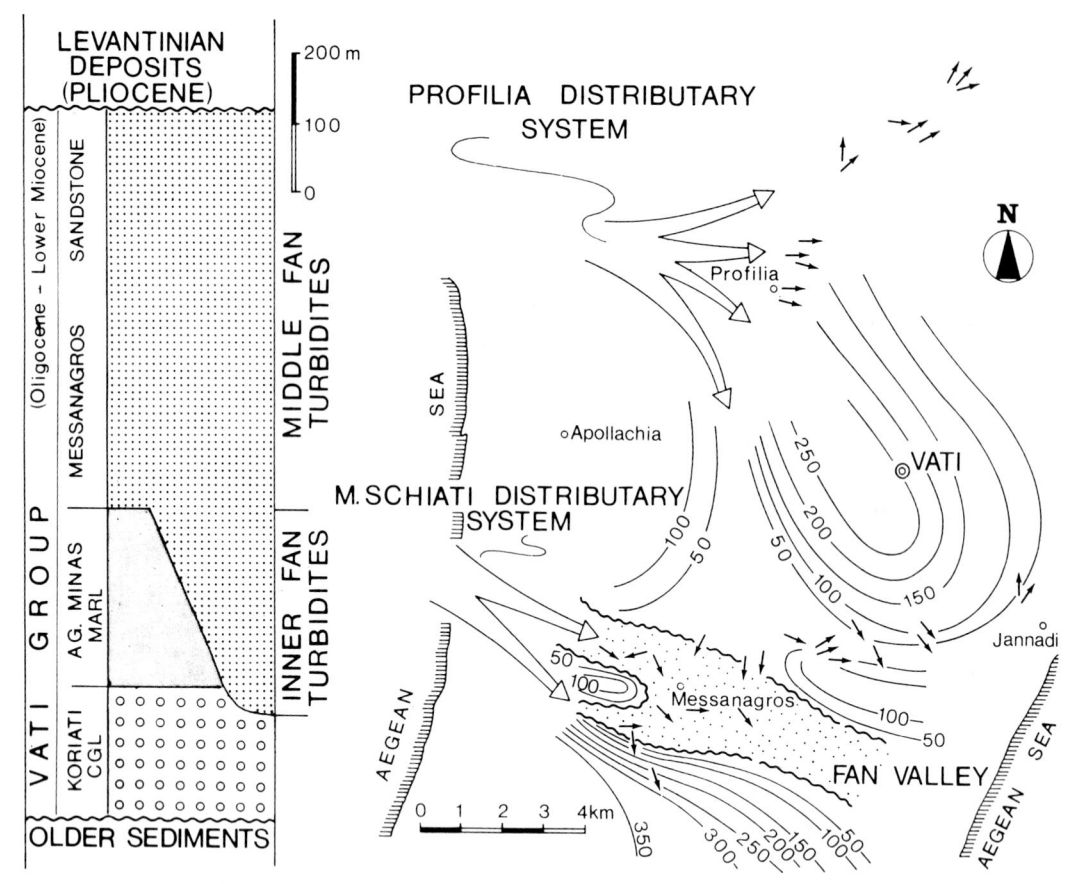

FIG. 3.—Stratigraphy and paleogeography of the Vati Group in south-central part of island of Rhodes, Greece. Paleogeographic sketch on right refers to time interval during which Agios Minas Marl and inner fan (submarine valley fill) Messanagros Sandstone were deposited contemporaneously. Isopach contours (in meters) refer to Agios Minas Marl, whereas paleocurrent directions refer to Messanagros Sandstone. Most directions were measured within middle fan deposits of Messanagros Sandstone and thus belong to sediments younger than those represented on map. Note also the remarkable dispersion of paleocurrent directions away from the two main distributary systems (modified from Mutti, 1969).

lent Agios Minas Marl provide data that allow a reasonably accurate environmental reconstruction based on facies characteristics and paleocurrent analysis. The Messanagros Sandstone and the Koriati Conglomerate are in direct contact (i.e., the Agios Minas Marl is absent) at the site of a large erosional feature in the Mount Schiati area, southwestern part of the island of Rhodes (see detailed geological map in Mutti, 1969). This erosional feature represents a submarine valley 1.5 km wide and at least 200 m deep. It has been filled with thick-bedded sandstones and conglomerates associated with abundant slump units. In the uppermost portion of the valley-fill sequence, interfingering of the coarse-grained sediments with the adjacent Agios Minas Marl shows that the submarine valley eventually became considerably less incised. This valley extends as a distinct feature eastward downslope for at least 8 km and gradually assumes the characteristics of a depositional fan valley as defined by Normark (1970). On the north side, the fan valley is bordered by a prominent levee of contemporaneously deposited Agios Minas sediments, and erosional contacts at the base of this portion of the valley are virtually absent. Moreover, the Agios Minas Marl contains many sandstone interbeds on both the northern and southern sides of the valley that probably represent overbank deposits. All these sedimentary features indicate an inner fan environment not unlike that described by Normark (1970) in his model for Recent deep-sea fans.

Turbidite sedimentation of the Vati Group was confined to the Mount Schiati fan valley during the whole of this period, whereas varying thicknesses of marl were deposited over the remainder of the basin. Middle fan deposits comprising the remaining bulk of the Messanagros Sandstone form a continuous stratigraphic cover over the lower sediments described above. Numerous paleocurrent directions indicate a radial pattern of development from the Profilia distributary system to the north of the Mount Schiati area. The main fan valley of this system has not been preserved. Repeated juxtaposition of channel-fill and interchannel sediments are characteristic features of these middle fan deposits. The channels are generally very broad, and their fill is a few tens of meters thick and consists mostly of coarser grained facies exhibiting well-defined thinning and fining upward cycles (Mutti, 1969, p. 1054, fig. 16). Basal units of channel-fill sequences commonly consist of rather well-packed pebbly to cobbly conglomerates and chaotic slump levels. These sediments make up broadly lenticular bodies as much as 10 m thick and rest on erosional surfaces cut into underlying finer grained deposits. The conglomerates are overlain by thick-bedded coarse-grained massive sandstones containing scattered pebbles and shale clasts throughout. The sandstones grade upward into alternating shales and thin-bedded, finer grained, and current-laminated sandstones and siltstones. Stratigraphic relationships indicate that the shales and the thin-bedded sandstones represent either the final stages of channel filling or interchannel deposition. Both channel-fill and interchannel sequences contain thick, parallel beds of somewhat graded pebbly mudstones. These pebbly mudstones have wide lateral continuity and were probably the result of rather mobile suspensions capable of spreading over a large part of the fan surface. The origin and significance of the abundant conglomeratic strata contained within the Messanagros Sandstone have been discussed elsewhere (Mutti, 1969, p. 1057). Large-scale truncation in the westernmost portion of the Mount Schiati fan valley, as well as pebble size and composition, clearly indicates that the clasts were derived from erosion of the underlying Koriati Conglomerate within the deeply incised fan valley and adjacent canyons.

The paleogeographic significance of the Messanagros Sandstone, as part of the general history of development of the area has been discussed by Mutti (1969) and by Mutti and others (1970). The sandstone was deposited at the base of a slope controlled by tension faulting in an area where nappes from eugeosynclinal basins previously had been piled on the local miogeosynclinal sequence. Messanagros and Koriati detritus, in fact, was wholly derived from the subaerial erosion of these nappes. This is one of the few circumstances in which both facies distribution and paleocurrent patterns concur in demonstrating a lateral origin for turbidite sediments within a geosynclinal zone.

The first outbuilding of the Mount Schiati fan valley, the subsequent retreat and northward shift of the main distributary system, and the consequent recessive deposition of middle fan sediments over a previous inner fan environment are expressions of the tectonic instability of the basin, particularly along its western edge.

Outer fan deposits.—Turbidite-facies associations formed of nonchannelized, broadly lenticular, coarse-grained, and thick-bedded sand bodies are considered as outer fan sediments (Mutti and Ghibaudo, 1972; Mutti and Ricci Lucchi, 1972) along with the enclosing, laterally persistent parallel beds of mudstone and thoroughly current-laminated fine-grained sandstones.

Sediments of these types are extremely widespread in ancient turbidite sequences in the northern Apennines, especially in those sequences deposited in Tertiary miogeosynclinal troughs, for example, the Macigno and Marnoso-arenacea formations and the Modino-Cervarola Group (Mutti and Ricci Lucchi, 1972). The geological setting of these sediments has been discussed in detail by Bortolotti and others (1970). A significant example of outer fan turbidites is provided by the Bobbio Formation, a thick Miocene sequence of the Modino-Cervarola Group cropping out in the core of a classic tectonic window 50 km south of Piacenza in the northwestern part of the northern Apennines (Mutti and Ghibaudo, 1972).

The stratigraphy of this formation is illustrated by figure 4. The lower member (Brugnello Shale) onlaps progressively southwest onto slope sediments of the Sanguineto complex and consists predominantly of silty mudstones containing thin, parallel interbeds of current-laminated fine-grained sandstone. Local groupings of these interbeds form distinctly sandier horizons than are generally encountered within this member; they are interpreted as fan-fringe deposits.

The San Salvatore Sandstone was deposited over the Brugnello Shale. The lower part of the San Salvatore Sandstone is shown in some detail in figure 5. Vertical subdivisions of the sequence are shown in terms of coarse- and fine-grained sediments. The coarse sediments consist of proximal (in sense of Walker, 1967) turbidite beds whose aggregate thickness ranges between 2 and 90 m. The internal picture includes thick-bedded sandstone and mudstone interbeds.

The sandstone beds are well graded, and the graded divisions may or may not be capped by thinner divisions of current-laminated finer grained sandstone and siltstone. The complete Bouma sequence (i.e., Ta-e) is present in some places. Finer grained sediments form strata that range between 2 and 50 m in thickness and consist primarily of distal turbidites, that is, mudstone enclosing thin, parallel beds of fine-grained thoroughly current-laminated sandstone and siltstone.

The coarse-grained strata shown on the left column of figure 5 are clearly nonchannelized sandstone bodies. A section at right angles to the paleocurrent direction (fig. 6) shows that these bodies are markedly lenticular and lens out within a few hundred meters. Detailed field mapping in a downcurrent direction reveals rapid thinning followed by total disappearance over a distance of 2 to 5 km. The lateral and downcurrent equivalents consist of finer grained thin-bedded sediments that are similar to those observed vertically in the measured section and to those forming the Brugnello Shale.

The coarse-grained sediments are here interpreted as virtually nonchannelized sand lobes within an overall prograding sequence in an outer fan environment. The sand lobes developed at the debouching terminus of a migrating channel system located higher on the fan surface. The obvious cyclic character of these sandstone bodies (fig. 5) and its possible paleoenvironmental significance are discussed in detail in the following section.

DISCUSSION AND CONCLUSIONS

The examples which have been discussed in the preceding sections indicate that readily observable and measurable sedimentary features can be used for recognizing and interpreting paleoenvironments for ancient deep-sea fan deposits. Basically, these features include the geometry of the coarse-grained sandstone bodies, the presence of channeling at the base of these bodies, and the type of finer grained sediments incorporating the sandstone bodies. In accord with the criteria set forth particularly by Mutti and Ricci Lucchi (1972), these features allow recognition of a broad depositional model by which ancient deep-sea fans can be subdivided into these three main portions: (1) an inner fan portion displaying large submarine valleys (lowermost reaches of submarine canyons or fan valleys), (2) a middle fan portion reflecting an extensively channelized area cut into an open fan surface, and (3) an outer fan portion reflecting an area virtually devoid of channels.

The model is suitable for practical use in field studies, particularly on a regional scale in which ancient deep-sea fans can be examined within the context of their depositional and stratigraphic relationships with the enclosing sediments. Examples of applications of this model to ancient basins, with particular reference to middle and outer fan deposits, have been given by Mutti and Ricci Lucchi (1972) and by Ricci Lucchi and Pialli (in press) from the northern and central Apennines respectively. These studies also have shown that one of the advantages of the model is that each of the three main facies associations that characterize ancient deep-sea fans is generally expressed by sediment thicknesses on the order of some hundreds of meters. This appears to be of primary importance in turbidite-facies analysis because these facies associations can be considered as mappable units. Moreover, their thickness allows use of conventional biostratigraphic techniques for establishing reliable time-correlation patterns on which paleogeographic reconstructions have to be based.

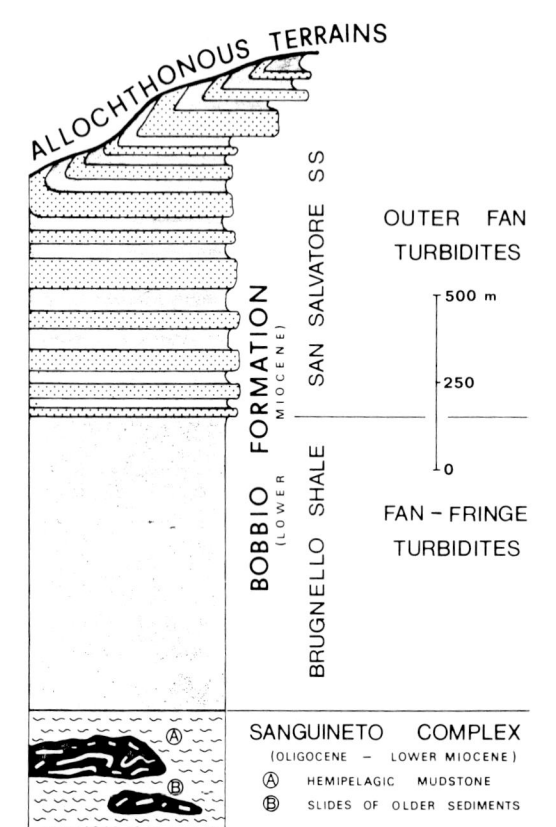

FIG. 4.—Stratigraphy and paleoenvironmental interpretation of Bobbio Formation in Trebbia valley, south of Piacenza, northern Apennines (modified from Mutti and Ghibaudo, 1972).

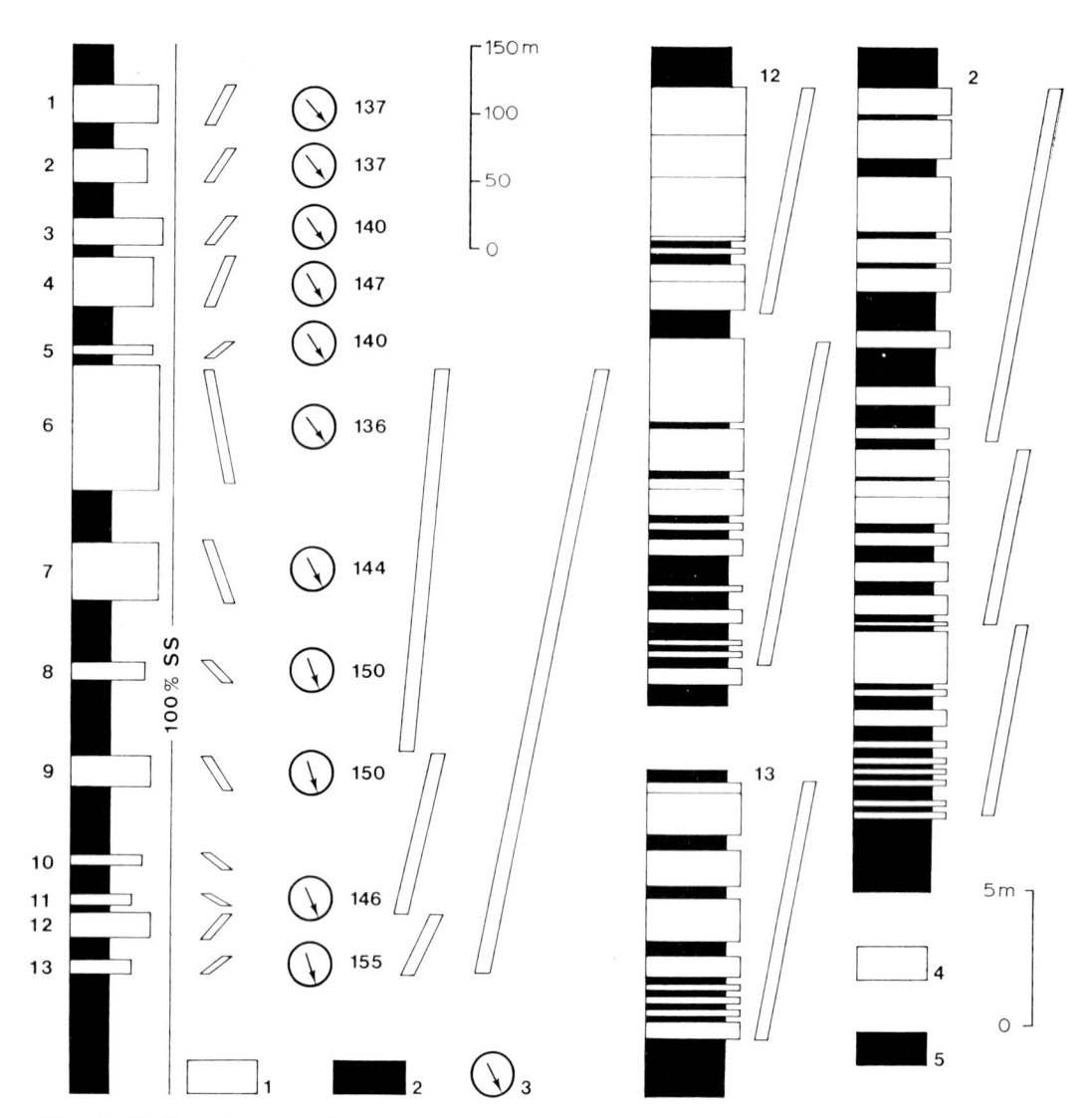

FIG. 5.—Cyclic and progradational patterns of outer fan turbidites of San Salvatore Sandstone (Bobbio Formation, northern Apennines). See text for discussion. *1,* Coarse-grained thick-bedded turbidites; *2,* finer grained thin-bedded turbidites; *3,* vector mean of paleocurrent directions from within each measured level; *4,* A-division; *5,* current-laminated divisions and homogenous shale. Diagonal bars show thickening upward versus thinning upward cycles (see fig. 7).

The model admittedly has some oversimplifications and discrepancies when compared in the context of the physiographic and depositional settings of Recent deep-sea fans. Various features of Recent fan sedimentation are well known in detail (e.g., Normark, 1970; Piper, 1970b; Haner, 1971). Their importance, however, is minor in comparison with the information that can be obtained from an examination of the much greater volume of sediment contained in ancient sequences. In fact, careful study of ancient deep-sea fan deposits allows a more satisfactory interpretation of sedimentary features than can usually be made from short cores penetrating thin veneers of Recent sediments. Moreover, physiographic criteria are mainly used for modern deep-sea fan zonation. Such criteria are largely inapplicable to ancient deposits because extensive three-dimensional exposures are rarely present and because it is virtually impossible to determine gradients of depositional paleosurfaces within a few degrees.

Obviously, models derived from the modern sea floor must be adapted to include criteria that

FIG. 6.—Exposures of Bobbio Formation (Miocene) along Trebbia valley, northern Apennines. Exposure is approximately normal to main paleocurrent trend. From base to top (left to right), section exposes thin-bedded fan-fringe deposits (Brugnello Shale) followed by outer fan sediments (San Salvatore Sandstone). Outer fan sediments include increasingly thicker sand bodies from left to right, which results from basinward progradation of coarse-grained sediments. Note how abruptly sandstone bodies lens out in a direction normal to the main paleocurrent trend, that is, uphill in this exposure.

are applicable to the objective field study of ancient sediments. For example, although channels of various sizes are reported from Recent outer fans (e.g., Shepard and others, 1969; Nelson and others, 1970; Haner, 1971), only minor channels, generally shallower than 2 m and likely related to braided distributary systems, can be considered as occasional features of ancient outer fan sediments (Mutti and Ghibaudo, 1972). Normark (this volume) points out that small distributary channels of comparable depth may be common on the lower reaches of all Recent deep-sea fans, though they cannot be resolved by conventional sounding techniques. In ancient sequences, the association of coarse-grained channel-fill sediments and enclosing finer grained and thin-bedded deposits is here thought to be restricted, for most practical purposes, to the middle fan environment.

The model that has been discussed above can be improved significantly so as to permit its application to ancient deep-sea fans, particularly in those examples for which paleoenvironmental reconstructions have to be attempted primarily on the basis of vertical sequence analysis.

As has been pointed out in the preceding sections, distinct depositional cycles occur in both the Messanagros Sandstone and in the Bobbio Formation. Although somewhat overlooked in the past, a cyclic arrangement of bed thickness, sand-shale ratio, internal sedimentary structures, and grain size is present in many turbidite sequences. Walker (1970) gave an extensive review of such cycles and discussed the possible origin of thinning and thickening upward cycles in terms of shifting points of supply on the platform. He (p. 242) concluded:

"It is very important that sequences of turbidites be analyzed for these minor cycles—it would not be at all surprising to find that they are very common in certain situations. If they are discovered, the art of basin analysis is made easier because a model must be sought to explain the sequence. Thick sequences of beds and long periods of time can therefore be considered simultaneously, rather than the present tendency to view each turbidite as an individual intruder into an area of quiet water deposition."

Detailed observations in the northern Apennine area have led Mutti and Ricci Lucchi (1972) to conclude that a cyclic arrangement of bed thickness and grain size is among the most typical features of ancient deep-sea fans; also, that middle fan channel-fill deposits characteristically display thinning upward cycles, whereas nonchannelized sandstone bodies in the outer fan have a tendency to develop thickening upward sequences. Whatever their origin may be, such cyclic deposits are an obvious feature of ancient deep-sea fans, and, therefore, they should be overlooked no longer.

Marked physiographic and depositional similarities among Recent and ancient deep-sea fans, alluvial fans, and certain deltaic complexes have been recognized by several workers (Jacka and others, 1968; Nelson and others, 1970; Normark, 1970; Haner, 1971). If such close comparison holds true, roughly similar depositional sequences should be found in the vertical successions of all these sediments as a result of an overall pattern, including a distributary system and a peripheral fringe of outbuilding deposits.

Comparing patterns of sedimentation in constructive (in sense of Fisher, 1969) delta systems and deep-sea fans has been attempted by Mutti and Ghibaudo (1972) in order to explain the cycles found in the San Salvatore Sandstone. Active deltaic outbuilding is represented typically by stream-mouth bars. These deposits are markedly lenticular in cross section and are enclosed by finer grained sediments. Basal channeling is completely absent, and distinct thickening and (or) coarsening upward sequences are clear evidence of progradational growth (e.g., Fisk, 1955, 1961; Coleman and Gagliano, 1965; Fisher and others, 1970). In delta systems advancing seaward, one stream-mouth bar sequence tends to be superimposed on another, forming larger scale progradational cycles, this cyclic pattern being controlled primarily by the rate of sedimentation and rate of subsidence and by lateral shifting of sites of sand deposition. These arguments have been discussed particularly by Curtis (1970). Examination of the section illustrated in figure 5 shows that the San Salvatore sequence may be subdivided locally into a number of depositional cycles having varying degrees of importance. A well-marked thickening upward cycle results from variations in the thickness of the coarse-grained intervals (level 13 through level 6). On the same basis, this major cycle can be subdivided into three smaller thickening upward units (levels 13–12, 11–9, and 8–6). Thus, the overall depositional history of the local San Salvatore section is one of progradation for the lowermost 550 m and of a following recessional phase that probably is attributable to lateral shifting of the main distributary system. The renewal of progradational details (levels 5 through 1) is obscured by outcrop discontinuity and tectonic complexities.

Within the more general thickening upward cycles described above are smaller thickening and thinning upward cycles as shown by variations in bed thickness and particularly in the thickness of the Bouma graded division. These cycles, which can be either simple or composite, are quite obvious in the field and in terms of bed thickness represent the internal organization of the individual sandstone bodies comprising the local San Salvatore sequence. Three thickening upward cycles are shown in the detailed columns on the right side of figure 5. Cycle 13 is simple; cycles 2 and 12 are composite (i.e., made up of minor superimposed thickening upward units). The sandstone bodies displaying these thickening upward cycles are also characterized by absence of basal channeling, lenticular cross-sectional geometry, gradual transition with the underlying finer grained sediments, and by commonness of minor channels and amalgamations at their tops. All these features compare favorably with those of stream-mouth bars. On this basis, Mutti and Ghibaudo (1972) concluded that the cycles of the San Salvatore Sandstone are indicative of progradational sand lobes developed at the debouching terminus of middle fan channels located higher on the fan surface.

Fining and (or) thinning upward cycles are well known from channel-fill sequences in alluvial and deltaic plains that result from vertical accretion following channel abandonment or from lateral accretion following channel migration. In Recent deep-sea fans, middle fan areas exhibit both meandering (Nelson and others, 1970; Haner, 1971) and braided (Normark, 1970) channel systems. Many channels are now abandoned in these middle fan areas due to shifting of the distributary systems along with rapid vertical accretion in the interchannel areas (Normark, 1970). Moreover, with the possible exception of the Ranzano example of inner fan sandstone discussed here, no conclusive evidence for lateral accretion has been reported in the literature from ancient channel-fill deposits in turbidite sequences. It is tempting, therefore, to suggest that the channel-fill, thinning upward cycles reported by Mutti and Ricci Lucchi (1972) from middle fan deposits of the northern Apennines and of Rhodes could be related primarily to processes of vertical accretion (aggradation) upon gradual abandonment of the channels.

From the above, vertical sequence analysis appears to result in better paleoenvironmental understanding of middle and outer fan sediments. The general depositional model for ancient deep-sea fans is thus enhanced. The resulting new model, shown diagrammatically in figure 7, emphasizes vertical accretion in channels and interchannel areas in the middle fan sandstone and progradation in the outer fan sandstone. Nevertheless, cycles of pure aggradation (thinning upward) or progradation (thickening upward) are rarely developed in ancient turbidite sequences. More frequently, an aggradational or progradational character can be inferred only on the basis of the general trend of composite cycles that reflect a complex depositional history of sandstone bodies.

No major discrepancy seems to exist between the model proposed here and that inferred by Normark (1970) for Recent deep-sea fans. Normark's suprafan is a *physiographic* feature that appears as a convex-upward bulge on longitudinal sections of many Recent deep-sea fans and exhibits numerous shallow braided channels in its inner portion. In ancient sequences this channelized portion probably would appear as a middle fan turbidite-facies association, that is, as channel and interchannel deposits, whereas the outer portion of the suprafan and extensive reaches of Normark's lower fan would display progradational sand lobes. At the present time, the fact that cyclical depositional sequences detected in ancient sequences have thicknesses on the order of several meters severely limits comparison of patterns of sedimentation in ancient and Recent deep-sea fans. Conventional coring devices cannot penetrate a complete cyclical sequence in modern deep-sea fans. Overcoming this problem will result in substantially better understanding of the depositional setting of both Recent and ancient deep-sea fans.

Size and shape of Recent fans vary considerably depending upon basin size and geometry, which in turn reflect structural setting (Nelson and Nilsen, this volume; Normark, this volume). The same parameters also must have been of primary importance in ancient basins, although available data on size of ancient deep-sea fans do not yet permit significant comparison. Different basin sizes and geometries, however, probably do not affect substantially the general pattern of sediment transport and depo-

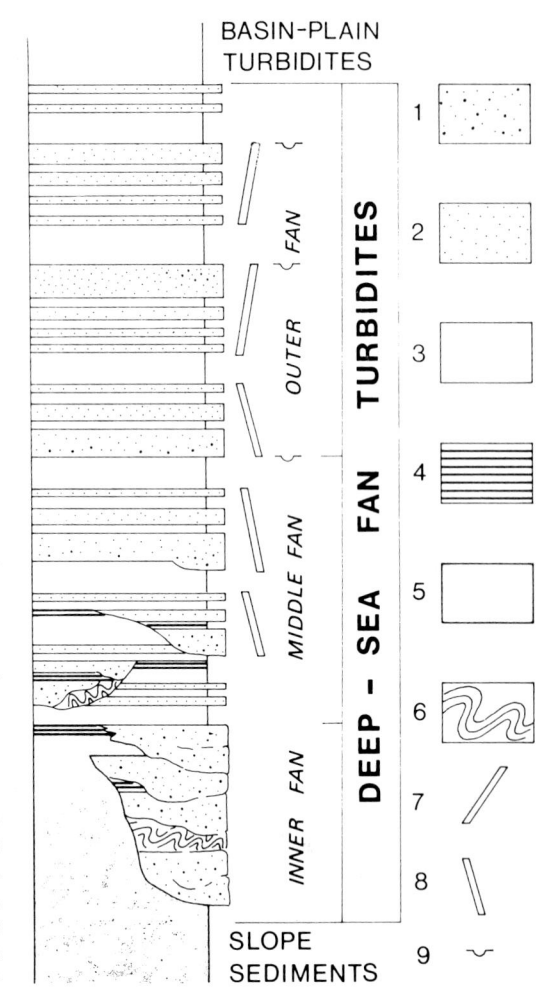

Fig. 7.—Depositional model for ancient deep-sea fan deposits. *1,* thick-bedded graded and(or) crudely laminated coarse-grained sandstone and conglomerate; *2,* thick-bedded graded sandstone and varying amounts of finer grained current-laminated deposits, including mudstone; *3,* mudstone and parallel bedded, thoroughly current-laminated sandstone and siltstone; *4,* thin, lenticular, and discontinuous beds of fine- to coarse-grained sandstone exhibiting grading and poorly developed current laminae; *5,* so-called "hemipelagic" mudstone; *6,* chaotic sediments; *7,* thickening upward cycle; *8,* thinning upward cycle; *9,* shallow channels (less than 2m). This model, modified and simplified from Mutti and Ricci Lucchi (1972), is based on recognition of several sedimentary facies that are only briefly discussed here. (See original reference.)

sition. If this is true, the depositional model presented in this paper can be applied regardless of scale differences, particularly in the example of vertical sequence analysis.

In the next several years, detailed paleogeographic reconstructions of ancient deep-sea fans in the circum-Mediterranean geosynclines will provide us with a great deal of information regarding the sedimentary evolution of the continental margin and ocean-basin portions of these geosynclines. This type of investigation is just beginning, however, so that results to this time are only preliminary. Nevertheless, one noteworthy example is that, at least in the northern Apennines, sediments of ancient deep-sea fans appear to characterize most of the Oligocene and Miocene basins, which developed in the miogeosynclinal zone (e.g., as represented by the Macigno and Marnoso-arenacea formations and the Modino-Cervarola Group) even though such sediments also developed in the Ligurian eugeosynclinal zone (e.g., as represented by the Monte Gottero Sandstone) (E. Mutti and F. Ricci Lucchi, studies in progress). The miogeosynclinal fans apparently were built up as relatively narrow, elongate features within narrow basins and were oriented roughly parallel to the main structural trends, that is, in a northwest-southeast direction. Facies distribution and paleocurrent directions indicate outbuilding of these fan systems toward the southeast and their transition in the same direction into more distal basin-plain turbidites (Ricci Lucchi and Pialli, in press; E. Mutti and F. Ricci Lucchi, studies in progress).

ACKNOWLEDGMENTS

The writer wishes to thank F. Ricci Lucchi for stimulating discussions and criticism pertinent to this paper. The manuscript was improved by critical commentary by R. H. Dott, Jr., and R. G. Walker. C. C. Daetwyler, who also reviewed an early draft of the manuscript, G. Ghibaudo, G. V. Middleton, D. J. Stanley, and G. Zanzucchi contributed during several field trips to the development of some of the ideas discussed in this paper. It is the writer's pleasure to acknowledge all this assistance even though full responsibility for expression remains with him. The field work was made possible by the generous financial support received from the Italian Consiglio Nazionale delle Ricerche, Rome. Drafting of illustrations was kindly done by A. Arrobbio.

REFERENCES

BASSET, D. A., AND WALTON, E. K., 1960, The Hell's Mouth Grits: Cambrian greywackes in St. Tudwal's Peninsula: Geol. Soc. London Jour.: v. 116, p. 85–110.
BOCCALETTI, M., ELTER, P., AND GUAZZONE, G., 1971, Plate tectonic models for the development of the western Alps and northern Apennines: Nature, v. 234, p. 108–111.
BORTOLOTTI, V., PASSERINI, P., SAGRI, M., AND SESTINI, G., 1970, The miogeosynclinal sequence, in SESTINI, G. (ed.), Development of the northern Apennines geosyncline: Sed. Geology v. 4, p. 341–444.
COLEMAN, J. M., AND GAGLIANO, S. M., 1965, Sedimentary structures: Mississippi River deltaic plain, in MIDDLETON, G. V. (ed.), Primary sedimentary structures and their hydrodynamic interpretation: Tulsa, Oklahoma, Soc. Econ. Paleontologists and Mineralogists Special Pub. 12, p. 133–148.
CURTIS, D. M., 1970, Miocene deltaic sedimentation, Louisiana Gulf Coast, in MORGAN, J. P. (ed.), Deltaic sedimentation, modern and ancient: ibid., Special Pub. 15, p. 293–312.
DZULYNSKI, S., KSIAZKIEWICZ, M., AND KUENEN, PH. H., 1959, Turbidites in flysch of the Polish Carpathian Mountains: Geol. Soc. America Bull., v. 70, p. 1089–1118.
ELTER, G., ELTER, P., STURANI, C., AND WEIDMANN, M., 1966, Sur la prolongation du domaine ligure de l'Apennin dans le Montferrat et les Alpes et sur l'origine de la Nappe de la Simme s.l. des Prealpes Romandes et Chablaisiennes: Archives Sci. Genève, v. 19, p. 279–377.
FISHER, W. L., 1969, Facies characterization of Gulf Coast Basin delta systems, with some Holocene analogues: Gulf Coast Assoc. Geol. Socs. Trans., v. 19, p. 239–261.
———, PROCTOR, C. V., GALLOWAY, W. E., AND NAGLE, J. S., 1970, Depositional systems in the Jackson Group of Texas. Their relationship to oil, gas, and uranium: ibid., v. 20, p. 234–261.
FISK, H. N., 1955, Sand facies of Recent Mississippi delta deposits: World Petroleum Cong. 4 (Rome, 1955), Proc., Sec. 1, p. 377–398.
———, 1961, Bar-finger sands of the Mississippi delta, in Geometry of sandstone bodies: Tulsa, Oklahoma, Am. Assoc. Petroleum Geologists, p. 29–52.
GHIBAUDO, G., AND MUTTI, E., 1973, Facies ed interpretazione paleoambientale delle Arenarie di Ranzano nei dintorni di Specchio (Val Pessola, Appennino Parmense): Soc. Geol. Italiana mem. 12, p. 251–265.
HANER, B. E., 1971, Morphology and sediments of Redondo submarine fan, southern California: Geol. Soc. America Bull., v. 82, p. 2413–2432.
HOORN, B. VAN, 1969, Submarine canyon and fan deposits in the Upper Cretaceous of the south-central Pyrenees, Spain: Geologie en Mijnb., v. 48, p. 67–72.
JACKA, A. D., BECK, R. H., ST. GERMAIN, L. C., AND HARRISON, S. C., 1968, Permian deep-sea fans of the Delaware Mountain Group (Guadalupian), Delaware Basin, in Guadalupian facies, Apache Mountain area, west Texas: Soc. Econ. Paleontologists and Mineralogists, Permian Basin Sec., Pub. 68–11, p. 49–90.
KELLING, G., AND WOOLLANDS, M. A., 1969, The stratigraphy and sedimentation of the Llandoverian rocks of the Rhayader District, in WOOD, A. (ed.), The pre-Cambrian and lower Paleozoic rocks of Wales: Swansea, Univ. Wales Press, p. 255–282.
MUTTI, E., 1969, Sedimentologia delle Arenarie di Messanagros (Oligocene-Aquitaniano) nell'isola di Rodi: Memorie Società Geologica Italiana, v. 8, p. 1027–1070.

―――, AND GHIBAUDO, G., 1972, Un esempio di torbiditi di conoide esterna: le Arenarie di San Salvatore (Formazione di Bobbio, Miocene) nell'Appennino di Piacenza: Accad. Sci. Torino Mem., Cl. Sci. Fis., Mat. e Nat., ser. 4, v. 16, 40 p.
―――, OROMBELLI, G., AND POZZI, R., 1970, Geological map of the island of Rhodes (Greece). Explanatory notes: Géol. Pays Helleniques Ann., v. 22, p. 77–226.
―――, AND RICCI LUCCHI, F., 1972, Le torbiditi dell'Appennino settentrionale: introduzione all'analisi di facies: Soc. Geol. Italiana Mem., v. 11, p. 161–199.
NELSON, C. H., CARLSON, P. R., BYRNE, G. V., AND TAU RO ALPHA, 1970, Development of the Astoria Canyon-fan physiography and comparison with similar systems: Marine Geology, v. 8, p. 259–291.
NORMARK, W. R., 1970, Growth patterns of deep-sea fans: Am. Assoc. Petroleum Geologists Bull., v. 54, p. 2170–2195.
―――, AND PIPER, D. J. W., 1969, Deep-sea fan-valleys, past and present: Geol. Soc. America Bull., v. 80, p. 1859–1866.
PIPER, D. J. W., 1970a, A Silurian deep sea fan deposit in western Ireland and its bearing on the nature of turbidity currents: Jour. Geology, v. 78, p. 509–522.
―――, 1970b, Transport and deposition of Holocene sediment on La Jolla deep sea fan, California: Marine Geology, v. 8, p. 221–227.
―――, AND NORMARK, W. R., 1971, Re-examination of a Miocene deep-sea fan and fan-valley, southern California: Geol. Soc. America Bull., v. 82, p. 1823–1830.
RICCI LUCCHI, F., 1969, Channelized deposits in the middle Miocene flysch of Romagna (Italy): Gior. Geologia, v. 36, p. 203–282.
―――, AND PIALLI, G., 1973, Apporti secondari della Marnoso-arenacea. 1: Torbiditi di conoide e pianura sottomarina a ENE di Perugia: Soc. Geol. Italiana Boll. 1973.
SESTINI, G., 1970, Sedimentation of the late geosynclinal stage, in SESTINI, G. (ed.), Development of the northern Apennines geosyncline: Sed. Geology, v. 4, p. 445–479.
SHEPARD, F. P., DILL, R. F., AND RAD, U. VON, 1969, Physiography and sedimentary processes of the La Jolla submarine fan and fan-valley: Am. Assoc. Petroleum Geologists Bull., v. 53, p. 390–420.
STANLEY, D. J., 1967, Comparing patterns of sedimentation in some modern and ancient submarine canyons: Earth and Planetary Sci. Letters, v. 3, p. 371–380.
―――, 1969, Sedimentation in slope and base-of-slope environments, in STANLEY, D. J. (ed.), The new concepts of continental margin sedimentation: Washington, Am. Geol. Inst., p. DJS-8-1-16.
―――, AND UNRUG, R., 1972, Submarine channel deposits, fluxoturbidites and other indicators of slope and base-of-slope environments in modern and ancient marine basins, in RIGBY, J. K., AND HAMBLIN, W. K. (eds.), Recognition of ancient sedimentary environments: Tulsa, Oklahoma, Soc. Econ. Paleontologists and Mineralogists Special Pub. 16, p. 287–340.
SULLWOLD, H. H., JR., 1961, Turbidites in oil exploration, in PETERSON, J. A., AND OSMOND, J. A. (eds.), Geometry of sandstone bodies, a symposium: Tulsa, Oklahoma, Am. Assoc. Petroleum Geologists, p. 63–81.
UNRUG, R., 1963, Istebna beds—a fluxoturbidite formation in the Carpathian Flysch: Soc. Géol. Pologne Ann., v. 33, p. 49–92.
WALKER, R. G., 1966, Shale Grit and Grindslow Shales: transition from turbidite to shallow water sediments in the Upper Carboniferous of northern England: Jour. Sed. Petrology, v. 36, p. 90–114.
―――, 1967, Turbidite sedimentary structures and their relationship to proximal and distal depositional environments: ibid., v. 37, p. 25–43.
―――, 1970, Review of the geometry and facies organization of turbidites and turbidite-bearing basins, in LAJOIE, J. (ed.), Flysch sedimentology in North America: Geol. Assoc. Canada Special Paper 7, p. 219–251.
―――, AND SUTTON, R. G., 1967, Quantitative analysis of turbidites in the Upper Devonian Sonyea Group, New York: Jour. Sed. Petrology, v. 37, p. 1012–1022.

ANCIENT SUBMARINE CANYONS AND FAN VALLEYS

J. H. McD. WHITAKER
University of Leicester, England

ABSTRACT

Many modern submarine canyons and deep-sea fans originated in pre-Pleistocene time. Similar submarine canyons, fans, and fan valleys are found in the geological record certainly as far back as the Precambrian. Criteria for recognizing ancient submarine channels include: (1) proved or inferred size comparable to modern canyons and fan valleys; (2) comparable geometry (e.g., high axial gradients diminishing seaward and steep wall slopes, some of which become vertical or overhanging); (3) similar locations (or submarine canyons) between shallow-marine (shelf) and deep-marine (basin) environments and (for fan valleys) incision into inferred deep-sea fans at the lower ends of canyons; (4) similarities in lithologies, grain sizes, and primary structures of the fills and their variations along the length and width of the canyons or valleys; (5) similarities in the observed or deduced processes of fast cutting and filling, together with clean-cut channel contacts, concave upward form, minor channels at the base of or within channel fill, compaction effects, and partial flushing out; and (6) similarities of multiple origin of faunas within the fill (indigenous, swept in from surrounding shelf areas, or derived from the canyon walls).

These criteria emerge from a study of all available data on ancient submarine canyons and fan valleys from 32 areas. The information is grouped, tabulated, and discussed under eight stratigraphic and geographic headings: (1) Lower Paleozoic of the Caledonian-Appalachian Geosyncline, (2) Carboniferous of the Pennine Basin, (3) Upper Paleozoic of the Variscan Geosyncline, (4) Permian of the Delaware Basin, (5) Mesozoic and Tertiary of the Tethyan Geosyncline, (6) Mesozoic and Tertiary of California, (7) Tertiary of the Gulf Coast Basin, and (8) other areas.

A study of these and future examples of ancient submarine canyons and fan valleys is important as an aid both in reconstructing the continental slopes and rises of geosynclinal and other basins and in the search for possible oil and gas traps. Further intensive studies of transitional areas between shelf and basin facies should reveal many more examples of ancient canyons and fan valleys.

INTRODUCTION

Two lines of evidence have led to acceptance of the antiquity of submarine canyons, deep-sea fans, and fan valleys: (1) evidence pointing to pre-Pleistocene and even early Tertiary origins for many present-day features and (2) evidence from the geological column of large filled channels that closely resemble modern examples in geometry and sedimentary evolution. Evidence from modern features is examined briefly here, and data from the geological column is presented in greater detail, including listing of well-documented ancient canyons and fan valleys, summarization of their characters, and discussion of their positions in ancient basins of deposition.

PRE-PLEISTOCENE ORIGINS FOR SOME MODERN SUBMARINE FEATURES

Recent work on present-day oceanic topographic features, such as continental shelves and slopes, submarine canyons, deep-sea fans, abyssal plains, and deep-sea channels, has led to the belief that many of these features may be older than at first suspected. For example, some deep-sea fans are so large that they may have originated long before the Pleistocene. Menard (1960) thought that the Delgada and Monterey Fans may date from pre-Pliocene time; Menard, Smith, and Pratt (1965) inferred that the Rhône Fan may have been forming since Oligocene time, and similar studies point to a long history of fan building and thus to a comparable antiquity for the submarine canyons that fed them. Daly (1936) realized that canyons may be preglacial, and there are many subsequent references to the antiquity of present-day canyons, such as by Fisk and McFarland (1955), Starke (1956), Bourcart (1958), Roberson (1964), Bourcart (1965), Martin and Emery (1967), Conolly (1968), and Von der Borch (1968).

FILLED CANYONS AND FAN VALLEYS IN THE GEOLOGICAL COLUMN

Hypothetical ancient canyons and fan valleys.—Increasing knowledge of present-day submarine canyons and deep fans together with their fan valleys, well summarized by Shepard and Dill (1966), led stratigraphers and sedimentologists to ask the question, "Where are the ancient submarine canyons and their fans?" Crowell (1952, p. 81), Shrock (1957, p. 1407), Kuenen (1958, p. 338), Knill (1959, p. 324), and others raised the matter in general terms, and more recently many workers on ancient basin environments have invoked lateral canyons through which sediment could be channelled from shallow to deeper water and spread out as submarine fans at the lower ends of the canyons.

Demonstrable ancient canyons and fan valleys.—Whilst for many authors the old canyons or fan valleys remained more or less hypothetical concepts, erosion or burial having precluded their discovery or revealed, at best, only fragmentary evidence, other workers have been able to locate a number of ancient submarine canyons and fan valleys and to demonstrate their geometry and the nature of their sedimentary fill. Examples from 32 areas (15 in Europe, 14 in North America, and 1 each in South America, Africa, and Australia) have been described in some detail, the first in 1948, the rest since 1954. Their degree of preservation varies considerably, from short stretches well exposed at outcrop to vast channel fills located entirely by subsurface methods. Outcrop studies usually give detailed information over limited parts of a channel, whereas subsurface data, less good for detailed study, tend to give overall dimensions more accurately. Both types of study may yield considerable data on channel fills. Present research is seeking to differentiate ancient canyons from fan valleys and to distinguish fan valleys on suprafan, inner, middle, and outer fans (see, for example, Nelson and Nilsen, this volume). The old canyons and fan valleys, although imperfectly known, may be matched in their dimensions, shape, and sedimentary histories with these various parts of modern canyon systems: canyon heads, the main, deeply cut canyons, and the fan valleys, which shallow, bifurcate, and migrate as they cross deep-sea fans.

The ages of ancient (pre-Pleistocene) submarine canyons and fan valleys range from Pliocene to Precambrian, and such features have known examples in every system except the Triassic and Devonian. With further search it is likely that canyons of these ages will be found in marine sequences.

CRITERIA FOR RECOGNIZING ANCIENT CANYONS
AND FAN VALLEYS

Before discussing the ancient examples in detail, the criteria used in deciding which channels in the geological column qualify as canyons and which as fan valleys must be stated on the basis of comparison with modern submarine canyons (Shepard and Dill, 1966) and fans and fan valleys (see recent studies by Normark and Piper, 1969; Shepard and others, 1969; Nelson and others, 1970; Normark, 1970; Curray and Moore, 1971; Haner, 1971; Normark and Nelson and Nilsen, this volume). The following six main criteria have proved useful for comparison, actual examples being indicated by table 1 entry number (in parentheses) or, if not in the table, by author reference:

(1) Proved or inferred comparable size to modern canyons and fan valleys in length, width, and depth; also, variation of width and depth along the length.

(2) Comparable geometry to modern canyons and fan valleys, such as high axial gradients diminishing seaward (6) (7) (10) and steep wall slopes, some of which become vertical or overhanging (1) (2) (5) (6) (7) (8) (11) (21) (27) (Bosellini, 1967, pl. 3). Tributaries joining canyons (20) (30) (32) and branching of fan-valley distributaries (10) (Mutti and Ghibaudo, 1972) (Mutti and Ricci Lucchi, 1972) and their migration (8) (14) have been noted, and occasionally a northern hemisphere left hook (14) (21) or a southern hemisphere right hook (after correcting for paleolatitude) (5) (6) (Schenk, 1970) have been found, which is consistent with the behavior of modern canyon and fan-valley systems.

(3) Similar positions of canyons on the slopes between shallow-marine (shelf) and deep-marine (basin) environments; fan valleys cut into inferred deep-sea fans at the lower ends of canyons. In favorable situations, inner (=upper, proximal), middle (=intermediate), and outer (=lower, distal) fan valleys and channels may be differentiated (8) (9) (10) (24) (28) (Mutti and Ghibaudo, 1972) (Mutti and Ricci Lucchi, 1972) (Mutti and Nelson and Nilsen, this volume).

(4) Similarities in lithologies, grain-size distributions, and primary structures of the fills and their variation along the length (proximal to distal) and width (axial to marginal) of the canyons or fan valleys. Characteristic features include the presence of olistoliths (11) (14) (21); boulder beds (4) (5) (6) (7) (10) (15) (16) (18) (27); cobble or pebble conglomerates and pebbly mudstones (3) (4) (5) (7) (8) (10) (13) (14) (16) (21) (22) (24) (25) (26) (27) (31), some having long-axis preferred orientation or imbrication (4) (6) (Unrug, 1963); breccias (1) (3) (10) (11), and, less commonly, sandstone spheroids, some of which have mud centers (13); and armored mud balls (5) (16) (Unrug, 1963), these clasts being set in a much finer grained matrix of sandstone or mudstone. The sandstones may be massive, amalgamated (8) (13) (17) (Stanley and Unrug, 1972), well bedded (some bedding slightly concave upward), or laminated (3) (4) (5) (6) (8) (9) (11) (13) (14) (26) (27) (28). Channel levees may be present (8) (10). Also characteristic are mass-flow phenomena, including slide conglomerates, slump folds (1) (2) (4) (6) (9) (13) (14) (26) (27) (28), and other channel-controlled primary structures, some being governed by remarkably consistent paleocurrents (6) (14) and others by downflank or downchannel paleoslopes (6) (13).

(5) Similarities in the observed or deduced pro-

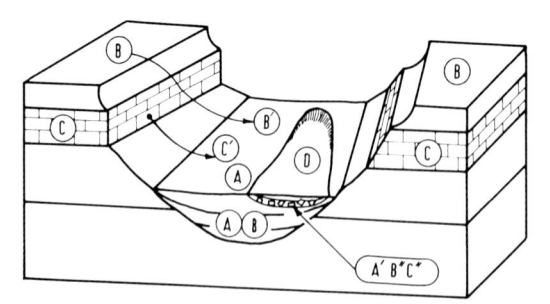

FIG. 1.—Channel fill bearing faunas of different origins. The letters A, B, C, D refer to fossil assemblages (commonly thanatocoenoses). A single suffix denotes derivation, a double suffix, rederived assemblages.

cesses of fast cutting and filling, some processes operating down active fault scarps (1) (4) (6) (15) (20) (28), on sedimentary cover tilted by buried faulting (4), or along active fault zones (6) (11) (29); clean-cut channel contacts that are concave upward (5) (6) (11) (13) (Bosellini, 1967), minor channels at the base of or within the main channel fill (3) (4) (5) (6) (14) (27), and compaction affecting the fill (6) (20) (23) (29), which itself may undergo partial flushing out (6).

(6) Similarities in the multiple origin of the faunas within the fill, expressed diagrammatically in figure 1: A, an *indigenous* deeper water fauna, which may give useful depth indications (e.g., foraminifera), and some of which may be channel dwellers only; B, a contemporary shallow-water fauna washed in from surrounding shelf areas to give a derived fauna B′ termed *exotic;* and C′, an older *remanié* fauna (terms from Craig and Hallam, 1963) derived from erosion of the canyon walls, sometimes recognizable from the abraded, limonitized, or encrusted nature of the derived fossils (4) (6) (12) (13) (16) (19) (24) (28) and figure 2 or from their presence in fallen blocks of wall lithology (6) (11). Rarely, a derived (A′) or a rederived (B″ or C″) fauna may become concentrated at the base of a minor channel within the canyon fill, the result of flushing-out processes, as in (6) and figures 1 and 2. The minor channel sediments may have a fauna, D, slightly younger than A and B ((6) and fig. 2). Downchannel currents may align elongate fossils parallel with the channel axis (6) (10), but if such fossils become lodged in ripplemark troughs, their orientation may be at right angles to the axis (10).

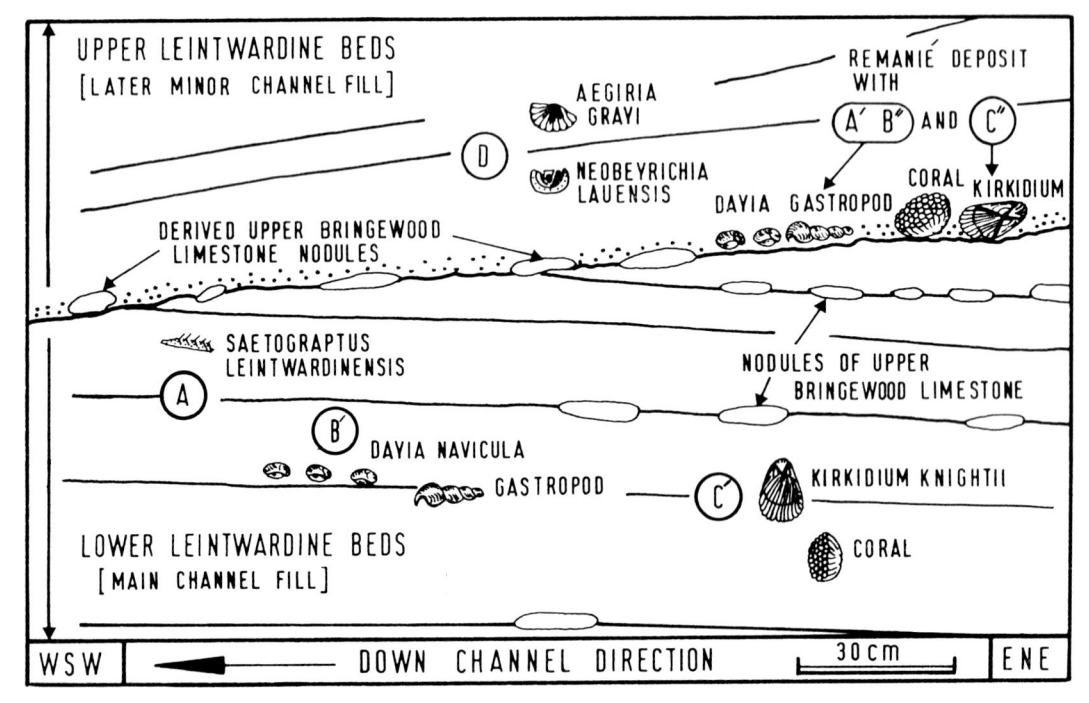

FIG. 2.—Derived and rederived faunas resulting from partial flushing out of channel fill, Todding Lane, Leintwardine, Welsh Borders (British National Grid Reference SO 417756). A, graptolites and indigenous fauna; B′, exotic current-oriented *Dayia navicula* and gastropods; C′, *remanié Kirkidium knightii* and corals derived from Upper Bringewood limestones of the canyon walls; A′B″, abraded and limonitized fossils (such as *Dayia navicula* and gastropods) in basal conglomerate of minor channel fill; C″, rederived *Kirkidium knightii* and corals, also abraded and limonitized; D, indigenous fauna of *Aegiria grayi* and *Neobeyrichia lauensis* (Upper Leintwardine Beds) slightly younger than the AB fauna (Lower Leintwardine Beds).

Differentiation between the lower ends of old submarine canyons and ancient deeply cut valleys on the inner fans is by no means easy; geometrical and spatial relationships are generally of most help. Canyons and inner fan valleys may be more readily distinguished from middle fan valleys, as the former are more deeply incised and contain more poorly bedded (Bouma *ae*), coarser, and more disturbed sediments than the middle fan valleys. These middle fan valleys, as well as the outer fan valleys, which may be wider but as little as 2 m deep, contain well-bedded complete *a-e* sequences. Fan valleys are flanked by dark silty overbank deposits characterized by *cde* sequences that are not found in association with canyons (Jacka, and others, 1968; Stanley and Unrug, 1972; Mutti, this volume; Nelson and Nilsen, this volume).

The criteria listed above exclude shelf channels (for these, see Passega, 1954, and Sedimentation Seminar, 1969) and the extensive and well-studied fluvial channels from many areas and horizons.

TABULATION OF DATA

The main dimensional and sedimentological data on all ancient canyons and fan valleys known to the author (excluding, for reasons of space, channels less than 15 m deep and those insufficiently described) are set out in table 1, so that comparisons and contrasts may be made readily. For a few entries, the writer had to deduce dimensions, trends, etc., from maps or sections. By comparing this table with the appendix in Shepard and Dill (1966), where data on modern canyons are tabulated, the reader may, if he wishes, seek the present-day equivalent to any ancient example.

The entries in table 1 are numbered consecutively and arranged in eight main groups. This arrangement conveniently deals with the older canyons and fan valleys first and the younger ones later. Within each group there is, where possible, a geographic arrangement, and if there are several papers on one channel or group of channels, these papers are grouped under a single entry number.

In the following section, points of special interest from a few selected canyons and fan valleys are discussed, especially for those examples that amplify diagnostic criteria listed earlier or that throw some light on the margin of the basin into which they funnelled sediment. "Strongly inferred" canyons are occasionally referred to, and personal communications also are included.

Lower Paleozoic channels of the Caledonian-Appalachian Geosyncline.—This well-studied geosyncline is widely known through the work of Charles Lapworth, O. T. Jones, W. J. Pugh, Marshall Kay, other pioneers, and, more recently in a plate-tectonics setting, by Wilson (1966), Dewey (1969), Bird and Dewey (1970), Ziegler (1970), Mitchell and Reading (1971), McKerrow and Ziegler (1971, 1972), Bird (this volume), and contributors to the 1969 symposium on the North Atlantic—geology and continental drift (Kay, 1969).

The canyons and fan valleys detailed in table 1, I, extended south and southeast down the proto-North American continental slope and west and northwest down the proto-European continental slope. The earliest canyons are of Ordovician age. These fed sediment (some of it being very coarse grained like the Cow Head Breccia) into a wide proto-Atlantic. The remarkable Cow Head Breccia (Kindle and Whittington, 1958) is interpreted by Dewey (1) (numbers in parentheses refer to entries in table 1) as a submarine fan at the lower end of a canyon. The base of the canyon, one steep wall, and its fill of limestone breccia and slumped argillaceous calcilutites are seen in a cliff exposure. The continental slope into which the canyon was cut may have been determined by the northeast-trending White Bay Fault zone. Later, the fan deposits were thrust over the canyon fill. Lock (1972, fig. 3) confirmed Dewey's views by a palinspastic reconstruction of western Newfoundland for the Middle Ordovician Epoch. Cambro-Ordovician canyons were postulated at various places along 2,000 km of strike in the northern Appalachians by Burke and Waterhouse (1973) on the basis of presence of resedimented carbonate breccias. They associated canyons in the stable miogeoclinal areas of North America with a long period of lowered sea level resulting from glaciation in the Sahara region. Canyons and fan valleys of similar age in Quebec were postulated by Hubert and others (1970), and basal Silurian channels in Newfoundland were proposed by Helwig and Sarpi (1969). In early Llandoverian (Early Silurian) time, a canyon now sited in New Brunswick led southward into a fan valley (2). Also trending south are the late Llandoverian canyons and fan valleys in north Connemara, Eire (3), and the Wenlockian canyons in south Scotland postulated by Warren (1962). As suggested by Schenk (1970), the early Paleozoic sediments of the Meguma Group, deposited on the European-African side of the proto-Atlantic but now exposed in Nova Scotia, were "funnelled north-northwestward down the paleoslope by gravity creep, slump, and turbidity currents through canyons recorded now by local anoma-

TABLE 1.—SUMMARY OF INFORMATION ON ANCIENT

Provincial Grouping	Entry number	Date of publication	Author	Location	Name of canyon (C) or fan valley (F)	Occurrence (surface or subsurface)	Straight (St) or sinuous (Si)	Length in km	Maximum width in km	Maximum axial gradient
I. Lower Paleozoic Channels of Caledonian-Appalachian Geosyncline	1	1969, pers. commun.	Dewey	Sops Arm, W White Bay, Newfoundland	C	Sur			0.03 seen; 0.06 inferred	
	2	1971	Hamilton-Smith	NW Brunswick	C in N, F in S	Sur and sub		35+		
	3	1969, 1970, 1972	Piper	Co. Galway, Eire	C and F (1969 fig. 6.11),	Sur			1	
					F (1969 fig. 6.30)	Sur			0.35	
	4	1967, 1972 1969	James James & James	W-central Wales	Numerous channels, C and F	Sur	St	Up to 7	Up to 3	8° on Corris-Elerch slope
	5	1969 and pers. commun.	Kelling & Woollands	Rhayader, central Wales	Caban Channel, C, and several smaller channels, C and F	Sur	Si	5	1.6	
	6	1962, 1963 1969	Whitaker Jones	Leintwardine, N Herefordshire, England	Marlow, Bagdon, Todding, Mocktree, Church Hill, and Tatteridge Channels, C	Sur	St	Up to 4	0.8+	175m/km, =10°
II. Carboniferous of Pennine Basin, England	7	1964, 1970 1969, 1972, pers. commun.	Sadler Simpson & Broadhurst	Castleton, Derbyshire, England	Winnats Reef Channel, ?C	Sur and sub	Si	0.4	0.4	300m/km, =16½°
	8	1966a, b	Walker	N Derbyshire, England	Grindslow Channels (17), C and F	Sur		3+	0.12 to 1	
III. Variscan Geosyncline	9	1966b	Walker	N Devon, England	Westward Ho! Channels (3), C	Sur			0.13	
		1969	Walker	do	Northam Channel, C	Sur			0.05+	
IV. Permian of Delaware Basin, Texas and New Mexico	10	1963	Pray & Stehli	Guadalupe Mountains, New Mexico and W Texas	Bone Spring Channel, C	Sur			0.1	
		1968	Jacka and others	do	Last Chance Canyon Channel, C	Sur	St	8+	3.2+	175m/m, =10°, decreasing to 35m/km, =2°
					W Dog Canyon-Cutoff Ridge Channel, C, continued as?:	Sur	St	12+ (2 and 3 together)	Fairly broad	
					Shumard and Bone Canyon Channel, C	Sur	St		do	
					Guadalupe Pass, Glover, and Lamar Canyons, W Chico Draw, Long Point, etc., Channels, F	Sur	St		0.4	

SUBMARINE CANYONS AND FAN VALLEYS

Geometry			Age		Regimen		
Downaxis direction or channel trend	Wall slopes	Maximum depth of axis below rims in meters	Beds cut by channel (inclusive)	Age of channel fill	Sedimentology of channel fill	Inferred site at time of formation	Inferred cutting and filling agents
	Up to 70°	30+	Arenigian (Lower Ordovician)		Limestone breccia, blocks up to 0.3 m, slumps in argillaceous calcilutites	Continental margin	
To S	To 90° and overhanging	50	Early Llandoverian (Carys Mills Fm.)	Early Llandoverian (Siegas Fm.)	Limestone-slate conglomerate in matrix of lithic wacke; clasts 0.6 to 1.5 m; slumps, pullapart, plastic deformation	Between Taconic folded region and Aroostook-Matapedia Trough	Sliding lenticules, turbulent high-density flow, wall collapse
To S	15–40°	100	Connemara Schists	Late Llandoverian (Gowlaun member)	Conglomerates and coarse sandstones	Steep margin of turbidite basin	In part subaerial?
To S	10–30°	100		Late Llandoverian	Conglomerates, fine breccias, and sandstones	Submarine fan fed by canyon	Turbidity currents
To W and NW	11°	115	Ashgillian (Ordovician)		Arenites, conglomerates, and a variety of mass-flow deposits; flow rolls adjacent to steep channel margins	Stepped basin floor having slope-lip and slope-base channels	Turbidity currents, wall collapse, slumping
To WNW in SW, to NNW in N	30° to vertical and overhanging	110	Llandoverian (Silurian)		Pebbly grits to cobble conglomerates containing 2 m-long angular mudstone blocks, some armoured and contorted; periods of finer grained sedimentation	On submarine slope, cutting obliquely (Caban Channel); others are fan channels	Gravity flows of fluxo-turbidities, undermining of walls
To WSW	35°, locally to 90°	183	Mid-Eltonian to early Lower Leintwardine (Ludlovian, Silurian)	Later Lower and Upper Leintwardine (Ludlovian, Silurian)	Laminated calcareous siltstones, boulder beds, slumps, ripple marks, many paleocurrent indicators such as skip, prod and groove casts; derived and rederived fossils	Continental shelf edge and slope	?Turbidity currents aided by slumping; cut and fill, $=1$ Ma
To ESE, then ENE	Steep	?70+	Lower Carboniferous (Mid D_1 = Mid B_2)		More or less *in situ* limestone boulders in calcareous matrix; fan of calcirudite	From shelf through back reef, down fore reef to basin	
To S and ESE inc	Gentle to 70°, Stepped	50	Namurian (Late Carboniferous), nine in Upper Shale Grit, one in Lower Shale Grit, and seven in Grindslow Shale		Sandstones (proximal turbidites), pebbly sandstones, mudstones	Inner fan at upcurrent edge of apron of basin turbidites On slope	Cut by fast underladen turbidity currents
	Less than 5°	14+	Westphalian (Late Carboniferous, Westward Ho! Fm.)		Two filled with turbidite sandstones, silty mudstones, and sandstone slump balls; one mudstone-filled channel complex	On slope (Haner, 1971, stated outer fan)	As above, or cut by permanent ocean currents or by slumping
		15	Westphalian (Northam Fm.)		Sandstones, siltstones	do	Cut in agitated water environment by turbidity currents
To E or SE	Sharp, concave upward	34	Leonardian (Permian)		Carbonate rudites, blocks up to 9×12×15 m; geopetal fabrics different in each block; matrix limestone, minor dolomite	Submarine slope between Diablo Platform and Delaware Basin	Submarine slides from shallow-water environment
To ESE	do	?460	Guadalupian (Permian) ?to Pennsylvanian	Guadalupian (Permian)	Very fine sandstone, giant foresets; siltstone and conglomeratic lime mudstone; many paleocurrent indicators; shallow-water fossils	Incised several km into northwestern shelf margin	High-density, high-velocity salinity currents, suction currents, sand flows, under-cutting and collapse of canyon walls
To SE	do	Deep	do	do	As above, downaxis thickening of formations	do	do
To SE	do	do	Leonardian and Guadalupian (Permian)		Conglomeratic mudstones, 2-m blocks derived from walls; contorted sandstones below	do	do
	do	15+ on proximal fan, less incised on distal fan	Guadalupian (Permian)		Proximal: minor flow units (Bouma a-d), conglomeratic mudflows, cross-bedded sandstones; intermediate: major flow units of cross-bedded sandstone followed by Bouma b-e intervals; distal: minor flow units	Cut into deep-sea fans in Delaware Basin	do

TABLE 1. (*Continued*)

Provincial Grouping	Entry number	Date of publication	Author	Location	Name of canyon (C) or fan valley (F)	Occurrence (surface or subsurface)	Straight (St) or sinuous (Si)	Length in km	Maximum width in km	Maximum axial gradient
		This vol.	Harms & Pray	do	Several, C or F?	Sur		Many km	1	
V. Mesozoic and Tertiary Channels of Tethyan Geosyncline	11	1969, 1970	Van Hoorn	S-central Pyrenees, (Esera), Spain	C and F	Sur	Si	10	2+	Steep
	12	1965	Schoeffler	Aquitaine, SW France	Le gouf de Capbreton fossile, C	Sub	Si	1.160+	20	
								2.80+	10	
								3.72+	10	
	13	1964, 1967	Stanley	SE France	Annot,	Sur	Si	8	2	
		1964	Stanley & Bouma		Contes, and	Sur	Si	6	3	
		1968	Stanley & Mutti		Menton Channels, C	Sur	Si	2	2	
		1972	Stanley & Unrug							
	14	1969	Ricci Lucchi	N Apennines, Italy	Several, F	Sur	St	To 105 km inferred	2	2°
	15	1968, pers. commun.	Sargent	Calabria, S Italy	C	Sur	St	6	2	
	16	1964	Marschalko	Klenov-Suchá Dolina, Czechoslovakia	C	Sur	St	10+	6 to 8	
		1967	Koráb, Leško & Marschalko							
	17	1969	Mutti	Island of Rhodes, Greece	Arnita, C	Sur	Si	Approx. 6	Approx. 1.5	
					Monte Schiati, C	Sur	Si	Approx. 5	Approx. 1.2	
	18	1960	Neev	S Coastal Plain, N Negev, Israel	1. Kurnub-Beersheba-Saad Channel, C	Sur and Sub	Si	100	4	14m/km, =1°
					2. Nahal-Shorek Channel, C					
VI. Mesozoic and Tertiary Channels of California	19	1959	Frick, Harding & Marianos	N Sacramento Valley	C	Sur	Si	64	Narrow	
	20	1965	Edmondson	Sacramento Valley	Meganos Channel C	Sur and sub	Si	80+	3.2 to 9.7	2° approx
		1967	Dickas & Payne							
	21	1972a	Lowe	Sacramento Valley	C and F	Sur	St	C approx. 8	C 2.5	
								F approx. 15	F to 3	
	22	1957	Almgren & Schlax	S Sacramento Valley	Markley Gorge, C	Sur and sub		96		
	23	1967	Martin & Emery	Monterey	Pajaro Gorge, C	Sub	Si	19	5.6	
	24	1963	Martin	Bakersfield	Rosedale Channel, C	Sub	Si	9.6 (perhaps 29)	1.1 in N to 2.1 in S	40m/km =2½°

	Geometry			Age		Regimen		
Downaxis direction or channel trend	Wall slopes	Maximum depth of axis below rims in meters	Beds cut by channel (inclusive)	Age of channel fill	Sedimentology of channel fill	Inferred site at time of formation	Inferred cutting and filling agents
To	Steep sided	40	Guadalupian (Permian)		Sandstones, upper-flow regime features	On slopes between shelf and deep basin floor	Density currents in relatively deep submarine environment
To ESE, To SW	To nearly vertical	385	Late Cretaceous (Aguas Salenz Fm.)	Santonian (Late Cretaceous, Campo Breccia Fm.)	In west: coarse limestone breccias and immense olistoliths (Triassic to Albian and Santonian) derived from nearby diapiric uplift to west; in east: limestone microbreccias, turbidites, blue marls	Steep slope leading from shallow shelf to California-type basins separated by swells	Sudden catastrophic event; fault-controlled subsidence to form small graben; slumping from walls
To, To WNW		500 to 1,000		Lower Eocene	Epineritic facies, pelagic microfauna flanked by shallower facies		
				Oligocene	Clayey and sandy marls		
				Miocene	Sands and detrital limestones		
To N, To NNW, To N, To, To app, To S and W		200+, 350+	Priabonian (Eocene, Marnes Bleues Fm.)	Late Eocene to early Oligocene (Annot Sandstone Fm.)	Massive wedge-shaped coarse sandstones, thickening downchannel, sandstone spheroids, thin silts, and clays; channel tongues of gravel in axes and draping flanks; thick slumps; reworked fossils from canyon walls	On relatively steep slopes, deeper basin to north	Slide and sand flows, rapid but intermittent sedimentary transport leading to unstable fills
To S and W	5 to 20°		Tortonian (middle Miocene, Marnoso-arenacea Fm.)		Thick-bedded sandstones and polymictic conglomerates, blocks several hundred cubic meters in size; fining upward; as channels filled, they migrated to NE	Zone between slope and fans	Very dense and fast turbidity currents, mass flows and slumps, slipping of clays on channel walls
To SE, To NW		250	?Devonian granites	Miocene and Pliocene	Coarse boulder beds	Submarine fault scarp	
To N and W, To		150 to 200+		Late Eocene	Conglomeratic flysch (coarse graded beds, 2-m boulders, slumps) and wildflysch (medium sandstones and coarse siltstones, thick slides); mud balls; fully marine (foraminifera) but reworked plant detritus	Cutting slopes of steep continental terraces	Fast-moving slide masses and turbidity currents
To E and SE, To W				Middle-late Oligocene and Aquitanian (Messanagros Sandstone)	Sands, some amalgamated, and some slide conglomerates	Shelf-basin transitional area	
	10° to 15°+	To 1,100 in W	Early Cretaceous to Oligocene	Neogene	1. Coarse clastics, gypsum, sandy shale, marls 2. Silts and clays, calcareous sandstones, large boulders	Subaerial river system subsequently drowned	Probably turbidity currents
S		610+	Upper Cretaceous	Eocene	Deltaic, but foraminifera show these beds are all submarine		
S and W	5 to 15°	614	Late Cretaceous to late Paleocene	Late Paleocene	95% silty shale, some shaly siltstone and glauconitic units, 35–60% compaction	Neritic to upper bathyal on foraminifera; fault control	Extensive slumping, turbidity currents (rapid cutting)
SW, SSW	Steepsided		Early Turonian to Albian sediments, Late Jurassic and older igneous and metamorphics	Turonian (Venaio Fm.)	Sandstones and pebbly mudstones, blocks up to 150 m long	Shelf-slope-rise area	Grain flow, mass movements, wall collapse
S		762	Late Cretaceous to late Eocene	Post-Eocene (Oligocene and ?Miocene)	Shales, sandstones, and conglomerates, lateral and vertical variation		
W		1,525	Late Cretaceous or earlier	Early middle Miocene	Reddish sandstones and siltstones, few fresh-water gastropods washed in	Cut subaerially on Elkhorn erosion surface, then submerged	
S	11° to 22°	366+	Middle Miocene	Middle Mohnian (late Miocene)	Mainly sandstones, some pebble conglomerates and siltstones, poorly sorted; fragments of shallow-water megafossils and mixed shallow and deep microfossils	Water depth probably greater than 400 m on foraminifera (Haner, 1971, stated inner fan)	Turbidity currents or gravity flows; cut and fill, ≈0.7 Ma

gular shale clasts, some of the shale clasts being contorted, others armored. The flat-lying shales cut by the gravity flows were undermined and downwarped and acquired steep dips.

Just before the complete closure of the proto-Atlantic at the end of Ludlovian (Late Silurian) time in the Welsh Borders (McKerrow and Ziegler, 1972), continental slopes steep enough to encourage the rapid cutting of six parallel canyon heads at Leintwardine still existed (6). These are well exposed on outcrop. They are short and narrow but were relatively deeply cut and have high axial gradients (10°) and steep, locally vertical walls. Channel margins (Whitaker, 1962, fig. 4) and channel bottoms clearly show the unconformable relations of channel fill resting on earlier Ludlovian sediments. Even though more than 180 m of the normal succession is cut out, little more than one graptolite zone (*Saetograptus leintwardinensis*) is involved, and the author estimates that canyon cutting and filling could have taken place within one million years even though some partial flushing out (fig. 2) might have delayed the filling of the channels.

The sedimentary fill and associated fauna have been studied in detail. Varied primary structures confirm the geometry of the canyon heads that has been deduced from mapping and direct observation (fig. 3). This diagram (fig. 3) shows the concave-up bedding, slumps, boulder beds, ripple, groove, skip, and prod marks characteristic of the channels, together with a rare *indigenous* fauna (italicized terms from Craig and Hallam, 1963) that is not found on the interchannel shelf areas. *Exotic* (derived from contemporaneous shelf environments and commonly current oriented), *remanié* (derived from channel walls), and rederived faunas are also found as illustrated in figure 2. Faunal complexity such as this within a few cubic meters of channel-fill sediments calls for careful study but can yield important evidence concerning the evolutionary history of a canyon or fan channel.

According to Jones' (1969) quantitative study of the paleoecology of the Leintwardine Beds, significant differences exist between the lithologies of the Leintwardine channels and the shelf regions and also between the faunas of the two environments. The unlaminated siltstones of the shelf were accumulated above wave base and have a predominantly benthonic fauna, whereas the channels accumulated finely laminated siltstones by gravity settling below wave base. Also, the channels contain finely disseminated shell material and abundant ostracodes and have a comparatively high percentage of nektobenthonic filter feeders and scavengers, all of which probably reflects a concentration of organic detritus moving from shelf to basin within the channels. Occasional spilling over of shelf sediments set up small turbidity currents in the canyon heads that carried along and occasionally oriented fossils derived from the shelf.

Work in progress by the author shows that the fill of Ludlovian age in the Tatteridge Channel is continued westward to form Brandon Hill (Earp and others, in press) and suggests that the boulder beds and pebbly mudstones around Lingen, which involve older (Wenlockian) strata, may in part be slide deposits that accumulated in the lower parts of more deeply cut channels similar to those at Leintwardine but situated farther south.

Carboniferous channels of the Pennine Basin, northern England.—So far, submarine canyons and fan valleys from the upper Paleozoic of Europe have proved to be scarce. A small canyon (first suspected by Shirley and Horsfield, 1940, p. 290) is cut into the Lower Carboniferous Limestone near Castleton in the Pennine region of northern England (Sadler, 1964). Most of the infilling sediments have been eroded away, leaving more or less *in situ* boulder beds (Simpson and Broadhurst, 1969) in and on the north side of the present Winnats Valley. Also, the so-called "Beach-Beds" remain at the lower end. Apparently, the present Carboniferous Limestone surface is an exhumed subaerial or submarine surface of pre-*Posidonia* age (Simpson and Broadhurst, 1969). The writer, following W. H. C. Ramsbottom's and B. J. Taylor's discussion of the Simpson and Broadhurst paper, prefers the concept of a submarine slope (at least 130 m deep according to Sadler, 1970, p. 286) on which the boulders accumulated as a submarine scree. That this slope was incised by the Winnats Channel is shown by the position of the boulder bed within it and on its northern margin. Nearby Cave Dale possibly is a second channel of similar age and character (F. M. Broadhurst and I. M. Simpson, personal commun., 1972).

The controversial Beach-Beds, which are composed predominantly of broken shells (mainly brachiopods) on the north side of the Winnats Channel and of more rounded shells on the south side, are interpreted as forming a small submarine fan that was partly redistributed toward the east by submarine currents in the direction of thinning (Sadler, 1964). Boring algae of shallow-water type penetrated the shells of the Beach-Beds, confirming the idea that the shells must have been transported down the channel from shallow water into water too

DISCOVERY OF CANYONS AND FAN VALLEYS

A study of publication dates of papers on ancient canyons and fan valleys reveals that they are being discovered at an increasing rate. Most of the United States, Venezuelan, Australian, and African examples were found during the search for oil and gas, whereas the European, Canadian, and Japanese features were discovered mainly by academic research. The present almost total lack of knowledge (at least on the part of the writer) of comparable features from the rest of the world may result from a lack of detailed mapping and drilling. Very detailed work in shelf-basin transition areas is certainly required to find them, even in regions apparently well known such as the Texas Permian or the Welsh Borderland Silurian, where channels have been found only recently. This close scrutiny is needed because the requisite shelf-to-basin transitional zones are of limited area and the areal extent of channel fills is even more limited. In addition, these zones may be lost through later erosion or may be hidden by deep sedimentary burial or orogenic thrusting of basin sequences over shelf sediments. The search is worth while, however, as coarse channelized deposits may be used for distinguishing canyon and fan-valley environments and as an aid in recognizing slope and base-of-slope environments in ancient marine basins (Stanley and Unrug, 1972). This will help us to interpret more fully the general paleogeographies of ancient geosynclines, for we are understanding more clearly that canyons have been the major conduits for introducing coarse clastic sediments, via deep-sea fans, into deep-water basins within ancient geosynclinal belts.

CONCLUSIONS

There is now ample evidence that submarine canyons and their fans, deeply scored by fan valleys, are not unique features of Pleistocene and Holocene time. Evidence obtained from detailed studies of modern submarine canyons and fans (including estimates of the rates of cutting of canyons and building of fans in comparison with the rates of denudation of source areas) has shown that many of them must have begun during pre-Pleistocene time, some of them well back in the Tertiary. Different lines of evidence, from careful study of surface exposures and detailed subsurface drilling, have revealed not only filled canyons and fan valleys of Tertiary age but much older examples in nearly every system as far back as the Precambrian. Often the evidence is fragmentary as these transitory features of limited area are readily lost by erosion or buried by later sedimentation or overthrusting. But sufficient data have now been collected from both ancient and modern submarine canyons and fan valleys to provide support for the uniformitarian principle that "The Present is the Key to the Past." Additionally, they support the equally useful doctrine that "The Past is the Key to the Present," for some show evidence for their complete evolutionary histories, such as one or more episodes of rapid downcutting followed by a period of channel filling that may be reversed from time to time by flushing out processes. The channel fills are composed of varied sediments (commonly very coarse grained) and show many different primary structures and complex faunal assemblages. For students of modern canyons and fan valleys, this knowledge of the complete histories of some ancient examples may prove to be a useful stimulant to their researches on the origins and probable histories of the present-day canyons and fan valleys.

This brief review (for a fuller survey, see Whitaker, in press) shows also that the majority of ancient canyons and fan valleys occur along the margins of actively developing geosynclines. Most authors compare their channels with the well-studied modern canyons and fan valleys off California. The nongeosynclinal Pennine and Delaware Basins, however, have canyons and fan valleys of character similar to the others, both modern and ancient. Clearly, it is not yet possible to differentiate ancient continental margins into various types on the basis of canyons and fan valleys alone.

ACKNOWLEDGMENTS

For guidance in the field, I am indebted to F. P. Shepard, R. F. Dill, and G. Kelling. For permission to quote from theses and personal communications, I wish to thank F. M. Broadhurst, R. P. Coats, J. F. Dewey, A. D. Jacka, D. M. D. James, M. D. Jones, G. Kelling, D. J. W. Piper, W. A. Pryor, L. Redwine, G. E. G. Sargent, and I. M. Simpson. I have also been helped by D. S. Gorsline, A. Hallam, J. Helwig, J. D. Hudson, C. Lewis, R. Marschalko, D. F. Merriam, E. Mutti, H. E. Sadler, J. Schoeffler, P. C. Sylvester-Bradley, and D. Vass. A Fulbright Travel Award and grants from the William Waldorf Astor Foundation and the University of Leicester materially assisted the field work and are gratefully acknowledged. Finally, I wish to thank R. H. Dott, Jr., for inviting me to attend the "Conference on Modern and Ancient Geosynclinal Sedimentation" and to contribute to this volume in honor of Marshall Kay.

REFERENCES

ALMGREN, A. A., AND SCHLAX, W. N., 1957, Post-Eocene age of "Markley Gorge" fill, Sacramento Valley, California: Am. Assoc. Petroleum Geologists Bull., v. 41, p. 326–330.
BARTOW, J. A., 1966, Deep submarine channel in upper Miocene, Orange County, California: Jour. Sed. Petrology, v. 36, p. 700–705.
BIRD, J. M., AND DEWEY, J. F., 1970, Lithosphere plate-continental margin tectonics and the evolution of the Appalachian Orogen: Geol. Soc. America, Bull., v. 81, p. 1031–1060.
BORNHAUSER, M., 1948, Possible ancient submarine canyon in southwestern Louisiana: Am. Assoc. Petroleum Geologists Bull., v. 32, p. 2287–2290.
———, 1960, Depositional and structural history of Northwest Hartburg Field, Newton County, Texas: ibid., v. 44, p. 458–470.
———, 1966, Marine unconformities in the northwestern Gulf Coast: Gulf Coast Assoc. Geol. Societies Trans., v. 16, p. 45–51.
BOSELLINI, A., 1967, Frane sottomarine nel Giurassico del Bellunese e del Friuli: Accad. Naz. Lincei, v. 43, p. 563–567.
BOURCART, J., 1958, Problèmes de géologie sous-marine: Paris, Masson, 123 p.
———, 1965, Les canyons sous-marine de l'extremité orientale des Pyrénées, in SEARS, M. (ed.), Progress in oceanography, 3: London, Pergamon Press, p. 63–69.
BOUROULLEC, J., AND DELOFFRE, R., 1972, Esquisse paléogéographique de l'Albien supérieur a l'Yprésien en Aquitaine: Soc. Natl. Pétroles Aquitaine Centre Rech. Pau Bull., v. 6, p. 263–287.
BURKE, K., 1972, Longshore drift, submarine canyons, and submarine fans in development of Niger delta: Am. Assoc. Petroleum Geologists Bull., v. 56, p. 1975–1983.
———, AND WATERHOUSE, J. B., 1973, Saharan glaciation dated in North America: Nature, v. 241, p. 267–268.
CONOLLY, J. R., 1968, Submarine canyons of the continental margin, east Bass Strait (Australia): Marine Geology, v. 6, p. 449–461.
CRAIG, G. Y., AND HALLAM, A., 1963, Size-frequency and growth-ring analyses of *Mytilus edulis* and *Cardium edule,* and their palaeoecological significance: Palaeontology, v. 6, p. 731–750.
CROWELL, J. C., 1952, Submarine canyons bordering central and southern California: Jour. Geology, v. 60, p. 58–83.
———, 1955, Directional-current structures from the Prealpine Flysch, Switzerland: Geol. Soc. America Bull., v. 66, p. 1351–1384.
CURRAY, J. R., AND MOORE, D. G., 1971, Growth of the Bengal deep-sea fan and denudation in the Himalayas: ibid., v. 82, p. 563–572.
DALY, R. A., 1936, Origin of submarine "canyons": Am. Jour. Sci., ser. 5, v. 31, p. 401–420.
DAVIES, D. K., 1972, Mineralogy, petrography and derivation of sands and silts of the continental slope, rise and abyssal plain of the Gulf of Mexico: Jour. Sed. Petrology, v. 42, p. 59–65.
DEWEY, J. F., 1969, Evolution of the Appalachian/Caledonian Orogen: Nature, v. 222, p. 124–129.
———, AND BIRD, J. M., 1970, Mountain belts and the new global tectonics: Jour. Geophys. Research, v. 75, p. 2625–2647.
DICKAS, A. B., AND PAYNE, J. L., 1967, Upper Paleocene buried channel in Sacramento Valley, California: Am. Assoc. Petroleum Geologists Bull., v. 51, p. 873–882.
DILL, R. F., 1964, Sedimentation and erosion in Scripps submarine canyon head, in MILLER, R. L. (ed.), Papers in marine geology (Shepard Commemorative Vol.): New York, Macmillan Co., p. 23–41.
DZULYNSKI, S., KSIAZKIEWICZ, M., AND KUENEN, PH. H., 1959, Turbidites in flysch of the Polish Carpathian Mountains: Geol. Soc. America Bull., v. 70, p. 1089–1118.
EARP, J. R., AND OTHERS, in press, Map and explanation, Leintwardine-Ludlow sheet: Great Britain Inst. Geol. Sci.
EDMONDSON, W. F., 1965, The Meganos Gorge of the southern Sacramento Valley: San Joaquin Geol. Soc. Selected Papers, no. 3, p. 36–51.
FISK, H. N., AND MCFARLAND, E., 1955, Late Quaternary deposits of the Mississippi River, in POLDERVAART, A. (ed.), Crust of the earth, a symposium: Geol. Soc. America Special Paper 62, p. 279–302.
FRICK, J. D., HARDING, T. P., AND MARIANOS, A. W., 1959, Eocene gorge in northern Sacramento Valley (abs.): Am. Assoc. Petroleum Geologists Bull., v. 43, p. 255.
GOHEEN, H. C., 1959, Sedimentation and structure of the *Planulina*-Abbeville trend, South Louisiana: Gulf Coast Assoc. Geol. Societies Trans., v. 9, p. 91–103.
HAMILTON-SMITH, T., 1971, A proximal-distal turbidite sequence and a probable submarine canyon in the Siegas Formation (early Llandovery) of northwestern New Brunswick: Jour. Sed. Petrology, v. 41, p. 752–762.
HANER, B. E., 1971, Morphology and sediments of Redondo Submarine Fan, southern California: Geol. Soc. America Bull., v. 82, p. 2413–2432.
HELWIG, J., AND SARPI, E., 1969, Plutonic-pebble conglomerates, New World Island, Newfoundland, and history of eugeosynclines, in KAY, G. M. (ed.), North Atlantic—geology and continental drift: Am. Assoc. Petroleum Geologists Mem. 12, p. 443–466.
HOYT, W. V., 1959, Erosional channel in the middle Wilcox near Yoakum, Lavaca County, Texas: Gulf Coast Assoc. Geol. Societies Trans., v. 9, p. 41–50.
HSÜ, K. J., 1959, Flute- and groove-casts in the Prealpine Flysch, Switzerland: Am. Jour. Sci., v. 257, p. 529–536.
———, 1971, Origin of the Alps and western Mediterranean: Nature, v. 233, p. 44–48.
———, 1972a, The concept of the geosyncline, yesterday and today: Leicester Lit. and Philos. Soc. Trans., v. 66, p. 26–48.
———, 1972b, When the Mediterranean dried up: Sci. American, v. 227, p. 26–36.
HUBERT, C., LAJOIE, J., AND LÉONARD, M. A., 1970, Deep sea sediments in the lower Paleozoic Québec Super-

group, *in* LAJOIE, J. (ed.), Flysch sedimentology in North America: Geol. Soc. Canada Special Paper 7, p. 103–125.
JACKA, A. D., AND OTHERS, 1968, Permian deep-sea fans of the Delaware Mountain Group (Guadalupian), Delaware Basin: Soc. Econ. Paleontologists and Mineralogists, Permian Basin Sec. Pub. 68–11, p. 49–90.
JAMES, D. M. D., 1967, Sedimentary studies in the Bala of central Wales (Ph.D. thesis): Swansea, Univ. Wales, 126 p.
———, 1972, Sedimentation across an intra-basinal slope: the Garnedd-wen Formation (Ashgillian), west central Wales: Sed. Geology, v. 7, p. 291–307.
———, AND JAMES, J., 1969, The influence of deep fractures on some areas of Ashgillian-Llandoverian sedimentation in Wales: Geol. Mag., v. 106, p. 562–582.
JONES, M. D., 1969, The palaeogeography and palaeoecology of the Leintwardine Beds of Leintwardine, Herefordshire (M.Sc. thesis): Leicester, England, Univ. Leicester, 67 p.
KAY, G. M. (ed.), 1969, North Atlantic—geology and continental drift: Am. Assoc. Petroleum Geologists Mem. 12, 1082 p.
KELLING, G., AND WOOLLANDS, M. A., 1969, The stratigraphy and sedimentation of the Llandoverian rocks of the Rhayader district, *in* WOOD, A. (ed.), The Pre-Cambrian and lower Palaeozoic rocks of Wales: Cardiff, Univ. Wales Press, p. 255–282.
KINDLE, C. H., AND WHITTINGTON, H. B., 1958, Stratigraphy of the Cow Head region, Newfoundland: Geol. Soc. America Bull., v. 69, p. 315–342.
KNILL, J. L., 1959, Axial and marginal sedimentation in geosynclinal basins: Jour. Sed. Petrology, v. 29, p. 317–325.
KOMAR, P. D., 1970, The competence of turbidity current flow: Geol. Soc. America Bull., v. 81, p. 1555–1562.
KORÁB, T., LEŠKO, B., AND MARSCHALKO, R., 1967, Inner-Carpathian and Outer Flysch of the East Slovakia: 23rd Internat. Geol. Cong., Prague, 1968, Guide to Excursion 17AC, 38 p.
KRUIT, C., BROUWER, J., AND EALEY, P., 1972, A deep-water sand fan in the Eocene Bay of Biscay: Nature, v. 240, p. 59–61.
KUENEN, PH. H., 1953, Origin and classification of submarine canyons: Geol. Soc. America Bull., v. 64, p. 1295–1314.
———, 1958, Problems concerning source and transportation of flysch sediments: Geol. en Mijnb., v. 20, p. 329–339.
LAPWORTH, H., 1900, The Silurian sequence of Rhayader: Geol. Soc. London Quart. Jour., v. 56, p. 67–137.
LOCK, B. E., 1972, Lower Paleozoic history of a critical area; eastern margin of the St. Lawrence Platform in White Bay, Newfoundland, Canada: 24th Internat. Geol. Cong., Montreal, sec. 6, p. 310–324.
LOWE, D. R., 1972a, Implications of three submarine mass-movement deposits, Cretaceous, Sacramento Valley, California: Jour. Sed. Petrology, v. 42, p. 89–101.
———, 1972b, Submarine canyon and slope channel sedimentation model as inferred from Upper Cretaceous deposits, western California: 24th Internat. Geol. Cong., Montreal, sec. 6, p. 75–81.
MCKERROW, W. S., AND ZIEGLER, A. M., 1971, The Lower Silurian paleogeography of New Brunswick and adjacent areas: Jour. Geology, v. 79, p. 635–646.
———, AND ———, 1972, Silurian paleogeographic development of the proto-Atlantic Ocean: 24th Internat. Geol. Cong., Montreal, sec. 6, p. 4–10.
MARSCHALKO, R., 1964, Sedimentary structures and paleocurrents in the marginal lithofacies of the central-Carpathian flysch, *in* BOUMA, A. H., AND BROUWER, A. (eds.), Turbidites: Amsterdam, Elsevier Pub. Co., Developments in sedimentology 3, p. 106–126.
MARTIN, B. D., 1963, Rosedale Channel—evidence for late Miocene submarine erosion in Great Valley of California: Am. Assoc. Petroleum Geologists Bull., v. 47, p. 441–456.
———, AND EMERY, K. O., 1967, Geology of Monterey Canyon, California: *ibid.*, v. 51, p. 2281–2304.
MENARD, H. W., 1960, Possible pre-Pleistocene deep-sea fans off central California: Geol. Soc. America Bull., v. 71, p. 1271–1278.
———, SMITH, S. M., AND PRATT, R. M., 1965, The Rhône deep-sea fan, *in* WHITTARD, W. F., AND BRADSHAW, R. (eds.), Submarine geology and geophysics: London, Butterworth and Co., p. 271–285.
MITCHELL, A. H., AND READING, H. G., 1971, Evolution of island arcs: Jour. Geology, v. 79, p. 253–284.
MUTTI, E., 1963, Confronto tra le direzioni d'apporto dei clastici entro il Macigno e il "Tongriano" dell'Appennino di Piacenza: Riv. Italiana Paleontologie e Stratigrafia, v. 69, no. 3, p. 235–258.
———, 1964, Schema paleogeografico del Paleogene dell'Appennino di Piacenza: *ibid.*, v. 70, p. 869–885.
———, 1969, Studi geologici sulle isole del Dodecaneso (Mare Egeo). X. Sedimentologia delle Arenarie di Messanagros (Oligocene-Aquitaniano) nell'isola di Rodi: Soc. Geol. Italiana Mem., v. 8, p. 1027–1070.
———, AND DE ROSA, E., 1968, Caratteri sedimentologici delle Arenarie di Ranzano e della formazione di Val Luretta nel basso Appennino di Piacenza: Riv. Italiana Paleontologia, v. 74, p. 71–120.
———, AND GHIBAUDO, G., 1972, Un esempio di torbiditi di conoide sottomarina esterna: le arenarie di San Salvatore (Formazione di Bobbio, Miocene) nell'Appennino di Piacenza: Accad. Sci. Torino Mem., Cl. Sci. Fis., Mat. e Nat., ser. 4a, no. 16, 40 p.
———, AND RICCI LUCCHI, F., 1972, Le torbiditi dell'Appennino Settentrionale: introduzione all'analisi di facies: Soc. Geol. Italiana Mem., v. 11, p. 161–199.
NASU, N., 1964, The provenance of the coarse sediments on the continental shelves and the trench slopes off the Japanese Pacific coast, *in* MILLER, R. L. (ed.), Papers in marine geology (Shepard Commemorative Vol.): New York, Macmillan Co., p. 65–101.
NEEV, D., 1960, A pre-Neogene erosion channel in the southern coastal plain of Israel: Israel Ministry Devel., Geol. Survey Bull. 25, Oil Div. Paper 7, 20 p.
NELSON, C. H., AND OTHERS, 1970, Development of the Astoria Canyon-Fan physiography and comparison with similar systems: Marine Geology, v. 8, p. 259–291.
NORMARK, W. R., 1970, Growth patterns of deep sea fans: Am. Assoc. Petroleum Geologists Bull., v. 54, p. 2170–2195.

———, AND PIPER, D. J. W., 1969, Deep-sea fan-valleys, past and present: Geol. Soc. America Bull., v. 80, p. 1859–1866.
PAINE, W. R., 1966, Stratigraphy and sedimentation of subsurface Hackberry wedge and associated beds of southwestern Louisiana. Gulf Coast Assoc. Geol. Societies Trans., v. 16, p. 261–274.
PASSEGA, R., 1954, Turbidity currents and petroleum exploration: Am. Assoc. Petroleum Geologists Bull., v. 38, p. 1871–1887.
PIPER, D. J. W., 1969, Silurian sediments in western Ireland (Ph.D. thesis): Cambridge, England, Univ. Cambridge, 213 p.
———, 1970, A Silurian deep sea fan deposit in western Ireland and its bearing on the nature of turbidity currents: Jour. Geology, v. 78, p. 509–522.
———, 1971, Sediments of the Middle Cambrian Burgess Shale, Canada: Lethaia, v. 5, p. 169–175.
———, 1972, Sedimentary environments and palaeogeography of the late Llandovery and earliest Wenlock of north Connemara, Ireland: Geol. Soc. London Quart. Jour., v. 128, p. 33–51.
———, AND NORMARK, W. R., 1971, Re-examination of a Miocene deep-sea fan and fan-valley, southern California: Geol. Soc. America Bull., v. 82, p. 1823–1830.
PRAY, L. C., AND STEHLI, F. G., 1963, Allochthonous origin, Bone Spring "patch reefs," west Texas (abs.): Geol. Soc. America Special Paper 73, p. 218–219.
RADOMSKI, A., 1961, On some sedimentological problems of the Swiss flysch series: Eclogae Geol. Helvetiae, v. 54, p. 451–459.
RICCI LUCCHI, F., 1969, Channelized deposits in the middle Miocene flysch of Romagna (Italy): Gior. Geologia., v. 36 (for 1968), p. 203–260.
ROBERSON, M. I., 1964, Continuous seismic profiler survey of Oceanographer, Gilbert and Lydonia submarine canyons, Georges Bank: Jour. Geophys. Research, v. 69, p. 4779–4789.
ROBINSON, F. M., 1964, Core tests, Simpson area, Alaska: U.S. Geol. Survey Prof. Paper 305-L, p. 645–730.
SADLER, H. E., 1964, The origin of the "Beach-Beds" in the Lower Carboniferous of Castleton, Derbyshire: Geol. Mag., v. 101, p. 360–372.
———, 1970, Boring algae in brachiopod shells from Lower Carboniferous (D$_1$) limestones in north Derbyshire, with special reference to the conditions of deposition: Mercian Geologist, v. 3, p. 283–290.
SCHENK, P. E., 1970, Meguma Group (lower Paleozoic), Nova Scotia, in LAJOIE, J. (ed.), Flysch sedimentology in North America: Geol. Soc. Canada Special Paper 7, p. 127–153.
SCHOEFFLER, J., 1965, Le "Gouf" de Capbreton, de l'Eocène inférieur a nos jours, in WHITTARD, W. F., AND BRADSHAW, R. (eds.), Submarine geology and geophysics: London, Butterworth and Co., p. 265–270.
SCOTT, K. M., 1966, Sedimentology and dispersal pattern of a Cretaceous flysch sequence, Patagonian Andes, southern Chile: Am. Assoc. Petroleum Geologists Bull., v. 50, p. 72–107.
SEDIMENTATION SEMINAR (Indiana Univ.), 1969, Bethel Sandstone of western Kentucky and south-central Indiana: Kentucky Geol. Survey, ser. 10, Rept. Inv. 11, p. 7–24.
SHEPARD, F. P., AND DILL, R. F., 1966, Submarine canyons and other sea valleys: Chicago, Rand McNally and Co., 381 p.
———, ———, AND VON RAD, U., 1969, Physiography and sedimentary processes of La Jolla submarine fan and fan-valley, California: Am. Assoc. Petroleum Geologists Bull., v. 53, p. 390–420.
SHIRLEY, J., AND HORSFIELD, E. L., 1940, The Carboniferous Limestone of the Castleton-Bradwell area, north Derbyshire: Geol. Soc. London Quart. Jour., v. 96, p. 271–299.
SHORT, K. C., AND STÄUBLE, A. J., 1967, Outline geology of Niger delta: Am. Assoc. Petroleum Geologists Bull., v. 51, p. 761–779.
SHROCK, R. R., 1957, New geological horizons: ibid., v. 41, p. 1403–1408.
SIMPSON, I. M., AND BROADHURST, F. M., 1969, A boulder bed at Treak Cliff, north Derbyshire: Yorkshire Geol. Soc. Proc., v. 37, p. 141–151.
SMITH, A. G., 1971, Alpine deformation and the oceanic areas of the Tethys, Mediterranean, and Atlantic: Geol. Soc. America Bull., v. 82, p. 2039–2070.
STANLEY, D. J., 1964, Large mudstone-nucleus sandstone spheroids in submarine channel deposits: Jour. Sed. Petrology, v. 34, p. 672–676.
———, 1967, Comparing patterns of sedimentation in some modern and ancient submarine canyons: Earth and Planetary Sci. Letters, v. 3, p. 371–380.
———, AND BOUMA, A. H., 1964, Methodology and paleogeographic interpretation of flysch formations: a summary of studies in the Maritime Alps, in BOUMA, A. H., AND BROUWER, A. (eds.), Turbidites: Amsterdam, Elsevier Pub. Co., Developments in Sedimentology 3, p. 34–64.
———, AND MUTTI, E., 1968, Sedimentological evidence for an emerged land mass in the Ligurian Sea during the Palaeogene: Nature, v. 218, p. 32–36.
———, AND UNRUG, R., 1972, Submarine channel deposits, fluxoturbidites and other indicators of slope and base-of-slope environments in modern and ancient marine basins, in RIGBY, J. K., AND HAMBLIN, W. K. (eds.), Recognition of ancient sedimentary environments: Soc. Econ. Paleontologists and Mineralogists Special Pub. 16, p. 287–340.
STARKE, G. W., 1956, Genesis and geologic antiquity of the Monterey Submarine Canyon (abs.): Geol. Soc. America Bull., v. 67, p. 1783.
SULLWOLD, H. H., 1960, Tarzana Fan, deep submarine fan of late Miocene age, Los Angeles County, California: ibid., v. 44, p. 433–457.
———, 1961, Turbidites in oil exploration, in PETERSON, J. A., AND OSMOND, J. C. (eds.), Geometry of sandstone bodies: Tulsa, Oklahoma, Am. Assoc. Petroleum Geologists, p. 63–81.
TRETTIN, H. P., 1970, Ordovician-Silurian flysch sedimentation in the axial trough of the Franklinian Geosyncline, northeastern Ellesmere Island, Arctic Canada, in LAJOIE, J., (ed.), Flysch sedimentology in North America: Geol. Soc. Canada Special Paper 7, p. 13–35.
UNRUG, R., 1963, Istebna Beds—a fluxoturbidity formation in the Carpathian flysch: Soc. Géol. Pologne Annales, v. 33, p. 49–92.

VAN HOORN, B., 1969, Submarine canyon and fan deposits in the Upper Cretaceous of the south-central Pyrenees, Spain: Geol. en Mijnb., v. 48, p. 67–72.
———, 1970, Sedimentology and palaeogeography of an Upper Cretaceous turbidite basin in the south-central Pyrenees, Spain: Leidse Geol. Meded., v. 45, p. 73–154.
VON DER BORCH, C. C., 1968, Southern Australian submarine canyons: their distribution and ages: Marine Geology, v. 6, p. 267–279.
WALKER, R. G., 1966a, Shale Grit and Grindslow Shales: transition from turbidite to shallow water sediments in the Upper Carboniferous of northern England. Jour. Sed. Petrology, v. 36, p. 90–114.
———, 1966b, Deep channels in turbidite-bearing formations: Am. Assoc. Petroleum Geologists Bull., v. 50, p. 1899–1917.
———, 1969, The juxtaposition of turbidite and shallow-water sediments: study of a regressive sequence in the Pennsylvanian of north Devon, England: Jour. Geology, v. 77, p. 125–143.
WARREN, P. T., 1962, The petrography, sedimentation and provenance of the Wenlock rocks near Hawick, Roxburghshire: Edinburgh Geol. Soc. Trans., v. 19, p. 225–255.
WEZEL, F. C., 1968, Osservazioni sui sedimenti dell'Oligocene-Miocene inferiore della Tunisia settentrionale: Soc. Geol. Italiana Mem., v. 7, p. 417–439.
WHITAKER, J. H. McD., 1962, The geology of the area around Leintwardine, Herefordshire: Geol. Soc. London Quart. Jour., v. 118, p. 319–351.
———, 1963, The geology of the area around Leintwardine, Herefordshire: discussion: *ibid.*, v. 119, p. 513–514.
———, in press, Submarine canyons and deep-sea fans, modern and ancient: Stroudsburg, Pennsylvania, Dowden, Hutchinson and Ross.
WILHELM, O., AND EWING, M., 1972, Geology and history of the Gulf of Mexico: Geol. Soc. America Bull., v. 83, p. 575–600.
WILSON, J. T., 1966, Did the Atlantic close and then re-open? Nature, v. 211, p. 676–681.
ZIEGLER, A. M., 1970, Geosynclinal development of the British Isles during the Silurian Period: Jour. Geology, v. 78, p. 445–479.

ANCIENT SUBMARINE CANYONS OF THE CARPATHIAN MIOGEOSYNCLINE

FRANTIŠEK PÍCHA[1]
Geological Survey of Czechoslovakia, Prague

ABSTRACT

Many submarine canyons and channels have been described in modern seas, but little conclusive evidence about their existence has been found in ancient rock series. Like the modern submarine canyons, the ancient counterparts were situated on unstable edges of continents. During their further geologic history, they usually underwent extensive tectonic and erosional destruction or were buried below younger sediments.

The western part of the Carpathian Flysch Belt in the territory of Czechoslovakia in central Europe is one of the convenient places where the critical zone between the platform and the former geosynclinal trough can be studied. Among the most interesting contributions of this investigation is the discovery of two large buried depressions described as Nesvačilka (N) and Vranovice (V) grabens (fig. 1). Their existence was proved both by geophysical measurements and drilling operations. The depressions, traditionally regarded as tectonic structures, however, show many similarities with modern submarine canyons.

The depressions are cut in Paleozoic and Mesozoic carbonate rocks covering the crystalline complexes of the Bohemian Massif. They are filled with Eocene and Oligocene deposits and overlain by Neogene sequences of the Carpathian foredeep. These autochthonous formations deposited on marginal sectors of the platform dip below the Carpathian flysch nappes that comprise Cretaceous and Paleogene miogeosynclinal series (fig. 1). The longitudinal axes of the depressions are oriented in a northwest-southeast direction perpendicular to the margin of the platform. The canyons probably join each other farther downdip to form one channel system not unlike the Scripps and La Jolla Canyons along the coast of California. At their upper ends, the depressions are surrounded by steep, high walls resembling the heads of modern canyons. The thickness of the sedimentary fill near the head of the Nesvačilka Depression exceeds one thousand meters. Even though this thickness does not wholly correspond to the original depth, it indicates the rugged relief of the structure. Downward the canyons become shallower and their walls less steep. The measured width of the depressions varies from about 2 km at their heads to as much as 7 km in their distal parts. Both structures have been followed for a distance of about 25 km, their further courses being hidden beneath 4- to 7-km-thick flysch and molasse sequences.

The Eocene and Oligocene sediments filling the depressions are composed predominantly of dark-brown calcareous silty shales rich in organic matter. They contain abundant planktonic microfauna, proving the marine origin of these sediments. The shales are intercalated with laminae and thin beds of siltstones and fine sandstones. The lamination is a common structure present elsewhere. Scour-and-fill structures of small size found in many sandstone beds indicate the activity of strong, erosional, bottom-seeking currents sweeping through the canyons. No turbidites showing the characteristic succession of internal structures have been identified, but, because of the small number of drill cores available, their presence cannot be excluded. Thick beds of massive sandstones and boulder conglomerates were found at the bottom of the Nesvačilka Depression.

Though the oldest deposits found inside the canyons are of late Eocene age, such huge structures should have originated much earlier. Redeposition of Cretaceous and Paleocene microfauna indicates that the canyons could have been active during the entire time of existence of the flysch miogeosyncline from Cretaceous to Oligocene. The early stage of canyon development was characterized by erosion and transportation of material but was followed during Oligocene time by sedimentation inside the canyons, which brought about the end of their development. The origin of submarine canyons has not been explained satisfactorily, though several hypotheses have been submitted. The Carpathian canyons, oriented parallel to the fault system of the platform, are believed to be of combined tectonic and erosional origin. They intersected the shelf and entered the former geosynclinal flysch trough, which, considering the morphology of the canyons, must have been at least one thousand meters deep.

There is an open question what role the canyons played in geosynclinal sedimentation. According to the commonly accepted paleogeographical concept, the flysch trough of the Western Carpathians was supplied predominantly from internal sources (cordilleras), while the platform yielded mostly only fine pelitic material. The distribution of sandy and shaly facies and the composition of clastics apparently support this hypothesis. Modern oceanographic explorations show, however, that canyon-derived sediments can be transported long distances before finally being deposited. In the San Diego Trough, for example, the coarser sediments accumulate at the distant oceanic side of the basin far from the mouth of supplying canyons along at the coast of California (Shepard, Dill, and Rad, 1969). Also, in the Carpathian Trough the sandy facies were not necessarily related to some nearby sources such as cordilleras. The potential existence of large submarine canyons at the continental side of the Carpathian flysch trough, serving as conduits for flysch sediments, therefore must be taken into consideration in any paleogeographical reconstruction.

[1] Present address, Department of Geology, Kuwait University, Kuwait.

REFERENCE

SHEPARD, F. P., DILL, R. F., AND RAD, ULRICH VON, 1969, Physiography and sedimentary processes of La Jolla submarine fan and fan-valley, California: Am. Assoc. Petrol. Geologists Bull. v. 53, p. 390–420.

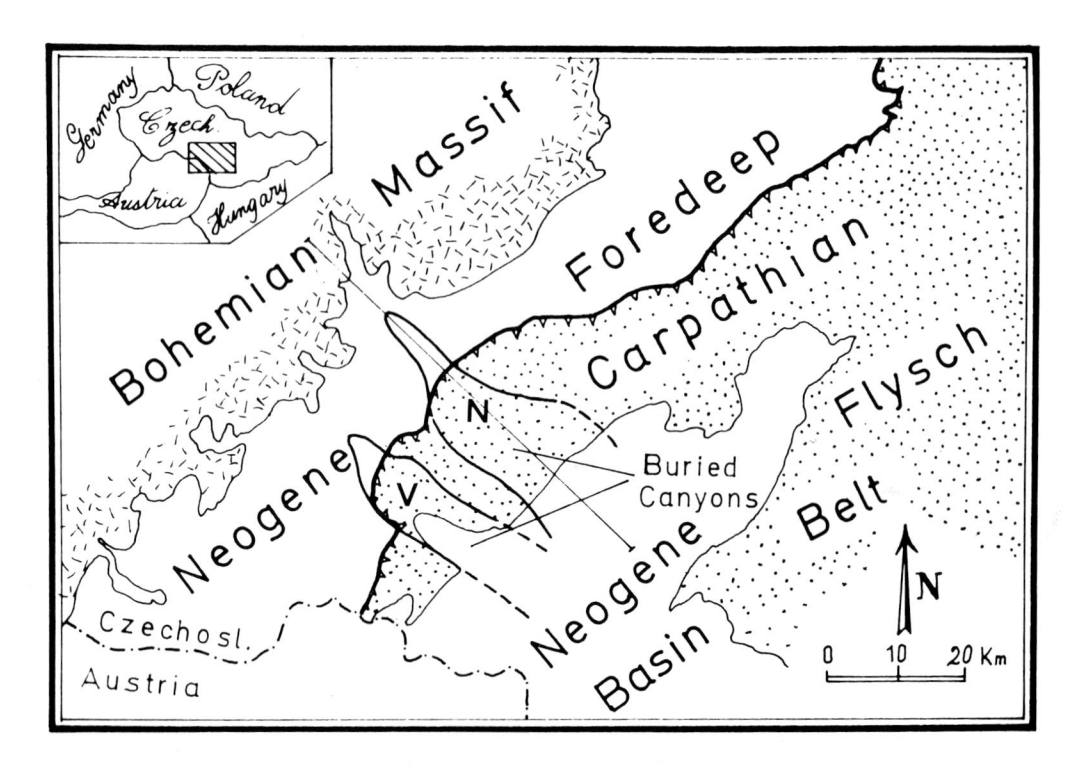

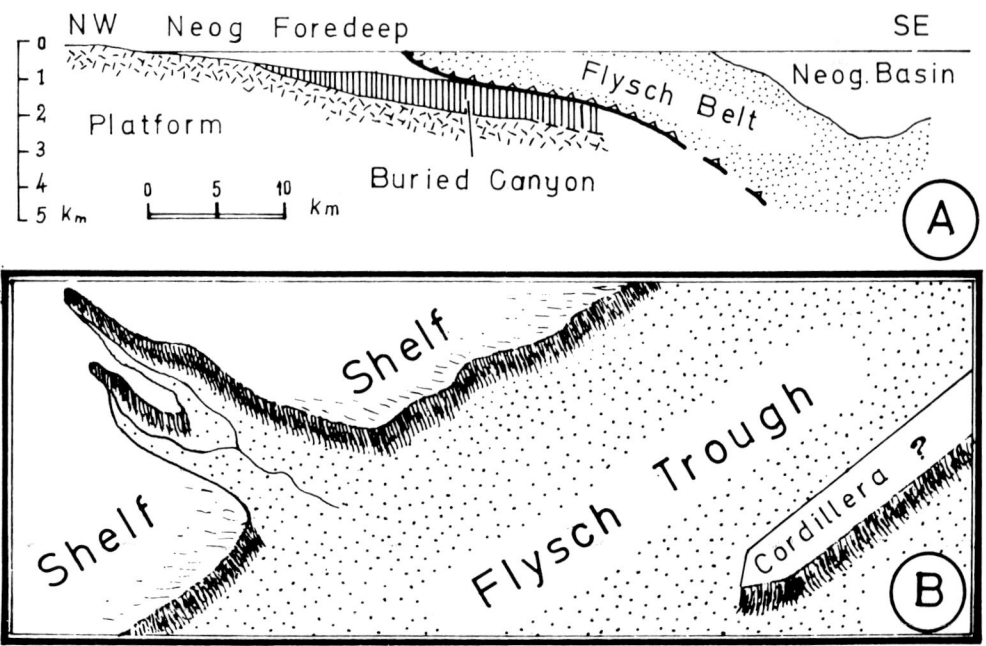

Fig. 1.—Map and section showing positions of buried canyons in Western Carpathians (A). Flysch belt, thrusted over the platform deposits, was originally deposited in the miogeosyncline situated southeastward as reconstructed in drawing B.

DEEP-SEA PELAGIC SEDIMENTS AND OPHIOLITE ASSEMBLAGES

THE ODD ROCKS OF MOUNTAIN BELTS

A. G. FISCHER
Princeton University, Princeton, New Jersey

ABSTRACT

The earth's great mountain belts contain sediments and volcanic complexes of kinds that are well known in the stable cratonic regions and of kinds that are exposed only (or almost only) in the mountain belts. The latter kinds include ophiolites, great sequences of pillow basalts and (or) andesites, bedded, generally radiolarian cherts, uniform shale sequences containing pelagic biotas and representing long time spans, pelagic carbonate sequences, great bodies of turbidite flysch, olistostromes, and metamorphic rocks that are not dealt with here. The sediments are commonly red colored.

For a century, explanations for these rocks have been sought along two different lines (paradigms): In one they are viewed as the products of mobile belts and orogeny—specifically, as the contents of a special generative trough or complex of troughs, the eugeosyncline of Hans Stille and of Marshall Kay. In the other, which we may term the oceanic model and which was stated most clearly by R. S. Dietz and J. C. Holden, they are viewed as the normal rocks of the oceans and oceanic margins brought to the mobile belts by plate subduction and incorporated into the continental margins by orogeny.

Progressive exploration of the oceans, first by dredging and coring and more lately by drilling, has shown that most of these rocks indeed correspond closely to the rocks of the great ocean floors. This seems to be particularly true when sediments of the same ages are compared.

Only the great andesite sequences and olistostromes remain as rock types linked to mobile belts or convergent plate margins. A good case can be made for a distant oceanic derivation of some ophiolite-sediment sequences in mountains (e.g., Elba), and therewith the eugeosynclinal model seems to have outlived its usefulness.

At the same time, a model substituting the main ocean basins for the eugeosyncline is even less realistic. Through the work of D. E. Karig and others, confirmed by legs 6 and 7 of the Deep Sea Drilling Project, it has become essentially certain that juvenile oceanic or semioceanic crust overlain by oceanic type sediment can grow in mobile belts to produce the interarc and marginal or small-ocean basins. Yet other sites of crustal growth are likely (e.g., Canary Islands). The ophiolites and sediments of such sites have a better chance of unmetamorphosed incorporation into mountains than does the main sea floor. The ophiolite complexes of Cyprus (Mamonia, Troodos) and of Kandahar, Afghanistan, should be viewed in this light.

Furthermore, the odd sediments of the mountain belts commonly form parts of otherwise normal or miogeosynclinal successions (e.g., Martinsburg flysch, Jurassic sediments of Northern Limestone Alps), and the great andesite sequences occur either on oceanic crust (Antilles) or on sial (Andes). Olistostromes are most likely trench generated.

Thus, in principle, the odd rocks of the mountain belts represent a wide range of generative settings, including crust-generating ocean ridges and their flanks, abyssal plains, trenches, volcanic arcs, tectonic welts, interarc basins, and the downwarped edges of continents. Tectonic and paleogeographic reconstructions will remain extremely uncertain until we have learned to recognize these constituents more clearly.

ALPINE, MEDITERRANEAN, AND CENTRAL ATLANTIC[1] MESOZOIC FACIES IN RELATION TO THE EARLY EVOLUTION OF THE TETHYS

DANIEL BERNOULLI AND HUGH C. JENKYNS
Universität Basel, Switzerland; University of Durham, Great Britain

ABSTRACT

The main Alpine-Mediterranean Mesozoic lithofacies, excluding flysch, are outlined and placed in their paleogeographic setting within the broader context of an evolving ocean basin. In the external (ophiolite-free) zones of the Alpine-Mediterranean orogen, Mesozoic pelagic facies almost invariably overlie kilometers-thick successions of Bahamian-type platform carbonates. Wherever the basement of these platform carbonates is exposed, it is continental, comprising low- to high-grade metamorphic rocks and granites. We suggest that the pelagic sediments of these zones were deposited on a deeply submerged continental margin of the Atlantic type. Palinspastic reconstructions of the central Alpine-Mediterranean area place the depositional setting of most of these pelagic facies on the southern continental margin of the Tethys. In this area supply of clastics and organic matter was minimal, thus encouraging pelagic conditions.

The earliest pelagic sediments of the Alpine-Mediterranean region are of Triassic age and comprise gray and red limestones or cherts that commonly are associated with volcanics. These sediments were deposited in embayments and basins between extensive carbonate platforms and reefs. During the Liassic Epoch, a phase of block faulting, probably related to rifting in the oceanic Tethys, destroyed many of these shallow-water sites, and pelagic conditions became more widespread. During the Jurassic Period a basin-swell morphology was produced by irregular subsidence of the different blocks. On submarine highs, or seamounts, the following stratigraphically condensed facies were developed: pisolitic ironstones, red biomicrites containing ferromaganese nodules and crusts, crinoid-pelagic bivalve-gastropod-ammonite biosparites, pelagic pelmicrites and micro-oncolitic sparites, and certain red, fine-grained, nodular limestones. In the neighboring basins, more expanded successions containing slumped blocks and turbidites accumulated; the basinal facies were developed as red, more clay-rich, nodular limestones, gray limestone-marl interbeds, radiolarian cherts, and white nannofossil limestones. The Cretaceous Period saw a smoothing of submarine topography and a general deepening of the water as the continental margin continued to founder. Deposition of varicolored marls and red and white coccolith limestones was widespread.

True ocean-floor lithofacies are represented by those rocks associated with, or lying upon, ophiolites. In the central Mediterranean area they comprise ophicalcites, umbers, radiolarites, white nannofossil limestones, and black shales. Their age is Jurassic and Cretaceous.

In the western central Atlantic pelagic facies of Late Jurassic and Cretaceous ages occur. These facies resemble both the continental margin and ocean-crust lithologies of the Alpine-Mediterranean Tethys. A section through the Mesozoic portion of this undeformed continental margin and ocean-basin complex comprising the Bahamas, the inter-platform straits, and oceanic realm illustrates a paleogeographic arrangement that strongly resembles the reconstructed section for the Alpine-Mediterranean Tethys. This resemblance illustrates the parallel evolution of these two now widely separated areas, so that they both can be considered as representatives of an east-west Mesozoic seaway, or Tethyan realm, that stretched from the Caribbean to Indonesia.

INTRODUCTION

Since publication of the papers of Fuchs (1883), Haug (1900), Steinmann (1905, 1925), and Heim (1924), certain European geologists have thought that among the Mesozoic sediments preserved in the Alps and Mediterranean region (figs. 1 and 2) were deep-sea deposits comparable to those documented by Murray and Renard (1891) in the reports of the *Challenger* expedition. The presence of deep-sea deposits on land necessarily cast doubt upon the concept of permanency of continents and ocean basins, a concept to which at that time most of the Anglo-American geological fraternity adhered. Because they contravened the current dogma, and possibly because they did not write in English, the central European workers made little headway against the mainstream of extra-Alpine geological thought. More recently, some workers of Alpine heritage (e.g., Trümpy, 1960, 1970; Garrison and Fischer, 1969) have once more championed the deep-water hypothesis; these authors have again stressed the great similarities between Alpine Mesozoic sediments and Recent abyssal deposits. Other recent work has, however, spotlighted evidence of photic deposition for certain of these pelagic facies (Szulczewski, 1963, 1966; Radwański and Szulczewski, 1965; Wendt, 1969a, 1970; Jenkyns, 1971a; Sturani, 1971; our fig. 3), and it is apparent that within the Mesozoic of the Alpine-Mediterranean orogen there is a suite of relatively shallow-water sediments that closely mimic deep-sea deposits. It is these pelagic facies, both the relatively shallow-water and more basinal types

---

[1] We have used "central Atlantic" to denote the oceanic region lying between Africa and North America, thus avoiding the more ambiguous "North Atlantic."

often associated with a characteristic fauna, that generally are referred to as "Alpine," "Mediterranean," or "Tethyan." We attribute them to both continental-margin and oceanic settings (fig. 2).

The Tethyan facies comprise a variety of lithologies, the most renowned of which are red nodular limestones and marls known as *Ammonitico Rosso* or *Knollenkalk*. It was to this kind of rock type that Arkell (Warman and Arkell, 1954) referred when he wrote the following on the Jurassic of western Sicily: "... all the Bathonian, Callovian and Oxfordian faunas occur in highly condensed deposits. Current text-books hardly prepare one for finding all these stages and substages in a few metres of rock in the middle of the 'Mediterranean Geosyncline.' None of these deposits is, in fact, in any way geosynclinal." Arkell, therefore, was inclined to dispense altogether with the term geosynclinal for these particular deposits. His comment underlines the fact that Tethyan facies do not match classical geosynclinal models.

To Trümpy (1960) certain of these Jurassic red limestones, together with the stratigraphically associated radiolarian cherts and white nannofossil limestones, represented a so-called "leptogeosynclinal" formation deposited during a preorogenic, sediment-starved bathyal lull (term from Goldring, 1961). In an earlier paper, Trümpy (1955) cited the Triassic pelagic Hallstatt facies of the Austrian Alps as a leptogeosynclinal deposit.

Aubouin (1965) wrote the following on the paleogeographic significance of Ammonitico Rosso: "... the *Ammonitico Rosso* facies may characterize certain zones, or parts of zones, during the *generative stage* or the *development stage* of the geosynclinal period. This is not represented in the orogenic stage of the same period, even less in the late- and post-geosynclinal periods."

Clearly the condensed nature of certain of these facies, reflecting reduced sedimentary supply and sometimes coupled with slow subsidence rates, has caused some authors to hesitate to label these rocks as geosynclinal (see discussion in Trümpy, 1955). Recognition of spatially related stratigraphic equivalents of some of these condensed facies, equivalents whose thicknesses range up to 4000 meters in the lower to middle Lias alone (Bernoulli, 1964), suggests that too much significance should not be placed on thinness as an objection to geosynclinal interpretation. Furthermore, the white nannofossil limestone may grade into terrigenous flysch, which is more typical of geosynclinal terranes (Durand-Delga, 1960; Blanchet and others, 1970; de Booy, 1969). There seems no doubt, therefore, that these Mesozoic pelagic facies of the Tethys can be considered characteristic of the early evolution of the Alpine-Mediterranean geosyncline.

We propose to describe the most important pelagic facies of the Tethyan continental margins and to outline their depositional setting and paleogeographical significance. Furthermore, we wish to point to those sediments associated with ophiolites as true representatives of ocean-floor facies. The distinction between these very similar sedimentary associations is most important. In the past they often have been confused because some lithologies occur in both settings, which must explain, in part, the longevity of the deep water-shallow water dispute.

Jurassic and Cretaceous pelagic sediments from JOIDES boreholes in the western central Atlantic (Leg XI) are here examined with reference to continental-margin and ocean-floor facies of the Tethyan geosyncline (fig. 2). The paleogeographic setting of these cores relative to the Atlantic margin in turn throws light on conditions that prevailed in the Mesozoic Tethys.

TECTONIC POSITION OF MESOZOIC FACIES IN THE ALPINE-MEDITERRANEAN REGION

The reconstruction of preorogenic paleogeography in the Alpine-Mediterranean belt depends primarily on the genetic significance of the ophiolite assemblages (fig. 1). That they represent slivers of oceanic crust and lithosphere, hinted at by Steinmann (1905), and later suggested by Hess (1965), has been confirmed by the work of Moores (1969), Moores and Vine (1971), and Bezzi and Piccardo (1971) on petrology and gross structure, and by Pearce and Cann (1971) and Bickle and Nisbet (1972) on trace elements. This view, which for the central Mediterranean area is now commonplace (Decandia and Elter, 1969; Laubscher, 1969), implies allochthony for the ophiolites in contrast to classical geosynclinal concepts (Aubouin 1965). Recent field work in the western Alps (Elter, 1971; Lemoine and others, 1970), the Apennines (Decandia and Elter, 1969), and in Greece (Bortolotti, Dal Piaz, and Passerini, 1969; Hynes and others, 1972; Zimmerman, 1972), has demonstrated that these ultramafic rocks have been emplaced tectonically on their substratum.

The Tethyan ophiolite zone can be followed from Turkey to Greece and northwestern Yugoslavia, where it is offset by a complex system of early to late Alpine strike-slip faults (fig. 1; Laubscher, 1971a). In the Eastern Alps it is covered by the north-thrusted Austro-Alpine nappes, but its remnants appear in tectonic windows. Ophiolites occur also in the south-Pennine zone of the Western Alps and in the

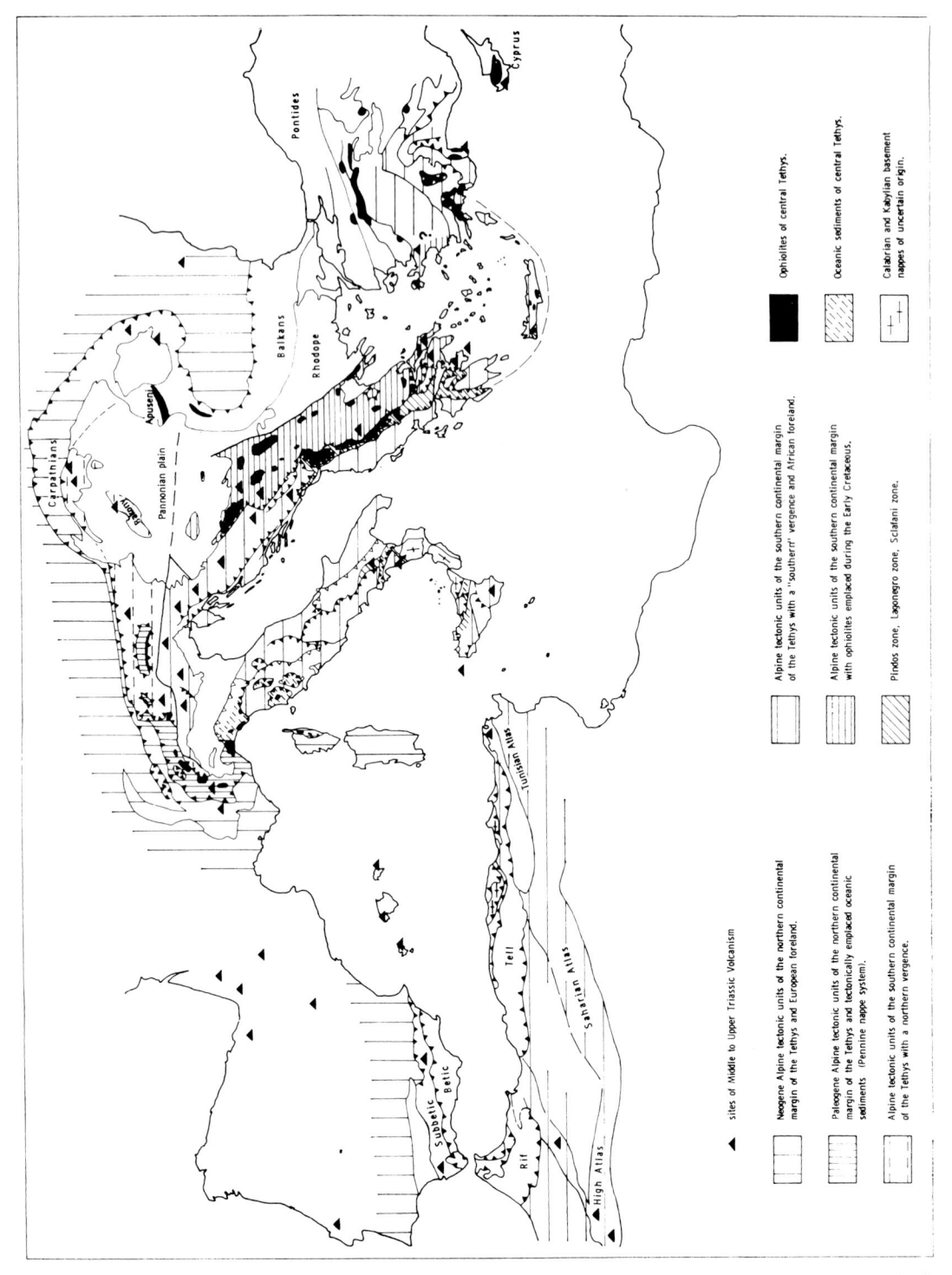

Fig. 1.—Alpine tectonic units of the Alpine-Mediterranean area and present-day distribution of ocean basin and continental-margin complexes. Note that many of the units formed originally on the *southern* margin of the Tethys, although they now lie on the northern side of the Mediterranean Sea. Mainly after Bernoulli and Laubscher (1972), Laubscher (1971a, 1971b). Data on Triassic volcanism from Arkell (1956), Beauseigneur and Rangheard (1967), Bellair (1948), Ćirić and Kamarata (1960), Cristofolini (1966), Cros and Lagny (1969), Dietrich and Scandone (1972), Giunta (personal commun.), Guernet (1965), Hoppe (1968), Jacobshagen (1972), Joja and others (1968), Leonardi and others (1967), Mahel, Buday, and others (1968), Marcoux (1970), Milch and Renz (1911), Pantò (1961), Paquet (1969), Rocci and Lapierre (1969), Roch (1950), Römermann (1968), Schneider (1964), Schönenberg (1971), Sellwood (personal commun.), Terry (1971), Trümpy (1960), and Winnock (1971).

Ligurian nappes of the northern Apennines. In the southern Apennines, olistoliths and pillow basalts associated with oceanic sediments testify to the former presence of ophiolites. West of Sicily the continuation of the Mesozoic oceanic realm is still questionable. Many of the ultramafic bodies in the western Mediterranean may in fact be intrusive representatives of abortive oceans rather than being tectonically emplaced slices (Loomis, 1972; Nicolas and Jackson, 1972).

Palinspastic reconstructions of the Alpine-Mediterranean place the external[2] (ophiolite-free) zones of the Taurides, Hellenides, Dinarides, and Apennines, the Austro-Alpine nappes, and the internal Carpathians on the southern continental margin of the Tethys. (For the complex kinematics see Laubscher, 1970, 1971a, 1971b; Bernoulli and Laubscher, 1972.) The Sicilian, north African, and Betic mountain chains are most probably the continuation of these peri-Adriatic zones, but the origin of crystalline basement nappes included in them cannot be established with certainty (fig. 1). The northern continental margin comprises the European foreland and the external zones of the Alpine-Carpathian orogen that are still connected to the margin (fig. 1). In this paper we shall limit ourselves primarily to an illustration of the sediments deposited on the former southern continental margin and in the adjacent Tethyan ocean basin.

During the different phases of Alpine orogeny, most if not all the Tethyan ocean had been eliminated, and the continental margins were greatly affected by folding and faulting. Some orogenic movements took place during, or even before, the Early Cretaceous (Mercier, 1966). Such movements along the southern margin of the Tethys are documented by the pre-Albian emplacement of the ophiolite nappe in eastern Greece and Yugoslavia (e.g., Moores, 1969; Bernoulli and Laubscher, 1972; Hynes and others, 1972). There is abundant circumstantial evidence, however, that coeval orogeny occurred elsewhere in the internal zones of the Alpine-Mediterranean chains (Laubscher, 1970). Later, during the Paleogene and the Neogene, new discordant tectonic zones developed externally, integrating new segments of the foreland into the orogen. Contrary to the classical geosynclinal concept of embryonic tectonics (Argand, 1916; Aubouin, 1965), these Alpine structures generally trend obliquely and in some examples at right angles to older paleogeographic units (Trümpy, 1960). Consequently, the Mesozoic pelagic sediments here described now occur in several different tectonic settings: (1) in sedimentary cover nappes detached from oceanic or continental basement, (2) as sedimentary cover of continental or oceanic basement nappes, (3) as olistoliths (slide blocks) or *klippes sédimentaires* (Broquet and others, 1966) associated with orogenic flysch deposits or as isolated fragments in tectonic mélanges, and (4) as fragments of autochthonous foreland that escaped the Miocene to Quaternary break-down of the present Mediterranean Sea (e.g., the Apulian Platform). Examples of different genetic sequences are illustrated in figure 2.

MESOZOIC FACIES OF THE ALPINE-MEDITERRANEAN TETHYS

Continental Margin Deposits

Carbonate-platform facies.—Underlying many of the pelagic sequences of the southern continental margin of the Tethys are thick successions of shallow-water limestones and dolomites, mostly of Triassic and Early Jurassic age (fig. 2). Such sequences comprise stromatolitic, oolitic, and pelletal sediments and commonly exhibit birdseye structure and shrinkage cracks.

---

[2] "External" and "internal" are relative terms used to describe the original positions of segments in an orogenic belt with respect to continent (foreland) and ocean. Here we use "external zones" to denote the foreland and more stable parts of the continental margin that were affected only by Tertiary nappe movements. "Internal zones" denotes the ocean-derived units of the orogen and the oceanward parts of the continental margin where ophiolites were emplaced during Cretaceous time.

FIG. 2.—Mesozoic genetic sequences of the Alpine-Mediterranean and western central Atlantic regions. *1*, Serra del Prete, Southern Limestone Apennines, Italy (from Sartoni and Crescenti, 1962); *2*, Pizzo Cefalone and Monte Portella, Gran Sasso d'Italia, Abruzzi, Italy (from Bernoulli, 1967; Crescenti, 1969; Crescenti and others, 1970; and personal observations); *3*, Ionian zone, western Greece, combination of sections: Kouklessi (Bernoulli and Renz, 1970), Terovo (Institut de Géologie, etc., 1966), and Arakhtos (Institut de Géologie, etc., 1966); *4*, Rocce Maranfusa, western Sicily (from Jenkyns, 1970b,c); *5*, Othrys, eastern Greece (from Hynes and others, 1972); *6*, Bracco unit, Ligurian Apennines (from Elter and others, 1966; Decandia and Elter, 1969); *7*, Borehole, Bahamas Oil Co. No. 1 Andros Island (from Spencer, 1967; Goodell and Garman, 1969; Paulus, 1972). *#98, #99A–101A, #105,* JOIDES boreholes (from Hollister, Ewing, and others, 1972).

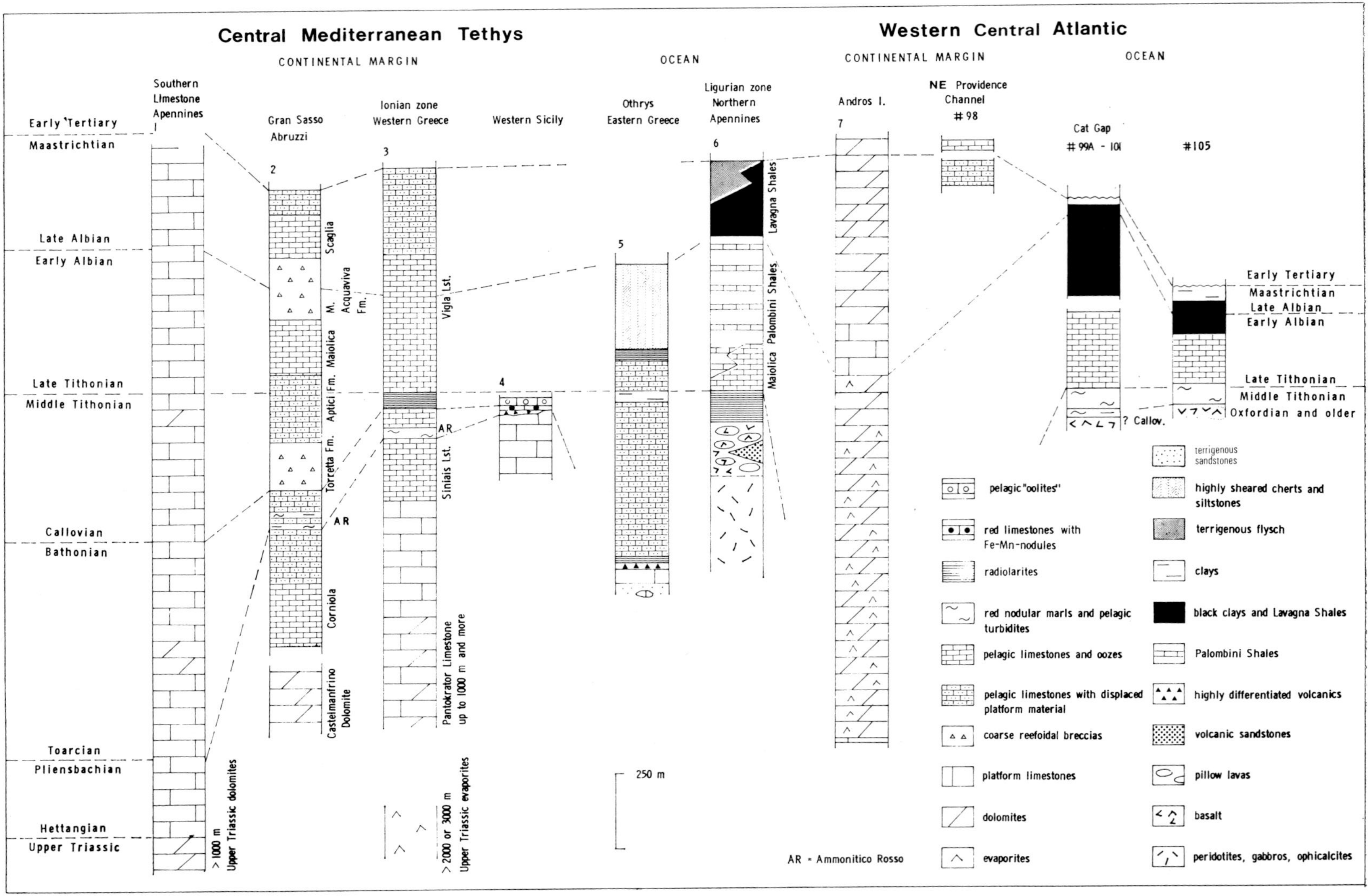

Central Mediterranean Tethys
CONTINENTAL MARGIN
OCEAN
Western Central Atlantic
CONTINENTAL MARGIN
OCEAN
Early Tertiary
Maastrichtian
Late Albian
Early Albian
Late Tithonian
Middle Tithonian
Callovian
Bathonian
Toarcian
Pliensbachian
Hettangian
Upper Triassic
Southern Limestone Apennines
1
> 1000 m
Upper Triassic dolomites
Gran Sasso Abruzzi
2
Scaglia
Acquaviva Fm.
M.
Maiolica
Aptici Fm.
Torretta Fm.
AR
Corniola
Castelmanfrino Dolomite
Ionian zone Western Greece
3
Vigla Lst.
AR
Siniais Lst.
Pantokrator Limestone up to 1000 m and more
> 2000 or 3000 m
Upper Triassic evaporites
Western Sicily
4
Othrys Eastern Greece
5
Ligurian zone Northern Apennines
6
Lavagna Shales
Maiolica
Palombini Shales
Andros I.
7
NE Providence Channel
98
Cat Gap
99A - 101
#105
? Callov.
Early Tertiary
Maastrichtian
Late Albian
Early Albian
Late Tithonian
Middle Tithonian
Oxfordian and older
250 m
AR = Ammonitico Rosso
pelagic "oolites"
red limestones with Fe-Mn-nodules
radiolarites
red nodular marls and pelagic turbidites
pelagic limestones and oozes
pelagic limestones with displaced platform material
coarse reefoidal breccias
platform limestones
dolomites
evaporites
terrigenous sandstones
highly sheared cherts and siltstones
terrigenous flysch
clays
black clays and Lavagna Shales
Palombini Shales
highly differentiated volcanics
volcanic sandstones
pillow lavas
basalt
peridotites, gabbros, ophicalcites

Gastropods and calcareous algae are common, bivalves occur sporadically, and ammonites are very rare indeed. Zones representing emergence are indicated by red and green clay-rich material that locally is associated with vadose pisolites (Bernoulli and Wagner, 1971), by more extensive bauxitic deposits (d'Argenio, 1970a), by black shales that probably are marsh deposits, and by coal seams. Fresh-water limestones occur at some levels. Detailed sedimentology of these facies has been described by Fischer (1964), d'Argenio (1966), Bosellini and Broglio Loriga (1971), and by Pialli (1971). Associated with these supratidal to shallow subtidal carbonate rocks are less extensive reef deposits containing abundant corals, sponges, hydrozoans, and calcareous algae (Agard and du Dresnay, 1965; Fabricius, 1966; Zankl, 1969). The pattern of sedimentation, both reef and nonreef, compares extremely well with that of the present-day Bahamas and Florida Bay region, and the overall depositional setting was that of a carbonate platform.

The lower, Triassic levels of these carbonate-platform successions may be evaporitic, and it is upon such a mobile substratum that the successions were often sheared off during orogeny. In some areas, however, the carbonate successions can be traced downward, in places through marine and fresh-water sandstones, to a crystalline basement. The basement is invariably continental, comprising low- to high-grade metamorphic rocks and granites (e.g., Southern Alps, Bernoulli, 1964; Toscanide Nappes, Bortolotti and others, 1970; Durmitor Zone, Rampnoux, 1970; Tatra Mountains, Kotański, 1961; eastern Sicily, Truillet, 1970).

The inception of this carbonate-platform pattern of sedimentation was generally during the Middle or Late Triassic. Rapid subsidence of the depositional site was balanced by massive carbonate production, and thicknesses of one to several kilometers were sometimes built up before there was a change to pelagic facies. In areas where carbonate platforms were not formed or did not persist for any length of time, basinal areas resulted by default and were the sites of some of the earliest pelagic sedimentation. Such areas were limited in extent, however, until the Early Jurassic, when there was widespread destruction of the carbonate-platform areas (see fig. 7).

Of all the Alpine-Mediterranean carbonate platforms, those in the peri-Adriatic region displayed the most remarkable longevity, persisting from the Early Triassic to the Late Cretaceous and Tertiary. These resulted in sedimentary piles as much as 7 kilometers thick (Herak and others, 1970). Their areal extent was reduced during the Mesozoic, but platform morphology and environment were maintained until finally overwhelmed by orogenic flysch sedimentation (d'Argenio and others, 1971). Although widespread Late Triassic to early Liassic carbonate platforms of the Alpine-Mediterranean Tethys duplicated many present-day Bahamian environments, the similarity of the peri-Adriatic platforms to the present-day Florida-Bahamas platform goes far beyond environment. A strong similarity exists, not only in terms of facies, but of gross geometry, time-space relationships, and subsidence rates (d'Argenio, 1970b; Bernoulli, 1972). This underscores the parallel evolution of the western central Atlantic and Alpine-Mediterranean regions.

Pelagic facies.—Many of the pelagic sequences of the southern continental margin were deposited in settings spatially connected with carbonate platforms. In basins bordering such platforms, coarse boulder beds and graded breccias were deposited that pass laterally into graded and laminated calcarenites to calcisiltites. Components of such turbidites include skeletal fragments of neritic organisms, calcareous algae, oncolites, pellets, and intraclasts (Aubouin, 1959a,b; Flügel and Pölsler, 1965; Bernoulli, 1967; Garrison, 1967). Replacement chert is common in the turbidites. These platform-margin deposits closely resemble pelagic and re-sedimented shallow-water facies of the basins surrounding Bahamian carbonate platforms. We restrict ourselves to description of the true pelagic facies of Triassic, Jurassic, and Cretaceous ages.

Triassic: Included within this discussion are the Triassic (mostly Carnian and Norian) pelagic sediments of the Pindos, Lagonegro, and Sclafani Basins of Greece and Yugoslavia, southern Italy, and Sicily respectively. The ultimate basement of these elongated basins is unknown. The deepest outcrops generally comprise volcanics with sandstones, siltstones, argillites and marls (e.g., Carnian "flysch," so-called, of central Sicily and Anisian "flysch" of southern Yugoslavia). Such "flysch" may contain olistoliths of Paleozoic limestones and possibly pene-contemporaneously displaced reef material. Examples are found in southern Italy, Sicily, Yugoslavia, and possibly in Rhodes (Scandone, 1967; Donzelli and Crescenti, 1970; Mascle, 1967; Kochansky-Dévidé, 1958; Orombelli and Pozzi, 1967).

To a large extent Triassic pelagic sediments are exact analogues of those developed in the Jurassic. Three Triassic facies (a through c) are briefly surveyed.

(a) Red biomicrites. Stratigraphically con-

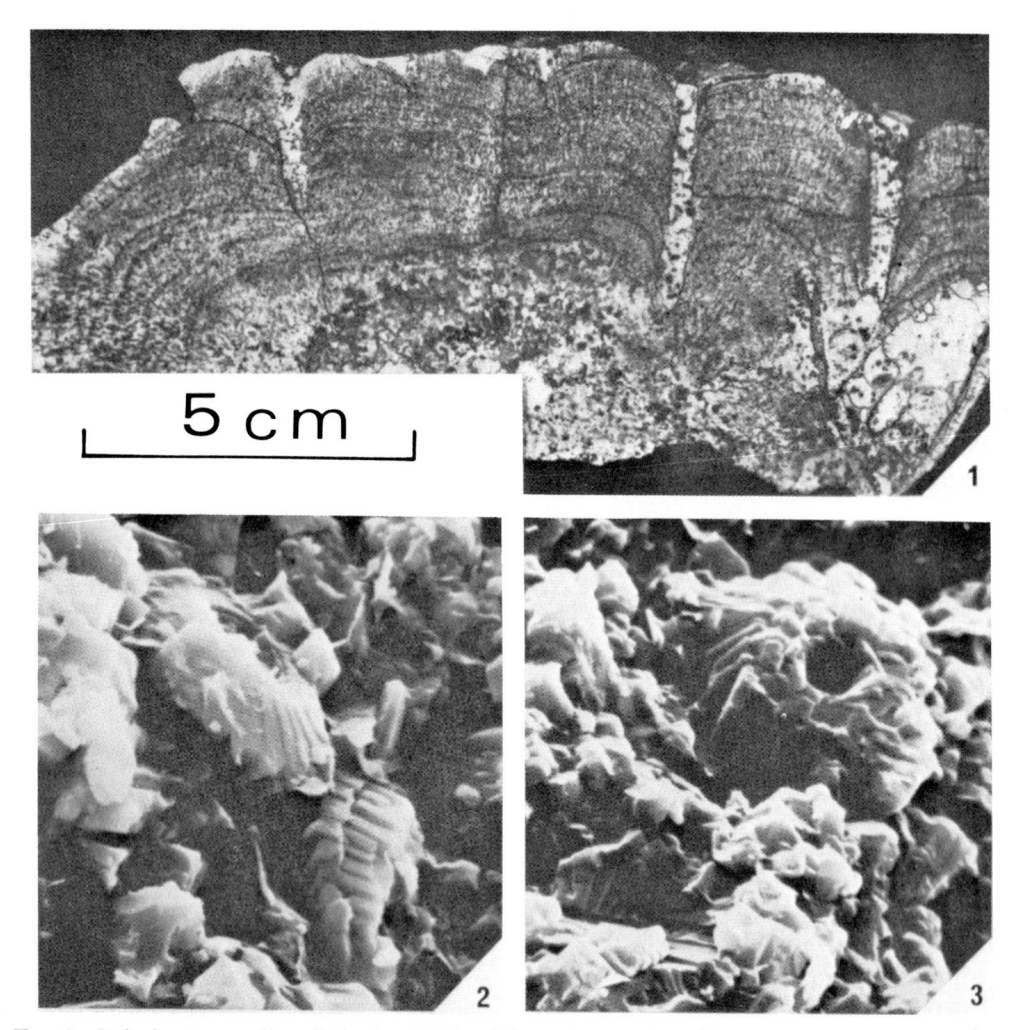

FIG. 3.—Pelagic stromatolite, Callovian, Villány Mountains, southern Hungary. *1,* Polished surface; note ammonite cross section at bottom right. *2,* Fine sediment trapped in the stromatolitic structure; note well-preserved coccoliths. *3,* Fine sediment caught between the stromatolitic columns; note coccoliths and their fragments. Figures 3-2 and 3 are scanning electron micrographs, gold-coated, ×6000.

densed red limestones occur in association with ferromanganese crusts or as marly nodular facies. Lenses of pelagic bivalves occur locally. These rocks are analogous to their Jurassic counterparts, containing similar faunas except that conodonts are present only in the Triassic rocks. Also, problematical nannofossils are present (Zankl ,1971).

Red limestones of this type have been interpreted as deposits of a starved basin surrounded by carbonate platforms (e.g., by Zankl, 1967; Schlager, 1969). Fischer (1964) suggested a deep-water environment; however, the interfingering of these pelagic facies with carbonate-platform deposits, together with the presence of boring algae, encrusting Foraminifera, and possibly algal stromatolites (cf. our fig. 3), suggest photic deposition (Wendt, 1969b; Krystyn and others, 1971). Consequently, Zankl (1971) suggested a water depth of 50 to 200 meters. As opposed to more expanded basinal facies, these red limestones probably were deposited on local sheltered highs bounded by normal faults.

These facies have been described in great detail from the type region of the Hallstatt Limestone in the Eastern Alps, Austria (see above references). They are also known, locally associated with volcanics, from the Carpathians, through the High Karst and Durmitor Zones of Yugoslavia, to the more internal zones of Greece (Patrulius, 1967; Ćirić, 1963; Jacobshagen, 1967).

(b) Light-colored limestones (biomicrites) and marls. These stratigraphically expanded facies comprise alternations of gray, pink, and yellow limestones and marls that commonly contain pelagic bivalves (*Daonella, Halobia*) and calcite-filled radiolarian molds (fig. 4–6); ammonites occur less commonly (Gemmellaro, 1904). Their basinal setting is clearly indicated by thickness greater than that of the condensed red limestones and by the presence of slumps and proximal to distal pelagic and carbonate-platform turbidites. Such facies occur in the Eastern Alps (Schlager, 1969) and in the Dinarides (Rampnoux, 1970) and are particularly widespread in the Pindos Zone of Greece, the Lagonegro Zone of southern Italy, and in the Sclafani Zone of Sicily (Aubouin, 1959a; Scandone, 1967; Broquet and others, 1966).

(c) Other facies. Radiolarites, black shales, and varicolored marls, commonly associated with volcanic sandstones, acid to basic tuffites, and basic lavas, occur in the Middle Triassic of the Southern Alps and can be traced through Yugoslavia, Greece, and western Turkey (Porphyrite-Hornstein Formation, Ćirić and Kamarata, 1960; fig. 4–5). In the Buchenstein and Wengen Formations of the Italian Dolomites, displaced carbonate-platform sediments and blocks occur in these facies (Leonardi and others, 1967).

Jurassic: Varied paleogeography ensured that the Jurassic Period was the time of maximum facies diversity in the Alpine-Mediterranean Tethys. Where pelagic conditions had been initiated during the Triassic, basinal environments generally resulted by Jurassic time, and radiolarites and displaced carbonate-platform material were deposited. Many of the Jurassic lithologies are gradational with one another, but nine (a through i) main divisions are recognizable.

(a) Red biomicrites containing less than 5 percent clay (fig. 5–1). This facies is perhaps the most stratigraphically condensed of all the Tethyan lithologies (Kondensationskalk of Hollmann, 1962) and is generally characterized by ferromanganese crusts and nodules (Jenkyns, 1970a; Wendt, 1970; Germann, 1971). An inclusive faunal list would comprise ammonites, aptychi, belemnites, pelagic bivalves, crinoid fragments, Foraminifera, ostracods, Radiolaria, fish teeth, some gastropods, and rare single corals. Aragonitic fossils commonly show evidence of sea-floor and postburial solution. Encrusting fauna occurs within the ferromanganese nodules and crusts. Calcareous nannofossils occur sporadically. Algal stromatolites and traces of boring algae are common in many condensed sequences (our fig. 3; Szulczewksi, 1966; Wendt, 1970; Jenkyns, 1971a; Sturani, 1971), indicating deposition in the photic zone. Much of this flora and fauna is dependent on hard substrates, which suggests early lithification of the sea bottom possibly due to precipitation of a high-magnesian calcite cement (Garrison and Fischer, 1969; Jenkyns, 1971a). The paleogeographic situation envisaged for such condensed red pelagic limestones is that of a fault-bounded swell or seamount (Jenkyns and Torrens, 1971; Castellarin, 1972; Galácz and Vörös, 1972) where ambient currents swept much of the sediment into neighboring basins.

This facies is widespread throughout the Alpine-Mediterranean region; it is perhaps best displayed in the Trento and Lake Garda region of north Italy (Hollmann, 1964; Sturani, 1964, 1971) and in western Sicily (Wendt, 1963; Jenkyns, 1971a).

(b) Pisolitic ironstones (ferruginous biomicrites). Ferruginous limestones containing goethitic, hematitic and chamositic pisoliths make up a stratigraphically condensed facies, which may be associated with ferromanganese crusts and nodules. The color of the limestone matrix varies through red and brown to green; char-

---

Fig. 4.—Thin sections showing Mesozoic pelagic microfacies of Alpine-Mediterranean region. *1*, Cretaceous-Tertiary boundary, ×10; Trento Zone, Southern Alps, Nago, Province of Trento, Italy (slide courtesy of K. Novbakht). Uppermost part of Scaglia Rossa Formation (Upper Cretaceous) is almost completely replaced by phosphatic micrite, but original texture is still recognizable; planktonic Foraminifera of early Maastrichtian age are replaced by iron hydroxides. Overlying Paleocene biomicrite shows some patchy replacement by collophane and contains reworked fragments of the phosphatized Upper Cretaceous limestone (white). *2*, Biomicrite containing calpionellids, ×50; Maiolica Formation, Lower Cretaceous, Pizzo Cefalone, Gran Sasso d'Italia, Central Apennines, Italy. *3*, Radiolarite showing silica-filled molds of Radiolaria in fine-grained matrix of silica and hematite, ×12; Upper Jurassic, Bracco Zone of Ligurian Apennines, west of Nascio, Province of Genova, Italy. *4*, Sponge-spicule biomicrite, ×12; Lower Jurassic, Prekarst (Kuči) Zone, Dinarides, north of Gacko, Yugoslavia. *5*, Radiolarite showing graded radiolarian-rich laminae and hematite-stained silica matrix, ×12; Ladinian, Porphyrite-Hornstein Formation, Budva (Pindos Zone), road from Budva to Bar, southern Yugoslavia. *6*, Radiolaria-pelagic bivalve biomicrite, ×12; Carnian-Norian, Scillato Formation, Sclafani Zone, Palazzo Adriano, Sicily.

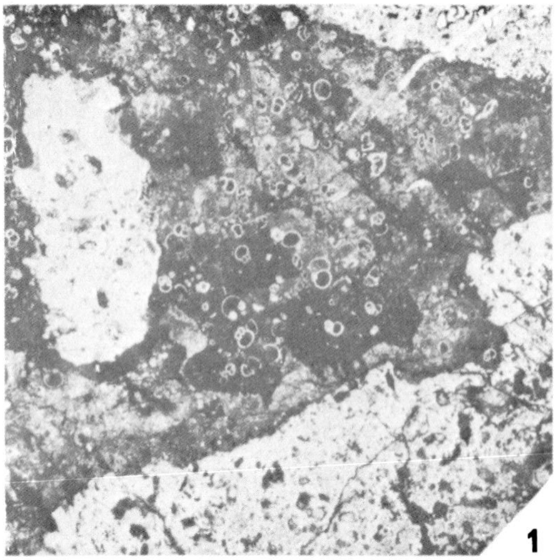
1

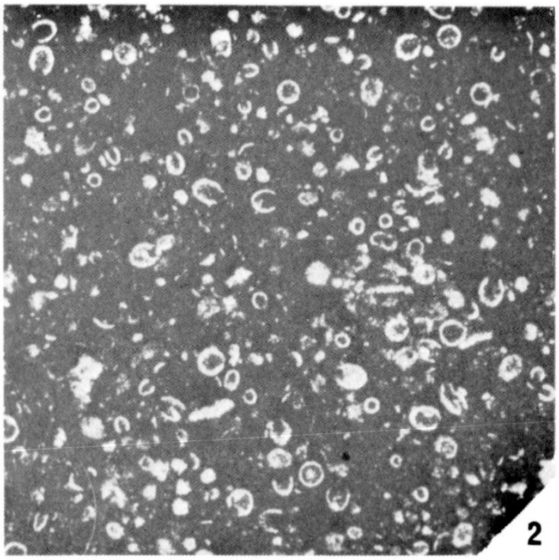
2

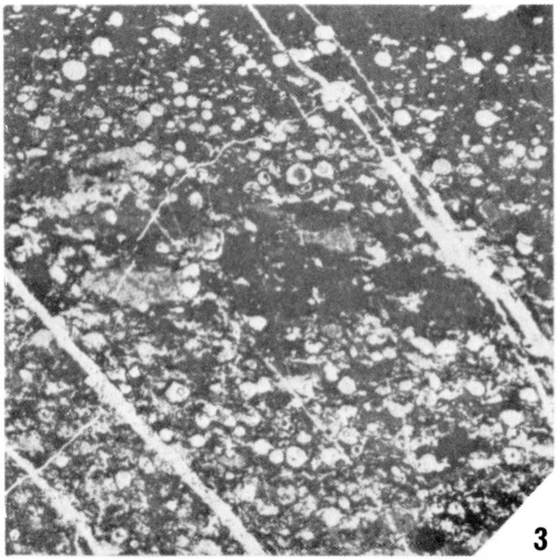
3

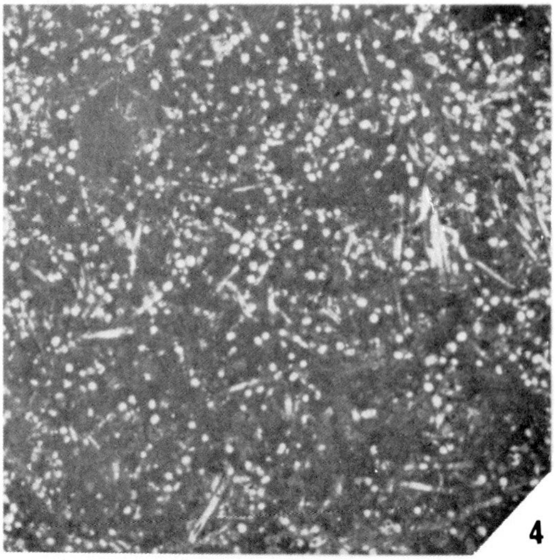
4

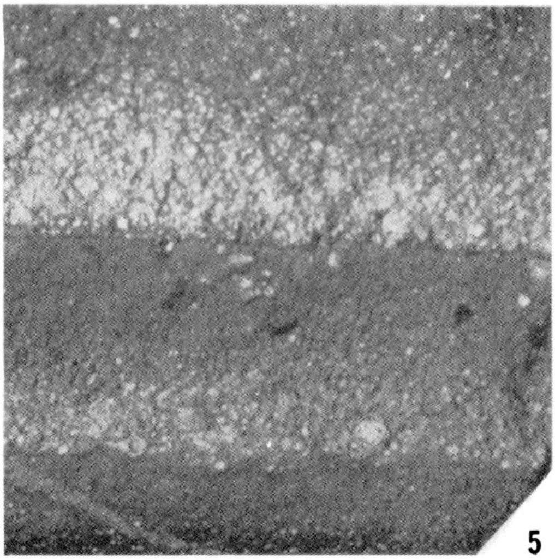
5

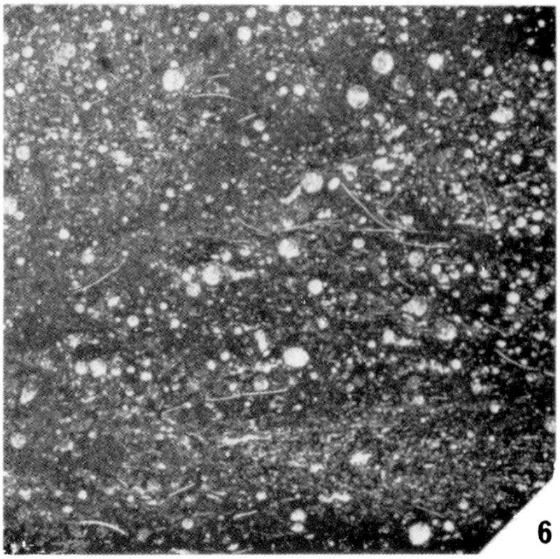
6

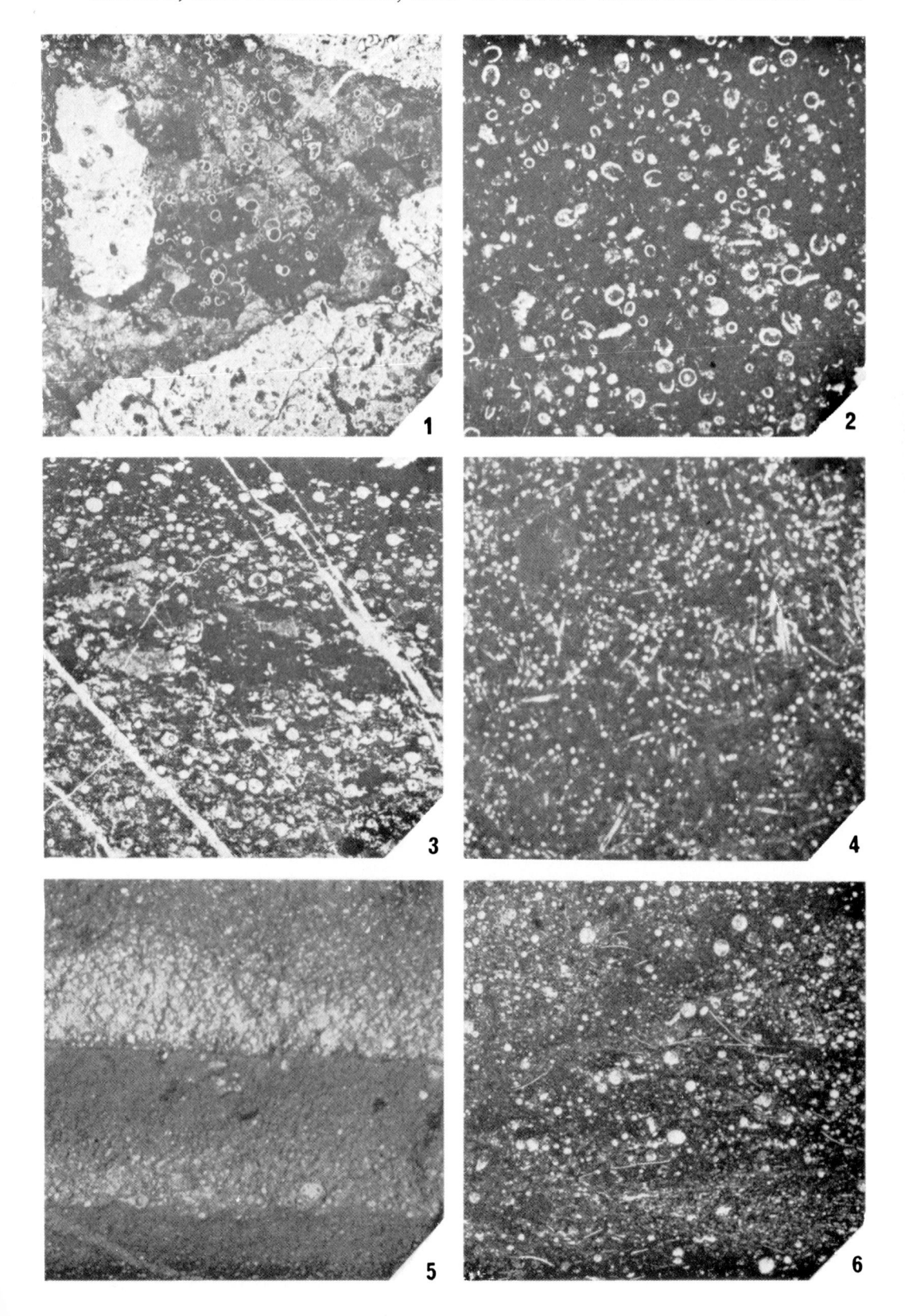

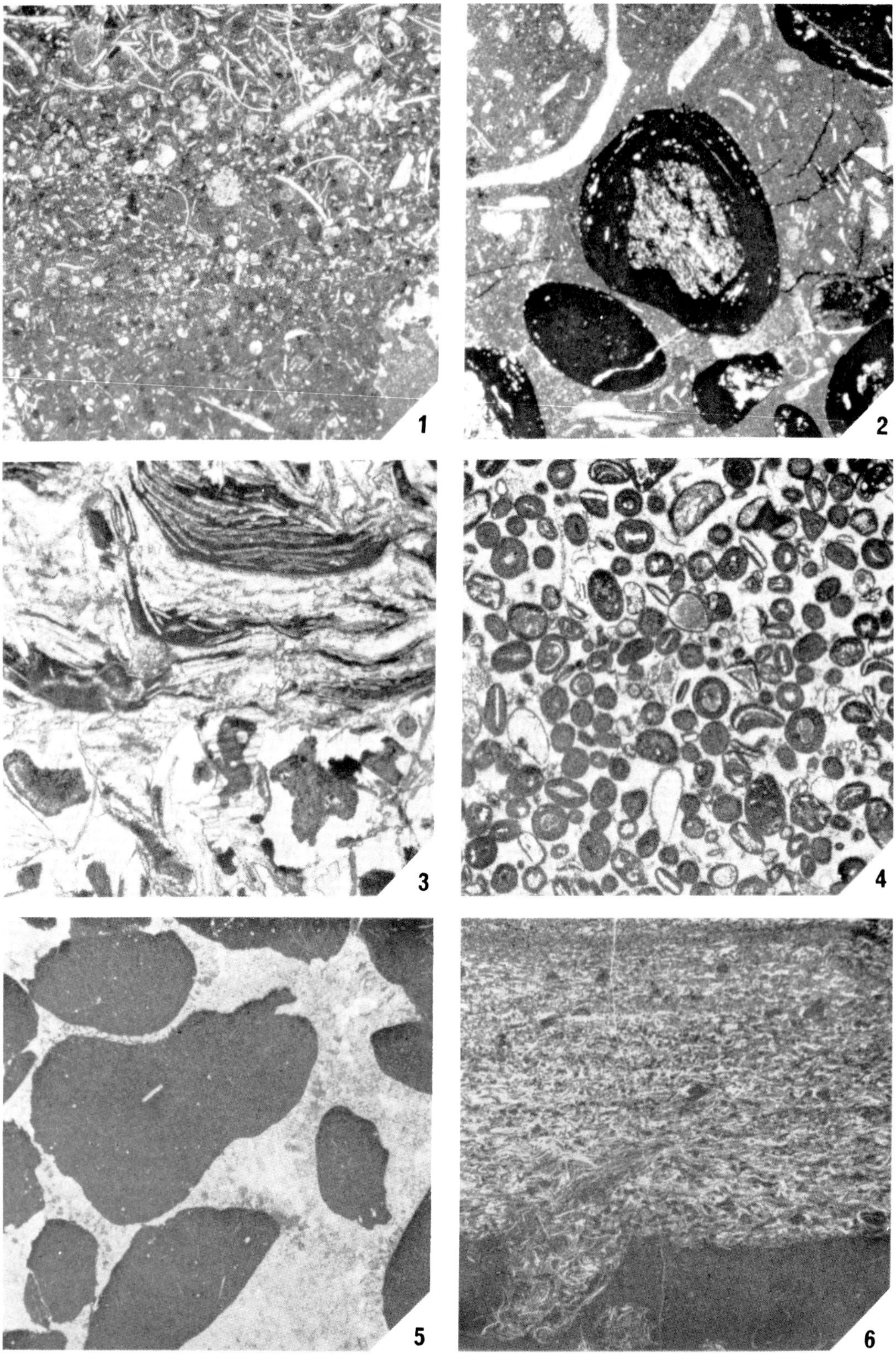
1
2
3
4
5
6

acteristically, the pisoliths are packed together in bundles. The associated fauna is similar to that in the red biomicrites described above; gastropods are generally more common, however, and fish teeth are particularly abundant. Traces of boring algae and fungi are common within the pisoliths and the fish teeth. This facies is also interpreted as a condensed deposit formed on a shallow seamount. The pisolitic facies of Toarcian age in western Sicily contains abundant fragments of sanidine trachyte (fig. 5-2) and can be interpreted as a volcanogenic ironstone (Jenkyns, 1970b).

Pisolitic ironstones are an extremely rare Tethyan lithofacies. Apart from western Sicily, they occur sporadically in the Spanish Subbetic (Geyer, 1967) and in the Venetian Alps (personal observations, 1970).

(c) Crinoidal and ammonite-gastropod-brachiopod-pelagic bivalve biosparites. These white to red sparite facies generally occur as lenses and may be overlain by, or grade into, the red condensed biomicrites. Crinoidal biosparites are widespread in the Lower Jurassic (e.g., Hierlatzkalk of the Eastern Alps, Austria). Pelagic bivalve biosparites, often referred to as *Posidonia alpina* lumachelles, are widespread in the Middle Jurassic but may be mixed with the crinoidal facies (fig. 5-3).

The ammonite-gastropod-brachiopod biosparites have more limited geographical distribution, and their paleoecological interpretation poses certain problems. Detailed studies of these facies by Sturani (1971) in the area of the Trento Swell in the Southern Alps has shown that their fauna is almost invariably of small size and recalls the so-called "dwarfed" or "stunted" forms of earlier authors. Sturani suggested that, although influenced to some extent by sorting, the diminutive size of the molluscs is a primary feature. The gastropods are mostly herbivorous, suggesting the former presence of algal meadows. These meadows may have harbored only small-sized ammonite species or the juveniles of such larger forms as the phylloceratids and lytoceratids. That algal meadows can act as biometric sieves is established for many recent environments (Jenkyns and Torrens, 1971; Sturani, 1971, and references therein).

The depositional milieu of all these calcarenites reflects turbulent conditions, and the characteristic lenslike forms of the calcarenites have been interpreted as submarine dunes or sand waves on current-swept seamounts (Jenkyns, 1971b). The pitted upper surfaces of some of these sand bodies have been interpreted as evidence of subaerial exposure and karst development; however, there is no general evidence of emergence.

The crinoidal and pelagic bivalve biosparites occur widely in the Alpine-Mediterranean region not only as seamount deposits but also as turbidites (fig. 5-6). The ammonite-brachiopod-gastropod biosparites, apart from those in the Trento Swell, occur in the Pieniny Klippen Belt, Polish Carpathians, in parts of Sicily, and in the Austrian Alps (Birkenmayer, 1963; Sturani, 1971).

(d) Pelagic pelletal and micro-oncolitic limestones (oncosparites and pelmicrites; fig. 5-4). These are white, cream, or pale-pink limestones that contain abundant ammonites, some brachiopods, and gastropods, and a rich microfauna including globigerines, the planktonic crinoid *Saccocoma*, and the spores of *Globochaete*. Radiolaria and calpionellids occur in some successions. This facies is generally confined to the Upper Jurassic and Lower Cretaceous. Electron-microscope study of bodies resembling ooliths has revealed coccoliths within their cortex (Jenkyns, 1972). This suggests that the so-called "ooliths" grew at least partially by contact adhesion of fine particles, so that they can be interpreted as small algal balls or oncolites displaying remarkable geometrical perfection. Depth of formation is uncertain, although it is certainly greater than that of true Bahamian-type ooliths. Facies maps (Jenkyns, 1972) demonstrate that this lithology was limited to structural highs.

FIG. 5.—Thin sections showing Mesozoic pelagic microfacies of Alpine-Mediterranean region. *1*, Radiolaria-pelagic bivalve biomicrite (Kondensationskalk), ×17; Bathonian, Monte Bonifato, western Sicily. *2*, Pisolitic ironstone containing large sanidine trachyte fragment in pisolith at center, ×17; Toarcian, Rocce Maranfusa, western Sicily. *3*, Crinoid-pelagic bivalve biosparite as part of sand wave, ×17; upper Bathonian to lower Oxfordian, Subbetic Zone, Cehegin, southern Spain. *4*, Pelagic "oolite", ×17; upper Tithonian, Rocce Maranfusa, western Sicily. *5*, Pebbles of pelagic limestone (biomicrite) containing thin-shelled bivalves and calcite-replaced Radiolaria in coarse calcite cement, ×4; pelagic cephalopod limestones, Upper Jurassic, Louros Valley, Epirus, western Greece. *6*, Pelagic turbidite consisting of graded skeletal and intraclastic calcarenite to calcisiltite intercalated with pelagic bivalve biomicrites, ×4; Pelagic Bivalve Limestone, Middle Jurassic, Ionian Zone, Kouklessi, western Greece.

These deposits are not as condensed as are the red ferromanganiferous biomicrites, and trapping of sediment by algae may have kept pace with subsidence of the seamount blocks.

Pelagic micro-oncolites have been recorded from western Sicily (Jenkyns, 1972), southern Hungary (Kaszap, 1963), Polish Carpathians (Lefeld and Radwánski, 1960), and from the Ligurian Briançonnais, Franco-Italian Alps (Royant and others, 1971). The San Vigilio Oolite (lower Middle Jurassic) of the Trento Swell (Venetian Alps) is comparable to these Upper Jurassic facies in some respects (Jenkyns, 1972).

(e) Red nodular and marly facies (biomicrites). This is the well-known Ammonitico Rosso or Knollenkalk containing more than 5 percent clay; with greater than 20 percent clay the rock takes on a generally marly appearance. The fauna includes typical Tethyan pelagic elements, but hard-substrate dwellers are notably absent. Ammonite molds are locally common; calcareous nannofossils are very rare. Algal stromatolites occur in some feebly nodular calcareous rocks, suggesting that deposition could have taken place as shallowly as in the photic zone; more probably, however, most of the facies was deposited in somewhat deeper water. Hollmann (1962) ascribed the origin of the nodular structure to the action of irregular sea-floor solution, producing limestone remnants in an insoluble marly residue. In contrast, Lucas (1955) suggested that early diagenetic segregation had taken place. Petrographic studies support the latter interpretation (Jenkyns, in preparation). Solution of aragonite and fine-grained calcite below the sediment-water interface probably supplied excess carbonate for the formation of nodules.

Two types of Ammonitico Rosso may be distinguished (Aubouin, 1964): a strongly calcareous facies gradational with the red condensed biomicrites and a more stratigraphically expanded marly facies. The lime-rich varieties were evidently deposited on a swell or seamount. Such rock types are well exposed in parts of the Southern Alps (Sturani, 1964). The marly facies, exposed in the Eastern Alps, the central Apennines, western Greece, and elsewhere, were formed in basinal settings. Such a depositional environment is established from the abundant intercalated turbidites (fig. 5-6) and from slump-folded complexes (Bernoulli, 1964, 1971; Bernoulli and Jenkyns, 1970).

The Ammonitico Rosso, although particularly common in the Jurassic System, may extend into the Cretaceous (e.g., Spanish Subbetic, Kuhry, 1972).

(f) Spar-cemented nodular limestones. Calcareous nodules of centimeter scale, cemented by coarse calcite, constitute a facies associated stratigraphically with red calcareous biomicrites and ferromanganese encrustations. This association with such condensed deposits suggests deposition on or at the edge of a seamount (Bernoulli and Jenkyns, 1970), although they may occur in more basinal settings. Their origin has been ascribed to resedimentation of an Ammonitico Rosso with the marly matrix being lost in transit (Garrison and Fischer, 1969; Hudson and Jenkyns, 1969; Jurgan, 1969). The void-filling cement is considered to have formed during a submarine sedimentary pause (Bernoulli and Jenkyns, 1970).

This is an uncommon facies. The most famous example is the so-called "Scheck" which occurs at Adnet near Salzburg, Austria. This facies has also been recorded by Bernoulli and Renz (1970) from western Greece (fig. 5-5).

(g) Gray limestone-marl interbeds (biomicrites). This facies, which has a variety of local names (e.g., Fleckenkalk, Medolo, Corniola, Siniais Limestone, or, if containing abundant chert, Hornsteinkalk), is particularly widespread in the Lower Jurassic and occurs also in younger sequences. Ammonites and belemnites are common to very rare and the microfauna includes Radiolaria, Foraminifera, ostracods, and sponge spicules. Sponge spicules may constitute such an important part of this rock type that the facies is referred to as Spongienkalk (fig. 4-4). *Chondrites* burrows are common. Calcareous nannofossils are locally abundant (Bernoulli and Jenkyns, 1970).

The expanded nature of this facies is illustrated by very thick successions (4,000 meters (Bernoulli, 1964) in the Liassic Generoso section of the Southern Alps) and by the widespread occurrence of slump-folded complexes and turbidites (Bernoulli, 1964; Bosellini, 1967b; Bernoulli and Jenkyns, 1970; Castellarin, 1972). This suggests deposition in a structural basin. That conditions were sometimes stagnant is shown by the sporadic occurrence of black sapropelic marls and iron-manganese carbonate deposits (documented from Hungary, the Carpathians, and the Eastern Alps by Szabó-Drubina, 1961; Andrusov, 1965; and Germann and Waldvogel, 1971).

This gray limestone-marl facies can be traced throughout the Alpine-Mediterranean region; also, in the northern continental margin deposits of the Tethys between the Briançonnais high and the oceanic Piemont Zone, where it is typically associated with marine breccias (Breccia Nappe, Gondran Zone, Lemoine, 1971). This particular facies was formerly included within the metamorphic Schistés lustrés.

(h) Radiolarites. The radiolarites are generally red, gray, or green and consist of more or less regular alternations of centimeter-thick chert layers and paper-thin siliceous argillites. The fauna includes a few sponge spicules and badly preserved coccoliths in some of the more calcareous varieties (Garrison and Fischer, 1969). A patina of ferromanganese oxides is common along cleavage planes.

Most radiolarites are Middle and Late Jurassic in age or younger; exact dating is not usually possible because of the lack of well-preserved diagnostic fossils. Such facies are uniformly interpreted as one of the deepest of the Tethyan lithofacies. Slumps and turbidites occur locally (Price and Nisbet, 1973; Schlager and Schlager, 1973). Most authors, following Cayeux (1924), draw a comparison with Recent radiolarian oozes, suggesting that radiolarites formed below the calcite-compensation depth as an insoluble siliceous residue. Cosmic spherules have been recovered from some Alpine radiolarites (Mutch and Garrison, 1967).

A particularly interesting radiolarite sequence is exposed in the Sclafani Zone of central Sicily; this ranges through the whole of the Jurassic and Cretaceous Systems and into the Eocene Series. Intercalated in this sequence are a series of shallow-water carbonate beds that led earlier authors to speak of repeated transgressions (Schmidt di Friedberg and others, 1960; Broquet and others, 1966). We interpret these beds as redeposited units, which are not, therefore, indicative of shallow-water radiolarite formation (personal observations, 1970; Scandone and others, 1972).

Radiolarites are widespread in the Alps and Mediterranean area, not only in continental margin settings, but also on Mesozoic oceanic crust (see below).

(i) White nannofossil limestones. White fine-grained limestones containing blue-black chert nodules were first deposited during the Tithonian, commonly directly overlying radiolarite. They extend into the Cretaceous (fig. 4–2). The abrupt lithological change from chert to limestone may be related to an evolutionary burst of the coccolithophorids (Garrison and Fisher, 1969; Hallam, 1971a). This facies is known by several local names such as Maiolica, Biancone, Lattimusa, Aptychenkalk, and Vigla Limestone. The fauna comprises rare ammonites, belemnites, aptychi, Radiolaria, and calpionellids. Very thin thin sections (Steinmann, 1925) and electron microscopy (Grunau and Studer, 1956; Garrison, 1967) reveal abundant coccolithophorids, particularly *Nannoconus*.

In the Eastern Alps this facies has shaley partings (Oberalm Beds) and contains abundant turbidites of carbonate-platform origin (Garrison, 1967). Redeposited pelagic sediments occur in this and in some other areas but are not common.

The white nannofossil limestone, together with the radiolarite, must constitute one of the deepest Jurassic lithologies of the Tethys; it also shares, again with the radiolarite, the distinction of being a sedimentary facies deposited on ocean crust. Of all the Tethyan Jurassic lithologies, it is perhaps the most widespread, occurring not only in areas that belonged to the deep sea and southern continental margin but also in structural units that unequivocally were derived from the northern continental margin (e.g., Helvetic Nappes and Sub-Alpine chains, Trümpy, 1960; Remane, 1967; Pieniny Klippen Belt, Carpathians, Birkenmayer, 1963).

Cretaceous: The most common pelagic facies of Cretaceous age in the Alpine-Mediterranean region is the white nannofossil limestone described above (fig. 4–2); a few other lithologies persisted from the Jurassic. Two other important facies (a and b) are described.

(a) Greenish-gray and variegated marls and marly limestones (biomicrites). This facies consists of well-bedded alternations of darker, more argillaceous laminated sediments and lighter, more calcareous sediments in which original structure has been modified by burrows (*Chondrites*, fucoids). It is known locally as Marne, a fucoidi or Scaglia variegata and is mostly of middle Cretaceous age. Colors vary from lilac to pink and from dark gray to light greenish gray. The sediments are biomicrites bearing planktonic Foraminifera and Radiolaria set in a fine-grained matrix composed of calcareous nannoplankton and various amounts of clay minerals (Dufour and Noël, 1970; Bernoulli, 1972). Replacement chert is present in some levels. Additionally, intercalations of dark-gray to black, strongly siliceous, carbonate-free argillites occur that are particularly rich in organic matter and Radiolaria. These may reflect episodes of euxinic conditions.

The Marne a fucoidi and Scaglia Variegata are distinguished from the underlying white nannofossil limestones mainly by virtue of their increased clay content. The clay minerals comprise illite and some montmorillonite (Porrenga *in* Bernoulli, 1972). The clay content could result from an increased terriginous influx during middle Cretaceous time contemporaneous with flysch deposition in the more internal zones, or with volcanic activity, or with bathymetric deepening. Deposition took place in a topographically subdued basin.

The Marne a fucoidi is widespread in the central Apennines; the Scaglia variegata is developed in the Southern Alps.

(b) Red to white biomicrites containing planktonic Foraminifera. Red, pink, and white, well-bedded, faintly nodular biomicrites are a widespread Upper Cretaceous facies. They contain abundant planktonic Foraminifera (*Globotruncana*) and calcite-replaced Radiolaria. The fine matrix consists of solution-welded and recrystallized nannoplankton (Farinacci, 1964; Bernoulli, 1972). Mostly referred to as Scaglia, this rock type resembles the white nannofossil limestones of Late Jurassic and Early Cretaceous ages. In some places there is stratigraphic continuity between the two facies (e.g., Vigla Limestone of western Greece, Institut de Géologie, etc., 1966). In western Greece, in the Pindos Zone, in the Apennines, and in parts of the Southern Alps and Sicily, this facies contains numerous intercalations of breccias, displaced blocks, and turbidites, all of carbonate-platform origin (Ferasin, 1958; Aubouin, 1959a,b; Luterbacher and Premoli Silva, 1964; Bortolotti and others, 1970). All this attests to a basinal environment.

In the area of the Trento Swell in north Italy (Southern Alps), the Scaglia is characterized by stratigraphic gaps and hardgrounds. Finely laminated ferruginous crusts occur along the nondepositional surfaces, and the limestones are highly phosphatized (fig. 4–1). Borings and tubes penetrating this limestone may be filled with younger pelagic sediment or lower Tertiary neritic deposits (Premoli Silva and Luterbacher, 1966). Several generations of mineralization are recognizable, and the effects of local reworking are demonstrated by the presence of breccias comprising mineral crusts and limestones. These breccias and phosphatized sediments are very similar to Tertiary rocks dredged from Caribbean seamounts and from the Agulhas Bank off South Africa (Marlowe, 1971; Parker and Siesser, 1972). An environment of fault-bounded pelagic high is thus indicated. Fault scarps along the edge of the Trento Swell that were active during the deposition of Scaglia limestones can still be reconstructed in great detail by careful mapping (Castellarin, 1972).

Similar limestones also occur both as stratigraphically condensed and expanded sequences along the northern continental margin (Briançonnais Ridge facies, Bourbon, 1971; Seewen Limestone, Bolli, 1944). Microfacies of the same type occur in the Carpathians (Mišík, 1966), Spain (Paquet, 1969), and elsewhere.

Reasons for prevalence of pelagic facies on the southern margin of Tethys.—Recognition that Tethyan pelagic limestones mostly belong to tectonic units that seem to have been derived from the southern continental margin of the Tethys (see above) suggests that their open-sea character and general lack of terrigenous detritus may be related to their specific paleogeographic situation.

Considering first the northern landmass of Eurasia, we envision an arid region of salt lakes that was flooded by a shallow shelf sea during Muschelkalk (Triassic) time (Wurster, 1968). By the beginning of Jurassic time, a variety of marine facies were being deposited, some of which, such as ironstones, bear witness to the proximity of river mouths and to moist, well-vegetated tropical landmasses (Hallam, 1966; Chowns, 1966; see also Porrenga, 1965). Jurassic ironstones are widespread, not only in Great Britain, but also in France, Germany, and parts of Switzerland (Cayeux, 1909; Bubenichek, 1961; Aldinger and Frank, 1944; Aldinger, 1957; Deverin, 1940, 1948). In Late Jurassic and Early Cretaceous time, the Wealden deltas spread across much of northern Europe, including parts of Spain (Arkell, 1956; Allen, 1967). The Wealden episode also must have brought significant amounts of sediment, organic matter, and nutrients to the shallow European seas. Pelagic conditions were extant during Middle and Late Jurassic time in some areas (e.g., Briançonnais, Pieniny Klippen Belt), probably owing to the sheltering effects of intervening troughs. Such conditions, however, became widespread over much of northern Europe during the Late Cretaceous Epoch with the deposition of chalk, which, in its upper levels, consists of a remarkably pure, coccolith-rich deposit (Black, 1953). It has been suggested that the surroundings of the chalk sea were bevelled arid plains (Bailey, 1924) that thus provided negligible amounts of terrigenous matter. This prompts us to examine the Mesozoic successions of North Africa to see if they shed any light on the reasons for widespread pelagic sedimentation on the southern margins of the Tethys.

There is no doubt that the Saharan Mesozoic is substantially different from coeval facies in northern Europe. Busson (1967, 1970) has described a variety of lithologies from southern Tunisia and Algeria that include recurring evaporites (Triassic, Jurassic, and Cretaceous), fluvial and aeolian sandstones, minor deltaic and lagoonal sequences, and widespread shallow-water limestones. Similar facies occur in parts of the Atlas Mountains of Morocco (du Dresnay, 1956, 1963). Fluvial and aeolian sands of Mesozoic age are present in Libya (Kallenbach, 1972) and in much of northeast Africa (Nubian Sandstone, Whiteman, 1970). The lack of ironstones in North Africa is particularly noteworthy; minor Jurassic examples are recorded by Lucas (1942) from the Algeria-Morocco frontier area. These circumstances suggest a dry desert landmass that was subject to occasional flash floods and was ringed by a broad coastal plain (sabka) on which evaporites and shallow-water carbonates were deposited.

We thus suggest that, in contrast to the northern Eurasian landmass, only a few sluggish or ephemeral rivers on the southern continental margin debouched into the sea. Supply of clastics and organic matter thus was minimal. Furthermore, the carbonate platforms on the southern margin of the Tethys may have acted as shields for the pelagic basins that transected them, particularly during Late Triassic and Early Jurassic time. Any encroaching clastics probably would have been incorporated in these rapidly deposited carbonate rocks. These circumstances must have enabled pelagic conditions to persist in moderate to great depths (say, 50 to 4,000 m) at the northern edge of the African landmass during much of Mesozoic time. In Tunisia, pelagic sediments possibly can be traced southward into lagoonal and lacustrine facies (Bonnefous, 1967).

Ocean-Floor Deposits

Introduction.—Distinguishing ocean-floor sedimentary facies from pelagic continental margin deposits is not easy. The only reliable guide to recognition of true oceanic sediments is an association with ophiolites. Generally, oceanic sediments comprise carbonate-poor lithologies, such as cherts and argillites, that presumably were deposited below the calcite-compensation depth; such facies, however, also can occur in deeply submerged continental margin settings. Nevertheless, calcareous sediments do occur on some ophiolitic masses. Such sediments may have formed on a ridge crest during an early phase of spreading, or their accumulation may be owed to depression of the calcite-compensation depth during a period of prolific calcareous nannoplankton production.

The earliest documented ocean-floor facies, if correctly interpreted, are marls, radiolarites, and *Halobia* limestones from the Upper Triassic of Turkey (Juteau, 1970; Marcoux, 1970). They occur in the Antalya Nappe complex, which consists of thrust sheets whose origin and relationship to the central Tethyan ophiolites is still ambiguous. Similar occurrences of *Daonella* limestones associated with pillow lavas have been described from northern Greece (Terry, 1971), but some uncertainty also exists here as to whether these volcanics are part of an ophiolite complex or a document of early rifting in the continental margin (e.g., Porphyrite-Hornstein Formation). Such Triassic rocks are rare, however; most of the undoubted ocean-floor facies are Jurassic or Cretaceous in age. The best preserved sequences are known from northern Greece (Vourinos, Moores, 1969; Zimmerman, 1972; Othrys, Hynes and others, 1972; our fig. 2), from the Ligurian Apennines (Decandia and Elter, 1969; Abbate and Sagri, 1970; our fig. 2), and from the Rumanian Carpathians (Burchfiel, in preparation), where they escaped Alpine metamorphism. Essentially the same association can be recognized in the Western Alps (Lemoine and others, 1970; Lemoine, 1971).

Facies.—Five facies are most important.

Ophicalcites: In the northern Apennines and elsewhere, the surface of the oceanic basement may be heavily brecciated where it consists of serpentinite or gabbro, the fissures being filled by alternating white sparry calcite and red or green crystalline limestone. Generally, serpentinites that are fragmented *in situ* and penetrated by complex fissures grade upward to matrix-rich breccias containing some radiolarite and lava fragments. These breccias, called ophicalcites, have been interpreted previously as consanguineous associations of ultramafic rocks and magmatic carbonatites (Bailey and McCallien, 1960). Both oxygen isotope data (Galli and Togliatti, 1965) and cement fabrics in the sparry calcite, however, indicate an exogenous origin (Abbate and others, 1972). The varicolored crystalline limestones could represent, therefore, the earliest oceanic sediments deposited above the calcite-compensation depth (see Thompson, 1972, for a possible modern equivalent). The negligible thickness of these limestones is probably due to deposition on a topographic ridge influenced by ambient currents.

Umbers: Brown iron-rich mudstones, of Late Cretaceous age, locally termed umbers, directly overlie the pillow lavas of the oceanic Troodos Massif of Cyprus. They are enriched in certain trace elements and have been interpreted as products of rapid precipitation resulting from interaction of magmatic fluids and sea water. In this respect they presumably are comparable with the spreading ridge sediments of the East Pacific Rise and with the iron-rich basal facies of the Atlantic and Pacific (Robertson and Hudson, 1973).

Radiolarites: Generally, the pillow lavas that form the uppermost member of ophiolite stratigraphy are overlain by radiolarites (fig. 6), but in some places radiolarites, fine-grained tuffites, graded volcanic sandstones, and gabbroic arkoses alternate (e.g., Monte Rossola near Levanto, northern Apennines, Abbate and others, 1972). The radiolarites are mostly regularly bedded, almost carbonate-free siliceous rocks having paper-thin intercalations of argillites. In thin section (fig. 4–3) radiolarian skeletons appear as pigment-free, silica-filled molds set in a very fine-grained, hematite-rich siliceous matrix. Although no stratigraphically diagnostic forms have yet been described from these rocks, a Middle to Late Jurassic age is indicated by the overlying white nannofossil limestone (*Calpionella Limestone*). Thicknesses of the radiolarites vary from 0 to 300 m, extreme differences in thickness being taken as evidence of sediment-ponding resulting from normal or transform faulting in the basement (Abbate and others, 1972).

The depositional environment of the radiolarites was certainly a deep ocean below the compensation depth for calcite. Apart from the Apennines, Jurassic radiolarites resting on oceanic pillow lavas or ophicalcites can be seen in the Western Alps (Lemoine and others, 1970), Greece (Moores, 1969), and elsewhere. In Corsica, radiolarites have been found associated with oolitic limestones (Jodot, 1933) that formerly were believed to indicate shallow water. Recent investigations, however, have shown the oolites to have been redeposited (Bortolotti and Passerini, in press; for other examples, see Grunau, 1965).

White nannofossil limestones and Palombini: White nannofossil limestones, which typically overlie Jurassic radiolarites, are lithologically indistinguishable from their continental margin equivalent (Maiolica) and, like the radiolarites described above, are Late Jurassic to Early Cretaceous in age. Together with the ophiolites and radiolarites, they have been reported from the northern and southern Apennines, the Western Alps, and from Greece (Decandia and Elter, 1969; Abbate and Sagri, 1970; Ogniben, 1969; Lemoine, 1971; and Moores, 1969).

In the northern Apennines, this facies passes laterally into shales intercalated with fine-grained dove-gray argillaceous limestones (Palombini). This lithology seems to be restricted to the oceanic environment. Limestones, which make up only a thin part of the formation, contain Radiolaria, calcareous nannofossils, some calpionellids, and, in higher parts of the succession, rare and poorly preserved planktonic Foraminifera. The most pronounced features of the Palombini are the hard, brownish-weathering siliceous zones along bedding planes. Equivalents of this facies occur in the Western Alps (Formation de la Replatte, Lemoine, 1971) and in Greece (personal observation, 1971).

Laterally, both the white nannofossil limestones and Palombini may pass into Early Cretaceous flysch (southern Apennines: Formazione del Frido, Santa Venere Formation; see Ogniben, 1969; Sicily: Monte Soro Flysch, Ogniben, 1963).

Black shales: Black argillites, known as Lavagna Shales in the Ligurian Apennines, are particularly widespread in the middle Cretaceous rocks. In addition, laminated siltstone and sandstone turbidites also are present, limestones and marls are subordinate, and planktonic Foraminifera are rare. The formation was probably deposited near or below the calcite-compensation depth under euxinic conditions and may represent the deepest bathymetry attained in the ocean basin. In the Alpine-Mediterranean Tethys, this lithology marks a turning point in paleogeographic evolution and the onset of orogenic movements. In the Ligurian Apennines, the upper part of the Lavagna Shales and their equivalents contain large olistrostomes and olistoliths of ophiolites (Elter and Raggi, 1965a, b) and are capped by an Upper Cretaceous flysch wedge. Oceanic black shales overlain by flysch also occur in the southern Apennines (Crete Nere Formation, Vezzani, 1968)

FIG. 6.—Jurassic radiolarites overlying a breccia composed of diabase and subordinate shale fragments (ophiolite suite); Bracco unit, Ligurian Apennines, Montaretto, Province of La Spezia, Italy.

and in the Western Alps (part of Schistes lustrés, Lemoine, 1971).

MESOZOIC FACIES OF THE WESTERN CENTRAL ATLANTIC

Introduction

The most complete record of undeformed Mesozoic pelagic facies on the ocean floor was recovered during Leg XI of the Deep Sea Drilling Project in the western central Atlantic. The following account is based mainly on this material, which was examined in detail by Bernoulli in 1972. Extensive illustration of the central Atlantic facies has been made by Hollister, Ewing, and others (1972); Windisch and others (1968); Ewing, Worzel, and others (1969); Peterson and others (1970); and by Hayes, Pimm, and others (1972). Schematic logs of relevant successions are given in Figure 2.

Continental Margin Deposits

Structural setting.—Site 98 of Leg XI is located in the Northeast Providence Channel between Great Abaco and New Providence Islands in the Bahamas. According to Uchupi and others (1971), the northwestern Bahama Islands and their intervening troughs belong to a continental margin floored by continental crust. (For an alternative interpretation see Dietz and others, 1970.)

Facies.—White to yellow nannoplankton ooze: The Upper Cretaceous sediments from the Northeast Providence Channel comprise indurated nannoplankton ooze or chalk containing planktonic Foraminifera and unidentified silt-sized skeletal matter of possible shallow-water origin. Intercalations of graded and laminated calcarenites and calcisiltites, typically composed of rounded to subangular shallow-water particles, also are present as well as large benthonic Foraminifera, rudistid or echinoderm fragments, and gravel-sized fossil debris and pieces of hard, vuggy skeletal limestones. The latter apparently have undergone meteoric-water diagenesis and subsequent introduction into the deeper marine environment as lithified fragments.

This sedimentary association matches well with the description of the younger, upper Tertiary to Recent sediments of Tongue of the Ocean and Northeast Providence Channel, so that a similar environment may be assumed (Paulus, 1972, and references therein). Inasmuch as the whole Bahamian area has undergone considerable subsidence since the Late Cretaceous, however, water depth at that time was much less and probably not more than a few hundred meters. (See discussions in Bernoulli, 1972; Paulus, 1972.) Finally, the Upper Cretaceous sediments of site 98 are very similar to those coeval Tethyan pelagic limestones that were deposited in the vicinity of carbonate platforms (Bernoulli, 1972).

Ocean Floor Deposits

Structural setting.—Basalt underlying oceanic sediments was reached by deep drilling in the Cat Gap area, off the Bahamian platforms, in the magnetically quiet zone (site 100) as well as near the boundary between the magnetically quiet zone and rough basement at the foot of the continental rise off New York (site 105). These basalts, which contain some sediment, are interpreted as the uppermost level of ocean crust. They correspond with the top of the acoustic basement that can be followed throughout the western central Atlantic basin (Hollister, Ewing, and others 1972).

Facies.—Green to gray argillaceous limestones: The oldest oceanic sediments are greenish-gray limestones of Oxfordian and possibly Callovian age. They only occur in the Cat Gap area (site 100), which, according to Pitman and Talwani (1972), belongs to the oldest part of the Atlantic. These sediments consist of relatively homogenous argillaceous micrites having some faint laminations, current bedding, and burrows. Although the limestones contain abundant carbonaceous matter and plant debris, deposition in a pelagic setting is indicated by calcareous nannoplankton and Radiolaria. The composition of the benthonic foraminiferal faunas suggests an upper bathyal environment (Luterbacher, 1972). This facies resembles somewhat the gray limestone-marl interbeds of Jurassic age in the Alpine-Mediterranean region.

Red argillaceous limestones: In the Cat Gap area, the green-gray argillaceous facies grades upwards into red, marly, slightly nodular limestones of Oxfordian to Tithonian age. Most of the sediments are very fine grained, comprising various mixtures of clay minerals (mainly montmorillonite), calcareous nannoplankton, and neomorphic micrite containing detrital minerals. Megafossils include aptychi, pelagic bivalves, and some echinoderm fragments; ammonites are very rare, occurring only as poorly preserved internal molds. Radiolaria are represented by calcite- or silica-filled spheres. The poor benthonic assemblages consist mainly of primitive agglutinating Foraminifera and deep-water ostracods. Planktonic microorganisms such as *Globochaete* and *Stomiosphaera* are common.

At site 105 the red marls and limestones exhibit evidence of pronounced mobilization and redeposition, which seems to be primarily related to the rugged surface of the volcanic basement. (See seismic profiles of Lancelot and others, 1972.) Resedimented layers typically comprise incomplete cycles of large-scale slump complexes and pebbly mudstones overlain by so-called "pelagic turbidites," which grade from calcarenites into homogenous marly mudstones. The redeposited beds are composed dominantly of penecontemporaneously displaced pelagic intraclasts or fossils and lumps of white indurated limestone containing small calcite-replaced ammonites and the foraminifer *Globigerina* occur rarely. *Globigerina* is not present elsewhere in the section, which suggests that the white limestone was deposited on topographic highs above the dissolution level for planktonic Foraminifera and then was displaced into local depressions. In the Cat Gap area, the basement surface appears smoother, and the overlying sediments contain fewer resedimented layers.

A deep bathyal environment above calcite-compensation depth seems to be indicated from the faunal composition. The whole lithological association is very similar to the basinal marly Ammonitico Rosso (Jurassic) of the Mediterranean, so that similar physiographic settings may be postulated for both the Mediterranean and Atlantic deposits even though they formed in different paleotectonic situations.

Cherts are very rare in the red argillaceous limestones, although at site 105 some silica occurs in replacement fabrics and as cement in a pelagic turbidite. At site 100 a few red chert layers contain silica-filled radiolarian molds set in a hematite-rich siliceous matrix. This could be considered as an intercalation of radiolarite.

White and gray chalky nannofossil limestones: At sites 101A and 105, the red argillaceous limestones grade upward into a well-bedded white to light-gray, intensely burrowed chalky limestones alternating with darker, organic rich, finely laminated layers that become abundant in the uppermost part of this sequence. Fossils include rare and badly preserved calpionellids, abundant calcite-filled radiolarian molds, and well-preserved aptychi. The upward disappearance of slumping and the transitional contact with the overlying black clays indicate gradual elimination of local sea-floor topography and progressive stagnation during general deepening of the basin.

This facies is late Tithonian to Hauterivian in age, and in the Cat Gap area it passes laterally into white nannoplankton oozes and chalks (see below). The similarity between the white and gray limestones and some coeval or slightly younger facies in the Mediterranean region is notable (Oberalm Beds, Marne a fucoidi).

White nannoplankton ooze and chalk: This lithology consists of homogenous and structureless, soft or slightly indurated nannoplankton ooze containing sparse nodules of gray or black replacement chert. The fauna consists of molds of Radiolaria, badly preserved calpionellids, primitive looking Foraminifera and occasional aptychi. At one locality (Leg I, site 4, Ewing, Worzel, and others, 1969), silicified lime turbidites contain grains that suggest an origin from Bahamian platforms.

The Upper Tithonian to Neocomian white nannoplankton oozes and chalks are obvious equivalents of the white nannofosil limestones (Maiolica), which are the most widespread Mesozoic lithology of the Tethys (Cuba, Cape Verde Islands, Alpine-Mediterranean Tethys, Indonesia; Colom, 1955, 1965). Most probably they were deposited in a deep bathyal environment. Similar lithologies occur also in the Upper Cretaceous (Turonian to lower Maastrichtian) of Cat Gap. They comprise white or cream nannoplankton marls and chalks and intercalated pebbly mudstones and graded calcarenites.

Black clay: In large areas of the western central Atlantic, middle Cretaceous sediments consist of black to greenish-gray or olive-green carbonaceous clays, some organic matter and zeolites, and subordinate quartz, feldspar, siderite, pyrite, and calcite. Abundant silty laminae are composed of zeolites, siderite, pyrite, and montmorillonite. The fauna includes rare siderite-filled molds of Radiolaria, fish remains, and both planktonic and simply structured arenaceous Foraminifera indicative of deep water (*Rhabdammina* fauna). This lithology was certainly deposited under stagnant conditions in a deep basin, and the lack of calcareous material was caused either by sinking below the carbonate-compensation depth or by dissolution resulting from high pH of the interstitial water (discussion in Lancelot and others, 1972).

The black clays find a parallel in the coeval dark clays of the eastern central Atlantic (e.g., site 135, Hayes, Pimm, and others, 1972) and in the black shales of the Ligurian Apennines (Lavagna Shales) and elsewhere in the Mediterranean region.

Black radiolarian mudstones: Black radiolarian mudstones and cherts of Albian and Cenomanian age were recorded east of the Bahamian platform at site 5, Leg I (Ewing, Worzel, and others, 1969). The color and limited carbonate content of this facies suggest episodic euxinic conditions on a sea bottom near the compensation depth. A parallel with some Tethyan radiolarites may be drawn.

Deep-sea clays: Varicolored, generally brown

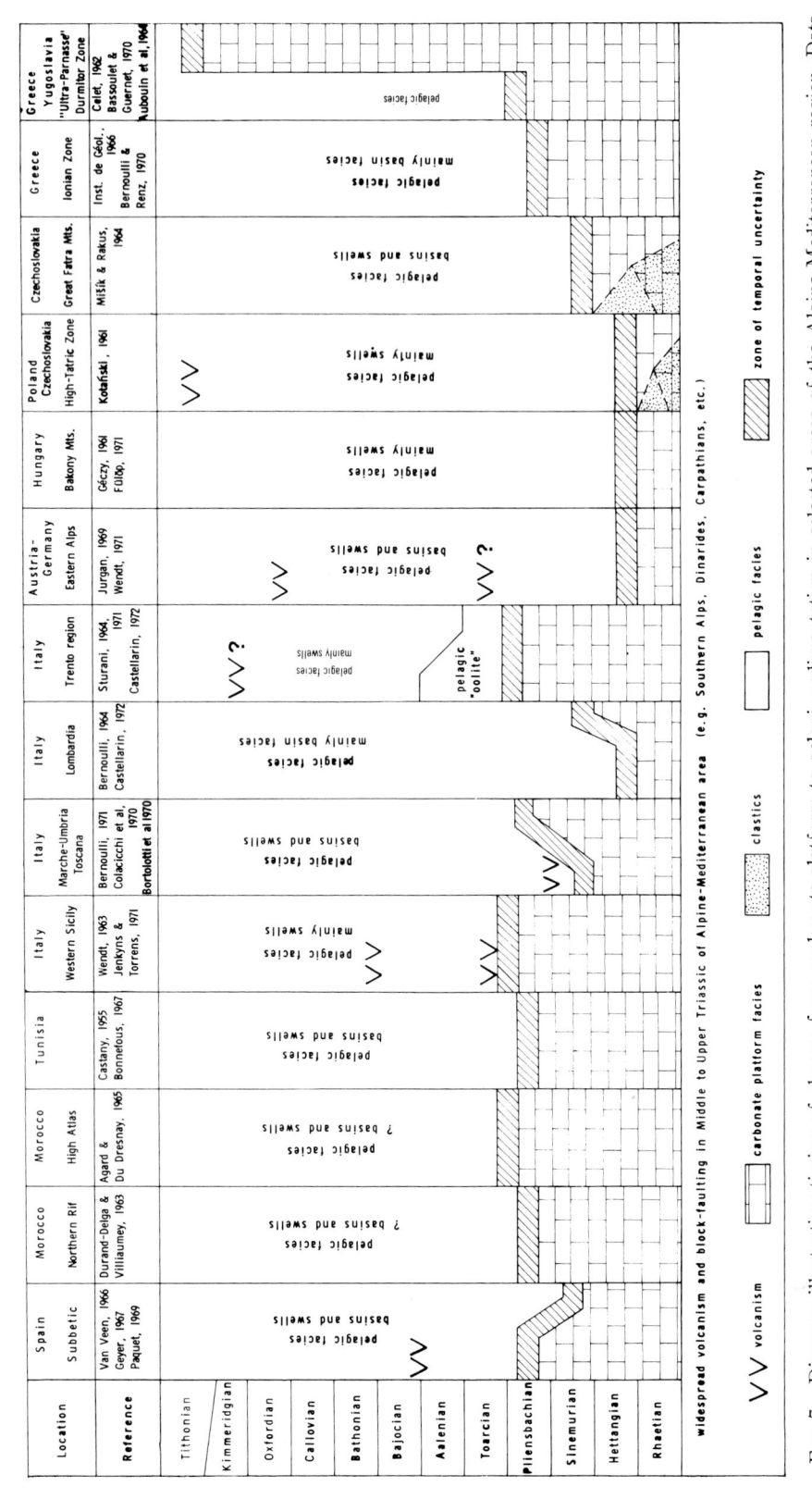

FIG. 7.—Diagram illustrating timing of change from carbonate-platform to pelagic sedimentation in selected areas of the Alpine-Mediterranean region. Data on Jurassic volcanism from Bernoulli and Peters (1970), Boccaletti and Manetti (1972), Fabiani (1926), Germann and Waldvogel (1971), Huckriede (1971), Jenkyns (1970b), Kotański and Radwański (1959), Paquet (1969), and Trevisan (1937).

to yellow zeolitic clays of Late Cretaceous age are widespread (Peterson and others, 1970; Hayes, Pimm, and others, 1972). Multicolored clays associated with abundant iron and manganese oxides and various kinds of sulphide mineralization are abundant at site 105. No definite equivalent of these lithologies occurs in the Alpine-Mediterranean region.

MESOZOIC EVOLUTION OF THE ALPINE-MEDITERRANEAN TETHYS AND WESTERN CENTRAL ATLANTIC

Alpine-Mediterranean Tethys

In Early Triassic time, the Alpine-Mediterranean region probably existed as a unified landmass produced by the Hercynian Orogeny. Sediments of the Lower Triassic include a preponderance of shallow-marine sandstones, some shales and continental red beds, and a few shallow-water carbonates. However, a pelagic limestone of Early Triassic age, recorded by Renz and Renz (1948) from the island of Chios in the Aegean Sea, and a coeval radiolarian chert, illustrated by Mišík (1966) from the Carpathians, suggest the formation of some deeper marine areas sheltered from clastics as early as this time. The Middle and Upper Triassic rocks generally record an increasing marine influence as shown by thick evaporites (Gill, 1965) followed by rapidly accumulating platform carbonates and reef deposits. Rates of subsidence and sedimentation were of the order of 100 mm/10^3 years (Garrison and Fischer, 1969; d'Argenio, 1970b). During this time, block faulting in certain of these shallow-water sites produced some topographically irregular basins in which pelagic limestones were deposited (figs. 7 and 8). These are recorded from the Middle Triassic (e.g., in the Eastern Alps, Schlager, 1967; in the Dolomites, Leonardi and others, 1967; Cros and Lagny, 1969; in Yugoslavia, Ćirić, 1963; in the Carpathians, Mišík, 1966), but they became even more widespread during the Late Triassic (e.g., the Hallstatt facies in the Eastern Alps, Schlager, 1969; in the Carpathians, Mišík, 1966; Patrulius, 1967; in Spain, Busnardo and others, 1969). Although some of these basins were rather deep at times—about a thousand meters can be inferred from geometrical relationships in the basinal Buchenstein Formation of the Dolomites—they were of limited extent and sometimes of short lifespan. More extensive basins, however, developed in Greece (Pindos Zone, Aubouin, 1959a), southern Italy (Lagonegro Zone, Scandone, 1967) and in central Sicily (Sclafani Zone, Broquet and others, 1966).

Tectonism, particularly during the Late Triassic, is documented by dated neptunian dikes[3] in some Alpine areas (Krystyn and others, 1971; Schöll and Wendt, 1971). Widespread volcanic activity, generally yielding felsic or mafic magmas, also took place, particularly in the Middle and Late Triassic Epochs (e.g., quartz keratophyres, Milch and Renz, 1911; Porphyrite-Hornstein Formation, Ćirić and Karamata, 1960). These magmatic events can be traced all around the Alpine-Mediterranean region (fig. 1) and must be considered as one of the most spectacular events in the early history of the Tethys. Lead and zinc ores resulted from certain of these volcanic-hydrothermal activities (Schneider, 1964). Ophiolites of possible Late Triassic age, recorded from the Antalya Nappes, Western Taurids, Turkey (Juteau, 1970), suggest that a small ocean was formed in the eastern Mediterranean at this time. In this context the Porphyrite-Hornstein Formation of the Yugoslavian Dinarides can be interpreted as the record of an abortive ocean, which ended towards the west in a number of troughs of the Pindos type.

Taking a broad view of the Alpine-Mediterranean region during Late Triassic time, we gain the impression of a continental area that was beginning to rift. This beginning is somewhat comparable with the Miocene development of the Red Sea, where the stratigraphic record of the continental margins shows a change from clastic facies to carbonate and evaporite facies representative of the beginning separation of Ethiopia and Arabia (Sestini, 1965; Hutchinson and Engels, 1970). Volcanics, both rhyolitic and basaltic, have been erupted in that area since the Miocene (Black and others, 1972). Potential lead-zinc ore bodies may be represented by the Red Sea geothermal brine deposits (Bischoff, 1969).

On the southern margin of the Tethys, block faulting was widespread during the Early Jurassic. It destroyed many of the preexisting carbonate platforms and reefs (fig. 7). Only limited volcanism accompanied this tectonic event

---

[3] Neptunian dikes (Spaltenfüllungen, filons sédimentaires) are fissures containing sediment whose orientations range from vertical to horizontal even within one dike. Cracking of a consolidated substrate and sediment filling of the resultant voids is the mechanism usually envisaged for their origin. The sedimentary fill commonly derives from above, and composite fills are common. In the Alpine-Mediterranean region, fracture of the substrate is clearly linked to extensional tectonics (Wendt, 1965, 1971a; Castellarin, 1972). In some areas the sediments and fauna of neptunian dikes and sills preserve a more continuous sedimentary and faunal record than that found in the normal stratigraphic succession, which may be punctuated by nonsequences (Wendt, 1971a).

(e.g., western Sicily, Jenkyns and Torrens, 1971; northern Apennines, Boccaletti and Manetti, 1972), but a thermal event of 180 million years ago in the deeper crust of the Southern Alps (Ivrea Zone) may be linked to this phase of block faulting and foundering (Laubscher, 1970). Neptunian dikes and sills, most of them penetrating platform carbonate rocks, were also formed throughout the Alpine-Mediterranean region during the Early Jurassic (Wiedenmayer, 1963; Agard and Du Dresnay, 1965; Wendt, 1965, 1971a; Castellarin, 1965; Jurgan, 1969). After this, pelagic conditions became widespread. Differential subsidence of the shallow-water carbonate blocks gave rise to a seamount-and-basin topography (fig. 8) that closely controlled the facies developments. Stagnant conditions prevailed in certain of the basins (Germann and Waldvogel, 1971). Similar Early Jurassic block faulting and differential subsidence has been established for the northern continental margin of the Tethys (Trümpy, 1960; Plancherel and Weidmann, 1972).

The relationship between block faulting and initial stages of rifting is clearly evident from a study of the Red Sea (Girdler, 1968; Hutchinson and Engels, 1970; Lowell and Genik, 1972). Similar Early Jurassic tectonism probably accompanied, or perhaps preceded, a major phase of sea-floor spreading in the oceanic Tethys (Hallam, 1971b; Smith, 1971). Although most of the block faulting took place during Early Jurassic time, it was undoubtedly diachronous as indicated by figure 7; nevertheless, focus was distinctly within the Pliensbachian-Toarcian interval. Magnetic anomalies and JOIDES results in the central Atlantic Ocean suggest a separation of Africa from North America some 180 million years ago (Pitman and Talwani, 1972), which time, according to the time scale of Harland and others (1964), falls within the Pliensbachian-Toarcian interval. Opening of this part of the Atlantic necessarily involved translational movement between Africa and Eurasia, so that a more or less synchronous opening of the central Atlantic and Alpine-Mediterranean Tethys may be assumed.

Fission-track dating of zircons from ophiolitic gabbros in the Ligurian Apennines has yielded ages of 162 ± 31 my and 167 ± 17 my (Bigazzi and others, 1972). This suggests the time of spreading as Early to Middle Jurassic, which is in accord with timing of the block faulting.

Extensional tectonics persisted throughout the Jurassic in many parts of the Tethys. Mafic volcanism has been recorded from the Middle Jurassic of western Sicily (Fabiani, 1926; Trevisan, 1937) and from the Subbetic of Spain (Paquet, 1969). Bajocian neptunian dikes have been recorded from the Carpathians by Kotański (1961), the Venetian Alps (Castellarin, 1965), and western Sicily (Wendt, 1965). Large amounts of basaltic lavas and tuffs were extruded during the Middle (?) to Late Jurassic along part of the continental margin-ocean boundary now exposed in Yugoslavia and eastern Greece (Diabase-Hornstein Formation, Ćirić and Kamarata, 1960).

The actual configuration of the oceanic Tethys during the Middle and Late Jurassic Epochs can only be surmised. In the western and central Mediterranean area, however, one elongated basin seems to be consistent with palinspastic reconstructions (Laubscher, 1970, 1971a, 1971b); a series of small oceans and microcontinents is an alternative model (Hsü, 1971; Smith, 1971).

The seamount-and-basin topography characterized much of the southern continental margin of the Tethys during the Middle Jurassic (fig. 8). Sinking of the seamounts led to progressive facies changes from biosparites and manganese deposits to red nodular limestones. By Late Jurassic time, water depth had increased and topographic influences decreased except for those seamounts on which so-called pelagic "oolites" were forming (fig. 8). Radiolarites, probably deposited below the calcite-compensation depth, were widespread until Tithonian time, when coccolith oozes in large quantities became the dominant deposits (Garrison and Fischer, 1969). In some areas, however, carbonate platforms persisted throughout the Jurassic (e.g., peri-Adriatic platforms, d'Argenio and others, 1971; eastern Madonie, north-central Sicily, Schmidt di Friedberg and others, 1960). The Plassen Limestone in the Eastern Alps (Fenninger, 1967; Fenninger and Holzer, 1972) also seems to represent an Upper Jurassic platform facies. Some block faulting of carbonate platforms took place in the Late Jurassic (e.g., Greece, Celet, 1962; our fig. 7), and neptunian dikes of this age occur in western Sicily, the Sonnwend Mountains, Austria, the central Apennines, and in the Venetian Alps (Wendt, 1965, 1969a, 1971a; Colacicchi and others, 1970; Castellarin, 1965). Traces of Oxfordian volcanism have been recorded from the Eastern Alps by Huckriede (1971), Kimmeridgian volcanism from the Venetian Alps by Bernoulli and Peters (1970), and Tithonian volcanism from the Carpathians by Kotański and Radwański (1959).

Flysch, apart from an anomalous occurrence in the Aalenian of the Pieniny Klippen Belt, Carpathians (see Birkenmayer and Pazdro, 1968), first appeared in the Tithonian (e.g.,

(Pitman and Talwani, 1972) possibly was accompanied by tectonic movements that initiated the partial disintegration of the western continental margin. The resulting Late Cretaceous environments in this area matched almost exactly coeval situations in the Alpine-Mediterranean region wherever orogenic sedimentation was still excluded.

The oceanic sediments of the western central Atlantic reveal a different story. Some Upper Jurassic red marls and limestones are resedimented, suggesting that parts of the ocean floor were topographically irregular. Indeed, in some areas differential relief of several hundred meters can still be established (Hollister, Ewing, and others, 1972). Deposition of these sediments on a midocean ridge seems probable. Thus, although the central Atlantic facies are lithologically very similar to the basinal Ammonitico Rosso of the Alpine-Mediterranean Jurassic, they were deposited on a different type of basement.

During the Late Jurassic and Early Cretaceous Epochs, the ocean basin of the western central Atlantic subsided deeply. Upward disappearance of slumping in Lower Cretaceous oceanic deposits indicates gradual elimination of sea-bottom topography by a sedimentary blanket. At this time widespread deposits of white nannofossil ooze were exactly matched by Alpine-Mediterranean oceanic and continental-margin counterpart (Maiolica), even with respect to paucity of resedimentation phenomena.

By middle Cretaceous time, the older parts of the ocean basin had sunk below compensation depth, and euxinic conditions prevailed during deposition of black shales. Comparable environments existed in parts of the Alpine-Mediterranean region, but here consumption of oceanic crust was already taking place, along with emplacement of ophiolite nappes. Elimination of this ocean was linked to a pronounced shift in tectonic rotation poles as the northernmost Atlantic began to open during middle to Late Cretaceous time (Pitman and Talwani, 1972). During the Late Cretaceous the western central Atlantic basin continued to subside. In its deeper parts multicolored clays were deposited. No Alpine-Mediterranean equivalent exists for these, and it seems, therefore, that by middle Cretaceous time the formerly parallel evolutionary trends of the Tethys and western central Atlantic were already diverging.

CONCLUSIONS

We have attempted to demonstrate the strong similarity between the Mesozoic histories of the southern continental margin-ocean basin of the Alpine-Mediterranean Tethys and the western central Atlantic. Such a similarity can be viewed in terms of paleotectonics, paleogeography, and facies. Many of the described lithologies can be traced from the Alps through the Himalayas, Indonesia, and back to the Caribbean. In this context the Mesozoic sequences of the Alpine-Mediterranean and western central Atlantic areas can both be considered as representatives of an east-west Mesozoic seaway, or Tethyan realm, that stretched from the Caribbean to Indonesia.

ACKNOWLEDGMENTS

Our studies of Mesozoic facies in the central Mediterranean area are part of a research program on the sedimentary and paleotectonic evolution of the southern Tethys at the Geological Institute of Basel University. Support of this program by the Swiss National Science Foundation (Grant 2.421.70) is gratefully acknowledged. Bernoulli is also very much indebted to the Deep Sea Drilling Project, and particularly to J. Ewing and C. Hollister, for the opportunity to work on the cores from Leg XI. Jenkyns gratefully acknowledges funds from the Royal Society, The Natural Environment Research Council of Great Britain, and the British Association for the Advancement of Science.

Both authors are very grateful to all colleagues who, by demonstration of field evidence and stimulating discussions, have contributed to this paper, particularly A. Castellarin, P. Elter, R. E. Garrison, Y. Lancelot, H. Laubscher, M. Lemoine, H. P. Luterbacher, I. Price and A. G. Smith. R. H. Dott, Jr. and D. S. Gorsline furnished helpful reviews.

Considerable help was received from C. Kapellos in preparing the photomicrographs and from U. Pfirter in drafting the text-figures.

REFERENCES

ABBATE, E., BORTOLOTTI, V., AND PASSERINI, P., 1972, Studies on mafic and ultramafic rocks. 2—Paleogeographic and tectonic considerations on the ultramafic belts in the Mediterranean area: Soc. geol. italiana Boll., v. 91, p. 239–282.
———, AND SAGRI, M., 1970, The eugeosynclinal sequences, *in* SESTINI, G. (ed.), Development of the Northern Apennines Geosyncline, Sed. Geology, v. 4, p. 251–340.
AGARD, J., AND DU DRESNAY, R., 1965, La région minéralisée du jbel Bou-Dahar, près Beno-Tajjite (Haut Atlas oriental): Étude géologique et métallogénique: Service géol. Maroc Notes Mém., no. 181, p. 135–152.
ALDINGER, H., 1957, Zur Entstehung der Eisenoolithe in Schwäbischen Jura: Deutsch. Geol. Gesell. Zeitschr., v. 109, p. 7–9.

―――― AND FRANK, M., 1944, Vorkommen und Entstehung der südwestdeutschen jurassischen Eisenerze: Neues Jahrb. Mineralogie, Geologie und Paläontologie, Abh., v. 88, p. 293–336.
ALLEN, P., 1967, Origin of the Hastings facies in north-western Europe: Geol. Assoc. Proc., v. 78, p. 27–105.
ANDRUSOV, D., 1965, Geologie der tschechoslowakischen Karpaten. II: Berlin, Akademie-Verlag, 443 p.
ARGAND, E., 1916, Sur l'arc des Alpes occidentales: Eclogae Geol. Helvetiae, v. 14, p. 145–204.
ARKELL, W. J., 1956, Jurassic geology of the world: Edinburgh and London, Oliver and Boyd Ltd., 806 p.
AUBOUIN, J., 1959 a, Contribution à l'étude géologique de la Grèce septentrionale: les confins de l'Epire et de la Thessalie: Géol. Pays helléniques Ann., v. 10, p. 1–403.
――――, 1959b, Granuloclassement vertical (graded bedding) et figures de courants (current marks) dans les calcaires purs: les brèches de flanc des sillons géosynclinaux: Soc. géol. France Bull., sér. 7, v. 1, p. 578–582.
――――, 1964, Réflexions sur le faciès "Ammonitico Rosso": *ibid.,* v. 6, p. 475–501.
――――, 1965, Geosynclines. Developments in geotectonics, v. 1: Amsterdam, Elsevier, 335 p.
――――, CADET, J., RAMPNOUX, J.-P., DUBAR, G., AND MARIE, P., 1964, A propos de l'âge de la série ophiolitique dans les Dinarides yougoslaves: la coupe de Mihajlovici aux confins de la Serbie et du Monténégro (région de Plevlja, Yougoslavie): Soc. géol. France Bull., sér. 7, v. 6, p. 107–112.
BAILEY, E. B., 1924, The desert shores of the Chalk Seas: Geol. Mag., v. 61, p. 102–116.
――――, AND McCALLIEN, W. J., 1960, Some aspects of the Steinmann Trinity: mainly chemical: Geol. Soc. London, Quart. Jour. v. 116, p. 365–395.
BALL, M. M., 1967, Tectonic control of the configuration of the Bahama Banks: Gulf Coast Assoc. Geol. Socs. Trans., v. 17, 265–267.
BASSOULLET, J. P., AND GUERNET, J. C., 1970, Le Trias et le Jurassique de la région des lacs de Thèbes (Béotie et Locride, Grèce): Rev. Micropaléontologie, v. 12, p. 209–217.
BEAUSEIGNEUR, C., AND RANGHEARD, Y., 1967, Contribution a l'étude des roches éruptives de l'ile d'ibiza (Baléares): Soc. géol. France Bull., sér. 7, v. 9, p. 221–224.
BELLAIR, P., 1948, Pétrographie et tectonique des massifs centraux dauphinois. I. le haut massif: Service Carte géol. France Mém., 345 p.
BERNOULLI, D., 1964, Zur Geologie des Monte Generoso (Lombardische Alpen): Geol. Karte Schweiz Beitr., new ser., v. 118, 134 p.
――――, 1967, Probleme der Sedimentation im Jura Westgriechenlands und des zentralen Apennin: Naturf. Gesell. Basel, Verh., v. 78, p. 35–54.
――――, 1971, Redeposited pelagic sediments in the Jurassic of the Central Mediterranean Area, *in,* VÉGH NEUBRANDT, E. (ed.), Colloque du Jurassique méditerranéen: Inst. Geol. Pub. Hungarici Ann., v. 54/2, p. 71–90.
――――, 1972, North Atlantic and Mediterranean Mesozoic facies: a comparison, *in* HOLLISTER, C. D. AND EWING, J. I., AND OTHERS, Initial reports Deep Sea Drilling Project, v. 11: Washington, D.C., U.S. Govt. Printing Office, p. 801–871.
――――, AND JENKYNS, H. C., 1970, A Jurassic basin: the Glasenbach Gorge, Salzburg, Austria: Geol. Bundesanstalt Wien Verh., 1970, p. 504–531.
――――, AND LAUBSCHER, H., 1972, The palinspastic problem of the Hellenides: Eclogae. Geol. Helvetiae, v. 65, p. 107–118.
――――, AND PETERS, TJ., 1970, Traces of rhyolitic-trachytic volcanism in the Upper Jurassic of the Southern Alps: *ibid.,* v. 63, p. 609–621.
――――, AND RENZ, O., 1970, Jurassic carbonate facies and new ammonite faunas from western Greece: *ibid.,* p. 573–607.
――――, AND WAGNER, C., 1971, Subaerial diagenesis and fossil caliche deposits in the Calcare Massiccio Formation (Lower Jurassic, central Apennines, Italy): Neues Jahrb. Geologie und Paläontologie, Abh., v. 138, p. 135–149.
BEZZI, A., AND PICCARDO, G. B., 1971, Structural features of the Ligurian ophiolites: petrologic evidence for the "oceanic" floor of the northern Apennines geosyncline; a contribution to the problem of the Alpine-type gabbro-peridotite associations: Soc. geol. italiana Mem, v. 10, p. 53–63.
BICKLE, M. J., AND NISBET, E., 1972, The oceanic affinities of some Alpine mafic rocks based on their Ti-Zr-Y contents: Geol. Soc. London Jour., v. 128, p. 267–271.
BIGAZZI, G., FERRARA, G., AND INNOCENTI, F., 1972, Fission track ages of gabbro from northern Apennines ophiolites: Earth and Planetary Sci. Letters, v. 14, p. 242–244.
BIRKENMAYER, K., 1963, Stratigraphy and palaeogeography of the Czorsztyn Series (Pieniny Klippen Belt, Carpathians) in Poland: Studia Geol. Polonica, v. 9, 380 p.
――――, AND PAZDRO, O., 1968, On the so-called "Sztolnia Beds" in the Pieniny Klippen Belt of Poland: Acta Geol. Polonica, v. 18, p. 325–365.
BISCHOFF, J. L., 1969, Red Sea geothermal brine deposits: their mineralogy, chemistry, and genesis, *in* DEGENS, E. T., AND ROSS, D. A. (eds.), Hot brines and Recent heavy metal deposits in the Red Sea: Berlin, Heidelberg, and New York, Springer Verlag, p. 368–401.
BLACK, M., 1953, The constitution of the Chalk: Geol. Soc. London Proc. 1949, p. lxxxi-lxxxvi.
BLACK, R., MORTON, W. H., AND VARET, J., 1972, New data on Afar tectonics: Nature, Phys. Sci. v. 240, p. 170–173.
BLANCHET, R., CADET, J. P., AND CHARVET, J., 1970, Sur l'existence d'unités intermédiaires entre la zone du Haut-Karst et l'unité du flysch bosniaque, en Yougoslavie: la sous-zone prékarstique: Soc. géol. France Bull., sér. 7, v. 12, p. 227–236.
――――, ――――, ――――, AND RAMPNOUX, J. P., 1969, Sur l'existence d'un important domaine de flysch tithonique-crétacé inférieur en Yougoslavie: l'unité du flysch bosniaque: *ibid.,* v. 11, p. 871–880.
BOCCALETTI, M., AND MANETTI, P., 1972, Traces of lower-middle Liassic volcanism in the crinoidal limestones of the Tuscan sequence in the Montemerano area (Grosseto, northern Apennines): Eclogae Geol. Helvetiae, v. 65, p. 119–129.

BOLLI, H., 1944, Zur Stratigraphie der Oberen Kreide in den höheren helvetischen Decken: *ibid.*, v. 37, p. 217–329.

BONNEFOUS, J., 1967, Jurassic stratigraphy of Tunisia: a tentative synthesis (northern and central Tunisia, Sahel and Chotts areas), *in* MARTIN, L. (ed.), Guidebook to the geology and history of Tunisia: Tripoli, Petroleum Explor. Soc. Libya, p. 109–130.

BOOY, DE T., 1969, Repeated disappearance of continental crust during the geological development of the western Mediterranean area: K. Nederlands Geol.-Mijnb. Genoot. Verh., v. 26, p. 79–103.

BORTOLOTTI, V., DAL PIAZ, G. V., AND PASSERINI, P., 1969, Ricerche sulle ofioliti delle catene alpine: 5 nuove osservazioni sul massiccio del Vourinos (Grecia): Soc. geol. italiana Bol., v. 88, p. 35–45.

———, AND PASSERINI, P., in press, Paleogeographic and tectonic considerations on the ultramafic belts in the Mediterranean area: a correction: Soc. geol. italiana Boll.

———, ———, SAGRI, M., AND SESTINI, G., 1970, The miogeosynclinal sequences, *in* SESTINI, G. (ed.), Development of the northern Apennines geosyncline: Sed. Geology, v. 4, p. 341–444.

BOSELLINI, A., 1967a, La tematica deposizionale della Dolomia Principale (Dolomiti e Prealpi Venete): Soc. geol. italiana Boll., v. 86, p. 133–169.

———, 1967b, Torbiditi carbonatiche nel Giurassico delle Giudicarie e loro significato geologico: Univ. Ferrara Ann., Sezione IX, Sci. geol. e paleont., v. 4, p. 101–115.

———, AND BROGLIO LORIGA, C., 1971, I Calcari Grigi di Rozzo (Giurassico Inferiore, Altipiano di Asiago) e loro inquadramento nella paleogeografia e nella evoluzione tettonosedimentaria delle Prealpi Venete: *ibid.*, v. 5, p. 1–61.

BOURBON, M., 1971, Structure et signification de quelques nodules ferrugineux, manganésifères et phosphatés liés aux lacunes de la série crétacée et paléocène briançonnaise: Acad. Sci. (Paris), Comptes Rendus, ser. D, v. 273, p. 2060–2062.

BROQUET, P., CAIRE, A., AND MASCLE, G., 1966, Structure et évolution de la Sicile occidentale (Madonies et Sicani): Soc. géol. France Bull. sér. 7, v. 8, p. 994–1013.

BRYANT, W. R., MEYERHOFF, A. A., BROWN, N. K., FURRER, M. A., PYLE, T. E., AND ANTOINE, T. W., 1969, Escarpments, reef trends and diapiric structures, eastern Gulf of Mexico: Am. Assoc. Petroleum Geologists Bull., v. 53, p. 2506–2542.

BUBENICHEK, L., 1961, Recherches sur la constitution et la répartition du minerais de fer dans l'Aalénien de Lorraine: Sciences Terre, v. 8, p. 5–204.

BUSNARDO, R., LINARES, A., AND MOUTERDE, R., 1969, Trias fossilifère à faciès pélagique près de Alhama de Granada (Andalousie): Acad. Sci. (Paris), Comptes Rendus, sér. D, v. 268, p. 1364–1367.

BUSSON, G., 1967, Le Mésozoïque saharien. Part I. Extrême-sud Tunisien: Centre Rech. Zones Arides, sér. géol., no. 8, 194 p.

———, 1970, Le Mésozoïque saharien. Part II. Essai de synthèse des données des sondages algéro-tunisiens: *ibid.*, no. 11, 811 p. (2 vols.).

CASTANY, G., 1955, Les extrusions jurassiques en Tunisie: Mines Géol. Tunis Ann., v. 14, p. 1–71.

———, 1956, Essai de synthèse géologique du territoire Tunisie-Sicile: *ibid.*, v. 16, p. 1–101.

CASTELLARIN, A., 1965, Filoni sedimentari nel Giurese di Loppio (Trentino meridionale): Gior. Geologia, ser. 2, v. 33, p. 527–546.

———, Evoluzione paleotettonica sinsedimentaria del limite tra "piattaforma veneta" e "bacino lombardo" a nord di Riva del Garda: *ibid.*, v. 38, p. 11–212.

CAYEUX, L., 1909, Les minérais du fer oolithique de France. I. Minerais de fers primaires: Paris, Imprimerie Nationale, 344 p.

———, 1924, La question des jaspes à radiolaires: Soc. géol France, Comptes Rendus 1924, p. 11–12.

CELET, P., 1962, Contribution à l'étude géologique du Parnasse-Kiona et d'une partie des régions méridionales de la Grèce continentale: Geol. Pays helléniques Ann., v. 13, p. 1–446.

CHOWNS, T. M., 1966, Depositional environment of the Cleveland Ironstone Series: Nature, v. 211, p. 1286–1287.

ĆIRIĆ, B., 1963, Le développment des Dinarides Yougoslaves pendant le cycle alpin: Soc. géol. France, Livre à la mémoire du Prof. P. Fallot, v. 2, p. 565–582.

———, AND KARAMATA, S., 1960, L'évolution du magmatisme dans le géosynclinal dinarique au Mésozoïque et au Cénozoïque: *ibid.*, ser. 7, v. 2, p. 376–380.

COLACICCHI, R., 1963, Geologia del territorio di Pachino (Sicilia meridionale): Geologica Romana, v. 2, p. 343–404.

———, PASSERI, L., AND PIALLI, G., 1970. Nuovi dati sul Giurese umbro-marchigiano ed ipotesi per un suo inquadramento regionale: Soc. geol. italiana Mem., v. 9, p. 839–874.

COLOM, G., 1955, Jurassic-Cretaceous pelagic sediments of the western Mediterranean zone and the Atlantic area: Micropaleontology, v. 1, p. 109–124.

———, 1965, Essai sur la biologie, la distribution géographique et stratigraphique des Tintinnoïdiens fossiles: Eclogae Geol. Helvetiae, v. 58, p. 319–334.

CRESCENTI, U., 1969, Stratigrafia della serie calcarea dal Lias al Miocene nella regione marchigiano-abruzzese (Parte 1—Descrizione delle serie stratigrafiche): Soc. geol. italiana Mem., v. 8, p. 155–204.

———, CROSTELLA, A., DONZELLI, G., AND RAFFI, G., 1969, Stratigrafia della serie calcarea dal Lias al Miocene nella regione marchigiano-abruzzese. (Parte 2—Litostratigrafia, biostratigrafia, paleogeografia): *ibid.*, p. 343–420.

CRISTOFOLINI, R., 1966, Le manifestazioni eruttive basiche del Trias superiore nel sottosuolo di Ragusa (Sicilia sud-orientale): Periodico Mineral., v. 35, p. 1–28.

CROS, P., AND LAGNY, P., 1969, Paléokarsts dans le Trias moyen et supérieur des Dolomites et des Alpes Carniques occidentales. Importance stratigraphique et paléogéographique: Sciences Terre, v. 14, p. 139–195.

D'ARGENIO, B., 1966, Le facies littorali mesozoiche nell' Appennino meridionale: Soc. nat. Napoli Boll., v. 75, p. 497–552.

———, 1970a, Central and southern Italy Cretaceous bauxites: Inst. Geol. Pub. Hungarici Ann., v. 54/3, p. 221–233.

———, 1970b, Evoluzione geotettonica comparata tra alcune piattaforme carbonatiche dei Mediterranei Europeo ed Americano: Accad. Pontaniana Atti, new ser., v. 20, p. 3–34.

———, Radoičić, R., and Sgrosso, I., 1971, A paleogeographic section through the Italo-Dinaric external zones during Jurassic and Cretaceous times: Nafta, Zagreb, v. 22, p. 195–207.

Decandia, F. A., and Elter, P., 1969, Riflessioni sul problema delle ofioliti nell'Appennino settentrionale (nota preliminare): Soc. tosc. Sci. nat. Mem., ser. A, v. 76, p. 1–9.

Deverin, L., 1940, Les minérais de fer oolithiques du Dogger des Alpes Suisses: Schweizerische Mineralog. und Petrog. Mitt., v. 20, p. 101–116.

———, 1948, Oolithes ferrugineuses des Alpes et du Jura: ibid., v. 28, p. 95–103.

Dietrich, D., and Scandone, P., 1972. The position of the basic and ultrabasic rocks in the tectonic units of the southern Apennines: Accad. Pontaniana Atti, new ser., v. 21, p. 1–15.

Dietz, R. S., Holden, J. C., and Sproll, W. P., 1970, Geotectonic evolution and subsidence of Bahama Platform: Geol. Soc. America Bull., v. 81, p. 1915–1928.

Donzelli, G., and Crescenti, U., 1970, Segnalazione di una microbiofacies permiana, probabilmente rimaneggiata, nella formazione di M. Facito (Lucania occidentale): Soc. naturalisti Napoli Boll., v. 79, p. 3–19.

Dresnay, R. du, 1956, Contribution à l'étude de la série détritique jurassico-crétacée dans le Haut-Atlas oriental: Service géol. Maroc Notes, v. 14, p. 9–32.

———, 1963, Données stratigraphiques complémentaires sur le Jurassique moyen des synclinaux d'El mers et de Skoura (Moyen-Atlas, Maroc): Soc. géol. France Bull., sér. 7, v. 5, p. 883–900.

Dufour, T., and Noël, D., 1970, Nannofossiles et constitution pétrographique de la "Maiolica," des "Schistes à fucoides" et de la "Scaglia rossa" d'Ombrie: Rev. Micropaléontologie, v. 13, p. 107–114.

Durand-Delga, M., 1960, Le sillon géosynclinal du Flysch tithonique-néocomien de la Méditerranée occidentale: Accad. naz. Lincei Rendiconti, cl. sci. fis., mat. e nat., ser. 8, v. 29, p. 579–585.

———, and Villiaumey, M., 1963, Sur la stratigraphie et la téctonique du groupe du Jebel Musa (Rif septentrional, Maroc): Soc. géol. France Bull., sér. 7, v. 5, p. 70–79.

Elter, G., 1971, Schistes lustrés et ophiolites de la zone piémontaise entre Orco et Doire Baltée (Alpes Graies). Hypothèses sur l'origine des ophiolites: Géologie Alpine, v. 47, p. 147–169.

———, Elter, P., Sturani, C., and Weidmann, M., 1966, Sur la prolongation du domaine ligure de l'Apennin dans le Monferrat et les Alpes et sur l'origine de la Nappe de la Simme s.l. des Préalpes romandes et chablaisiennes: Archives Sci., Genève, v. 19/3, p. 279–377.

Elter, P., 1960, I lineamenti tettonici dell'Appennino a nord-ovest delle Apuane: Soc. geol. italiana Boll., v. 79/2, p. 273–312.

———, and Raggi, G., 1965a, Osservazioni preliminari sulla posizione delle ofioliti nella zona di Zignago (La Spezia): ibid., v. 84/3, p. 303–322.

———, and ———, 1965b, Tentativo di interpretazione delle brecce ofiolitiche cretacee in relazione con movimenti orogenetici nell'Appennino ligure: ibid., v. 84/5, p. 1–12.

Emery, K. O., Uchupi, E., Phillips, J. D., Bowin, C. O., Bunce, E. T., and Knott, S. T., 1970, Continental rise off eastern North America: Am. Assoc. Petroleum Geologists Bull. v. 54, p. 44–108.

Ewing, M., Worzel, J., and others, 1969, Initial reports of the Deep Sea Drilling Project, v. 1.: Washington D.C., U.S. Gov. Printing Office, 672 p.

Fabiani, R., 1926, Scoperta di un apparato eruttivo del Giurese medio in Sicilia: Assoc. Mineralog. Siciliana Boll., v. 2, p. 3–12.

Fabricius, F., 1966, Beckensedimentation und Riffbildung an der Wende Trias/Jura in den Bayrisch-Tiroler Kalkalpen: Internat. Sed. Petrology Ser.: Leiden, Netherlands, Brill, v. 9, 143 p.

Farinacci, A., 1964, Microrganismi dei calcari "Maiolica" e "Scaglia" osservati al microscopio elettronico (Nannoconi e Coccolithophoridi): Soc. paleont. italiana Boll., v. 3, p. 172–181.

Fenninger, A., 1967, Riffentwicklung im oberostalpinen Malm: Geol. Rundschau, v. 56, p. 171–185.

———, and Holzer, H. L., 1970, Fazies und Paläogeographie des oberostalpinen Malm: Geol. Gesell. Wien Mitt., v. 63, p. 52–141.

Ferasin, F., 1958. Il "Complesso di Scogliera" cretaceo del Veneto centro-orientale: 1st. Geol. Mineralog. Univ. Padova Mem., v. 21, p. 1–54.

Fischer, A. G., 1964, The Lofer Cyclothems of the Alpine Triassic, in, Merriam, D. W. (ed.), Symposium on cyclic sedimentation: Kansas Geol. Survey Bull., v. 169, p. 107–149.

Flügel, H., and Pölsler, P., 1965, Lithogenetische Analyse der Barmstein-Kalkbank B_2 nordwestlich von St. Koloman bei Hallein (Tithonium, Salzburg): Neues Jahrb. Geologie und Paläontologie, Monatsh., 1965, p. 513–527.

Fuchs, T., 1883, Welche Ablagerungen haben wir als Tiefseebildungen zu betrachten?: neues Jahrb. Mineralogie, Geologie u. Paläontologie, Beil. Bd., v. 2, p. 487–584.

Fülöp, J., 1971, Les formations Jurassiques de la Hongrie, in Végh-Neubrandt, E. (ed.), Colloque du Jurassique méditerranéen: Inst. Geol. Pub. Hung. Ann., v. 54/2, p. 31–46.

Galácz, A., and Vörös, A., 1972, Jurassic history of the Bakony Mountains and interpretation of principal lithological phenomena: Földtani Közlöny (Geol. Soc. Hungary Bull.), v. 102, p. 122–135.

Galli, M., and Togliatti, V., 1965, Ricerche petrografiche sulla formazione ofiolitica dell'Appennino ligure. Il Rosso di Levanto. Nuovo Contributo: Mus. Civ. Storia Nat. Genova Ann., v. 75, p. 359–381.

Garrison, R. E., 1967, Pelagic limestones of the Oberalm Beds (Upper Jurassic-Lower Cretaceous), Austrian Alps: Bull. Canadian Petroleum Geology, v. 15, p. 21–49.

———, and Fischer, A. G., 1969, Deep-water limestones and radiolarites of the Alpine Jurassic, in, Friedman, G. M. (ed.), Depositional environments in carbonate rocks, a symposium: Tulsa, Oklahoma, Soc. Econ. Paleontologists and Mineralogists, Special Pub. 14, p. 20–56.

GÉCZY, B., 1961, Die jurassische Schichtreihe des Tuzkoves-Grabens von Bakonycsernye: Inst. Geol. Pub. Hungarici Ann., v. 49, p. 507–563.
GEMMELLARO, G. G., 1904, I cefalopodi del Trias superiore della regione occidentale della Sicilia: Gior. Sci. Nat. Econ. Palermo., v. 24, p. 1–319.
GERMANN, K., 1971, Mangan-Eisen-führende Knollen und Krusten in jurassischen Rotkalken der Nördlichen Kalkalpen: Neues Jahrb. Geologie und Paläontologie, Monatsh., 1971, p. 133–156.
———, AND WALDVOGEL, F., 1971, Mineralparagenesen und Metallgehalte der "Manganschiefer" (unteres Toarcian) in den Allgäu-Schichten der Allgäuer und Lechtaler Alpen: ibid., Abh., v. 139, p. 316–345.
GEYER, O. F., 1967, Zur faziellen Entwicklung des subbetischen Juras in Südspanien: Geol. Rundschau, v. 56, p. 973–992.
GILL, W. D., 1965, The Mediterranean Basin, in Salt basins around Africa: London, Inst. Petroleum, p. 101–111.
GIRDLER, R. W., 1968, Drifting and rifting of Africa: Nature, v. 217, p. 1102–1105.
GOLDRING, R., 1961, The Bathyal Lull: Upper Devonian and Lower Carboniferous sedimentation in the Variscan Geosyncline, in COE, K. (ed.), Some aspects of the Variscan fold belt.: Manchester, England, Manchester Univ. Press, p. 75–91.
GOODELL, H. G., AND GARMAN, R. K., 1969, Carbonate geochemistry of Superior deep test well, Andros Island, Bahamas: Am. Assoc. Petroleum Geologists Bull., v. 53, p. 513–536.
GRUNAU, H. R., 1965, Radiolarian cherts and associated rocks in space and time: Eclogae Geol. Helvetiae, v. 58, p. 157–208.
———, AND STUDER, H., 1956, Elektronenmikroskopische Untersuchungen an Bianconekalken des Südtessins: Experientia, v. 12, p. 141–150.
GUERNET, C., 1965, Formations éruptives antéjurassiques en Eubée moyenne (Grèce): Soc. géol. France Bull., sér. 7, v. 8, p. 56–58.
HALLAM, A., 1966, Depositional environment of British Liassic ironstones considered in the context of their facies relationships: Nature, v. 209, p. 1306–1309.
———, 1971a, Evaluation of bathymetric criteria for the Mediterranean Jurassic, in VÉGH-NEUBRANDT, E. (ed.), Colloque du Jurassique méditerranéen: Inst. Geol. Pub. Hungarici Ann., v. 54/2, p. 63–70.
———, 1971b, Mesozoic geology and the opening of the North Atlantic: Jour. Geology v. 79, p. 129–157.
HARLAND, W. B., SMITH, A. G., AND WILCOCK, B., 1964, The Phanerozoic time scale, a symposium: Geol. Soc. London Quart. Jour., v. 120 (supp.), 458 p.
HAUG, E., 1900, Les géosynclinaux et les aires continentales. Contribution à l'étude des régressions et des transgressions marines: Soc. géol. France Bull., sér. 3, v. 28, p. 617–711.
HAYES, D. E., PIMM, A. C., AND OTHERS, 1972, Initial reports of the Deep Sea Drilling Project: Washington, D.C., U.S. Govt. Printing Office, v. 14, 975 p.
HEIM, A., 1924, Über submarine Denudation und chemische Sedimente: Geol. Rundschau, v. 15, p. 1–47.
HERAK, M., POLŠAK, I., GUSIĆ, I., AND BABIĆ, LJ., 1970, Dynamische und räumliche Sedimentationsbedingungen der mesozoischen Karbonatgesteine im Dinarischen Karstgebiet: Geol. Bundesanstalt Wien Verh., 1970, p. 637–643.
HESS, H. H., 1965, Mid-ocean ridges and tectonics of the sea floor, in Colston Papers, Proc. 17th Symposium Colston Research Soc., Univ. Bristol, Sci. Pub.: London, Butterworths and Co., p. 317–333.
HOLLISTER, C. D., EWING, J. I., AND OTHERS, 1972, Initial reports of the Deep Sea Drilling Project: Washington, D.C., U.S. Govt. Printing Office, v. 11, 1077 p.
HOLLMANN, R., 1962, Über Subsolution und die "Knollenkalke" des Calcare Ammonitico Rosso Superiore im Monte Baldo: Neues Jahrb. Geologie und Paläontologie, Monatsh., 1962, p. 163–179.
———, 1964, Subsolutions-Fragmente (Zur Biostratinomie der Ammonoidea im Malm des Monte Baldo/Norditalien): ibid., Abh., v. 119, p. 22–82.
HOPPE, P., 1968, Stratigraphie und Tektonik der Berge um Grazalema (S-W Spanien): Geol. Jahrb., v. 86, p. 267–338.
HSÜ, K. J., 1971, Origin of the Alps and western Mediterranean: Nature, v. 233, p. 44–48.
HUCKRIEDE, R., 1971, Rhyncholithen-Anreicherung (Oxfordium) an der Basis des älteren Radiolarits der Salzburger Kalkalpen: Geologica et Palaeontologica, v. 5, p. 131–147.
HUDSON, J. D., AND JENKYNS, H. C., 1969, Conglomerates in the Adnet Limestone of Adnet (Austria) and the origin of the "Scheck": Neues Jahrb. Geologie und Paläontologie, Monatsh., 1969, p. 552–558.
HUTCHINSON, R. W., AND ENGELS, G. G., 1970, Tectonic significance of regional geology and evaporite lithofacies in north-eastern Ethiopa: Royal Soc. (London) Phil. Trans, ser. A, v. 267, p. 313–329.
HYNES, A. J., NISBET, E. G., SMITH, A. G., WELLAND, M. J. P., AND REX, D. C., 1972, Spreading and emplacement ages of some ophiolites in the Othris region, eastern central Greece: Deutsch. Geol. Gesell., Zeitschr., v. 123, p. 445–468.
INSTITUT DE GÉOLOGIE ET RECHERCHES DU SOUS-SOL, ATHÈNES ET INSTITUT FRANÇAIS DU PÉTROLE-MISSION GRÈCE, 1966, Étude géologique de l'Epire (Grèce nord-occidentale): Paris, Editions Technip and Inst. Français Pétrole, 306 p.
JACOBSHAGEN, V., 1967, Cephalopoden-Stratigraphie der Hallstätter Kalke am Asklepieion von Epidauros (Argolis, Griechenland): Geologica et Palaeontologica, v. 1, p. 13–33.
———, 1972, Die Trias der mittleren Öst-Ägäis und ihre paläogeographischen Beziehungen innerhalb der Helleniden: Deutsch. Geol. Gesell. Zeitschr., v. 123, p. 445–454.
JENKYNS, H. C., 1970a, Fossil manganese nodules from the west Sicilian Jurassic: Eclogae Geol. Helvetiae, v. 63, p. 741–774.
———, 1970b, Submarine volcanism and the Toarcian iron pisolites of western Sicily: ibid., v. 63, p. 549–572.
———, 1970c, The Jurassic of western Sicily, in ALVAREZ, W., GOHRBANDT, K. H. A. (eds.), Geology and history of Sicily: Tripoli, Petroleum Explor. Soc. Libya, p. 245–254.
———, 1971a, The genesis of condensed sequences in the Tethyan Jurassic: Lethaia, v. 4, p. 327–352.

——, 1971b, Speculations on the genesis of crinoidal limestones in the Tethyan Jurassic: Geol. Rundschau, v. 60, p. 471–488.
——, 1972, Pelagic "oolites" from the Tethyan Jurassic: Jour. Geology, v. 80, p. 21–33.
——, AND TORRENS, H. S., 1971, Palaeogeographic evolution of Jurassic seamounts in western Sicily, *in* VÉGH-NEUBRANDT, E. (ed.), Colloque du Jurassique méditeranéen: Inst. Geol. Pub. Hung. Ann., v. 54/2, p. 91–104.
JODOT, P., 1933, Notes de pétrographie sédimentaire sur la Corse: Soc. géol. France Bull., sér. 5, v. 3, p. 767–798.
JOJA, T., ALEXANDRESCU, G., BERCIA, I., AND MUTIHAC, V., 1968, Note explicative pour la Carte Géologique de la Roumanie, sheet 5, Radauti: Inst. Géol. Roumanie, 63 p.
JURGAN, H., 1969, Sedimentologie des Lias der Berchtesgadener Kalkalpen: Geol. Rundschau, v. 58, p. 464–501.
JUTEAU, T., 1970, Pétrogénèse des ophiolites des nappes d'Antalya (Taurus lycien oriental, Turquie). Leur liaison avec une phase d'expansion océanique active au Trias supérieur: Sci. Terre v. 15, p. 265–288.
KALLENBACH, H., 1972, Beiträge zur Sedimentologie des kontinentalen Mesozoikums am Westrand des Murzukbeckens (Libyen): Geol. Rundschau, v. 61, p. 302–322.
KASZAP, A., 1963, Investigations on the microfacies of the Malm beds of the Villány Mountains: Univ. Sci. Budapestinensis Rolando Eötvös Nominatae Ann., Sec. Geol., v. 6, p. 47–57.
KOCHANSKY-DÉVIDÉ, V., 1957, Die Neoschwagerinenfaunen der südlichen Crna Gora (Montenegro): Geol. Vjesnik, v. 11, p. 45–76.
KOTAŃSKI, Z., 1961, Tectonogénèse et réconstitution de la paléogéographie de la zone haut-tatrique dans les Tatras: Acta Geol. Polonica, v. 11, p. 187–476.
——, AND RADWAŃSKI, A., 1959, High-tatric Tithonian in Osobita region, its fauna with Pygope diphya and products of the volcanoes (western Tatra (Mts.): *ibid.*, v. 9, p. 519–534.
KRYSTYN, L., SCHÄFFER, G., AND SCHLAGER, W., 1971, Über die Fossil-Lagerstätten in den triadischen Hallstätter Kalken der Ostalpen: Neues Jahrb. Geologie und Paläontologie, Abh., v. 137, p. 284–304.
KUHRY, B., 1972, Stratigraphy and micropaleontology of the Lower Cretaceous in the Subbetic south of Caravaca (Province of Murcia, SE Spain). I and II: K. Nederlands. Akad. Wetensch. Proc., ser. B, v. 75, p. 194–222.
LANCELOT, Y., HATHAWAY, T. C., AND HOLLISTER, C. D., 1972, Lithology of sediments from the western North Atlantic, leg IX, Deep Sea Drilling Project, *in* HOLLISTER, C. D., EWING, J. I., AND OTHERS, Initial reports Deep Sea Drilling Project: Washington, D.C., U.S. Govt. Printing Office, v. 11, p. 901–949.
LAUBSCHER, H. P., 1969, Mountain building: Tectonophysics, v. 7, p. 551–563.
——, 1970, Bewegung und Wärme in der alpinen Orogenese: Schweizerische Mineralog. and Petrog. Mitt. v. 50, p. 503–534.
——, 1971a, Das Alpen-Dinariden-Problem und die Palinspastik der südlichen Tethys: Geol. Rundschau, v. 60, p. 813–833.
——, 1971b, The large-scale kinematics of the Western Alps and the northern Apennines and its palinspastic implications: Am. Jour. Sci., v. 271, p. 193–226.
LAUGHTON, A. S., BERGGREN, W. A., AND OTHERS, 1972, Initial reports of the Deep Sea Drilling Project: Washington, D.C., U.S. Govt. Printing Office, v. 12, 1243 p.
LEFELD, J., AND RADWAŃSKI, A., 1960, Les crinoides planctoniques Saccocoma Agassiz dans le Malm et le Néocomien haut-tatrique des Tatras Polonaises: Acta Geol. Polonica, v. 10, p. 593–610.
LEMOINE, M., 1971, Données nouvelles sur la série du Gondran près Briançon (Alpes Cottiennes). Réflexions sur les problemes stratigraphique et paléogéographique de la zone piémontaise: Geologie Alpine, v. 47, p. 181–201.
——, STEEN, D., AND VUAGNAT, M., 1970, Sur le problème stratigraphique des ophiolites piémontaises et des roches sédimentaires associées: observations dans le massif de Chabrière en haute Ubaye (Basses-Alpes, France): Soc. Phys. et Histoire Nat. Genève, Comptes Rendus Séances, new ser., v. 5, p. 44–59.
LEONARDI, P., AND OTHERS, Le Dolomiti. Geologia dei monti tra Isarco e Piave: Trento, Italy, Consiglio Naz. Ricerche, 2 vols., 1019 p.
LOOMIS, T. P., 1972, Diapiric emplacement of the Ronda high-temperature ultramafic intrusion, southern Spain: Geol. Soc. America Bull., v. 83, p. 2475–2496.
LOWELL, J. D., AND GENIK, G. J., 1972, Sea-floor spreading and structural evolution of the Red Sea: Am. Assoc. Petroleum Geologists Bull., v. 56, p. 247–259.
LUCAS, G., 1942, Description géologique et pétrographique des Montes de Ghar Rhouban et du Sidi el Abed (frontière algéro-marocaine): Service Carte géol Algerie Bull., 2 sér., no. 16, 2 vols., 538 p.
——, 1955, Caractères pétrographiques des calcaires noduleux, à faciès *ammonitico rosso*, de la région méditerranéenne: Acad. Sci. (Paris), Comptes Rendus, v. 240, p. 1909–1911.
LUTERBACHER, H. P., 1972, Foraminifera from the Lower Cretaceous and Upper Jurassic of the northwestern Atlantic *in* HOLLISTER, C. D., EWING, J. I., AND OTHERS, Initial reports of Deep Sea Drilling Project: Washington, D.C., U.S. Govt. Printing Office, v. 11, p. 561–593.
——, AND PREMOLI SILVA, I., 1964, Biostratigrafia del limite Cretaceo-Terziario nell'Appennino Centrale: Rivista italiana Paleontologia, v. 70, p. 67–128.
MAHEL, M., BUDAY, T., AND OTHERS, 1968, Regional geology of Czechoslovakia. Part II. The west Carpathians: Prague, Academia, 723 p.
MARCOUX, J., 1970, Age Carnien de termes effusifs du cortège ophiolitique des nappes d'Antalya (Taurus lycien oriental, Turquie): Acad. Sci. (Paris), Comptes Rendus, sér. D, v. 271, p. 285–287.
MARLOWE, J., 1971, Dolomite, phosphorite and carbonate diagenesis on a Caribbean seamount: Jour. Sed. Petrology, v. 41, p. 809–827.
MARTINIS, B., AND PIERI, M., 1964, Alcune notizie sulla formazione evaporitica del Triassico superiore nel l'Italia centrale e meridionale: Soc. geol. italiana Mem., v. 4, p. 649–678.
MASCLE, G., 1967, Remarques stratigraphiques et structurales sur la région de Palazzo-Adriano, monts Sicani (Sicile). Soc. géol. France Bull., sér. 7, v. 9, p. 104–110.

MERCIER, J., 1966, Paléogéographie, orogenèse, métamorphisme et magmatisme des zones internes des Hellénides en Macédoine (Grèce) : vue d'ensemble : *ibid.*, v. 8, p. 1020–1049.
MILCH, L., AND RENZ, C., 1911, Über griechische Quarzkeratophyre: Neues Jahrb. Mineralogie, Geologie und Paläontologie, v. 31, p. 496–534.
MIŠÍK, M., 1966 Microfacies of the Mesozoic and Tertiary limestones of the west Carpathians: Slov. Akad. Vied., Bratislava, 269 p.
———, AND RAKUS, M., 1964, Bemerkungen zu räumlichen Beziehungen des Lias und zur Paläogeographie des Mesozoikum in der Grossen Fatra: Sborník Geol. Vied., Západné Karpaty, Rad ZK, v. 1, p. 159–199.
MOORES, E. M., 1969, Petrology and structure of the Vourinos ophiolite complex, northern Greece: Geol. Soc. America Special Paper, 118, 73 p.
———, AND VINE, F. J., 1971. The Troodos Massif, Cyprus, and other ophiolites as oceanic crust: evaluation and implications: Royal Soc. (London) Philos. Trans., ser A, v. 268, p. 443–466
MURRAY, J., AND RENARD, A. F., 1891, Report on deep-sea deposits based on specimens collected during the voyage of H.M.S. *Challenger* in the years 1872 to 1876, *in* "*Challenger* Reports:" Edinburgh, Her Majesty's Stationery Office, 525 p.
MUTCH, T. A., AND GARRISON, R. E., 1967, Determination of sedimentation rates by magnetic spherule abundances: Jour. Sed. Petrology, v. 37, p. 1139–1146.
NICOLAS, A., AND JACKSON, E. D., 1972, Répartition en deux provinces des péridotites des chaines alpines loreant la Méditerranée: implications géotectoniques: Schweizerische Mineralog. und Petrog. Mitt., v. 52, p. 479–495.
OGNIBEN, L., 1963, Stratigraphie tectono-sédimentaire de la Sicile: Soc. géol. France, Livre à la mémoire du Prof. P. Fallot, v. 2, p. 203–216.
———, 1969, Schema introduttivo alla geologia del confine calabro-lucano: Soc. geol. italiana Mem., v. 8, p. 453–763.
OROMBELLI, G., AND POZZI, R., 1967, Studi geologici sulle isole del Dodecaneso (Mare Egeo). V. Il Mesozoico nell'isola di Rodi (Grecia) : Riv. italiana Paleontologia, v. 73, p. 409–536.
PANTÒ, L., 1961, Le magmatisme mésozoïque en Hongrie: Inst. Geol. Pub. Hung. Ann., v. 49, p. 979–997.
PAQUET, J., 1969, Etude géologique de l'ouest de la province de Murcie (Espagne) : Soc géol. France Mém., new ser., v. 48, no. 111, 270 p.
PARKER, R. J., AND SIESSER, W. G., 1972, Petrology and origin of some phosphorites from the South African continental margin: Jour. Sed. Petrology, v. 42, p. 434–440.
PATRULIUS, D., 1967, Le Trias des Carpates orientales de Roumaine: Geol. Sbornik, v. 18, p. 233–244.
PAULUS, F. J., 1972, The geology of site 98 and the Bahama platform, *in* HOLLISTER, C. D., EWING, J. I., AND OTHERS, Initial reports of the Deep Sea Drilling Project: Washington, D.C., U.S. Govt. Printing Office, v. 11, p. 877–897.
PEARCE, J. A., AND CANN, J. R., 1971, Ophiolite origin investigated by discriminant analysis using Ti, Zr and Y: Earth and Planetary Sci. Letters, v. 12, p. 339–349.
PETERSON, M. N. A., AND OTHERS, 1970, Initial reports of the Deep Sea Drilling Project: Washington, D.C., U.S. Govt. Printing Office, v. 2, 501 p.
PIALLI, G., 1971, Facies di piana cotidale nel Calcare Massiccio dell'Appennino umbro-marchigiano: Soc. geol. italiana Boll., v. 90, p. 481–507.
PITMAN, W. C., AND TALWANI, M., 1972, Sea-floor spreading in the North Atlantic: Geol. Soc. America Bull., v. 83, p. 619–646.
PLANCHEREL, R., AND WEIDMANN, M., 1972, La zone anticlinale complexe de la Tinière (Préalpes médianes vaudoises) : Eclogae Geol. Helvetiae, v. 65, p. 75–92.
PORRENGA, D. H., 1965, Chamosite in Recent sediments of the Niger and Orinoco deltas: Geologie en Mijn., v. 44, p. 400–403.
PREMOLI SILVA, I., AND LUTERBACHER, H. P., 1966. The Cretaceous-Tertiary boundary in the Southern Alps (Italy) : Riv. italiana Palaeontologia, v. 72, p. 1183–1266.
PRICE, I., AND NISBET, E. G., in press, Bedded cherts as ocean-ridge derived turbidites: Nature.
RADWAŃSKI, A., AND SZULCZEWSKI, M., 1965, Jurassic stromatolites of the Villány Mountains (southern Hungary) : Univ. Sci. Budapestinensis, Rolando Eötvös Ann., Sec. Geol., v. 9, p. 87–107.
RAMPNOUX, J. P., 1970, Sur la géologie du Sandjak: mise en évidence de la nappe du Pešter (confins serbo-monténégrins, Yougoslavie) : Soc. géol. France Bull., sér. 7, v. 11, p. 881–893.
REMANE, J., 1967, Note préliminaire sur la paléogéographie du Tithonique des chaînes subalpines: *ibid.*, v. 8, p. 446–453.
RENZ, C., 1955, Die vorneogene Stratigraphie der normalsedimentären Formationen Griechenlands: Inst. Geol. Subsurface Research, Athens, 637 p.
———, AND RENZ, O., 1948, Eine untertriadische Ammonitenfauna von der griechischen insel Chios: Schweizerische Paläont. Abh., v. 66, p. 3–98.
ROBERTSON, A. H. F., AND HUDSON, J. D., 1973, Cyprus umbers: chemical precipitates on a Tethyan ocean ridge: Earth and Planetary Sci. Letters, v. 18, p. 93–101.
ROCCI, G., AND LAPIERRE, H., 1969, Étude comparative des diverses manifestations du volcanisme préorogenique au sud de Chypre: Schweizerische Mineralog. und Petrog. Mitt., v. 49, p. 31–46.
ROCH, E., 1950, Histoire stratigraphique du Maroc: Service géol. Maroc Notes, Mem. 80, 435 p.
RÖMERMANN, H., 1968, Geologie von Hydra: Geologica et Palaeontologica, v. 2, p. 163–171.
RONA, P. A., 1970, Comparison of continental margins of eastern North America at Cape Hatteras and Northwestern Africa at Cap Blanc: Am. Assoc. Petroleum Geologists Bull., v. 54, p. 129–157.
ROYANT, G., RIOULT, M., AND LANTEAUME, M., 1970, Horizon stromatolithique à la base du Crétacé supérieur dans le Briançonnais ligure: Soc. géol. France Bull., sér. 7, v. 12, p. 372–374.
SARTONI, S., AND CRESCENTI, U., 1962, Ricerche biostratigrafiche nel Mesozoico dell'Appennino meridionale: Gior. Geologia, ser. 2, v. 29, p. 161–302.

Scandone, P., 1967, Studi di geologia lucana: la serie calcareo-silico-marnosa e i suoi rapporti con l'Appennino calcareo: Soc. naturalisti Napoli Boll., v. 76, p. 3–175.

———, Radoičić, R., Giunta, G., and Liguori, V., 1972, Sul significato delle Dolomie Fanusi e dei calcari ad ellipsactinie nella Sicilia settentrionale: Riv. Mineraria siciliana, v. 23, p. 51–61.

Schlager, W., 1967, Hallstätter und Dachsteinkalk-Fazies am Gosaukamm und die Vorstellung ortsgebundener Hallstätter Zonen in den Ostalpen: Geol. Bundesanstalt Wien Verh., 1967, p. 50–70.

———, 1969, Das Zusammenwirken von Sedimentation und Bruchtektonik in den triadischen Hallstätterkalken der Ostalpen: Geol. Rundschau, v. 59, p. 289–308.

———, and Schlager, M., 1973, Clastic sediments associated with radiolarites (Taugboden-Schichten, Upper Jurassic, Eastern Alps): Sedimentology, v. 20, p. 65–89.

Schmidt di Friedberg, P., Barbieri, F., and Giannini, G., 1960, La geologia del gruppo montuoso delle Madonie (Sicilia centro-settentrionale): Servzio geol. Italia Boll., v. 81, p. 73–140.

Schneider, H.-J., 1964, Facies differentiation and controlling factors for the depositional lead-zinc concentrations in the Ladinian Geosyncline of the Eastern Alps, in Amstutz, G. C. (ed.), Sedimentology and ore genesis, Developments in sedimentology: Amsterdam, Elsevier, v. 2, p. 29–45.

Schöll, W. U., and Wendt, J., 1971, Obertriadische und jurassische Spaltenfüllungen im Steinernen Meer (Nördliche Kalkalpen): Neues Jahrb. Geologie und Paläontologie, Abh., v. 139, p. 82–98.

Scholle, P. A., 1971a, Sedimentology of fine-grained deep-water carbonate turbidites, Monte Antola Flysch (Upper Cretaceous), northern Apennines, Italy: Geol. Soc. America Bull., v. 82, p. 629–658.

———, 1971b, Diagenesis of deep-water carbonate turbidites, Upper Cretaceous Monte Antola Flysch, Northern Apennines, Italy: Jour. Sed. Petrology, v. 41, p. 233–250.

Schönenberg, R., 1971, Einführung in die Geologie Europas: Freiburg, West Germany, Verlag Rombach, 300 p.

Sestini, J., 1965, Cenozoic stratigraphy and depositional history, Red Sea Coast, Sudan: Am. Assoc. Petroleum Geologists Bull., v. 49, p. 1453–1472.

Sheridan, R. E., Drake, C. L., Nafe, J. E., and Hennion, J., 1966, Seismic refraction study of continental margin east of Florida: ibid., v. 50, p. 1972–1991.

Smith, A. G., 1971, Alpine deformation and the oceanic areas of the Tethys, Mediterranean, and Atlantic: Geol. Soc. America Bull., v. 82, p. 2039–2070.

Spencer, M., 1967, Bahama deep test: Am. Assoc. Petroleum Geologists Bull., v. 51, p. 263–268.

Steinmann, G., 1905, Geologische Beobachtungen in den Alpen. II. Die Schardtsche Überfaltungstheorie und die geologische Bedeutung der Tiefseeabsätze und der ophiolithischen Massengesteine: Naturf. Gesell. Freiburg Ber., v. 16, p. 18–67.

———, 1925, Gibt es fossile Tiefseeablagerungen von erdgeschichtlicher Bedeutung?: Geol. Rundschau, v. 16, p. 435–468.

Sturani, C., 1964, La successione delle faune ad ammoniti nelle formazioni mediogiurassiche delle Prealpi venete occidentali: Univ. Padova, Ist. Geologia e Mineralogia Mem., v. 24, p. 1–63.

———, 1971, Ammonites and stratigraphy of the "Posidonia alpina" Beds of the Venetian Alps (Middle Jurassic, mainly Bajocian): ibid., v. 28, 190 p.

Szabó-Drubina, M., 1961, Die liassischen Manganlagerstätten vom Bakony: Inst. Geol. Pub. Hung. Ann., v. 49, p. 1171–1179.

Szulczewski, M., 1963, Stromatolites from the high-tatric Bathonian of the Tatra Mountains: Acta Geol. Polonica, v. 13, p. 125–141.

———, 1966, Jurassic stromatolites of Poland: ibid., v. 18, p. 1–99.

Terry, J., 1971, Sur l'âge triasique de laves associées à la nappe ophiolitique du Pinde septentrional (Epire et Macédoine, Grèce): Soc. géol. France, Comptes Rendus Séances, p. 384–385.

Thompson, G., 1972, A geochemical study of some lithified carbonate sediments from the deep sea. Geochim. et Cosmochim. Acta, v. 36, p. 1237–1253.

Trevisan, L., 1937, Scoperta di formazioni basaltiche e piroclastiche presso Vicari (Palermo) e osservazioni sui fossili baiociani contenuti nei tufi: Soc. geol. italiana Boll., v. 56, p. 441–452.

Trümpy, R., 1955, Wechselbeziehungen zwischen Palaeogeographie und Deckenbau: Naturf. Gesell. Zürich Vierteljahrssch., v. 100, p. 217–231.

———, 1960, Paleotectonic evolution of the Central and Western Alps: Geol. Soc. America Bull., v. 71, p. 843–908.

———, 1970, Stratigraphy in mountain belts: Geol. Soc. London Quart. Jour., v. 126, p. 293–318.

Truillet, R., 1970, The geology of the eastern Peloritani Mountains of Sicily, in Alvarez, W., and Gohrbandt, K. H. A. (eds.), Geology and history of Sicily: Tripoli, Petroleum Explor. Soc., Libya, p. 171–185.

Uchupi, E., Milliman, J. D., Luyendyk, B. P., Bowin, C. O., and Emery, K. O., 1971, Structure and origin of southeastern Bahamas: Am. Assoc. Petroleum Geologists Bull. v. 55, p. 687–704.

Veen, G. W., van, 1966, Note on a Jurassic-Cretaceous section in the Subbetic SW of Caravaca (Province of Murcia, Spain): Geologie en Mijnb, v. 45, p. 391–397.

Vezzani, L., 1968, La Formazione del Frido (Neocomiano-Aptiano) tra il Pollino ed il Sinni (Lucania): Geologica Romana, v. 8, p. 129–176.

Warman, H. R., and Arkell, W. J., 1954, A review of the Jurassic of western Sicily based on new ammonite faunas: Geol. Soc. London Quart. Jour., v. 110, p. 267–282.

Wendt, J., 1963, Stratigraphisch-paläontologische Untersuchungen im Dogger Westsiziliens: Soc. paleont. italiana Boll., v. 2, p. 57–145.

———, 1965, Synsedimentäre Bruchtektonik im Jura Westsiziliens: Neues Jahrb. Geologie und Paläontologie, Monatsh., 1965, p. 286–311.

———, 1969a, Stratigraphie und Paläogeographie des Roten Jurakalks im Sonnwendgebirge (Tirol, Österreich): ibid., Abh., v. 132, p. 219–238.

———, 1969b, Foraminiferen "Riffe" im karnischen Hallstätter Kalk des Feuerkogels (Steiermark, Österreich) : Paläont. Zeitschr., v. 43, p. 177–193.

———, 1970, Stratigraphische Kondensation in triadischen und jurassischen Cephalopodenkalken der Tethys: Neues Jahrb. Geologie und Paläontologie, Monath., 1970, p. 433–448.

———, 1971a, Genese und Fauna submariner sedimentärer Spaltenfüllungen im mediterranen Jura: Palaeontographica, v. 136, p. 122–192.

———, 1971b, Geologia del Monte Erice (Provincia di Trapani, Sicilia occidentale) : Geol. Romana, v. 10, p. 53–76.

WEZEL, F. C., 1970, Interpretazione dinamica della "eugeosinclinale meso-mediterranea": Riv. Mineraria Siciliana, v. 21, p. 187–198.

WHITEMAN, A. J., 1970, Nubian Group: origin and status: Am. Assoc. Petroleum Geologists Bull., v. 54, p. 522–526.

WIEDENMAYER, F., 1963, Obere Trias bis mittlerer Lias zwischen Saltrio und Tremona (Lombardische Alpen) : Eclogae Geol. Helvetiae, v. 56, p. 529–640.

WINDISCH, C. C., LEYDEN, R. J., WORZEL, J. L., SAITO, T., AND EWING, J., 1968, Investigation of Horizon Beta: Science, v. 162, p. 1473–1479.

WINNOCK, E., 1971, Géologie succinte du bassin d'Aquitaine (Contribution à l'histoire du Golfe du Gascogne), *in* Institut Français du Pétrole and Centre Nationale pour l'Exploitation des Océans, Histoire structurale du Golfe du Gascogne: Paris, Editions Technip, v. 1, p. IV. 1–1—IV. 1–30.

WORZEL, J. L., BRYANT, W., AND OTHERS, 1973, Initial reports of the Deep Sea Drilling Project: Washington, D.C., U.S. Govt. Printing Office, v. 10, 748 p.

WURSTER, P., 1968, Paläogeographie der deutschen Trias und die paläogeographische Orientierung der Lettenkohle in Südwestdeutschland: Eclogae Geol. Helvetiae, v. 61, p. 157–166.

ZANKL, H., 1967, Die Karbonatsedimente der Obertrias in den nördlichen Kalkalpen: Geol. Rundschau, v. 56, p. 128–139.

———, 1969, Der Hohe Göll. Aufbau und Lebensbild eines Dachsteinkalk-Riffes in der Obertrias der nördlichen Kalkalpen: Senckenbergische Naturf. Gesell. Abh., no. 519, p. 1–123.

———, 1971, Upper Triassic carbonate facies in the Northern Limestone Alps, *in* MÜLLER, G. (ed.), Sedimentology of parts of central Europe, Guidebook to excursions of VIII Internat. Sed. Cong., Frankfurt, West Germany, Kramer, p. 147–185.

ZIMMERMAN, J., JR., 1972, Emplacement of the Vourinos ophiolitic complex, northern Greece, *in* SHAGAM, R., AND OTHERS (eds.), Studies in earth and space sciences, a Memoir in honor of H. H. Hess: Geol. Soc. America Mem. 132, p. 225–239.

PALEOZOIC AND MESOZOIC GEOSYNCLINAL VOLCANISM IN THE JAPANESE ISLANDS AND ASSOCIATED CHERT SEDIMENTATION

KAMETOSHI KANMERA
Kyushu University, Japan

ABSTRACT

The main part of the Japanese Islands consists of the middle and late Paleozoic and the early Mesozoic eugeosynclinal sequences containing voluminous volcanic beds. The change through time and space of the nature of the volcanic rocks and the lithologic assemblages of the Paleozoic sequences suggest that they were formed in an ancient marginal or interarc basin and on the outer flank of a migrating ridge. The basin is presumed to have been initiated by a volcanotectonic rift zone that was associated with silicic tuffs and flows and coralline limestones during Middle Silurian to Middle Devonian time. This was followed by the crustal extension that resulted in the separation of the Kurosegawa-Abukuma-South Kitakami insular ridge from the Hida belt to form a marginal basin during Late Devonian to Middle Permian time. Thick eugeosynclinal deposits, including mafic volcanics and chert, accumulated concomitantly in the spreading basins.

In the early stage (Late Devonian to Early Carboniferous) of basin spreading, alkaline basalt was produced in association with a large amount of gabbroic rocks and serpentinite, which are assumed to be an oceanic basin crust created along the basin axes. In the later stage (Late Carboniferous to middle Permian) flow rocks and pyroclastics of mainly alkaline, partly tholeiitic basalt dominated. They are conformably intercalated in clastic sediments, and they are arranged in an alternating en echelon pattern of a few to several linear volcanic zones. Two zones of ophiolitic assemblages, the Mikabu Green Rocks (Early Permian) and the Yakuno Basic Rocks (Middle Permian), are interpreted to have occurred as short-lived volcanic rises within the marginal basin.

The Triassic and Cretaceous mafic volcanic zones respectively were remarkably stepped southward along with the shifting of the main sedimentary basin to the outer side. This shifting was almost coeval with the orogenic events in the main sedimentary basin during the preceding Paleozoic ages.

The mafic volcanic rocks are commonly accompanied by bedded chert, and a rhythmic interlayering of tuff and chert is most characteristic. The cherts consist essentially of Radiolaria, some sponge spicules, their fragments, and a small amount of aphanitic material, probably volcanic clay. Their sedimentary features and intimate association with tuff suggest that chert beds accumulated rapidly because of an enormous supply of siliceous organisms. These organisms perished episodically in great masses when explosive submarine eruptions at depths not greater than 500 m resulted in heated water columns rising to the surface and spreading detrimentally among the plankton populations.

INTRODUCTION

Recent advances in plate tectonic theory have provided an actualistic basis for interpreting geologic evolution of ancient orogenic belts, and the nature and setting of ancient geosynclinal rock masses within island arcs may be specified by analogy with modern tectonic and depositional features in marginal basins and along younger island arc systems.

The Japanese Islands exhibit an older and longer geologic history than any other island arc, and the main part is made up of the middle and upper Paleozoic and Mesozoic systems of eugeosynclinal rock assemblages containing many beds of volcanic rocks and bedded chert. A recent remarkable advance has been made in knowledge of the geochemistry of volcanic rocks in these geosynclinal sequences (e.g., Hashimoto and others, 1970; Sugisaki and Tanaka, 1971; Sugisaki and others, 1972; Suzuki and others, 1971), and the distribution in time and space of volcanic rocks has also been examined (Kanmera, 1971).

This paper interprets the tectonic and sedimentary framework of the Paleozoic and Mesozoic geosynclines of Japan in relation to the change in time and space of the nature of the volcanic rocks. Some comments are also given on the implication of the submarine volcanism for the origin of bedded chert.

GEOSYNCLINAL VOLCANISM AND EVOLUTION OF THE PALEOZOIC GEOSYNCLINE

Paleozoic rocks cover the larger part of the Japanese Islands (fig. 1) and the oldest deposits dated by fossils belong to the Middle Silurian Series. Paleozoic deposits consist entirely of eugeosynclinal assemblages typified by sandstone of the wacke suite, slate, volcanic rocks, chert, and subordinate lenticular limestone and intraformational conglomerate. Nowhere within Japan are miogeosynclinal or miogeoclinal Paleozoic sediments recognized.

The general nature in time and space of Paleozoic and Triassic volcanic episodes is shown in figure 2. Volcanism of the Paleozoic geosyncline was not confined to such limited times as it was in European and eastern Ameri-

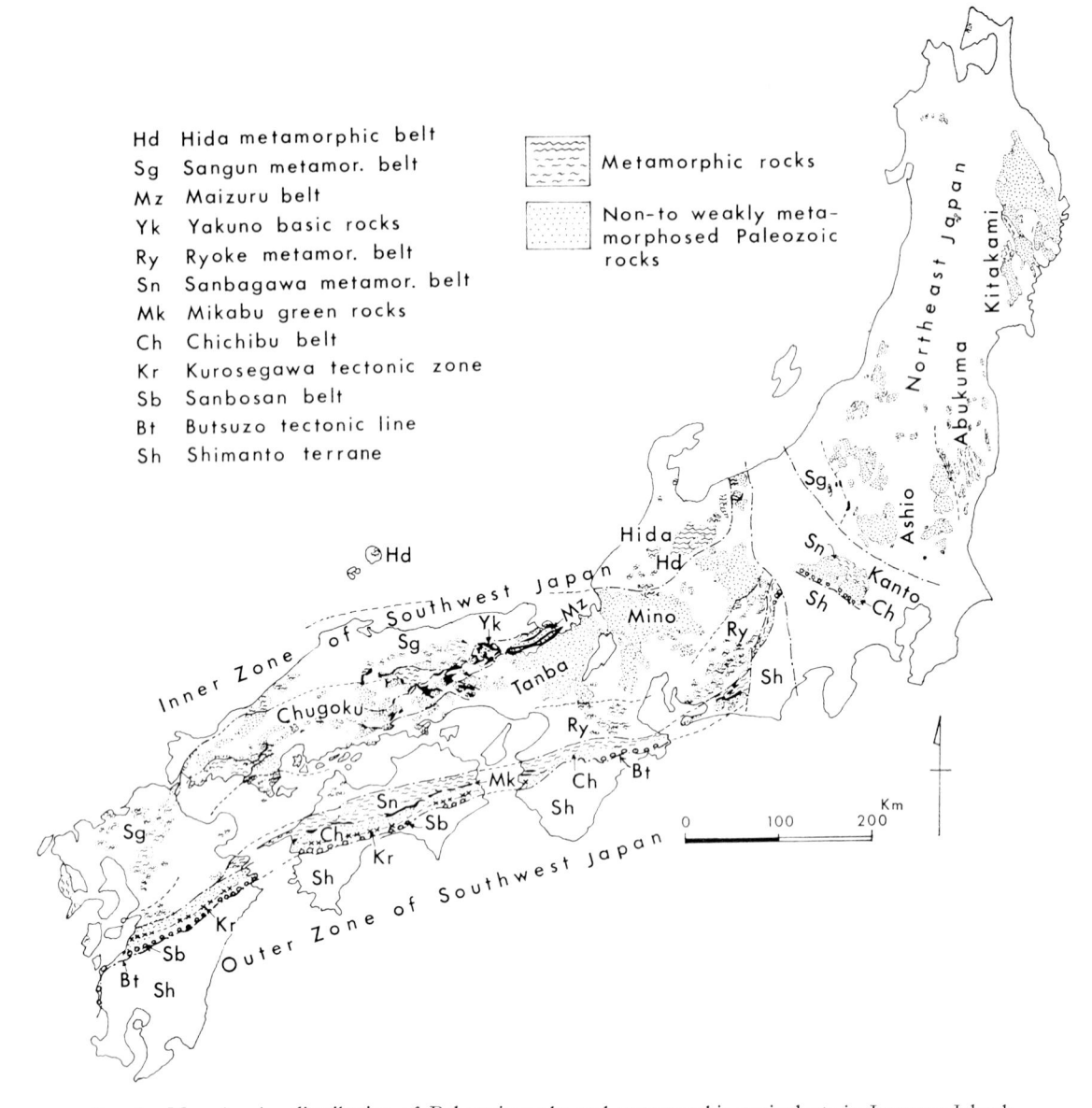

Hd Hida metamorphic belt
Sg Sangun metamor. belt
Mz Maizuru belt
Yk Yakuno basic rocks
Ry Ryoke metamor. belt
Sn Sanbagawa metamor. belt
Mk Mikabu green rocks
Ch Chichibu belt
Kr Kurosegawa tectonic zone
Sb Sanbosan belt
Bt Butsuzo tectonic line
Sh Shimanto terrane

FIG. 1.—Map showing distribution of Paleozoic rocks and metamorphic equivalents in Japanese Islands.

can geosynclines. The products are mostly non-ophiolitic sequences that mostly lack ultramafic rocks.

Silurian and Devonian.—The Silurian rocks are distributed in southern Kitakami of northeast Japan and in the Kurosegawa tectonic zone of the Outer Zone of southwest Japan. They consist of coralline limestone and andesitic tuff and shale. The former existence of the Silurian coralline limestone in the Hida Mountains in the Inner Zone of southwest Japan is suggested by cobbles in the Permian conglomerates in the southerly adjacent Mino Mountains. Devonian rocks are found not only in the three areas mentioned above but also in the Abukuma Mountains, which belong to the same continuous structural domain as southern Kitakami. In the southern Kitakami-Abukuma and Kurosegawa, these rocks, except for those of Late Devonian age, consist of dacitic or rhyolitic lava and tuffs, of volcaniclastic sandstone, and of shale containing some brachiopods; in the Hida Mountains they consist of coralline limestone, of andesitic to dacitic lava and tuffs, and of

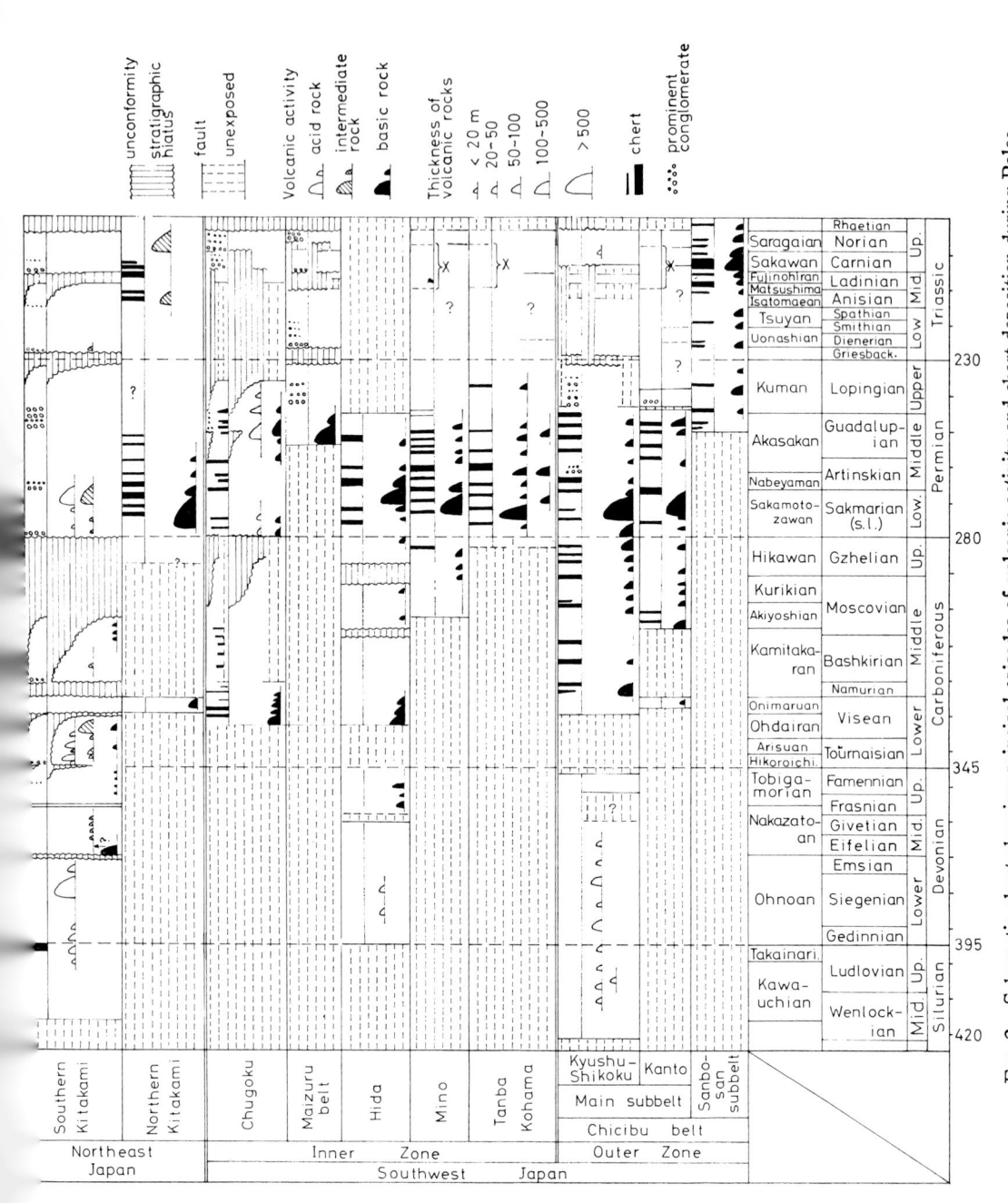

FIG. 2.—Schematic chart showing principal episodes of volcanic activity and chert deposition during Paleozoic and Triassic time in Japan. Part of sequence indicated by X includes some chert beds containing Middle and (or) Upper Triassic conodonts but is not yet stratigraphically differentiated.

volcaniclastic sandstone and shale.

Thus the early stage of the Paleozoic geosyncline is characterized by shallow marine deposits and andesitic to rhyolitic volcanic rocks that suggest their emplacement in an tectonic belt that must have had a continental type crust. The question remains, however, as to whether they are the deposits of two separated belts, as they are separated at present, or whether they belong to one and the same tectonic domain. This problem requires consideration of the basement of the Paleozoic geosyncline.

Along with the Siluro-Devonian rocks are high-grade metamorphic rocks consisting mainly of garnet-bearing amphibolite, biotite schist, mica gneiss, and others. In the Kurosegawa and the Abukuma, granites are also present. These rocks occur along the tectonic zones as squeezed-out or thrust-up bodies associated everywhere with sheared serpentinite and are in tectonic contact with rocks of the Silurian and Devonian Systems, which are generally non-metamorphosed. The radiometric age data for these rocks show a remarkable concentration in the range of 350 to 370 my, although some are as great as 420 my (Shibata and others, 1970; Ishizaka, 1972; Kawano and Ueda, 1967; Maruyama, 1972). These metamorphic and plutonic features of Early to Middle or Late Devonian age are probably related to the intense rhyolitic and andesitic volcanism mentioned above. The rocks under consideration are at present separated into two zones, but the close similarity in distribution, rock assemblages, metamorphic facies, and simultaneity of the metamorphic and plutonic events suggest that they formed primarily along one and the same volcanoplutonic orogen or magmatic arc.

Much controversy exists on the basement of the Paleozoic geosyncline of Japan. Several workers (e.g., Minato and others, 1965; Minato, 1968; Ichikawa and others, 1970; Sugisaki and others, 1972) considered that the Paleozoic geosyncline was formed initially on the stable or inactive older Precambrian continental crust that was extended from the Asiatic continent, and they believed that the Hida metamorphic belt occupying the innermost of southwest Japan represents an exposed part of the Precambrian basement. The reason, all circumstantial, for this inference has been discussed in detail by Sugisaki and others (1972). An alternative interpretation is that the Hida gneiss and schist are metamorphic equivalents of the upper Paleozoic sediments produced by Late Permian to Middle Triassic orogenesis (Kobayashi, 1941; Miyashira, 1961, 1967). The radiometric age data obtained by several workers indicate that the Hida metamorphics have undergone polymetamorphism, which includes, besides the main events of the Late Permian and Early Triassic and the Early Jurassic, the Middle to Late Devonian phase (Shibata and others, 1970). In any event, the basement of the Paleozoic geosyncline, even if Precambrian rocks were not a part, was not a stable one but an active volcanoplutonic belt at least in the initial stage of development of the geosyncline. It is not celar, however, that this magmatic belt was standing on the edge of the Asian continent or that it was a volcanoplutonic arc itself.

This volcanoplutonic belt is thought to have been separated into two ridge systems (the Hida belt in the inner side and the southern Kitakami-Abukuma and the Kurosegawa insular belt in the outer) during latest Devonian time to Early Carboniferous time and that the Carboniferous and Permian rocks were deposited in the widening trough between the two ridges and on the outer flank of the outer ridge. This is inferred from the sudden change of volcanic character during those times to the mafic suites (fig. 2). On the western flank of the southern Kitakami is a glaucophanitic metamorphic complex (Motai Group), primarily comprised of basalt, diabase, and gabbro of the alkaline suite (Kanisawa, 1971) and subordinate amounts of tuff, sandstone, slate, and chert. This complex is associated with a large quantity of serpentinite. The group interdigitates with the uppermost Devonian argillite beds in which basaltic volcanics first appeared. Fairly large bodies of serpentinite also occur on the western flank of the Abukuma massif in close association with weakly metamorphosed sedimentary rocks of Devonian lithologic aspect. On the opposite side, that is, in the Sangun metamorphic (also glaucophanitic) zone and its northeastern extension, the Hida marginal zone, of the northern flank of the Paleozoic terrane, large masses of serpentinite associated with gabbro and diabase also are present. The Sangun Group is considered to be of Early Carboniferous and possibly Late Devonian age on the basis of its structural relation to other dated Paleozoic rocks. It exhibits a eugeosynclinal facies of thick mafic flow rocks and tuffs and some cherts. The serpentinite and gabbroic bodies appear to be mostly in thrust or partly in intrusive relationship with the Sangun Group, but their emplacement apparently was prior to the metamorphism of Late Permian to Early Triassic age. In addition to serpentinite in the southern margin of the Hida metamorphic belt, thick sill-like bodies of diabase and basaltic flows and tuffs are present at intercalations in the Upper Devonian Series. The ultramafic and mafic rocks, including volcanic beds intercalated in sedimentary sequences of these two flank areas of the main Paleozoic terrane, are considered to reflect the opening of an oceanic basin in association with generation of oceanic type crust, on which immediately the Upper Devonian and Lower Carboniferous sediment aprons were formed dominantly by the supply of volcano-plutonic debris (Mikami, 1971) from the two-ridge system mentioned above.

The existence of the Abukuma-Kitakami insular massif is suggested by several differences between this area and other parts of the upper Paleozoic terrane. These differences are observed as follows: (1) in the rock assemblage and its sedimentary facies and deformation pattern, as exemplified by the Carboniferous andesitic and dacitic volcanics that belong predominantly to a rock series different from that of other parts of the terrane (Sugisaki and Tanaka, 1971), (2) in the lack of volcanics and chert in the Permian, (3) in the Early Permian cyclic sediments of lagoonal to very shallow sea facies, and (4) in a much weaker and different type of deformation of rocks. Influence of the Kurosegawa is indicated by the Middle and Upper Permian intraformational conglomerates and coarse-grained sandstones that occur only along the Kurosegawa zone and thin laterally both northward and southward into the coeval formations of the muddy and cherty facies (Kanmera, 1961). These observations indicate that along the Kurosegawa zone there was an uplifted source area from which coarse material was supplied. Miyashiro (1972) and Horikoshi (1972) inferred the Kurosegawa zone as an ancient Benioff zone, but this view is not borne out by the information mentioned above. The conglomerates contain abundant cobbles of granitic rocks and other kinds of plutonic and effusive rocks of felsic to mafic composition, some low-grade metamorphic rocks, and subordinate sedimentary rocks. Granite-bearing conglomerates of the same ages are also present in the southern Kitakami and Abukuma. Kano (1967, 1971) concluded from detailed petrographic studies of Permian conglomerates of the Kitakami, Hida and Kurosegawa that the main provenance was a "volcano-plutonic formation," which seems to have been related intimately to Siluro-Devonian volcanism.

Carboniferous and Permian.—Between the two zones of the Siluro-Devonian rocks and the basement complex, thick sedimentary piles occupy a belt about 200 km wide. They contain abundant volcanic rocks and chert of Carboniferous and Permian age and subordinate Triassic rocks. In contrast to the Silurian and Lower and Middle Devonian volcanic rocks mentioned above, the Carboniferous and Permian rocks are mostly of mafic composition and occur as basaltic and diabasic lava, pillow breccia, tuff breccia, and tuff, except for those rocks of the southern Kitakami and Abukuma that are mentioned above. Gabbroic rocks as thick bodies are mostly not associated with them. Very small amounts of ultramafic rocks are present in the Sanbagawa and Mikabu zones.

The Carboniferous rocks are very much restricted in exposure, but insofar as known the Visean basaltic rocks are distributed in the northern part of the Inner Zone of southwest Japan. Their lower limit cannot be determined, but in the Mino Mountains to the south of Hida is a volcanic series more than 800 m thick under Visean limestone (Igo, 1964). Evidence of late Early to Late Carboniferous volcanism is most prominent in the northern Chichibu belt immediately north of the Kurosegawa tectonic zone.

Most of the Carboniferous and Permian volcanic rocks are conformably intercalated as lenticular bodies in clastic sediments and do not show any particular deformation pattern different from the over- and underlying clastic beds. Nor do they show such tectonic style as mélanges. Thus, they are hardly assignable to fragments of ophiolitic sequences that were generated at an oceanic ridge and subsequently conveyed to a convergent plate juncture, but rather to *in situ* emplacement.

The volcanic rocks occur as linear chains or zones. The alignment of volcanic chains and the facies distribution of the Carboniferous and Permian sediments show an en echelon pattern. They trend obliquely (northeast-southwest) rather than parallel to the distribution of the Siluro-Devonian rocks, although some later modification by left- or right-lateral displacement along the Median Tectonic Line must have been added. This pattern is best exemplified by the Early Permian volcanics, which represent the greatest volcanic episode in the Paleozoic and attain as much as 500 to 1,500 m of thickness in some places. They occur in at least five zones (fig. 3), of which three of the inner side are arranged en echelon and trend obliquely to the outer ones and to the Siluro-Devonian volcanic zone. These volcanic zones are less than 20 km in width, and the coeval strata on both sides of them are represented by clastic rocks. This distribution pattern may be comparable with that of the basaltic sea mounts in the Andaman Basin and of the volcanic archipelago along the inside of the Andaman-Nichobar island arc (Rodolfo, 1969) and with the subparallel disposition of basaltic ridges (axial highs) in the interarc basin between the frontal and the third arc of the Mariana and Tonga-Kermadec island arc systems (Karig, 1970, 1971a). The distribution

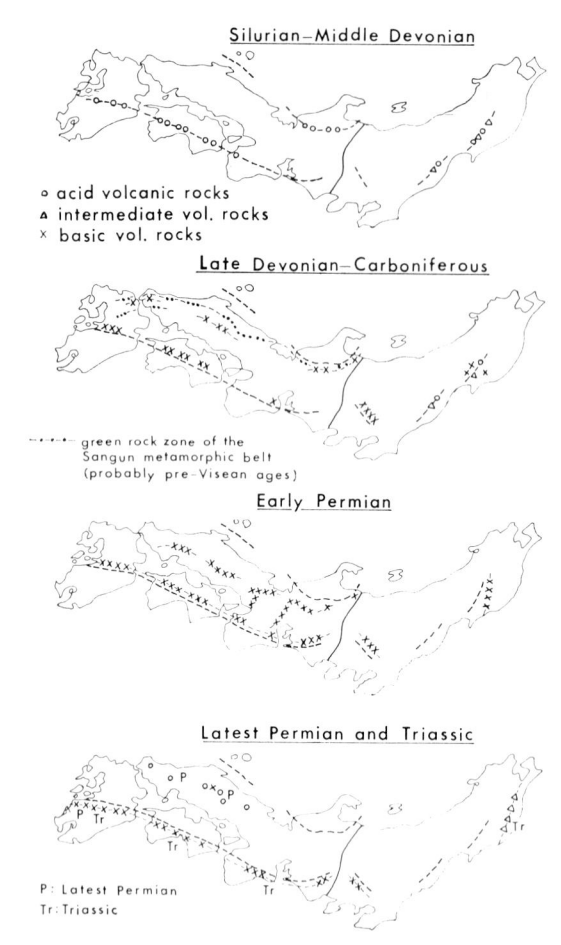

Fig. 3.—Distribution maps of prominent volcanic zones for some Paleozoic and Triassic stages in Japan.

in time and space of the volcanic rocks suggests that the trough in which the Carboniferous and Early Permian rocks were deposited was a site of crustal extension as in an interarc or marginal basin. The disposition of the Early Permian volcanic rocks suggests that the crustal extension was directed southeastward. The rifting and gradual widening of the sedimentary basin during Carboniferous and Permian time are also suggested by the apparent inward accretion of younger sediments.

The most prominent Early Permian volcanism is manifested by the Mikabu Green Rocks distributed in a narrow strip along the boundary between the Sambagawa crystalline schist belt and the southern adjacent Chichibu belt that consists mainly of weakly metamorphosed upper Paleozoic rocks (fig. 1). They display a layered sequence comprising, in the upper part, basaltic lava, pillow breccia, tuff breccia, tuff, and some intercalated chert beds; in the lower part, diabase and thick layered gabbros and in some places subsequent intrusives of quartz diorite and serpentinite (Suzuki and others, 1971, 1972, Iwasaki, 1969). The basaltic rocks are mostly tholeiite but include subordinate alkali basalt and lack high-alkali basalt (Sugisaki and others, 1972). Thus, the Mikabu Green Rocks display a sequence very similar to the ophiolite suite, which is currently considered as the intact slabs of old oceanic crust, and the basaltic members are compared in chemical composition with abyssal basalt of the present oceanic ridge.

Ernst (1972) referred the Mikabu zone to the site of convergent crust. Miyashiro (1972) also suggested that the Kiyomizu tectonic zone, which lies at about 5 km north of the Mikabu zone, might be a relict Benioff zone. The Mikabu Green Rocks are underlain conformably, however, by thick slate and sandstone probably of Late Carboniferous age. To the north, they change laterally into volcanic beds of the Sanbagawa southern marginal zone, and, to the south, they change into the volcanic beds of the northern zone of the Chichibu belt. These beds in turn constitute the northern and the southern wing respectively of an anticline whose axis is along the Mikabu zone. Thus, the rocks of the two zones comprise a structural and stratigraphic continuity through the Mikabu zone. No tectonic mélange or imbricated sheared sheets suggestive of a subduction zone are found along or near the Mikabu zone. Considered from the above-mentioned facts, the Mikabu rocks are the products of submarine volcanic eruption of and intrusion along a short-lived volcanic rise generated within a marginal basin behind the migrating ridge mentioned above.

Middle Permian volcanism became generally much less prominent than Early Permian volvanism but it is represented in the Chugoku belt by rhyolite lavas and pyroclastics in addition to some alkali basalt and in the Maizuru belt (fig. 1) by basaltic and diabasic lavas and subordinate tuffs. Both these belts are in the inner zone of the Paleozoic terrane. The Maizuru belt constitutes a tectonic zone that obliquely delimits the southeastern margin of the Chugoku belt and the Sangun metamorphic belt and that is characterized by steeply imbricated zonal distribution of the upper Lower to Upper Permian Maizuru Group and of the Triassic strata. This zone is also characterized by the tectonic intrusions of mafic rocks, called the Yukuno Basic Rocks, which occur within the belt and along its northern and southern boundaries. Coarse-grained cataclastic hornblende metagabbro (K-Ar ages dated at 241 to 278 my) is the main constituent of the rocks and is intermingled with or prograded into gneissic hornblende gabbro and amphibolite on the one hand and into the Permian diabase, metabasalt, and tuffs in the lower part of the Maizuru Group mentioned above. Diorite and trondjemite are also subsidiarily associated in intrusive relationships with the mafic rocks, and apparent xenoliths of mica schist (Rb-Sr ages dated at 269 to 332 my) are common. The tectonic implication of these intricate rocks is not clear, but they are thought to represent a tectonic wedge of the short-lived mid-Permian volcanic rise.

In the Late Permian Epoch, volcanism decreased markedly. Only the rhyolite flows and tuffs of the Chugoku belt (fig. 3) are a notable volcanic manifestation in the main geosynclinal basin, and the mafic volcanic belt was shifted southward to the Sambosan belt mentioned below. The Chugoku belt and the Hida marginal belt were subjected to tectonic movements in Late Permian to Early Triassic time, which are exemplified as follows: (1) by metamorphism of the Sangun zone (dated at 260 to 270 my); (2) by formation of the Upper Permian chaotic limestone conglomerates and development of erosional unconformities in the limestone mound areas, such as at Akiyoshi, Atetsu, and Taishaku in the northern part of the Chugoku belt; and (3) by the remarkable unconformity between the strongly folded Permian rocks and the Sangun metamorphic rocks (below) and by the molasse-type, mainly deltaic Upper Triassic formations consisting of conglomerate and sandstone with shale and coal seams (above). Thus, the northernmost region of the Paleozoic geosynclinal basin became a scene of compression with felsic volcanism, high P/T metamorphism, and intermittent uplift in Late Permian to Early Triassic time. At the same time, the Maizuru belt was the site of thick sedimentation along the frontal margin of the above-mentioned region. The Upper Permian Maizuru Group rests conformably on the mafic volcanic rocks mentioned above and consists of sandstone and shale containing sporadic intercalated conglomerate.

The Lower Triassic sequence shows a facies change from coarse sediments comprising submarine-fan conglomerate and sandstone in the northwestern side of the belt to finer sediments mainly of shale in the southeastern side (Nakazawa, 1958). This signifies that the Maizuru belt in Late Permian to Early Triassic time was a site of convergent plate juncture, as suggested by Horikoshi (1972), between the uplifting areas (including the Chugoku belt that existed to the north) and the main geosynclinal belt to the south.

The Upper Permian and Triassic rocks in most parts of Japan, except for those of the Sambosan zone mentioned below, are composed of clastic sediments and some limestones (fig. 4). Recently, conodonts and bivalves of Middle and Upper Triassic aspect have been found in a series of alternating sandstone and chert (which had been considered as Paleozoic in age) in several places in the Tamba, Mino,

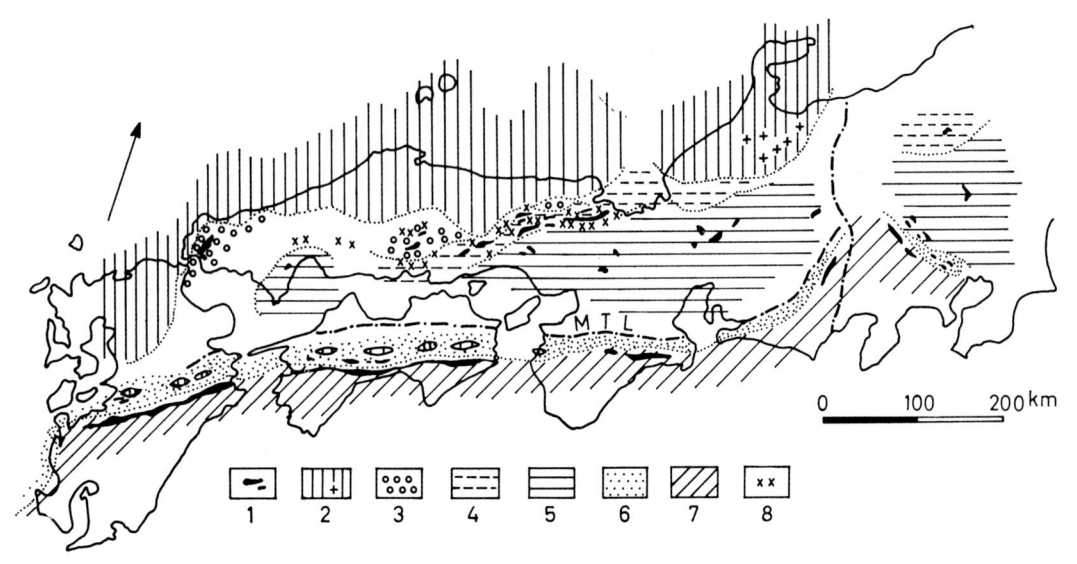

FIG. 4.—Interpretation of sedimentary framework of Triassic rocks in southwest Japan. 1, Outcrop of Triassic rocks; 2, inferred terrane of Paleozoic rocks, metamorphic equivalents, and granites; 4, very coarse nonmarine sediments dominating facies; 5, marine sandstone-shale facies; 6, chert-turbidite sandstone facies; 7, mafic volcanics, chert, and micritic limestone facies; 8, outcrop of Yakuno Basic Rocks in Mailzura belt.

Kanto, and Ashio Mountains. Although existence of the Lower Triassic has not yet been confirmed in these places and nearby areas, neither distinct lithologic difference nor physical break has been recognized between the Triassic mentioned above and the apparent Permian. Thus, the Tamba-Mino-Ashio belt very probably was maintained as a sedimentary trough of eugeosynclinal facies at least until Late Triassic time.

MESOZOIC GEOSYNCLINAL VOLCANISM IN WESTERN JAPAN

Triassic.—The Triassic volcanic rocks are found only in the southernmost zone of the Chichibu terrane of southwest Japan, which is called the Sambosan zone (figs. 1 and 3). The zone is about 5 km in maximum width but extends at least 1,000 km from Kyushu to the Kanto Mountains, although in part it is tectonically discontinuous. It is bordered on the south by the Butsuzo thrust fault that dips to the north against the Shimanto terrane, which consists of late Mesozoic and early Tertiary sediments. The best display of rocks in the zone is seen along the middle course of the Kuma River in southern Kyushu, where it is a 1,300-m thick conformable series ranging from early Late Permian to Late Triassic in age. There it consists mainly of basaltic tuff breccia, pillow breccia, and tuff. Included are many beds of chert and micritic limestone containing planktonic faunas, subordinate intercalated wedge beds of slate and some thin lenticular beds of sandstone, and some autocannibalistic chert and limestone conglomerates. The volcanic rocks are most prominent in the Upper Triassic Series. The great abundance of pyroclastics and marked amygdaloidal nature of lavas and breccias suggest the frequency of the explosive submarine eruptions that are mentioned on a following page.

This group is the most representative sediment of deep-sea pelagic aspect in the entire geologic columns of Japan. The limestones are dominantly biomicrite and contain variable amounts of Radiolaria, extremely thin-shelled bivalves, and foraminifers. They are very similar to the Upper Jurassic and Cretaceous micritic limestones of the Alpine Geosyncline of the Mediterranean region. (For details of litho- and biofacies of the limestones, see Kanmera, 1969.)

This group shows a remarkable lithologic difference from the coeval strata that are zonally distributed to the north. The lithofacies of the Triassic in the Inner Zone and their stratigraphic relationships with the Paleozoic systems have already been described. The Triassic of the northern and central zones of the Chichibu terrane, which is juxtaposed with the Sambosan zone, is of a shallow neritic facies. The lower part consists of limestone rich in mollusc shells and some beds of gray sandstone and shale; the middle and upper part consists of gray sandstone and shale and some lentils of coarse oolite and arenaceous oolitic limestone. These Triassic rocks are devoid of volcanic sediments except for a local thin bed of fine-grained, probably air-fall felsic tuff of Carnian age. Siliceous sediments are entirely lacking. In contrast, the Sambosan zone was a scene of a remarkable submarine mafic volcanic outburst during Triassic time, particularly during Late Triassic time.

The juxtaposition of these remarkably different facies, shelf and volcanic-pelagic, is similar to that of the Great Valley and Franciscan Sequences in California, which are respectively described as miogeosynclinal or arc-trench gap sediments on the basis of lack of volcanic rocks (e.g., Dickinson, 1972) and either as offshore ocean-floor equivalents having oceanic substrates (e.g., Hamilton, 1969) or as a trench-type eugeosynclinal prism of subducted mélanges and tectonites (Dickinson, 1972). Notable differences are seen, however, in the lack of ultramafic and coarse mafic rocks in the Sambosan belt and in the lack of a major thrust between the two belts.

Thus, the volcanic zone of Triassic age was shifted far to the south of those of Carboniferous and Permian age. This may suggest generation of a new subduction zone in the south of the Sambosan belt during Triassic time as the result of generation of a new lithospheric plate. Such a plate could have formed from conjugation of two, inner and outer arc ridges with sedimentary fills in the interarc basin, which were at least 5,000 m thick and were deformed and metamorphosed in part, and with the accreted plutonic crust beneath the sedimentary piles in the interarc geosynclinal basin.

Jurassic and Cretaceous.—No intense volcanism is known for the Jurassic Period in southwest Japan except as represented by a few thin beds of fine-grained felsic tuff of Late Jurassic age in the Hida Mountains. Cretaceous volcanism is manifested in two belts, in the Inner Zone and in the Shimanto terrane of the Outer Zone, the latter occupying a broad belt about 100 km wide and at least 1,000 km long in the southernmost part (Pacific side) of southwest Japan and in the Kanto Mountains. The Inner Zone belt is characterized by an enormous amount of andesitic and rhyodacitic pyroclastic flows that were erupted subaerially and by extensive granitic intrusions. The sedimentary rocks are terrestrial.

In contrast to the Inner Zone, the Shimanto terrane was the site of thick marine sedimentation dominated by alternating sandstone and shale of the flysch facies of Cretaceous to early Tertiary age. The submarine volcanic episodes of tholeiitic basalt occurred mostly during the early stage of development of this sedimentary basin, corresponding to intense volcanoplutonic activity and upheaval in the inner side and to the southward shifting of the main site of sedimentation to the outer side.

The Shimanto terrane is divided by a thrust fault into two belts, the northern belt comprising the Cretaceous sediments called the Shimanto (or Shimantogawa) Group and the southern belt comprising the Paleogene formations. The Shimanto Group is referred by Dickinson (1971) to the trench assemblage. The Cretaceous and the Paleogene each attain more than 10,000 m of thickness.

The mafic volcanic material and chert are in formations consisting mainly of shale and some thin beds of fine-grained sandstone, but they are very rare or entirely lacking from formations that are prevalently sandstone or alternating sandstone and shale of the flysch facies. At least three shale units contain volcanic rocks and chert in the Shimanto Group. One occupies the lowest part of the group, and the other two are conformably intercalated in the thick flysch formations (Imai and others, 1971). Each unit is some hundred meters thick.

The volcanic rocks tend to occur in swarms consisting of 10 to 15 beds either overlying one another successively or intercalated with other rocks (Shiida and others, 1971). These swarms are scattered within the same formation. All this probably means that the volcanic eruptions took place nearly continuously or intermittently at certain volcanic centers distributed widely and that the eruptions continued to form swarmed volcanic sequences during a considerable period of time.

The volcanic rocks of the Shimanto Group consist of basaltic and diabasic lava and tuffs of varying grades of coarseness. Gabbroic rocks may be locally associated, but no ultramafic rocks have been found. These rocks are mostly less than 100 m in thickness, rarely 150 m or more, and are conformable with the clastic rocks. Systematic petrochemical studies of these rocks have not been achieved, but available data (Shiida and others, 1971; Sugisaki and others, 1972) demonstrate that the basalts have low potassium content and are chemically similar to abyssal basalt. They mostly are referred to high-alkali tholeiite, but some are alkali basalt.

CONCLUDING REMARKS ON THE PALEOZOIC AND MESOZOIC GEOSYNCLINAL VOLCANISM

The tectonic and sedimentary sequences in young active island arcs, such as the Marianas, Tonga-Kermadec, and New Hebrides (Karig, 1970, 1971a, b), have been depicted to begin with generation of a volcanotectonic rift zone accompanied by silicic magmatic intrusion and extrusion. This lead to development of an interarc basin that was produced by an extensional process involving movement of the frontal arc away from the third arc. In this gradually opening and deepening interarc basin, new oceanic crust was created along with some subparallel or en echelon axial volcanic rises, and sedimentary filling followed or was concomittant. The marginal basins lying behind the volcanic chains of island arc systems are thought to have formed by a similar extensional process (Karig, 1971b) as is well exemplified by the Andaman-Nichobar basin (Rodolfo, 1969).

The Paleozoic sequence in Japan shows a remarkable change in sedimentary facies and volcanic rocks from Silurian to Middle Devonian reefoid limestone, volcaniclastic shallow-marine sediments, and rhyolitic-andesitic volcanic rocks to Upper Devonian to mid-Permian assemblages containing many beds of radiolarian chert and voluminous mainly alkaline, partly tholeiitic basalt. This change reflects changes in tectonic and depositional setting, which seem to compare well with the sequence of events in the active island arcs or in the marginal basins mentioned above. The occurrence of alkaline and tholeiitic basaltic rocks at many stratigraphic levels and in many areas in the Carboniferous and Permian sequences suggests that the sediments of these ages are most probably allied with the deposits in the relict interarc or marginal basin. This basin was generated during latest Devonian time and gradually opened at least from Carboniferous to mid-Permian time as the result of separation of a ridge (Kurosegawa-Abukuma-South Kitakami zone) from the Hida belt. Furthermore, Carboniferous and Permian sediments also include those deposited on the outer flank of the migrating ridge. The gradual widening of the Carboniferous and Permian basin is suggested by the progressive inward accretion of younger sediments and of volcanic zones against the older rocks along the marginal zone. The mafic volcanic rocks in the Paleozoic sequence, including the Mikabu and Maizuru Green Rocks, which are associated with gabbroic and ultramafic rocks, are interpreted to be short-lived volcanic rises within the widening basin.

The tectonic implication of the remarkable southward shifting of the volcanic zone in Triassic and Early Cretaceous times cannot be understood fully, but Triassic shifting was coeval with upheaval of the Late Permian Sangun metamorphic belt along the inner side of the Paleozoic terrane and Early Cretaceous shifting was coeval with enormous felsic magmatism and upheaval of the main Paleozoic terrane. These upheavals of the inner side may have caused extension and thinning or rifting of the crust on the outer side to give rise to submarine volcanism. That these volcanic rocks are characteristically associated with deep-pelagic sediments is noteworthy.

CHERT SEDIMENTATION ASSOCIATED WITH SUBMARINE VOLCANISM

Modern siliceous sediments are almost entirely of organic origin, being composed of siliceous skeletons of mainly diatoms, Radiolaria, and siliceous sponges. A very minor amount of non-biogenically precipitated opal forms in special lagoons (Peterson and Borch, 1965) and hypersaline lakes (Eugster, 1969, Surdam and others, 1972), but all available data on the solubility and behavior of silica in solution (Krauskopf, 1959; Siever, 1962; and others) indicate that chemical precipitation of silica cannot occur in normal sea water because of undersaturation of dissolved silica with respect to amorphous silica.

The origin of bedded chert has long been a subject of controversy. On the basis of data obtained from my study of some Paleozoic and Mesozoic cherts of Kyushu and Shikoku, as well as those from published reports on other siliceous rocks of Phanerozoic age, I was once lead to the five following conclusions regarding the constituents and origin of chert (Kanmera, 1968): (1) Most cherts, normally occurring in bedded form, are primarily sediments of siliceous organic tests and their fragments. (2) Nodular chert in limestone is a replacement products formed by early diagenetic processes in carbonate sediments that contain disseminated siliceous organic remains. (3) The conversion of amorphous silica to quartz or chalcedony often destroys largely or almost completely the original structure of organic opaline tests and the primary sedimentary textures, resulting in apparently nonfossiliferous equigranular microcrystalline chert. (4) Remarkable epidiagenetic recrystallization and deformation make the bedded structure of normal chert obscure and finally give rise to a massive body of chert. (5) Massive chert of another type also is known, and, although of minor occurrences, it seems to be related to chemical precipitation from silica-rich solutions under very restricted special conditions.

The last-mentioned type is presumed to represent such an environment as crater lakes, where silica-rich fumaroles assure that the lake water is saturated with silica. Significant revision of the conclusions mentioned above probably are not needed on the basis of knowledge obtained since that time, and hydroflouric acid etching as exploited and advanced by Imoto and Takahashi (1972) and Saito (1972) has revealed clearly that the cherts, even those in which biogenic elements are hardly visible by ordinary thin-section methods, are composed of siliceous organic skeletons and their fragments.

Volcanic rocks of eugeosynclinal sequences are commonly accompanied by bedded chert. Cherts of the Paleozoic (fig. 1) and the Mesozoic of Japan are similarly closely associated with mafic volcanic rocks but rarely with felsic and intermediate rocks (fig. 1).

Then, why are the bedded cherts associated so intimately with mafic volcanic rocks? One interpretation is that the cherts originally were ocean-floor pelagic siliceous oozes deposited very slowly and directly upon the oceanic volcanic basement generated from oceanic ridges. This view is not tenable for the Paleozoic and Mesozoic cherts of Japan because the volcanic rocks of those ages cannot be referred to such oceanic basement as was described in foregoing pages. On the other hand, Bailey and others (1964) and some other persons have considered that bedded cherts represent chemical precipitation of silica that was released from volcanic material by rapid reaction between hot magmatic material and sea water in the deep sea at the time of eruption. This inorganic chemical precipitation hypothesis is also untenable in view of recently acquired knowledge on the nature of submarine basaltic eruptions, which, in turn, suggests that chemical reaction between the two substances in most examples proceeds slowly and continuously after chilling of the surface of the erupted material (Moore, 1965, 1966; Iijima and Hay, 1967). Most bedded cherts associated with volcanic rocks are intercalated with or rest on tuff beds but rarely directly overlie flow rocks. Such pyroclastic rocks are the products of explosive eruptions, which, in turn, are not likely to happen at depths greater than 500 m (Nayudu, 1964; McBirney, 1963; Moore, 1965, 1966; Moore and Fiske, 1969). Volcanic activity on the deep sea floor probably gives rise to quiet extrusion of lava flows on the floor so as to form a sea mound of flow rocks having hyaloclastite on their surfaces that is owed to the chilling effect of cold water, or such activity gives rise to sills intruded into soft sediments.

Cherts of the Shimantogawa Group are most revealing of petrographic features in relation to other kinds of rocks in Japan because of their relatively undeformed condition. So far as I have observed, they are all associated with basaltic volcanic rocks, especially with lapilli to fine-grained tuff in an interbedded or superposed relationship. Rhythmic interlayering is most characteristic. Even for those cherts that seemingly are intercalated directly in slate, very thin tuffaceous beds or partings are seen at the base or between chert layers, and by following the beds in the field we can confirm that the cherts are at the same horizon as the volcanic beds. The most common circumstance is that tuff breccia and lapilli tuff, commonly with intercalated flow beds, change laterally into finer grained tuff having fine siliceous layers or lenticular bands, a fraction of a millimeter thick; further, they change to interbedded chert and fine-grained tuff and finally to bedded chert having thin tuff partings. Similar changes are also seen vertically in the volcanic chert sequences. This stratigraphic sequence is so distinct that it can be used for determining the stratigraphic top of beds, which is consistent with the sequence recognized in sedimentary structures of clastic rocks.

Cherts are colored generally reddish to greenish, matching the color of interlayered tuffs. They are generally less than 20 m thick and rarely attain 30 m. Most are thin bedded, each layer generally being 1 to 5 cm thick.

Thin sections of bedded cherts and siliceous bands or laminae in tuff beds show that they are composed of abundant radiolarian tests and of subordinate aphanitic matrix. Microscopic observation of surfaces of the same specimens etched by hydroflouric acid reveals additional details of the texture of the matrix as well as of the biogenic skeletal constituents. That is, radiolarian tests standing out in contrast to the matrix show delicate mesh structure of the tests and fine radial spines, and the matrix is made up largely of fragments of siliceous tests and spines and of a minor amount of unidentifiable fine dust, probably clay of volcanic origin. No detrital grains of quartz or other minerals have been found.

Interlayered tuffs in chert layers commonly show an upward fining in grain size, which suggests deposition from suspension of volcanic material derived from submarine eruption. They often change into cherts through tuffaceous beds containing disseminated radiolarian remains. Similar upward fining in a chert layer with respect to the size of radiolarian tests is occasionally observed, although more detailed examinations are needed in order to be certain of this.

As stated above, the chert layers are mostly interlayered with tuffs, and even those that apparently overly the flow rocks also have thin partings of fine-grained tuff. This intimate relation of cherts and volcanic rocks and the extreme scantiness of siliceous organic remains in the overlying and underlying shales or sandstones indicate that accumulation of biogenic material that formed chert beds was rapid and contemporaneous with the submarine volcanism but was abruptly impoverished shortly after cessation of submarine volcanic activity. The enormous episodic supply of planktonic siliceous organisms is probably owed to mass perishments when explosive submarine eruptions resulted in rising and spreading of heated water columns and in associated water agitation. These eruptions took place at depths not greater than 500 m. The rapid deposition of chert is also suggested by the lack of detrital quartz and other minerals in the chert and volcanic rocks. Furthermore, although it is difficult to identify taxonomically the Radiolaria contained in cherts, preliminary examination reveals that their assemblage and generic ratios are almost the same throughout one chert formation as they are in a single chert layer, but these indices often are distinguished easily for different chert formations. The above-mentioned features suggest that the chert beds of the Shimanto Group must be of rapid deposition rather than of such slow accumulation as are radiolarian oozes in the modern deep ocean.

ACKNOWLEDGMENTS

I wish to express my sincere gratitude to R. H. Dott, Jr., of the University of Wisconsin, for his invitation to present this paper and for his review and to T. Matsumoto, of the Kyushu University, for his stimulating discussions on these and related matters.

REFERENCES

BAILEY, E. H., IRWIN, W. P., AND JONES, D. L., 1964, Franciscan and related rocks and their significance in the geology of western California: California Div. Mines Bull. 183, p. 1–177.

DICKINSON, W. R., 1971, Clastic sedimentary sequences deposited in shelf, slope, and trough settings between magmatic arcs and associated trenches: Pacific Geology, v. 3, p. 15–30.

———, 1972, Evidence for plate-tectonic regimes in the rock record: Am. Jour. Sci., v. 272, p. 551–576.

ERNST, W. G., 1972, Possible Permian oceanic crust and palte juncture in central Shikoku, Japan: Tectonophysics, v. 15, p. 233–239.
EUGSTER, H. P., 1969, Inorganic bedded cherts from the Magadi area, Kenya: Contr. Mineralogy and Petrology, v. 22, p. 1–31.
HAMILTON, W., 1969, Mesozoic California and underflow of Pacific mantle: Geol. Soc. America Bull., v. 80, p. 2409–2430.
HASHIMOTO, M., KASHIMA, N., AND SAITO, Y., 1970, Chemical composition of Paleozoic greenstones from two areas of southwest Japan: Geol. Soc. Japan Jour., v. 76, p. 463–376.
HORIKOSHI, E., 1972, Orogenic belts and plates of Japan: Tokyo, Kagaku, Iwanami Shoten, v. 42, p. 665–673 (in Japanese).
ICHIKAWA, K., FUJITA, Y., AND SHIMAZU, M., 1970, Geologic development of the Japanese Islands: Tokyo, Tsukuzi Shokan, 232 p. (In Japanese).
IGO, H., 1964, On the occurrence of *Goniatites* (s. s.) from the Hida Massif, central Japan: Paleont. Soc. Japan Trans. and Proc., new ser., no. 54, p. 234–238.
IIJIMA, A., AND HAY, R. L., 1967, Nature and origin of zeolitic palagonite tuffs on Oahu, Hawaii: Geol. Soc. Japan Jour., v. 73, p. 573–590 (in Japanese with English abstract).
IMAI, I., TERAOKA, Y., AND OKUMURA, K., 1971, Geologic structure and metamorphic zonation of the northeastern part of the Shimanto terrane in Kyushu, Japan: *ibid.*, v. 77, p. 207–220 (in Japanese with English abstract).
IMOTO, N., AND TAKAHASHI, A., in press (1972), Microstructure of chert: Jour. Marine Geology, v. 8.
ISHIZAKA, K., 1972, Rb-Sr dating on the igneous and metamorphic rocks of the Kurosegawa tectonic zone: Geol. Soc. Japan Jour., v. 78, p. 596–575 (in Japanese with English abstract).
IWASAKI, M., 1969, The basic metamorphic rocks at the boundary between the Sambagawa metamorphic belt and the Chichibu unmetamorphic Paleozoic sediments: Geol. Soc. Japan Mem. 4, p. 41–50 (in Japanese with English abstract).
KANISAWA, S., 1971, Metamorphic rocks of the southwestern part of the Kitakami mountainland, Japan: Tohoku Univ. Sci. Repts., ser. 3, v. 9, p. 155–198.
KANMERA, K., 1961, Middle Permian Kozaki Formation: Kyushu Univ. Fac. Sci. Sci. Repts., v. 5, p. 196–215 (in Japanese with English abstract).
———, 1968, On some sedimentary rocks associated with geosynclinal volcanic rocks: Geol. Soc. Japan Mem. 1, p. 23–32 (in Japanese with English abstract).
———, 1969, Litho- and bio-facies of Permo-Triassic geosynclinal limestone of the Sambosan belt in southern Kyushu: Paleont. Soc. Japan Special Paper 14, p. 13–39.
———, 1971, Paleozoic and early Mesozoic volcanicity in Japan: Geol. Soc. Japan Mem. 6, p. 97–110 (in Japanese with English abstract).
KANO, H., 1967, On the Usuginu granitic rocks in Kyushu, Japan: Akita Univ. Jour. Min. Coll., ser. A, v. 4, p. 1–37.
———, 1971, Studies on the Usuginu conglomerates in the Kitakami Mountains: Geol. Soc. Japan Jour., v. 77, p. 415–440 (in Japanese with English abstract).
KARIG, D. E., 1970, Ridges and basins of the Tonga-Kermadec island arc system: Jour. Geophys. Research, v. 75, p. 239–254.
———, 1971a, Structural history of the Mariana island arc system: Geol. Soc. America Bull., v. 82, p. 323–344.
———, 1971b, Origin and development of marginal basins in the western Pacific: Jour. Geophys. Research, v. 76, p. 2542–2561.
KAWANO, Y., AND UEDA, Y., 1967, K-Ar dating on the igneous rocks in Japan: Japanese Assoc. Mineralogy, Petrology and Econ. Geology Jour., v. 65, p. 247–264 (in Japanese with English abstract).
KOBAYASHI, T., 1941, The Sakawa orogenic cycle and its bearing on the origin of the Japanese Islands: Tokyo Imp. Univ. Fac. Sci. Jour., ser. 2, v. 5, p. 219–578.
KRAUSKOPF, K. B., 1959, The geochemistry of silica in sedimentary environments, *in* IRELAND, H. A. (ed.), Silica in sediments: Soc. Econ. Paleontologists and Mineralogists Special Pub. 7, p. 4–19.
McBIRNEY, A. R., 1963, Factors governing the nature of submarine volcanism: Bull. Volcanology, v. 26, p. 455–469.
MARUYAMA, T., 1972, Rb-Sr whole rock ages of granitic and metamorphic rocks of the Gozaisho-Takanuki area, Abukuma Mountains: Mimeographed report of studies of the basements of Japan, no. 3, p. 14–20 (in Japanese).
MIKAMI, T., 1971, Preliminary notes of the Paleozoic sandstones from the Hikoroichi area, south Kitakami Mountains: Mem. Geol. Soc. Japan Mem. 6, p. 33–37 (in Japanese with English abstract).
MINATO, M., 1968, Basement complex and Paleozoic orogeny in Japan: Pacific Geology, v. 1, p. 85–95.
MINATO, M., GORAI, M., AND FUNAHASHI, M. (eds.), 1965, The geologic development of the Japanese Islands: Tokyo, Tsukiji Shokan, 442 p.
MIYASHIRO, A., 1961, Evolution of metamorphic belts: Jour. Petrology, v. 2, p. 277–311.
———, 1967, Orogeny, regional metamorphism and magmatism in the Japanese Islands: Dansk Geol. Fören. Meddel., v. 17, p 390–446.
———, 1972, Metamorphism and related magmatism in plate tectonics: Am. Jour. Sci., v. 272, p. 629–656.
MOORE, J. G., 1965, Petrology of deep-sea basalt near Hawaii: *ibid.*, v. 263, p. 40–52.
———, 1966, Rate of palagonitization of submarine basalt adjacent to Hawaii: U.S. Geol. Survey Prof. Paper 550-D, p. 163–171.
MOORE, J. G., AND FISKE, R. S., 1969, Volcanic substructure inferred from dredge samples and ocean-bottom photographs, Hawaii: Geol. Soc. America Bull., v. 80, p. 1191–1202.
NAKAZAWA, K., 1958, The Triassic System in the Maizuru zone, southwest Japan: Kyoto Univ. Coll. Sci. Mem., ser. B, v. 24, p. 265–313.
NAYUDU, Y. R., 1964, Palagonite tuff (hyaloclastites) and the products of post-eruptive processes: Bull. Volcanology, v. 27, p. 391–410.

Peterson, M. N. A., and Borch, C. C. von der, 1965, Chert: Modern inorganic deposition in a carbonate precipitating locality: Science, v. 149 (3961), p. 1501–1503.
Rodolfo, K. S., 1969, Bathymetry and marine geology of the Andaman Basin, and tectonic implications for southeast Asia: Geol. Soc. America Bull., v. 80, p. 1203–1230.
Saito, Y., 1972, Origin of chert: Nat. Sci. and Museum, v. 39, p. 173–178 (in Japanese).
Shibata, K., Nozawa, T., and Wanless, R. K., 1970, Rb-Sr geochronology of the Hida metamorphic belt, Japan: Canadian Jour. Earth Sci., v. 7, p. 1383–1401.
Shiida, I., and others, 1971, Greenstones of the Cretaceous Hitakagawa belt of the Shimanto terrain in the Totsukawa area, Nara Prefecture, central Japan: Geol. Soc. Japan Mem. 6, p. 137–149 (in Japanese with English abstract).
Siever, R., 1962, Silica solubility 0–200°C and the diagenesis of siliceous sediments: Jour. Geology, v. 61, p. 127–149.
Sugisaki, R., and Tanaka, T., 1971, Magma types of volcanic rocks and crustal history in the Japanese pre-Cenozoic geosynclines: Tectonophysics, v. 12, p. 393–413.
Sugisaki, R., and others, 1972, Late Paleozoic geosynclinal basalt and tectonism in the Japanese Islands: Tectonophysics, v. 14, p. 35–56.
Surdam, R. C., Eugster, H. P., and Mariner, R. H., 1972, Magadi-type chert in Jurassic and Eocene to Pleistocene rocks, Wyoming: Geol. Soc. America Bull., v. 83, p. 2261–2266.
Suzuki, T., Sugisaki, R., and Tanaka, T., 1971, Geosynclinal igneous activity of the Mikabu Green Rocks of Ozu City, Ehime Prefecture: Geol. Soc., Japan Mem. 6, p. 121–136 (in Japanese with English abstract).
Suzuki, T., and others, 1972, Geosynclinal volcanism of the Mikabu Green-Rocks in the Okuki area, western Shikoku, Japan: Japanese Assoc. Mineralogy, Petrology and Econ. Geology Jour., v. 67, p. 177–192.

PALEOZOIC MARGINAL OCEAN BASIN-VOLCANIC ARC SYSTEMS IN THE CORDILLERAN FOLDBELT

MICHAEL CHURKIN, JR.
U.S. Geological Survey, Menlo Park, California

ABSTRACT

The Paleozoic stratigraphic-structural belts of the Cordilleran foldbelt from Alaska to Nevada and California can be explained by a succession of marginal ocean basins opening and closing behind volcanic arcs. Cambrian through Devonian rocks of the foldbelt are divided into (1) a carbonate rock and quartzite belt (miogeosyncline), a continental shelf and upper continental slope deposit, (2) a graptolitic shale and chert belt (inner part of eugeosyncline), a marginal ocean-basin deposit developed behind a volcanic arc, and (3) a volcanic rock and graywacke belt (outer part of eugeosyncline), a volcanic arc deposit.

The volcanic rock and graywacke belt had a long history of volcanic island development, which began in northern California no later than the Late Ordovician Epoch and in southeastern Alaska and western Washington during the Precambrian or Cambrian Period. Initially, rocks of this belt formed a primitive crust on oceanic basement and through a long process of volcanism, erosion and sedimentation, metamorphism, and plutonism developed a crust of more continental character. In this interpretation, the outer eugeosynclinal rocks are not continental fragments of Asia or parts of oceanic features of unknown provenance that collided with North America. Instead, they are parautochthonous deposits of volcanic island chains and interarc basins developed outward from the North American continental margin as it faced a proto-Pacific basin. Deep-sea pelagic deposits, condensed sections of graptolitic shale and chert, apparently accumulated on basaltic basement in newly formed marginal ocean basins behind the volcanic chains as they migrated from the continent. In the Late Devonian and Mississippian, these marginal ocean basins closed, perhaps by collision with the frontal volcanic arcs, and the deep-sea pelagic sediments were thrust eastward onto the continental shelf, uplifted, and eroded to provide thick wedges of coarse clastic sediments.

In the Pennsylvanian and Permian Periods, the collapsed earlier Paleozoic marginal ocean basins were rifted, and a new system of basins, floored apparently by new crust, developed behind segments of the older volcanic arc complexes. This late Paleozoic basin in the latitude of Nevada was closing during the Late Permian and Early Triassic, and its rocks were again thrust eastward as during the Late Devonian and Mississippian orogeny. This succession of migrations of Paleozoic volcanic arc complexes away from the North American continent and creation of new basins, presumably with oceanic crust behind the frontal arcs, is closely analogous with the tectonic history of the western Pacific, where most of the world's marginal seas are now concentrated.

INTRODUCTION

This paper compares the geologic record of the early development of the continental margin of western North America with present-day continental margin features of the Pacific. Special emphasis is given to comparisons between the Paleozoic geology of northern California, Nevada, Idaho, and southeastern Alaska and parts of the western Pacific, such comparisons being enhanced by the results of deep-sea drilling. The entire margin of western North America is treated on the basis of the best known Paleozoic sequences illustrating each tectonic province. Thus, the wide latitudinal gaps in the presented data produce, at best, greatly generalized composite sections and introduce some inaccuracies in correlative details.

Paleozoic rocks of the Cordilleran foldbelt have been traced nearly continuously from Alaska south into the southern parts of California, Nevada, and northern Mexico (King, 1966) (fig. 1). The northern end of the Cordilleran Geosyncline trends west across southern Alaska to the edge of the Bering Sea, and its continuation reappears in the Koryak Mountains and in other parts of northeast U.S.S.R. (Churkin, 1970). Along its length, the foldbelt is exposed in a series of arcuate segments (King, 1966) whose structure, I believe, generally reflects an original arcuate and segmented paleogeography like that of the volcanic arcs and marginal ocean basins along the modern west Pacific margin.

Cambrian through Devonian rocks of the foldbelt can be subdivided conveniently into three stratigraphic-structural belts that can in turn be related to specific continental margin features (fig. 2). From the continental interior progressively outward these belts are:

(1) Carbonate and quartzite belt (miogeosyncline), the deposits of which are comparable with those forming on continental shelves.

(2) Graptolitic shale and chert belt (inner part of eugeosyncline), the deposits of which are comparable with deep-sea pelagic deposits forming in ocean deeps.

(3) Volcanic and graywacke belt (outer part of eugeosyncline), the deposits of which are comparable with those forming on or in the vicinity of volcanic arcs.

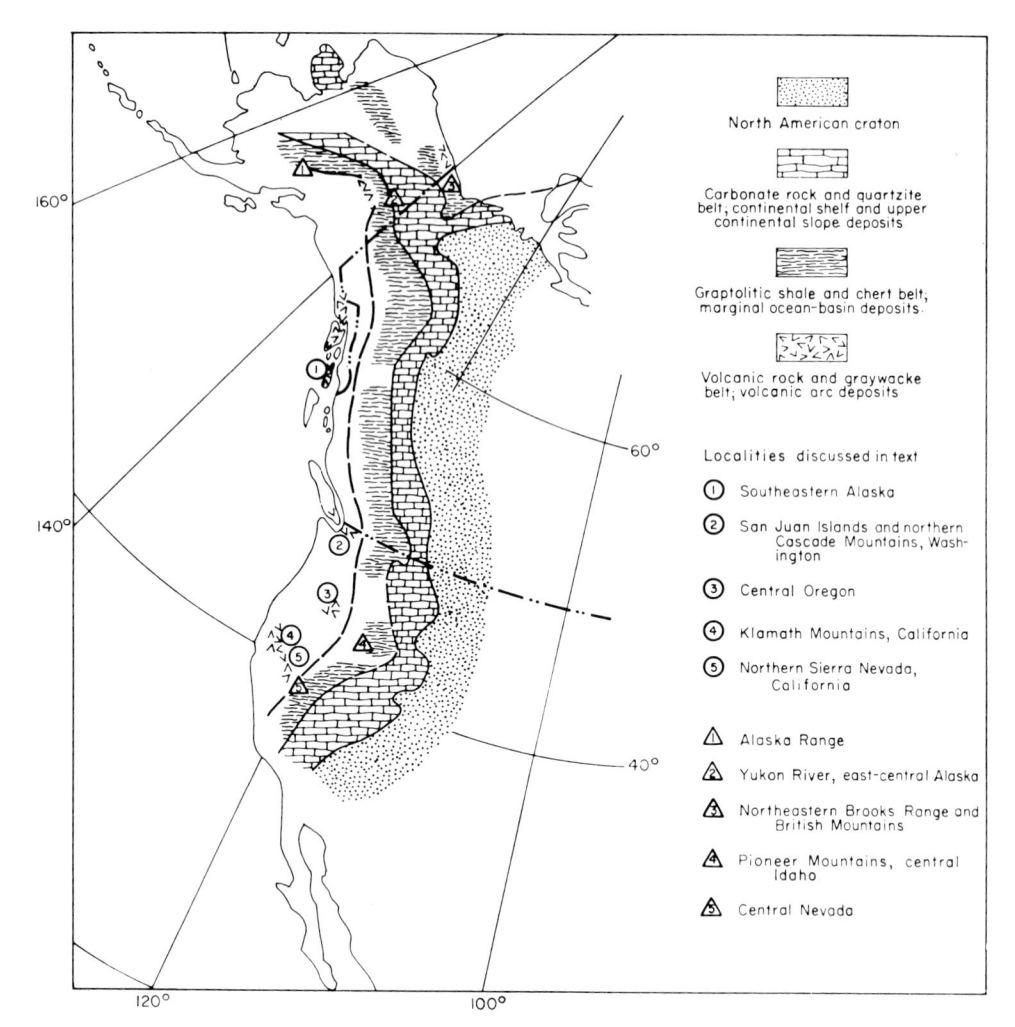

FIG. 1.—Major lower Paleozoic stratigraphic belts within the Cordilleran foldbelt and localities discussed in text.

Acknowledgment.—Publication has been authorized by the Director, U.S. Geological Survey.

CARBONATE AND QUARTZITE BELT—A CONTINENTAL SHELF DEPOSIT

A predominantly carbonate-rock sequence of Cambrian through Devonian age forms a broad wedge-shaped deposit mantling the North American continental nucleus. Westward, away from the continental interior, these carbonate rocks thicken markedly and are referred to as the Millard Belt or Millard Miogeosyncline (Kay, 1951). The deposits of this belt lie on a thick sequence of late Precambrian strata that in turn rest on an earlier Precambrian crystalline basement to the east (Stewart, 1972). The limestone and dolomite that form the major part of the Paleozoic section yield much sedimentological and biological evidence showing that they originated in shallow marine waters, including the intertidal and supratidal zones, like those covering modern continental shelves. Most of the rocks, where not recrystallized or dolomitized, have abundant shelly fossil fragments, and extensive coral reefs are known, particularly along the outer edge of the belt (Winterer and Murphy, 1960). Several widespread quartzites and related regional unconformities within the Cambrian and Ordovician parts of the section mark episodes of marine transgression and regression of the North American continent (Sloss and others, 1949). The quartzites, made of well-rounded and frosted, extremely pure quartz sand grains, can be traced to sources in the continental interior. The maturity of the quartzites and their widespread distribution indicate that they are strandline sand deposits that were progressively joined laterally to produce their broad, tabular shape. As marine transgression of the craton continued, bioclastic carbonate sediments were produced across a wide shelf offshore of the beach sand; meanwhile, coral reefs grew along the outer edge of the shelf. The environment of deposition was apparently a broad continental shelf adjacent to a low-standing continental landmass of mature physiography. The Bahama Banks off of Florida and the Queensland Plateau off the east coast of Australia are approximate modern-day analogues.

On their westward side, the relatively pure carbonate rocks thin abruptly and grade into a much narrower belt of argillaceous rock, limestone, dolomite, and chert, which are best known in central Nevada (Roberts and others, 1958; Kay and Crawford, 1964; McKee, in press). The stratigraphic relations of this belt seem to be transitional between the purer carbonate rock facies on its continental side and the graptolitic shale and chert belt on its west side. Details of stratigraphy in the transitional zone, however, are imperfectly known because the rocks are involved in eastward directed imbricate thrusts. The transitional rocks by their thinness and the presence of major disconformities in the section have been interpreted as having formed around islands that interrupted the otherwise continuous seas extending from the carbonate-quartzite shelf into a deep basin on the west (Kay and Crawford, 1964). I believe that the relatively condensed sections of the transitional belt may also be interpreted as the result of submarine erosion and slumping along an abrupt shelf break that separated the predominantly carbonate sediments of the outer shelf from the predominantly argillaceous sediments of the continental slope and ocean basin farther west.

GRAPTOLITIC SHALE AND CHERT BELT—A MARGINAL OCEAN BASIN DEPOSIT

The graptolitic shale and chert belt forms the inner zone of the Cordilleran Geosyncline (fig. 1). The shales, despite their structural complexity and generally uniform lithology, have yielded numerous graptolite collections in their Ordovician, Silurian, and Lower Devonian parts that, in association with marker beds such as quartzites, identify sequences that can be correlated along the length of the Cordil-

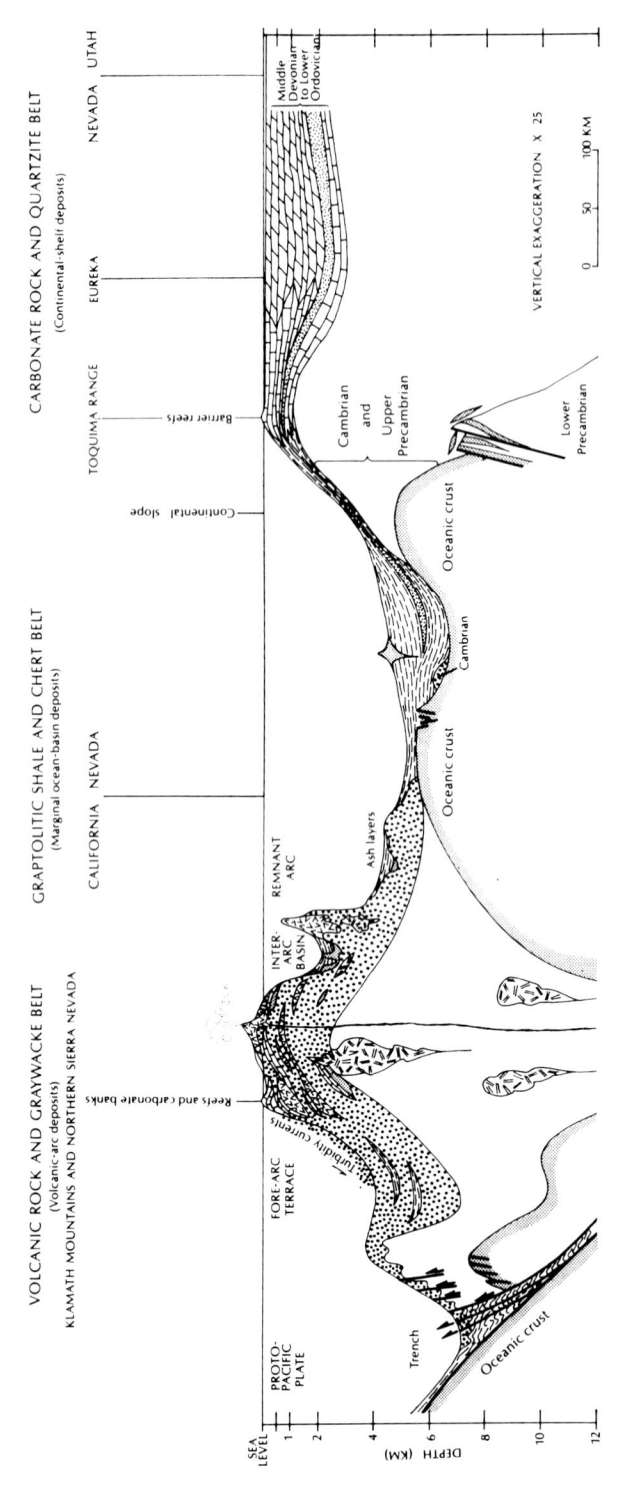

FIG. 2.—Middle Paleozoic reconstruction of western margin of North America showing relation of stratigraphic belts to the continental margin. (See figure 5 for explanation of lithology patterns.)

lera (fig. 3). Where stratigraphic sections have been measured, the largest part of these is composed of argillaceous rocks, siltstone, and chert (Ketner, 1969). Quartzite locally forms a large part of the section; greenstone, limestone, and sandstone generally are minor constituents.

In the northern part of the graptolitic shale and chert belt, there are areas in east-central Alaska and in the Yukon Territory where graptolitic shales notably lacking lavas and quartzites overlie earliest Ordovician and Cambrian carbonate strata. These thin sequences of graptolitic shale interfinger with limestones of the Yukon shelf, which separate the Cordilleran graptolitic shale and chert belt from a similar belt exposed along the edge of the Canada Basin part of the Arctic Ocean (Churkin, 1969). In most other places, the graptolite shales have been thrust eastward onto the carbonate rock and quartzite belt.

Along the outer side of the graptolitic shale and chert belt, there are no known areas of pre-Ordovician crystalline basement. In several isolated places in west-central parts of Nevada, however, Cambrian rocks occur in thrust slices. These include the Scott Canyon Formation, mainly of chert, argillite, and greenstone, which resembles nearby Ordovician graptolitic rocks, and the Harmony Formation, a sandstone and grit characterized by a high percentage of detrital feldspar, mica, and quartz indicating a granitic source (Roberts and others, 1958).

At some places within the graptolitic shale and chert belt, minor amounts of pillow basalt are interlayered with the shale and chert. These are found at the base or in lower parts of the sections (Kay and Crawford, 1964; Churkin and Kay, 1967; Riva, 1970) and are unknown in Silurian and Devonian rocks of similar facies (fig. 3). These facts suggest that the basalts represent proximity to a basaltic basement as is true for some of the basalts found in the bottoms of holes drilled in the Deep Sea Drilling Project. Also, good evidence of an early Precambrian crystalline basement west of the Utah-Nevada boundary is lacking. Lead isotope ratios (R. Zartman, personal commun., 1973) suggest that the crust in the northern part of the Basin and Range Province west of the Utah-Nevada line is unlike continental crust farther east. Instead, the crust in the western part of the Basin and Range Province may have a basaltic basement and a crust not unlike that of ocean basins. Furthermore, Sr 87/86 isotopes in Mesozoic granitic rocks indicate that the crust that underlaid the western part of the Basin and Range Province was intermediate in character between oceanic crust that was under northern California and continental crust in the eastern Basin and Range Province (Kistler and Peterman, in press). Thus, the graptolitic shale and chert belt in the western Basin and Range Province, its site of deposition, appears to have had a crustal structure like that of a marginal ocean basin filled with sediment, and it is bordered on the west in the northern Sierra Nevada Mountains by a crust not unlike that of a magmatic arc, which it apparently was at various times during the Paleozoic Era and early part of the Mesozoic Era.

Thicknesses of the graptolitic shale originally were estimated to be several times those of the carbonate rock and quartzite belt farther east. Reconstructions of the Cordilleran Geosyncline thus depicted the geosyncline as a westward thickening wedge (Kay, 1951; Roberts and others, 1958). Detailed mapping, however, has shown that the structure within the graptolitic shale and chert belt is very complex and that most large blocks are not only internally folded but also imbricately sliced by low-angle thrust faults (Kay and Crawford, 1964; Gilluly and Gates, 1965). In several places, bed-by-bed collecting of graptolites coupled with structural mapping of marker beds has established local stratigraphic successions (Churkin, 1963; Kay and Crawford, 1964; Churkin and Kay, 1967; Riva, 1970). These successions have relatively lesser thicknesses (figs. 2 and 3) than carbonate and quartzite facies representing the same geologic time farther east. The condensed nature of the graptolitic shale sections is not peculiar to this region but seems to be the rule in other geosynclinal areas—British Isles (Toghill, 1968), Canadian Arctic Isles (Trettin, 1971), and Alaska (Churkin, Carter, and Eberlein, 1971), to name a few. When the condensed aspect of the graptolitic shale and chert belt is taken into account, the reconstruction (fig. 2) shows a close similarity to sedimentary profiles along modern continental margins. Comparison of sedimentation rates obtained from various modern oceanic provinces (Scientific Staff, DSDP legs 7, 21, 1969, 1972) with thicknesses of graptolitic shales and cherts suggests that the general accumulation rate (<10 m/my) of abyssal clay most nearly approximates that of the graptolitic shale. Where thick quartzites are present in the graptolitic shale, they are confined to narrow biostratigraphic intervals and thus represent only a fraction of the geologic time represented by an equal thickness of shale or chert (fig. 3). Example of this are in the Antler Park area, central Nevada, where massive quartzite units form about 600 m of a 1,500-m total section (Roberts, 1964), and in the Independence Range, northeastern Nevada, where

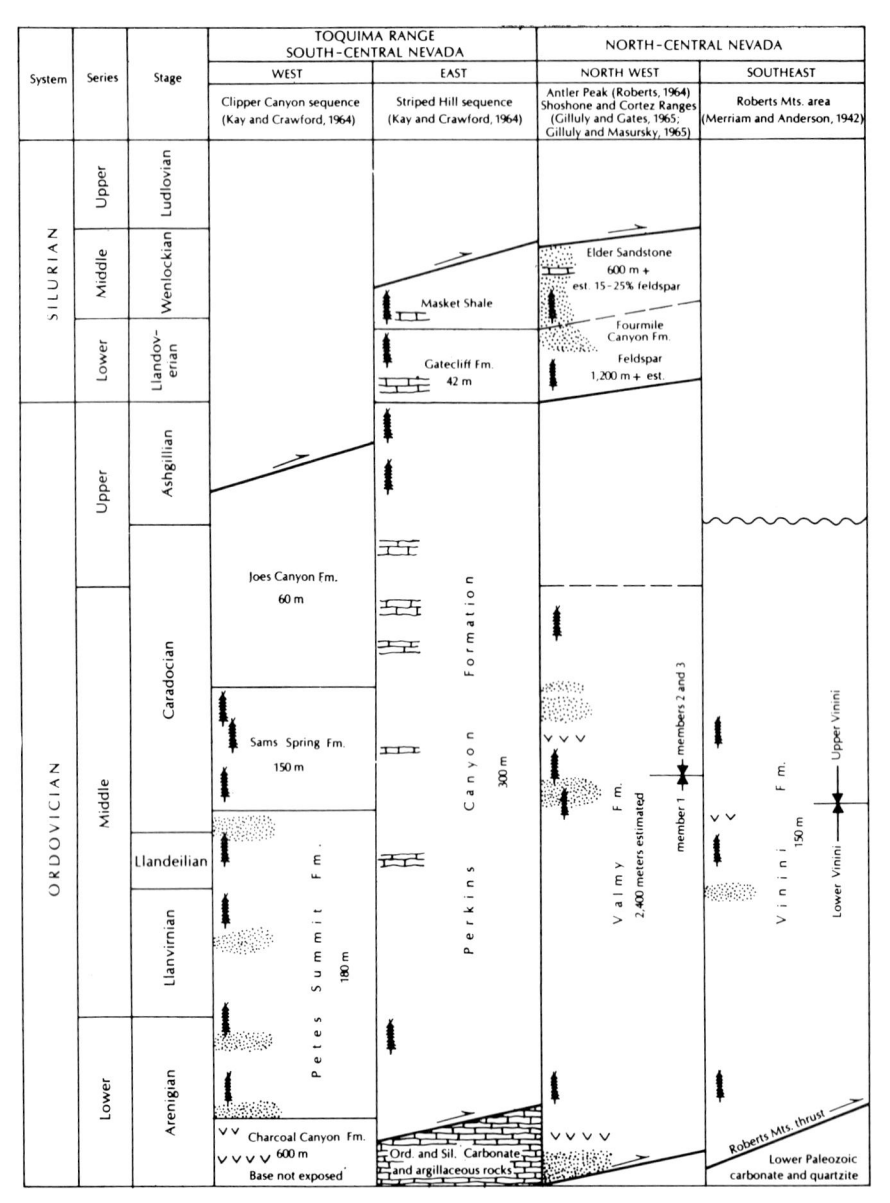

FIG. 3.—Correlation of sequences within the graptolitic shale and chert belt of Cordilleran Geosyncline. Stippled pattern, sandstone; brick pattern, limestone; simulated graptolite symbol, grapolite zones; V symbols, basaltic lavas.

massive quartzite units form about 360 m of an 810-m total section (Churkin and Kay, 1967).

Chemical analyses of the argillaceous rocks of the graptolitic shale and chert belt show 75 to 90 percent of SiO_2, which is a much higher content than that of most shales, modern deep-sea clays, or even of radiolarian oozes. The close resemblance of the composition of Ordovician argillaceous rocks in this belt with that of acid to intermediate volcanic rocks elsewhere (Ketner, 1969, fig. 6) suggests that ash from explosive volcanoes (described next from the volcanic rock and graywacke belt adjacent to the west) may have enriched the argillaceous sediments in silica. Relic pumice shards, "porcelaneous grains that appear to be devitrified volcanic glass," and 15 to 25 percent feldspar in silt-size grains in the Elder Sandstone of Silurian age provide evidence that ash reached the graptolitic shale area (Gilluly and Gates, 1965, p. 35, 36; Gilluly and Masursky, 1965, p. 58). A similarly high percentage of pyroclastic(?) feldspar grains is noted for siltstone of approximately coeval age at Trail Creek, Idaho (Churkin, 1963) and for siltstone within the Slaven Chert of Devonian age (Gilluly and Gates, 1965, p. 38). A modern analogue is the deep-sea sediments rich in airfall ash that are forming in the modern marginal ocean basins and interarc basins (e.g., Lau, South Fiji, Caledonia, and Coral) of the western Pacific (Scientific Staff DSDP, leg 21, 1972). The original amount of fine-grained volcanic detritus in the Paleozoic rocks, however, is difficult to estimate because unstable glassy volcanic fragments this old would be virtually recrystallized and now could be represented by much of the microcrystalline quartz and chert in the section.

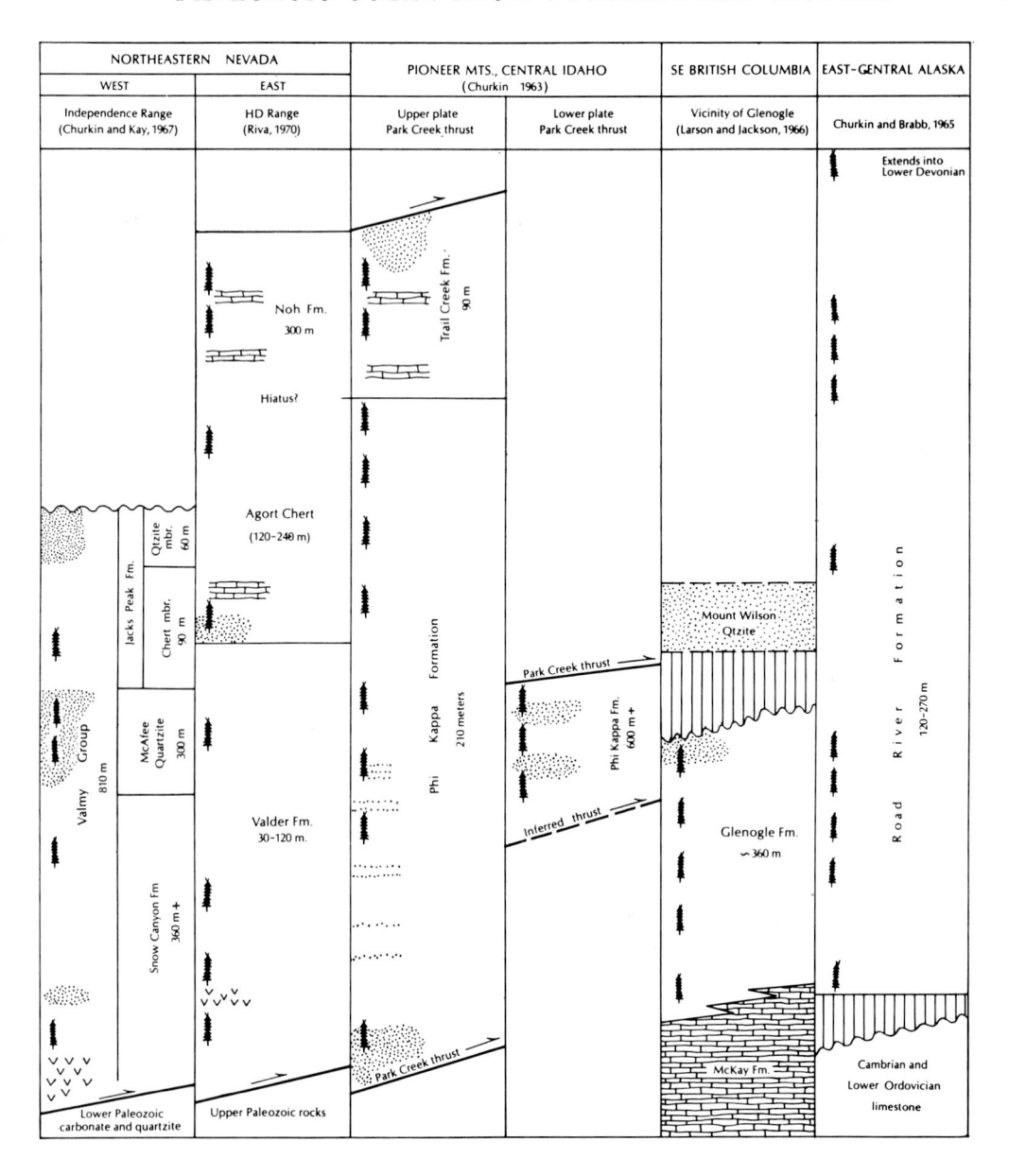

The depth of water and environment of deposition of graptolitic shales have long been a subject of speculation (Bulman, 1970). The Black Sea and other landlocked basins with restricted circulation and reducing bottom environment poor in life have been suggested as modern analogues to the environment in which graptolitic shale was deposited. The graptolite faunas found in the Cordilleran belt, however, are characterized not only by species of a Pacific faunal province but also by many species of the Arctic-Atlantic province. Application of the Black Sea model to explain the Cordillera with its mixture of graptolites from different faunal provinces would require a chain of Mediterranean-like seas circling the earth. Instead, it is more reasonable to think of the graptolitic shale and chert belt as having formed in a series of interconnected marginal ocean basins. The next section of this paper shows that volcanic islands rimmed the outer edge of the graptolitic shale belt and acted as barriers to oceanic currents and thus could have restricted the bottom-water circulation in the marginal ocean basins and helped to produce a reducing environment unfavorable to bottom-living or burrowing organisms.

The association of chitinophosphatic or siliceous fossils (phyllocarid crustaceans, linguloid brachiopods, sponge spicules, and radiolarians) with graptolites in the dark shales, instead of an association with thick-shelled calcareous fossils, suggests a very deep water environment of deposition below the calcium carbonate compensation depth (in modern seas, 3,500 to 5,000 meters in depth).

The vesicularity and bulk density of pillow lavas interlayered with the argillaceous sediments can give, as in modern examples (Moore, 1965), an independent measure of water depth. Samples from progressively deeper water have higher specific gravity and contain fewer and smaller vesicles. Preliminary results of a study of vesicularity of the Ordovician pillow lavas from the graptolite shale belt, when compared with modern lavas, suggest depths greater than 500 m for extrusion of the submarine lavas. In southeastern Alaska, where pillow lavas are interbedded with various types of rocks, those submarine lavas that are closely associated with graptolitic shale and chert sequences also suggest water depths greater than 500 m by their poorly developed vesicularity. In contrast, the more vesicular pillow lavas that are closely associated with thick massive reef and reef-breccia limestones and other shallow-water sediments suggest much shallower water, generally less than a few tens of meters deep.

The graptolitic shales and chert, although representing relatively deeper water deposits than the carbonate section farther east, were not necessarily everywhere formed at abyssal depths. For example, trilobites and brachiopods that are normally interpreted as shallow-water fossils occur in limestone that fills pockets in greenstone at one locality in the Shoshone Range of Nevada (Gilluly and Gates, 1965, p. 24). Complexly cross-bedded clastic limestone also found in the volcanic section there records strong current action, perhaps related to a volcanic submarine high such as a volcanic ridge or seamount. A volcanic plug of coarsely crystalline gabbro, which is surrounded by pillow lava and breccia that grade outward into fine-grained tuff and shale, may also be related to relatively shallow water volcanic activity (Roberts, 1964, p. A19). The series of active and inactive interarc basins separated by volcanic ridges and highs behind the active Tonga-Kermadec frontal arc-trench system illustrate the large number of volcanic submarine features possible in the development of marginal ocean basins behind migrating arcs (Karig, 1971; Scientific Staff, DSDP, leg 21, 1972).

Layered barite is another minor lithology in the graptolitic shale and chert belt that may have far-reaching significance in determining the condition of sedimentation within the belt. The barite is interbedded with cherts of Ordovician and Devonian age (Shawe and others, 1969; Poole and others, 1968). Similar marine barite deposits in a deep-water pelagic sediment section have been reported to be forming on the East Pacific Rise as a consequence of volcanic exhalations (Arrhenius and Bonatti, 1965; Bostrom and Peterson, 1969). Other marine barite mineralizations in areas of high heat flow, the Afar and Red Sea Rifts, for example, also appear to be associated with active oceanic rifts (Bonatti and others, 1972). By analogy, the bedded barite in the pelagic sequences of the Great Basin may be evidence of high heat flow during early Paleozoic time and of related oceanic volcanism and sea-floor spreading.

The quartzites in the graptolitic shale and chert belt vary in detail from massive, cliff-forming, nearly structureless bodies up to several hundreds of feet thick to thin beds of generally finer grained sandstone that in places are graded and interlaminated with shale. The quartzites are composed of frosted well-rounded quartz sand grains of characteristically bimodal size. These quartzites very closely resemble the Eureka and Kinnikinic Quartzites interbedded with carbonate rocks farther east. However, the quartzites in the graptolitic shale and chert sequences, according to Ketner (1966), apparently are slightly coarser grained and less well sorted than the quartzites in the carbonate rock and quartzite belt farther east. Ketner therefore attributed their origin to an unknown source, probably on the opposite side of the geosyncline. There are no sandstones or other rocks sufficiently rich in quartz, however, to produce the quartzites from the volcanic rock and graywacke belt along the outer edge of the palinspastically restored graptolite shale and chert belt. There are a few quartz-bearing sandstones of Ordovician and Silurian ages in the volcanic rock and graywacke belt in the northern Klamath Mountains (Churkin and Langenheim, 1960, p. 264), but these are only minor and impure quartz sandstone units. In southeastern Alaska a rare quartzo-feldspathic sandstone in a graywacke and mudstone section appears to have developed locally from quartz-bearing lavas (Eberlein and Churkin, 1970, p. 12). On the other hand, pre-Ordovician rifting conceivably could have carried westward a yet undiscovered continental fragment including Cambrian or Precambrian quartzite that may have served as a western source (fig. 5A).

Another important consideration in establishing the source of the sand is that the thickest beds of quartzite in the graptolitic shale and chert belt are interbedded with graptolitic shales of about Middle Ordovician (early Caradocian) age, which is approximately the age of the mineralogically identical Eureka (Ross, 1964) and Kinnikinic Quartzites (Hobbs and others, 1968) in the carbonate rock and quartzite belt farther east. All these units are western facies of the well-known St. Peter Sandstone of the Craton.

Besides these Middle Ordovician quartzites, there are some less conspicuous Early Ordovician quartzites in the graptolitic shale and chert belt. Because comparable quartzites were thought to be unknown in rocks of the shelf facies that are older than the Eureka, a more westerly source for the quartz sand was postulated (Gilluly and Gates, 1965). However, some quartzites are recognized in the earlier Ordovician sections of central Nevada (quartzite below the Copenhagen Formation; Ross, 1964), western Utah (Swan Peak Quartzite; Hintze, 1951), and central Idaho (Clayton Mine Quartzite; Hobbs and others, 1968).

Available data suggest that most of the quartzites in the graptolitic shale and chert belt are multicycled mature sands derived from the North American craton and carried off a sharp shelf break down submarine canyons by slumping and turbidity currents into the graptolitic shale and chert basins. The general lack of primary sedimentary structures, especially graded bedding considered characteristic of turbidites, can be explained by the textural maturity of the sand grains already developed before the sands were transported into the deep oceanic basin. In places, cyclical interlaminations of the quartzite with argillaceous rocks (Churkin, 1963) and sole markings (Gilluly and Masursky, 1965; Kay, 1966) have been reported. All these current features are found to some degree in deep-sea turbidites, for example, those in the Coral Sea abyssal plain at a water depth of 4,650 m (Scientific Staff, DSDP leg 21, 1972). The commonly thick quartzites of central Idaho, in part recycled from deep erosion of the eastern shelf quartzites, suggest the existence of a major delta-submarine canyon and deep-sea fan system in Idaho that could have conveyed the sand into the graptolitic shale and chert basin, whence it was transported south along the basin axis.

VOLCANIC ROCK AND GRAYWACKE BELT—AN
ISLAND ARC DEPOSIT

Paleozoic sections composed mainly of coarse volcaniclastic sedimentary rocks and rich in lavas interlayered with reef limestones are known from many widely separated localities along the rim of the northern Pacific (Churkin, 1970). Along the northeast side of the Pacific Ocean basin (fig. 1), they form the outer margin of the Cordilleran Geosyncline and presumably continue to the northwest side of the Pacific, where they occur in the Koryak Mountains and in a few isolated places within the northeast U.S.S.R. (Krasniy, 1966) and in Japan (Minato and others, 1965).

These rocks have many characteristics in common with modern deposits, both marine and nonmarine, forming in volcanic island areas (table 1). The oldest and most complete section of this facies is exposed in southeastern Alaska (Brew and others, 1966; Eberlein and Churkin, 1970; Ovenshine and Webster, 1970). Here, thick sequences of graywacke, polymictic conglomerate, submarine lava, and volcaniclastic rocks closely associated with biogenic limestone were deposited in several parts of this region during most of Ordovician, Silurian, Devonian, Carboniferous, and Permian time (fig. 4).

Most of the lavas are basaltic or andesitic, but they have a wide range in composition and include rhyolite (Brew and others, 1966; Brew, 1968; Eberlein and Churkin, 1970). Pillow structure characterizes many of the basalt flows, and in numerous places there is a progressive fragmentation from pillow breccias into aquagene tuff. In many places the volcaniclastic rocks are cemented thoroughly by limestone and interbedded with strata containing marine fossils (Eberlein and Churkin, 1970). In contrast, rhyolitic tuffs, although in places cyclically interlayered with marine limestones, probably represent subaerial ash flows that were sufficiently hot at the time of deposition to give the rock a flow-banded and welded texture (G. D. Eberlein and M. Churkin, Jr., unpub. data). In gen-

TABLE 1.—MAJOR FEATURES OF THE LOWER PALEOZOIC CORDILLERAN GEOSYNCLINE

Feature	Eugeosyncline — Volcanic rock and graywacke belt	Eugeosyncline — Graptolitic shale and chert belt	Miogeosyncline — Carbonate-rock and quartzite belt

Sedimentary rocks	Volcaniclastic graywacke-mudstone conglomerate including turbidites
	Minor chert and shale
	Biogenic limestones:
	  reef and reef breccia
	  brachiopod-pelecypod, shell-bank deposits
	  intertidal algal mats
	Oolitic limestone and lime mudstone
		Graptolitic shale
		Bedded chert
		Quartzite
		Minor intraformational chert grit
			Limestone and dolomite; mainly bioclastic carbonate rocks including reef and related deposits
			Quartzite
			Limy argillaceous rocks

Sedimentary environment	Extremely variable water depths related to faulting and volcanic activity resulting in abrupt facies changes Coarse clastic rocks (partly nonmarine) and reef, lagoon, and intertidal limestones outline volcanic islands Turbidites with minor pelagic layers developed in deeper water fore-arc terraces and interarc troughs	Deep-water pelagic sediments deposited mainly below carbonate-compensation depth Interbedded quartzitic turbidites Minor intraformational microbreccia on submarine highs Subaerial volcanic ash present but difficult to identify	Shallow-marine largely biogenic carbonate sediments Strandline and coastal plain clastics Relatively gradual facies changes except around reefs
Volcanic rocks	Abundant basaltic-andesitic breccia and tuff including carbonate-rich aquagene tuff Pillow basalt and abundant broken pillow breccia Structureless submarine and subaerial lava Rare rhyolitic welded tuff	Pillow basalt rare except in lower parts of sequence Very rare broken pillow breccia and tuff	None
Syntectonic intrusive rocks	Some large plutonic masses ranging widely in composition from granite to gabbro, rarely ultramafic Abundant dikes, sills, and small porphyritic masses	Virtually none except for rare mafic bodies	None
Oldest fossiliferous rocks and basement	Southeast Alaska: oldest strata are Early Ordovician volcaniclastic and volcanic rocks Underlying metamorphosed volcanic arc facies of Cambrian or Precambrian age Klamath Mountains: Upper Ordovician and Silurian strata overlie Ordovician(?) ultramafic sheet	Early Ordovician pillow lavas, chert, and graptolitic shale In a few places similar Cambrian rocks Pillow lavas in lowest parts of sequences suggest approach to basaltic basement Outer part of belt in thrust slices has no sign of basement Inner part in places rests on Cambrian limestone and quartzite	Thick Precambrian strata in outer part of belt, including older Precambrian crystalline rocks toward the continental interior
Type of crust underlying initial deposits	Oceanic or transitional where metavolcanic arc basement	Mainly oceanic in outer part Transitional in inner part	Continental
Modern tectonic examples	Volcanic island arcs and associated troughs	Marginal ocean basins behind island arcs including lower continental slope	Continental shelf and coastal plain, including upper continental slope

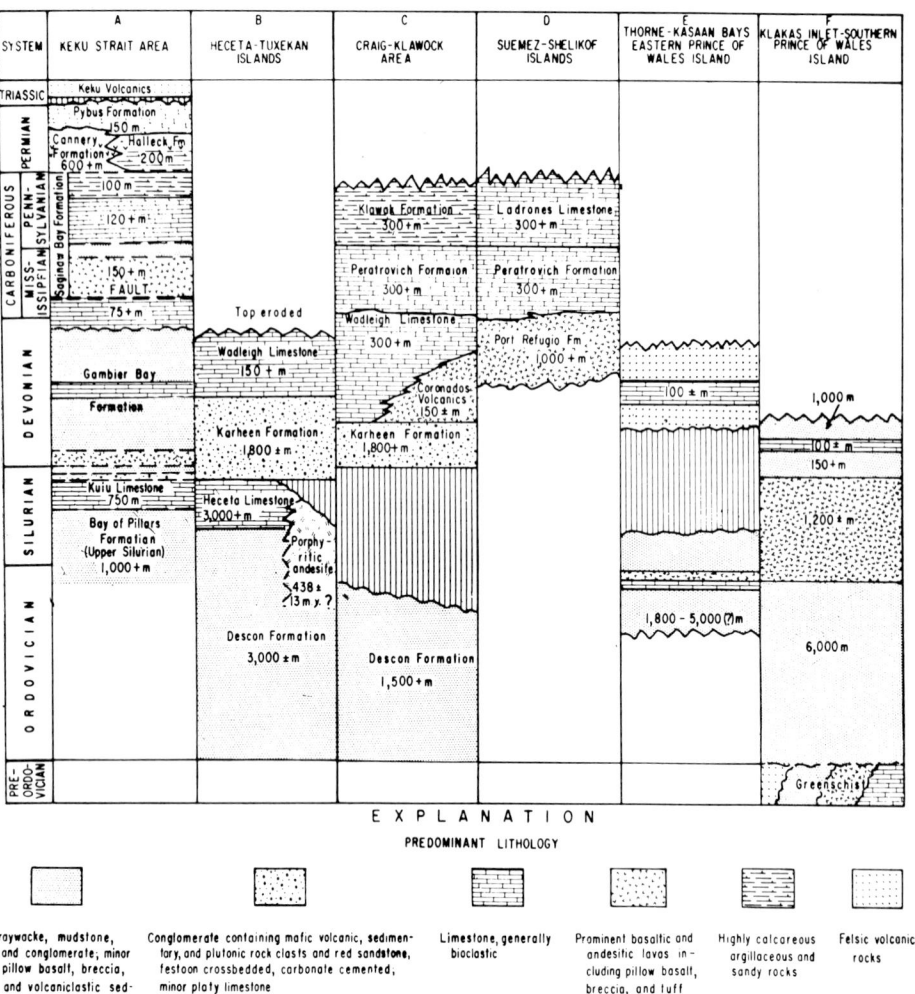

FIG. 4.—Correlation chart of major rock types of Paleozoic age in southeastern Alaska showing abrupt lateral changes in stratigraphy (after Churkin, 1973). Sections oriented from Keku Strait on the north to southern Prince of Wales Island on the south, a distance of 250 km.

eral, the wide range in composition and texture of the volcanic rocks supports their correlation with the type of complex volcanism associated with modern volcanic areas.

Closely associated with the lavas are several types of generally structureless bodies of crystalline igneous rocks. The more coarsely crystalline rocks range in composition generally from monzonite-syenite to granodiorite-diorite but include gabbro and some ultramafic rocks. Their contacts with Paleozoic sedimentary and volcanic rocks are crosscuting and intrusive. Some of the granitic rocks, however, give isotopic dates that are Ordovician (Lanphere and others, 1964) and Devonian (Lanphere and others, 1965), not widely different from the Paleozoic rocks they intrude.

Besides definitely plutonic rocks, massive and brecciated fine-grained igneous rocks rich in phenocrysts of feldspar, pyroxene, and amphibole are closely associated with compositionally similar submarine lavas and volcaniclastic rocks of early Paleozoic age. These porphyritic rocks form steep-sided, dome-shaped bodies not unlike the more crystalline plutonic rocks but in a few places are interbedded with sedimentary rocks. One of these porphyries, forming Staney Cone and Kogish Moutains, Prince of Wales Island, southeastern Alaska, is interbedded with a lens of limestone containing Silurian corals. A potassium-argon isotope date of 438 ± 13 my obtained on amphiboles from the porphyry is consistent with the paleontologic dating (Churkin and Eberlein, in press). The plutonic rocks, isolated areas between the widely separated regions.

LATE PALEOZOIC SEDIMENTATION

The graptolitic shale and chert belt was formed over a relatively long period spanning at least the Early Ordovician through most of the Devonian—about 150 million years. In the Late Devonian Epoch, this cycle of nearly continuous sedimentation was sharply interrupted by the Antler Orogeny that deformed and uplifted the belt (Dott, 1955; Roberts and others, 1958). An enormous volume of conglomerate and sandstone composed mainly of chert and argillite detritus forms a clastic wedge deposit that lies unconformably on the graptolitic shale and chert belt and extends progressively eastward with less disconformity and finer grain size across the miogeosyncline.

In the Carboniferous and Permian Systems, varying amounts of volcanic rocks associated with deep-water argillaceous and chert sequences lithically similar to those in the older graptolitic shales indicate that oceanic conditions again developed along the axis of the Cordilleran belt. The Havallah Sequence (Roberts, 1964; Silberling, in press) and the Schoonover Formation of Fagan (1962) were formed in the southern part of the belt. Farther north the Cache Creek Group in British Columbia (Monger and Ross, 1971), the Pennsylvanian and Permian rocks of southern Alaska (Bond, 1971; Richter and Jones, 1971), and the Yukon-Koyukuk Basin sequence in western Alaska (Patton, 1973) also suggest oceanic conditions in places, including volcanic arc activity. The Havallah Sequence is allochthonous, and its basement is uncertain, although ultramafic rocks occur along its thrust border (Speed, 1971). The other units, however, rest on ultramafic rock basement that in places sems to show an ophiolitic succession that can be interpreted as oceanic crust. As an exception, the Schoonover Formation has a basal conglomerate that rests unconformably on Ordovician graptolitic shale and chert (Churkin and Kay, 1967). During the Late Permian and Early Triassic Epochs, deposition of the Havallah Sequence, a marginal ocean basin deposit in the latitude of Nevada, was terminated by the Sonoma Orogeny, and the Havallah rocks were again thrust east as in the earlier Antler Orogeny (Silberling, in press).

PALEOZOIC TECTONICS

Individual segments of the Cordilleran foldbelt have been separately fitted into conceptual models embracing plate-tectonic theory (Moores, 1970; Burchfiel and Davis, 1972; Monger and others, 1972; Stewart, 1972). The reconstructions in this paper, although grossly resembling some of the conceptual models (Burchfiel and Davis, 1972, figs. 2A,B; Monger and others, 1972, figs. 4 and 6), differ widely in detail. This discussion is intended to emphasize how various stratigraphic and structural details within the Cordillera can be related to specific continental margin features influenced in various degrees by the processes of sea-floor spreading.

The thick carbonate rock and quartzite belt that rimmed the continent and thinned into platform deposits of the continental interior (craton) shows little sign of tectonic unrest—only the effects of epeirogenic uplift and subsidence. If there was any compressive interaction of the western continental margin of North America with an oceanic plate farther west, the effects on Cambrian through Middle Devonian sedimentation on the continental shelf were shielded by the tectonic activity and resulting deposits of the volcanic arc system along its outer margin (figs. 2 and 5A-C).

The graptolitic shale and chert belt, which is interpreted as having formed in a marginal ocean basin environment created by oceanic rifting and spreading behind a magmatic arc, also shows little sign of tectonic unrest during its sedimentation. The stratigraphic record is a condensed section of pelagic sediments deposited on deep-water pillow lavas.

The volcanic rock and graywacke belt (outer eugeosynclinal belt), in contrast to the graptolitic shale and chert belt and the carbonate rock and quartzite belt, had a complex record of violent volcanism, plutonism, and uplift that can be related to volcanic arc tectonics.

Preliminary reconstructions of the sequence of major tectonic events are illustrated in figure 5 and are noted under the next five headings to show their possible influences on the major stratigraphic belts of the Cordilleran geosyncline.

Early to Middle Ordovician.—Rifting of the continental margin of western North America in the late Precambrian (Stewart, 1972) and migration of pre-Ordovician arc complexes away from the continent created marginal ocean basins (fig. 5A). This reconstruction is based on evidence in southeastern Alaska and in the San Juan Islands and north Cascade Mountains region indicating that Early Ordovician through Devonian arc sequences were deposited, respectively, on still older, highly deformed, intruded, and metamorphosed arc complexes. Much farther south, in Nevada, feldspathic sandstone and grit of Cambrian age were apparently derived from a granitic source that may represent a remnant of an orogenically active continental margin or rifted volcanic arc left behind as a new frontal arc moved away from the continent during pre-Ordovician time.

Deep-sea pelagic sediments accumulated apparently on a basaltic basement. Layered barite may represent some of the initial deposits related to a process of rifting and submarine volcanism as portrayed in figure 5A.

Late Ordovician.—This reconstruction (fig. 5B) is based on the fact that the oldest fossiliferous strata in the Klamath Mountains and northern Sierra Nevada Mountains are Late Ordovician and Silurian and that they represent a volcanic arc assemblage that may have built up directly on oceanic crust as did the Aleutian Ridge.

Late Ordovician through Middle Devonian.—This was a long, relatively static period when the outer volcanic arc system shielded the continental margin of western North America and its marginal ocean basin from deformation by a proto-Pacific plate that moved against the arc (fig. 5B,C).

Late Devonian and Mississippian.—In the Late Devonian Epoch, the cycle of nearly continuous sedimentation for about 150 my in the graptolitic shale and chert belt was sharply interrupted by the Antler Orogeny that deformed and uplifted the belt (Dott, 1955; Roberts and others, 1958) (fig. 5D). An enormous volume of conglomerate and sandstone composed mainly of chert and argillaceous detritus forms a clastic wedge deposit that lies unconformably on the belt and extends progressively eastward with diminishing disconformity and grain size across the miogeosyncline.

The reconstruction shows the closure of the marginal ocean basin by collision with its frontal arc and uplift of pelagic sediments by imbricate underslicing. The Roberts Mountains thrust fault is shown as a result of post-uplift gravity sliding (Roberts and others, 1958; Kay and Crawford, 1964).

Pennsylvanian and Permian.—In the Pennsylvanian and Permian Periods, renewed rifting and formation of new oceanic crust within the collapsed earlier Paleozoic marginal ocean basins led to the deposition of deep-water argillaceous and chert sequences similar to those of the older graptolitic shales along the axis of the Cordilleran belt in Nevada (Roberts, 1964; Speed, 1971; Fagan, 1962), in British Columbia (Monger and Ross, 1971), in southern Alaska (Bond, 1971; Richter and Jones, 1971), and in western Alaska (Patton, 1973).

The reconstruction (fig. 5E) shows renewed rifting within the collapsed earlier Paleozoic marginal basin sequences and development of another marginal ocean basin-volcanic arc system, thus nearly repeating the tectonic events of pre-Ordovician time—a tectonic history resembling that in the southwestern Pacific (Karig, 1971).

If the Paleozoic stratigraphic belts that developed concentrically around the proto-Pacific basin are interpreted as accretions of volcanic arc and marginal ocean basin deposits to the North American craton, then it should be possible to see changes in crustal structure that are transitional from purely oceanic type to purely continental type. In fact, there is a striking similarity in crustal structure between a line of section from the Pacific margin across the northern Great Basin to the Colorado Plateaus and across the present-day Aleutian volcanic arc-marginal ocean basin system (M. Churkin, Jr., unpub. data). The Great Basin, with its thin and well-layered crust, closely resembles the transitional crust of a marginal ocean basin filled with sediment and is bordered on the west by a crust not unlike that of an island arc. Thus, some of the major features of the crustal structure of the region, although affected by Cenozoic high heat flow, low-velocity mantle, and Basin and Range structure, may be inherited from Paleozoic time when the region was the site of a succession of marginal ocean basins.

CONCLUSIONS

The Paleozoic stratigraphic and structural belts of the Cordilleran foldbelt can be explained by a succession of marginal ocean basins developing behind volcanic arcs. The volcanic rock and graywacke belt, the outer part of the eugeosyncline, indicates a long history of volcanic island development, which started in Precambrian or Cambrian time in southeastern Alaska and in Early Ordovician time at the latitude of northern California. Initially, rocks of this belt formed a primitive crust on oceanic basement and, through a long process of volcanism, erosion, sedimentation, metamorphism, and plutonism, developed a crust of more continental character. The outer eugeosynclinal rocks in this interpretation are not considered continental fragments of Asia or parts of oceanic features of unknown provenance that collided with North America. Instead, they are considered parautochthonous deposits of volcanic island chains and interarc basins developed outward from the North American continental margin as it faced a proto-Pacific basin. Deep-sea

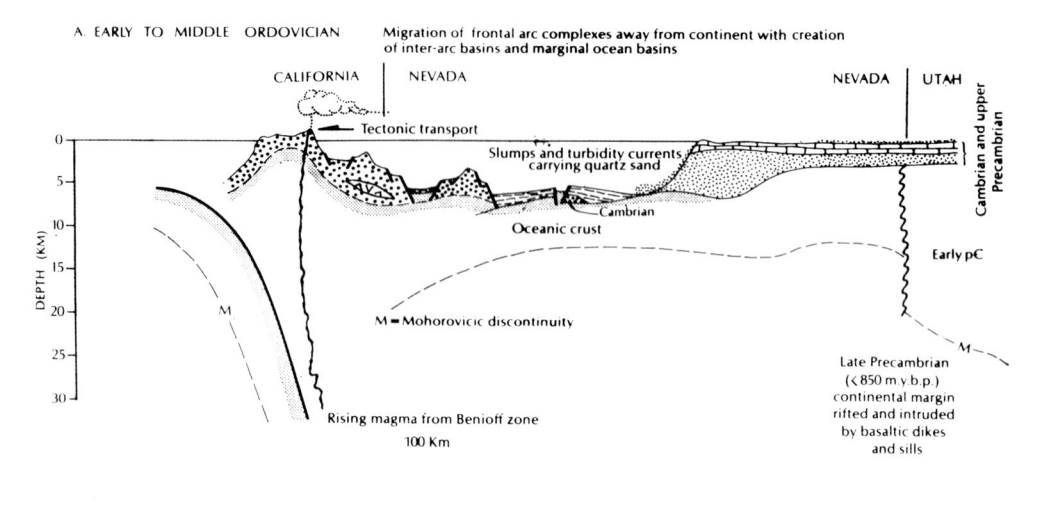

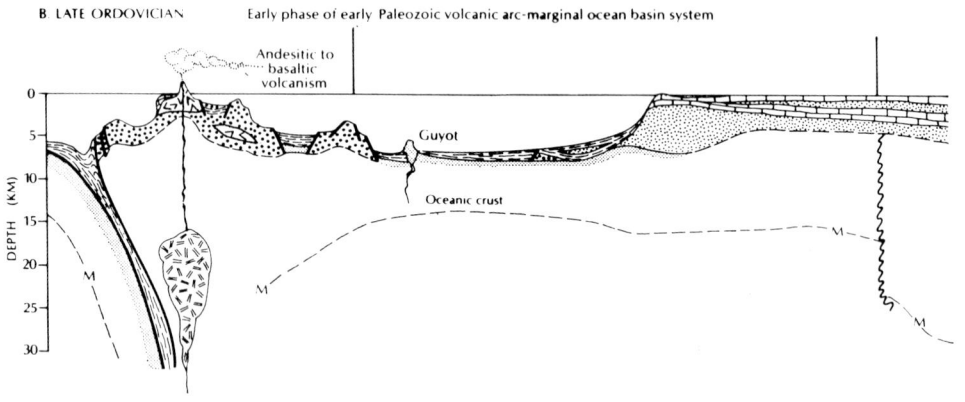

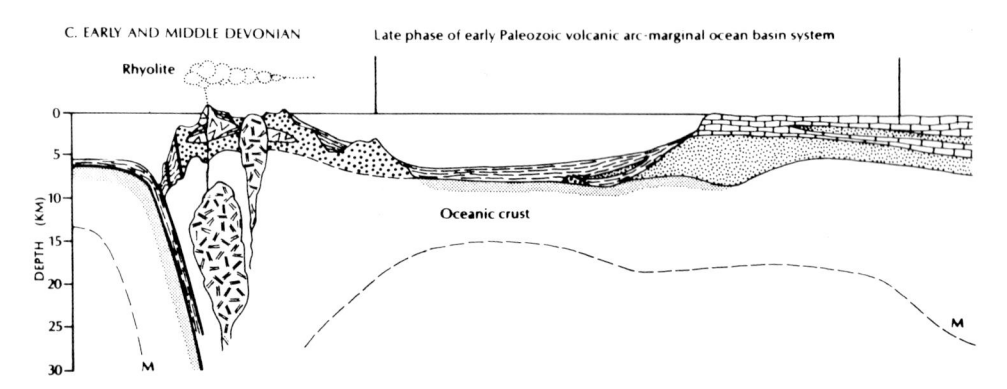

FIG. 5.—Preliminary reconstructions of the major stages in the Paleozoic development of the western margin of North America (see text for explanation).

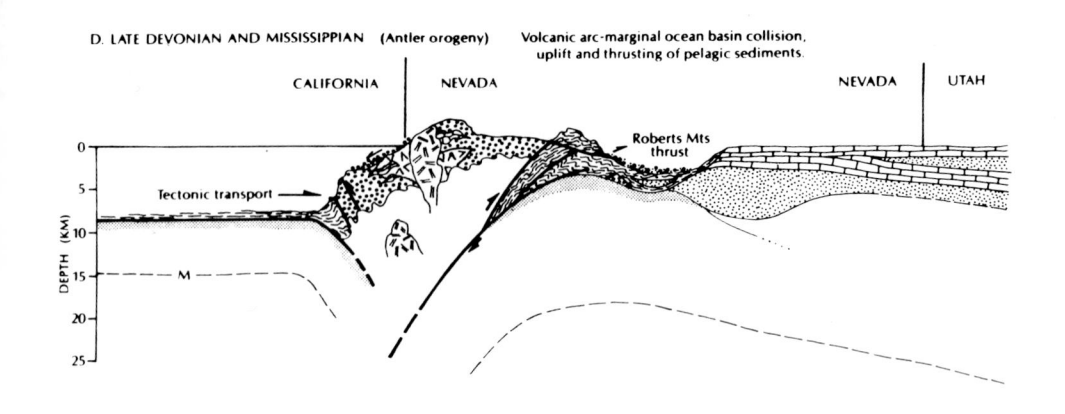

PALEOZOIC OCEAN BASIN-VOLCANIC ARC SYSTEMS

D. LATE DEVONIAN AND MISSISSIPPIAN (Antler orogeny) — Volcanic arc-marginal ocean basin collision, uplift and thrusting of pelagic sediments.

CALIFORNIA | NEVADA | NEVADA | UTAH

Tectonic transport

Roberts Mts thrust

M

DEPTH (KM)

E. PENNSYLVANIAN AND PERMIAN — Renewed rifting and creation of new volcanic arc-marginal ocean basin system

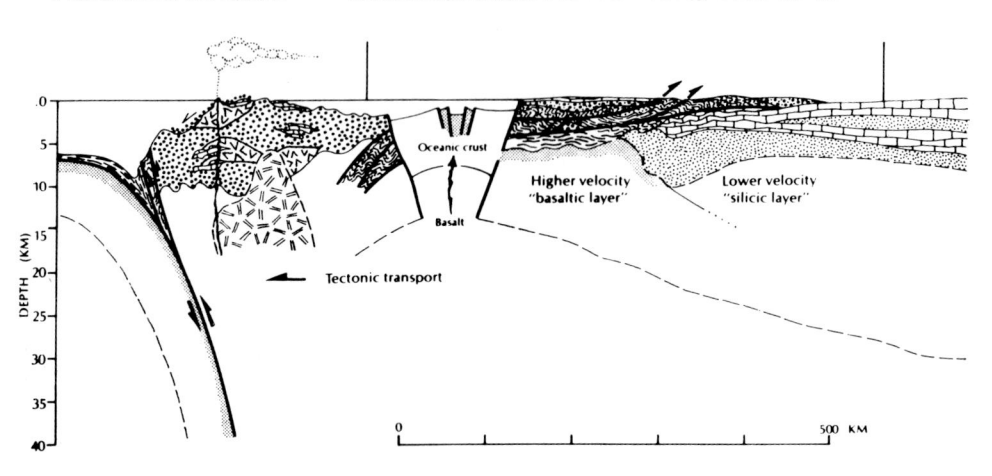

Oceanic crust

Higher velocity "basaltic layer"

Lower velocity "silicic layer"

Basalt

Tectonic transport

0   500 KM

DEPTH (KM)

EXPLANATION

Volcanic graywacke, mudstone, and conglomerate

Plutonic rocks

Volcanic breccia, tuff, and structureless porphyritic lava

Dolomite

Limestone

Quartzite

Pelagic sedimentary rocks, mainly graptolitic shale and chert

Basaltic lava, in places has pillow structure

pelagic deposits accumulated in newly formed marginal ocean basins behind the volcanic chains as they migrated from the continent. In the Late Devonian and Mississippian, these marginal ocean basins closed, perhaps by collision with the frontal volcanic arcs, and the deep-sea pelagic sediments were thrust east onto the continental shelf, uplifted, and eroded to provide thick wedges of coarse clastic sediments.

In the Pennsylvanian and Permian, the collapsed earlier Paleozoic marginal ocean basins were rifted, and a new system of basins floored apparently by new crust developed behind segments of the older volcanic arc complexes. This migration of Paleozoic volcanic arc complexes away from the North American continent and creation of new basins presumably with oceanic crust behind the frontal arcs is closely analogous with the tectonic history of the western Pacific, where most of the world's marginal seas are now concentrated.

REFERENCES

ARRHENIUS, GUSTAF, AND BONATTI, ENRICO, 1965, Neptunism and vulcanism in the ocean, in SEARS, MARY (ed.) : Oxford, England, Pergamon Press, Progress in oceanography, v. 3, p. 7–22.

BERG, H. C., 1972, Thrust faults, Annette-Gravina area, southeastern Alaska, in Geological Survey research 1972 : U.S. Geol. Survey Prof. Paper 800-C, p. 79–83.

BONATTI, E., AND OTHERS, 1972, Iron-manganese-barium deposit from the northern Afar Rift (Ethiopia) : Econ. Geology, v. 67, p. 717–730.

BOND, GERARD, 1971, Early Permian sedimentation and paleogeography in east central Alaska Range, and regional implications (abs.), in Second internat. symposium on Arctic geology, San Francisco : Am. Assoc. Petroleum Geologists, Pacific Sec., Program Abs., p. 8.

BOSTROM, K., AND PETERSON, M. N. A., 1969, The origin of aluminum-poor ferromanganoan sediments in areas of high heat flow on the East Pacific Rise: Marine Geology, v. 7, p. 427–447.

BREW, D. A., 1968, The role of volcanism in post-Carboniferous tectonics of southeastern Alaska and nearby regions, North America: 23rd Internat. Geol. Cong., Prague, Proc., v. 2, p. 107–121.

———, LONEY, R. A., AND MUFFLER, L. J. P., 1966, Tectonic history of southeastern Alaska, in A symposium on tectonic history and mineral deposits of the western Cordillera in British Columbia and neighboring parts of the United States: Canadian Inst. Mining and Metallurgy Special Vol. 8, p. 149–170.

BULMAN, O. M. B., 1970, Graptolithina with sections on Enteropneusta and Pterobranchia, in TEICHERT, CURT (ed.), Treatise on invertebrate paleontology, pt. 5, 2d ed.: Boulder, Colorado, Geol. Soc. America and Univ. Kansas Press, 163 p.

BURCHFIEL, B. C., AND DAVIS, G. A., 1972, Structural framework and evolution of the southern part of the Cordilleran Orogen, western United States: Am. Jour. Sci., v. 272, p. 97–118.

CHURKIN, MICHAEL, JR., 1963, Graptolite beds in thrust plates of central Idaho and their correlation with sequences in Nevada: Am. Assoc. Petroleum Geologists Bull., v. 47, p. 1611–1623.

———, 1969, Paleozoic tectonic history of the Arctic basin north of Alaska: Science, v. 165, p. 549–555.

———, 1970, Fold belts of Alaska and Siberia and drift between North America and Asia, in ADKINSON, W. L., AND BROSGÉ, M. M. (eds.), Proceedings of the geological seminar on the North Slope of Alaska: Los Angeles, California, Am. Assoc. Petroleum Geologists, Pacific Sec., p. G1–G17.

———, 1973, Paleozoic and Precambrian rocks of Alaska and their role in its structural evolution: U.S. Geol. Survey Prof. Paper 740, 64 p.

———, AND BRABB, E. E., 1965, Ordovician, Silurian and Devonian biostratigraphy of east-central Alaska : Am. Assoc. Petroleum Geologists Bull., v. 49, p. 172–185.

———, CARTER, CLAIRE, AND EBERLEIN, G. D., 1971, Graptolite succession across the Ordovician-Silurian boundary in southeastern Alaska: Geol. Soc. London Quart. Jour., v. 126, p. 319–330.

———, AND EBERLEIN, G. D., in press, Geologic map of the Craig C-4 quadrangle (scale, 1:63,360) : U.S. Geol. Survey Geol. Quad. Map.

———, AND KAY, MARSHALL, 1967, Graptolite-bearing Ordovician siliceous and volcanic rocks, northern Independence Range, Nevada: Geol. Soc. America Bull., v. 78, p. 651–668.

———, AND LANGENHEIM, R. L., JR., 1960, Silurian strata of the Klamath Mountains, California: Am. Jour. Sci., v. 258, p. 258–273.

DANNER, W. R., 1966, Limestone resources of western Washington: Washington Div. Mines and Geology Bull. 52, p. 474.

———, 1970, Paleontologic and stratigraphic evidence for and against sea floor spreading and opening and closing oceans in the Pacific Northwest (abs.) : Geol. Soc. America Abs. with Programs, v. 2, p. 84–85.

DEWEY, J. F., AND BIRD, J. M., 1971, Origin and emplacement of the ophiolite suite, in Appalachian ophiolites in Newfoundland: Jour. Geophys. Research, v. 76, p. 3179–3206.

DOTT, R. H., JR., 1955, Pennsylvanian stratigraphy of Elko and northern Diamond Ranges, northeastern Nevada : Am. Assoc. Petroleum Geologists Bull., v. 39, p. 2211–2305.

EBERLEIN, G. D., AND CHURKIN, MICHAEL, JR., 1968, Paleozoic mobile belt along the continental margin, southeastern Alaska (abs.) : 23rd Internat. Geol. Cong., Prague, Abs., p. 45.

———, AND ———, 1970, Paleozoic stratigraphy in the northwest coastal area of Prince of Wales Island, southeastern Alaska: U.S. Geol. Survey Bull. 1284, 67 p.

FAGAN, J. J., 1962, Carboniferous cherts, turbidites and volcanic rocks in the northern Independence Range, Nevada : Geol. Soc. America Bull., v. 73, p. 595–612.

GILLULY, JAMES, AND GATES, OLCOTT, 1965, Tectonic and igneous geology of the northern Shoshone Range, Nevada: U.S. Geol. Survey Prof. Paper 465, 153 p.

———, AND MASURSKY, HAROLD, 1965, Geology of the Cortez Quadrangle, Nevada: U.S. Geol. Survey Bull. 1175, 117 p.
HINTZE, L. F., 1951, Lower Ordovician detailed stratigraphic sections for western Utah: Utah Geol. and Mineralog. Survey Bull. 39, 99 p.
HOBBS, S. W., HAYS, W. H., AND ROSS, R. J., JR., 1968, The Kinnikinic Quartzite of central Idaho—Redefinition and subdivision: U.S. Geol. Survey Bull. 1254-J, p. 1–22.
IRWIN, W. P., 1966, Geology of the Klamath Mountains province, in BAILEY, E. H. (ed.), Geology of northern California: California Div. Mines and Geology Bull. 190, p. 19–38.
———, AND BATH, G. D., 1962, Magnetic anomalies and ultramafic rock in northern California: U.S. Geol. Survey Prof. Paper 450-B, p. 65–67.
JONES, D. L., IRWIN, W. P., AND OVENSHINE, A. T., 1972, Southeastern Alaska—a displaced continental fragment?, in Geological Survey research 1972: U.S. Geol. Survey Prof. Paper 800-B, p. 211–217.
KARIG, D. E., 1971, Origin and development of marginal basins in the western Pacific: Jour. Geophys. Research, v. 76, p. 2542–2561.
KAY, MARSHALL, 1951, North American geosynclines: Geol. Soc. America Mem. 48, 143 p.
———, 1966, Comparison of the lower Paleozoic volcanic and non-volcanic belts in Nevada and Newfoundland: Canadian Petroleum Geologists Bull., v. 14, p. 579–599.
———, AND CRAWFORD, J. P., 1964, Paleozoic facies from the miogeosynclinal to the eugeosynclinal belt in thrust slices, central Nevada: Geol. Soc. America Bull., v. 75, p. 424–454.
KETNER, K. B., 1966, Comparison of Ordovician eugeosynclinal and miogeosynclinal quartzites of the Cordilleran Geosyncline: U.S. Geol. Survey Prof. Paper 550-C, p. 54–60.
———, 1969, Ordovician bedded chert, argillite, and shale of the Cordilleran eugeosyncline in Nevada and Idaho: ibid., Prof. Paper 650-B, p. 23–34.
KING, P. B., 1966, The North American Cordillera, in A symposium on tectonic history and mineral deposits of the western Cordillera in British Columbia and neighboring parts of the United States: Canadian Inst. Mining and Metallurgy Special Vol. 8, p. 1–25.
KISTLER, R. W., AND PETERMAN, Z. E., in press, Variations in Sr, Rb, K, Na and initial Sr 87/86 in Mesozoic granitic rocks and intruded wall-rocks in California: Geol. Soc. America Bull.
KRASNIY, L. I. (ed.), 1966, Geologicheskoe stroenie severo-zapadnoi chasti tikhookeanskogo podvizhnogo poyasa (Geological structure of the northwestern part of the Pacific mobile belt): Ministerstvo geologii SSSR, Vses. Nauchno-issled. Geol. Inst., Nedra, Moscow, 516 p.
LA FEHR, T. R., 1966, Gravity in the eastern Klamath Mountains, California: Geol. Soc. America Bull., v. 77, p. 1177–1190.
LANPHERE, M. A., IRWIN, W. P., AND HOTZ, P. E., 1968, Isotopic age of the Nevadan Orogeny and older plutonic and metamorphic events in the Klamath Mountains, California: Geol. Soc. America Bull., v. 79, p. 1027–1052.
———, LONEY, R. A., AND BREW, D. A., 1965, Potassium-argon ages of some plutonic rocks, Tenakee area, Chichagof Island, southeastern Alaska: U.S. Geol. Survey Prof. Paper 525-B, p. 108–111.
———, MACKEVETT, E. M., JR., AND STERN, T. W., 1964, Potassium-argon and lead-alpha ages of plutonic rocks, Bokan Mountain area, Alaska: Science, v. 145, p. 705–707.
LARSON, M. L., AND JACKSON, D. E., 1966. Biostratigraphy of the Glenogle Formation (Ordovician) near Glenogle, British Columbia: Bull. Canadian Petroleum Geology, v. 14, p. 486–503.
MCKEE, E. H., in press, Geology of the northern part of the Toquima Range, Nevada: U.S. Geol. Survey Prof. Paper.
MATTINSON, J. M., 1972, Ages of zircons from the northern Cascade Mountains, Washington: Geol. Soc. America Bull., v. 83, p. 3769–3784.
———, AND HOPSON, C. A., 1972, Paleozoic ages of rocks from ophiolitic complexes in Washington and northern California (abs.): Am. Geophys. Union Trans., Ann. Mtg. EOS, v. 53, no. 4, p. 543.
MERRIAM, C. W., AND BERTHIAUME, S. A., 1943, Late Paleozoic formations of central Oregon: Geol. Soc. America Bull., v. 54, p. 145–171.
MINATO, M., GORAI, M., AND HUNAHASHI, M., 1965, The geologic development of the Japanese Islands: Tokyo, Tsukiji Shokan, 442 p.
MONGER, J. W. H., AND ROSS, C. A., 1971. Distribution of fusulinaceans in the western Canadian Cordillera: Canadian Jour. Earth Sci., v. 8, p. 259–278.
———, SOUTHER, J. G., AND GABRIELSE, H., 1972, Evolution of the Canadian Cordillera; a plate-tectonic model: Am. Jour. Sci., v. 272, p. 577–602.
MOORE, J. G., 1965, Petrology of deep-sea basalt near Hawaii: ibid., v. 263, p. 40–52.
MOORES, ELDRIDGE, 1970, Ultramafics and orogeny, with models of the U.S. Cordillera and the Tethys: Nature, v. 228, p. 837–842.
MUFFLER, L. J. P., 1967, Stratigraphy of the Keku Islets and neighboring parts of Kuku and Kupreanof Islands, southeastern Alaska: U.S. Geol. Survey Bull. 1241-C, 52 p.
OVENSHINE, A. T., 1972, Tidal origin of parts of the Karheen Formation (Lower Devonian), southeastern Alaska: Univ. Miami, Florida, Tidal deposits—Workshop and Conf., p. 73–78.
———, AND BREW, D. A., 1972, Separation and history of the Chatham Strait Fault, southeast Alaska, North America: 24th Internat. Geol. Cong., Montreal, Proc., sec. 3, p. 245–254.
———, EBERLIN, G. D., AND CHURKIN, MICHAEL, JR., 1969, Paleotectonic significance of a Silurian-Devonian clastic wedge, southeastern Alaska (abs.): Eugene, Oregon, Geol. Soc. America, Cordilleran Sec. Ann. Mtg. Program, pt. 3, p. 51.
———, AND WEBSTER, G. D., 1970, Age and stratigraphy of the Heceta Limestone in northern Sea Otter Sound, southeastern Alaska: U.S. Geol. Survey Prof. Paper 700-C, p. 170–174.
PATTON, W. W., JR., 1973, Reconnaissance geology of the northern Yukon-Koyukuk province, Alaska: ibid., Prof. Paper 774-A, p. 1–17.

POOLE, F. G., BROBST, D. A., AND SHAWE, D. R., 1968, Sedimentary origin of bedded barite in central Nevada (abs.) : Geol. Soc. America Abs. with Programs, v. 1, p. 242.

RICHTER, D. H., AND JONES, D. L., 1971, Structure and stratigraphy of eastern Alaska Range, Alaska (abs.), *in* Second internat. symposium on Arctic geology, San Francisco: Tulsa, Oklahoma, Am. Assoc. Petroleum Geologists Program Abs., p. 45.

RIVA, JOHN, 1970, Thrusted Paleozoic rocks in the northern and central HD Range, northeastern Nevada: Geol. Soc. America Bull., v. 81, p. 2689–2716.

ROBERTS, R. J., 1964, Stratigraphy and structure of the Antler Peak Quadrangle, Humboldt and Lander Counties, Nevada: U.S. Geol. Survey Prof. Paper 459-A, 93 p.

———, AND OTHERS, 1958, Paleozoic rocks of north-central Nevada: Am. Assoc. Petroleum Geologists Bull., v. 42, p. 2813–2857.

ROHR, D., AND BOUCOT, A. J., 1971, Northern California (Klamath Mountains) pre-Late Silurian igneous complex (abs.) : Riverside, California, Geol. Soc. America, Cordilleran Sec., Ann. Mtg. Program, p. 186.

ROSS, R. J., JR., 1964, Relations of Middle Ordovician time and rock units in Basin Ranges, western United States: Am. Assoc. Petroleum Geologists Bull., v. 48, p. 1526–1554.

SCIENTIFIC STAFF, DEEP SEA DRILLING PROJECT, 1969, Deep sea drilling project, leg 7: Geotimes, v. 14, p. 12–13.

———, 1972, *Glomar Challenger* down under: Geotimes, v. 17, p. 14–16.

SHAWE, D. R., POOLE, F. G., AND BROBST, D. A., 1969, Newly discovered bedded barite deposits in east Northumberland Canyon, Nye County, Nevada: Econ. Geology, v. 64, p. 245–254.

SILBERLING, N. J., in press, Geologic events during Permo-Triassic time along the Pacific margin of the United States, *in* LOGAN, A., AND HILLS, L. U. (eds.), International Permian-Triassic conference volume: Calgary, Alberta, Alberta Soc. Petroleum Geologists.

SLOSS, L. L., KRUMBEIN, W. C., AND DAPPLES, E. C., 1949, Integrated facies analysis: Geol. Soc. America Mem. 39, p. 91–124.

SPEED, R. C., 1971, Permo-Triassic continental margin tectonics in western Nevada (abs.) : *ibid.*, Abs. with Programs, v. 3, p. 197.

STEWART, J. H., 1972, Initial deposits in the Cordilleran geosyncline; evidence of a Late Precambrian (<850 m.y.) continental separation: *ibid.*, Bull., v. 83, p. 1345–1360.

TOGHILL, PETER, 1968, The graptolite assemblages and zones of the Binkhill Shales (Lower Silurian) at Dobb's Linn: Palaeontology, v. 11, p. 654–668.

TRETTIN, H. P., 1971, Geology of lower Paleozoic formations, Hazen Plateau and southern Grant Land Mountains, Ellesmere Island, Arctic Archipelago: Geol. Survey Canada Bull. 203, 134 p.

WILSON, J. T., 1968, Static or mobile earth; the current scientific revolution: Am. Philos. Soc. Proc., v. 112, no. 5, p. 309–320.

WINTERER, E. L., AND MURPHY, M. A., 1960, Silurian reef complex and associated facies, central Nevada: Jour. Geology, v. 68, p. 117–139.

DEPOSITS IN MAGMATIC ARC AND TRENCH SYSTEMS

SEDIMENTARY SEQUENCE IN MODERN PACIFIC TRENCHES AND THE DEFORMED CIRCUM-PACIFIC EUGEOSYNCLINE

D. W. SCHOLL AND M. S. MARLOW

U.S. Geological Survey, Menlo Park, California

ABSTRACT

Nearly 30,000 km of trench floor fringe the western, northern, and eastern sides of the Pacific Ocean. About 55 percent of this contains less than 400 m of sediment. The remainder comprises trenches with deposits as thick as 2,500 m. Most of the empty or nearly empty trenches border the island arcs of the western Pacific, and include Mesozoic to late Cenozoic pelagic or hemipelagic beds. Although the Panama Trench (?) and the southern part of the Middle America Trench contain 400 to 600 m of pelagic and hemipelagic deposits, the remaining eastern Pacific trenches (i.e., southern part of the Peru-Chile Trench and the Washington-Oregon Trench) and the central and eastern segments of the northern (Aleutian) trench are filled or nearly so with a thick (1,000 to 2,000 m) two-layer sedimentary sequence. Off South America the older unit (200 to 250 m thick) is typically composed of landward dipping pelagic and hemipelagic beds of Tertiary age, whereas in the northwestern Pacific the lower layer is a hemiterrigenous unit composed of pelitic turbidites interlayered with pelagic beds. In both regions the upper unit is a wedge-shaped body of flat-lying terrigenous turbidites that thins seaward over the landward dipping older beds. Only about one million years old, the wedge formed in response to the low sea levels and to increased continental erosion that accompanied episodes of expansive continental glaciation. Similar wedges may have formed only infrequently during the Mesozoic and Tertiary.

In the recent past, many geologists have speculated that the Mesozoic eugeosynclinal rocks flanking much of the Pacific represent raised and deformed trench fillings. However, modern trenches are not of geosynclinal proportions as they are narrow troughs incapable of accumulating a section of terrigenous turbidites thicker than about 3 km. Also, because the terrigenous wedge of existing trenches may be a unique body of Pleistocene age, there is reason to believe that Mesozoic trenches, like most modern ones, may have been chiefly the site of pelagic and hemipelagic sedimentation. In recent years it has become known that trenches are not static downwarps but mark the junction of rapidly converging lithospheric plates. Hence, many engeosynclinal masses are now thought to represent tectonically offscraped and partially subducted Mesozoic trench deposits. This model greatly compounds the problem of the presumed missing pelagic deposits. During the 100 my or so involved in the formation of the typical ensimatic eugeosynclinal assemblage, the volume of pelagic and hemipelagic deposits that would have been tectonically injected into the trench by the incoming oceanic plate is itself nearly of geosynclinal proportion. The pelagic debris should be especially concentrated in the mélange units. These units, however, as well as the remainder of the Pacific eugeosynclinal rocks, are composed chiefly of terrigenous detritus.

If the earth is not expanding, we reason that the paucity of oceanic offscrapings in Pacific fold belts means that the greater part of the sediment reaching a trench, whether pelagic or terrigenous, whether deposited there or carried there on the back of an oceanic plate, is subducted beneath the adjacent insular or continental crust. This conclusion also means that the bulk of the terrigenous masses of the Pacific eugeosynclines are not deformed trench deposits.

INTRODUCTION

One of the more dramatic consequences of sea-floor spreading and plate tectonics is the sweeping of oceanic pelagic deposits into trenches. Here these ocean floor beds must mix tectonically with the more terrigenous and pyroclastic sediments shed from an adjacent island arc or continental region. This realization has led several investigators to speculate that the deformed Mesozoic rocks of Pacific fold belts that overlie simatic crust may represent partially subducted trench deposits scraped off a descending oceanic plate[1] (W. Hamilton, 1969; Dewey and Bird, 1970a; Hsü, 1971; Dickinson, 1971a,b,c). The geologic implication is that a trench is part of some sort of ensimatic geosyncline, most likely a type of eugeosyncline (Dewey and Bird, 1970b).

The likelihood that major parts of the Pacific fold belt, especially segments peripheral to the northern and southwestern Pacific, are tectonically offscraped and partially subducted morphosed rock. Offscraped deposits are strata and rocks that have been skimmed off the upper part of the oceanic lithosphere and folded against and partially thrust beneath the base of the continental crust. Offscraped deposits retain much of their original formational characteristics, but they may have an imprint of high P/T mineralization due to partial subduction.

[1] Subducted deposits are regarded as strata and rock deeply injected below the crust that can return to the surface of the earth only as igneous or highly meta-

trench deposits is especially appealing to North American and Japanese geologists. The subduction process seems to provide the heretofore elusive mechanism explaining the formation of the perplexing mélange units that are structurally interwoven into the largely terrigenous rocks of the Franciscan Sequence (Page, 1970, 1972a,b; Hsü, 1971). Also, the outer of the paired Pacific-type metamorphic belts of Miyashiro (1961) is readily attributable to the consequence of a deeply underthrust trench sequence subsequently brought into view by uplift and erosion (Ernst, 1965, 1970; Ernst and others, 1970).

Despite these appealing arguments, a study of the sequence of sedimentary deposits occurring in modern Pacific trenches does not support the idea that the circumpacific fold belt includes substantial amounts of uplifted trench deposits or tectonic scrapings. One difficulty lies in the discovery that, although modern trenches may contain a thick (as much as 2,000 m) seaward thinning or wedge-shaped body of terrigenous turbidite beds, this section is a late Cenozoic deposit of the glacial period that may not have been duplicated often in Mesozoic and Tertiary trenches. Moreover, as noted by Mitchell and Reading (1971), the typical sedimentary fill of a trench includes a sequence of pelagic, hemipelagic, or pelitic hemiterrigenous deposits. These are types of sedimentary deposits that are poorly represented volumetrically in the dominantly terrigenous graywacke beds and mafic volcanic rocks common to the fold belt (Bailey and others, 1964; Burk, 1965, in press; Bailey and Blake, 1969; Bogdanov, 1969, 1970; Ernst and others, 1970; Barbat, 1971; Avdeiko, 1971; Moore, 1972; Landis and Bishop, 1972). Also, nearly the entire area of the Pacific crust has been renewed at least once since the Early Cretaceous (Larson and Chase, 1972). In the course of this renewal, many thousands of kilometers of oceanic plate necessarily underthrust the perimeters of the Pacific (Larson and Pitman, 1972). During the Mesozoic Era, the volume of oceanic debris that should have been offscraped in trenches would form a very noticeable part of the circum-pacific mountains if it were present there.

In the following text we briefly describe the sedimentary sequence filling modern Pacific trenches, comment on the origin of these deposits, discuss their apparent rarity in the uplifted ensimatic eugeosynclinal accumulations fringing much of the northern and southwestern Pacific, and suggest a solution to the problem of the missing oceanic debris.

Definition and recognition of sedimentary units.—Drawing substantially from the literature but also on personal communications and our unpublished records, we have assembled in table 1 information about the type and thickness of depositional units occurring in the major Pacific trenches. These units, terrigenous, hemiterrigenous, pelagic, and hemipelagic, are defined in this table. Estimates of the volume of the Pleistocene turbidite wedge are also given. Age assignments are based either on information gathered at DSDP (Deep Sea Drilling Project) sites or on the age of the crust underlying the trench estimated from magnetic chronology data.

Criteria for recognizing the depositional units underlying the trench floor are based chiefly on their acoustical properties and depositional geometry (Ewing, Ewing, and others, 1968; Ewing and others, 1969; E. Hamilton, 1967), but a number of DSDP sites provide direct lithologic information. Terrigenous beds deposited by nepheloid layers (Ewing and Connary, 1970) or the slow settling of finely comminuted mineral matter shed from a nearby source could form a thick depositional body over a trench floor that would be difficult to distinguish on reflection records from normal pelagic beds. Thus, some of the pelagic or hemipelagic trench units we identified may in fact be substantially composed of beds of fine (clay-size) detritus derived from a nearby continental or island arc source. Because these terrigenous beds would form a shale or claystone section, their inadvertent inclusion in the pelagic and hemipelagic units noted in table 1 does not affect the intended implication of the terms terrigenous and hemiterrigenous to mean those deposits that could form graywacke beds, which are the rocks most similar to those constituting the bulk of the circumpacific eugeosynclinal assemblages.

SEDIMENTARY SEQUENCE IN
PACIFIC TRENCHES

General description.—Trenches, although not occurring everywhere, border nearly 30,000 km of the margins of the Pacific Ocean (fig. 1; Fisher and Hess, 1963; Menard, 1964; Hayes and Ewing, 1970). Although bathymetrically unrevealed or poorly revealed, trenches as structural features also occur along the western side of South America between the southern tip of Chile and about lat 37°S, seaward of Panama(?), also off Oregon and Washington, and perhaps western Canada and southeastern Alaska farther to the north. The trenches in these areas have been filled with sedimentary deposits.

As structural features, the major Pacific

trenches are similar (Fisher and Hess, 1963; Worzel, 1965; Hayes and Ewing, 1970); their sedimentary fills, however, are not. For example, the *eastern* trenches bordering North and South America typically contain a two-layer sequence. The upper layer is a wedge-shaped body of terrigenous turbidites that thins seaward over a lower depositional unit of chiefly hemipelagic or hemiterrigenous beds. These can be traced seaward to pelagic or abyssal plain deposits (fig. 2). In contrast, the *western* trenches, those bordering Asia and the western and southwestern Pacific Island arcs, are either empty or characterized by a relatively thin (100 to 400 m) blanket of predominantly pelagic or hemipelagic deposits (fig. 2). Transitional between what may be called these eastern and western types of sedimentary sequences is that of the northern or connecting Aleutian Trench, which is dominated by terrigenous and hemiterrigenous units to the east but pelagic and hemipelagic units to the west.

A brief description of the sedimentary deposits of the major Pacific trenches is given below, and all pertinent references are listed in table 1.

Eastern trenches.—From south to north, the eastern Pacific is bordered by the Peru-Chile, Panama(?), Middle America, and the Washington-Oregon Trenches (fig. 1). South of about lat 37°S the Peru-Chile Trench is completely filled. As much as 2,000 m of sedimentary deposits underlie the inner edge of the trench near the base of the continental slope. Overflow may extend seaward more than one hundred kilometers. The upper half to two-thirds is typically constructed of flat-lying or slightly landward dipping turbidite deposits that wedge out seaward over older deposits (fig. 3). These older beds dip more steeply landward and are roughly parallel to the underlying surface of the basaltic rocks forming the upper part of the oceanic crust. The depositional geometry and acoustic characteristics of the lower sedimentary section indicate that north of about lat 40°S it is a pelagic or hemipelagic unit. Farther south, possibly to the tip of South America, the lower unit appears to include many turbidite beds and perhaps should be classified as hemipelagic if not hemiterrigenous.

North of lat 37°S the trench is only partially filled, but nonetheless it locally contains as much as 2,000 m of terrigenous turbidite beds. Underlying pelagic beds are only a few hundred meters thick. North of lat 37°S, off the Atacama Desert, the trench is without a significant turbidite or pelagic section. This in part reflects the great aridity of the adjacent drainages.

Very little information is available concerning the type of sedimentary deposits in the northern part of the Peru-Chile Trench. In many areas the trench is apparently occupied mainly by a pelagic or hemipelagic unit, although a turbidite section exists locally (Kulm and others, 1972). Somewhat greater thicknesses of hemipelagic deposits partially fill the complicated troughs referred to as the Panama Trench in table 1. The nearly contiguous Middle America Trench to the north also contains principally pelagic or hemipelagic deposits; off Mexico (north of about lat 17°N) a turbidite section generally less than 500 m in thickness overlies several hundred meters of hemipelagic or pelagic deposits.

Seaward of northern California, Oregon, Washington, and southwestern Canada (possibly as far north as lat 50 to 55°; Shor, 1966), a completely filled trench lies at the base of the continental margin. North of about lat 40°N the trench sequence consists of a thick (500 to 1000 m) section of horizontal to subhorizontal terrigenous turbidite beds overlying an equally thick hemiterrigenous section (fig. 4). Beneath the Astoria Fan off the Columbia River the upper turbidite unit is only about 1.0 my old (DSDP site 174, von Huene, Kulum and others, 1971; Kulm, von Huene, and others, 1973). In this area the underlying predominantly Pliocene section is composed principally of clay-rich distal turbidite beds of an older abyssal plain sequence, that is, the typical hemipelagic or hemiterrigenous unit is absent beneath the Astoria Fan.

Northern (Aleutian) Trench.—In the eastern part of the Aleutian Trench the upper turbidite wedge is less than 1.0 my old, whereas the underlying landward dipping hemiterrigenous unit, an abyssal plain sequence, began to form in the early Neogene (fig. 3; DSDP site 180, Kulm, von Huene, and others, 1973). The central segment of the Aleutian Trench lies south of the gracefully arcuate Aleutian Ridge. Between long 170°W and 170°E, a distance of about 1,500 km, the bulk of the trench fill is a thick (1,500 to 2,000 m) wedge-shaped mass of turbidite beds (fig. 5, profile B6). These deposits were supplied by westward axial flow of turbidity currents from the eastern trench segment south of mainland Alaska. Without this flow, the eastern trench would have filled with turbidites and overflowed to the south. A thin (about 200 m thick) pelagic unit lies beneath the turbidites of the central segment. West of about long 170°E, the lower pelagic section thickens relative to that of the overlying turbidite wedge (fig. 4, profile B36). The pelagic unit correlates with pelagic and hemipelagic beds cored at DSDP site 192. These deposits range in age

TABLE 1.—SEDIMENTARY DEPOSITS OF THE PRINCIPAL PACIFIC TRENCHES

Trench name and approximate limits	Length (km)	Depositional units[1]	Average maximum thickness (km)	Age[2]	Average width (km)	Volume[3] (km³)	Remarks and data sources
EASTERN PACIFIC Washington-Oregon lat 40–52(?)°N	1,300	terrig. turb.	1.1	Pleistocene	200	140,000	Upper turbidite section includes large fans of Cascadia Basin that fill trench and extend westward toward Juan de Fuca Ridge; Shor (1966), Ewing, Ewing, and others (1968), Silver, (1969), Hayes and Ewing (1970), Dehlinger and others (1970), von Huene, Kulm, and others (1971), Silver (1971, 1972), Tiffin and others (1972), McManus and others (1972) and Kulm, von Huene, and others (1973)
		hemiterrig.	1.3	LC			
Middle America lat 17–21°N	1,000	terrig. turb.	0.5	Pleistocene	10	2,500	Fisher (1961), Ross and Shor (1965), Ewing, Ewing and others (1968), Hayes and Ewing (1970), Van Andel and others (1971) and Ross (1971)
		pel.	0.2	M-LC			
lat 9–17°N	2,000	pel.	0.4	M-LC			
Panama lat 2°S–7°N	1,200	hemipel.+pel.	0.7	M-LC			A complex trough or trenchlike structure at base of continental margin, designated here as the Panama Trench, contains as much as 1,500 m of largely hemipelagic deposits; Ewing, Ewing, and others (1968) and Van Andel and others (1971)
Peru-Chile lat 2°S–20°S	2,500	hemipel.+pel.+turb.	0.2(?)	E-LC			Very little information is available about sedimentary fill of this segment of trench; Ewing, Ewing, and others (1968), Van Andel and others (1971), and Kulm and others (1972)
lat 20°S–27°S	800	Nil					This segment of trench is seaward of Atacama Desert and typically contains less than about 50 m of sedimentary debris; Fisher and Raitt (1962), Scholl and others (1968), Scholl, Christensen, and others (1970), Hayes and Ewing (1970) and Scholl and von Huene (1970)
lat 27°S–37°S	1,100	terrig. turb.	0.8	Pleistocene	20	9,000	Ewing, Ewing, and others (1968), Scholl and others (1968), Scholl, Christensen, and others (1970), Scholl and von Huene (1970), Hayes and Ewing (1970), and Hayes and others (1972)
		pel.+hemipel.	0.2	E-LC			
lat 37°S–56°S	2,250	terrig. turb.	1.1	Pleistocene	150	185,000	
		pel.+hemipel. hemiterrig.(?)	0.5	E-LC			
NORTHERN PACIFIC Aleutian long 145–170°W	1,900	terrig. turb.	0.7	Pleistocene	20	13,000	Lower section of hemiterrigenous beds are abyssal plain deposits downwarped into trench floor; von Huene and Shor (1969), von Huene, Kulm, and others (1971), von Huene (1972), Kulm, von Huene, and others, (1973), and E. Hamilton (1973)
		hemiterrig.	0.8	E-LC			
long 170°W–170°E	1,500	terrig. turb.	1.5	Pleistocene	15	17,000	Ewing, Ewing, and others (1968), Hayes and Ewing (1970), Holmes and others (1972), Marlow and others (1973), Marlow, Scholl, and others (1973), Scholl, Buffington, and Marlow (in press), and Scholl and Marlow (in press)
		pel.	0.2	LK-LC			
long 164°–170°E	500	terrig. turb.	0.4	Pleistocene	5	500	Buffington (1973)
		pel.	1.0	LK-LC			

TABLE 1.—(Continued)

Trench name and approximate limits	Length (km)	Typical sedimentary sequence			Late Cenozoic turbidite wedge		Remarks and data sources
		Depositional units[1]	Average maximum thickness (km)	Age[2]	Average width (km)	Volume[3] (km³)	
WESTERN PACIFIC Kamchatka-Kuril lat 41–55°N	2,100	pel.+hemipel.	0.4	EK-LC			Turbidites of unknown thickness occur in small restricted basins; Buffington (1973) and E. L. Hamilton (1972, written commun.)
Japan-Bonin lat 55–24°N	1,900	pel.+hemipel.	0.4	Jr(?)-LC			Pockets of turbidites occur locally along trench axis; Ludwig and others (1966), Ewing, Ewing, and others (1968), Hayes and Ewing (1970), and Reynolds (1970)
Mariana-Yap-Palau lat 5–24°N	2,100	pel.	<0.1(?)	Tr(?)-LC			Very little information is available on sedimentary deposits of these trenches; where available, reflection profiles reveal a virtually empty trench; seaward of trench, at DSDP site 59, about 130 m of pelagic clay, ash, and some chert were penetrated without reaching basalt; Fischer and others (1970), Hayes and Ewing (1970), and Karig (1971)
Ryuku lat 23–30°N	1,200	Nil					Although many profiles reveal no significant fill, near Taiwan, and probably also southern Japan, are terrigenous turbidites and hemipelagic deposits as much as 600 m thick; Emery and others (1969) and Wageman and others (1970)
Philippine lat 5–15°N	1,100	Nil					Hayes and Ewing (1970) and J. Ewing (personal commun., 1972)
Malaita lat 5–10°S	1,000	pel. turbs.	0.4	LC			Upper pelagic turbidite wedge is not found everywhere and in many areas is replaced by large slide or slump masses; Kroenke (1972)
		pel.	0.6	LK-LC			
Vityas lat 9–12°S	500	pel. turbs.(?)	0.5	LC(?)			Hayes and Ewing (1970)
Tonga-Kermadec lat 5–37°S	2,700	Nil					Seaward of trench, at DSDP site 204, about 100 m of pelagic and volcanogenic sediment was penetrated without reaching basement; Mesozoic deposits may have been reached; Fisher and Engel (1969), Ewing and others (1969), Hayes and Ewing, (1970), and Burns, Andrews, and others (1972)
Hikurang lat 37–42°S	600	terrig. turbs.	1.0	LC			Houtz and others (1967) and Ewing and others (1969)
		hemipel. or hemiterrig.	1.2	Jr-Lc(?)			

[1] Terrig.=terrigenous; turb.=turbidite; hemiterrig.=hemiterrigenous; pel.=pelagic; hemipel.=hemipelaigc; terrig. turb./pel means a depositional unit of terrigenous turbidites overlies a unit of pelagic deposits. As used in this paper, these units are defined as follows: (1) *Terrigenous unit:* a sequence of sedimentary beds formed chiefly (i.e., more than about 75 percent) by the accumulation of solid mineral detritus derived through erosion of relatively nearby continental or insular terrane. (2) *Hemiterrigenous unit:* a terrigenous unit that includes a volumetrically large (but less than 50 percent) component of either distinct pelagic beds or interspersed pelagic debris. (3) *Pelagic units:* a sequence of beds composed of particulate or chemically precipitated matter derived chiefly from the water column by either organic (biopelagic) or inorganic processes or by the settling of finely comminuted detritus derived from the continents generally or from a laterally broad portion of an adjacent continent. (4) *Hemipelagic unit:* a pelagic unit that includes a volumetrically large (but less than 50 percent) component of terrigenous beds or interspersed pelagic detritus derived from a nearby terrane.

[2] EC, MC, and LC =early, middle, and late Cenozoic, respectively; EK and LK =Early and Late Cretaceous, respectively; Jr = Jurassic; and Tr =Triassic.

[3] Computed as one-half average thickness times average width and length of trench fill.

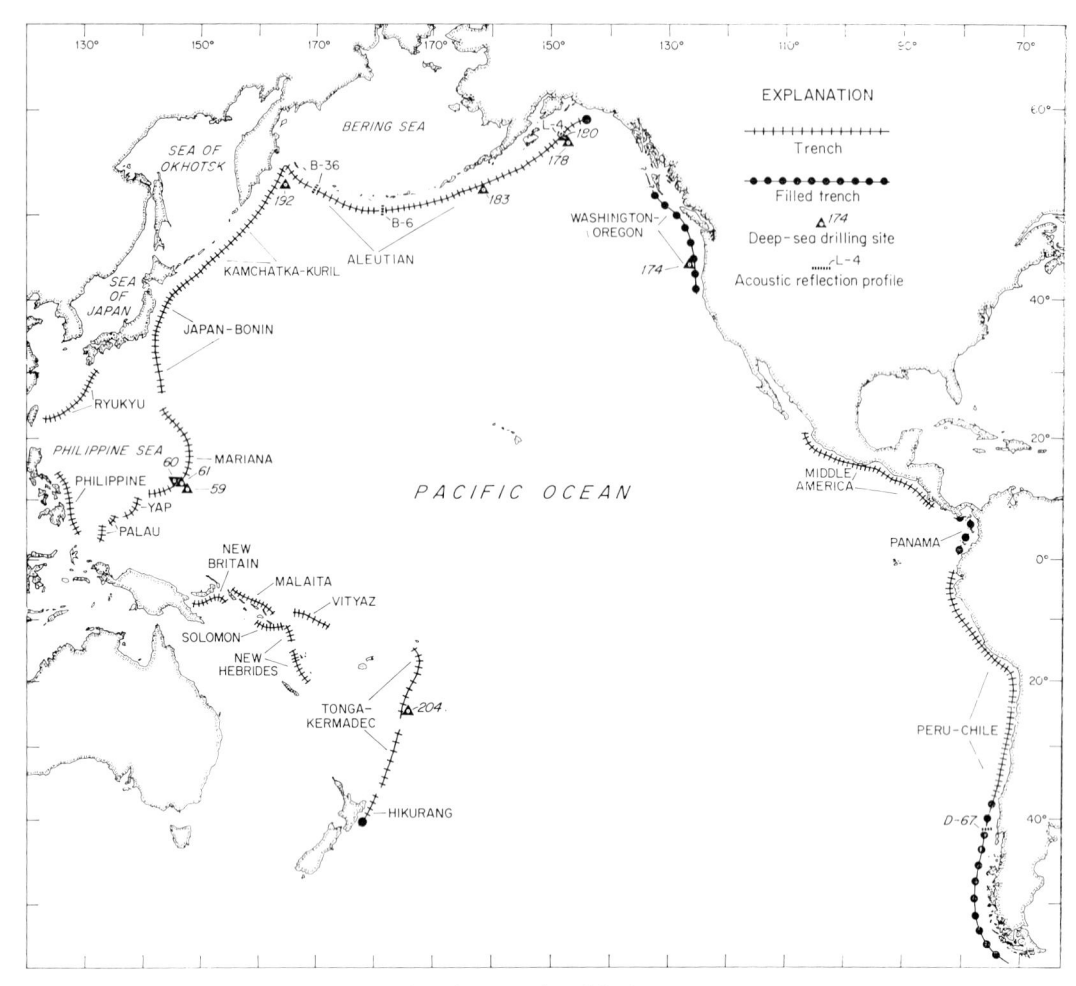

Fig. 1.—Index chart showing location of major trenches fringing the Pacific, DSDP sites where trench or trench-related deposits have been sampled, and line of reflection profiles shown on figures 3, 4, 5, and 6.

from Late Cretaceous to Holocene (Fig. 1; Scholl, Creager, and others, 1971; Creager, Scholl, and others, 1973). In the vicinity of its confluence with the Kamchatka Trench, the pelagic-hemipelagic section of the western part of the Aleutian Trench is about 2,500 m thick; an overlying turbidite section has not been observed.

Western trenches.—In the Northern Hemisphere, the great deeps along the western side of the Pacific form a crudely S-shaped trough reaching from northern Kamchatka (lat 55°N) to Melanesia (lat 5°N). The principal segments are the Kuril-Kamchatka (lat 55–41°N), Japan-Bonin (lat 41–24°N), Mariana (lat 24–10°N) and Yap-Palau (lat 10–5°N) Trenches (fig. 1, table 1). Other trenches occur west of or inside the sweeping arcs of these trenches, for example, the Ryukyu and the Philippine Trenches. These deep troughs occur along the western side of the Philippine Sea and, like the New Britain, New Hebrides, and Solomon Trenches of the Melanesian area, are seperated from the outer or more typical Pacific trenches by island archipelagos (fig. 1). South of the Melanesian region, and considerably (5,500 km) to the east of the Yap-Palau Trench, the Tonga-Kemadec Trench extends southward (lat 5–37°S) to the Hikurangi Trench, which lies east of northeastern New Zealand. Pelagic beds dominate the western trenches (fig. 2; Mitchell and Reading, 1971).

Pelagic and hemipelagic deposits in the Kamchatka Trench are remarkably thick and may exceed 2,000 m. A wedge of terrigenous turbidites is not characteristic of the Kamchatka Trench. However, a turbidite unit of local extent has been found off the southern tip of

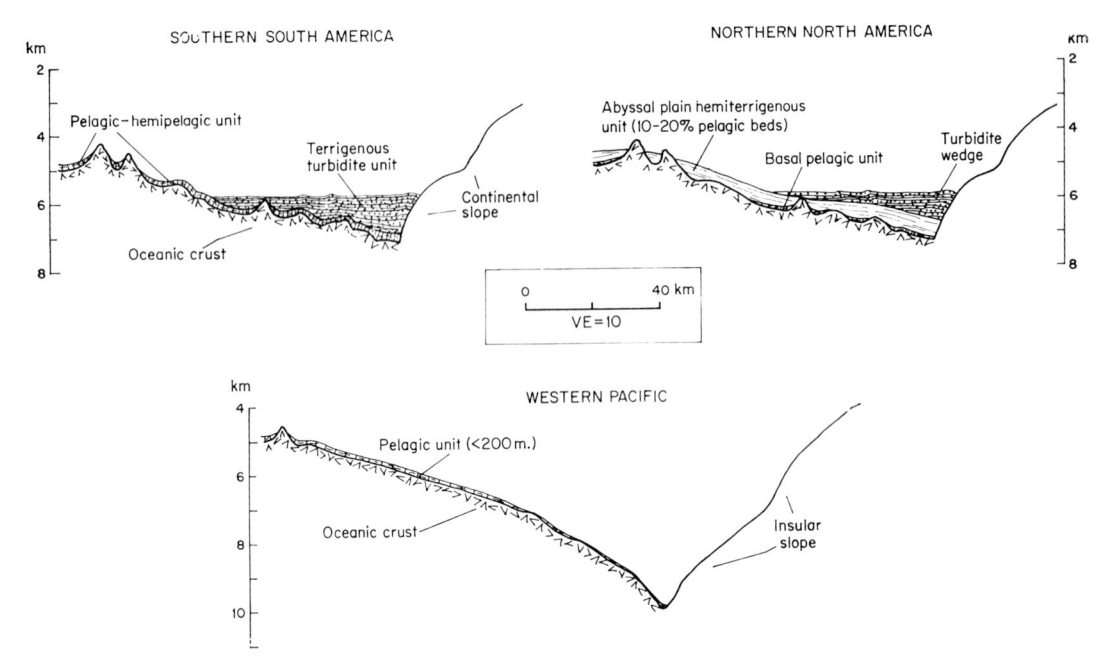

Fig. 2.—Schematic drawings showing stratigraphic relation of major depositional units in typical North American, South American (lat 27–40°S), and western Pacific trenches.

Kamchatka. Farther to the south (lat 41–50°N), pelagic deposits as much as 400 m thick are the principal depositional unit of the Kuril Trench. In the Japan Trench, pockets of terrigenous turbidites as thick as 1,000 m occur in restricted areas. Nonetheless, as is apparently true of most western Pacific trenches, the Japan-Bonin Trench is characteristically burdened with a pelagic depositional unit averaging about 300 or 400 m in thickness.

The inner Ryukyu and Philippine Trenches skirting the western margin of the Philippine Sea are generally devoid of a significant (i.e., >100 m) sedimentary fill. However, near Taiwan, and probably also near southern Japan, the typically sediment-empty Ryukyu Trench is floored by as much as 600 m of turbidite beds.

Continuing southward (lat 15–24°N), the Mariana-Yap-Palau trench system, although poorly investigated, appears to lack a significant trench fill, a circumstance also indicated by its great depth (including the ocean's greatest deep, 11,022 m, southwest of Guam). Sampling at DSDP sites 59, 60, and 61 indicates that any significant thickness of sediment that may overlie the trench floor is likely to be composed of pelagic or volcanogenic debris (Fischer and others, 1970; Fischer, Heezen, and others, 1971; Winterer, Riedel, and others, 1969, 1971). Magnetic anomalies seaward of the Mariana Trench suggest that any sediment beneath its floor may be the oldest (Triassic?) occurring anywhere in the Pacific (Larson and Chase, 1972).

In the Melanesian area, Pacific trenches are absent or poorly formed. The best known of these trenches, New Britain, New Hebrides, and Solomon (not listed in table 1), are inter-arc structures isolated depositionally, structurally, and in part tectonically from the Pacific basin. Limited information about them indicates they are virtually empty of sedimentary debris (Ewing and others, 1969; Karig and Mammerickx, 1972; J. Ewing, personal commun., 1972).

Although a deep (viz., >6,000 m) trench is absent east of the Solomon Islands (lat 5–10°S), a shallower trough, designated in table 1 as the Malaita Trench, does exist. The trench floor in many areas is underlain by an upper wedge-shape mass of flat-lying beds as much as 1,000 m thick. The wedge overlies an equally thick or thicker landward dipping section. Both units are composed principally of pelagic beds; however, the upper wedge-shaped unit is a sequence of pelagic turbidites formed through the remobilization of previously deposited pelagic beds. Presumably the narrow, flattish floor of the more easterly but nearby Vityaz Trench (lat 9–12°S), is also underlain by a few hundred meters of pelagic turbidites.

Unfortunately, information on the sedimentary fill of the 2,700-km-long Tonga-Kermadec

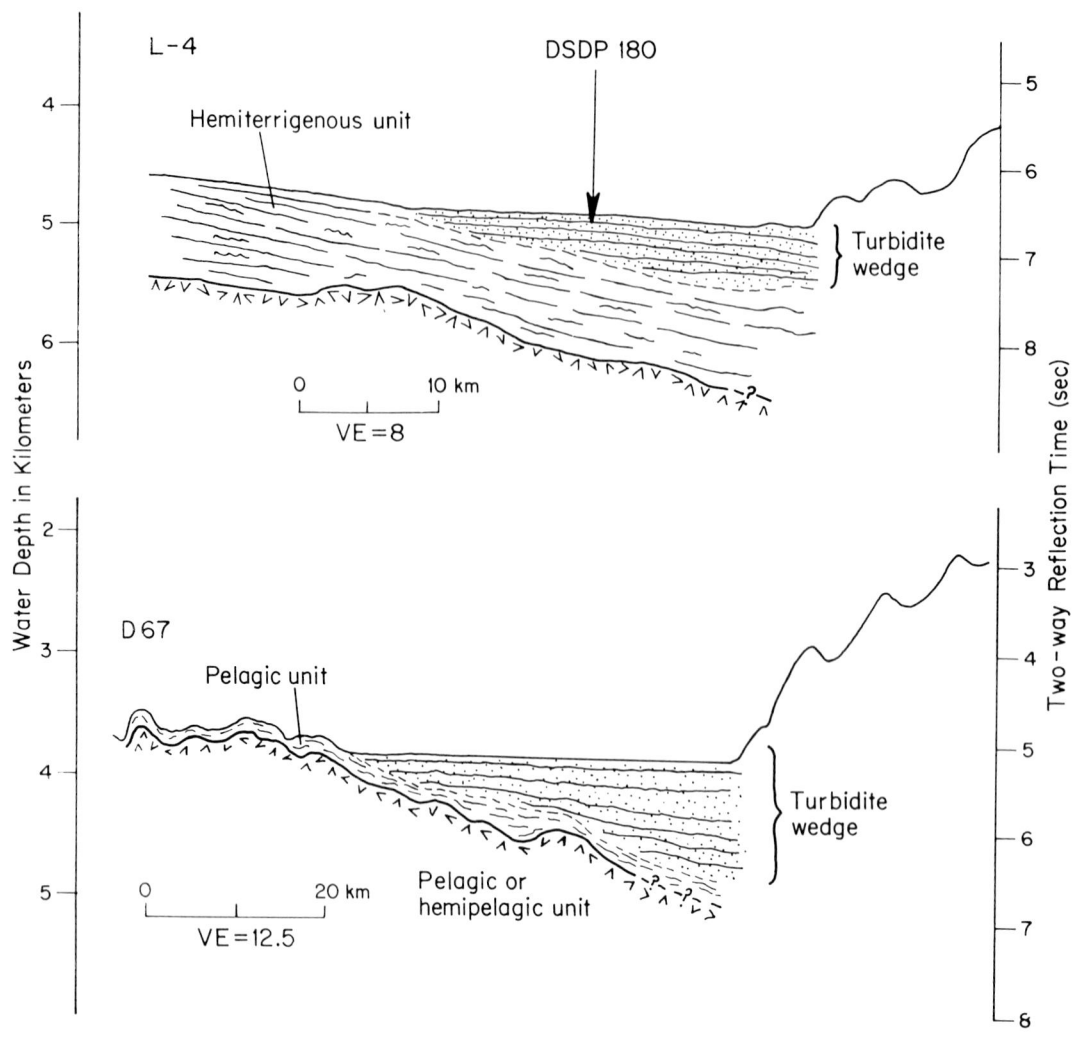

FIG. 3.—Diagrammatic drawings of acoustic reflection profiles that cross southern part of central segment of Peru-Chile Trench (D67) and eastern part of Aleutian Trench (L4, adapted from von Huene, 1972).

Trench is sketchy. It appears to be devoid of a significant fill, although it probably contains a thin (100 m?) layer of pelagic beds. However, near New Zealand, its linear continuation, the Hikurangi Trench (lat 37–42°S), picks up a thick (1,000 m) terrigenous turbidite fill overlying an older hemipelagic or hemiterrigenous section.

DISCUSSION[2]

Origin of trench sequence.—In the Aleutian and the northern and southern trenches of the eastern Pacific, rapid sedimentation did not be-

[2] Several papers relevant to this discussion and to table 1 have appeared since this paper was written. They include: Ludwig, W. J., Den, N., and Marauchi, S., 1973, Seismic reflection measurements of southwest

gin until late Cenozoic time. At this time turbidity currents began to convey the first of the hundreds of thousands of cubic kilometers of terrigenous debris that form their upper wedge-

Japan margin: Jour. Geophys. Research, v. 78, p. 2508–2516; Moore, D. G., 1973, Plate-edge deformation and crustal growth, Gulf of California structural province: Geol. Soc. America Bull., v. 84, p. 1883–1906; Piper, D. J. W., von Huene, R., and Duncan, J. R., 1973, Late Quaternary sedimentation in the active eastern Aleutian Trench: Geology, v. 1, p. 19–22; and Yoshii, T., Ludwig, W. J., and others, 1973, Structure of southwest Japan margin off Shikoku: Jour. Geophys. Research, v. 78, p. 2517–2525. Other papers that are pertinent to this discussion but that are not cited here include papers listed in the references section: Marlow, Scholl, and Buffington (1973); and Scholl, Buffington, and Marlow (in press); and Kulm, von Huene, and others (1973).

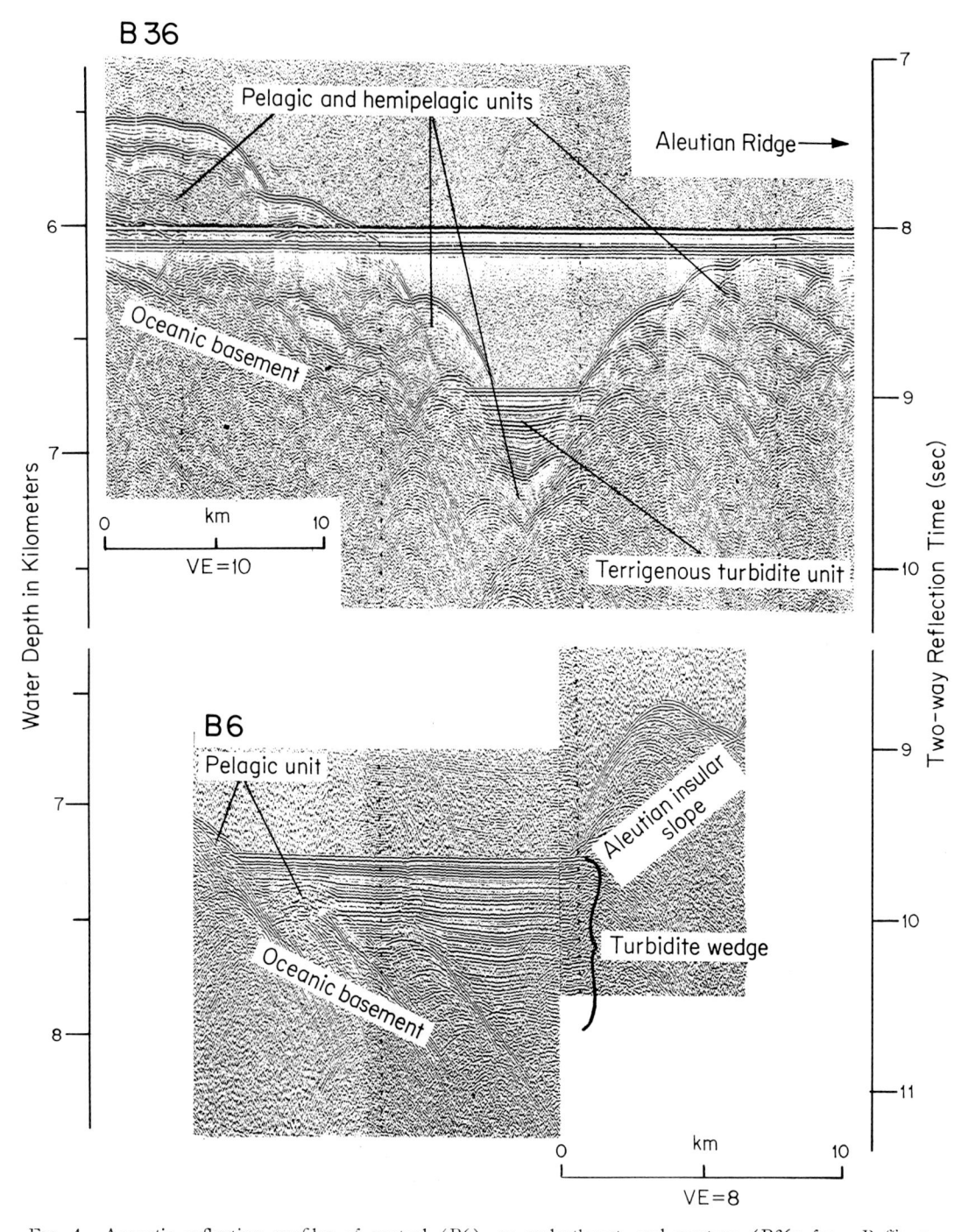

Fig. 4.—Acoustic reflection profiles of central (B6), or underthrust, and western (B36; from Buffington, 1973), or nonunderthrust, segments of Aleutian Trench.

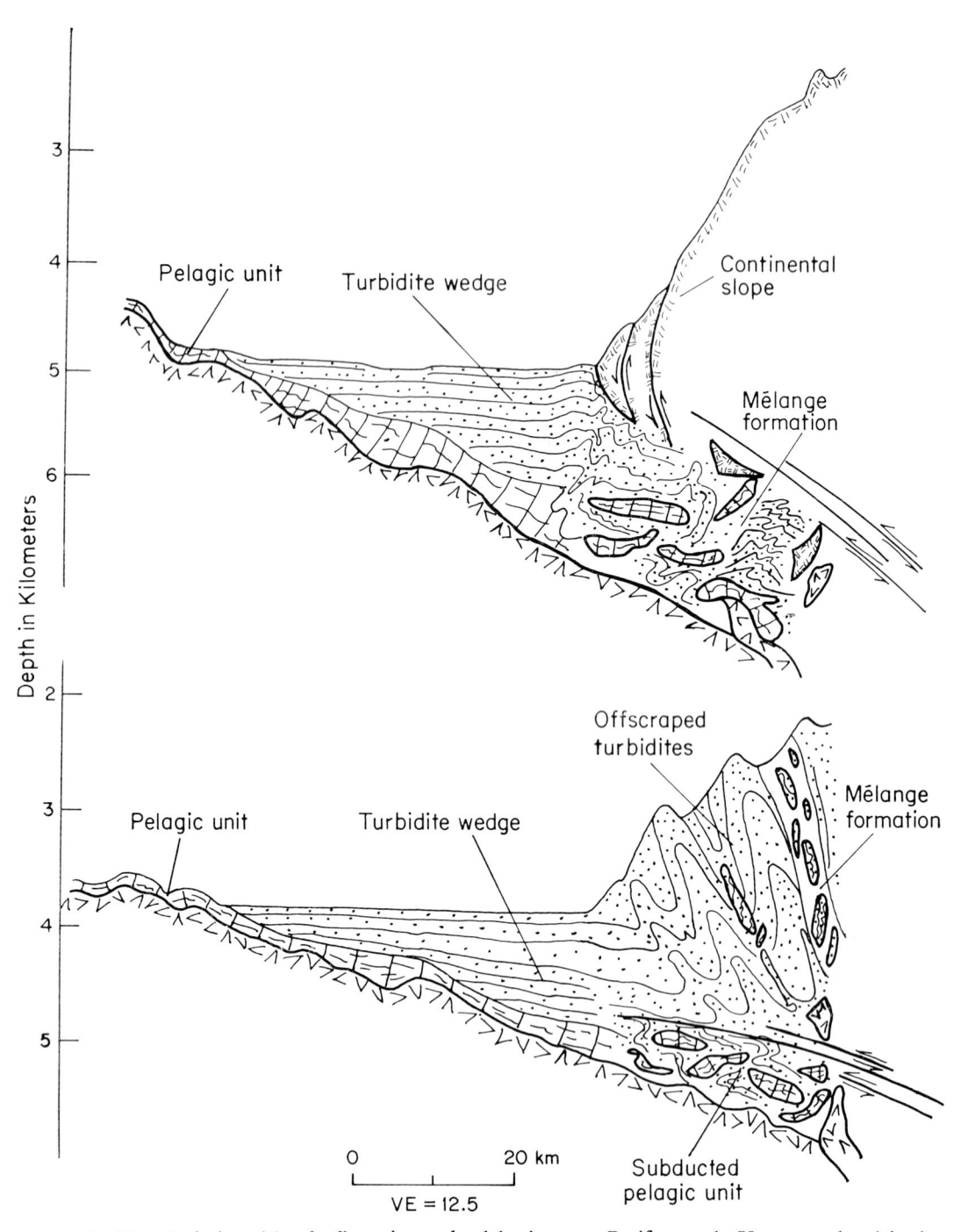

FIG. 5.—Hypothetical models of offscraping and subduction at a Pacific trench. Upper panel, subduction of entire trench sequence to form a mélange of terrigenous and pelagic blocks together with scraps of oceanic and continental crust. Lower panel, a possible mechanism for selective offscraping of terrigenous deposits and subduction of pelagic beds.

shaped sedimentary bodies (table 1; figs. 2, 3, 4). Deep-sea drilling indicates that rapid filling began only about one million years ago (von Huene, Kulm, and others, 1971; Kulm, von Huene, and others, 1973). We relate the spectacular accumulation of the terrigenous turbidites to the inception of expansive continental glaciation, which reached southward to midcontinent latitudes beginning about 1.2 million years ago (Kent and others, 1971; Izett and others, 1971; Hopkins, 1972). During episodes of expansive continental glaciation, sea level was greatly lowered (100 to 200 m), and rivers, burdened with ice-eroded detritus, were able to bypass the continental shelf and empty along the top of the continental slope. Sediment dumped here was conveyed via canyon-guided turbidity currents to the abyssal floors of flanking trenches (Scholl and others, 1968; Scholl, Christensen, and others, 1970). Figure 1 and table 1 show that only those trenches adjacent to extensively glaciated coastal mountains or adjacent but more interior areas, were filled to a point of seaward overflow.

Prior to about 1.2 my ago, episodes of late Cenozoic glaciation were far less extensive, being generally limited to alpine ice fields. In the North American Cordillera and the South American Andes, these fields contributed newly eroded debris, but the volume of ice on the continents was probably insufficient to cause emergence of the continental shelves. As a consequence, we suppose that terrigenous detritus was contributed less voluminously to the deep sea floor prior to about the middle Pleistocene (e.g., 1.2 my ago). This model of sediment distribution is highly significant because it means that the existing turbidite wedge may be considered as a special sedimentary body of great size that formed only in middle and late Pleistocene time. Accordingly, a thick wedge of trench turbidites may not have been characteristic of older trenches. This conclusion is in accord with Menard's (1964, p. 113-116) reasoning that, except where special conditions prevail, rapid terrigenous filling of trenches is unlikely.

Because plate subduction must involve trench deposits, we must consider the possibility that the existing wedge is, in fact, typical of older, now subducted or offscraped, wedges of Tertiary age. We think that pre-Pleistocene wedges existed rarely, or were small (i.e., a few hundred meters thick), because they are not found in the virtually nonunderthrust western segment of the Aleutian Trench and the southern segment of the Peru-Chile Trench (table 1). As these trench segments have been free of active plate subduction during at least the last 5 to 10 my (Pitman and Hayes, 1968; Herron nad Hayes, 1969; Atwater, 1970; Herron, 1972), their sedimentary sequence should be representative of a long-term infilling of a static or nearly static trench. The terrigenous wedge in these trench segments is restricted to the upper part of their two-layer sedimentary sequence (fig. 4, profile 36). It is therefore reasonable to conclude that the similarly positioned wedge in actively underthrust trench segments represents a special depositional body—one related to repeated episodes of accelerated continental erosion and lowered sea level during approximately the last 1.2 my. The history of late Neogene sedimentation over the abyssal floor of the Bering Sea (north of the Aleutian Trench) seems to confirm this, for the rate of accumulation of sandy and silty turbidites markedly increased in the middle Pleistocene. Before this time pelagic or pelitic terrigenous sedimentation dominated (Creager, Scholl, and others, in press).

Much less is known about the history of sedimentation in the western Pacific trenches. However, findings at DSDP sites and the magnetic anomaly chronology of the western Pacific indicate that the dominantly pelagic deposits of these trenches may be as old as early Mesozoic (Fischer and others, 1970; Fischer, Heezen, and others, 1971; Larson and Chase, 1972). Considering that the western trenches typically flank island arcs rather than continents, it is not surprising that the upper terrigenous turbidite wedge in these trenches is poorly developed. However, it is important to note that the wedge is also virtually absent in the Kamchatka Trench, which lies adjacent to a well-drained and extensively glaciated terrane. The large Cenozoic basin underlying what has been called the Kamchatka Terrace (Chase and others, 1970), situated midway down the continental slope, intercepted much of the coarse terrigenous detritus shed toward the trench during the Pleistocene Epoch (Buffington, 1973).

Tectonic control over trench filling is an important factor to emphasize. For example, although turbidites of terrigenous detritus have been reaching the abyssal floor of the Gulf of Alaska since at least early Eocene time (E. Hamilton, 1967, 1973; Scholl and Creager, 1973), far less would have been shed to the gulf in Miocene through early Pleistocene time had there not been orogenic uplift of the Pacific margin to form partially glaciated mountains in the Neogene (Denton and Armstrong, 1969). Trench filling in the Pleistocene to form the turbidite wedges of the Washington-Oregon and Aleutian Trenches would probably also have been far less rapid if adjacent coastal mountains capable of nourishing extensive ice fields had not formed.

Mountains formed by uplifted trench deposits.

—Modern eastern Pacific trenches are narrow structures, typically 50 to 100 km wide (measured between the defining 6,000-m contours), and they cannot accumulate a sedimentary section thicker than about 2 to 3 km without losing additional sediment to the adjacent sea floor. In themselves, these figures are not geosynclinal in proportion. Simply uplifted, a typical trench bordering North or South America would form a coastal mountain range underlain by oceanic crust thinly mantled by two distinguishable lithologic units, a 200- to 500-m thick section of older pelagic to hemiterrigenous deposits and as much as 2,000 m of overlying younger terrigenous turbidites. The older unit would have been deposited over a span of time ranging from 10 to as much as 100 times that of the turbidite section. In contrast, if a Panamanian, Central American, or Western Pacific trench were uplifted, the mountain thus formed would be constructed of oceanic crust overlain by a hemipelagic or pelagic section ranging from a few hundred meters to perhaps 2,000 m in thickness. Interbedded extrusive rocks would be rare, although tuffaceous and pyroclastic and turbidite beds would be common.

Without specifying the exact structural setting of the depositional trough (or troughs) of what may be called the Mesozoic eugeosynclines of the Pacific, it is nonetheless clear that they were deep-water (bathyal depths) areas dominated by terrigenous sedimentation (Bailey and others, 1964; Burk, 1965; Matsumoto, 1967; Bogdanov, 1969, 1970; Avdeiko, 1971; Landis and Bishop, 1972; Plafker, 1972; Moore, 1972). Because most modern trenches are floored chiefly by pelagic and hemipelagic deposits (table 1), because the turbidite wedge in many of them is probably a special feature related to expansive continental glaciation, and because glaciation of any significant extent is unknown in the Mesozoic Era (Crowell and Frakes, 1970), there is scant reason for supposing that trenches were the principal site of coarse terrigenous sedimentation in the Mesozoic. More likely, most segments of Mesozoic and Tertiary trenches probably received a more normal amount of terrigenous detritus that mixed with equal or slightly lesser amounts of pelagic debris. Pelagic sediment would have reached the Mesozoic and Tertiary trenches as they do modern ones, either directly by *in situ* sedimentation or indirectly by tectonic transportation to the trench on the back of an underthrusting oceanic plate. Except where local masses of turbidites had formed, possibly associated with the outfall of a major river, an uplifted Mesozoic or Tertiary trench sequence would therefore have probably been characterized by abundant pelagic and hemipelagic units.

Trench deposits and circum-Pacific eugeosynclinal rocks.—Uplifted thick piles of crumpled graywacke, micrograywacke, volcanically derived deposits, and mafic lava of late Paleozoic to early Tertiary age flank the margins of the Pacific. Many of these largely Mesozoic accumulations, especially those fringing the northern and southwestern Pacific, are presumed to have formed on oceanic crust; they commonly also contain ultramafic bodies and terranes of high P/T or blueschist metamorphism. Bogdanov (1969) referred to those ensimatic eugeosynclinal assemblages as thalassogeosynclines. It has been hypothesized that the deformed eugeosynclinal rocks may represent either one or more uplifted trench sequences (Menard and Dietz, 1951; Burk, 1965, 1973; Bogdanov, 1969; Avdeiko, 1971; Moore, 1972) or an uplifted tectonic complex (mélange) formed by the vertical and horizontal stacking of many offscraped and partially subducted trench sequences (Seyfert, 1969; Mitchell and Reading, 1969, 1971; Dewey and Bird, 1970a,b; Dickinson, 1971a,b,c; Plafker, 1972; Page, 1972a,b). The type tectonic or mélange eugeosyncline is represented by the Franciscan rocks of California (Hsü, 1971).

All the deformed and partially metamorphosed circum-Pacific eugeosynclinal masses, whether ensimatic or not, are dominated by terrigenous and volcanically-derived deposits (Bogdanov, 1969). Pelagic deposits, such as sequences of cherty micro- and nannofossil limestone, bedded chert, finely particulate or zeolitic abyssal shale, and manganese nodules, are uncommon in or absent from most areas. Chert or siliceous beds and limestone do occur interbedded with terrigenous deposits and in close association with extrusive lava. This is also true of contemporaneous rocks deposited on oceanic crust landward of the recognized inner limits of the Franciscan Sequence (Bailey and others, 1970; Page, 1972a). Bogdanov (1969) further pointed out that the siliceous rocks tend to be concentrated near the base of the Pacific eugeosynclinal assemblage. These basal units are probably deep-water pelagic and hemipelagic deposits; in some areas they could be true trench deposits. The critical relation is that these lower pelagic beds are depositionally overlain by younger and much thicker terrigenous deposits.

The great thickness, 15 to 30 km, of the overlying graywacke and related terrigenous and volcanically-derived deposits that form the bulk of the eugeosyncline could not have formed in a static (i.e., nonunderthrust) trench. Modern trenches cannot accumulate a thickness greater than about 3 km without loosing additional turbidites by overflow to the adjacent oceanic basin. Additional thickening requires that a continental rise prism (Drake and others, 1959) or apron (Wang, 1972) form depositionally above the filled trench. As Dietz (1963) emphasized, marginal eugeosynclinal rocks may be deformed rise prisms. Alternatively, by the formation of a younger volcanic arc seaward of the trench, thickening can be accomplished by transforming the trench into a type of inner arc basin. Avdeiko (1971) effectively used this model to account for the progressive accretion of eugeosynclinal complexes (Gnibidenko and others, in press) to the early Mesozoic continental margin of eastern Siberia (Kamchatka-Koryak region). In each case, the trench sequence, or initial deep-water trough sequence, would be located at the bottom of the eugeosynclinal pile and would constitute only a small fraction of its bulk.

As separately emphasized by Dewey and Bird (1970a,b), Hsü (1971), and Dickinson (1971a,b,c), and Page (1972a,b), plate consumption beneath trenches requires a new consideration of the type as well as of the structural arrangement of sedimentary units that might be expected to form there. The trench is presumably a tectonic mixing trough of sediment normally deposited there and of sediment conveyed there by the underthrusting oceanic plate. Mixing should take place at the back of the trench and along its tectonizing throat, so to speak, extending deeply beneath the adjacent continent or island arc (i.e., along the inclined Benioff or subduction zone; fig. 5). A combination of offscraping and subduction could conceivably stack trench fill upon trench fill beneath the base of the continental margin. If uplifted, such a tectonostratigraphic assemblage would form a mountain belt underlain by a great thickness of jumbled terrigenous and pelagic units and fragments of oceanic crust. If pelagic and terrigenous blocks were not tectonically separated from each other, such that one type was preferentially uplifted over the other (fig. 5, lower panel), then the relative proportions of each unit would presumably depend upon the average rate of terrigenous sedimentation in and adjacent to the trench and the average rate of tectonic injection of pelagic and hemipelagic debris scraped off the oceanic plate (fig. 5, upper panel).

The approximate amount of pelagic or hemipelagic deposits injected into the trench can be estimated in several ways. One assumes that Mesozoic trenches were underthrust at the average rate taking place beneath modern trenches, which we estimate to be between 5 and 10 cm/yr (Le Pichon, 1968; Isacks and others, 1968). Taking the average life of a circumpacific eugeosyncline to be 100 my, then during this time from 5,000 to 10,000 km of oceanic plate would have been consumed beneath it. Judging by the present distribution of pelagic and hemipelagic deposits on the Pacific crust (Ewing, Ewing, and others, 1968; Ewing and others, 1969), the Mesozoic plate seaward of its trench should have been covered by similar deposits at least 200 m and possibly as much as 500 m thick. Over a 100-my period, approximately 2.5×10^6 km^3 of incoming pelagic and hemipelagic deposits could be subducted or offscraped along a 1000-km trench segment.

Another method bases the calculations on the former position of Pacific spreading centers and plates during the Mesozoic. For example, the reconstructions of Larson and Chase (1972) and Larson and Pitman (1972) require that during the last 120 my about 7,000 km of oceanic plate have underthrust most Pacific margins. The average convergence rate has therefore been about 6 cm/yr, which corresponds to about 2×10^6 km^3 of offscraped or subducted pelagic and hemipelagic deposits for each 1,000-km segment of trench. An equivalent mass should have been offscraped or subducted beneath most Mesozoic trenches.

The volume of terrigenous and volcanic rock in a 1,000-km long segment of a typical Pacific eugeosyncline can be estimated by taking an average width of 200 km and a thickness of 20 km. The corresponding volume is 4×10^6 km^3. If the typical eugeosyncline is an assemblage of deposits that either depositionally or tectonically reached a Pacific trench, then at least one-third (a factor that takes compaction effects into account) of the uplifted tectonic mélange should be blocks and pieces of pelagic offscrapings. Deep-sea drilling in the Pacific indicates that the bulk of the Mesozoic pelagic debris would have been biopelagic limestone (Davies and Supko, 1973). To our knowledge, no circum-Pacific eugeosynclinal assemblage, whether ensimatic or not, or whether including mélange terranes or not, comprises more than a small amount of oceanic pelagic offscrapings.

The paucity of exposed oceanic deposits around the periphery of the Pacific has been long known and was for many years given as a strong argument for the stability of oceanic and continental areas (Kuenen, 1950, p. 123). This is not to say that pelagic deposits do not occur; for example, they are found on New Caledonia (Lillie and Brothers, 1970), the Solomon Islands (P. Coleman, 1966; Kroenke, 1972), the New Hebrides (Mitchell and Reading, 1971; Karig and Mammerickx, 1972), and also on New Guinea (Davies and Smith, 1971; Coleman, 1971). Some of these chiefly siliceous and calcareous beds may overlie obducted slabs of oceanic crust (R. Coleman, 1971), and some may be uplifted trench deposits, but, with the possible exception of the New Caledonian rocks

(M. C. Blake, personal commun., 1972), none appears to be a tectonically emplaced mass of trench offscrapings.

CONCLUDING REMARKS

If the coastal mountains fringing the Pacific are virtually devoid of oceanic deposits, what, then, has happened to the tens of millions of cubic kilometers of pelagic and hemipelagic deposits that should have accumulated depositionally and tectonically in Mesozoic and Cenozoic trenches, and what was the depositional environment of the graywacke-dominated circum-Pacific eugeosynclinal assemblage?

If the world is expanding (Hilgenberg, 1969; Rickard, 1969; Carey, 1970; Kolchanov, 1971; Jordan, 1971), then these questions are largely rhetorical. Ocean debris would not be expected in great quantities in eugeosynclinal belts because crustal underthrusting along Benioff zones would be limited in extent, and the largely graywacke and volcanic deposits of eugeosynclinal accumulation can be ascribed to uplifted rise prisms or aprons or to filled inner arc basins (i.e., Sea of Japan) that may or may not have a deeply buried trench sequence as a depositional root (Rickard, 1969).

On a nonexpanding earth, the only apparent way to rid virtually an entire Pacific basin of pelagic deposits is to insert them deeply beneath its insular or continental margins via subtrench subduction (Scholl and Marlow, 1972). Gilluly and others (1970) and Gilluly (1971), basing their arguments principally on missing terms in a terrigenous sediment budget, have also argued that much if not all sediment reaching the deep-sea floor is subducted beneath continents and island arcs. Terrigenous turbidites that may have reached the trench floor would presumably have been similarly disposed. Graywacke and associated volcanic rocks of the Pacific eugeosyncline must therefore have accumulated in a basin or a series of basins landward or upslope of the trench (Scholl and Marlow, 1972; in press). Following the implications of Matsumoto's (1967) models, the Aleutian Terrace, that is, part of the arc-trench gap of Dickinson (1971a,b,c) separating the volcanic arc and the trench, would be part of a modern eugeosynclinal area. The Aleutian Terrace is underlain by at least 7 to 8 km of volcanic rich terrigenous and hemiterrigenous deposits of Tertiary age (Scholl and Creager, 1973; Grow, 1973). The depositional sequence probably rests on an igneous basement, possibly a detached slab of oceanic crust. Even thicker accumulations of sediment, including huge slump masses, underlie the large midslope terrace (see bathymetric charts of Chase and others, 1970) landward of the Kamchatka Trench (Buffington, 1973). This trench lacks a Pleistocene turbidite wedge because of the efficacy of the Kamchatka Terrace as a sediment trap.

Other modern examples of Mesozoic eugeosynclines might be inner arc basins, such as the Shikoku Basin-Nankai Trough-Tosa Terrace and Shichito-Iwo Jima Ridge area (Hilde and others, 1969). This complex area, separated and isolated tectonically from the Pacific plate by the Japan-Bonin Trench, has trapped large amounts of terrigenous debris. Chai (1972) employed a similar model, although he called upon a seaward dipping Benioff zone (extending below a trench located inside or continentward of a reverse island arc) to account for the Cenozoic mélanges that underlie eastern Taiwan. We are also mindful of the possibility that exact structural and depositional analogs of the late Paleozoic and early Tertiary Pacific eugeosyncline may not exist today (Matsumoto, 1967). These geosynclines were formed, deformed, and metamorphosed mainly during the early phases of continental drift. Most of the depositional basins now flanking the Pacific are post-Cretaceous in age and were formed seaward of, and long after, the initial consequences of drift had been imprinted on the periphery of the Pacific.

Our preferred upslope location of the principal accumulation area for the typical Pacific eugeosyncline of Mesozoic age should not be construed to mean that trenches, in part or totally, could not have been either the principal tectonic assembly area or the depositional trough of a specific segment of a Pacific eugeosyncline. If the fans and abyssal plains of the Gulf of Alaska and the Cascadia Basin off Washington and Oregon were swept into their adjacent trenches, as the reasoning of Silver (1969, 1971, 1972), von Huene (1972), and Plafker (1972) implies, the resulting offscrapings and partial subduction products could form a geosynclinal pile greatly dominated by terrigenous beds. Deep-sea drilling (site 178, fig. 1) and reflection records indicate that at least 15 percent of the Cenozoic abyssal plain deposits of the Gulf of Alaska are pelagic or hemipelagic beds (von Huene, personal commun., 1972). It is therefore not likely that the present gulf contains sufficient turbidite beds to form a typical Pacific eugeosyncline by offscraping at a marginal trench.

Besides requiring that a rich supply of erosional detritus reach the deep-sea floor, the formation of a eugeosynclinal assemblage by the offscraping of abyssal plain deposits or a turbidite wedge also requires that the incoming plate

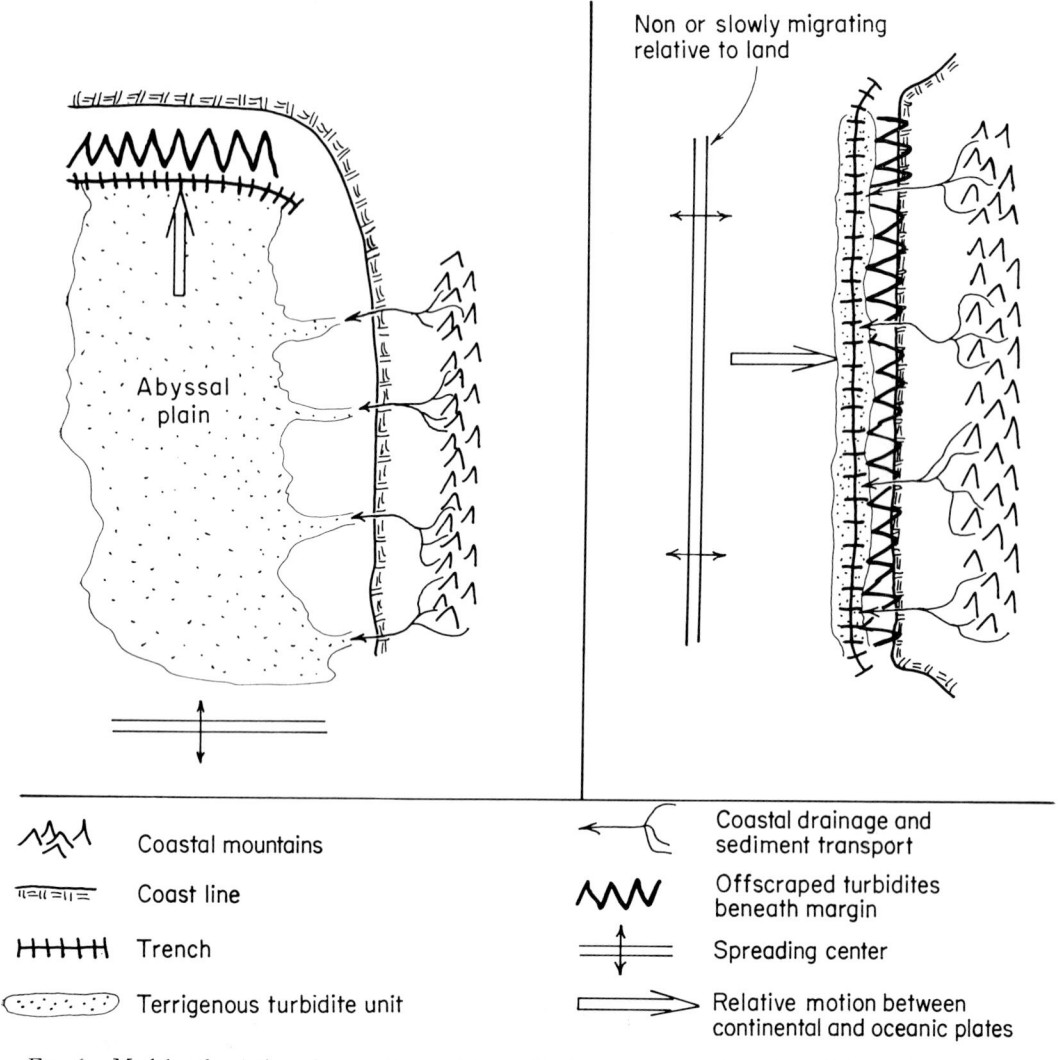

Fig. 6.—Models of relative plate motion, sediment discharge, and trenches that might result in formation of a graywacke-dominated eugeosyncline by offscraping of abyssal plain deposits (left panel) or a trench-filling turbidite wedge (right panel).

be free of a thick layer of pelagic deposits older than the turbidites. This condition is met if a stable or slowly migrating spreading center is located near the trench throughout most of the time the geosynclinal rocks are being assembled (Chipping, 1971; fig. 6, right panel). It is also met if the portion of the plate underlying the abyssal plain has traveled generally parallel to and near a continent since its formation at a spreading center (fig. 6, left panel). In neither circumstance would the subducted oceanic crust have accumulated a thick blanket. The crust of the northwest Pacific appears to have fulfilled many of these conditions since the early Neogene (Atwater, 1970; Dehlinger and others, 1970).

The reconstructions of Mesozoic plate boundaries and spreading centers by Larson and Chase (1972) and Larson and Pitman (1972) indicate that the above conditions were met rarely around the periphery of the Pacific during the Mesozoic Era. In fact, the perspective gained by viewing the entire Pacific basin and its friging trenches as an interacting system of tectonic and sedimentological elements forces the observer to realize that between 50 and 100 million km^3 of pelagic and hemipelagic deposits have been injected into the circum-Pacific trenches during the past 200 my. As a consequence, the low concentration of pelagic debris in circum-Pacific mountains stringently discourages any generalizing that the typical eugeosynclinal pile is composed of trench offscrapings.

A possibility remains that, because most of the pelagic and hemipelagic beds are located near the bottom of the trench fill, the overlying terrigenous beds would be tectonically skimmed off and thereby selectively preserved (fig. 5, lower panel). And, accordingly, the pelagic units would be selectively lost via subduction. Two factors argue against selective subduction. First, selective offscraping of terrigenous debris to form a eugeosynclinal mass requires that a thick steady-state turbidite wedge fill the Mesozoic trenches, a likelihood that is not supported by the history of sedimentation in Cenozoic trenches. And second, a thickness of pelagic and hemipelagic beds equivalent to the typical turbidite wedge (500 to 2,000 m) must have been swept into many segments of Mesozoic trenches. Where this took place geosynclinal masses of pelagic offscraping should have formed, yet none are known.

We believe that the distribution and volumes of sedimentary units in modern trenches and in deformed Pacific mountain belts signify that the deposits of many eugeosynclines did not directly interact with those of an oceanic plate in the bowels of a trench. We prefer to place the principal zone of accumulation in a more landward or upslope basin, such as an inner arc or reverse arc basin, a midslope terrace, or the continental or insular slope generally (Scholl and Marlow, 1972). Subsequent or contemporaneous deformation and partial subduction of these basinal deposits could readily account for the peculiar mélange terranes and high-pressure metamorphism characteristic of many circum-Pacific eugeosynclines.

Finally, granting that, if nothing else, a geosyncline is the site of deposition of a thick sequence of sedimentary or volcanic deposits (Kay, 1951; King, 1969), we can find little reason to believe that, in themselves, ancient or modern trenches were or are geosynclines. They may form part of a geosynclinal system of sediment-receiving basins separated by intervening ridges (Gnibidenko and Shashkin, 1970), but trenches are not the whole system, and, in our opinion, their deposits and tectonic accumulations form only a small part of the rocks exposed on island arcs and continental margins.

ACKNOWLEDGMENTS

We wish to extend our thanks and gratitude to Roland von Huene, Eli A. Silver, Edgar H. Bailey, David L. Jones, and Parke D. Snavely Jr., U.S. Geological Survey, for helpful discussions concerning some of the ideas presented in this paper. Our manuscript has also benefited significantly from critical readings by Benjamin M. Page, Stanford University; M. Clark Blake, U.S. Geological Survey; John V. Byrne, Oregon State University; R. H. Dott, Jr., University of Wisconsin; and Edwin C. Buffington, Naval Undersea Center, San Diego. Buffington also kindly gave us permission to publish the acoustic reflection profile B36 (fig. 5) and, prior to their publication, allowed us to incorporate some of the results of his investigation into the depositional histories of the western part of the Aleutian Trench and the connecting Kamchatka Trench (Buffington, 1973). John Ewing, Lamont-Doherty Geological Observatory, was equally kind and allowed us to study unpublished reflection records of many Pacific trenches.

Publication has been authorized by the Director, U.S. Geological Survey.

REFERENCES

ATWATER, T., 1970, Implications of plate tectonics for the Cenozoic tectonic evolution of western North America: Geol. Soc. America Bull., v. 81, p. 3511–3536.
AVDEIKO, G. P., 1971, Evolution of geosynclines on Kamchatka: Pacific Geology, v. 3, p. 1–14.
BAILEY, E. H., AND BLAKE, M. C., JR., 1969, Tectonic development of western California during the late Mesozoic: Geotectonics, no. 3, p. 148–154, and no. 4, p. 225–230.
BAILEY, E. H., IRWIN, W. P., AND JONES, D. L., 1964, Franciscan and related rocks, and their significance in the geology of western California: California Div. Mines and Geology Bull. 183, 177 p.
———, ———, AND JONES, D. L., 1970, On-land Mesozoic oceanic crust in California Coast Ranges: U.S. Geol. Survey Prof. Paper 700-C, p. 70–81.
BARBAT, W. F., 1971, Megatectonics of the Coast Ranges, California: Geol. Soc. America Bull., v. 82, p. 1541–1562.
BOGDANOV, N. A., 1969, Thalassogeosynclines of the Circumpacific ring: Geotectonics, no. 3, p. 141–147.
———, 1970, Nekotorye osobennosti tektoniki vostoka Koryakskogo Nagor'ya [Some unusual features in the tectonics of the east of the Kryak Uplands]: Akad. Nauk SSSR, Doklady v. 192, p. 607–610.
BUFFINGTON, E. C., 1973, The Aleutian-Kamchatka Trench convergence and investigations of lithospheric plate interaction in the light of modern geotectonic theories (thesis): Los Angeles, Univ. Southern California, Dept. Geol. Sci.
BURK, C. A., 1965, Geology of the Alaska Peninsula—Island arc and continental margin: Geol. Soc. America Mem. 99, 250 p.

———, 1973, Uplifted eugeosynclines and continental margins: *ibid.*, Hess vol., Mem. 132, p. 75–85.
BURNS, R. E., ANDREWS, J. E., AND OTHERS, 1972, Deep Sea Drilling Project leg 21: Geotimes, May, p. 14–16.
CAREY, S. W., 1970, Australia, New Guinea and Melanesia in the current revolution in the concepts of the evolution of the earth: Search, v. 1, p. 178–187.
CHAI, B. H. T., 1972, Structure and tectonic evolution of Taiwan: Am. Jour. Sci., v. 272, p. 389–422.
CHASE, T. E., MENARD, H. W., AND MAMMERICKX, J., 1970, Bathymetry of the North Pacific: Scripps Inst. Oceanography and Inst. Marine Res. Tech. Rept. Ser. TR-7, 10 sheets.
CHIPPING, D. H., 1971, Paleoenvironmental significance of chert in the Franciscan Formation of western California: Geol. Soc. America Bull., v. 82, p. 1707–1712.
COLEMAN, P. J., 1966, The Solomon Islands as an island arc: Nature, v. 211, p. 1249–1251.
COLEMAN, R. G., 1971, Plate tectonic emplacement of upper mantle peridotites along continental edges: Jour. Geophys. Research, v. 76, p. 1212–1222.
CREAGER, J. S., SCHOLL, D. W., AND OTHERS, 1973, Initial reports of the Deep Sea Drilling Project, v. 19: Washington, D.C., U.S. Govt. Printing Office.
CROWELL, J. C., AND FRAKES, L. A., 1970, Phanerozoic glaciation and the cause of ice ages: Am. Jour. Sci., v. 268, p. 193–224.
DAVIES, H. L., AND SMITH, I. E., 1971, Geology of eastern Papua: Geol. Soc. America Bull., v. 82, p. 3299–3312.
DAVIES, T. A., AND SUPKO, P. R., 1973, Oceanic sediment and diagenesis: Jour. Sed. Petrology, v. 43, p. 381–390.
DEHLINGER, P., AND OTHERS, 1970, Northeast Pacific structure, *in* MAXWELL, A. E. (ed.), New concepts of sea floor evolution, pt. 2, Regional observations concept: New York, Wiley-Interscience Publishers, The sea, v. 4, p. 133–189.
DENTON, G. H., AND ARMSTRONG, R. L., 1969, Miocene-Pliocene glaciations in southern Alaska: Am. Jour. Sci., v. 267, p. 1121–1142.
DEWEY, J. F., AND BIRD, J. M., 1970a, Mountain belts and the new global tectonics: Jour. Geophys. Research, v. 75, p. 2625–2647.
———, 1970b, Plate tectonics and geosynclines: Tectonophysics, v. 10, p. 625–638.
DICKINSON, W. R., 1971a, Plate tectonics in geologic history: Science, v. 174, p. 107–113.
———, 1971b, Clastic sedimentary sequences deposited in shelf, slopes and trough settings between magmatic arcs and associated trenches: Pacific Geology, v. 3, p. 15–30.
———, 1971c, Plate tectonics models of geosynclines: Earth and Planetary Sci. Letters, v. 10, p. 165–174.
DIETZ, R. S., 1963, Collapsing continental rises: an actualistic concept of geosynclines and mountain building: Jour. Geology, v. 7, p. 314–333.
DRAKE, C. L., EWING, M., AND SUTTON, J., 1959, Continental margins and geosynclines—the east coast of North America north of Cape Hatteras: *in* AHRENS, L. H., AND OTHERS (eds.), Physics and chemistry of the earth, v. 3: London, Pergamon Press, p. 110–198.
EMERY, K. O., AND OTHERS, 1969, Geological structure and some water characteristics of the east China Sea and the Yellow Sea: ECAFE Tech. Bull., v. 2, p. 3–43.
ERNST, W. G., 1965, Mineral parageneses in Franciscan metamorphic rocks, Panoche Pass, California: Geol. Soc. America Bull., v. 76, p. 879–914.
———, 1970, Tectonic contact between the Franciscan mélange and the Great Valley Sequence—crustal expression of a late Mesozoic Benioff zone: Jour. Geophys. Research, v. 75, p. 868–901.
ERNST, W. G., AND OTHERS, 1970, Comparative study of low-grade metamorphism in the California Coast Ranges and the outer metamorphic belt of Japan: Geol. Soc. America Mem. 124, 276 p.
EWING, J., EWING, M., AND OTHERS, 1968, North Pacific sediment layers measured by seismic profiling, *in* KNOPOFF, L., DRAKE, C. L., AND HART, P. J. (eds.), The crust and upper mantle of the Pacific area: Am. Geophys. Union Geophys. Mem. 12, p. 147–173.
EWING, M., AND CONNARY, S. D., 1970, Nepheloid layer in the North Pacific: *in* HAYS, J. D., AND OTHERS (ed.), Geological investigations of the North Pacific, Geol. Soc. America Mem. 126, p. 41–82.
———, HOUTZ, R., AND EWING, J., 1969, South Pacific sediment distribution: Jour. Geophys. Research, v. 74, p. 2477–2511.
FISCHER, A. G., AND OTHERS, 1970, Geological history of the western North Pacific: Science, v. 168, p. 1210–1214.
———, HEEZEN, B. C., AND OTHERS, 1971, Initial reports of the Deep Sea Drilling Project, v. 6: Washington, D.C., U.S. Govt. Printing Office, 1329 p.
FISHER, R. L., 1961, Middle America Trench: topography and structure: Geol. Soc. America Bull., v. 72, p. 703–720.
———, AND ENGEL, C. G., 1969, Ultramafic and basaltic rocks dredged from the nearshore flank of the Tonga Trench: *ibid.*, v. 80, p. 1373–1378.
———, AND HESS, H. H., 1963, Trenches: *in* HILL, M. N. (ed.), The earth beneath the sea history: New York, Wiley-Interscience Publishers, The sea, v. 3, p. 411–436.
———, AND RAITT, R. W., 1962, Topography and structure of the Peru-Chile Trench: Deep-Sea Research, v. 9, p. 423–443.
GILLULY, J., 1971, Plate tectonics and magmatic evolution: Geol. Soc. America Bull., v. 82, p. 2383–2396.
———, REED, J. C., JR., AND CADY, W. M., 1970, Sedimentary volumes and their significance: *ibid.*, v. 81, p. 353–376.
GNIBIDENKO, H. S., AND SHASHKIN, K. S., 1970, Basic principles of geosynclinal theory: Tectonophysics, v. 9, p. 5–13.
———, AND OTHERS, in press, Geology and deep structure of Kamchatka: Pacific Geology.
GROW, J. A. 1973, Crustal and upper mantle structure of the central Aleutian Arc: Geol. Soc. America Bull., v. 84, p. 2169–2192.

HAMILTON, E. L., 1967, Marine geology of abyssal plains in the Gulf of Alaska: Jour. Geophys. Research, v. 72, p. 4189–4213.
———, 1973, Marine geology of the Aleutian abyssal plain: Marine Geology, v. 14, p. 295–325.
HAMILTON, W., 1969, Mesozoic California and the underflow of Pacific mantle: Geol. Soc. America Bull., v. 80, p. 2409–2430.
HAYES, D. E., AND EWING, M., 1970, Pacific boundary structure, in MAXWELL, A. E. (ed.), New concepts of sea floor evolution, pt. 2, Regional observations concepts, The seas, v. 4: New York, Wiley-Interscience Publishers, p. 29–72.
———, AND OTHERS, 1972, Seismic reflection profiles, Pt. C, in EWING, M. (ed.), Preliminary report of volume 22, *USNS Eltanin*, cruises 28–32, March 1967-March 1968, Tech. Rept. CU-1-72: Washington, D.C., National Science Foundation, p. 147–200.
HERRON, E. M., 1972, Sea-floor spreading and the Cenozoic history of the east-central Pacific: Geol. Soc. America Bull., v. 83, p. 1671–1692.
———, AND HAYES, D. E., 1969, A geophysical study of the Chile Ridge: Earth and Planetary Sci. Letters, v. 6, p. 77–83.
HILDE, T. W. C., WAGEMAN, J. M., AND HAMMOND, W. T., 1969, The structure of Tosa Terrace and Nankai Trough off southeastern Japan: Deep-Sea Research, v. 16, p. 67–75.
HILGENBERG, O. C., 1969, Earth expansion, deep-sea trenches, and the inclination of shelf-sea floors: Neues Jahrb. Geologie u. Palaontologie, Monatsh., p. 138–145.
HOLMES, M. L., VON HUENE, R. E., AND MCMANUS, D. A., 1972, Seismic reflection evidence supporting underthrusting beneath the Aleutian arc near Amchitka Island: Jour. Geophys. Research, v. 77, p. 959–964.
HOPKINS, D. M., 1972, Changes in oceanic circulation and late Cenozoic cold climates: 24th Internat. Geol. Cong., Montreal, 1972, p. 370.
HOUTZ, R., AND OTHERS, 1967, Seismic reflection profiles of the New Zealand Plateau: Jour. Geophys. Research, v. 72, p. 4713–4730.
HSÜ, K. J., 1971, Franciscan mélange as a model for eugeosynclinal sedimentation and underthrusting tectonics: *ibid.*, v. 76, p. 1162–1170.
ISACKS, B., OLIVER, J., AND SYKES, L. R., 1968, Seismology and the new global tectonics: *ibid.*, v. 73, p. 5855–5899.
IZETT, G. A., AND OTHERS, 1971, Evidence for two pearlette-like ash beds in Nebraska and adjoining areas: Geol. Soc. America Abs. with Programs, v. 3, p. 610.
JORDAN, P., 1971, The expanding earth: Oxford, England, Pergamon Press, 202 p.
KARIG, D. E., 1971, Site surveys in the Mariana area (SCAN IV), in FISCHER, A. G., HEEZEN, B. C., AND OTHERS (eds.), Initial reports of the Deep Sea Drilling Project, v. 6: Washington, D.C., U.S. Govt. Printing Office, p. 681–689.
———, AND MAMMERICKX, J., 1972, Tectonic framework of the New Hebrides Island Arc: Marine Geology, v. 12, p. 187–205.
KAY, M., 1951, North American geosynclines: Geol. Soc. America Mem. 48, 143 p.
KENT, D., OPDYKE, N., AND EWING, M., 1971, Pattern of climate change in the North Pacific using ice-rafted detritus as a climatic indicator: *ibid.*, Bull., v. 82, p. 2741–2754.
KING, P. B., 1969, The tectonics of North America—a discussion to accompany the tectonic map of North America (scale 1:5,000,000): U.S. Geol. Survey Prof. Paper 628, 95 p.
KOLCHANOV, V. P., 1971, O. C. Hilgenberg's paleogeographic representation of an expanding earth: Geotectonics, no. 4, p. 252–259.
KROENKE, L. W., 1972, Geology of the Ontong Java Plateau: Hawaii Inst. Geophysics Rept. HIG-72-5, 119 p.
KUENEN, PH. H., 1950, Marine Geology: New York, John Wiley and Sons, Inc., 568 p.
KULM, L. D., AND OTHERS, 1972, Deformation in the Peru Trench: Geol. Soc. America Abs. with Programs, v. 4, p. 570.
KULM, L. D., VON HUENE, R., AND OTHERS, 1973, Initial report of the Deep Sea Drilling Project, v. 18: Washington, D.C., U.S. Govt. Printing Office, p. 1077.
LANDIS, C. A., AND BISHOP, D. G., 1972, Plate tectonics and regional stratigraphic-metamorphic relations in the southern part of the New Zealand Geosyncline: Geol. Soc. America Bull., v. 83, p. 2267–2284.
LARSON, R. L., AND CHASE, C. G., 1972, Late Mesozoic evolution of the western Pacific Ocean: *ibid.*, p. 3627–3644.
———, AND PITMAN, W. C. III, 1972, World-wide correlation of Mesozoic magnetic anomalies, and its implications: *ibid.*, p. 3645–3662.
LE PICHON, X., 1968, Sea-floor spreading and continental drift: Jour. Geophys. Research, v. 72, p. 3661–3697.
LILLIE, A. R., AND BROTHERS, R. N., 1970, The geology of new Caledonia: New Zealand Jour. Geology and Geophysics, v. 13, p. 145–183.
LISTER, C. R. B., 1971, Tectonic movement in the Chile Trench: Science, v. 173, p. 719–722.
LUDWIG, W. J., AND OTHERS, 1966, Sediments and structure of the Japan Trench: Jour. Geophys. Research, v. 71, p. 2121–2137.
MARLOW, M. S., SCHOLL, D. W., AND BUFFINGTON, E. C., 1973, Discussion of paper by M. L. Holmes, R. E. von Huene, and D. A. McManus, Seismic reflection evidence supporting underthrusting beneath the Aleutian Arc near Amchitka Island: *ibid*, p. 3517–3522.
———, ———, AND OTHERS, 1973, Tectonic history of the central Aleutian Arc: Geol. Soc. America Bull., v. 84, p. 1555–1574.
MATSUMOTO, T., 1967, Fundamental problems in the circum-Pacific orogenesis: Tectonophysics, v. 4, p. 595–613.
MCMANUS, D. A., AND OTHERS, 1972, Late Quaternary tectonics, northern end of Juan de Fuca Ridge (northeast Pacific): Marine Geology, v. 12, p. 141–164.
MENARD, W. H., 1964, Marine Geology of the Pacific: New York, McGraw-Hill Book Co., 271 p.

———, AND DIETZ, R. S., 1951, Submarine geology of the Gulf of Alaska: Geol. Soc. America Bull., v. 62, p. 1263–1285.
MITCHELL, A. H., AND READING, H. G., 1969, Continental margins, geosynclines, and ocean floor spreading: Jour. Geology, v. 77, p. 629–646.
———, AND ———, 1971, Evolution of island arcs: ibid., v. 79, p. 253–284.
MIYASHIRO, A., 1961, Evolution of metamorphic belts: Jour. Petrology, v. 2, p. 277–311.
MOORE, J. C., 1972, Uplifted trench sediments: southwestern Alaska-Bering shelf edge: Science, v. 175, p. 1103–1105.
PAGE, B. M., 1970, Sur-Nacimiento Fault zone of California—continental margin tectonics: Geol. Soc. America Bull., v. 81, p. 667–690.
———, 1972a, Oceanic crust and mantle fragment in subduction complex near San Luis Obispo, California: ibid., v. 83, p. 957–972.
———, 1972b, Cannibalism in Franciscan mélanges of California: Geol. Soc. America Abs. with Programs, v. 4, p. 620.
PITMAN, W. C., III, AND HAYES, D. E., 1968, Sea-floor spreading in the Gulf of Alaska: Jour. Geophys. Research, v. 73, p. 6571–6580.
PLAFKER, G., 1972, Alaskan earthquake of 1964 and Chilean earthquake of 1960: implications for arc tectonics: ibid., v. 77, p. 901–925.
REYNOLDS, L., 1970, Oceanographic cruise report, northwest Pacific, July-August, 1970: Washington, D.C., U.S. Naval Oceanographic Office OCR 931003, 6 p.
RICKARD, M. J., 1969, Relief of curvature on expansion—a possible mechanism of geosynclinal formation and orogensis: Tectonophysics, v. 8, p. 129–144.
ROSS, D. A., 1971, Sediments of the northern Middle America Trench: Geol. Soc. America Bull., v. 82, p. 303–322.
———, AND SHOR, G. G., JR., 1965, Reflection profiles across the Middle America Trench: Jour. Geophys. Research, v. 70, p. 5551–5572.
SCHOLL, D. W., BUFFINGTON, E. C., AND MARLOW, M. S., in press, Plate tectonics and the structural evolution of the Aleutian-Bering Sea region—solutions and complications, in FORBES, R. B. (ed.), The geophysics and geology of the Bering Sea region: Geol. Soc. America Special Paper 151.
———, CHRISTENSEN, M. N., AND OTHERS, 1970, Peru-Chile Trench sediments and sea-floor spreading: ibid., Bull., v. 81, p. 1339–1360.
———, AND CREAGER, J. S., 1973, Geologic synthesis of leg 19 (DSDP) results: far North Pacific, Aleutian Ridge and Bering Sea, in CREAGER, J. S., AND SCHOLL, D. W., AND OTHERS (eds.), Initial reports of the Deep Sea Drilling Project, v. 19: Washington, D.C., U.S. Govt. Printing Office.
———, ———, AND OTHERS, 1971, Deep Sea Drilling Project leg 19: Geotimes, v. 16, no. 11, p. 12–15.
———, AND MARLOW, M. S., 1972, Ancient trench deposits and global tectonics: a different interpretation: Geol. Soc. America Abs. with Programs, v. 4, p. 232–233.
———, AND ———, in press, Global tectonics and the sediments of modern and ancient trenches: some different interpretations, in KAHLE, C. F., AND MEYERHOFF, A. A. (eds.), Sea floor spreading, other interpretations: Tulsa, Oklahoma, Am. Assoc. Petroleum Geologists.
———, AND VON HUENE, R., 1970, Comments on paper by R. L. Chase and E. T. Bunce, 'Underthrusting of the eastern margin of the Antilles by the floor of the western north Atlantic Ocean, and origin of the Barbados Ridge': Jour. Geophys. Research, v. 75, p. 488–490.
———, ———, AND RIDLON, J. B., 1968, Spreading of the ocean floor—undeformed sediments in the Peru-Chile Trench: Science, v. 159, p. 869–871.
SEYFERT, C. K., 1969, Undeformed sediments in oceanic trenches with sea-floor spreading: Nature, v. 222, p. 70.
SHOR, G. G., JR., 1966, Continental margins and island arc of western North America, in POOLE, W. H. (ed), Continental margins and island arcs: Geol. Survey Canada Paper 66-15, p. 216–222.
SILVER, E. A., 1969, Late Cenozoic underthrusting of the continental margin off northernmost California: Science, v. 166, p. 1265–1266.
———, 1971, Transitional tectonics and late Cenozoic structure of the continental margin off northernmost California: Geol. Soc. America Bull., v. 82, p. 1–22.
———, 1972, Pleistocene tectonic accretion of the continental slope off Washington: Marine Geology, v. 13, p. 239–249.
TIFFIN, D. L., CAMERON, B. E. B., AND MURRAY, J. W., 1972, Tectonics and depositional history of the continental margin off Vancouver Island, British Columbia: Canadian Jour. Earth Sci., v. 9, p. 280–296.
VAN ANDEL, T. H., AND OTHERS, 1971, Tectonics of the Panama Basin, eastern equatorial Pacific: Geol. Soc. America Bull., v. 82, p. 1489–1580.
VON HUENE, R., 1972, Structure of the continental margin and tectonism at the eastern Aleutian Trench: ibid., v. 83, p. 3613–3626.
———, KULM, L. D., AND OTHERS, 1971, Deep Sea Drilling Project leg 18: Geotimes, v. 16, no. 10, p. 12–15.
———, AND SHOR, G. G., JR., 1969, The structure and tectonic history of the eastern Aleutian Trench: Geol. Soc. America Bull., v. 80, p. 1889–1902.
WAGEMAN, J. M., HILDE, T. W. C., AND EMERY, K. O., 1970, Structural framework of east China Sea and Yellow Sea: Am. Assoc. Petroleum Geologists Bull., v. 54, p. 1611–1643.
WANG, C. S., 1972, Geosynclines in the new global tectonics: Geol. Soc. America Bull., v. 83, p. 2105–2110.
WINTERER, E. L., AND RIEDEL, W. R., AND OTHERS, 1969, Deep Sea Drilling Project leg 7: Geotimes, v. 14, no. 12, p. 12–14.
———, ———, AND OTHERS, 1971, Initial reports of the Deep Sea Drilling Project, v. 7: Washington, D.C., U.S. Govt. Printing Office, p. 1.
WORZEL, J. L., 1965, Deep structures of coastal margins and mid-oceanic ridges, in WHITTARD, W. F., AND BRADSHAW, R. (eds.), Submarine geology and geophysics, 17th Symposium, Colston Research Soc.: London, Butterworth & Co. (Publishers) Ltd., p. 335–361.

CENOZOIC SEDIMENTARY FRAMEWORK OF THE GORDA-JUAN DE FUCA PLATE AND ADJACENT CONTINENTAL MARGIN—A REVIEW

L. D. KULM AND G. A. FOWLER

Oregon State University, Corvallis; University of Wisconsin—Parkside, Kenosha

ABSTRACT

Earlier studies show that the area of western Oregon and Washington and the adjacent continental margin was the site of a large geosyncline during Cenozoic time. The marine sandstone and siltstone and locally interbeded volcanic rocks reach a maximum thickness of about 8 km. Late Cenozoic continental shelf and upper slope deposits consist primarily of siltstone and mudstone and subordinate arkosic and lithic wackes. Lower slope deposits are characterized by lithic, arkosic, or volcanic sand turbidites and mudstones.

During late Eocene time, the geosyncline was subjected to intense tectonism. The seaward part of the basin probably was destroyed by subduction in middle to late Cenozoic time. As the Coast Range was uplifted late in the Oligocene Epoch, deposition shifted westward into the structural basins in the vicinity of the present continental shelf. Several periods of deposition followed by uplift and erosion occurred on the continental shelf. At least two regional unconformities, late middle Miocene and Pliocene-Pleistocene, are interpreted from seismic reflection records of the Oregon shelf. The earlier unconformity probably is related to the worldwide change in lithospheric plate motion 10 million years ago; the later one to small-plate tectonics within the Gorda-Juan de Fuca plate.

Seismic reflection profiling in the Cascadia and Gorda Basins shows that a thick turbidite wedge occurs at the base of the continental slope in an elongate basement low between Vancouver Island and the Mendocino Fracture Zone. This wedge thins westward toward the Gorda and Juan de Fuca Ridges and reaches a maximum thickness of 2.5 km at the base of the continental slope in the vicinity of the Columbia River. The configuration of the basement depression and its associated sedimentary fill are similar to that found in the Aleutian and Peru-Chile Trenches.

Two large submarine fans, Astoria and Nitinat, occupy a large portion of the Gorda-Juan de Fuca plate and consist of middle to late Pleistocene turbidites. Sand turbidites of the Astoria Fan form a discordant contact with the underlying, landward dipping silt turbidites of the abyssal plain. The fan sediments were deposited in a trenchlike depression at the base of the continental slope. With continued deposition, the fan sediments prograded westward as the depression filled.

Silt and sand turbidites, which comprise the bulk of the deposits of the Gorda-Juan de Fuca plate, are intercalated with thin hemipelagic and pelagic deposits. These turbidites are characterized by lithic, arkosic, and volcanic sands. Paleoenvironmental analyses suggest that the Pliocene and Pleistocene abyssal-plain turbidites of the Gorda-Juan de Fuca plate are being uplifted and accreted to the lower continental slope. A maximum uplift of approximately 1,200 m has occurred on the Oregon continental margin since the late Miocene.

Although both the Cenozoic geosyncline and Cascadia Basin are dominated by terrigenous deposits, there are significant differences in the periodicity of volcanic activity and the configuration of the sedimentary bodies.

INTRODUCTION

Seismic reflection profiling and deep-sea drilling show that the northeastern Pacific Ocean is an area of relatively thick Cenozoic sedimentary deposits (e.g., Ewing and others, 1968; Kulm, von Huene, and others, 1973). The bulk of these deposits are terrigenous turbidites (e.g., Horn and others, 1970; von Huene and others, 1971; Kulm, von Huene and others, 1973), which were derived ultimately from the drainage basins of the adjacent continent. The thickest sedimentary accumulations are found in the Tertiary basins onshore, on the continental margin, and on the sea floor adjacent to the margin (Shor and others, 1968; von Huene, Lathram, and Reimnitz, 1971). A thick clastic wedge of sediment may partially or completely fill the trenches that lie at the base of the continental slope (Scholl and Marlow, this volume).

The primary objective of this paper is to determine the nature of the sedimentary framework of the Gorda-Juan de Fuca plate and the adjacent continental margin (fig. 1). This geologic setting contrasts with that described for the Cenozoic geosyncline of western Oregon and Washington. We emphasize the Oregon continental margin and Cascadia Basin because the most complete sedimentary and tectonic interpretation can be developed for these features.

THE OCEANIC CRUSTAL SETTING

Early in the Cenozoic Era, a north-south, elongate depositional basin occupied the present location of the Coast Range of Oregon and Washington, the Olympic Mountains, and the adjacent continental margin (e.g., Snavely and Wagner, 1963; Snavely and others, 1964; Snavely and others, 1968; Braislin and others, 1971) (fig. 1). More than 7 to 8 km of Ceno-

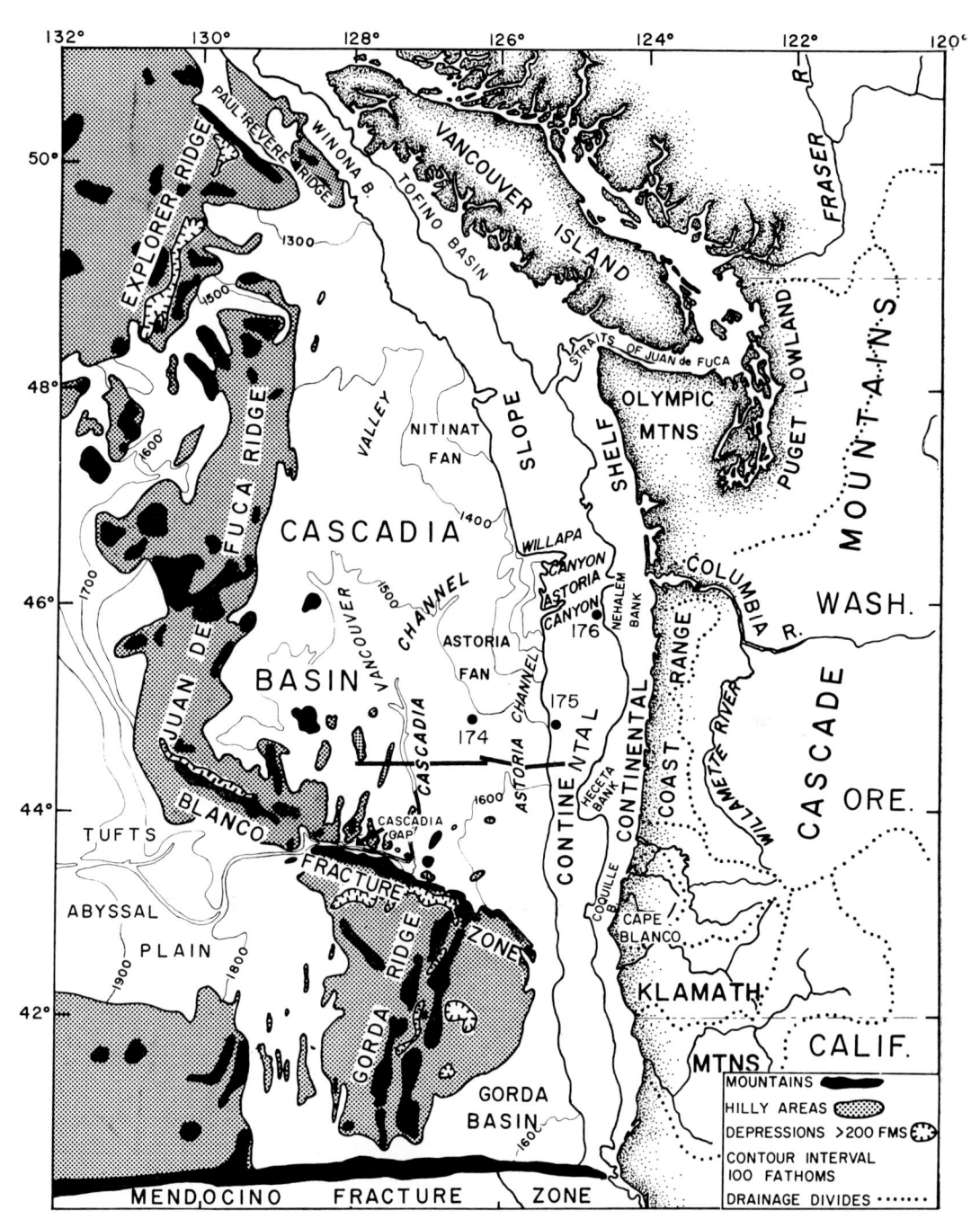

FIG. 1.—Map showing Gorda-Juan de Fuca plate (Cascadia and Gorda Basins), adjacent continental margin, and northwestern United States. Note locations of Deep Sea Drilling sites 174, 175, and 176 in Cascadia Basin and on Oregon margin.

zoic marine sedimentary and volcanic rocks accumulated in the area. Such accumulations of thick marine sedimentary rocks associated with pillow lavas generally are considered to be eugeosynclinal deposits.

In early Eocene time, the basin extended from Vancouver Island in the north to the Klamath Mountains in the south, a distance of 640 km (Snavely and Wagner, 1963). Young lava flows of the Cascade Range cover the eastern boundary; the western or seaward boundary cannot be defined and may have been subducted. A thick unit of early to middle Eocene age and consisting of tholeiitic basaltic pillows and breccia was extruded within the basin; locally a younger alkalic basaltic unit formed seamounts and islands (Snavely and others, 1968). This volcanic sequence may reach a thickness of 20,000 feet (6 km) near centers of volcanic activity, although the base of the sequence is not exposed and seismic data on crustal layering in the Coast Range are scarce. Basalts found in the lower unit of the Siletz River Volcanic Series in the central part of the Oregon Coast Range and in one of the Siletz River correlative units, the Crescent Formation of western Washington, have chemical compositions similar to oceanic tholeiitic basalts (Snavely and others, 1968). Additional analyses of trace and major elements of the Crescent Formation confirms the oceanic basalt affinities (Glassy, 1973).

Seismic refraction studies (Shor and others, 1968) show that an oceanic crustal velocity of 6.6 km/sec occurs below a depth of 10 km on the continental shelf off central Oregon (fig. 2, station G9-10) and northern California. Cenozoic sedimentary and volcanic rocks probably

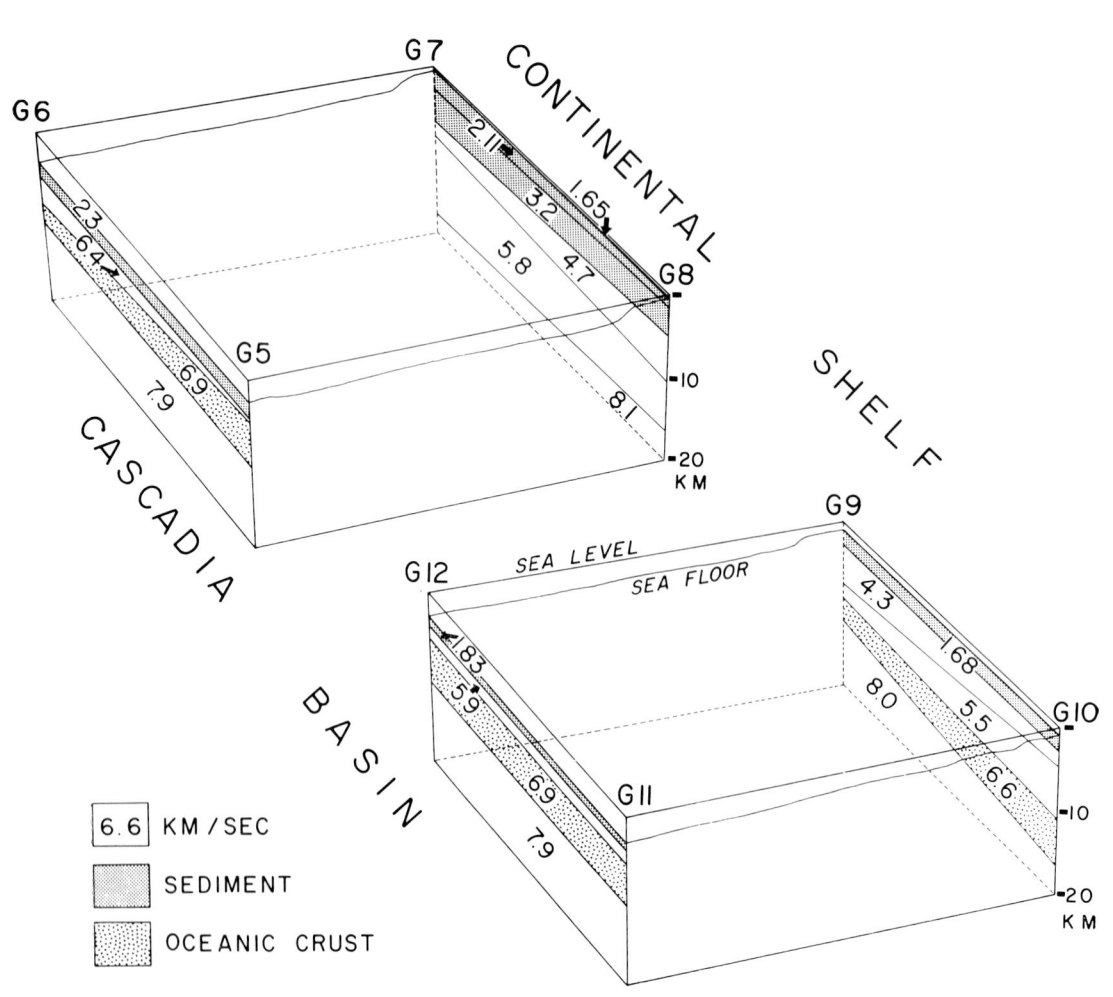

Fig. 2.—Seismic refraction sections in Cascadia Basin and adjacent continental shelf (data from Shor and others, 1968). Upper block is located off northern Oregon and southern Washington; lower block off central Oregon. Horizontal scale variable; vertical scale as indicated.

account for most of the upper 10 km. Relatively shallow mantle depths of 14 km (lat 44°30′ N) and 17 km (lat 46°30′ N) (fig. 2, stations G7-8 and G9-10) were found beneath the continental shelf (Shor and others, 1968). Mantle depths ranging from 16 km (Berg and others, 1966) to less than 20 km (Thiruvathukal and others, 1970; Dehlinger and others, 1971) were determined for western Oregon, whereas a mantle depth of about 40 km was found in eastern Oregon (Dehlinger and others, 1968). Tatel and Tuve (1955) showed a relatively shallow mantle depth of 19 km between Seattle and the coastal mountains and a much greater one, 30 km, east of Seattle. Seismic refraction studies by Johnson and Couch (1970) show that the crustal thickness increases from north to south and decreases toward the west in western Washington. On the basis of the 1965 Seattle earthquake, McKenzie and Julian (1971) suggested that a high-velocity layer dips beneath southwestern Canada and the northwestern United States.

In the Cascadia Basin, at the base of the continental slope, the mantle is at a depth of 10 to 11 km (fig. 2, stations G5-6 and G11-12), which is typical for normal oceanic crust. Because of the relatively thick sediments in Cascadia Basin, however, the area should exhibit a deep mantle, indicating regional compensation (Shor and others, 1968). According to Shor and others (1968), the shallow mantle may be related to the low-density upper mantle that exists beneath the Juan de Fuca Ridge and Cascadia Basin (Dehlinger and others, 1968, 1971). This low-density mantle apparently extends beneath the continental margin (shelf and slope) and has velocities similar to those found at the base of the continental slope in the Cascadia Basin.

Magnetic anomalies (Vine, 1968; Atwater and Menard, 1970) indicate that the oceanic basement near the continental slope is no older than late Miocene (8 million years) in the Cascadia Basin and early Pliocene (4 million years) in the Gorda Basin (fig. 3). Deep Sea Drilling Project (DSDP) site 174 is positioned over the 8-million-year anomaly; lower Pliocene sediments (4 to 5 million years) were recovered near acoustic basement (von Huene and others, 1971; Kulm, von Huene, and others, 1973). The asymmetrical magnetic anomalies about the spreading Gorda and Juan de Fuca Ridges suggests that a substantial portion of the Gorda-Juan de Fuca plate (Farallon plate) has been subducted beneath the North American plate (Silver, 1969b; Atwater, 1970).

SEDIMENTARY FRAMEWORK

The geologic history of the Cenozoic geosyncline of western Oregon and Washington was first summarized by Snavely and Wagner (1963). The material presented here on the geosyncline is summarized chiefly from their work. Reference is also made to the studies of Snavely and others (1964), Dott (1966), Lovell (1969), Braislin and others (1971), and Tiffin and others (1972). The sedimentary history of the Gorda-Juan de Fuca plate is known largely through the regional studies of Duncan (1968), Nelson (1968), McManus and others (1970), Carson (1971), von Huene and others (1971), and Kulm, von Huene, and others, 1973.

Geosyncline and Continental Margin

Early Cenozoic.—In early Eocene time, the thick volcanic sequence described above became intertongued complexly with marine siltstone and basaltic sandstone. Numerous volcanoes apparently contributed basaltic debris to the section. Continuental sediments were deposited along the northeastern edge of the geosyncline in Washington, and marine siltstone and sandstone were deposited to the west in the offshore regions.

Uplift occurred in the Klamath Mountains to the south and in the Vancouver Island area to the north in middle Eocene time. Immature sediments were eroded from the uplifted ancestral Klamath Mountains, deposited in deltas along the shore, and carried northward into the basin by turbidity currents. The Tyee Formation, which consists of arkosic, lithic, feldspathic, and volcanic wacke (fig. 4E, accounts for as much as 3 km of rhythmically bedded section in the geosyncline (Snavely and others, 1964; Lovell, 1969). Foraminifera in these turbidites suggest middle bathyal to deeper environments. As relief in the source areas diminished, silt and clay were deposited in the geosyncline (Snavely and others, 1964).

The geosyncline was separated into several smaller basins due to tectonism during middle late Eocene time (Snavely and Wagner, 1963). A major unconformity truncates older strata on the central Oregon shelf (Braislin and others, 1971, fig. 3). Volcanism was renewed (within the trough and on the eastern side), the volcanics intertonguing laterally with marine siltstone. Marine deposition continued in most of western Washington throughout the Eocene Epoch. Shale, sandstone, and conglomerate of late Eocene to Oligocene age are exposed along the west coast of Vancouver Island (Tiffin and others, 1972). Middle bathyal depths are indicated for these deposits, which are about 7 to 8 km thick on the continental shelf.

Middle Cenozoic.—Uplift occurred along the southern margin of the trough in the early Oligocene, restricting the size of the basin of

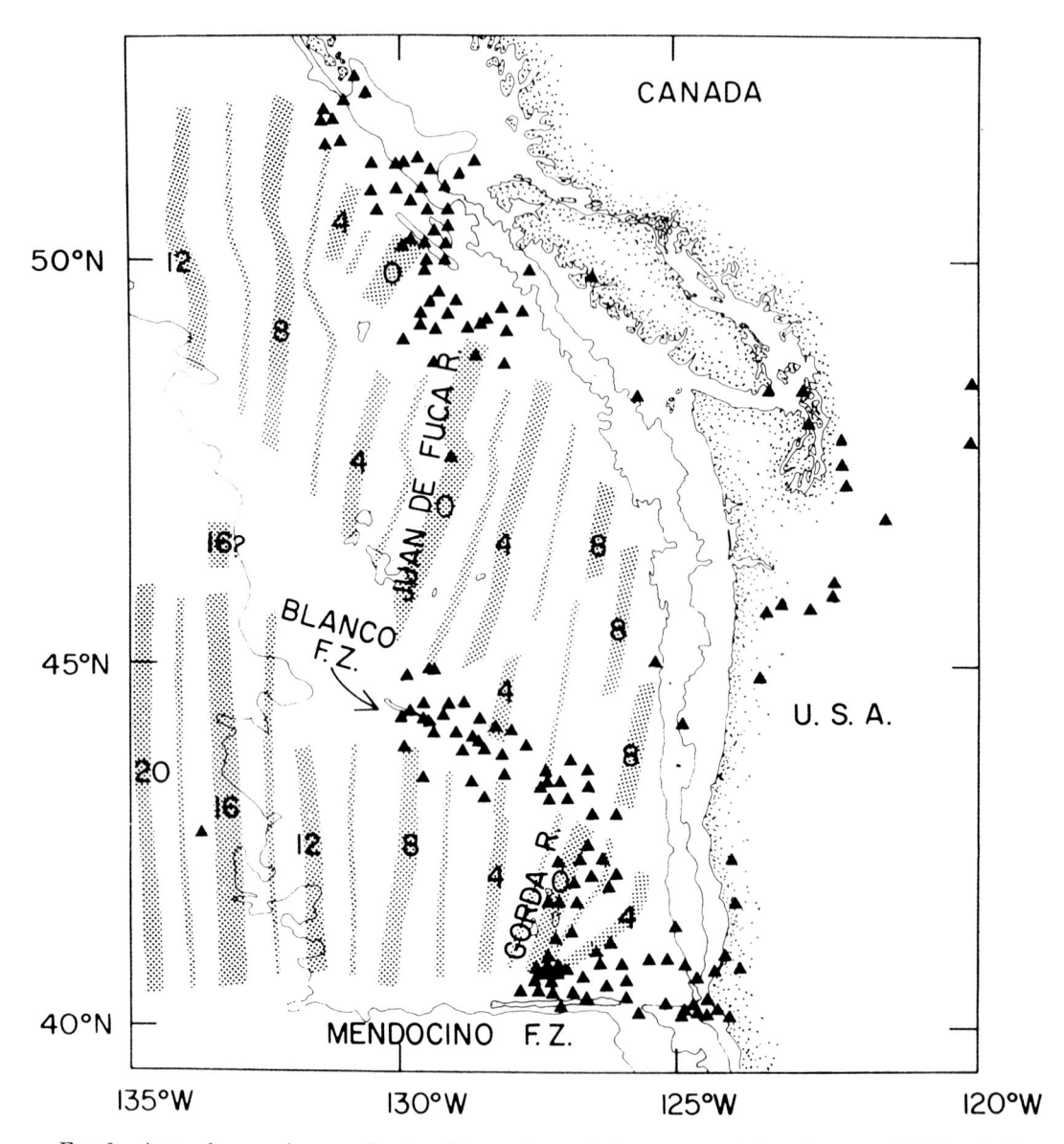

FIG. 3.—Ages of magnetic anomalies in millions of years before present (after Atwater, 1970, fig. 10A). Triangles represent earthquake epicenters.

FIG. 4.—Classification of sediments and sedimentary rocks (after Williams and others, 1954) on Oregon margin and in Cascadia Basin. A, Continental shelf sands off Oregon and southern Washington (Bushell, 1964; White, 1970; Spigai, 1971); B, Cascadia Basin sand turbidies (Duncan and Kulm, 1970; Carson, 1971); C, Mesozoic and Cenozoic rocks from Klamath Mountains (after Dott, 1965, fig. 3); D, Pliocene and Pleistocene sandstones from southern Oregon shelf and slope (Spigai, 1971); E, Tyee Formation (Eocene) from Oregon Coast Range (Snavely and others, 1964, fig. 6).

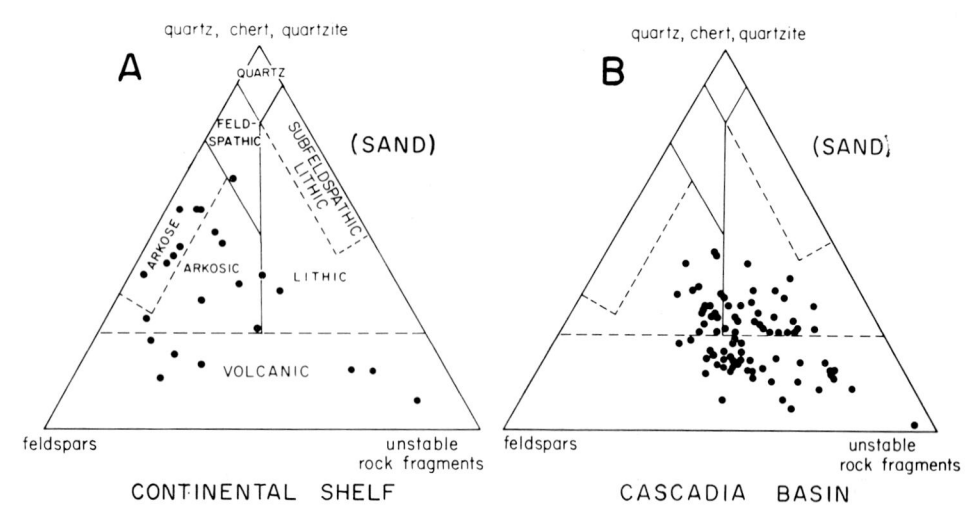

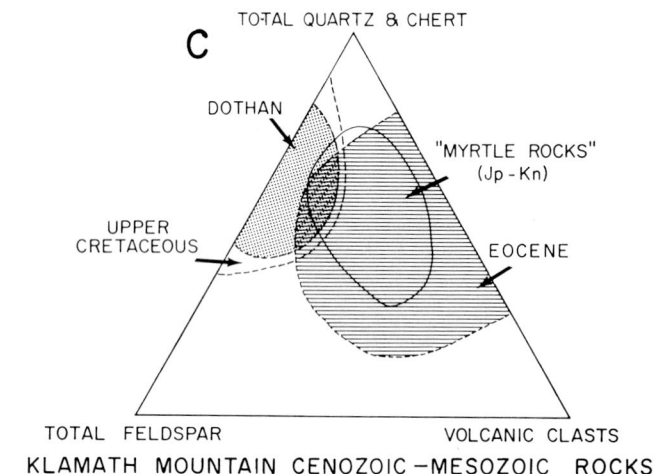

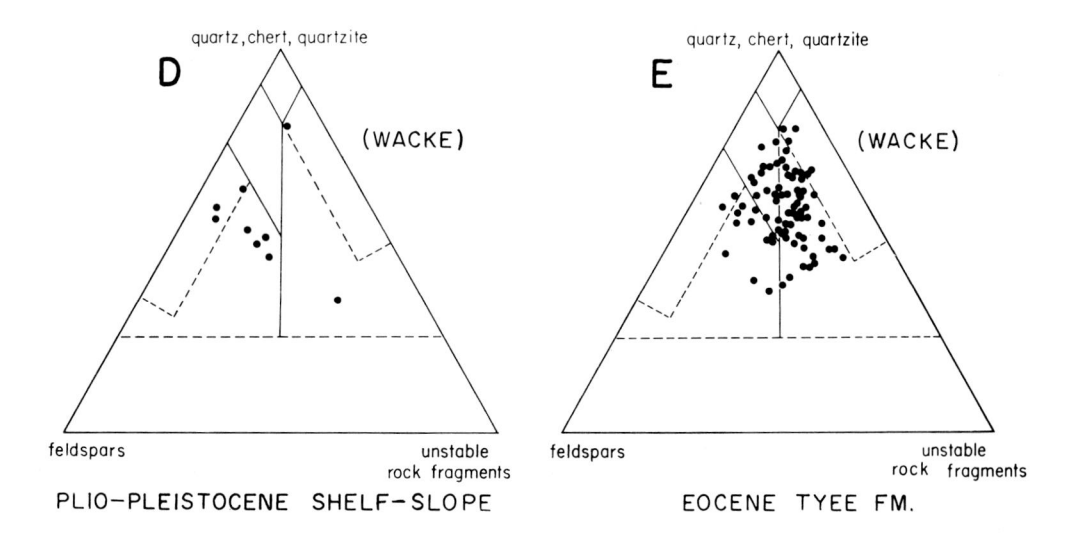

A CONTINENTAL SHELF

B CASCADIA BASIN

C KLAMATH MOUNTAIN CENOZOIC-MESOZOIC ROCKS

D PLIO-PLEISTOCENE SHELF-SLOPE

E EOCENE TYEE FM.

marine deposition (Snavely and Wagner, 1963). Large volumes of volcanic debris were deposited along with the marine sandstone and siltstone, indicating substantial volcanism in the ancestral Cascade Range to the east. On the north side of the present Olympic Mountains, more than 3 km of coarse to fine clastic debris was deposited but only minor amounts of volcanic material.

The Coast Range was uplifted, and thick grabbroic sills were emplaced in the late Oligocene. In Oregon, marine deposition was limited mostly to the area of the present continental shelf and western side of the Coast Range. Structural basins along the western flanks of the Coast Range and in the vicinity of the present continental shelf were the sites of continued marine deposition. Organic-rich siltstone and mudstone were deposited in marine basins on the Oregon margin during early Miocene time (Braislin and others, 1971). Relatively thick deposits of near-shore sandstone and siltstone characterize the upper lower and middle parts of the Miocene Series.

A period of uplift and erosion produced a major angular unconformity on the central Oregon cnotinental shelf in the middle Miocene. This unconformity is indicated in the seismic reflection records of the stratigraphic section across Heceta Bank (fig. 5, profile E) and in the offshore well drilled off Depoe Bay (Braislin and others, 1971, fig. 4). A middle to late Miocene unconformity is also present in western Washington (Weaver, 1937; Fowler, 1965), which suggests that the region was deformed on a broad front. Middle Miocene basalt from centers of volcanism near the present coastline of central and northern Oregon (Snavely and others, 1973) produced the large positive magnetic anomalies (Emilia and others, 1968) and the strongly reflecting acoustic basement (fig. 5, profile A) that occurs locally beneath the inner continental shelf.

Post-middle Miocene.—The thickest and most widespread marine deposits in the basins on the Oregon margin are late Miocene in age. Exploratory offshore wells (Braislin and others, 1971) and more than 200 rock samples collected from outcrops on the shelf by Oregon State University personnel show that the upper Miocene deposits are chiefly mudstones (sandy siltstone and claystone) and fine-grained sandstone (Muehlberg, 1971). Some of these deposits are diatomaceous (Fowler and others, 1971). Benthonic foraminifera indicate a bathyal environment of deposition. Miocene radiolarian and diatomaceous mudstone in the Tofino Basin off Vancouver Island were also deposited in bathyal environments (Tiffin and others, 1972). Another significant period of uplift and erosion is indicated in the Tofino Basin by a prominent early Pliocene unconformity.

Pliocene deposits are widespread on the continental shelf and consist mainly of massive siltstone and claystone. These thick deposits are exposed on the uplifted submarine banks (Nehalem and Heceta Banks, fig. 5, profiles A and E, respectively) on the Oregon shelf and occur in the shelf synclines and the anticlinal structures on the continental slope (fig. 5, profiles A and D). Pliocene deposits are at least 1.5 km thick on the shelf (Braislin and others, 1971) and may be thicker on the slope to judge from seismic reflection profiles.

Coarse-grained arkosic and lithic wackes (fig. 4D) of probable Pliocene and Pleistocene age have been dredged from the upper continental slope off southern Oregon (Spigai, 1971). These deposits are a more shallow-water facies than are the Pliocene siltstones recovered from the central and northern Oregon margin, and they were derived from the Mesozoic rocks of the Klamath Mountains (Spigai, 1971; compare C and D of fig. 4).

Seismic reflection records show that as much as 0.5 km of Pleistocene siltstone and sandstone may be present in the synclines above the youngest unconformity on the continental shelf (fig. 5, profile A). These rocks probably form the bulk of the sediment ponded on the continental slope.

Pliocene sand and silt turbidites and mudstones were dredged from a steep escarpment on the lower continental slope off central Oregon (fig. 5, profile D). These rocks have benthonic foraminiferal assemblages and lithologies characteristic of adjacent abyssal plain deposits, which indicates they have been uplifted from 500 to 1,200 meters above the adjacent abyssal plain.

Samples of Pleistocene silty clays and interbedded sand and silt turbidites in the lower part of the 233-m section were recovered from DSDP site 175 on the lower continental slope off Oregon (Kulm, von Huene, and others, 1973). The turbidites are similar to the abyssal plain turbidites found below the Astoria Fan at site 174. These lower slope deposits were uplifted 200 to 700 m above the adjacent abyssal plain 0.3 to 0.4 million years ago.

Gorda-Juan de Fuca Plate

Sedimentary wedge.—A thick wedge of sediment occurs at the base of the continental slope in an elongate basement depression that extends from the Mendocino Fracture Zone in the south to Paul Revere Ridge in the north (fig. 6). The wedge appears to be thickest near the apices of

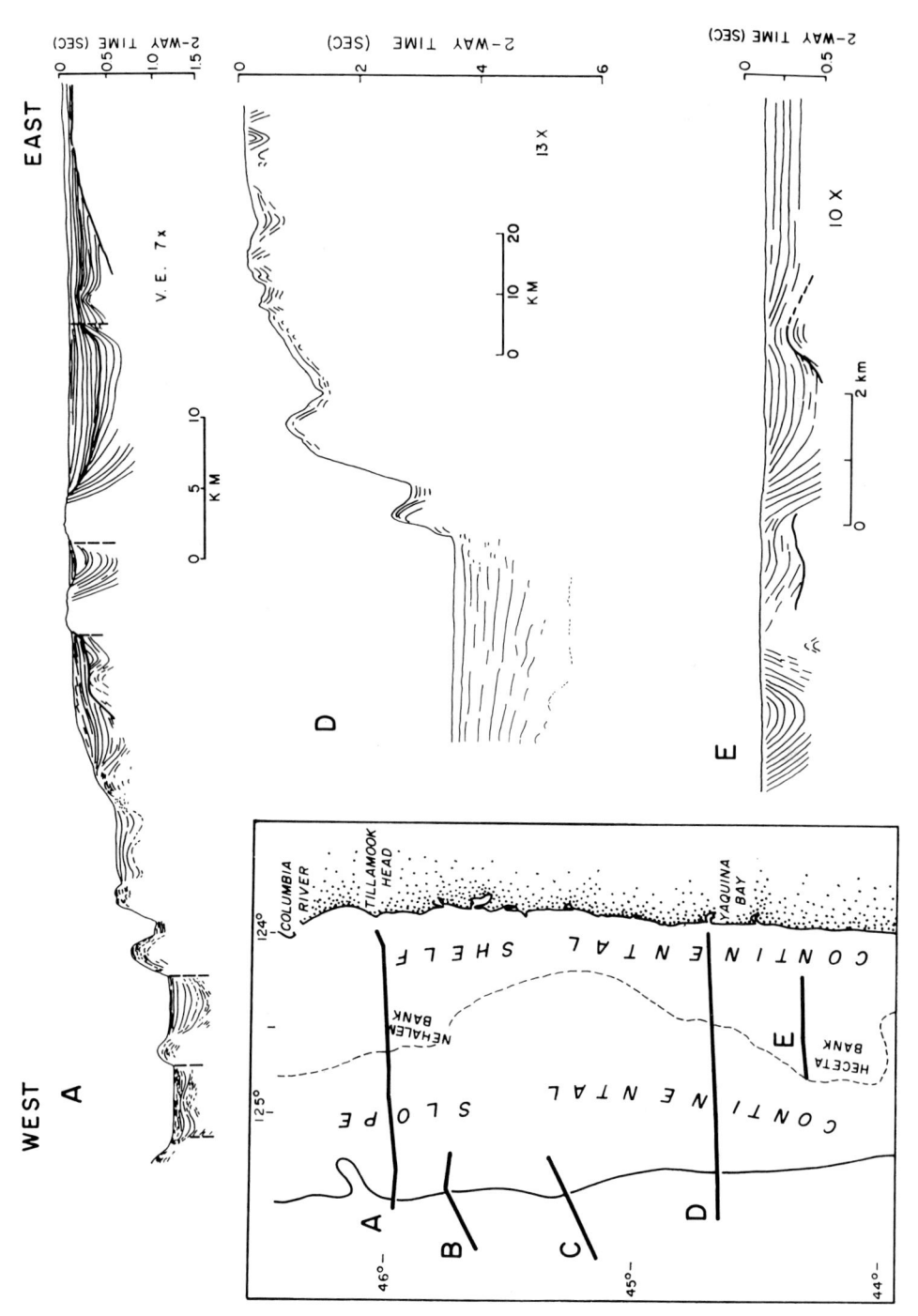

FIG. 5.—Line drawings of seismic reflection profiles from central and northern Oregon continental margin.

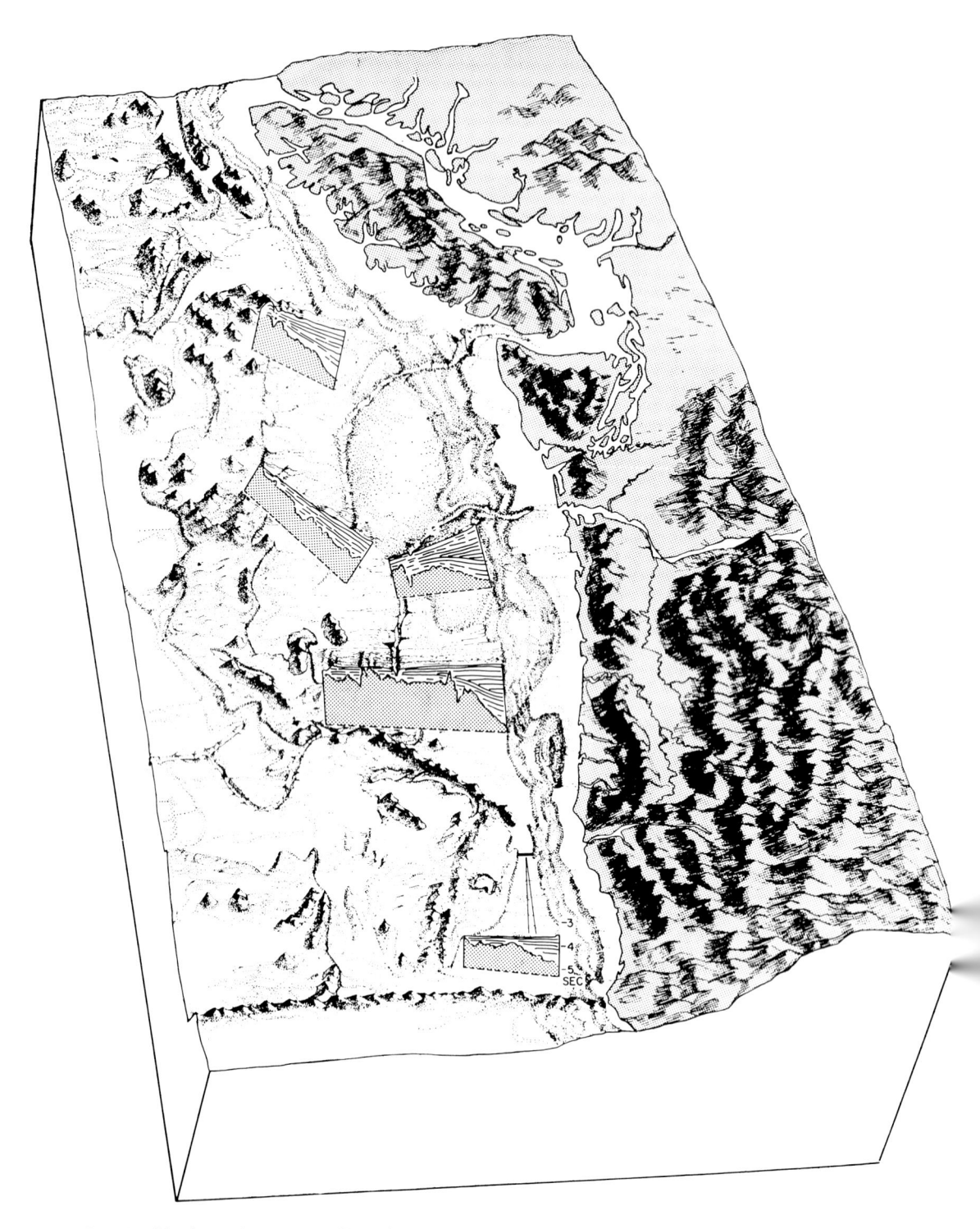

Fig. 6.—Physiographic map of Cascadia and Gorda Basins and surrounding features. Diagram drawn to scale according to bathymetry. Superimposed seismic reflection profiles drawn schematically to scale. Northern profile opposite Straits of Juan de Fuca from Ewing and others (1968, fig. 7), southern profile from Silver (1969a, fig. 7), and remaining profiles from Oregon State University collection.

the Astoria and Nitinat submarine fans and to thin westward away from these fans and toward the flank of Juan de Fuca Ridge and southward toward Gorda Ridge. Near the apex of the Astoria Fan, seismic refraction data show 2.5 km of sediment having a velocity of 2.3 km/sec (fig. 2, G5-6 and G11-12) (Shor and others, 1968). A similarly thick section of sediment is found on the Nitinat Fan (Ewing and others, 1968) and in the Winona Basin (Tiffin and others, 1972), which lies between Paul Revere Ridge and the edge of the continental shelf near the northern end of Vancouver Island (fig. 1). To the south, in the Gorda Basin, the thickest sediment accumulation occurs in an L-shaped trough along the eastern side (adjacent to the continental slope) and along the southern side (Silver, 1971a). The submarine fan, located in the southeastern part of the Gorda Basin and associated with the Eel Canyon, may be the thickest sedimentary section in the basin, but it was not penetrated acoustically (Silver, 1971a). Refraction data indicate about 1 km of sediment fill in the basin (Shor and others, 1968).

In general, a weakly reflective or nearly transparent acoustic layer is indicated immediately above acoustic basement in the several reflection profiles across the Cascadia and Gorda Basins (figs. 6 and 7). As the base of the continental slope off central Oregon is approached, this layer becomes more strongly reflective above acoustic basement.

Astoria Fan.—A seismic discontinuity occurs within the sedimentary wedge described in the vicinity of the Astoria Fan (Kulm and others, 1973). It can be traced from the eastern wall of Cascadia Channel to the base of the Oregon continental slope (figs. 6 and 7), where the discordance is not as apparent. Reflectors below the discontinuity dip toward the continental margin, and those above form a seaward transgressive sequence of Astoria Fan deposits, which is thickest at the base of the continental slope and which thins to the west and south. A complex sedimentary sequence is seen within the fan deposits that may have been produced by the shifting distributary channels on the Astoria Fan. A similar discontinuity was observed beneath the Nitinat Fan (Carson, 1971).

Pliocene and Pleistocene turbidites dominate the sedimentary section recovered from DSDP site 174 on the distal portion of the Astoria Fan (fig. 1) (Kulm, von Huene, and others, 1973). The fine- to medium-sand turbidites of the Astoria Fan overlie a thick section of silt turbidites that are typical of abyssal plain environments in the Cascadia Basin (Duncan, 1968). The fan deposits are middle to late Pleistocene (0.4 to 1.0 my) in age (von Huene and Kulm, 1973). Late Pleistocene sand turbidites also are common on the upper and middle parts of the Astoria Fan (Nelson, 1968) and no doubt account for the bulk of the deposits above the angular contact that separates the fan deposits from the underlying abyssal deposits.

Basin deposits.—From deep-sea drilling (Kulm, von Huene, and others, 1973) and seisfic reflection profiling (McManus and others, 1972; this study), it is clear that silt turbidites underlie most of the eastern side of the Cascadia Basin and probably a large part of the western side of the basin. Sand and silt turbidites are found in the western and northern parts of the basin close to the flank of the Juan de Fuca Ridge (Duncan, 1968; Carson, 1971). In fact, the northern end of the ridge near Vancouver Island has a thick sequence of turbidites in the axial valley (McManus and others, 1972). The coarse-grained components of the late Pleistocene and Holocene deposits in the Cascadia and Gorda Basins are principally lithic, arkosic, and volcanic sands (fig. 4B). These young, immature sediments consist mostly of such unstable grains as feldspar and rock fragments.

Terrigenous sedimentation rates were high (100 m/my) and masked the biogenic components throughout most of the Cascadia Basin (Duncan, 1968, even on the eastern flank of the Juan de Fuca Ridge. A muddy limestone was recovered from the bottom of the hole at site 174 (Kulm, von Huene, and others, 1973). This deposit probably formed on the spreading Juan de Fuca Ridge and was later covered by silt turbidites as the east flank of the ridge subsided. These early Pliocene pelagic and hemipelagic sediments contain clay minerals derived from submarine sources (Hayes, 1973).

Silt turbidites may produce the weakly reflective to nearly transparent layering effect seen in the seismic reflection records (fig. 7). It is doubtful that a thick, truly pelagic deposit exists beneath the thick part of the sediment wedge as Hayes and Ewing (1971) suggested. Some pelagic sediments may underlie the Pleistocene turbidites of the western part of the Cascadia Basin (Duncan, 1968), especially high on the flank of the Juan de Fuca Ridge, but apparently they are not very extensive. The poorly layered or transparent effect seen in the reflection records across the Gorda Basin (Silver, 1971a) may represent silt turbidites. If the Gorda Ridge has been in its present position, or nearly so, for the past 4 million years (the age of the basement near the base of the continental slope, fig. 3), we would not expect significant amounts of pelagic sediments to underlie the turbidites.

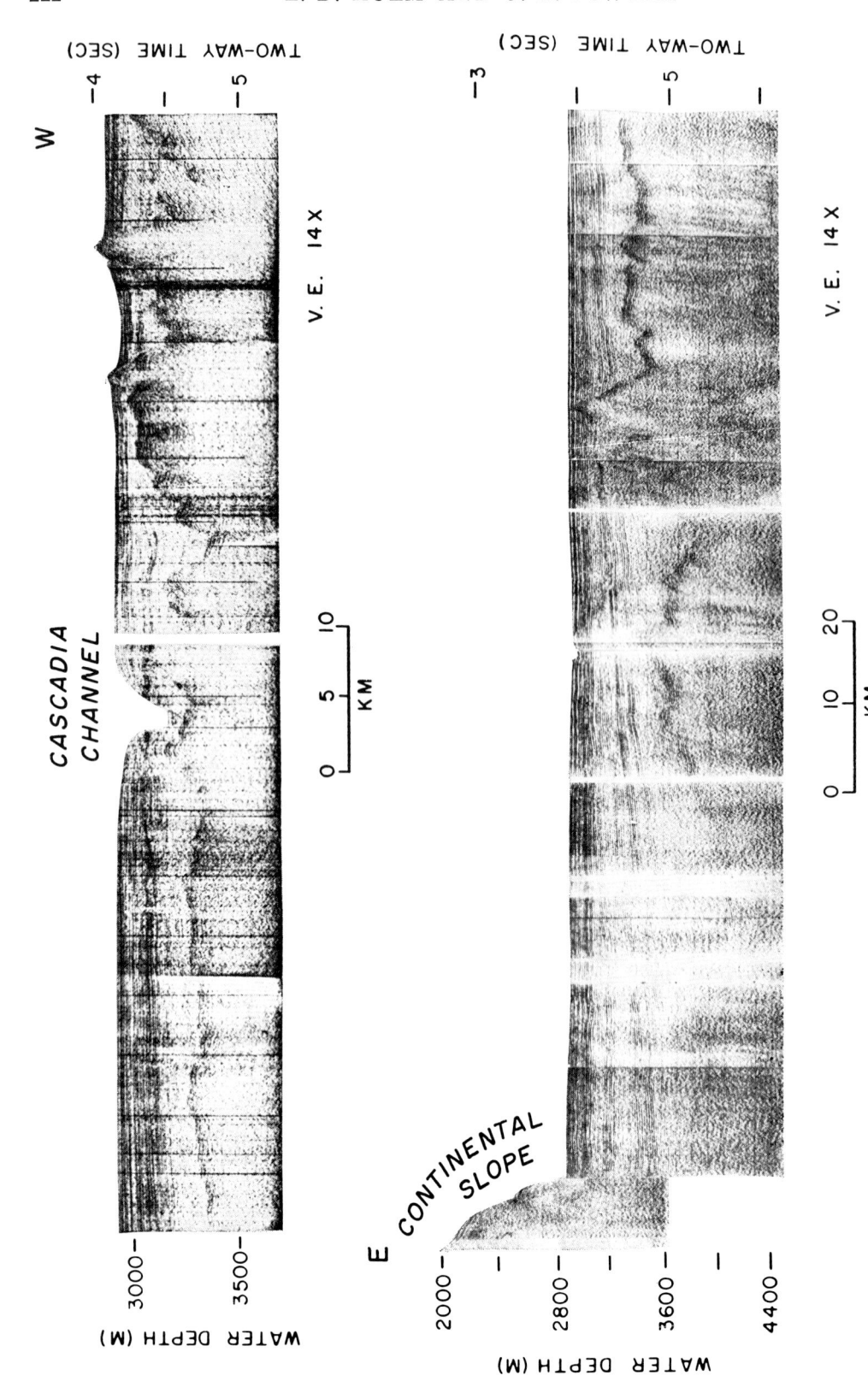

FIG. 7.—Seismic reflection profile across western Cascadia Basin and distal part of Astoria Fan (see fig. 1, thick solid line, for location).

Sediment dispersal.—Sediments originating on the continental margin are transported through the Cascadia Basin by turbidity currents flowing through Cascadia Channel (Duncan and Kulm, 1970; Griggs and Kulm, 1970); they pass through Cascadia Gap in the Blanco Fracture Zone and onto the Tufts Abyssal Plain to the southwest (fig. 1). Thick Columbia River sands were cored in a sediment-filled depression on the south side of the Blanco Fracture Zone on the Tufts Abyssal Plain (Duncan and Kulm, 1970). Cascadia Channel has served as an avenue of sediment transport since Pliocene time (Griggs and Kulm, in press), and apparently large volumes of clastic sediment have bypassed the Cascadia Basin.

Sediments of the Gorda Basin have been derived from the adjacent Klamath Mountains and Columbia River, principally during glacial periods when sands were transported down Astoria Canyon, across the Astoria Fan through fan channels, and then southward into the basin (Duncan and Kulm, 1970). Deep-sea drilling in the Escanaba rough, the median valley of Gorda Ridge, revealed a 390-m section of fine- to coarse-grained Pleistocene turbidites (McManus and others, 1970). Heavy mineral studies (Vallier, 1970) of these sands show that they were derived from both the Columbia River and Klamath Mountain drainage basins. Columbia River sands must have been transported through the Gorda Basin, around the south end of the Gorda Ridge, and into the trough. Semiconsolidated clayey sandstone, derived from the same two sources, occurs higher on the inner walls and has been uplifted 1,000 m above the floor of the trough by means of block faulting during the past 1 million years (Fowler and Kulm, 1970). The distribution of turbidites indicates that the Cascadia and Gorda Basins have been connected with respect to sediment transport during much of the Pleistocene Epoch, if not longer.

CONTINENTAL MARGIN STRUCTURE AND TECTONICS

Published studies of the shallow structure of the continental margin (e.g., Ewing and others, 1968; Bales and Kulm, 1969; Kulm and Bales, 1969; Carson and McManus, 1969; Bennett, 1969; Grim and Bennett, 1969; Silver, 1969a, 1971b, 1972; Kulm and Fowler, 1971; Fowler and Kulm, 1971; Wissmann, 1971; Carson, 1971; McManus and others, 1972; Tiffin and others, 1972) and the unpublished data of Oregon State University form the data base for the study described here. The shallow structure was determined mainly through the use of seismic records made in this area. Gravity data (Dehlinger and others, 1971) and magnetics (Emilia and others, 1968) are also used in the interpretation of the structure.

Continental Shelf

The continental shelf off northern California, Oregon, Washington, and Vancouver Island is characterized by anticlines and unconformities that indicate uplift and erosion and by synclines that are filled with thick sedimentary deposits (e.g., fig. 5). Diapirs pierce the youngest shelf deposits off Vancouver (Tiffin and others, 1972), Washington (Bennett, 1969; Grim and Bennett, 1969), and possibly off northern Oregon but appear to be most prominent north of the Columbia River. The diapir cores probably consist of hard mudstone that pierce thick lower density deposits. Elongate, large free-air gravity anomalies yielding negative values to 70 mgal on the shelf suggest deep sediment-filled basins (Dehlinger and others, 1971). Seismic refraction measurements (Shor and others, 1968) show a probable sediment thickness (velocities less than 4 km/sec) of about 6 km in the vicinity of the Columbia River (fig. 2) and much thinner sections near Heceta Bank (2 km) and Eureka, California (0.6 to 1.1 km). Rocks having seismic velocities intermediate between sediments and oceanic crust occur beneath the sediments and may be older, more indurated sedimentary rocks. Both seismic reflection and refraction data show the presence locally of a structural high beneath the outer edge of the shelf. Mantle depths beneath the outer shelf are also quite shallow (Shor and others, 1968).

The shallow structure of the Oregon continental shelf has been studied in detail by personnel of Oregon State University. More than 8,000 km of seismic reflection profiles are used to delineate the structure of the region. This shelf is characterized by a series of uplifted and faulted submarine banks (e.g., Heceta and Nehalem Banks, fig. 5), which occur near the outer edge, and by shallow synclinal basins on the middle and inner shelf. A broad synclinal basin is present on the inner shelf off northern Oregon, and smaller basins predominate to the south (Kulm and Fowler, 1971). At least two prominent unconformities are indicated from some seismic reflection profiles across the shelf (Kulm and Fowler, 1971). The youngest unconformable surface (Pliocene-Pleistocene) can be traced southward from the vicinity of Nehalem Bank to Heceta Bank and to the vicinity of Coquille Bank (fig. 1). Sediments beneath the unconformity range in age from early to late Pliocene in the north to late Miocene in the south. Rocks of probable Eocene and Oligocene age are also truncated by this surface near the shoreline. Pleistocene sediments occur above the unconformity. This unconformable surface was drilled (DSDP site 176) on the eastern flank of Nehalem Bank. Pleistocene sediments were found to overlie Pliocene shales, which were deposited in water at least 500 m deeper than present depth at the site (von Huene and others, 1971; Kulm, von Huene, and others, 1973). A period of regional uplift and erosion is inferred from the widespread Pliocene-Pleistocene unconformity and from the varying ages of sediments truncated by the unconformity.

An older unconformity is present deeper in the sedimentary section (fig. 5, profile E, west end), particularly on the central Oregon shelf. A few, late Miocene rock samples were collected above this surface but not below it. Braislin and others (1971) reported a significant, middle to late Miocene unconformity from a well drilled on the central Oregon shelf. An unconformity of pre-late Miocene age is also present on shore in western Washington (Fowler, 1965) and possibly in the circum-Pacific generally (Dott, 1969). This suggests a major period of uplift and erosion.

As discussed earlier, two prominent unconformities, (1) a post-Oligocene to pre-Miocene and (2) post-early Pliocene, are present on the continental shelf off Vancouver Island and indicate major periods of uplift and erosion, but they appear to be older than their possible counterparts on the Oregon shelf.

Rocks ranging in age from middle Miocene to Pleistocene were recovered from the folded and faulted submarine banks on the shelf; the oldest sedimentary rocks are found on Stonewall and Heceta Banks. On the basis of paleodepth determinations, a maximum uplift of 900 meters has been calculated for late Miocene and early Pliocene strata (Fowler, 1966; Byrne and others, 1966) and progressively smaller amounts of uplift for younger strata. Some downwarp is also noted on the outer shelf and upper slope.

Continental Slope

Folded and faulted structures are common on the continental slope. In some areas, sediment has been ponded behind anticlines. Marginal plateaus interrupt the slope gradient off northern California (Silver, 1969a), southern Oregon (Spigai, 1971), and near the southern end of Vancouver Island (Tiffin and others, 1972). Steep escarpments also are common on the continental slope, and downslope movement of sedimentary blocks takes place on these faulted surfaces (Tiffin and others, 1972).

A series of broad anticlinal folds and intervening synclinal basins or troughs form the lower continental slope off northern Oregon (fig. 5A); a similar set of structures exists off Washington (Silver, 1972), where the same morphology dominates. Off northern Oregon, some sediment ponds display essentially flat-lying sediments, whereas others show a deeper, underlying folded structure having a discordant contact between the two sedimentary sequences. Some of these anticlinal folds form small ridges within the basins.

It is difficult to ascertain from the profiles whether the deeper folded structures represent a preorogenic sedimentary sequence followed by post-orogenic sediment ponding or whether tectonism is continuing but is masked by the high sedimentation rates of late Pleistocene time. Continued tectonism is more likely in view of the results obtained from DSDP site 175, which was drilled in a narrow synclinal trough on the lowermost part of the continental slope. The deposits drilled here probably were uplifted above the adjacent abyssal plain during middle to late Pleistocene time (Kulm, von Huene, and others, 1973). As discussed previously, Pliocene abyssal plain turbidites were dredged from the steep scarps of these structures (fig. 5D), and the benthonic foraminiferal assemblages indicate that they have been uplifted as much as 1,200 meters. Other evidence of possible late Pleistocene deformation is seen in the reflection profiles across the continental slope-abyssal plain interface (fig. 8). In most reflection profiles across this boundary, there is little return from the seismic signal and the nature of the contact is uncertain. However, in one profile (fig. 8C), the seismic reflectors can be traced continuously from the Astoria Fan onto the continental slope. Smaller folds occur near the base of the slope, which suggests that the deformation may be continuing and new structures may be forming in the abyssal plain and accreting immediately adjacent to the lower slope. Silver (1972) inferred that similarly folded structures on the lowermost slope are younger than folded structures higher on the slope.

COMPARISON OF CASCADIA BASIN AND CENOZOIC GEOSYNCLINE

Depositional Basin

The present configuration (length and width) of the Cascadia Basin is generally similar to that of the early and middle Eocene geosyncline in western Oregon and Washington. The size and shape of the Cascadia Basin changed, however, throughout Cenozoic time because of the subduction of the oceanic plate along North America (Atwater, 1970). In cross section, the Cascadia Basin is characterized by a wedge-shaped sedimentary deposit that is similar to the deposits of late Cenozoic trenches. Cenozoic trench deposits have limited widths and thicknesses and wedge-shaped cross sections (Scholl and Marlow, this volume). In contrast, the early and middle Eocene geosyncline apparently was characterized by lens-shaped deposits.

Although a substantial amount of turbidite sediment has bypassed the Cascadia Basin via transport through Cascadia Channel, sediments are ponding at the southern end of the Cascadia Basin and filling its western part (Griggs and Kulm, in press). If subduction and downwarping of the oceanic plate ceased, the Gorda and Cascadia Basins would fill with approximately 4 km of sediment or to the height of topographic barriers, that is, the Blanco Fracture Zone and spreading ridges, that border the western and southern parts of the two basins (fig. 1). Cascadia Channel would still continue to funnel sediment from the basin through abyssal gaps in the Blanco Fracture Zone. The deposits of the two basins, nevertheless, would not reach the 7 to 8 km of sedimentary accumulation of the Cenozoic geosyncline without substantial downwarping of the ocean plate.

The 3-km thick, rhythmically bedded sandstone sequence (Tyee Formation) was deposited in the southern part of the Eocene geosyncline and interfingers with tuffaceous siltstone and volcanic material north of lat 45° N (Snavely and Wagner, 1963). These middle Eocene turbidites buried the preexisting volcanic highs as the Pliocene and Pleistocene turbidites did in the Cascadia Basin (fig. 7; McManus and others, 1972; Griggs and Kulm, in press). A relatively thin (perhaps 50 to 100 m) very calcareous hemipelagic or pelagic deposit underlies the Pliocene silt turbidites of the Cascadia and Gorda Basins and probably covers oceanic layer 2. Deposits of this nature are not readily apparent in the early Eocene sediments of the geosyncline, although chert beds and calcareous siltstone are found locally in the section (Snavely and Wagner, 1963). Benthonic foraminifera suggest that the Tyee Formation was deposited in water depths ranging from 1,500 m to possibly more than 2,000 m. The Pliocene and Pleistocene turbidites of the Cascadia Basin were deposited in water depths ranging from 2,500 m to more than 3,000 m. They contain an excellent paleobathymetric indicator species, *Melonis pompilioides,* which is presently found at depths greater than 2,200 or 2,300 meters (G. A. Fowler, unpublished research; Kulm, von Huene, and others, 1973).

The bulk of the sedimentary deposits both in the geosyncline and on the Gorda-Juan de Fuca plate is terrigenous and was derived from various sources. During Cenozoic time, the Klamath Mountains served as a source of sediment for portions of the geosyncline, for the Oregon continental margin, and for the Cascadia and Gorda Basins. As indicated by the apparent paucity of truly pelagic deposits (biogenic material or brown clay) on the plate and in the geosyncline, the depositional basins have always been within a few hundred kilometers of their terrigenous sediment sources. However, diatomaceous matter (biogenic opal) is common in the Miocene and Pliocene deposits at DSDP site 173 on the upper part of the Delgada Fan off northern California (Kulm, von Huene, and others, 1973) and in the Miocene mudstones on the Oregon shelf (Fowler and others, 1971). This indicates that coastal upwelling has been an important process during late Cenozoic time. During periods of increased rates of terrigenous sedimentation (e.g., Pleistocene), the biogenic content of the sediments was markedly reduced by dilution with terrigenous detritus (Kulm, von Huene, and others, 1973).

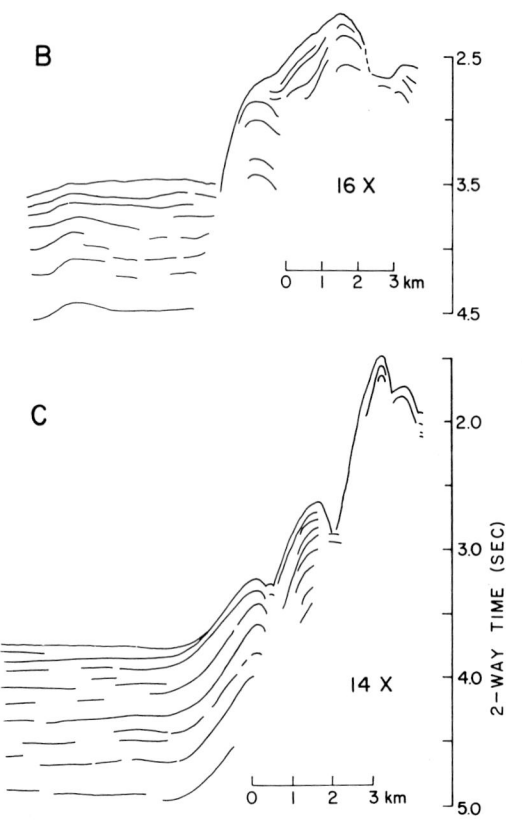

FIG. 8.—Seismic reflection profiles across interface of abyssal plain and continental slope off northern and central Oregon (from Kulm and others, 1973, figs. 5, 8) (see fig. 5 for locations of B and C profile lines.)

Volcanic Activity

The geosyncline was characterized by several periods of volcanism. Volcanic activity continued locally in two areas in the geosyncline from early to middle Eocene time; it was renewed on a larger scale during middle late Eocene time as the geosyncline was separated into smaller basins by tectonism (Snavely and Wagner, 1963). Thick gabbroic sills were emplaced during the Oligocene Epoch as the Coast Range was uplifted, and centers of volcanism developed near the present coastline during middle Miocene time.

Piston coring suggests that volcanic activity on the Gorda-Juan de Fuca plate has been limited to the crestal portions of the Gorda and Juan de Fuca Ridges. Cobb Seamount, which is located near the crest of the Juan de Fuca Ridge, has erupted several times during the past 3 to 4 my, producing volcanic ash layers in the surrounding sediments (Scheidegger, 1973). Ash from Mount Mazama occurs within the Holocene sediments of the Cascadia Basin, but it was carried into the basin by turbidity currents that originated on the continental margin (Nelson and others, 1968).

Neither seismic reflection profiling in or deep-sea drilling has produced evidence of large-scale volcanic activity (intrusives or lava flows) within the sediments of the Cascadia Basin. The clay mineral composition of early Pliocene deposits at site 174 suggests, however, that these deposits were derived in part from submarine volcanic sources (Hayes, 1973).

Although no direct evidence of midplate volcanism has been found yet on the older oceanic crust of the Gorda-Juan de Fuca plate, the possibility must be considered in view of recent studies. Luyendyke (1970) suggested that midplate volcanism, probably lava flows, may have occurred in the northeast Pacific. An 8.6-my-old tholeiitic basalt ridge discovered in the axis of the Peru Trench represents a piece of block-faulted oceanic crust from the underthrusting Nazca plate (Kulm and others, 1972 and 1973). Because the late Miocene tholeiitic basalt is younger than the Eocene oceanic crust at the eastern edge of the Nazca plate, midplate volcanism is suggested. If there is midplate volcanism on the Gorda-Juan de Fuca plate, it is probably associated with the basement faults (see fig. 3, left-lateral offset in magnetic anomalies) and is of limited extent.

Tectonism

Portions of the Gorda-Juan de Fuca plate and the continental margin are undergoing deformation. Most of the deformation is at the eastern boundary of the Gorda-Juan de Fuca plate, although large faults are present in the Gorda and Cascadia Basins (Silver, 1971b, 1972). The sediments of the plate are being folded against the lower continental slope, creating a series of ridges or a ridge-and-trough topography (Silver, 1969a, 1972; Kulm, von Huene, and others, 1973; Kulm and others, 1973). The deep-water silt and sand turbidites of the abyssal plain are being uplifted as much as 1,200 m and accreted to the continental margin.

The entire continental margin (shelf and slope) opposite the Gorda-Juan de Fuca plate is marked by folded, faulted, and truncated structures. Deformation is most intense on the uplifted submarine banks on the outer continental shelf. Several angular unconformities are present on the shelf, the most prominent and widespread being those of middle Miocene and Pliocene to Pleistocene in age. These unconformities truncate strata of different ages and lithologies and indicate major periods of uplift and erosion.

A widespread middle to late Miocene unconformity on the continental shelf may be related to the well-documented change in plate motion that took place 10 my ago (Le Pichon, 1968; Morgan, 1968; Ewing and others, 1968). This was the approximate time of significant uplift and erosion on the continental margin.. East-west compression and greater net subduction at the boundary of the Farallon and American plates probably caused the deformation observed on the margin opposite the oceanic plate.

The youngest unconformity on the Oregon continental shelf (Pliocene-Pleistocene) is probably less widespread than the older unconformities. It may be related to the internal breakup of the Gorda-Juan de Fuca plate into smaller plates, which results in greater net subduction of the small plate (Cascadia) opposite the Oregon margin (Silver, 1971c).

CONCLUSIONS

Previous studies indicate that the present Gorda-Juan de Fuca plate and the Cenozoic geosyncline are underlain by an oceanic crust of similar chemical composition. The plate has a typical oceanic crustal thickness (10 km), however, whereas the geosyncline has a crustal thickness (20 km) that is intermediate between that of the ocean and continents. Seven or 8 km of sediment and volcanics characterize the geosyncline, whereas only 2.5 km of sediment fills the trenchlike depression along the eastern edge of the plate.

Both the Gorda-Juan de Fuca plate and the geosyncline are dominated by terrigenous deposits that include turbidites, siltstone, and mudstone. These sediments are mineralogically immature and are classified as arkosic, lithic, feldspathic, and volcanic sands. Paleodepth determinations suggest that the middle Eocene turbidites were deposited in water depths nearly as great as those presently found in the Cascadia Basin. Pelagic deposits, highly calcareous siltstone, chert beds, and diatomaceous siltstone may be locally important in the deposits of the continental margin and geosyncline, whereas a relatively thin muddy limestone probably underlies the turbidites on the plate. Both depositional basins have been influenced by terrigenous sedimentation from nearby continental sources throughout most of their history.

Volcanism was important in the geosyncline, but apparently it became increasingly limited to the crestal portions of the spreading Gorda and Juan de Fuca Ridges. Although debris from local submarine volcanism was incorporated in the early Pliocene hemipelagic and pelagic deposits underlying the Cascadia Basin, no evidence of lava flows or intrusives within the basin deposits has been found. If midplate volcanism was important, its evidence cannot be recognized by means of the present techniques.

Widespread angular unconformities within the geosyncline and the continental margin deposits record several significant periods of tectonism. The late middle Miocene and Pliocene-Pleistocene unconformities on the margin may be related to changes in plate motion, which caused an increase in the rates of subduction along the continental margin. The middle to late Pleistocene sediments of the Cascadia Basin are being uplifted more than 1,000 m and accreted to the lower continental slope. Late Cenozoic uplift is also effective over much of the outer continental shelf and upper slope off Oregon.

ACKNOWLEDGMENTS

We thank R. Couch, R. H. Dott, Jr., D. A. McManus, K. F. Scheidegger, and P. D. Snavely, Jr., for their critical review of this manuscript. The views of the authors are not necessarily the same as some of the reviewers, and the authors take the sole responsibility for the contents of this paper.

Studies of the Oregon continental margin were sponsored by the U.S. Geological Survey, Office of Marine Geology. The National Science Foundation and the Office of Naval Research supported the unpublished work in the Cascadia Basin.

REFERENCES

ATWATER, T., 1970, Implications of plate tectonics for the Cenozoic tectonic evolution of western North America: Geol. Soc. America Bull., v. 81, p. 3513–3535.
———, AND MENARD, H. W., 1970, Magnetic lineations in the northeast Pacific: Earth and Planetary Sci. Letters, v. 7, p. 445–450.
BALES, W. E., AND KULM, L. D., 1969, Structure of the continental shelf off southern Oregon: Am. Assoc. Petroleum Geologists Bull., v. 53, p. 471.
BENNETT, L. C., 1969, Structural studies of the continental shelf off Washington (abs.): Am. Geophys. Union Trans., v. 50, no. 2, p. 63.
BERG, J. W., JR., AND OTHERS, 1966, Crustal refraction profile, Oregon Coast Range: Seismol. Soc. America Bull., v. 56, p. 1357–1362.
BRAISLIN, D. B., HASTINGS, D. D., AND SNAVELY, P. D., JR., 1971, Petroleum potential of western Oregon and Washington and adjacent continental margin: Am. Assoc. Petroleum Geologists Mem. 15, p. 229–238.
BUSHNELL, D. C., 1964, Continental shelf sediments in the vicinity of Newport, Oregon (master's thesis): Corvallis, Oregon State Univ.
BYRNE, J. V., FOWLER, G. A., AND MALONEY, N. M., 1966, Uplift of the continental margin and possible continental accretion off Oregon: Science, v. 154, p. 1654–1656.
CARSON, B., 1971, Stratigraphy and depositional history of Quaternary sediments in northern Cascadia Basin and Juan de Fuca abyssal plain, northeast Pacific Ocean (Ph.D. thesis): Seattle, Univ. Washington.
———, AND MCMANUS, D. A., 1969, Seismic reflection profiles across Juan de Fuca Canyon: Jour. Geophys. Research, v. 74, p. 1052–1060.
DEHLINGER, P., COUCH, R. W., AND GEMPERLE, M., 1968, Continental and oceanic structure from the Oregon coast westward across the Juan de Fuca Ridge: Canadian Jour. Earth Sci., v. 5, p. 1079–1090.
———, ———, MCMANUS, D. A., AND GEMPERLE, M., 1971, Northeast Pacific structure, in MAXWELL, A. E. (ed.), The sea: New York, John Wiley and Sons, v. 4, pt. 2, p. 133–189.
DOTT, R. H., JR., 1965, Mesozoic-Cenozoic tectonic history of the southwestern Oregon coast in relation to cordilleran orogenesis: Jour. Geophys. Research, v. 70, p. 4687–4704.
———, 1966, Eocene deltaic sedimentation at Coos Bay, Oregon: Jour. Geology, v. 74, p. 373–420.
———, 1969, Circum-Pacific late Cenozoic structural rejuvenation: implications for Sea floor spreading: Science, v. 166, p. 874–876.
DUNCAN, J. R., 1968, Late Pleistocene and post-glacial sedimentation and stratigraphy of deep-sea environments off Oregon (Ph.D. thesis): Corvallis, Oregon State Univ.
———, AND KULM, L. D., 1970, Mineralogy, provenance, and dispersal history of late Quaternary deep-sea sands in Cascadia Basin and Blanco Fracture Zone off Oregon: Jour. Sed. Petrology, v. 40, p. 874–887.
EMILIA, D. A., BERG, J. W., JR., AND BALES, W. E., 1968, Magnetic anomalies off the northwest coast of the United States: Geol. Soc. America Bull., v. 79, p. 1053–1061.

Ewing, J., Ewing, M., Aitken, T., and Ludwig, W. J., 1968, North Pacific sediment layers measured by seismic profiling, in Knopoff, L., Drake, C. L., and Hart, P. J. (eds.), The crust and upper mantle of the Pacific area: Am. Geophys. Union Geophys. Mon. 12, p. 147–173.

Fowler, G. A., 1965, The stratigraphy, foraminifera and paleoecology of the Montesano Formation, Grays Harbor County, Washington (Ph.D. thesis): Los Angeles, Univ. Southern California.

———, 1966, Notes on late Tertiary foraminifera from off the central coast of Oregon: The Ore Bin, v. 28, p. 53–60.

———, and Kulm, L. D., 1970, Foraminiferal and sedimentological evidence for uplift of the deep-sea floor, Gorda Rise, northeastern Pacific: Jour. Marine Research, v. 28, p. 321–329.

———, and ———, 1971, Late Cenozoic stratigraphy of the Oregon continental margin in relation to plate tectonics (abs.): Geol. America Abs. with Program, v. 3, p. 570–571.

———, Orr, W. N., and Kulm, L. D., 1971, An upper Miocene diatomaceous rock unit on the Oregon continental shelf: Jour. Geology, v. 79, p. 603–608.

Glassy, W. E., 1973, Part I. Geochemistry, metamorphism and tectonic history of the Crescent Volcanic Sequence, Olympic Peninsula, Washington. Part II. Phase equilibria in the prehnite-pumpellyite facies (Ph.D. thesis): Seattle, Univ. Washington.

Griggs, G. B., and Kulm, L. D., 1970, Sedimentation in Cascadia deep-sea channel: Geol. Soc. America Bull., v. 81, p. 1361–1384.

———, and ———, in press, Origin and development of Cascadia deep-sea channel: Jour. Geophys. Research.

Grim, M. S., and Bennett, L. C., 1969, Shallow sub-bottom geology of the Washington shelf off Grays Harbor (abs.): Am. Geophys. Union Trans., v. 50, p. 63–64.

Hayes, D. E., and Ewing, M., 1971, Pacific boundary structure, in Maxwell, A. E. (ed.), The sea: New York, John Wiley and Sons, v. 4, pt. 2, p. 29–72.

Hayes, J. B., Clay petrology of mudstones, in Kulm, L. D., von Huene, R., and others, 1973, Initial reports of the Deep Sea Drilling Project: Washington, D.C., U.S. Govt. Printing Office, v. 18, p. 903–914.

Horn, D. R., Horn, B. M., and Delach, M. N., 1970, Sedimentary provinces of the North Pacific, in Hays, J. D. (ed.), Geological Investigations of the North Pacific, Geol. Soc. America Mem. 126, p. 1–21.

Huene, R. von, and Kulm, L. D., 1973, Tectonic summary of leg 18, in Kulm, L. D., and Huene, R. von (eds.), Initial reports of the Deep Sea Drilling Project: Washington D.C., U.S. Govt. Printing Office, v. 18, p. 961–976.

———, Lathram, E. H., and Reimnitz, E., 1971, Possible petroleum resources of offshore Pacific-margin Tertiary basin, Alaska: Am. Assoc. Petroleum Geologists Mem. 15, p. 136–152.

———, and others, 1971, Deep Sea Drilling Project, Leg. 18; Geotimes, v. 16, p. 12–15.

Johnson, S. H., and Couch, R. W., 1970, Crustal structure in the north Cascade Mountains of Washington and British Columbia from seismic refraction measurements: Seismol. Soc. America Bull., v. 60, p. 1259–1269.

Kulm, L. D., and Bales, W. E., 1969, Shallow structure and sedimentation of upper continental slope off southern and central Oregon: a preliminary investigation (abs.): Am. Assoc. Petroleum Geologists Bull., v. 53, p. 472.

———, and Fowler, G. A., 1971, Shallow structural elements of the Oregon continental margin within a plate tectonic framework (abs.): Geol. Soc. America Abs. with Program, v. 3, p. 628.

———, and others, 1972, Deformation in the Peru Trench (abs.): ibid., v. 4, p. 570.

———, Prince, R. A., and Snavely, P. D., Jr., 1973, Site survey of the northern Oregon continental margin and Astoria Fan, in Kulm, L. D., von Huene, R., and others, Initial reports of the Deep Sea Drilling Project: Washington, D.C., U.S. Govt. Printing Office, v. 18, p. 979–987.

———, and others, 1973, Tholeiitic basalt ridge in the Peru Trench: Geology, v. 1, p. 11–14.

———, von Huene, R., and others, 1973, Initial reports of the Deep Sea Drilling Project: Washington, D.C., U.S. Govt. Printing Office, v. 18, 1077 p.

Le Pichon, X., 1968, Sea-floor spreading and continental drift: Jour. Geophys. Research, v. 73, p. 3661–3697.

Lovell, J. P. B., 1969, Tyee Formation: undeformed turbidites and their lateral equivalent: mineralogy and paleogeography: Geol. Soc. America Bull., v. 80, p. 9–21.

Luyendyk, B. P., 1970, Origin and history of abyssal hills in the northeast Pacific Ocean: ibid., v. 81, p. 2237–2260.

McKenzie, D. P., and Julian, B., 1971, The puget Sound, Washington earthquake and the mantle structure beneath the northwestern United States: ibid., v. 82, p. 3519–3524.

McManus, D. A., and others, 1970, Initial reports of the Deep Sea Drilling Project: Washington, D.C., U.S. Govt. Printing Office, v. 5, 827 p.

———, and others, 1972, Late Quaternary tectonics, northern end of Juan de Fuca Ridge (northeast Pacific): Marine Geology, v. 12, p. 141–164.

Morgan, W. J., 1968, Rises, trenches, great faults, and crustal blocks: Jour. Geophys. Research, v. 76, p. 1959–1982.

Muehlberg, G. E., 1971, Structure and stratigraphy of Tertiary and Quaternary strata, Heceta Bank, central Oregon shelf (master's thesis): Corvallis, Oregon State Univ.

Nelson, C. H., 1968, Marine geology of Astoria deep-sea fan (Ph.D. thesis): ibid.

———, and others, 1968, Mazama ash in the northeastern Pacific: Science, v. 161, p. 46–49.

Scheidegger, K. F., 1973, Volcanic ash layers in deep-sea sediments and their petrological significance: Earth and Planetary Sci. Letters, v. 17, p. 397–407.

Shor, G. G., Jr., and others, 1968, Seismic refraction studies off Oregon and northern California: Jour. Geophys. Research, v. 73, p. 2175–2194.

Silver, E. A., 1969a, Structure of the continental margin off northern California, north of the Gorda Escarpment (Ph.D. thesis): San Diego, Univ. California.

———, 1969b, Late Cenozoic underthrusting of the continental margin off northernmost California: Science, v. 166, p. 1265–1266.

———, 1971a, Tectonics of the Mendocino triple junction: Geol. Soc. America Bull., v. 82, p. 2965–2978.
———, 1971b, Transitional tectonics and late Cenozoic structure of the continental margin off northernmost California: *ibid.*, p. 1–22.
———, 1971c, Small plate tectonics in the northeastern Pacific: *ibid.*, p. 3491–3496.
———, 1972, Pleistocene tectonic accretion of the continental slope off Washington: Marine Geology, v. 13, p. 239–249.
SNAVELY, P. D., JR., AND WAGNER, H. C., 1963, Tertiary geologic history of western Oregon and Washington: Washington Div. Mines and Geology Rept. Inv. 22, 25 p.
———, ———, AND MACLEOD, N. S., 1964, Rhythmicbedded eugeosynclinal deposits of the Tyee Formation, Oregon Coast Range: Kansas Geol. Survey Bull. 169, p. 461–480.
———, MACLEOD, N. S., AND WAGNER, H. C., 1968, Tholeiitic and alkalic basalts of the Eocene Siletz River volcanics, Oregon Coast Range: Am. Jour. Sci., v. 266, p. 454–481.
———, ———, AND ———, 1973, Miocene tholeiitic basalts of coastal Oregon and Washington and their relation to coeval basalts of the Columbia Plateau: Geol. Soc. America Bull., v. 84, p. 387–424.
SPIGAI, J. J., 1971, Marine geology of the continental margin off southern Oregon (Ph.D. thesis): Corvallis, Oregon State Univ.
TATEL, H. E., AND TUVE, M. A., 1955, Seismic exploration of a continental crust, *in* POLDERVAART, A. (ed.), Crust of the earth (a symposium): Geol. Soc. America Special Paper 62, p. 35–50.
THIRUVANTHUKAL, J. V., BERG, J. W., JR., AND HEINRICHS, D. F., 1970, Regional gravity of Oregon: *ibid.*, Bull. v. 81, p. 725–738.
TIFFIN, D. L., CAMERON, B. E. B., AND MURRAY, J. W., 1972. Tectonics and depositional history of the continental margin off Vancouver Island, British Columbia: Canadian Jour. Earth Sci., v. 9, p. 280–296.
VALLIER, T. L., 1970, The mineralogy of some turbidite sands from sites 32 and 35, *in* MCMANUS, D. A., and others, Initial reports of the Deep Sea Drilling Project: Washington, D.C., U.S. Govt. Printing Office, v. 5, p. 535–539.
VINE, F. J., 1968, Magnetic anomalies associated with mid-ocean ridges, *in* PHINNEY, R. A. (ed.), The history of the earth's crust: Princeton, New Jersey, Princeton Univ. Press, p. 73–89.
WEAVER, C. E., 1937, Tertiary stratigraphy of western Washington and northwestern Oregon: Seattle, Washington Univ. Pub. Geology, v. 4, 266 p.
WHITE, S. M., 1970, Mineralogy and geochemistry of continental shelf sediments off the Washington-Oregon coast: Jour. Sed. Petrology, v. 40, p. 38–54.
WILLIAMS, H., TURNER, F. J., AND GILBERT, C. M., 1954, Petrography; an introduction to the study of rocks in thin sections: San Francisco, Freeman and Co., 406 p.
WISSMANN, G., 1971, Marine geophysical studies, offshore Cape Flattery, Washington and the implications for regional tectonics (master's thesis): Seattle, Univ. Washington.

SEDIMENTATION WITHIN AND BESIDE ANCIENT AND MODERN MAGMATIC ARCS

WILLIAM R. DICKINSON
Stanford University, Stanford, California

ABSTRACT

Rocks formed along magmatic arcs at convergent plate junctures are important components of eugeosynclinal terranes. Arc complexes include not only surficial volcanic accumulations and plutonic intrusions emplaced in the metamorphic roots of the arcs. Also important are clastic strata composed of debris dispersed from the igneous axes of the arcs by eruption or erosion and deposited in basins and troughs within or beside the arc trends. Magmatic arcs include both marginal volcanic chains along or near the edges of continental landmasses and volcanic island chains flanked on the side away from the trench either by shallow epicontinental seas or by deep oceanic areas that commonly were formed as interarc basins. Intra-arc basins and troughs are largely fault-controlled depressions, although environments of deposition range from varied terrestrial and shallow marine settings to deep marine waters where turbidites are dominant. Coarse air-fall and ash-flow pyroclastic strata, pillow breccias and aquagene tuffs, and subaqueous ash-flow deposits are distinctive facies common for intra-arc sequences. Thick prisms and wedges of clastic strata commonly accumulate along or near the flanks of magmatic arcs. On the trench side, elongate traps for sediment prisms within the arc-trench gaps include terrestrial valleys and sloping plains, subsiding shelves and slopes, and deep bathymetric benches and troughs. On the other side, in the rear of the arc, clastic wedges may be spread either as turbidite aprons built into a deep interarc basin lying beyond a rifted arc rear and bounded on the far side by a remnant arc or as riverine and deltaic plains built into a foreland basin lying beyond an arc-rear thrust belt and bounded on the far side by a cratonic mass. The history of salient arc-trench systems suggests that evolution through time typically involves coordinate widening of the arc-trench gap, the magmatic arc structure itself, and the interarc basin or thrust belt behind the arc.

ARCS AND GEOSYNCLINES

The term *magmatic arc* refers here to the curvilinear volcanic belts that stand roughly parallel to the subduction zones of trenches where oceanic lithosphere is consumed. The arc volcanism is typified by commonly explosive eruptions of intermediate magmas to form stratified piles of interbedded lavas and volcaniclastic rocks. The bulk composition is characteristically andesitic or dacitic, although basaltic lavas or rhyolitic ignimbrites may be abundant or even dominant locally. Presumably, the genesis of the magmas is related to the descent of slabs of lithosphere into the mantle beneath the arcs along sloping paths marked by the inclined seismic zones located at intermediate focal depths beneath the arcs. Geographically, magmatic arcs include both island chains, along which parts of the volcanoes are submerged below sea level, and chains of entirely subaerial volcanoes built near continental margins. By inference, the igneous rocks of magmatic arcs include not only the volcanic rocks erupted at the surface, but also cogenetic intrusive rocks injected into the crust beneath the volcanic belt. These underlying igneous bodies probably include large granitic plutons, with gabbroic and dioritic associates, as well as small hypabyssal dikes and sills (e.g., Dickinson, 1970). The host rocks for the intrusions include foundered lower levels of the overlying volcanic pile of the arc itself (Hamilton and Myers, 1967). Other wall rocks may represent remnants either of oceanic or continental crust older than any igneous rocks of the magmatic arc, or they may represent remnants of ophiolitic mélanges and underthrust slabs accreted by subduction to the flank of the arc structure during the timespan of magmatic activity. A complete arc structure is thus a volcanic and plutonic complex or orogen that includes: (a) the volcanic and volcaniclastic cover, (b) igneous roots formed by plutonic intrusions, and (c) metamorphic rocks incorporated into the roots surrounding the intrusions.

Kay (1944, 1947, 1951) showed plainly that classic eugeosynclinal terranes include voluminous assemblages of strata that represent ancient island arcs, but the terms "magmatic arc" and "eugeosyncline" are not synonymous. "Eugeosynclinal" implies deep subsidence in a belt having some kind of volcanism, but "magmatic arc" implies a certain type of volcanism regardless of amount of subsidence. On the one hand, the volcanic and volcaniclastic sequences of some magmatic arcs on continental margins may be thin terrestrial sequences for which the term "eugeosynclinal" may seem inappropriate. On the other hand, eugeosynclinal assemblages include not only magmatic arc complexes, but also oceanic sequences of pelagites and turbidites overlying ophiolitic slabs, as well as ophiolitic mélanges derived from oceanic sequences by deformation and metamorphism in subduction zones associated with trenches. As Hamilton (1966) indicated, the petrology of the associated volcanic strata can be used to distinguish between trench-type and arc-type eugeosynclinal complexes, although metasomatic alterations of the volcanic rocks and inherent ambiguities in the compositions of basalts from the two settings may frustrate clear distinctions in some cases. Additional criteria (Monger and Ross, 1971) include: (a) the abundance of volcaniclastic strata in magmatic arc complexes as opposed to the dominance of lava flows in sections of old oceanic crust; (b) the common occurrence of local unconformities within arc-type sequences; and (c) the characteristic occurrence, in ocean-floor sequences, of ribbon cherts and other pelagite-argillite strata undiluted by the floods of clastic debris common in arc-type sequences.

In this paper, I discuss depositional sites and processes within magmatic arcs, between arcs and trenches, and immediately behind arc structures. I do not treat the associated trench sites and processes, but this omission should not obscure the fact that the arc-trench system as a whole is an integrated tectonic entity.

TYPES OF MAGMATIC ARCS

Magmatic arcs include two broad structural classes, intraoceanic and continental margin (figs. 1, 2). Intraoceanic arc structures stand as elongate dividers of relatively thick crust sepa-

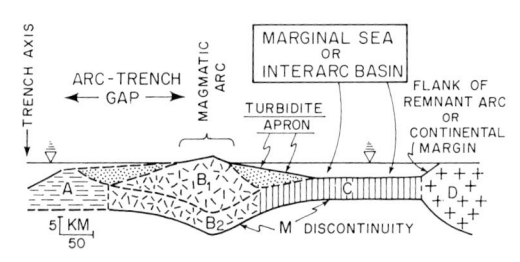

FIG. 1.—Diagrammatic transverse crustal section though idealized intraoceanic magmatic arc: A interpreted here as mélanges and associated slabs of oceanic ophiolites and turbidites of subduction zone, a progressively growing crustal element beneath trench slope break (Dickinson, 1973) at top of inner wall of trench; B is arc structure; C is oceanic crust; and D is edge of continental crust or semicontinental crust of a remnant arc. Arc structure is shown with upper and lower parts: B_1 is volcanic and plutonic igneous and metaigneous rocks produced by arc magmatism, together with derivative sedimentary and metasedimentary strata of foundered intra-arc basins; and B_2 is metamorphosed oceanic crust, possibly also intruded or partly melted, lying beneath welt built by arc magmatism. Stipples indicate prisms of derivative sedimentary strata flanking the arc.

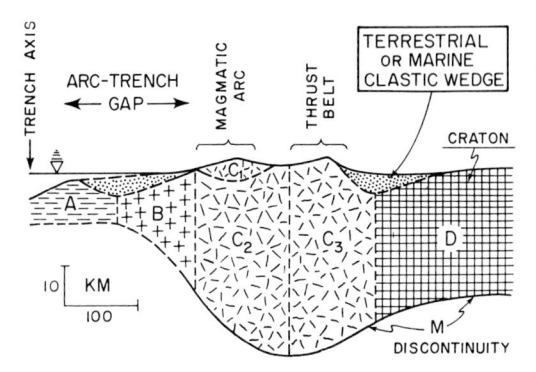

FIG. 2.—Diagrammatic transverse crustal section through an idealized continental-margin arc: A and stippled sedimentary prisms as in figure 1; B is transitional crust at ocean-continent interface; C is arc structure with thick root; and D is continental crust of cratonic interior. Arc structure is shown in three parts: C_1 is volcanic and volcaniclastic cover; C_2 is metamorphosed and perhaps partly melted continental crust injected by plutons of magmatic arc; and C_3 is tectonically thickened(?) and partly metamorphosed(?) continental crust beneath back-arc thrust belt. The clastic wedge behind the arc is built into a pericratonic or foreland basin.

rating broad areas underlain by thinner oceanic crust in the regions to either side, regions, that is, beyond the trench and behind the arc. Continental-margin arc structures, having crust of continental thickness in the region behind the arc, are built either along the edge of old sialic basement or upon accretionary prisms of mélange belts and other oceanic elements stacked tectonically against a continental margin. The contrast between the crustal structure in backarc areas of intraoceanic and continental-margin arcs implies fundamentally different tectonic histories for the two cases. Some intraoceanic arcs reach intraoceanic positions only as a result of the growth of the ocean basin of a marginal sea in the backarc area during the timespan of magmatic activity in the arc (e.g., Karig, 1970, 1971, 1972; Packham and Falvey, 1971). Backarc spreading of this type, including bodily migration of the arc structure away from the continental block, is precluded during the tectonic history of arc structures that remain welded to the edges of continental blocks. The common occurrence of contractional thrust belts in the backarc areas of continental-margin arcs, as in Sumatra (Van Bemmelen, 1949, p. 23–25) and the central Andes (Sonnenberg, 1963; Ham and Herrera, 1963), suggests the opposite tendency for the arc structure to be jammed against the continental block in the backarc area. In other instances, arc-trench systems may be initiated across oceanic regions, rather than along continental margins, but the relative importance of this process for the development of intraoceanic arcs, as opposed to lateral migration of the arc structure away from a continental margin, is not yet resolved.

Intraoceanic arcs include two main structural variants, those in which the crust beneath the arc has ordinary continental thickness and those in which the crust is thinner. In the first case, island landmasses are large and crustal thicknesses reach 30 to 35 km, as in Japan (Sugimura, 1968). The upper magmatic levels of the Japanese arc structure are likely underlain by a lower tier of sialic basement representing remnants of a sliver that was detached from the margin of the nearby continental block as migration of the arc structure accompanied the formation of the Sea of Japan (Matsuda and Uyeda, 1971). In the second case, the arc structure may lack any elements of sialic crust older than the igneous rocks of the arc. Markhinin (1968) has argued that all the sialic crust in the Kurile Islands, where the total crustal thickness is 15 to 25 km, was constructed by arc magmatism within roughly 100 my. A similar interpretation seems feasible for other largely submerged island arcs like Tonga-Kermadec and the Marianas, where total crustal thicknesses scarcely exceed 15 km (Shor and others, 1971, fig. 3; Murauchi and others, 1968, fig. 3). Unfortunately, thin crust beneath an intraoceanic arc does not preclude evolution of the arc-trench system by migration from a position marginal to a continental block. The opening of marginal seas behind migrating arcs apparently can proceed in distinct episodes involving the successive splitting of arc structures to form a series of discrete interarc basins that separate remnant arcs calved in sequence from the rear of the evolving frontal arc (Karig, 1972). The final intraoceanic arc developed by this process need contain no substance of the ancestral arc in existence at the start of the chain of migratory events.

Continental-margin arcs include two main geographic variants, those in which the volcanic chain forms islands and those in which the volcanic chain is a contiguous part of the continental landmass. In the first case, the backarc area is occupied by shallow seas like the Sunda Shelf behind Sumatra and Java. In the second case, the backarc area is occupied by terrestrial lowlands like the Amazon Basin behind the Andes. The crustal thickness of 50 to 75 km beneath the central Andes is much greater than under cratonic regions, and the excess crustal volume may be a rough measure of the magmatic materials added by volcanic and plutonic processes along the arc trend (James, 1971, figs. 6–7, p. 3340).

Volcaniclastic debris is the common component in sedimentary detritus shed from highlands extending along all types of magmatic arcs. Sediment dispersed widely as airborne ash of rhyolitic or dacitic composition is significant, but volcaniclastic sediment dispersed by surficial reworking of more localized but easily eroded pyroclastic deposits of andesitic or dacitic composition may be more important volumetrically. Erosion of andesitic or basaltic lavas is probably subordinate, for arc eruptions are commonly explosive and the flanks of arc volcanoes are typically mantled at any given time by a loose blanket of ash and ashy colluvium (e.g., Hay 1959a). In any case, the volume of volcaniclastic debris dispersed from sources along the arc trend can be greater than the net volume of the volcanoes standing along the arc trend at any given time.

Uplifts may permit erosion to bite deep enough into the volcanic and plutonic complex of the arc to expose granitic plutons from which quartzose and feldspathic sand can be derived (Dickinson, 1970). This effect is probably most common for continental-margin arcs having deep roots capable of supporting an isostatic rise of the arc structure, and the effect may be suppressed for intraoceanic arcs underlain by thin crust. Other supplemental sources of sediment derived from the arc itself or closely related tectonic elements include metamorphic rocks that encase the plutons of the arc roots, uplifted mélanges of accretionary subduction zones on the trench side of the arc, and sedimentary or metamorphic rocks uplifted in the thrust belts that lie behind some arc structures.

ARC SEDIMENTATION

Sedimentary strata that are largely but not exclusively of volcanic derivation and represent a variety of depositional environments occur within magmatic arc assemblages. Near volcanic centers, sediments are interstratified with both lavas and pyroclastic rocks, but volcaniclastic and related sediments may dominate columns that are hundreds or thousands of meters thick even though they are only a few kilometers or tens of kilometers from eruptive centers. For example, in the northern Sierra Nevada the Sierra Buttes and Taylor Formations of the metamorphosed upper Paleozoic pyroclastic sequence of McMath (1966) grade within 50 km along strike from more than 5 km of mainly lavas and coarse pyroclastics near Gold Lake to less than 2.5 km of mainly volcaniclastic turbidites near Indian Creek.

Three genetic types of facies can be recognized within arc assemblages: (1) central facies of eruptive centers bounded in special cases by transitional proximal facies marking a fringe of distinctive environments, such as biogenic reefs on the flanks of partly submerged volcanoes; (2) dispersal facies as aprons or blankets of sediment deposited by transport systems moving debris away from arc sources; and (3) basinal facies deposited in terminal sites of accumulation within arc structures. All three types of facies can be uplifted and eroded during the evolution of a magmatic arc, and unconformities which do not reflect any fundamental change in tectonic patterns are common within intra-arc sequences. On the other hand, basinal facies are not the only ones preserved, for bulk subsidence of the arc structure during its evolution may allow central and dispersal facies to founder and be covered by younger strata. Completed arc assemblages consist of superimposed volcanic and volcaniclastic lenses, each representing a packet of intertonguing facies related to successive eruptive centers (e.g., Schau, 1970). In some young sequences, the concentric arrangement of intra-arc facies in ovate patterns enclosing the central facies can be observed (e.g., Dickinson, 1968), but in most orogenic belts, where beds dip steeply, only a cross-sectional display of the facies gradations along one transect can be seen (e.g., Donnelly, 1966, fig. 6).

The character of central and proximal facies reflects chiefly modes of eruption, which are governed primarily by the explosiveness of the magmas and by the position of the surficial vents with respect to sea level. On both counts, continental-margin arcs and intraoceanic arcs present two possible extremes of behavior. For continental-margin arcs, most eruptions are subaerial, the proportion of silicic ignimbrites erupted can be appreciable, and typical stratovolcanoes are towering mountains having thick piedmont aprons of slurry-flood breccias (Swanson, 1966) and broad fans of reworked ash (Hay, 1959b). Deep dissection may expose plutonic components of the arc complex in close proximity to eroding volcanoes even as eruptions continue (Cobbing and Pitcher, 1972). For intraoceanic arcs, however, many eruptions may be submarine, the proportion of mafic lavas erupted can be high, and typical stratovolcanoes present only their tips as islands, which are fringed either by reefs built up from their flanks or by narrow shelves notched into their flanks. Pillow lavas, pillow breccias, and aquagene tuffs (Carlisle, 1963) may be more abundant than lava flows and vitroclastic airborne ash deposits, and submarine ash flows (Fiske and Matsuda, 1964), which are so-called "pyroturbidites," may form significant portions of volcaniclastic accumulations. For migratory intraoceanic arcs that have crustal underpinnings of continental aspect and for partly drowned continental-margin arcs backed by epicontinental seas, varied mixtures of these diverse lithologic types may occur among the central and proximal facies.

Dispersal facies of mainly volcaniclastic rocks may include, at one extreme, fluviatile suites spread from chains and clusters of subaerial volcanoes and, at the other extreme, turbidite suites spread on the submarine slopes of island or underwater volcanoes. Transit of well-sorted and well-rounded sediment across shelves of varying dimensions, or progradation of deltaic complexes across shallow seas, may also form links in the dispersal chain.

Distal equivalents of dispersal facies include diverse basinal facies of intra-arc depressions, as well as the strata deposited in arc-trench gaps and backarc areas discussed later. Intra-arc sedimentary basins are commonly elongate parallel to the trend of an arc and apparently are bounded by extensional faults. Modern examples of the wide range of intra-arc sedimentary accumulations include: (a) a kilometer or more of late Cenozoic marine strata, whose base is unknown, that is locally present on the tilted crest of the submerged Tonga Ridge and within the local Tofua Trough adjacent to part of the active volcanic chain of Tonga (Karig, 1970, fig. 4); (b) several kilometers of late Cenozoic marine sedimentary beds capped by lacustrine deposits in each of several basins on Honshu, where downwarping was accompanied by complementary uplift of intervening highlands upon which volcanoes were concentrated (Matsuda and others, 1967); and (c) many kilometers of intermontane continental redbeds interstratified with Cenozoic volcanic rocks beneath the altiplano of Peru and Bolivia (James, 1971). The association of intra-arc sedimentation with crustal extension to form grabenlike troughs emphasizes the general lack of strong contractional deformation across arc structures during magmatic activity and suggests further that some intra-arc depressions may be incipient interarc basins (e.g., Karig and Mammerickx, 1972).

The potential for preservation of intra-arc sedimentary and volcanic sequences varies for different types of arcs. The potential is low for continental-margin arcs that have deep roots supporting a subaerial mountain range and intermontane basins standing high above sea level. In the late Mesozoic magmatic arc system of western North America (Hamilton, 1969), the portion of the arc assemblage that has survived the termination of magmatic activity is mainly granitic plutons in the deeply eroded roots of

TABLE 1.—SPECULATIVE LIST OF DOMINANT DEPOSITIONAL ENVIRONMENTS AND STRATAL FACIES IN VARIED GEOSYNCLINAL SETTINGS WITHIN AND BESIDE MAGMATIC ARCS

Setting	Crust is oceanic or semioceanic (<15–20 km)	Crust is continental or semicontinental (>15–20 km)
Volcanic fields	Andesite and more mafic lavas and pyroclastics dominant and common pillow lavas and pillow breccias in central facies; biogenic reefs common as proximal facies on flanks of partly submerged cones	Andesite and more silicic subaerial pyroclastics dominant and common ignimbrites and pyroturbidites in central facies; fluvial breccias common as slurry-flood aprons and fanglomerates on flanks of high cones
Intra-arc basins	Volcaniclastic turbidites and marine tuffs in deep troughs flanked by narrow belts of local shelf facies and proximal turbidites on steep submarine slopes	Volcaniclastic redbeds and subaerial tuffs in grabens having lacustrine bolson fills; also, unstable shelf facies, conglomerates and local unconformities
Arc-trench gaps	Turbidites deposited in deep troughs as wedges of subsea fans displaying evidence of transverse paleocurrents and as prisms of fill showing evidence of longitudinal paleocurrents parallel to arc trend	Fluviatile, deltaic, and prodeltaic beds of progradational coastal plain complexes built upon subsiding shelfs open to the sea or within partly enclosed embayments
Backarc basins	Turbidite aprons built into deep interarc basins of marginal seas or into open oceans	Piedmont or deltaic wedges built into interior seaways or riverine lowlands of shallow foreland basins

the arc, which have been exposed by wholesale erosion of the cover rocks. In California, Cretaceous plutons along the trend of the arc were emplaced into older late Paleozoic and early Mesozoic eugeosynclinal complexes in the north, into Paleozoic miogeoclinal rocks in the middle, and into Precambrian basement rocks on the south. For some similar arcs, the surficial cover may be partly preserved in associations of subaerial volcanic rocks and mollaselike continental beds, as Mossakovskiy (1972) has argued for a postulated late Paleozoic arc along the southern margin of Eurasia.

Island arc assemblages, however, form eugeosynclinal prisms of stratified rock. Subsidence during volcanism can lead to net basinal development along the arc trend (Berg and others, 1972).

Stratigraphic thicknesses of upper Cenozoic marine volcanic and volcaniclastic strata exposed in the currently active Kurile island arc approach 10 km (Markhinin, 1968), and exposures of similar sequences of lower Cenozoic strata in Fiji, which is currently inactive but in an isolated position far from major land areas, probably exceed 10 km in combined thickness (Dickinson, 1967). Island arc assemblages in the rock record include: (1) those like the suite in the Greater Antilles, built on oceanic crust during the late Mesozoic and early Cenozoic (Donnelly, 1964), and (2) those like the middle Mesozoic suite in western North America, built on a thin wedge or newly accreted belt of crust near a continental margin and separated from the craton of the continental interior by shallow marine waters (Stanley and others, 1971). In Puerto Rico, the Antillean arc assemblage of intertonguing volcanic and volcaniclastic facies, including turbidites and cut by partly contemporaneous plutons, has an estimated stratigraphic thickness of roughly 7.5 km accumulated within a span of about 75 my (Berryhill and others, 1960; Mattson, 1966; Khudoley and Meyerhoff, 1971). In British Columbia, local sections of the middle Mesozoic arc suite are equally thick (e.g., Schau, 1970), and partly contermporary plutons also occur (Tipper, 1959). Detailed facies relations in an apparently intraoceanic island arc have been described by Mitchell (1970) for the Miocene Matanui Group of Malekula in the New Hebrides, and those in a probable continental-margin island arc have been described by Schau (1970) for the Upper Triassic Nicola Group in central British Columbia. Table 1 is a summary of inferred rock types to be expected in association with different types of magmatic arcs.

ARC-TRENCH GAP SEDIMENTATION

The distance between the bathymetric axis of the trench and the volcanic spine of the arc varies from 100 to 300 km in typical modern arc-trench systems. The inner wall of the trench is a steep declivity, along which unconsolidated and undeformed sediments occur only in local patches or pockets. The top of the inner wall of the trench is marked by a break in slope called the *trench slope break* (Dickinson, in press), which in some cases is at the crest or outer shoulder of a submerged bathymetric ridge or an emergent insular ridge. The trench slope break is commonly located above the crest of a structural barrier in the subsurface, which may be detected as acoustic basement on continuous seismic profiler records. The crest of this subsurface feature is interpreted here as the edge of deformed strata in the subduction zone related to the trench. The trench slope break thus marks the threshold or outer limit of the distinct morphotectonic belt that I have called the arc-trench gap (Dickinson, 1971). The inner limit of the arc-trench gap is the volcanic or magmatic front at the flank of the arc. The width of the arc-trench gap is 50 to 250 km in different cases, and the varied bathymetry and topography within this belt include a number of common configurations that imply the presence of traps for sediment shed from the arc (fig. 3).

Some arc-trench gaps include more than one of the elements illustrated. For example, in Japan both an uplifted belt (fig. 3b) exposing the massifs of the coastal ranges (Matsuda and others, 1967) and an outer sedimented submarine slope (fig. 3c; Ludwig and others, 1966) occur between the volcanic range of Honshu and the Japan Trench offshore. Other combinations occur elsewhere. Wherever basins filled with sediment exist, they apparently formed by initial downwarping or continual subsidence within the arc-trench gap, coupled with relative uplift of tectonically thickened materials in the subduction zone beneath the trench slope break. Seismic reflection profiles of a number of arc-trench gaps show the presence of undeformed or gently folded sedimentary strata too thick to display a clear acoustic base within a depth of 1 to 2.5 km beneath the bottom. Such sequences are most apparent beneath troughs (Ludwig and others, 1967), deep benches (Marlow, 1971) or plains (Karig, 1970), and beneath some shelves (Von Huene and Shor, 1969). Transverse slopes (Karig, 1971; Ludwig and others, 1966) and some other shelving areas (Scholl and others, 1970) are commonly masked by only a thin mantle of sediment.

The different possible configurations of potential depositional surfaces within arc-trench gaps imply that all the following types of strata may accumulate there together with varying proportions of airborne ash: (1) fluviatile-deltaic-shoreline complexes of subsiding coasts or depressions, (2) shallow-marine sediments of unstable shelves, (3) slope-covering turbidites showing evidence of transverse paleocurrents, and (4) trough-filling turbidites showing evidence of longitudinal paleocurrents. Complex interplay between the rates of the tectonic processes responsible for the formation of the basins and the rate of sediment delivery to the arc-trench gap may allow the configurations of basins to change in irregular fashion during the evolution of an arc-trench system. For example, for a given elevation of the threshold and a given rate of subsidence within the basin, varying rates of sediment delivery from the arc

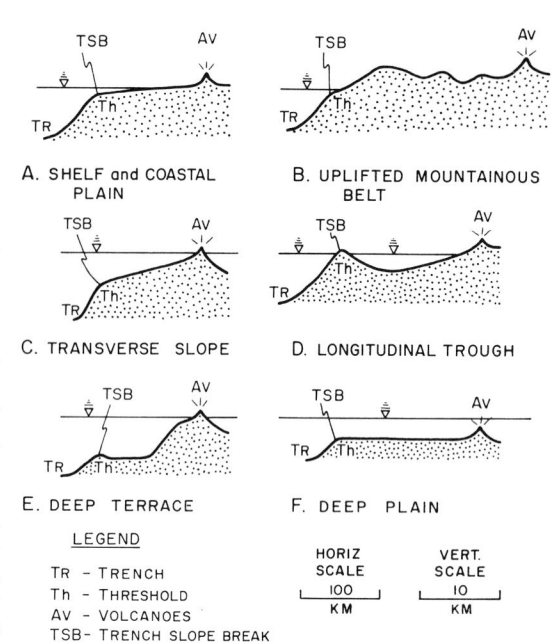

FIG. 3. Generalized sketches of bathymetric-topographic configurations within modern arc-trench gaps. Tectonic elements shown include TR, bathymetric axis of trench; AV, line of arc volcanoes at volcanic or magmatic front marking inner extremity of arc-trench gap; and TSB, trench slope break at outer threshold (Th) of arc-trench gap. Arc-trench systems depicted include: A B, Central American and Mexican segments of Middle America (King, 1969); C, typical parts of Tonga-Kermadec (Karig, 1970) and Marianas (Karig, 1971); D, Sumatra-Java segment of Sunda with emergent Mentawai Islands; E, central Aleutians (Marlow, 1971); and F, Raukumara Plain (Karig, 1970, fig. 5) northeast of New Zealand.

might allow the configuration of the basin to fluctuate with time. During times when sediment delivery was too slight to keep the basin full, a deep trough might develop. During times when sediment delivery was great enough to overfill the basin, the top of the sediment pile might form a broad shelf or bench. If the elevation of the trench slope break also fluctuated with time, the so-called sediment-full configuration could vary from a shallow shelf to a deep bench. The potential interplay between tectonics and sedimentation is well shown in the Eocene Series of coastal Oregon as described by Dott (1966). The progradational deltaic complex of the Coaledo Formation near the continental margin advanced into a basin that had previously received graded turbidite sequences. The depositional region presumably lay within an arc-trench gap west of the provenance, which was a volcanic chain that had orogenic foundations representing a continental-margin magmatic arc. Uplifts within arc-trench gaps may serve in other cases as addiitonal sources of sediment and may include islands along the trench slope break at the threshold of the gap.

The following diagnostic tectonic relations of arc-trench gap sequences can be inferred: (1) less deformed or metamorphosed than coeval arc and trench assemblages on either side of the gap, (2) contact with rocks of the subduction zone being mainly a syndepositional thrust zone but possibly onlapping locally, and (3) contact with igneous rocks of the arc being partly gradational to intra-arc volcaniclastic strata and partly onlapping across eroded volcanic and plutonic rocks and, also, syndepositional normal faults locally along the arc side of the gap.

ARC-TRENCH GAP SEQUENCES

In the circum-Pacific region, where old eroded arc-trench assemblages still face open ocean, uplifted arc-trench gap sequences occur as elongate belts of sedimentary strata, typically in great asymmetric synclinoria between more deformed and metamorphosed tectonic elements. Ideally, coeval mélange belts of subduction zones representing trench assemblages occur faulted against the steep limbs, and coeval batholithic or metavolcanic belts representing magmatic arc assemblages lie depositionally beneath and beyond the gentle limbs. A well-studied example is the Great Valley Sequence of late Mesozoic age deposited in California between the site of deformation of the Franciscan assemblage and the site of intrusion of the Sierra Nevada Batholith (Dickinson and Rich, 1972). The site of deposition evidently spanned the transition zone between oceanic and continental crust of the time, for the west limb of the so-called "megasyncline" containing outcrops and subcrops of the sequence includes an ophiolitic sequence at its base, whereas the east limb rests upon sialic crystalline rocks. The strata include 12 to 15 km of trough-filling turbidites showing mainly longitudinal paleocurrent structures, although subsea fan complexes displaying evidence of transverse paleocurrents were apparently also built into the trough from the arc side. Toward the flank of the arc to the east, thinner shelf facies rest depositionally, in part by onlap and in part probably by overlap across bounding fault scarps, on granitic and metamorphic rocks of the Sierra Nevada foothills. On the west, deep trough facies of the sequence and the underlying ophiolite slab are underthrust by mélanges and related rocks of the Franciscan Assemblage. Radiometric and paleontologic means of dating combine to show that deposition and deformation of most of the Franciscan Assemblage, deposition of all of the Great Valley Sequence, and intrusion of most of the Sierra Nevada Batholith took place during the same timespan of about 75 my. Details of the stratigraphy and sedimentary petrology of the Great Valley Sequence indicate that the provenance was the Sierran arc, and that the petrologic and tectonic evolution of the source terrane was reflected by episodes of deformation along the margin of the trough and by secular changes in the character of the detritus deposited in the arc-trench gap.

Within orogenic belts formed by continental collisions, arc-trench gap sequences deposited prior to collision are caught within broad terranes of deformed rocks adjacent to suture belts. Few such sequences have been well described, but one example may be the lower Paleozoic section of the Midland Valley in Scotland as described by Dewey (1971). The sequence was deposited across the transition between deformed Dalradian basement of sialic rocks to the north and deformed wedges of inferred oceanic crust and mantle in the Ballantrae Complex to the south. The granitic and metamorphic rocks of the Northern Highlands are taken to be the deeply eroded roots of the associated magmatic arc, whose front may have coincided roughly with the Highland Boundary Fault. Sediment delivery to the Midland Valley arc-trench gap was first only from high-rank metamorphic and plutonic sources in the inferred arc to the north but later also from low-rank metamorphic and sedimentary sources lying in the direction of the Southern Uplands. This southern provenance is interpreted as deformed sedimentary sequences and perhaps as intraoceanic arc structures of an oceanic plate that was exposed in tectonic uplifts associated with the suturing of the continental Midland Platform of England against the terranes to the north at the close of the Caledonide cycle (Dewey, 1969).

BACKARC SEDIMENTATION

Sedimentation behind intraoceanic arcs includes the spread of pyroclastic materials over wide areas and the construction of turbidite wedges (fig. 1) built into deep water from the rear of the arc structures (Karig, 1970, 1971). Where interarc basins and marginal seas form behind migratory arcs, the rear flanks of the migrating frontal arcs are typically compound fault scarps leading directly down from the volcanic chains to the floors of the oceanic basins (e.g., Karig, 1970, fig. 5; Karig, 1971, figs. 2-3). The rear flanks of remnant arcs commonly carry a turbidite cover that is starved by separation from the arc source at the time of opening of the interarc basin. If marginal spreading seas

Considerable disagreement remains as to the interpretative aspects and, in particular, the sediment transport mechanisms and the paleogeographic-tectonic framework in which flysch accumulated. That turbidity currents deposited most of the graded sandstone units (Kuenen and Migliorini, 1950) is widely accepted, although the role of turbidity currents is still questioned by some workers. In recent years, arguments have been presented in favor of perpetual vibrations (Oulianoff, 1960), oscillatory currents (Grossgeym, 1972), and bottom (Hubert, 1964) and contour-following currents (Heezen and others, 1966). Closer examination of modern sedimentation patterns in deep-marine base-of-slope environments suggests that the effect of both turbidity currents and bottom currents significantly influence deposition of both sands and muds in basins (Stanley, 1969, 1970; Piper, 1970, 1971). Exciting problems relating to flysch sedimentation await solution. Among others are the origin of pelitic sequences, conglomerates, and slumps; redefinition and significance of proximality-distality (Walker, 1967) and textural makeup as related to submarine topography; relation of mineralogical composition including maturity of sandstones and conglomerates as related to possible source terrains; the regional distribution of primary sedimentary and biogenic structures as related to dispersal patterns.

However, there are aspects even more basic to the understanding of the origin of flysch in general, and its relation to geosynclinal sedimentation—the principal theme of this symposium. These relate to the paleogeographic-tectonic framework in which this facies accumulates. Alpine workers first, and subsequently others, have allowed rather specific genetic structural-environmental implications to become an integral part of the definition of flysch. For instance, Tercier (1947, p. 181) stated:

> "Du point de vue paléogéographique, on doit le considerer comme un faciès propre à des bassins marins, tantôt étroits, parfois assez larges, le plus souvent moyennement profonds, du type des bassins d'archipels . . . C'est par le jeu réciproque de surrections maintenues à la marge des bassins d'archipels et de sédimentation puissante dans le bassin lui-même que des mers primitivement profondes vont finalement être peu à peu comblées. Et c'est en ce sens que le Flysch apparaît bien comme un faciès de fermeture de géosynclinaux."

While geographical age is ignored, geologists have considered flysch marine sequences to be preparoxysmal (Kuenen, 1958), early synorogenic or cataorogenic (Trümpy, 1960), or related to advanced orogenic development (de Raaf, 1968). Hsü (1970, p. 9) visualized an alpine-type origin "with a tectonic setting, and sedimentological features similar to the Alpine Flysch in its more typical setting." A recent trend is to define the term even more flexibly, that is, as a thick succession that has "been largely deposited by turbidity currents or mass flow in a deep water environment, within a geosynclinal belt" (Reading, 1972, p. 60). One might add that the "deep-water" origin, another aspect that crept into the definition in the mid-1900's (see arguments in Kuenen, 1967) is being relaxed, although few would subscribe to the very shallow water-neritic school (Mangin, 1962).

It appears that the concept of flysch sedimentation is once again in a state of flux, this time as a result of developing theories of sea-floor spreading and global tectonics (compare Dietz and Holden, 1966; Mitchell and Reading, 1969; Dewey and Bird, 1970). It is now popular to define marine sedimentary and volcanic sequences in terms of accumulations on mobile lithospheric plate margins. In this context, the relation of flysch sedimentation to concurrent tectonic activity is actually a significant factor. The present trend is to distinguish geosynclinal sedimentation—and this would include flysch deposition—at junctures where two lithospheric plates are moving apart (*divergent* case), or where two plates are moving laterally without substantial divergence (*strike-slip*) case or towards each other (*convergent* case) (Reading, 1972, p. 61).

Seismic-coring surveys reveal that the majority of depressions on the sea floor comprise at least a reasonable thickness of alternating sequences of unconsolidated turbidites and hemipelagic or pelagic units. In fact, the growing body of marine geological literature shows that a very considerable portion of the world's submarine surface is covered by turbidite-rich and associated sequences that appear flyschlike. Are all these sequences flysch? It has to be recognized that the term connotes genetic aspects as well as purely petrological-lithological characteristics to many workers. Although some workers believe that considerations of tectonic setting need not be included in the definition (Hsü, 1970, p. 8), few geologists would readily call all turbidite-bearing sequences flysch. At this stage, it is advisable to exercise caution in equating modern turbidite-filled lows with flysch basins of the ancient rock record; some restrictions and specific criteria should be applied if the term is to retain any meaning to geologists working on land and at sea.

For instance, one can seriously question the value of automatically defining as flysch (at least as the term is used by most land geologists) any deep-marine sequence simply because it presents some of the lithological aspects of flysch. As Hsü (1970) noted, sedimentological characteristics describe a flysch but do not define it. An example in point is the terrigenous wedge forming the continental rise off the relatively stable eastern North American margin (detailed discussion in Stanley, 1970); sediments forming this wedge, lithologically, are flyschlike but accumulate under conditions considerably different from those postulated settings in which well-known flysch sequences were deposited in the Alps, Carpathians, and other mobile belts.

Although beset as it is with semantic complications, there is little likelihood that the term will be abandoned. The field geologist, after all, finds it useful and is keenly interested in trying to understand the conditions under which the deposit he is mapping accumulated. Thus, I believe that there is some value in focusing on selected sectors of the modern sea floor that are presently receiving gravitative deposits in geographic-tectonic-sedimentary settings that are reasonable analogs of those in which some classical flysch types accumulated in the past. This paper considers sedimentation in one possible modern characteristic area of Quaternary flysch deposition, the mobile compressive setting of the Hellenic Arc in the eastern Mediterranean (fig. 1).

MOBILE SETTING OF THE HELLENIC ARC

Hersey (1965) and Ryan and others (1970) have indicated that the eastern Mediterranean is a region where flysch deposition is likely to occur, and Hsü and Schlanger (1970, figs. 1 and 2) and Hsü (1972) showed a number of geographic-tectonic analogies between Paleocene Alpine Flysch basins of Switzerland (Schlieren and Gurnigel Basins) and modern deep basins in the Aegean-Levantine region. Focus on this area is warranted in large part because of its particular geographic configuration and associated seismicity (Galanopoulos, 1968; Barazangi and Dorman, 1969; Lort, 1971; Papazachos and Comninakis, 1971; Comninakis and Papazachos, 1972) and volcanism (Ninkovich and Hays, 1972).

The principal seismic trends follow the Aegean arc and Hellenic Trench system. The Mediterranean Ridge, forming the southern boundary, is a broad (150 km wide) sinuous (1,600 km long) topographic high (depth between 1,300 to 3,200 m along the crest; average, 2,000 to 2,500 m) between the Ionian Sea and the area west of Cyprus (19° to 32°E long). The highly irregular surface (small-scale blockiness) is due to faulting according to Belderson and others (1972). Fault plane solutions of earthquakes show that the axis of compression is horizontal and perpendicular to the ridge (fig. 2) and that the axis of tension dips steeply, suggesting underthrusting of the Aegean Arc by the African lithospheric plate (McKenzie, 1970; Comninakis and Papazachos, 1972). This area, the most active in the Mediterranean, is at present designated as a convergent plate junction (see Le Pichon, 1908; Mitchell and Reading, 1969, fig. 2C; Dewey and Bird, 1970, figs. 13 and 14). Geophysical studies comprising gravity field, magnetics, seismicity, regional heat flow, and seismic reflection profiles were summarized by Rabinowitz and Ryan (1970) and by others. These authors also detailed the components of the east-west oriented island arc structure (convex margin facing south), which include from south to north: the Mediterranean Ridge, deep-trench system (including parallel Pliny and Strabo Trenches), an exposed sedimentary arc, an internal trough, and an interior andesitic volcanic arc (Ninkovich and Hays, 1972). Several rugged topographic highs (Strabo, Ptolemy, and Anaximander Mountains; Emery and others, 1966) lie between the Mediterranean Ridge and the exposed islands of Crete and Rhodes; the southern escarpment of these mountains, the Mediterranean Wall, is steep and relatively linear and continuous (Goncharov and Mikhailov, 1963).

Seismic profiles (see Hersey, 1965; Ryan and others, 1970; Wong and others, 1971) amply attest to the tectonic deformation of sea-floor features; faulting, offset of regional trends, and folded and tilted basin plains and highs. Trench walls appear to coincide with fault surfaces. Thick sequences (0.1 to >1.0 km) of Pliocene and Quaternary unconsolidated sequences, as well as pre-Pliocene units, are offset; movement along faults in this region exceeds 1 mm per year according to Rabinowitz and Ryan (1970). This rate (100 cm/1000 years) of offset exceeds the rate of sedimentation in most of this region. An exception is the region of high deposition on the Nile Cone south of the ridge (Wong and others, 1971).

Of particular interest is the interpretation of the Mediterranean Ridge and arc structure shown to result from imbrication of the crust (Ryan and others, 1970) produced by northwest-to-southeast-directed compression (fig. 2). In this model, the ridge is shown to contain intrabasinal sedimentary nappes, and its form and thickness are responses to underthrusting of the crust beneath the island arc. Seismic profiles indicate that the intensity of fracturing and folding, greatest along the island arc trend, de-

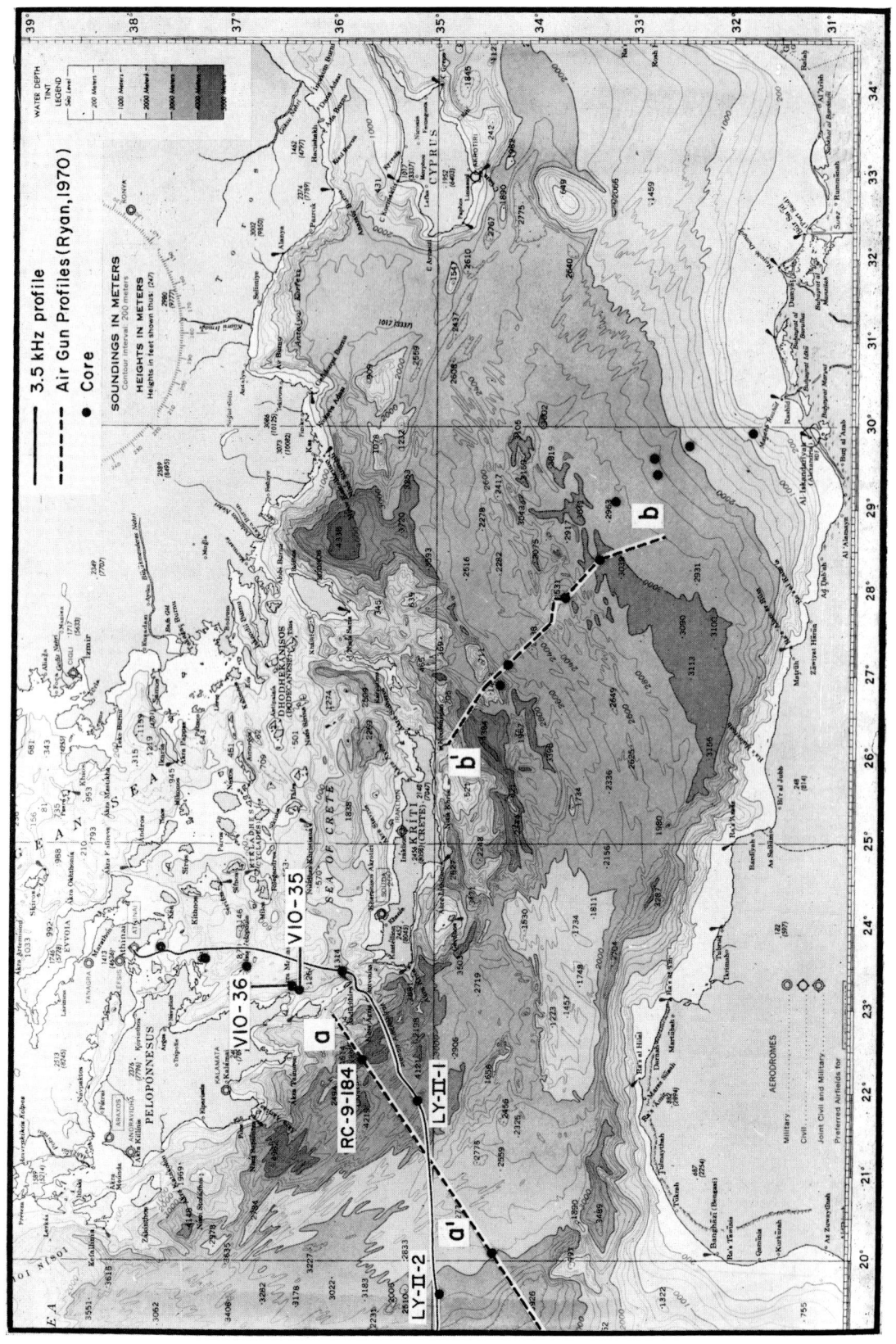

FIG. 1.—Chart of eastern Mediterranean (section of N.O. Chart 310) showing area discussed in text. Dots are cores (see figs. 7 and 8), dashed lines are air-gun profiles (fig. 5), and solid lines are 3.5-kHz profiles (fig. 9).

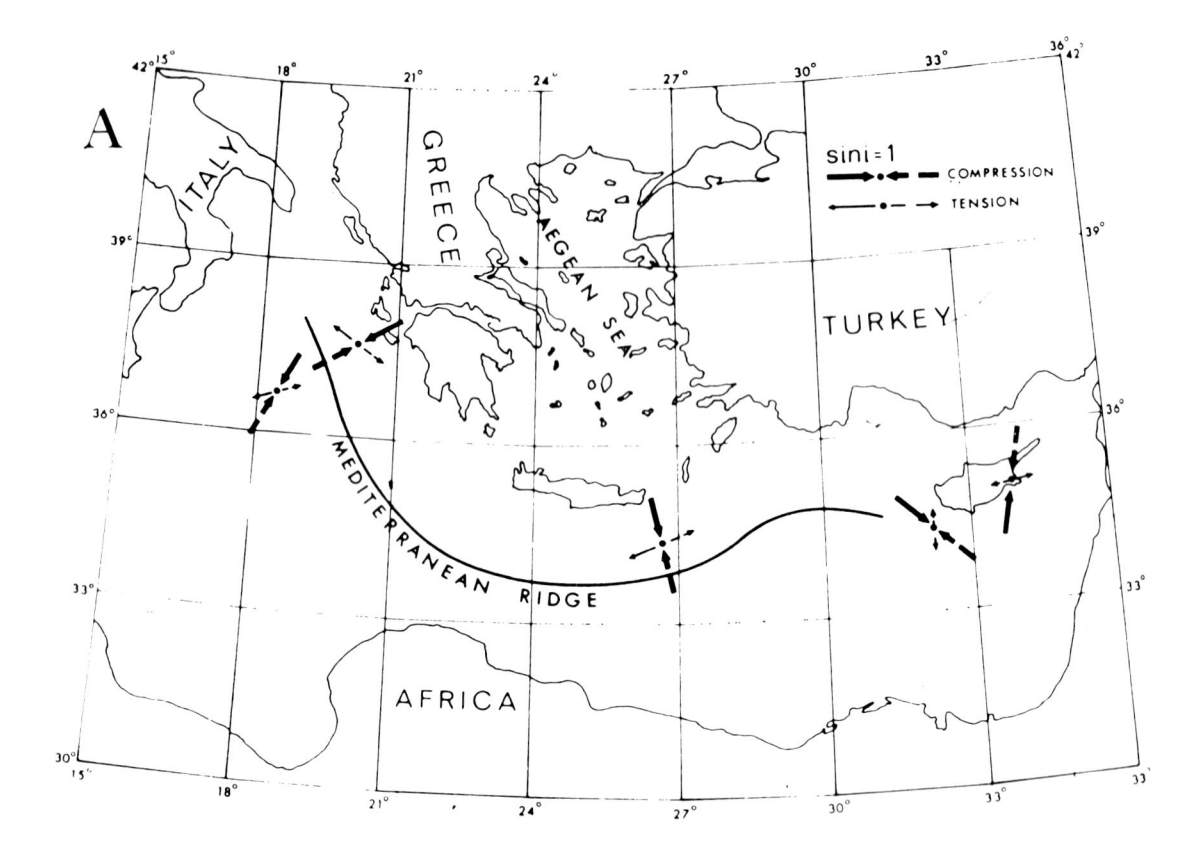

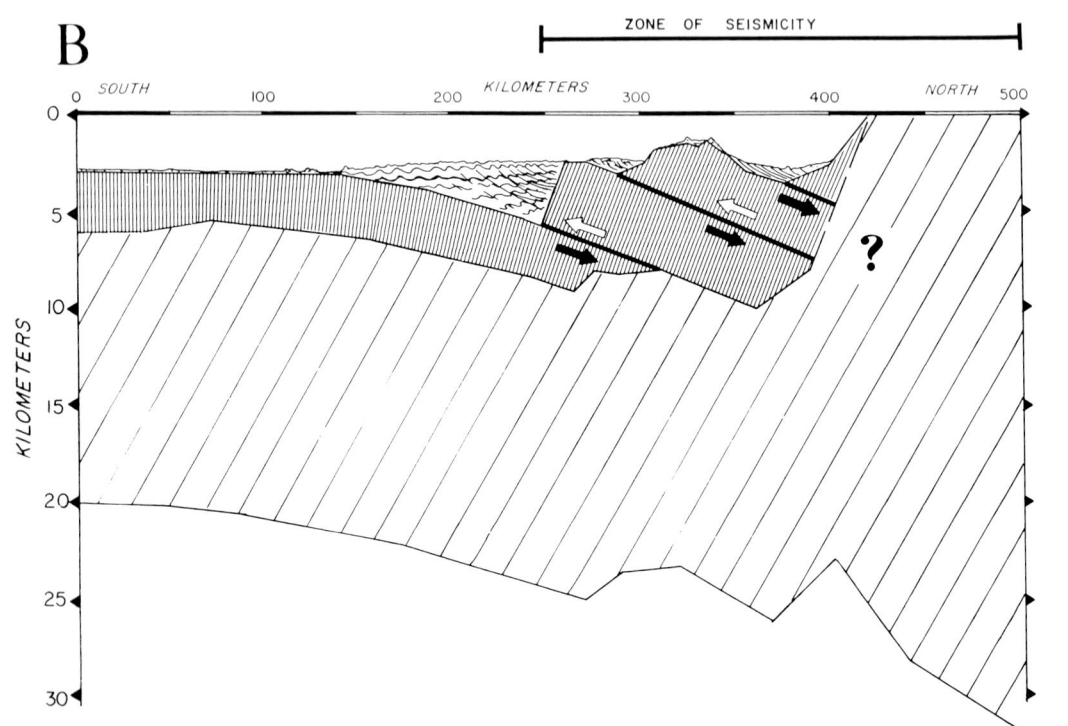

FIG. 2.—A, Directions of stress components in eastern Mediterranean (after Comninakis and Papazachos, 1972). B, Hypothetical crustal model showing crustal shortening by underthrusting and décollement between lower part of Nile Cone and Turkey (after Rabinowitz and Ryan, 1970).

creases southward toward the Nile Cone. Furthermore, the age of this deformation probably is greater toward the north and younger toward the Nile Cone. The trenches on the convex side of the Hellenic Arc are believed to owe their origin to underthrusting (McKenzie, 1970; Hsü, 1972), and this hypothesis is supported by recent D/V *Glomar Challenger* drilling.

Both the time-transgressive deformational aspect and the island arc configuration call to mind the classical model for flysch development in the Alpine Geosyncline, that is, a compressional flysch foredeep in front of a rising cordillera as postulated by Argand (1916). Modern studies (Trümpy, 1960; Aubouin, 1965; and others) also suggest that sedimentation results from the erosion of a zone situated to the interior of that receiving the sediment, and that the "flysch facies is developed throughout the region between the cordillera producing the source material and the neighboring (more external) furrow" (Aubouin, 1965, p. 139). Hellenic Arc deposits, like those of the Alpine Flysch, show that the influence of contemporaneous tectonic movements and sedimentation is locally transgressive on folds and minor thrust faults. Furthermore, active volcanoes are distributed along the arc structure in the Aegean Sea. The addition of volcanic ashes has been periodic and widespread (e.g., the influence of Santorini discussed by Ninkovich and Heezen, 1965; Ryan, 1972; and Bonatti and others, 1972). The association of active volcanoes and earthquakes lends support to the contention that the Hellenic area can be considered a geosyncline in development.

GENERAL GEOGRAPHIC SETTING OF THE HELLENIC REGION

Seismicity and volcanism are only two of the principal parameters affecting sedimentation in a geosynclinal setting. But what about the depositional sites and the effects of climate and oceanic circulation? As a result of recent intense tectonic activity, the Hellenic Arc displays the most complex and rugged land and submarine topography in the Mediterranean region (fig. 3). Space precludes a detailing of morphologic features, and the reader is referred to discussions of bathymetry in Emery and others (1966), Ryan and others (1970), Maley and Johnson (1971), and Wong and others (1971).

The classic flysch geographic setting as normally envisioned, high emerging cordilleras bounding deep depressions, is present. Much of the Levantine Sea floor lies below 2,000 m, whereas the Aegean Sea is considerably shallower (that is, the most part less than 1,000 m in depth with the exception of several deep basins north of Crete). The coastal plains of Greece and Turkey are generally narrow; in many areas mountains rise out of the sea, and these are bordered by a generally narrow shelf 5 to 15 km wide in the western part of the Aegean Sea and in the Ionian Sea. Shelf width is considerably more variable on island platforms such as in the region southeast of Athens (Cyclades Plateau), the northern Aegean, and locally off the Turkish coast. Sills topographically separate the Aegean Sea from the Levantine Sea (Caso Strait, Scarpano Strait, etc.).

Sea-floor slopes in the Levantine Sea commonly exceed 5°; an exception is the gentle slope (1–2°) on the fore-set beds of the Nile Cone (profiles in Emery and others, 1966; fig. 5C herein). Particularly steep margins occur on the Mediterranean Ridge margins (>5°) and trench walls (locally to 40°) of the Hellenic Trench (to 5,000 m) area west of Greece, including the Pliny (to > 4,000 m) and Strabo (deeper than 3,000 m) trenches south and southeast of Crete. Submerged highs such as the Ptolemy and Anaximander Mountains are bordered by slopes interpreted as probable high-angle faults (Wong and others, 1971).

Submarine canyons trending west and south dissect much of the margin off western Greece (Fig. 3), and, although few have been examined in detail, there appear to be some large canyons south of Crete (Shepard and Dill, 1966) as well as north. Mapping has shown that in many instances, canyons head in reentries on the shelf margin and upper slope and descend to the basins; canyons also debouch normal to trench axes.

A chart showing the location and configuration of the major flat or slightly tilted basin plains and trenches is shown in figure 4. The linear orientation of the majority of basins clearly parallels the predominant structural trends of this highly active region. This diagram illustrates the point that less than 10 percent of the Hellenic Arc region is floored with trenches and basin plains. (Some of these can correctly be termed abyssal plains.) Also noteworthy is the fact that basin plains behind the exposed island arc in the southern Aegean (Sea of Crete) are generally shallower, several times larger, and, for the most part, not as distinctly elongate as those seaward of the island arc. Their depths range from less than 400 to 2,509 m; their areas range from 10 km² to 10,000 km². (The majority are considerably larger than 400 km².) Depressions south of the Peloponnesus-Crete-Rhodes arc include a number of long and very narrow (1–3 km wide) trenches, which, for the most part, are small

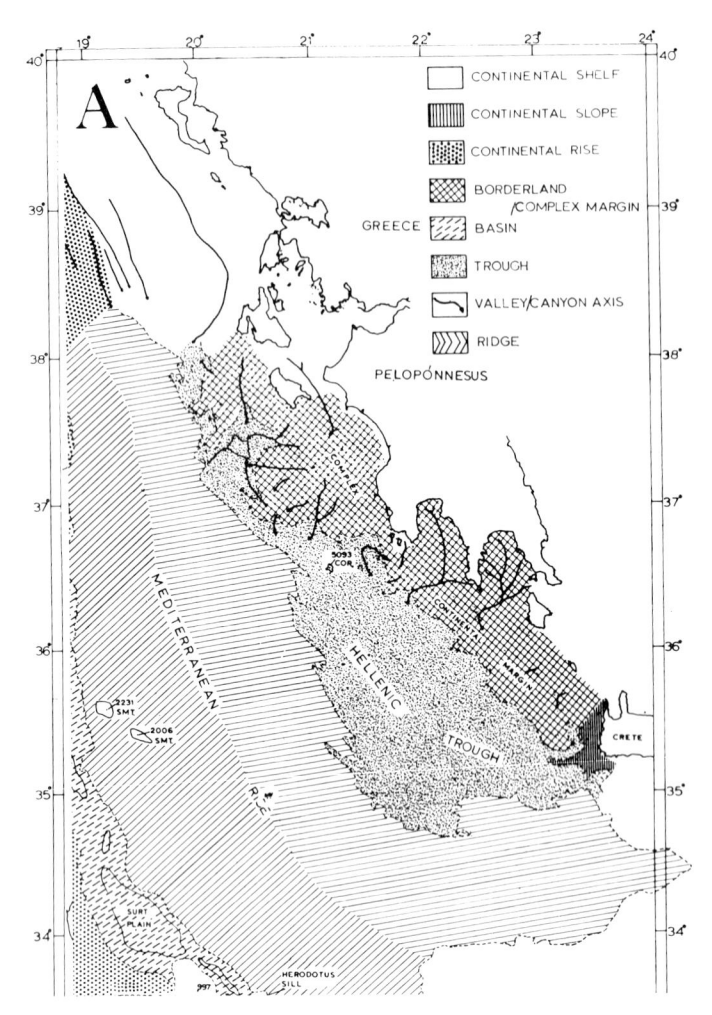

Fig. 3-A.—Physiographic diagrams of eastern Mediterranean showing topographic features discussed in text (after Carter and others, 1972). A, Ionian Sea west of Greece; B, Hellenic Arc and Levantine Sea.

(25 to 400 km²) and deep (1,500 to 5,000 m).

Only some of the larger basins on the highly dissected (fault-blocked) ridge proper are shown; these range from less than 10 km² to 120 km². By far the largest depression in the Levantine Sea is the Herodotus Plain, lying south of the ridge and northwest of the Nile Cone (Rosetta Fan). Its abyssal plain is about 1,200 km long, averages about 18 km in width, and is about 3,100 m deep in its deeper parts.

The normal temperature is about 10° to 21°C in January and 21° to 27° in July. Sedimentation is affected by seasonal shifting of wind patterns, which intensify fall, winter, and early spring wave activity, and by storms that in turn affect coastal erosion. The predominant sources of sediment, however, are rivers whose seasonal flow is controlled to a large degree by winter rains and snow melt. Rainfall in the Hellenic Arc is not uniform. Mountains act as barriers so that western Greece receives much heavier rainfall than does eastern Greece. A belt receiving from 1,000 to 1,500 mm of rainfall per year lies west of the Ionian Islands (Semple, 1931). Most of the Aegean receives about 500 to 750 mm per year, whereas the regions south of Crete and between Cyrenaica and the Nile receive less than 500 mm per year. As in most of the Mediterranean region, much of this rain is received during the colder, winter months. The summer is nearly rainless (less than 50 mm during the months of June, July, and August), and the streams are often dry, or at least considerably reduced, during the summer droughts. Periods of snow melt in spring help charge streams, and this sometimes contributes to local flooding.

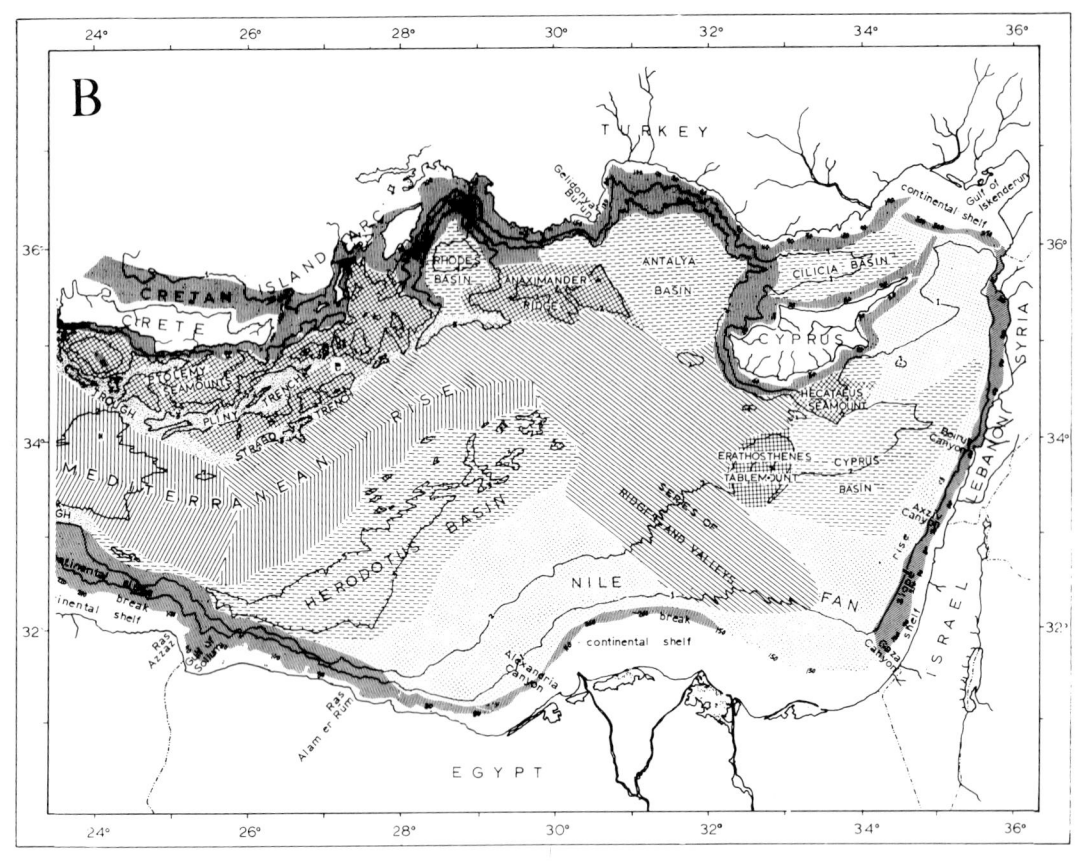

Fig. 3-B.

Some important rivers flow into the Aegean Sea from the Turkish margin, including the Meriç Ergene, Bakir, Gediz, Küçük Menderes, and Büyük Menderes. The following rivers flow into the Aegean from the Greek side: Néstos, Strimón, Vardar, Aliákmon, Piniós, and Sperkhiós. The shorter Evrótas, Ladhon, Mórnos, Akhelóüs, Arakhthos, Thiamis, and others drain the Peloponnesus and Pindus Mountains and flow west to the Ionian Sea. Short rivers also drain the southern margin of Turkey and flow south carrying sediment to the Antalyan margin. No rivers of any significance flow off the African margin except, of course, the Nile, which is by far the single most important source of sediment in the eastern Mediterranean, carrying a load from 57 million tons per year (estimated by Shukri, 1950) to 120 million tons per year (estimated by Holeman, 1968). Much of the North African margin is also a most important source of windblown sediment, some of which presumably finds its way northward as a result of the circulating water masses.

Violent autumn storms scour the thin covering of earth from the steep declivities, thus accelerating denudation of soil and resulting in rapid transfer of sediment downslope to the coastal plains and tideless sea beyond. Rivers of small to medium size are confined in deep ravines and traverse the narrow lowlands, bringing eroded *rendzina* soils and brown calcareous and red-brown sediment to the coast. Although there have been climatic changes during historic time, man in large part is responsible for a remarkably increased rate of erosion during the past three to four thousand years. The limestone soils, which he once deforested, did not recover. Furthermore, the onslaught of grazing by goats, sheep, and cattle, as well as of burning and irrigation, accelerated denudation, which progressed rapidly even during antiquity. The resulting acceleration of siltation during the past few thousand years is amply recorded by archaeological surveys of the coastal region; the fact that these recent man-induced changes are reflected in the upper sediment sections of cores collected offshore is noteworthy.

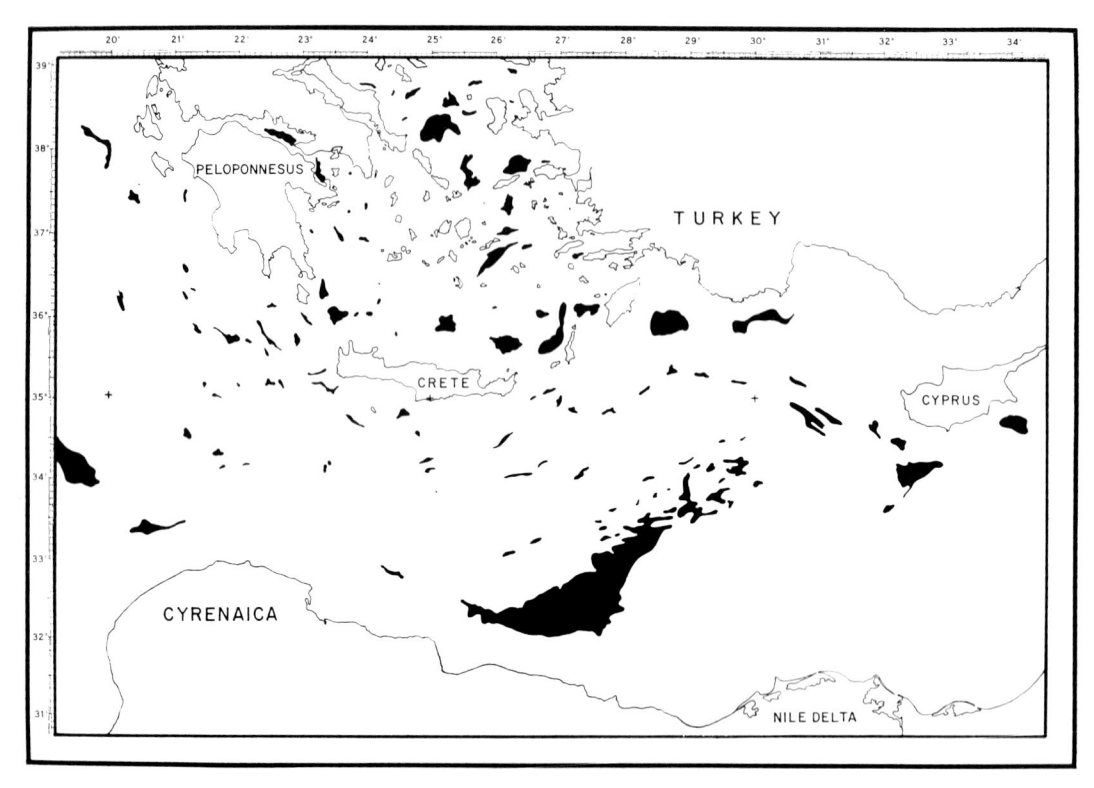

FIG. 4.—Chart showing major sedimentary ponds (in black). Note trend of narrow trenches between island arc and Mediterranean Ridge that parallel major structural trends (see fig. 2). Largest depression off Rosetta Fan (Nile Cone) is Herodotus abyssal plain.

Surface-water circulation initially plays the most important role in the transfer of fine sediments carried to the coast by rivers and of sediment eroded along the coast. This circulation pattern is generally counterclockwise in the Aegean Sea; water entering from the Levantine Sea moves northward along the margin of Turkey, and eventually turns west and south (Nielsen, 1912). Presumably some sediments are deposited in the Sea of Marmara, which acts as a settling trap between the Aegean and Black Seas. It is important to recall that rainfall is considerably higher in the Black Sea region (more than 2,500 mm/year) and northernmost part of the Aegean Sea. Some eroded sediments in this region reach the Mediterranean via surface-density overflows across the Dardanelles-Bosporus sills. Fines fed into the Aegean eventually are moved out between Crete and the Peloponnesus. A predominant southward flow of intermediate water takes place over the sill between Crete and Rhodes, whereas deep-water outflow also is postulated between Crete and the Peloponnesus (Venkatarathnam and Ryan, 1971; Lacombe and Tchernia, 1972; Miller, 1972). Off western Greece in the eastern part of the Ionian Sea, surface, intermediate, and deep water probably all flow northwest in the Hellenic Trench region. Surface circulation in the Levantine Sea is predominantly counterclockwise, and clay mineral studies (Venkatarathnam and Ryan, 1971) show that Nile sediments, windblown North African dust, and sediment introduced on the Israel and Lebanon margin are all carried northward.

SEDIMENTATION IN PONDS

The role of enclosed basins as effective sediment traps in the Mediterranean as elsewhere has been demonstrated amply on numerous continous seismic (air gun, sparker, etc.) profiles (Ryan and others, 1970; Wong and others, 1971; and others). As was emphasized in the preceding section, the topography of the Hellenic Arc is highly complex and the distribution of depressions is patchy; the regional pattern of the total sediment thickness is very irregular. The unconsolidated section is, to a large degree, directly related to morphology, that is, the greatest thicknesses are found in enclosed depressions, whereas the thinnest deposits generally are present on steeper slopes (fig. 6). The total

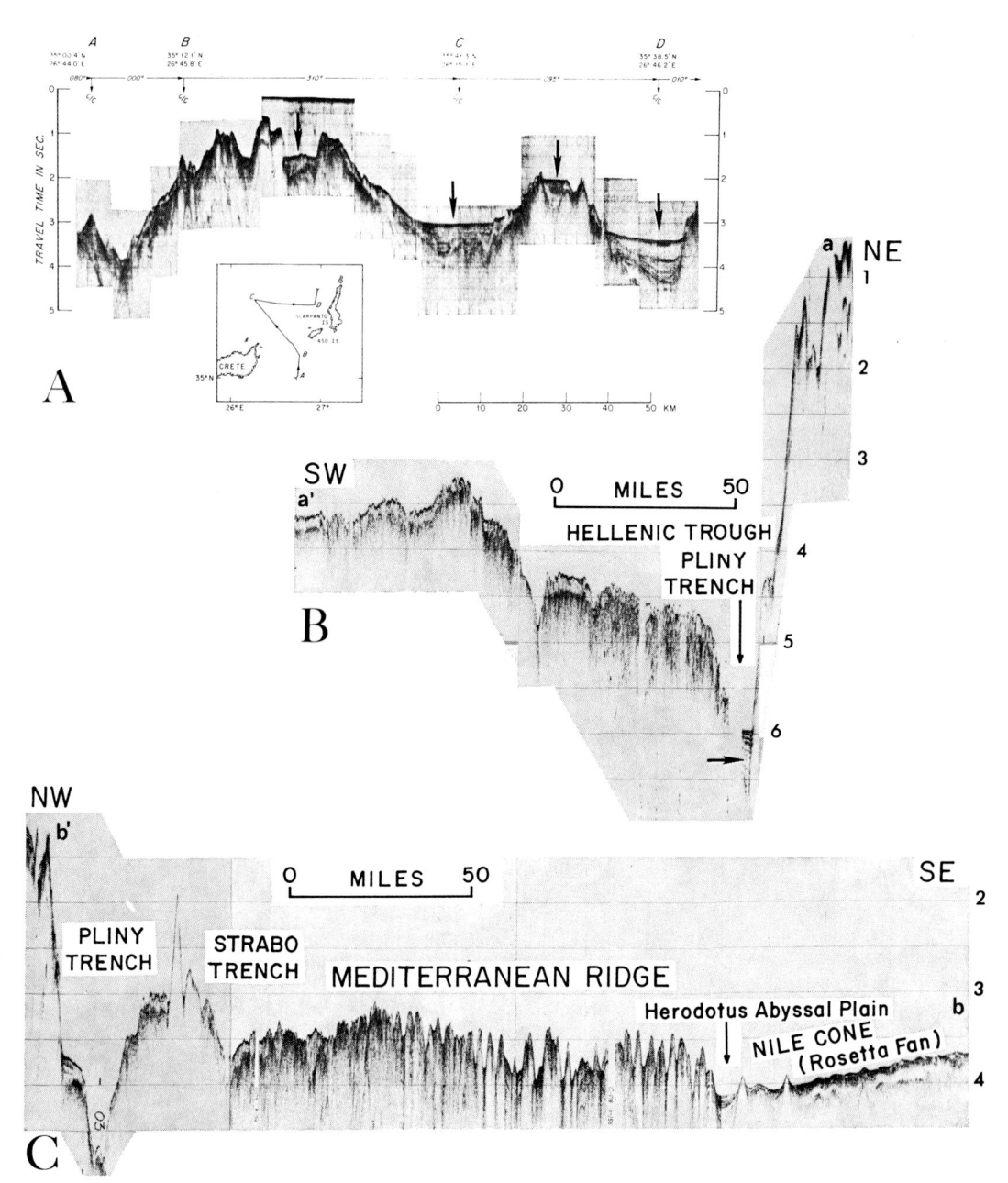

FIG. 5.—A, Seismic reflection profile along traverse east of Crete crossing Pliny Trench (between A and B), Caso Strait sill and basins west of Scarpanto Island in Sea of Crete (inner trough). Travel time in seconds. Arrows show thick pods of sediment in depressions as well as some perched on topographic highs attesting to very recent (neotectonics) offset by faulting (after Wong and others, 1971). B, Air-gun reflection profile along traverse crossing Hellenic Trench (see fig. 1) showing almost one second of penetration (two-way travel time). Note steep trench wall (>20°), narrow trench floor sloping toward inner trench wall, and dissected (fault-blocked) Mediterranean Ridge to WSW. Depth of trench floor is about 4,100 m. A core, Conrad 9–184, collected in trench along this crossing is shown in figure 8. C, Air-gun Conrad 9 profile across Nile Cone, narrow northeast sector of Herodotus abyssal plain, Mediterranean Ridge and, Strabo and Pliny Trenches. Note remarkable regional differences in intensity of tectonic deformation and thick wedge of unconsolidated Pliocene and Quaternary sediment on the Rosetta Fan (B and C after Ryan and others, 1970).

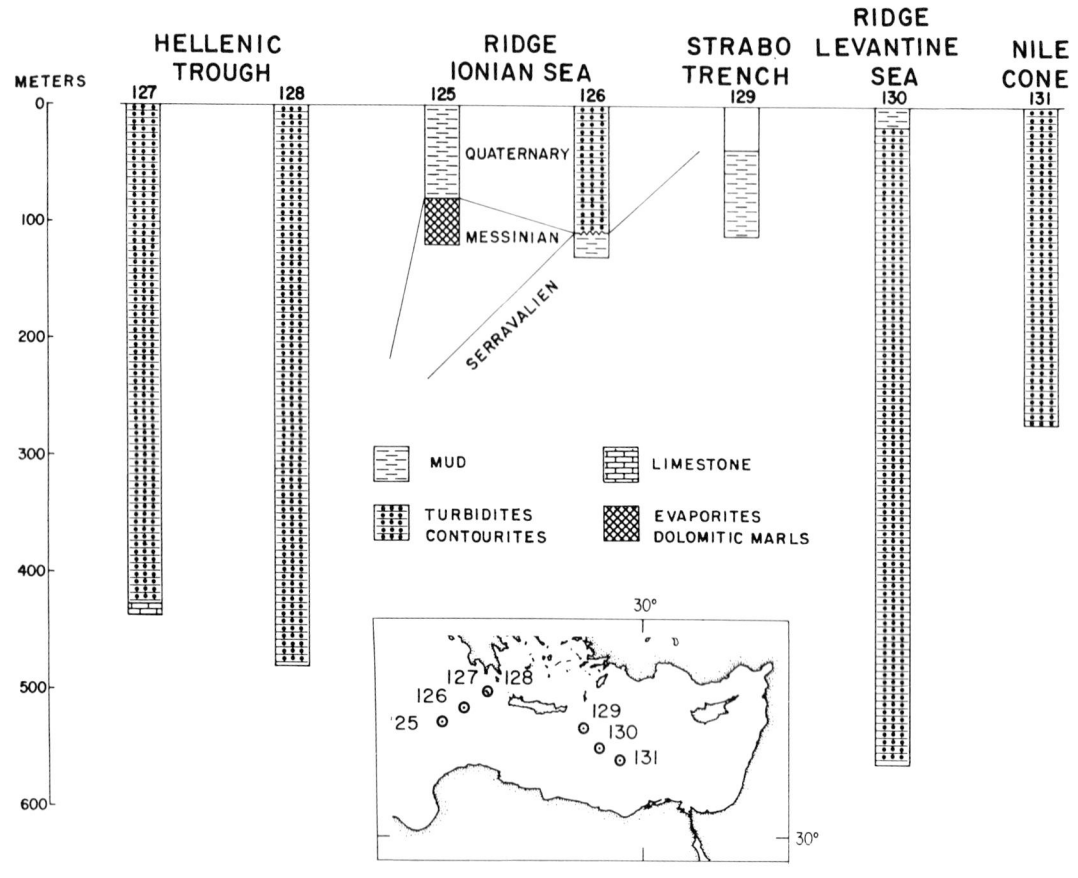

FIG. 6.—JOIDES leg 13 drill sites 125–131 in study area. Sites 127 and 128 in Hellenic Trough are located near air-gun profiles shown in figure 5B (after Nesteroff and others, 1972).

thickness of the unconsolidated section (Pliocene and Quaternary for the most part) is regionally scant (<0.25 to 0.5 km) according to Wong and Zarudzki (1969). Greatest accumulations (0.5 to 1.0 km or more fill former trench floors and depressions. There are some exceptions in that tectonic offset has left some thick isolated pockets draped on steep slopes or perched on topographic highs (fig. 5A). Almost all seismic profiles reveal that Hellenic Arc sedimentation has not kept up with tectonic activity as shown by the fact that some basin-plain floors, although flat, are not horizontal and by the fact that strata are tilted or offset as indicated by reflectors in trenches and basins (figs. 5, 9). Furthermore, the pods of strata appear to sag, attesting to concurrent foundering of the depression, or to uplift of the trench and basin margins, or to both during sedimentation.

Recent drilling by the D/V *Glomar Challenger* on its leg 13 recovered about 500 m of Quaternary terrigenous deposits and resedimented carbonate oozes from the Hellenic Trench (fig. 6). Near the trench wall this unconsolidated fill is underlain by bedrock (JOIDES site 127), probably the inner trench wall of contorted rock (tectonic mélange?) of Early Cretaceous age (Ryan and others, 1970; Nesteroff and others, 1972; Hsü, 1972). Estimates of rates of terrigenous sedimentation range from <10 to well more than 30 cm per 1,000 years, according to Mellis (1954) and others; pelagic oozes accumulate much more slowly, at about 3 cm per 1,000 years, according to Nesteroff and others (1973). Intense earthquake activity coupled with relatively small size and topographic isolation of trench floors and basin plains have resulted in rapid ponding. It is worth noting that the average estimate of sedimentation in nine ancient geosynclines ranges from 15 to 17 cm per 1,000 years (Kuenen, 1950). This rate approximates that of the Pliocene and Quaternary fill, considering that there has been a twofold to threefold reduction in thickness during lithification of these ancient rocks (that is, a rate of 30 to 45 cm/1,000 years prior to compaction and consolidation).

The thickness in trenches (see fig. 5B) ranges from almost no sediment to more than 1 km (Wong and others, 1971). Access to sediments and rates of compression of the trench floor modifies the sediment fill. The effective role of canyons as byways for the downslope movement of both coarse and fine sediment in certain sectors of the Hellenic Trench can be demonstrated. One such area is a small deep pond (about 5,100 m) southwest of the Peloponnesus (lat 36°03'N, long 21°05'E) termed the *Fosse Ouest* by French authors. This low, mapped by Hersey (1965) and Goncharov and Mikhailov (1966), appears as a small irregularly shaped basin plain bounded on the south by relatively steep slopes, locally exceeding 35°, and small steplike escarpments. A considerably more gentle topography comprising a valley and fan complex (fan area, 10 km²) is present along the northern margin. A series of submersible dives into this, the deepest recorded sector of the Mediterranean, has revealed gravitative flows from the adjacent steep margin onto the basin floor together with a classic depositional cone at the base of the slope. During two dives, Pareyn (1968, fig. 7) collected evidence for mudflows and mass movement as interpreted from undulating sea-floor surfaces, large cracks indicative of allochthonous-slab movement, and from the very irregular distribution of stiff clays alternating with zones of extremely soft oozes. Although he tended to dispute the importance of turbidity currents in this particular basin, Pareyn actually described a turbid density current that was triggered when his submersible struck the sea floor. In sum, metastable water-laden muds on the margins apparently are set into motion periodically by the frequent earthquakes affecting this region; failure results in the transfer of sediment into the basin, not from the shallow platform, but from the slope proper. The importance of these gravitative processes in flysch sedimentation and the presence of the fan-and-valley complexes in most turbidite basins preserved in the rock record are well established (Mutti and Ricci-Lucchi, 1972; Stanley and Unrug, 1972; Mutti, this volume).

Superimposed on the mass gravity-transport mechanisms is the role of the circulatory system alluded to in an earlier section. Mass movement of water undoubtedly modifies the sediment-distribution patterns as is attested by the long-distance transportation of volcanic ashes (Ninkovich and Heezen, 1965) and by wind- and river-derived sediment, as well as by pelagic organisms, for all these are found not only in basins but on isolated topographic highs as well, including the Mediterranean Ridge. A detailed study of a flyschlike basin in the Alboran Sea, western Mediterranean, has shown how bottom currents are effective in eroding sediment on basin margins and in redepositing them downslope on the basin plains (Bartolini and others, 1972; Huang and Stanley, 1972). Unfortunately, few data on bottom-current regime and velocities exist for the zone just above the sea floor in the deep areas of the Hellenic Arc; water flowing above the sills from the Aegean Sea toward the Levantine Sea however, probably is capable of eroding and redepositing fines. South of Crete, bottom water moves toward the west (Miller, 1972, fig. 1). Although the presence of contour-following currents is suspected, just how much of the thick sections of mud in this region are contourites (Nesteroff and others, 1972), so to speak, is still questionable.

One must use extreme caution in extrapolating existing oceanographic conditions to the past, for there can be little doubt that the present circulatory system is a recent phenomenon. That water movement was different in the Pliocene and Pleistocene Epochs is indicated by the dark layers (sapropels) present in almost all cores (fig. 7). These layers reflect circulatory restriction and stratification of the water masses unlike the presently observed conditions (Wüst, 1961). Climatic changes in the Pleistocene, that is, in precipitation and temperature, modified evaporation and river discharge to the levels we now know. During eustatic lower stands, the sills at the Strait of Gibraltar (Huang and others, 1972), at the Straits of Sicily, and those of the Dardanelles and Bosporus (Ryan, 1972) promoted density stratification and the resulting accumulation of pyrite, organic matter, diatoms, and pteropods in the dark layers (Olausson, 1961).

Dives to the sea floor have shown sediment surfaces to be highly modified by the activity of benthonic organisms (Drach, 1968; Pérès, 1968; Picard, 1968). Cores also show bioturbation (fig. 8). The generally quiet appearance of the Hellenic Trench floor, however, gives a somewhat misleading impression as to the processes that account for the bulk of the thick depositional wedges. The sum of data indicates that, in addition to slumping, most of the material in depressions accumulates from turbidity currents and the deposition of pelagic organisms. This is revealed in standard piston cores (fig. 8 herein; also, Emery and others, 1966; Ryan and others, 1970; and Horn and others, 1972) as

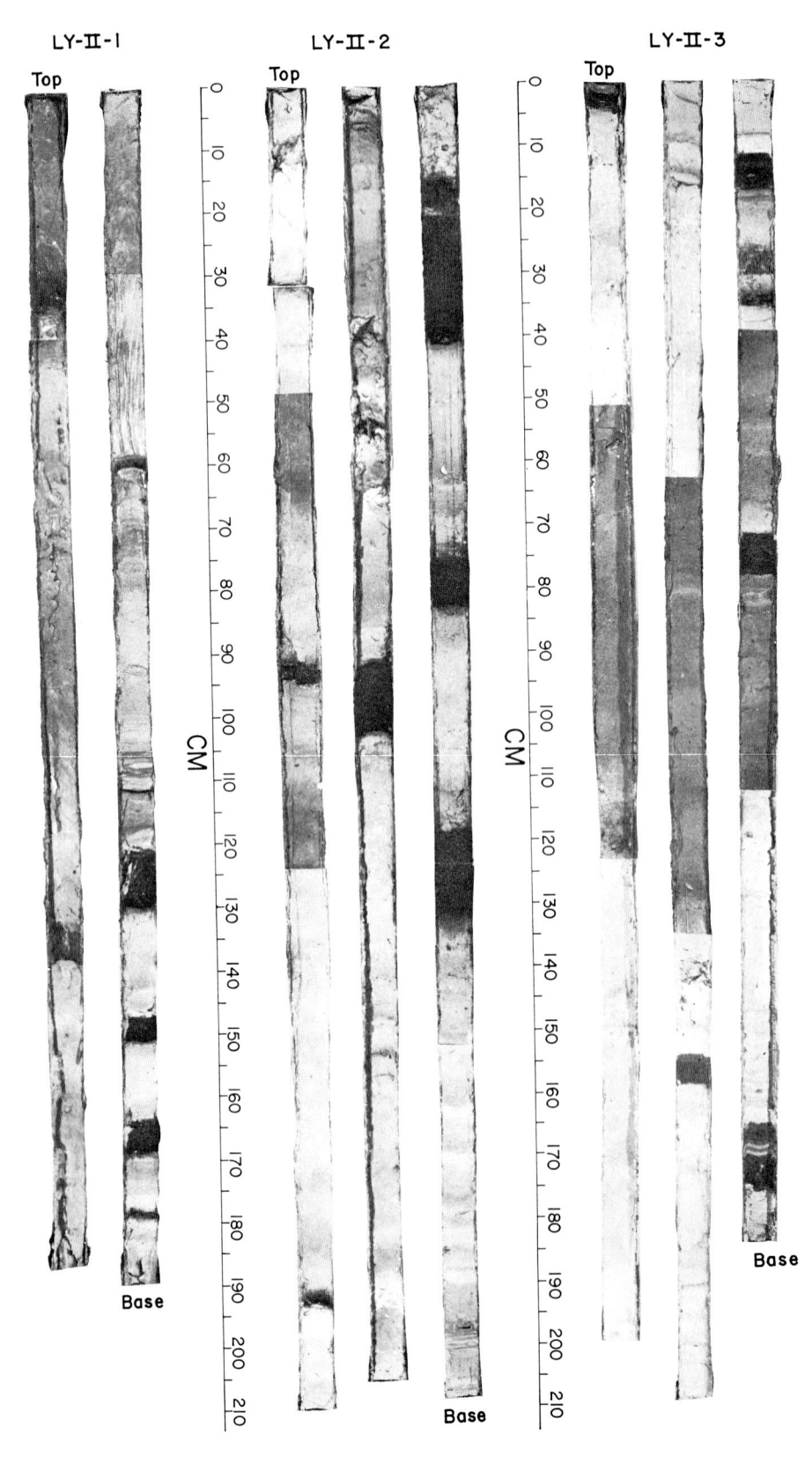

FIG. 7.—Piston cores (*LYNCH II* cruise, 1972) collected along 3.5-kHz line (see fig. 1) showing sapropels (dark layers), pelagic oozes, and hemipelagic mud. Core lithologies, and sapropels in particular, reflect fluctuations of stratification of water mass in Pleistocene and early Holocene time and a circulatory system unlike that presently observed in the Mediterranean.

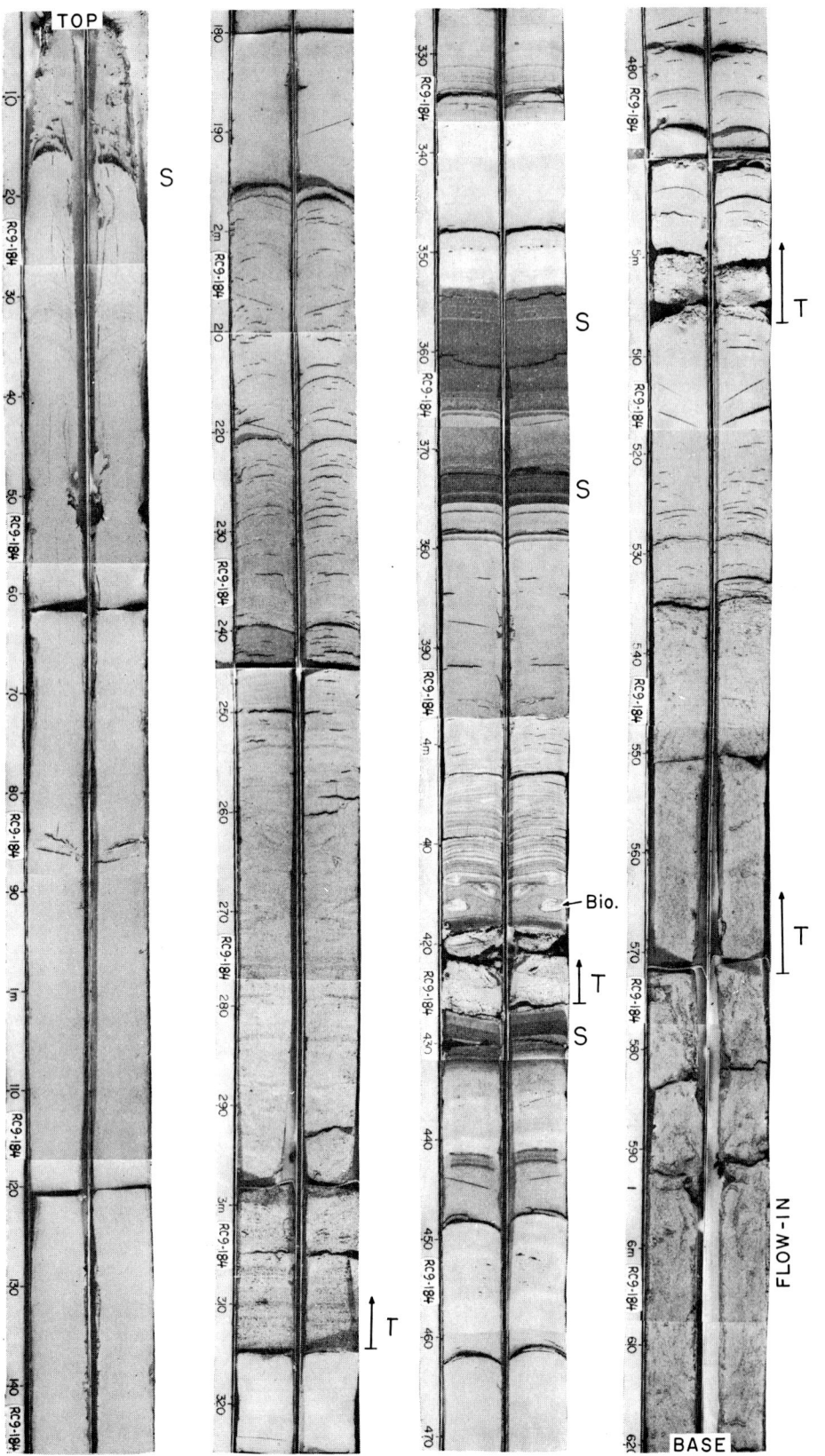

FIG. 8.—Core R69-184 (*Conrad 9* cruise, courtesy of Lamont-Doherty Core Laboratory) collected on trench floor of Hellenic Trough along air-gun profile a-a' (see figs. 1 and 5B). Note regular alternation of graded turbidite sands (T) and turbiditic muds, hemipelagic fines, and sapropels (S). Turbidite at depth of 420 cm probably reworked (Bio) by benthic organisms.

well as in cores from the recent JOIDES leg 13 (Ryan and others, 1973). Furthermore, the role of earthquakes as triggering mechanisms of gravity flows (turbidity currents, slumps, grain flows, etc.) is documented by recorded breaks of telegraph cables in the Gulf of Corinth (Heezen and others, 1966) and in the Ionian Sea (Ryan and Heezen, 1965). Neglected to date is an evaluation of the role of turbidity currents (and turbid layer flows, Moore, 1969) in the downslope transport of mud of the textural type that dominates the Pleistocene and Holocene sediment section as recorded by cores. Clearly, mud turbidites (Stanley, 1969; Piper, 1970) are often observed to be more common than the classic well-studied graded sand turbidites (fig. 8, T). Graded silt and clay sequences are usually overlooked in regional reconnaissance surveys, for their distinction from pelagic units requires special petrologic techniques (Hall and Stanley, 1973).

A continuous track line of 3.5-kHz profiles (fig. 9) penetrating 30 to 60 m of section (subparallel to the *Conrad 9* air-gun profiles described by Ryan and others, 1970) along a Hellenic Trench-Mediterranean Ridge traverse

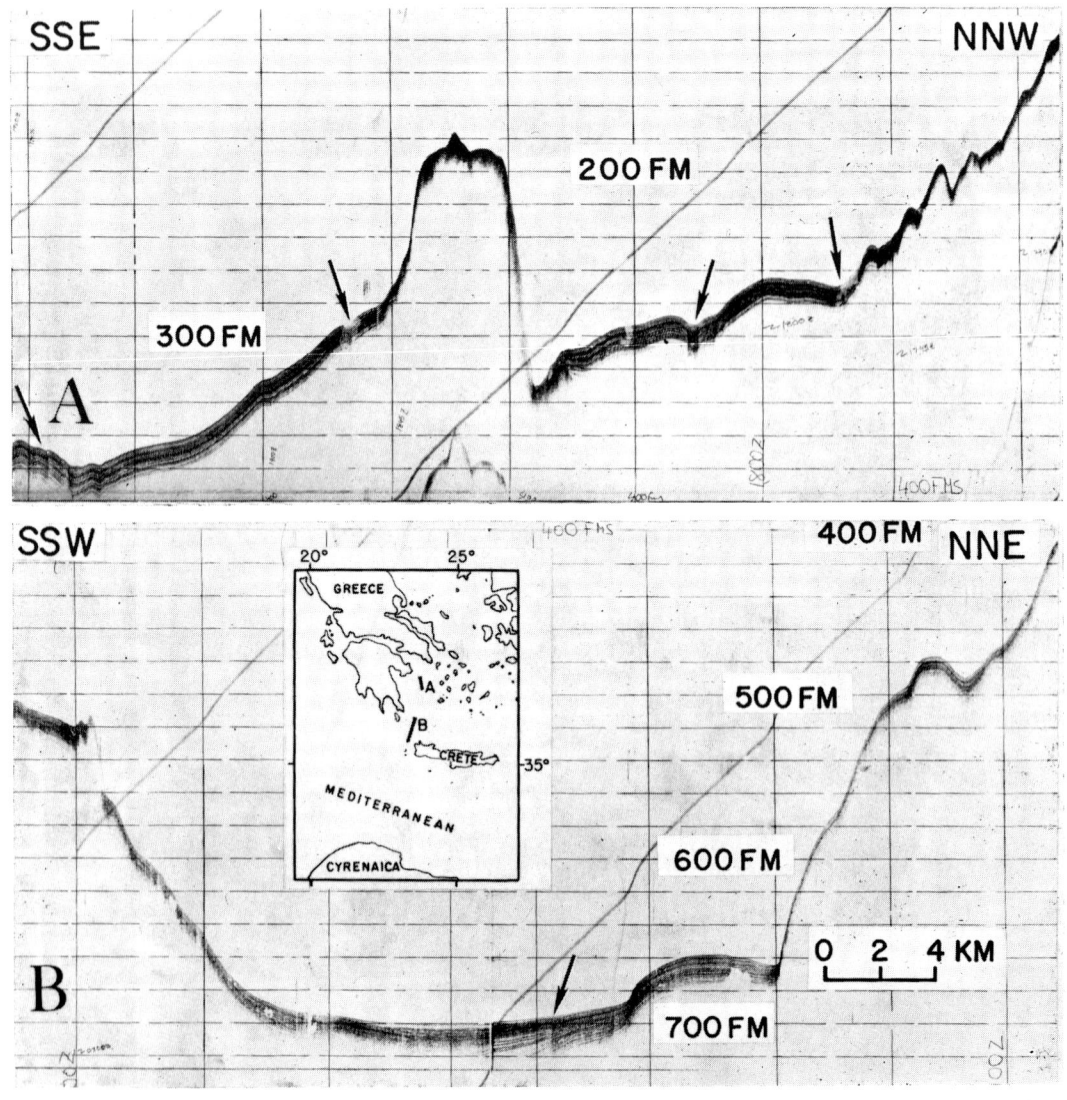

FIG. 9.—3.5 kHz profiles across sectors of the inner arc displaying remarkable resolution of upper 30 to 45 m of Quaternary section. Reflectors show considerable lateral continuity as well as neotectonic structural offset (arrows) affecting even uppermost strata.

shows the rhythmic nature of unconsolidated sections and the relatively long distance extension of reflectors. That many of these reflectors are thin sandy turbidites is borne out by an examination of cores V10-35, V10-36 (fig. 1) and RC-9-184 (fig. 8) collected along these seismic lines. As in other areas of the Mediterranean, for example, in the Alboran Sea (Bartolini and others, 1972), Tyrrhenian Sea (Ryan and others, 1965), and the Ionian Sea (Ryan and Heezen, 1965), these reflectors show that strata, although thin (generally < 50 cm), are continuous across much or all the floor of a depression. All the high-resolution profiles for the Sea of Crete and the Hellenic Arc show sharp offsets of reflectors (see arrows in fig. 9) that once again indicate the continuous activity of the substrate and adjacent margins in this seismically active region.

CONCLUSIONS

We return to the original question, Is there modern flysch? The very high resolution 3.5-kHz profiles, coupled with core data, reveal sequences of unconsolidated mud and sand in the Hellenic Arc that are comparable, at least petrologically, to mud-rich flysch formations exposed on land. Additional support for this interpretation is provided by the associated structural, geographic, and stratigraphic characteristics that many workers feel are implicit in the term flysch. In this respect, Hsü and Schlanger (1971, fig. 1) have compared the Paleocene Schlieren Basin with the Aegean Sea, the Gurnigel Trench with the Pliny and Strabo Trenches, the *Couches Rouges* with the Herodotus abyssal plain, and the Simmen Fan with the Nile Cone. An overall evaluation does indicate that this area presents all the prerequisites of one type of modern flysch in a compressive setting. On the basis of petrologic and facies characteristics and of comparison with ancient flysch in the rock record, I find enough similarities to designate as flysch the turbidite-rich deposits described here.

As stated earlier in this review, there is generally good agreement, at least sedimentologically, as to what lithological varieties are comprised in this type of deposit, how these units were deposited, and how to go about studying their geometry and facies organization (Walker, 1970). With new techniques, examination of the modern record now provides a fresh approach to an old and complicated problem and perhaps a means to interpret some of the features observed in the field where tectonics and lack of outcrop control is so often a limiting factor.

Although the area covered by the Hellenic Arc is negligible when compared to the world oceans, the diversity of conditions within even this small region is remarkable, and it should be taken into account when considering ancient time-equivalent flysch basins along any single regional trend. An example in point is the upper Eocene and lowermost Oligocene flysch (Champsaur, Annot, Taveyannaz, Val d'Illiez, etc.) in the Dauphinois-Ultradauphinois zone of the French Alps, which is a belt about 300 km long. Total thicknesses, sand-shale ratios, slump-turbidite-nonturbidite ratios, mineralogy (including zones of volcanic versus nonvolcanic content), paleocurrent directions, and degree of syndepositional tectonic and volcanic influence are variable along the belt, and each unit is distinguished as a separate formation. The area occupied by each of these formations is impressive to the geologist in the field, but, in reality, the arcuate belt along which they crop out represents only a fraction of the length of the Hellenic Arc. That is, each unit's length is about equal to the length of that part of the Hellenic Arc along Crete or of that part that is west of the Peloponnesus. Attention should also be called to the fact that the length of the present Hellenic Arc, which extends from west of Greece to Rhodes and southwestern Turkey, is roughly equal to that of the Alpine Arc from the Riviera through the French, Swiss, and Austrian Alps and to the Czechoslovakian border. Each of the flysch units of equivalent age in this Alpine belt presents a unique set of facies characteristics, and each is thus mappable as a distinct depositional unit. For the same reasons, it can be argued that the ponds of Pliocene and Quaternary sand and mud units in the Hellenic Arc are different enough from one another to be mapped as separate formations like those in the rock record.

Summarizing major observations, we find that turbidite-filled depressions occur on both sides of the Hellenic island arc. Even though the general setting of the area selected here is one of compression, there is ample evidence of distension in terms of both foundering of the basin and trench basement and of grabenlike structures and normal fault offsets in the stratified sequences. Basins of the inner arc have received a greater proportion of terrestrial (mostly fluvial) and volcanic input, are considerably larger, less markedly elongate, and generally much shallower than those south of the island arc. Although generally deep, some flysch-type fills are accumulating, in fact, at depths of 200 to 300 m. It appears that the very deep water connotation, an aspect recently incorporated in the definition of flysch, should be carefully appraised; alternating sequences of graded turbidites and muds occur on slopes and on basin plains at bathyal as well as abyssal depth.

Distance to land is by no means a critical factor as far as sedimentation rates are concerned. More important is access to sediments transported by rivers, circulating water masses, sediment-carrying bottom currents, and such gravity-influenced mechanisms as turbidity currents, turbid layer flows, grain flows, and slumps. Furthermore, such circumstances as the presence of submarine canyons, natural byways channelizing materials into ponds, and seasonal influences affecting winter and spring river floods or blooms of pelagic organisms need to be considered in the explanation of basin-to-basin differences in total thicknesses and rates of sedimentation. Of all factors, however, tectonics is the major agent. Tectonics controls the position, topographic isolation, and orientation of each depression, such control resulting, for instance, in the distinct alignment of the system comprised of the Hellenic, Strabo, and Pliny Trenches on the convex margin of the arc. Further, tectonics trigger the earthquakes that result in downslope transfer of the bulk of terrigenous sediment.

The core and seismic reconnaissance surveys show that gravitative processes clearly dominate and account for the largest volume of base-of-slope deposits. Deep-sea fans are best developed where sedimentation rates are either high or at least in excess of tectonic offset. The fills inevitably have higher percentages of mud than sand, and detailed petrologic methods are needed to distinguish mud turbidites from so-called contourites and pelagic strata, for all three types are undoubtedly present. Individual layers, some of them sand turbidites, can be traced across the length of some basins and in some instances bundles of well-stratified, alternating sand and mud layers can be traced over tens of kilometers.

Regional variations in seismicity have produced differences in thicknesses, attitude of bedding, degree of neotectonic structures, and in mineralogy north and south of the Mediterranean Ridge. The fill of trenches and basin plains adjacent to the island arc call to mind some of the classic Alpine Flysch, including the *flysch marneaux* and shaly facies of the Simmental. The composition of sands in the Hellenic Arc, including volcanic and some metamorphic and lithic fractions tend to be dirtier and regionally more varied than those on the Nile Cone and Herodotus Plain. These Nile and Herodotus deposits comprise thick silt-enriched facies that bear some similarity to the distal facies of the thick Tertiary *flysch gréseux* facies such as in the Nummidian and Annot units exposed in the western circum-Mediterranean region.

Lest the reader be misled, he should remember that the intricate details of sedimentation patterns in the Hellenic Arc are only beginning to be understood and that no complete analysis of even a single basin has been published. Although the Hellenic Arc admittedly is not unique, few other areas exist where the interaction of contemporary sedimentation and tectonics can be so well demonstrated. Here then, is an ideal natural laboratory in which to pursue the problem of flysch sedimentation.

ACKNOWLEDGMENTS

Cores and bathymetric and 3.5-kHz records serving as the basis for this paper were collected on the April 1972 cruise of the USNS *Lynch,* and I thank the U.S. Navy for use of the equipment as well as for ship-time support. I also express my appreciation to the administration of Lamont-Doherty Geological Observatory Core Laboratory for use of supplemental core material and photographs, to W.B.F. Ryan, also of Lamont-Doherty, for two *Conrad 9* airgun profiles, and to R. Zarudzki, CNR, Bologna, for permission to use one sparker profile. Drafts of this paper were read by S. Dzulynski, Jagellonian University, G. Kelling, University of Wales, F. McCoy, Smithsonian Institution, E. Mutti, University of Turin, and R. H. Dott, Jr., University of Wisconsin, and their suggestions have been helpful. Financial support for the Mediterranean study was provided by Smithsonian Institution Foundation grants FY-72:427235 and FY-73:430035.

REFERENCES

Argand, E., 1916, Sur l'arc des Alpes Occidentales: Ecologae Geol. Helvetiae, v. 14, p. 145–191.
Aubouin, J., 1965, Geosynclines: developments in geotectonics: Amsterdam, Elsevier Publishing Company, 335 p.
Barazangi, M., and Dorman, J., 1969, World seismicity maps compiled from ESSA Coast and Geodetic Survey, epicenter data, 1961–1967: Seismol. Soc. America Bull., v. 59, p. 369–380.
Bartolini, C., Gehin, C., and Stanley, D. J., 1972, Morphology and recent sediments of the Western Alboran Basin in the Mediterranean Sea: Marine Geology, v. 13, p. 159–224.
Belderson, R. H., Kenyon, N. H., and Stride, A. H., 1972, Comparison between narrow-beam and conventional echo-soundings from the Mediterranean Ridge: *ibid.,* v. 12, p. m11–m15.
Bertrand, M., 1897, Structures des alpes françaises et récurrence de certain faciès sédimentaires: 6th Cong. Internat. Géol., 1894, Compte Rendu, p. 163–177.

Bonatti, E., and others, 1972, Submarine iron deposits from the Mediterranean Sea, in Stanley, D. J. (ed.), The Mediterranean Sea: a natural sedimentation laboratory: Stroudsburg, Pennsylvania, Dowden, Hutchinson and Ross, p. 701–710.
Carter, T. G., and others, 1972, A new bathymetric chart and physiography of the Mediterranean Sea: ibid., p. 19–23.
Comninakis, P. E., and Papazachos, B. C., 1972, Seismicity of the eastern Mediterranean and some tectonic features of the Mediterranean Ridge: Geol. Soc. America Bull., v. 83, p. 1093–1102.
Dewey, J. F., and Bird, J. M., 1970, Mountain belts and the new global tectonics: Jour. Geophys. Research, v. 75, p. 2625–2647.
Dietz, R. S., and Holden, J. C., 1966, Deep-sea deposits in but not on the continents: Am. Assoc. Petroleum Geologists Bull., v. 50, p. 351–362.
Dott, R. H., Jr., 1963, Dynamics of subaqueous gravity depositional processes: ibid., v. 47, p. 104–128.
Drach, P., 1968, Observations topographiques et biologiques effectuées dans une fosse de la Mer Ionienne au S.-W. de l'ile de Sapientza, in Résultats scientifique des Campagnes du Bathyscaphe Archimède (Grèce, 1965): Annales Inst. Océanogr., new ser., v. 46, p. 35–40.
Dzulynski, S., and Walton, E. K., 1965, Sedimentary features of flysch and greywackes; Developments in sedimentology 7: Amsterdam, Elsevier Publishing Company, 274 p.
Emery, K. O., Heezen, B. C., and Allan, T. D., 1966, Bathymetry of the eastern Mediterranean Sea: Deep-Sea Research, v. 13, p. 173–192.
Galanopoulos, A. G., 1968, The earthquake activity in the physiographic provinces of the eastern Mediterranean Sea: Natl. Observatory Athens Seismol. Inst., Sci. Progress Rept. 11, 15 p.
Goncharov, V. P., and Mikhaylov, O. V., 1963, New data concerning the topography of the Mediterranean Sea bottom: Okeanologiya, v. 3, p. 1056–1601.
Grossgeym, V. A., 1972, Structure and conditions of formation of flysch: Geotectonics, no. 1, p. 22–25.
Hall, B. A., and Stanley, D. J., 1973, Levee-bounded submarine base-of-slope channels in the Lower Devonian Seboomook Formation, northern Maine: Geol. Soc. America Bull., v. 84, p. 2101–2110.
Heezen, B. C., Ewing, M., and Johnson, L. G., 1966, The Gulf of Corinth floor: Deep-Sea Research, v. 13, p. 381–411.
———, Hollister, C. D., and Ruddiman, W. F., 1966, Shaping of the continental rise by deep geostrophic contour currents: Science, v. 152, p. 502–508.
Hersey, J. B., 1965, Sediment ponding in the deep sea: Geol. Soc. America Bull., v. 76, p. 1251–1260.
Holeman, J. N., 1968, The sediment yield of major rivers of the world: Water Res. Research, v. 4, p. 737–741.
Horn, D. R., Ewing, J. I., and Ewing, M., 1972, Graded-bed sequences emplaced by turbidity currents north of 20°N in the Pacific, Atlantic and Mediterranean: Sedimentology, v. 18, p. 247–275.
Hsü, K. J., 1970, The meaning of the word flysch—a short historical search, in Lajoie, J. (ed.), Flysch sedimentology in North America: Geol. Assoc. Canada Special Paper 7, 272 p.
———, 1972, Alpine Flysch in a Mediterranean setting: 24th Internat. Geol. Cong., Sec. 6, p. 67–74.
———, and Schlanger, S. O., 1971, Ultrahelvetic flysch sedimentation and deformation related to plate tectonics: Geol. Soc. America Bull., v. 82, p. 1207–1218.
Huang, T.-C., and Stanley, D. J., 1972, Western Alboran Sea: sediment dispersal, ponding and reversal of currents, in Stanley, D. J. (ed.), The Mediterranean Sea: a natural sedimentation laboratory: Stroudsburg, Pennsylvania, Dowden, Hutchinson and Ross, p. 521–559.
———, Stanley, D. J., and Stuckenrath, R., 1972, Sedimentological evidence for current reversal at the Strait of Gibraltar: Marine Technology Soc. Jour., v. 6, p. 25–33.
Hubert, J. F., 1964, Textural evidence for deposition of many western North Atlantic deep-sea sands by ocean bottom currents rather than turbidity currents: Jour. Geology, v. 72, p. 757–785.
International Geological Congress, 1972, The comparative stratigraphy and sedimentology of flysch basins: Sec. 6, Stratigraphy and Sedimentology, p. 59–117.
International Sedimentological Congress (5th), 1959, Bassins détritiques: molasse, flysch, houiller et autres: Eclogae Geol. Helvetiae, v. 51, p. 827–1172.
Kuenen, Ph. H., 1950, Marine Geology: New York, John Wiley and Sons, Inc., 568 p.
———, 1958, Problems concerning source and transportation of flysch sediments: Geol. en Mijnb., ser. 20e, p. 329–339.
———, 1967, Emplacement of flysch-type sand beds: Sedimentology, v. 9, p. 203–243.
———, and Migliorini, C. I., 1950, Turbidity currents as a cause of graded bedding: Jour. Geology, v. 58, p. 91–127.
Lacombe, H., and Tchernia, P., 1972, Caractères hydrologiques et circulation des eaux en Méditerranée, in Stanley, D. J. (ed.), The Mediterranean Sea: a natural sedimentation laboratory: Stroudsburg, Pennsylvania, Dowden, Hutchinson and Ross, p. 25–36.
Lajoie, J. (ed.), 1970, Flysch sedimentology in North America: Geol. Assoc. Canada Special Paper 7, 272 p.
Le Pichon, X., 1968, Sea-floor spreading and continental drift: Jour. Geophys. Research, v. 73, p. 3661–3697.
Lort, J. M., 1971, The tectonics of the eastern Mediterranean: Rev. Geophysics and Space Physics, v. 9, p. 189–216.
McKenzie, D. P., 1970, Plate tectonics of the Mediterranean region: Nature, v. 226, p. 239–243.
Maley, T. S., and Johnson, L. G., 1971, Morphology and structure of the Aegean Sea: Deep-Sea Research, v. 18, p. 109–122.
Mangin, J. P., 1962, Le flysch, sédiment climatique?: Séances Soc. géol. France Compte Rendu Sommaire, ser. 2, p. 34–37.
Mellis, O., 1954, Volcanic ash-horizons in deep-sea sediments from the eastern Mediterranean Deep-Sea Research, v. 2, p. 89–92.
Miller, A. R., 1972, Speculations concerning bottom circulation in the Mediterranean Sea, in Stanley, D. J. (ed.), The Mediterranean Sea: a natural sedimentation laboratory: Stroudsburg, Pennsylvania, Dowden, Hutchinson and Ross, p. 37–42.

MITCHELL, A. H., AND READING, H. G., 1969, Continental margins, geosynclines, and ocean floor spreading: Jour. Geology, v. 77, p. 629–646.

MOORE, D. G., 1969, Reflection profiling studies of the California continental borderland. Structure and Quaternary turbidite basins: Geol. Soc. America Special Paper 107, 142 p.

MUTTI, E., AND RICCI-LUCCHI, F., 1972, Le torbiditi dell' Appennino settentrionale: introduzione all' analisi di facies: Soc. Geol. Italiana Mem., v. 11, p. 161–199.

NESTEROFF, W. D., AND OTHERS, 1972, Evolution de la sédimentation pendant le néogène en Méditerranée d'après les Forages Joides-DSDP, in STANLEY, D. J. (ed.), The Mediterranean Sea: a natural sedimentation laboratory: Stroudsburg, Pennsylvania, Dowden, Hutchinson and Ross, p. 47–62.

———, WEZEL, F. C., AND PAUTOT, G., 1973, Summary of lithostratigraphic findings and problems, in RYAN, W. B. F., HSÜ, K. J., AND OTHERS, Initial reports of the Deep Sea Drilling Project: Washington, D.C., U.S. Govt. Printing Office, v. 13, p. 1021–1040.

NIELSEN, J. N., 1912, Hydrography of the Mediterranean and adjacent waters. Danish Oceanogr. Exped. 1908–1910, Rept. v. 1, p. 77–191.

NINKOVICH, D., AND HAYS, J. D., 1972, Mediterranean island arcs and origin of high potash volcanoes: Earth and Planetary Sci. Letters, v. 16, p. 331–345.

NINKOVICH, D., AND HEEZEN, B. C., 1965, Santorini tephra, in Colston papers, v. 17, 17th Symposium Colston Research Soc. Proc.: London, Butterworth & Co. (Publishers), Ltd., p. 413–453.

OLAUSSON, E., 1961, Studies of deep-sea cores: Reports of the Swedish Deep-Sea Expedition, 1947–1948, Rept. 8, Swedish Nat. Sci. Research Council, p. 3–391.

OULIANOFF, N., 1960, Compaction, déplacement et granoclassement des sédiments: Internat. Geol. Cong., Scandinavia, Sec. 10, Submarine Geology, p. 54–58.

PAPAZACHOS, B. C., AND COMNINAKIS, P. E., 1971, Geophysical and tectonic features of the Aegean Arc: Jour. Geophys. Research, v. 76, p. 8517–8533.

PAREYN, C., 1968, Observations géologiques et sédimentologiques dans la fosse ouest de la mer Ionienne, in, Résultats scientifique des Campagnes du Bathyscaphe Archimède (Grèce, 1965): Annals Inst. Océanogr., new ser., v. 46, p. 6–69.

PÉRÈS, J. M., 1968, Observations efféctuées à bord du bathyscaphe Archimède dans la fosse située au S.-W. de l'île de Sapientza (Mer Ionienne): ibid., p. 41–46.

PICARD, J., 1968, Observations biologiques efféctuées à bord du bathyscaphe Archimède, dans l'une des fosses situées dans le Sud du Cap Matapan: ibid., p. 47–51.

PIPER, D. J. W., 1970, Transport and deposition of Holocene sediment in La Jolla deep-sea fan, California: Marine Geology, v. 8, p. 211–227.

———, 1972, Turbidite origin of some laminated mudstones: Geol. Mag., v. 109, p. 115–126.

RAAF, J. F. M., DE, 1968, Turbidites et associations sédimentaires apparentées. I, II: K. Nederlandsch Akad. Wetensch.-Amsterdam Proc., Ser., B, v. 71, p. 1–23.

RABINOWITZ, P. D., AND RYAN, W. B. F., 1970, Gravity anomalies and crustal shortening in the eastern Mediterranean: Tectonophysics, v. 10, p. 585–608.

READING, H. G., 1972, Global tectonics and the genesis of flysch successions: 24th Internat. Geol. Cong., Sec. 6, Stratigraphy and Sedimentology, p. 59–66.

RYAN, W. B. F., 1972, Stratigraphy of the late Quaternary sediments in the eastern Mediterranean, in STANLEY, D. J. (ed.), The Mediterranean Sea: a natural sedimentation laboratory: Stroudsburg, Pennsylvania, Dowden, Hutchinson and Ross, p. 149–169.

———, AND HEEZEN, B. C., 1965, Ionian Sea submarine canyons and the 1908 Messina turbidity current: Geol. Soc. America Bull., v. 76, p. 915–932.

———, AND OTHERS, 1970, The tectonics and geology of the Mediterranean Sea, in MAXWELL, A. E. (ed.), The sea, v. 4, Pt. 2: New York, Wiley-Interscience, p. 387–492.

———, VENKATARATHNAM, K., AND WEZEL, F. C., 1973, Mineralogical composition of the Nile cone, Mediterranean Ridge, and Strabo Trench sandstones and clays, in RYAN, W. B. F., HSÜ, K. J., AND OTHERS, Initial reports of the Deep-Sea Drilling Project: Washington, D.C., U.S. Govt. Printing Office, v. 13, p. 731–746.

———, WORKUM, F., AND HERSEY, J. B., 1965, Sediments on the Tyrrhenian abyssal plain: Geol. Soc. America Bull., v. 76, p. 1261–1282.

SEMPLE, E. C., 1931, The geography of the Mediterranean region—its relation to ancient history: New York, AMS Press, 737 p.

SHEPARD, F. P., AND DILL, R. F., 1966, Submarine canyons and other sea valleys: Chicago, Rand McNally & Co., 381 p.

SHUKRI, N. M., 1950, The mineralogy of some Nile sediments: Geol. Soc. London Quart. Jour., v. 105, p. 511–534; v. 106, p. 466–467.

STANLEY, D. J. (ed.), 1969, The new concepts of continental margin sedimentation: Washington, D.C., Am. Geol. Inst., 400 p.

———, 1970, Flyschoid sedimentation on the outer Atlantic margin off northeast North America: Geol. Assoc. Canada Special Paper 7, p. 179–210.

———, GEHIN, C. E., AND BARTOLINI, C., 1970, Flysch-type sedimentation in the Alboran Sea, western Mediterranean: Nature, v. 228, p. 979–983.

———, AND UNRUG, R., 1972, Submarine channel deposits, fluxoturbidites and other indicators of slope and base-of-slope environments in modern and ancient marine basins, in RIGBY, J., AND HAMBLIN, W. (eds.), Recognition of ancient sedimentary environments: Soc. Econ. Paleontologists and Mineralogists Special Pub. 16, p. 287–340.

STUDER, B., 1827, Remarques gèognostiques sur quelques parties de la chaîne septentrionale des Alpes: Annales Sci. naturales Paris, v. 11, p. 1–47.

———, 1848, Sur la véritable signification du nom de flysch: Swcheizerische Nature. Gesell. Verh., v. 33, p. 32–35.

TERCIER, J., 1947, Le flysch dans la sédimentation alpine: Ecologae Geol. Helvetiae, v. 40, p. 164–198.
TRÜMPY, R., 1960, Paleotectonic evolution of the Central and Western Alps: Geol. Soc. America Bull., v. 71, p. 843–908.
VENKATARATHNAM, K., AND RYAN, W. B. F., 1971, Dispersal patterns of clay minerals in the sediments of the eastern Mediterranean Sea: Marine Geology, v. 11, p. 261–282.
WALKER, R. G., 1967, Turbidite sedimentary structures and their relationship to proximal and distal depositional environments: Jour. Sed. Petrology, v. 37, p. 25–43.
WALKER, R. G., 1970, Review of the geometry and facies organization of turbidities and turbite-bearing basins: Geol. Assoc. Canada Special Paper 7, p. 219–251.
WONG, H-K., AND ZARUDZKI, E. F. K., 1969, Thickness of unconsolidated sediments in the Eastern Mediterranean Sea: Geol. Soc. America Bull., v. 80, p. 2611–2614.
———, AND OTHERS, 1971, Some geophysical profiles in the eastern Mediterranean: *ibid.*, v. 82, p. 91–100.
WÜST, G., 1961, On the vertical circulation of the Mediterranean Sea: Jour. Geophys. Research, v. 66, p. 3261–3271.

SUCCESSOR BASIN ASSEMBLAGES

NORTHERN ALPINE MOLASSE AND SIMILAR CENOZOIC SEQUENCES OF SOUTHERN EUROPE

F. B. VAN HOUTEN
Princeton University, Princeton, New Jersey

ABSTRACT

Cenozoic molasse in the northern Alpine and Pyrenean Aquitaine and Ebro foredeeps consists of late orogenic clastic wedges initiated by the first major uplift of the mountain belt and associated thrusting. Deposits in these basins, tens of kilometers wide and hundreds long, grade upward from autochthonous flysch, through a lower thin-bedded mudstone without turbidites, into an upper transitional paralic sequence with minor proximal conglomerates. The succeeding nonmarine to paralic molasse dominated the foredeeps and graded laterally into persistent marine deposits. During accumulation of molasse, sedimentation kept pace with subsidence, producing a surface near sea level that was susceptible to exchange of marine and nonmarine conditions. At times of reduced sediment supply, shallow marine transgressions spread across the lowlands.

Molasse is heterogeneous and lenticular. Its distinctive proximal, nonmarine facies, several thousands of meters thick, consists mainly of fanglomerates (nagelfluhen) and repetitious sequences of fining-upward fluvial cycles. Conglomeratic deposits, commonly exceeding 1000 m in thickness, are characterized by closely packed, rather well rounded clasts. Fluvial cycles comprise immature lithic and feldspathic sandstone and drab to variegated mudstone with thin lignite and coal, caliche, freshwater limestone, and evaporite. In the subordinate marine facies rather persistent, well-sorted, locally conglomeratic sandstone predominates. The composition of conglomerate clasts and sandstone grains has provided basic data about the location and progressive stripping of source areas, the emplacement of younger nappes, and the dispersal of sediments within the foredeep.

Although uplift of extensive source areas initiated and dominated the relatively undeformed molasse, older proximal molasse was overridden by nappes and folded and faulted in a zone a few tens of kilometers wide. Moreover, uplift of the mountains and emplacement of nappes were episodic and shifted along the range, as both nappes and foredeep encroached farther onto the craton. In the end the molasse was reduced to nearly half its original width.

Molasse accumulation persisted for a few tens of million years, with a maximum average preservation rate of about 400 m/my and as low as 150 to 200 m/my overall. There were two or three times of pronounced uplift and accentuated filling at intervals of about 10 to 15 my. Regional elevation and erosion of both mountain belts and foredeeps ended the molasse phase.

INTRODUCTION

The focus of this review is on late orogenic uplift[1] and thrusting, foredeep[2] subsidence, and accumulation of molasse, as significant products of orogeny. The tectofacies molasse comprises those thick clastic wedges[3] that accumulated in basins along cratonic margins and record major upheaval and denudation of orogens in their principal contribution of detritus to the mobile belt.

The problem of defining and delimiting molasse is much like that pertaining to the flysch (for example, Arkhipov, 1971; Beaudoin and others, 1970; Hsü, 1970; Perriaux and others, 1970), except that the molasse is an ordinary kind of sediment deposited under common conditions on the continents. Its complexities are more familiar, so they are the more difficult to synthesize. Concentration on small-scale environmental details within clastic wedges has often obscured a broader concern with their tectonic framework.

During the 18th century, a drab building stone in the Swiss foreland was called molasse. Studer (1825) first described these Cenozoic deposits in some detail and applied the name to the entire sequence (p. 72). In 1894 Bertrand (1897) pointed out that the molasse facies is characteristic of other mountain systems; then Haug (1900, p. 641–642) extended it to all mobile belts. In contrast, Cayeux (1929, p. 166; see

---
[1] A descriptive term for deformation that produced an extensive upland source area. It may be largely vertical movement that prevailed after the main orogenic phase as in the Central and Eastern Alps according to Milne (1969) and Cliff and others (1971); or it may be an important component of large-scale gravitational spreading accompanying upwelling of the hot, metamorphic infrastructure as in the southern Canadian Rockies according to Price (1971).

[2] In the widely accepted sense used on the International Tectonic Map of Europe (Bogdanoff and others, 1964); it is essentially the same as Kay's (1951) exogeosyncline.

[3] King (1959, p. 57) restricted the general term to deposits in foredeeps. It is much more usefully applied to any thick, nonmarine to shallow-marine product of deformation.

Bersier, 1959, p. 856–857) limited "molasse," or "mollasse," to calcareous, feldspathic sandstones like the Aquitanian sandstone of Lausanne, and this restricted usage, without reference to tectonic framework, was followed rather widely in France and adjacent Switzerland. For example, Crouzel (1957, p. 34–36, 217–218) referred to both "mollasse" sequences and "mollasse" (sandstone) in the Aquitaine Basin, and Bersier (1945, 1959) used "molasse" for a rock type and a major facies in western Switzerland. No good is gained by perpetuating the limited application of the term to a particular rock type, however.

I have attempted to summarize the characteristic sedimentary and tectonic frameworks of Cenozoic molasse in the northern Alpine and Pyrenean Aquitaine and Ebro Basins (fig. 1). An encouraging consistency emerges despite several significant differences. The definition that results aids in making meaningful comparisons of this important tectofacies.

ACKNOWLEDGMENTS

National Science Foundation grants, GA-380 and GA-14847, as well as Princeton University, supported parts of this study. It is a pleasure to record my thanks to Professor H. P. Laubscher for a most profitable and pleasant six months' association with the Geological and Paleontological Institute, University of Basel, and to acknowledge the help of D. Bernoulli (Basel), A. Matter (Bern), R. Trümpy (Zürich), H. Schmidt-Thome (München), J. Schoeffler (SNPA, Pau), and C. Puigdefabregas (Jaca). E. Dapples and R. Dott, Jr., reviewed the manuscript critically.

NORTHERN ALPINE FOREDEEP

History.—Heim's long review of the Swiss Molasse[4] (1919, p. 39–196; esp. p. 43–72) reflected nearly a century's progress after Studer's (1825) pioneer report. Since then many studies in Switzerland and southern Germany have presented abundant details. Outstanding recent examples are those by Briel (1962), Füchtbauer (1964), Matter (1964), and Gasser (1966, 1968). Good descriptions also accompany geologic maps of Switzerland and Bavaria (for example, Matter, 1970). Only a few, however, focused on the relation of the Molasse to the orogenic development of the Alps, even though the growth of the Alps is mirrored in its foreland sediments (Richter, 1927; Cadisch, 1928; Baumberger, 1934). Subsurface petroleum exploration in southern Germany and adjacent Switzerland, especially since 1945, did stimulate more concern with the framework and regional development of the Molasse (Büchi and others, 1965; Füchtbauer, 1967; Janoschek, 1963; Schmidt-Thomé, 1963; Wiedenmeyer, 1950).

General.—The Molasse north of the Central Alps (Lake Geneva to the Rhine) and Eastern Alps (the Rhine to the Danube) accumulated in a cratonic basin about 700 km long and 50 to 140 km wide that bulged northward between large Hercynian blocks (fig. 1, Molasse Basin). During early and middle Oligocene time, deposits in the foredeep graded upward from the autochthonous deep-water Flysch, through a slope facies (tonmergel), and into a paralic sequence (Lower Marine Molasse), as its axis was displaced toward the craton (see Hagn, 1960, fig. 7).

The succeding, characteristically coarse-grained, mainly nonmarine Molasse ranges in age from late Oligocene to late Miocene-early Pliocene and is as much as 3000 to 6000 m thick. In eastern Switzerland and adjacent Germany, for example, the proximal Molasse is about 4500 m thick, whereas 60 kilometers to the north distal deposits are less than 1800 m thick (Füchtbauer, 1967, fig. 7). On its northern margin the Molasse contains minor debris from cratonic sources and interfingers with marine deposits of the Rhine Graben and the Jura district. These near sea-level detrital deposits reflect active uplift of the orogen, high-angle faulting of the basement beneath the foredeep, and northward translation of nappes that began several million years after the main, early Eocene to early Oligocene orogeny (Clark and Jäger, 1969; Milne, 1969; Hunziker, 1970; Hsü and Schlanger, 1971).

The northern Alpine Molasse comprises two major and five minor coarsening-upwards megacycles several to many hundred meters thick (Füchtbauer, 1967) that record phases of Alpine deformation (fig. 2). The Lower Freshwater and Upper Freshwater Molasse (figs. 2 and 3) are separated by the mid-Miocene paralic to shallow marine Upper Marine Molasse which resulted from reduced influx of detritus and extensive marine flooding of the foredeep. Early in the Molasse episode nonmarine deposits graded eastward into paralic and marine facies, whereas the later marine and succeeding nonmarine Molasse were dispersed westward (fig. 3).

Contrasting proximal-distal and nonmarine-marine deposits have significantly different sedimentary textures and structures, making it difficult to delimit features of the Molasse succinctly. Of all its facies, the most distinctive comprises the thick fanglomerates (nagelfluhen)

---

[4] Capitalized when applied to type deposits in the Northern Alps. This procedure is consistent with Hsü's (1970, p. 7–8) recommendation that "flysch" be capitalized for appropriate Alpine formations.

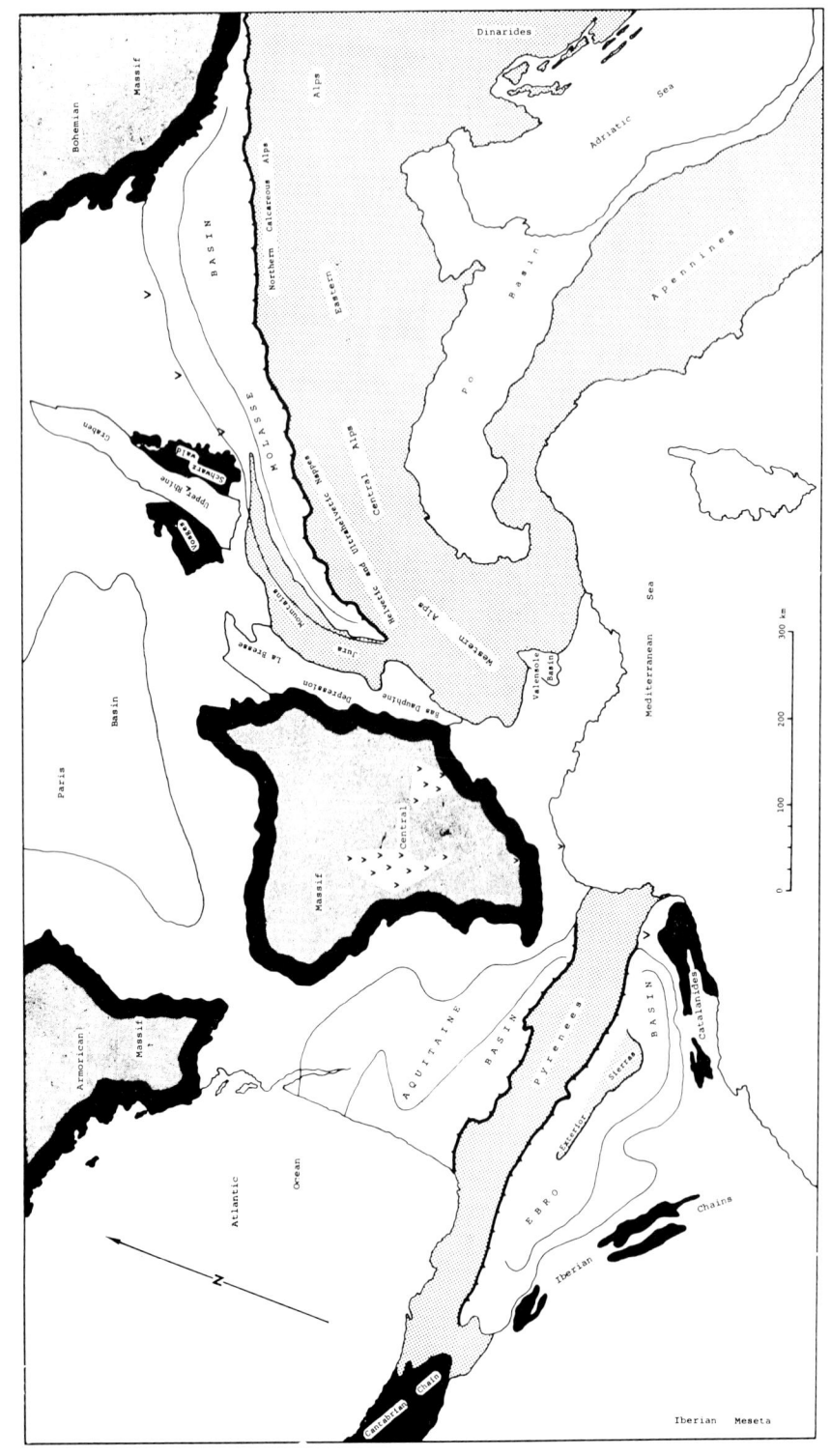

FIG. 1.—Principal tectonic elements of part of western Europe. Cenozoic mountain belts, stippled; Hercynian blocks on craton, black and gray; V, late Cenozoic volcanoes. Northern Alpine (Molasse), Aquitaine, and Ebro Basins depicted by form lines and heavy barbed orogenic margins.

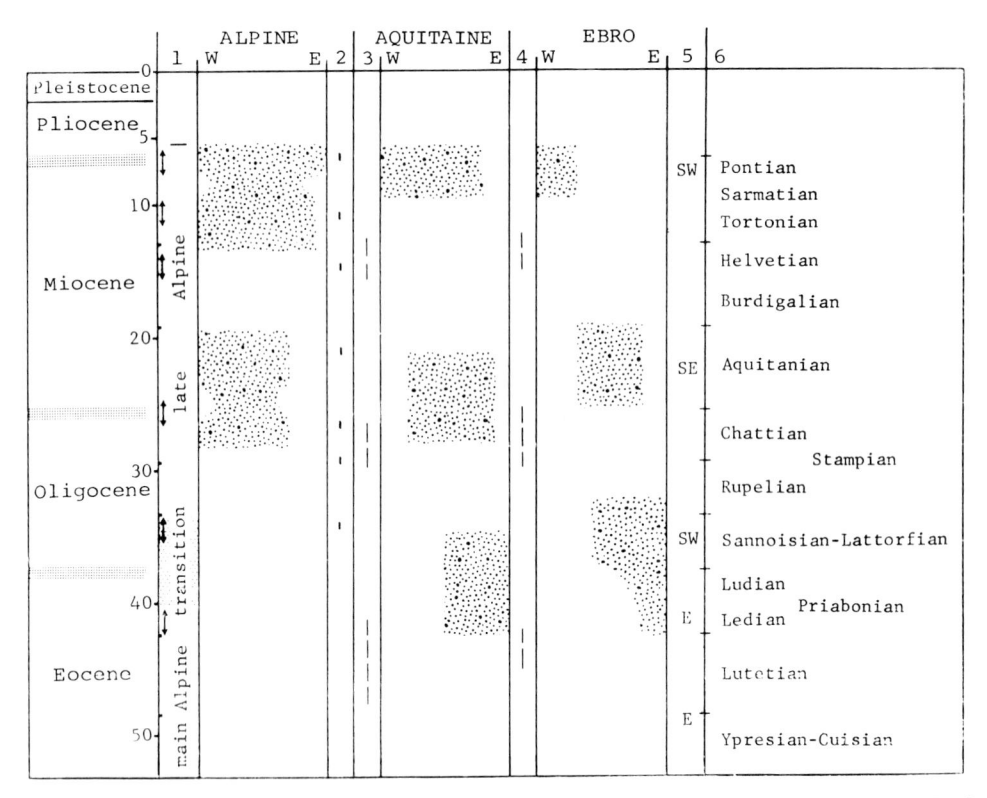

FIG. 2.—Stratigraphic distribution of thick Cenozoic conglomerates in northern Alpine part of Basin and in Aquitaine and Ebro Basins. Ranges of conglomerates and orogenic phases depend in part on the absolute time span assigned to Cenozoic epochs and on the assumption that Aquitanian and Pontian stages are Miocene in age. 1, Main and late Alpine orogeny (Milne, 1969). Arrow-pointed lines are Stille's Alpine phases of deformation (Richter, 1927; Breyer, 1960). 2, Post-Flysch, coarsening-upwards megacycles (Füchtbauer, 1967). 3, North Pyrenean orogenic phases (Schoeffler, 1971a). 4, South Pyrenean orogenic phases (Riba, 1955; Soler and Puigdefabregas, 1970; Henry and others, 1971). 5, Conglomerates from cratonic blocks (Ferrer, and others, 1968; Gross, 1968; Riba, 1955). 6, European Cenozoic stages.

and their laterally and distally equivalent fining-upward fluvial cycles. These provide the best known record of the size, shape, and distribution of ancient alluvial fans and of proximal sedimentation in a foredeep. As many as eight or nine large fans have been recognized along a 400-km mountain front (fig. 3). Although they varied in size, most of them were 20 to 40 km wide, as much as 1000 m thick, and they spread northward at least 50 km from their upland sources.

Lower Marine Molasse[5]—Early in mid-Oligo-

[5] The Lower Marine Molasse is commonly considered to comprise the middle Oligocene (Rupelian) tonmergel and paralic sandstone and its minor influx of alpine conglomerate (Trümpy, 1960; Rutsch, 1962; Füchtbauer, 1964). It has also been extended to include the widespread early Oligocene (Sannoisian, Lattorfian) Flysch that is as much as 1000 m thick, because its central, upper regressive (brackish) facies contains the first minor conglomeratic detritus from the Alps (Schmidt-Thomé, 1963; Füchtbauer, 1967).

cene (Rupelian; fig. 2) time, the foredeep was filled with tonmergel several hundred meters thick in the narrower western part and as much as 1000 m in the east, pointing to more active subsidence at the eastern end of the foredeep (Füchtbauer, 1964). The succeeding paralic sandstones (Gres de Carriers and Horwerplatten, a few tens of meters thick in the west; Bausteinschichten, a few hundred meters thick in the east) are quartzose lithic (calcite-dolomite) arenites eroded mostly from terranes of Flysch, Prealps and Northern Calcareous Alps. Associated minor conglomerates had a similar calcareous source and reflect incipient uplift and nappe emplacement of the late Alpine orogenic phase (fig. 2).

In Val d'Illiez in southwestern Switzerland (fig. 3), the transitional Lower Marine Molasse that conformably overlies the autochthonous Flysch, is exposed beneath overriding nappes (Trümpy, 1960, p. 880; Hsü, 1970, p. 6); the

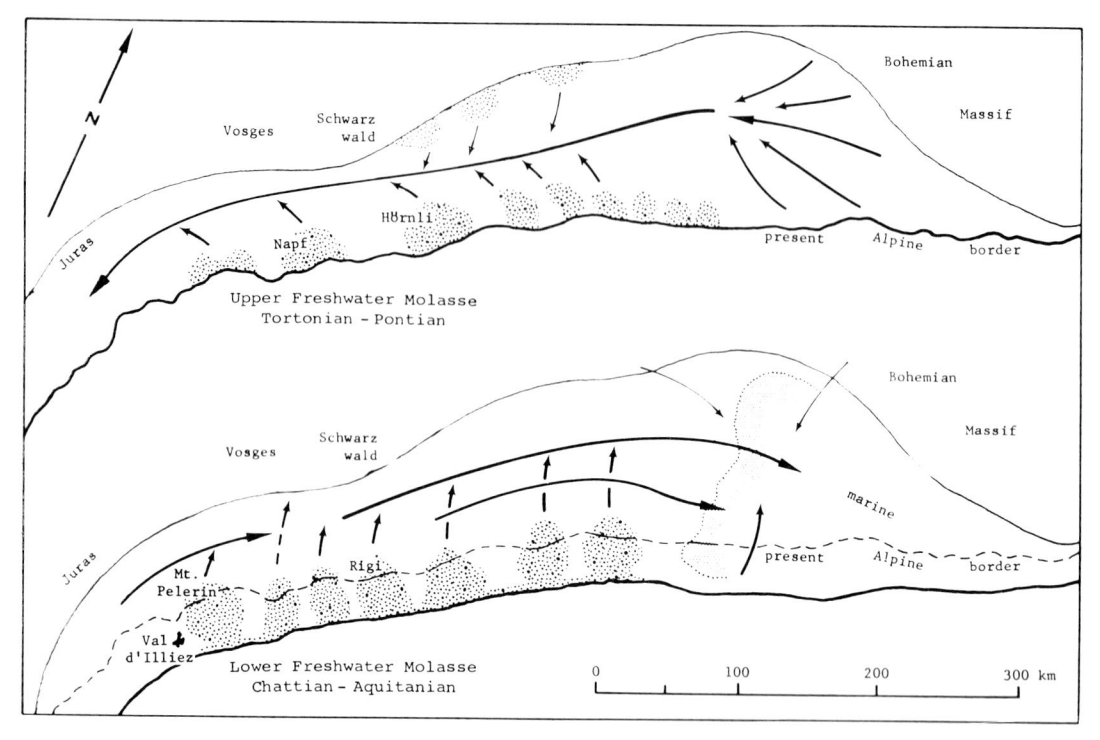

FIG. 3.—Alluvial fans and dispersal patterns in northern Alpine foredeep during accumulation of Lower Freshwater (Chattian-Aquitanian) and Upper Freshwater (Tortonian-Pontian) Molasse. After Füchtbauer 1967.

gradational succession is also present at least 120 km to the northeast (Gasser, 1968). These sites afford a clue to the general position of the southern margin of the foredeep in mid-Oligocene time. Farther east, in southern Germany and adjacent Austria, the Lower Marine Molasse equivalents, now overridden by Flysch and Helvetic nappes, as well as by the Northern Calcareous Alps, are unconformable on fractured cratonic Mesozoic strata and their metamorphic basement (Angenheister and others, 1972, p. 363-364).

Lower Freshwater Molasse.—During late Oligocene and early Miocene time the Eastern Alps remained relatively subdued, whereas the Central Alps were an actively uplifted source area. The episode began with accumulation of the Bunte Molasse (variegated nonmarine sandstone and mudstone) in the central and western parts of the foredeep. Then a fringe of eight or nine huge alluvial fans more than 1,000 m thick (fig. 3), as in Mt. Pelerin and Rigi nagelfluhen, dominated the proximal facies (Füchtbauer, 1964; Gasser, 1966, 1968). These commonly comprise successive fining-upward cycles, 10 to 20 m thick, with conspicuous beds of conglomerate several meters thick. The base of each is marked by scour-and-fill structures as much as 50 cm deep. Local truncation of superposed cycles has produced amalgamated conglomeratic units more than 50 m thick. Laterally and distally the fanglomerates succeeded and graded into variegated fining-upward fluvial cycles. Many of these are only a few meters thick and in the upper part include several centimeters of lignite and associated freshwater limestone (Bersier, 1945). Sediments transported beyond the fans were swept eastward toward a persisting marine embayment in southeastern Germany and adjacent Austria (fig. 3). Brackish to marine cyclothems that accumulated there (Lensch, 1961; Stephan, 1965) range up to a few tens of meters thick and contain thin beds of coal rarely more than 10 cm thick but with as many as 38 beds in 400 m of section.

Early in Chattian time major subsidence along the foredeep had shifted westward to the central part, where carbonate clasts derived from younger thrust sheets predominated. After a minor westward marine transgression in very early Miocene time, major subsidence shifted still farther westward in Switzerland. Renewed uplift and deep denudation produced extensive Aquitainian conglomerates characterized by relatively more clasts of metamorphic rocks in the western and central areas. Equivalent sandstones are rather immature calcareous feldspathic arenites like the Granitic Molasse and the Molasse of Lausanne. These Aquitanian sandstones occur in well-developed, fining-upward fluvial cycles as much as 15 m thick. Many of the beds of sandstone, normally 4 to 7 m thick, have been amalgamated to form multistory sand bodies (Bersier, 1959).

Upper Marine Molasse.—With waning of uplift and detrital influx in middle Miocene time, the northern Alpine foredeep was flooded by an extensive seaway bordered by a few small fans. The Upper Marine Molasse that resulted is not more than 1,000 m thick throughout most of its extent. At its eastern end, where subsidence was greater, equivalent marine marl (schliermergel) is more than 2,000 m thick, however.

Middle Miocene deposits in the west are predominantly well sorted arenites in well-bedded, persistent units with conspicuous directional features (Briel, 1962; Van der Linden, 1963). In contrast to the poorly developed bedding features in nonmarine molasse, conspicuous bedding, both horizontal bedding and cross bedding, characterizes this sandy marine facies. Asymmetrical giant ripples with wave lengths of several meters and megaflaser bedding with ripple lenses generally less than a meter long were formed in spits and offshore bars. Many local channels are filled with giant ripple deposits containing scattered small pebbles.

The older Burdigalian paralic to marine sediments include a distinctive glauconitic sandstone that has been quarried in the vicinity of Bern and Luzerne. Locally, it occurs in amalgamated units as much as 45 m thick. Later Helvetian marine sediments contain little glauconite, and their well-rounded proximal conglomerates contain increased amounts of crystalline clasts.

Upper Freshwater Molasse.—Upheaval of the Central and Eastern Alps in late Miocene to early Pliocene time was accompanied by northward movement of Helvetic and associated nappes, which overrode the foredeep a few tens of kilometers. Older proximal molasse was folded and faulted beneath the nappes as well as in the adjacent subalpine zone, as the axis of the foredeep shifted tens of kilometers to the north. In the end, exposed molasse was reduced by as much as half its original width.

Along the length of the displaced foredeep (fig. 3), coarse Alpine detritus filled the lowland with great Tortonian and Pontian alluvial fans of the foreland Molasse. Some are more than 1500 m thick, as at Napf and Hörnli (Matter, 1964, 1970). These coarse deposits, like those of the Lower Freshwater Molasse, contain abundant clasts of sedimentary rocks and are arranged in fining-upwards cycles 10 to 20 m thick with beds of conglomerate several meters thick. Amalgamation of conglomerates is common.

With increased uplift of the Eastern Alps, clasts from the Northern Calcareous Alps dominated the eastern fanglomerates (Janoschek, 1963, p. 334), and axial distribution of sediments shifted to the west, spreading eastern detritus across the Jura District (fig. 3). A minor amount of sediment was also derived from local cratonic sources, and ash was spread southward from volcanoes along the northern border of the basin (fig. 1).

Accumulation of the Upper Freshwater Molasse probably continued into Pliocene time, filling the foredeep as much as 1,000 m above sea level. Deformation of the Juras resulted in folding of distal molasse in that area. Then, later Plio-Pleistocene regional uplift terminated deposition, and deep erosion destroyed much of the latest record.

AQUITAINE BASIN

The Aquitaine Basin (fig. 1), between the Pyrenees on the south, the Massif Central on the northeast, and the Atlantic continental margin on the west (see Cateras, 1964, fig. 90, 91) has a southern foredeep, now about 50 km wide and nearly 300 km long, that is closed at the eastern end. From middle Eocene (late Lutetian) to late Miocene (Pontian) time, molasse accumulated along most of the length of the trough, accruing a thickness of 2,000 to 3,000 m in the east and 1,000 to 1,500 m in the west (Denizot, 1957, paleogeographic maps 1–12; Schoeffler, 1971a).

Although the conglomeratic molasse in the foredeep has long been known, relatively little has been written about it; lack of good outcrops prevents easy reconstruction. In contrast, much attention has been devoted to the stratigraphy and paleontology of the predominantly marine Cenozoic deposits in the western and northwestern parts of the basin (Gignoux, 1955, p. 502–510, 575–576). Significant exceptions include brief descriptions of the major conglomeratic units by Bergounioux and Crouzel (1951), a detailed sedimentologic analysis of the fluvial Miocene molasse by Crouzel (1957), reviews of the relations between the conglomerates and Pyrenean tectonics by Schoeffler (1971a, 1971b), and a succinct summary of the geological evolution of the basin by Winnock (1971).

Submergence of the Aquitaine Basin in early Eocene time (fig. 4) resulted in accumulation of calcareous flysch in the foredeep, together with

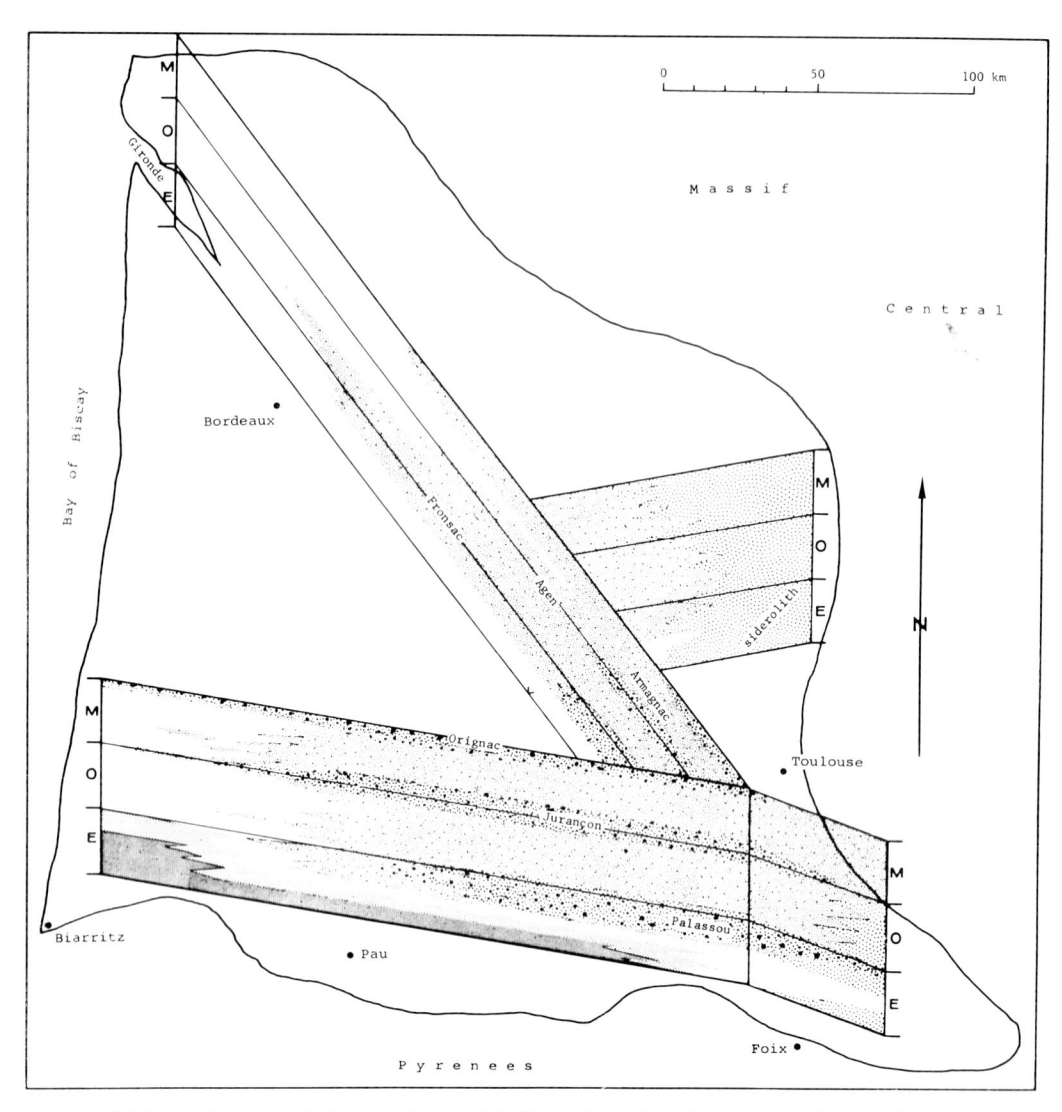

Fig. 4.—Major sedimentary facies associated with Cenozoic molasse in Aquitaine Basin. Flysch, dark, very fine stipple; blue marl, fine stipple; transitional and distal shallow marine facies, no pattern; nonmarine to paralic molasse and its conglomerates, mixed dots; siderolith and detritus from the craton, coarse uniform stipple.

widespread, shallow-water calcareous marine deposits and associated nonmarine sandstone and local conglomerate along the northeastern cratonic border. Individual beds of sandstone have commonly been called molasse (mollasse), following Cayeux (1929), but this is a confusing and unnecesary practice. In early middle Eocene time the molasse episode was heralded by accumulation of calcareous blue marl in the foredeep (Eardley and White, 1947), followed by a transitional shallow marine to paralic facies (Schoeffler, 1971b). Along the northeastern margin of the gulf siderolith formed on the exposed craton (fig. 4).

By late middle Eocene time (late Lutetian, Bonnard and others, 1958; Ypresian-Lutetian, Schoeffler, 1971b) deformation produced the first important Cenozoic uplift of the Pyrenees. Erosion of the northern half of its narrow upland, about 25 km wide and 300 km long, initiated a molasse facies that was dominated by three major conglomerates and their distal sandstone and mudstone arranged in fining-upwards fluvial cycles. During the three periods of active uplift and deformation, Pyrenean detritus flooded the basin (Bergounioux and Crouzel,

1951); in alternate, less active episodes, detritus from the Massif Central became more important (Vatan, 1950). As the molasse facies accumulated, nonmarine deposits spread slowly westward. But throughout the episode marine conditions persisted along the western margin of the basin (fig. 4).

The earliest and principal Cenozoic uplift and thrusting of the Pyrenees, beginning in middle Eocene and dominant in late Eocene time (Schoeffler, 1971a; Winnock, 1971), destroyed the southeastern gulf and produced great alluvial fans of the Palassou Conglomerate in the eastern half of the foredeep (fig. 4). This tectonic facies accumulated conformably on the preceding transitional deposits, reaching a thickness of 1,000 to 1,500 m. Distally to the northwest it interfingered with several tongues of marine sandstone and limestone. The fanglomerates consist largely of clasts of Cretaceous and early Cenozoic calcareous rocks. The equivalent, distal Lower Fronsac Molasse is composed mainly of calcareous lithic arenite and associated mudstone that extended northeastward across an earlier Eocene siderolith. At the western end of the foredeep, 1,200 m of middle Eocene calcareous flysch was succeeded by about 1,000 m of late Eocene blue marl.

By middle Oligocene time, lagoonal and shallow marine conditions were somewhat more widespread. Nevertheless, continuing effects of the Pyrenean uplift were recorded in the nonmarine conditions to the southeast, as in the Upper Fronsac Molasse, and in an increased detrital component in the nummulitic facies in the southwestern embayment.

A second period of uplift, thrusting, and associated major regression occurred in late Oligocene and early Miocene (Stampian-Aquitanian) time. As a result, the Palassou Conglomerate was folded and overridden by nappes while the widespread Jurançon Conglomerate filled the foredeep and displaced marine marl and a transitional facies far to the west. This conglomerate, about 600 m thick, contains more clasts of igneous and metamorphic rocks, recording deep dissection of the Pyrenees. Tongues of conglomeratic sandstone extending across the basin form the Molasse of Agen. These sandstones and the overlying middle Miocene Molasse of Armagnac comprise the fining-upwards fluvial cycles described in detail by Crouzel (1957, p. 213-224).

In middle Miocene time (Burdigalian-Helvetian), marine flooding was again rather widespread in the northern part of the basin, as it was in the perialpine depression to the east. During this interval minor deformation occurred in the sub-Pyrenean zone, producing folds in older strata and gently tilting the Jurançon Conglomerate.

Then, in latest Miocene (Pontian) time, a third important uplift of the range produced a 160-m-thick conglomerate and the associated Molasse of Orignac. These deposits records the last important differential movement in the Pyrenees. Later regional uplift of both range and basin terminated molasse deposition and initiated the present episode of degradation.

EBRO BASIN

The Ebro Basin south of the Pyrenees (fig. 1) displays one of the finest molasse sequences in Europe. Deposits of its foredeep now are more than 400 km long and as much as 75 km wide. The broader cratonic basin extends to the Hercynian Iberian chains to the southwest and is closed along its southeastern side by the coastal Catalanides.

From late Eocene to late Miocene time, molasse, as much as 3000 m thick, was supplied largely by the narrow Pyrenean upland shared with the Aquitaine Basin (see Cateras, 1964, fig. 91). During this interval, the center of conglomeratic nonmarine deposition shifted slowly westward along the axis of the foredeep, dispacing a marine embayment. Moreover, the bordering Hercynian blocks, more than in the Northern Alpine Basin and in the Aquitaine Basin, contributed significant amounts of detritus (fig. 5), and an evaporite facies accumulated in the interior of the basin.

No comprehensive report has been published on this important example of molasse. A general review of the relation of the Pyrenees to the Bay of Biscay (Mattauer and Seguret, 1971) provides a framework for the Late Cretaceous-early Cenozoic development of the Pyrenees (Henry and others, 1971). Short papers by Riba (1955, 1967), Rosell and Riba (1966), and Soler and Puigdefabregas (1970) describe local sequences and indicate some of their relations to the evolution of the Pyrenees. Characteristics of the transition from marine to paralic facies in the eastern part of the foredeep have been outlined by Ferrer and others (1968), Van Eden (1970), and Ferrer and Others (1971) in the central part by Puigdefabregas (1972) and in the west by Mangin and Rat (1960), and Feuillee and Rat (1971).

In early Eocene time (fig. 5) marine flooding spread eastward along the Ebro foredeep. This extensive transgression was the site of an early and middle Eocene calcareous flysch 4,000 to 5,000 m thick, and derived largely from the craton to the east and southeast. Correlative deposits in the eastern part of the foredeep include blue marl, paralic deposits, and a wedge of con-

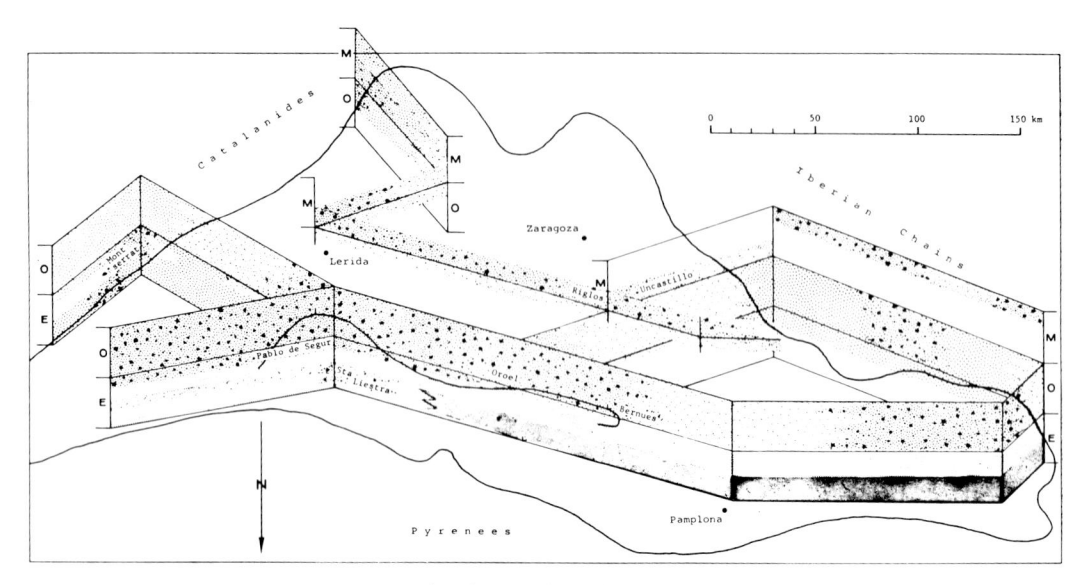

FIG. 5.—Major sedimentary facies associated with Cenozoic molasse in Ebro Basin. Flysch, dark, very fine stipple; blue marl, fine stipple; transitional facies, no pattern; nonmarine to paralic molasse and its conglomerates, mixed dots; detritus from cratonic blocks, coarse uniform stipple and mixed dots (conglomerates). Note that north is down.

glomerate, about 600 m thick, shed from an eastern Catalanid upland (Ferrer and others, 1968). At about the same time incipient uplift of the easternmost end of the Pyrenees produced the local Santa Liestra Conglomerate, which is about 500 m thick (fig. 5; see Ferrer and others, 1971; Henry and others, 1971; Van Eden, 1970). Then, a second, slightly younger uplift of the eastern Catalanides shed the Montserrat Conglomerate, as much as 1,200 m thick, which prograded into a transgressive marine facies in the foredeep.

The earliest extensive uplift of the Pyrenees and its associated thrusting (second phase, Soler and Puigdefabregas, 1970) began in the east in early late Eocene (post-Lutetian, pre-Ludian) time. This produced the 2,600-m-thick Pablo de Segur Conglomerate (Rosell and Riba, 1966; Nagtegaal, 1966) and equivalent fluvial deposits (Artes Formation) throughout the eastern end of the basin. In the central part of the foredeep as much as 1,500 m of blue marl and 600 m of successive transitional facies, derived from the southeast, accumulated in late Eocene time. Farther west, near Pamplona, the transitional facies persisted into early Oligocene time. Here it is characterized by some crude graded bedding, sole marks, abundant ripple marks, burrowing, saltcasts, minor gypsum, and tracks of wading birds (Mangin and Rat, 1960; de Raaf, 1964; de Raaf and others, 1965). Similar assemblages of structures and trace fossils occur in both nonmarine and paralic facies well within other molasse sequences (Beaudoin and Gigot, 1971; de Clercq and Holst, 1971).

Coarse detritus from the initial Pyrenean uplift accumulated in the central part of the foredeep in latest Eocene time, and by Oligocene time it prevailed along the length of the basin. These deposits consist of thick proximal conglomerates (Oroel), many hundreds of meters thick, and variegated fining-upwards fluvial cycles (Bernues) and their interior and western evaporite facies (fig. 5). During this interval an uplifted cratonic block on the southwestern border produced a wedge of conglomerate and distal fluvial deposits (Riba, 1955). As yet, the detailed succession and accurate dating of these Oligocene facies are not known. Consequently, proximal conglomerate has been portrayed arbitrarily as accumulating throughout the Oligocene Epoch (fig. 5).

Latest Oligocene to early Miocene uplift and deformation of the Pyrenees (third phase, Soler and Puigdefabregas, 1970) produced nappes of the Exterior Sierras and folded and faulted older molasse in the sub-Pyrenean zone. It also created a new influx of thick conglomerate (Riglos) and an associated distal facies (Uncastillo) that accumulated in the central and western part of the basin south of the Exterior Sierras. Concomitantly, an uplifted Catalanid block shed coarse conglomerate in the southeastern corner of the basin (Gross, 1968). As the supply of detritus waned, an interior evaporitic facies become more widespread.

In late Miocene time uplift of the western Pyrenees and along the southwestern cratonic border (Riba, 1955) produced another flood of conglomerates. No detailed information about the relations of the several Miocene conglomerates has been published; apparently the late Miocene activity occurred only in the western part of the basin.

COMPARISONS

Note of some differences among the examples reviewed is useful in attempting to define and delimit the molasse and its framework. Although each example was the product of late Alpine orogeny (fig. 2), the Pyrenean examples record important deformation and uplift several million years before the first major conglomerate accumulated in the northern Alpine foredeep. Consequently, the Pyrenean molasse episodes, measured by their conglomerates, lasted several million years longer. Yet, in spite of this, these sequences are thinner and less extensive than their northern Alpine counterparts.

Swiss-Bavarian molasse records a reversal of dispersal along the foredeep, from eastward in late Oligocene-early Miocene time to westward in late Miocene and early Pliocene time (fig. 3). In the Pyrenean basins, dispersal of detritus and shifting of source areas were continually westward.

Both the Alps and the Pyrenees were deformed and uplifted again in late Miocene time, but the conglomerates produced in the northern Alps greatly exceeded those in the Aquitaine and Ebro Basins in thickness and extent, suggesting an earlier waning of differential uplift in the Pyrenees.

Although late Miocene and Pliocene volcanism occurred on the craton just north of the Molasse Basin, east and northeast of the Aquitaine Basin, and at the east end of the Ebro Basin (fig. 1), only the Alpine Molasse contains layers of tuff and bentonite. Apparently the Pyrenean basins were not in a position to be showered with ash from their nearby volcanics.

Unlike the Alps, the Pyrenean core was not intensely metamorphosed in early Cenozoic time. At best, there was low-grade metamorphism of Cretaceous and early Cenozoic deposits on the orogenic side of the main North Pyrenean Fault, which produced schistosity and recrystallization of clay minerals to micas (Mattauer, 1968).

The Aquitaine Basin was unique in opening into an ocean basin and in having had a persistent marine facies during the entire molasse episode.

The Ebro Basin was unique in being bordered by active Hercynian blocks that shed significant amounts of conglomerate and in having had prominent evaporite deposition in the interior of the basin.

DEFINITION

The Cenozoic Molasse deposited in the Northern Alpine Foredeep, and its counterparts in the Pyrenean Aquitaine and Ebro Basins are prototypes of the tectofacies molasse. These late orogenic clastic wedges on cratonic margins were initiated by the first significant uplift of the range and associated emplacement of nappes. Molasse sedimentation was preceded by geosynclinical marine deposits comprising flysch, an overlying thin-bedded mudstone or marl, and a transitional paralic facies. The predominantly nonmarine to paralic molassic wedge with conspicuous proximal conglomerates and fining-upwards fluvial cycles graded laterally and distally into shallow marine deposits. During the molasse episode uplift was renewed, basement of the foredeep was fractured, nappes encroached onto the craton, basinfill was deformed along its orogenic border, and the foredeep migrated distally.

A late orogenic framework is fundamental to the deposition of molasse. In the Central Alps, for example, "probably the most important aspect of the [orogenic history] is that the main Alpine/late Alpine transition in the deeper levels of the Alpine orogene coincided temporally with the Flysch/Molasse[6] transition at the surface" (Milne, 1969, p. 111).

The main Alpine phase, involving compressive development of basement nappes, ended in the Central Alps in late Eocene to early Oligocene time (see Sutton, 1965, p. 30-35; Hunziker, 1970, p. 156). The late Alpine phase (fig. 2), marked by metamorphism, intrusion of granite, and major uplift and associated thrusting, was initiated during an early Oligocene transition. By late Oligocene time upheaval and erosion prevailed in the core zone (Clark and Jäger, 1969), as ultra-Helvetic and South Helvetic nappes of early Cenozoic strata slid northward over yet the undeformed late Eocene and early Oligocene North Helvetic Flysch formations (Hsü and Schlanger, 1971). This pile of supracrustal sheets not only contributed much of the detritus that overwhelmed the foredeep, but it may also have isostatically accentuated the peripheral depression in front of the encroaching range (see Price, 1971). Crustal shortening and uplift continued with interruption during the Miocene Epoch (Milne, 1969, fig. 1; Laubscher,

---

[6] In a general sense only. "Flysch" and "molasse" are not chronostratigraphic terms; locally their time spans overlapped in each of the examples reviewed.

1970, fig. 5). Then, in later Miocene time, Helvetic nappes spread northward over proximal molasse, producing a new source area that supplied abundant coarse detritus to the foredeep until early Pliocene time.

Geochronological data from the Eastern Alps also point to a correlation between orogenic events and the Molasse. According to Cliff and others (1971, p. 245–247, 255) emplacement of the East Alpine thrust sheets was completed about 55 my ago, followed by metamorphism during the Eocene Epoch (55–35 my). Uplift of the orogen lagged somewhat in mid-Cenozoic time, as several thousand meters of marine schliermergel accumulated at the east end of the foredeep. Marked uplift about 15 m.y. ago produced the first flood of coarse detritus from the Northern Calcareous Alps. These Late Miocene deposits not only filled the trough to the north, but they were dispersed for to the west as well.

BOUNDARY PROBLEM

As with most gradational stratigraphic sequences, selecting a boundary between flysch and its succeeding molasse involves a rather arbitrary procedure (see Hsü, 1970, p. 6). Nevertheless, it can be critical to such general measures as when the molasse began, its correlation with orogenic events, its duration, and its overall thickness. Traditionally, the transitional deposits in central and western parts of the Alpine foredeep have been assigned to the Lower Marine Molasse but without agreement as to the lower limits.

Many geologists have placed the base of the Lower Marine Molasse at the widespread influx of Alpine detritus in the middle Oligocene Bausteinschichten and equivalent sandstones. This selection correlates with current ideas about Central Alpine orogenesis. Füchtbauer (1967), on the other hand, has concluded that the Lower Marine Molasse began in early Oligocene time when a minor amount of Alpine detritus reached the foredeep (paralic Deutenhausenerschichten), while flysch still accumulated in the west and tonmergel prevailed in the east. Within this expanded range of Molasse, he has identified seven megacycles (fig. 2, column 2) and has correlated them with specific tectonic phases recognized by many geologists (Richter, 1927; Breyer, 1960). The earliest of these phases, the Pyrenean, occurred in late Eocene time (fig. 2, column 1).

As an alternative, all the Lower Marine Molasse might better be assigned to a new, post-Flysch transitional formation. The Molasse, redefined, would then focus on the distinctive widespread fanglomerates and associated alluvial plain deposits and would be a stratigraphic unit more easily delimited and compared with other late orogenic sequences. This suggested solution, like the other schemes, has its drawbacks.

SOME MEASURES

Each molasse reviewed records a distal migration of the axis of its foredeep. In the Molasse Basin, the axis shifted northward about 60 km from Oligocene to Pliocene time at an average rate of about 2 mm/y (Schmidt-Thome, 1963, p. 443).

Molasse episodes lasted a few tens of million years, accentuated by two or three pulses of orogenic activity and concomitant increase in conglomerate input, at intervals of 10 to 15 my (fig. 2).

Northern Alpine data suggest a maximum average preservation rate of about 400 m/my and a minimum overall of 150 to 200 m/my.

Very general estimates of the average amount of upland denudation required to yield the preserved molasse in these foredeeps (not the total amount eroded from associated source areas) can be derived as shown in table 1. These estimates are based on the reduced, exposed width of the molasse and disregard reworking of older, deformed proximal molasse. Presumably the effects would roughly cancel one another. Moreover, for so general an estimate, no allowance has been made for volume increases from source rock to derived sediments. The similar amount of vertical stripping estimated for the three examples, in spite of the much greater volume of

TABLE 1.—ESTIMATES OF UPLAND DENUDATION AND AMOUNTS OF PRESERVED MOLASSE

Measure	Basin		
	Alpine	Aquitaine	Ebro
Duration (my)	25	35	35
Average thickness (km)	2½	¾–1	1
Areal extent (km)	100×650	100×300	100×400
Volume (km³)	162,500	22,500–30,000	40,000
Upland area (km)	75×650[1]	25×300	25×400
Average stripping (km)	3¾	3–4[2]	4

[1] Baumberger (1934) estimated that a high southern divide lay 75 to 100 km south of the Alpine structural front in early Miocene time with subordinate watersheds 40 to 50 km south of the front. At this time there was no comparable high relief in the Eastern Alps.
[2] Crouzel (1957, p. 242) computed 4.5-km average stripping from data recorded in the central part of the foredeep.

the Alpine Molasse (Aquitaine—×1, Ebro—×1½, Alpine—×6), reflects the very narrow Pyrenean source areas. Local maximum stripping presumably was several times the estimated average, suggesting that at least 7 to 10 km must have been eroded from parts of the source areas during these molasse episodes.

Geochronologic and heat-flow data from the Central Alps (Clark and Jäger, 1969) yield a denudation rate of 0.4 to 1 mm/y, implying a maximum local stripping of 10 to 25 km during 25 my of Molasse accumulation. The difference between the estimated maximum denudation and vertical stripping recorded in the Molasse represents that portion of the total erosion product not preserved in the Molasse Basin. Most of this 30- to 60-percent excess probably bypassed the northern Alpine part of the perialpine foredeep. Some may also have geen diverted southward to the Po Basin (fig. 1).

REFERENCES

ANGENHEISTER, H., BÖGEL, H., GEBRANDE, H., GIESE, P., AND SCHMIDT-THOMÉ, P., 1972, Recent investigations of surficial and deeper crustal structures of the Eastern and Southern Alps: Geol. Rundschau, v. 61, p. 349–395.

ARKHIPOV, I. V., 1971, Contrast between flysch and nonflysch deposits: Internat. Geol. Rev., v. 14, N. 7, p. 720–728.

BAUMBERGER, E., 1934, Die Molasse des schweizerischen Mittellandes and Juragebietes: Geol. Führer Schweiz; Schweizerische Geol. Gesell., pt. 1, p. 57–75.

BEAUDOIN, B., AND GIGOT, P., 1971, Figures de courant et trace de pattes d'oiseaux associees dans la molasse Miocene de Digne, Basses Alpes (France): Sedimentology, v. 17, p. 241–248.

———, ———, AND HACCARD, D., 1970, Flysch et molasse, approche sedimentologique: Soc. géol. France Bull., ser. 7, v. 12, no. 4, p. 664–672.

BERGOUNIOUX, F. M., AND CROUZEL, F., 1951, Sur las nature, l'age et l'extension de quelques poudingues nord-Pyrenees: 3rd Internat. Cong. Sedimentology Proc., p. 67–80.

BERSIER, A., 1945, Sedimentation molassique; variations laterales et horizons continus a l'Oligocene: Eclogae Geol. Helvetiae, v. 38, p. 452–458.

———, 1959, Sequences detritiques et divagations fluviales: ibid., v. 51, p. 854–893.

BERTRAND, M., 1897, Structure des Alpes francaises et recurrence de certains facies sedimentaires: Internat. Geol. Cong., 1894; Compte Rendu, p. 161–177.

BOGDANOFF, A. A., MOURATOV, M. V., AND SCHATSCKY, N. S. (eds.), 1964, Tectonics of Europe; explanatory note to the international tectonic map of Europe, 1962: Moscow, Nauk i Zhizn, 360 p.

BONNARD, E., AND OTHERS, 1958, The Aquitanian Basin, southwest France, in WEEKS, L. G. (ed.), Habitat of oil: Tulsa, Oklahoma, Am. Assoc. Petroleum Geologists, p. 1091–1122.

BREYER, F., 1960, Die orogenen Phasen der gefalteten Molasse, des Helvetikums und des Flysches im westlichen Bayern und in Vorarlberg: Deutsche Akad. Wiss. Berlin Abh., Kl. 3, no. 1 (Ernst Kraus Festschr.), p. 95–98.

BRIEL, A., 1962, Geologie de la region de Lucens (Broye): Ecolgae Geol. Helvetiae, v. 55, p. 189–274.

BÜCHI, U. P., WIENER, G., AND HOFMANN, F., 1965, Neue Erkenntnisse im Molassebecken auf Grund von Erdöltiefbohrungen in der Zentral- und Ostschweiz: ibid., v. 58, p. 87–108.

CADISCH, J., 1928, Das Werden der Alpen im Spiegel der vorland Sedimentation: Geol. Rundschau, v. 19, p. 105–119.

CATERAS, M., 1964, Pyrenees, in BOGDONOFF, A. A., AND OTHERS, Tectonics of Europe: Moscow, Nauka (Nedva), p. 216–221.

CAYEUX, L., 1929, Les roches sedimentaires de France, roches siliceuses. Service Carte géol. France Mem., 774 p.

CLARK, S. P., AND JÄGER, E., 1969, Denuduation rate in the Alps from geochronologic and heat flow data: Am. Jour. Sci., v. 267, p. 1143–1160.

DE CLERCQ, S. W. G., AND HOLST, H. K. H., 1971, Footprints of birds and sedimentary structures from subalpine Molasse near Flühli (Canton of Luzern): Eclogae Geol. Helvetiae, v. 64, p. 63–69.

CLIFF, R. A., AND OTHERS, 1971, Structural, metamorphic and geochronological studies in the Reisseck and sou.hern Ankogel Groups, the Eastern Alps: Aust. Geol. Bundesanst., Jahrbuch, v. 114, p. 121–2/2.

CROUZEL, F., 1957, Le Miocene continental du bassin d'Aquitaine: Service Carte géol. France, Bull. 248, v. 54 (1956), 264 p.

DENIZOT, G. (ed.), 1957, France, Belgique, Pays-Bas, Luxembourg; Tertiare: Lexique Strat. Internat. v. 1, pt. 4a, vii + 217 p.

EARDLEY, A. J., AND WHITE, M. G., 1947, Flysch and molasse: Geol. Soc. Amer. Bull., v. 58, p. 979–990.

FERRER, J., ROSELL, J., AND REGUANT, S., 1968, Sintesis lithostratigraphica del Paleogeno del borde oriental de la depression del Ebro: Acta Geol. Hispanica, Anno 3, no. 3, p. 2–4.

———, AND OTHERS, 1971, El Paleogeno marino de la region de Tremp (Cataluña): 1st Cong. Hispanica-Luso-Am. Geologia Econ., sec. 1, v. 2, p. 813–827.

FEUILLEE, P., AND RAT, P., 1971, Structures et paleogeographies Pyreneo-Cantabrique: Inst. Français Pétrole Pub., Coll. Colloques et Semenaires 22, v. 2, paper 5.1, p. 1–48.

FÜCHTBAUER, H., 1964, Sedimentpetrographische Untersuchungen in der älteren Molasse nördlich der Alpin: Eclogae Geol. Helvetiae, v. 57, p. 157–298.

———, 1967, Die Sandsteine in der Molasse nördlich der Alpen: Geol. Rundschau, v. 56, p. 266–300.

GASSER, U., 1966, Sedimentologische Untersuchungen in der äusseren Zone der subalpinen Molasse des Entlebuchs (Kt. Luzern): Eclogae Geol. Helvetiae, v. 59, p. 723–772.

———, 1968, Die innere Zone der subalpinen Molasse des Entlebuchs (Kt. Luzern), Geologie und Sedimentologie: *ibid.*, v. 61, p. 229–319.
GIGNOUX, M., 1955, Stratigraphic geology: San Francisco, Freeman & Co., 682 p.
GROSS, G., 1968, Das Tertiär in südwestlichen Ebro-Becken: Neues Jahrbuch Geologie und Paläontologie Abh., v. 131, p. 23–32.
HAGN, H., 1960, Die stratigraphischen, paläogeographischen und tektonischen Beziehungen zwischen Molasse und Helvetikum im östlichen Oberbayern: Geologica Bavarica, no. 44, p. 3–208.
HAUG, E., 1900, Les geosynclinaux et les aires continentales: Contribution a l'étude des transgressions et des regressions marines: Soc. géol. France Bull., ser. 3, v. 28, p. 617–711.
HEIM, A., 1919, Geologie der Schweiz: Leipzig, Fauchnitz, v. 1, 704 p.
HENRY, J., LANUSSE, R., AND VILLANOVA, M., 1971, Evolution du domaine marin Pyreneen du Senonien Superieur a l'Eocene Inferieur: Inst. Francais Pétrole Pub., Coll. Colloques et Semenaires 22, v. 1, paper 4.7, p. 1–7.
HSÜ, K. J., 1970, The meaning of the word flysch—a short historical search: Geol. Assoc. Canada, Special Paper 7, p. 1–11.
———, AND SCHLANGER, S. O., 1971, Ultrahelvetic Flysch sedimentation and deformation related to plate tectonics: Geol. Soc. America Bull., v. 82, p. 1207–1218.
HUNSIKER, J. C., 1970, Polymetamorphism in the Monte Rosa, Western Alps: Eclogae Geol. Helvetiae, v. 63, p. 151–161.
JANOSCHEK, R., 1963, Das Tertiär in Österreich: Geol. Gesell. Wien Mitt., v. 56, p. 319–360.
KAY, M., 1951, North American geosynclines. Geol. Soc. America Mem. 18, 143 p.
KING, P. B., 1959, The evolution of North America: Princeton, New Jersey, Princeton Univ. Press, 190 p.
LAUBSCHER, H. P., 1970, Bewegung und Wärme in der alpinen Orogenese: Schweizerische Mineralog. und Petrogr. Mitt., v. 50, p. 565–596.
LENSCH, G., 1961. Stratigraphie, Fazies und Kleintektonik der Kohleführenden in der bayerischen Faltenmolasse: Geologica Bavarica no. 46, p. 1–52.
MANGIN, J. P., AND RAT, P., 1960, L'evolution post-Hercynienne entre Asturies et Aragon (Espagne), *in* DELGA, M. D. (ed.), Livre a la mem. Professeur Paul Fallot: Soc. géol. France Mem., v. 1, p. 333–349.
MATTAUER, M., 1968, Les traits structuraux essentiels de la chaine Pyreneene: Rev. géographie phys. et géologie dynamique, v. 10, p. 3–12.
———, AND SEGURET, M., 1971, Le relations entre la chaine des Pyrenees et le Golfe de Gascogne: Inst. Français Pétrole Pub., Coll. Colloques et Semenaires 22, v. 1, paper 5.4, p. 1–24.
MATTER, A, 1964, Sedimentologische Untersuchungen im östlichen Napfgebiet: Eclogae Geol. Helvetiae, v. 57, p. 315–428.
———, 1970, Molasse *in* HOTTINGER, L., AND OTHERS, Explanation of Geologic Atlas der Schweiz: 1093 Hörnli, p. 7–20.
MILNE, A. G., 1969, On the orogenic history of the Central Alps: Jour. Geology, v. 77, p. 108–112.
NAGTEGAAL, P. J. C., 1966, Scour-and-fill structures from a fluvial piedmont environment: Geologie et Mijnb., v. 46, p. 342–354.
PERRIAUX, J., AND OTHERS, 1970, Flysch et molasse: Géol. Soc. France, Synthesis of general meeting, June 15, 1970.
PRICE, R. A., 1971, Gravitational sliding and the foreland thrust and fold belt of the North American Cordillera: Geol. Soc. America Bull., v. 82, p. 1133–1138.
PUIGDEFABREGAS, C., 1972, Caracterizacion de estructuras de marea en el Eoceno Medio de la Sierra el Guara (Huesca): Pirineos, v. 104, p. 5–13.
RAAF, J. F. M. DE, 1964, The occurrence of flute casts and pseudomorphs after salt crystals in the Oligocene "Gres à ripplemarks" of the southern Pyrenees, *in* BOUMA, A. H., AND BROUWER, A. (eds.), Turbidites: Amsterdam, Elsevier, p. 192–198.
———, BEETS, C., AND KORTENBOUT VAN DER SLUIJS, G., 1965, Lower Oligocene bird-tracks from northern Spain: Nature, v. 207, p. 146–148.
RIBA, O., 1955, Sur le type de sedimentation du Tertiare continental de la partie ouest du bassin de l'Ebre: Geol. Rundschau, v. 43, p. 363–71.
———, 1967, Resultados de un estudie sobre el Terciario continental de la parte este de la depresion central Catalana: Acta Geol. Hispanica, v. 2, no. 1, p. 1–6.
RICHTER, M., 1927, Molasse und Alpen: Deutsch. Geol. Gesell. Zeitschr., v. 79, p. 124–135.
ROSELL, J., AND RIBA, O., 1966, Nota sobre la disposicion sedimentaria de los conglomerados de Pobla de Segur: 5th Internat. Cong. Study Pyrenees; Inst. Estud. Pirenees, p. 3–16.
RUTSCH, R. F., 1962, Zur Palogeographie der subalpinen Unteren Meeresmolasse (Rupelien) der Schweiz: Assoc. suisse Géologues et Ingénieurs Pétrole Bull., v. 28, no. 74, p. 27–32; no. 75, p. 13–24.
SCHMIDT-THOMÉ, P., 1963, Le bassin de la molasse d'Allemagne du sud, *in* DELGA, M.D., (ed.): Soc. géol. France Mem., v. 2, p. 431–452.
SCHOEFFLER, J., 1971a, Étude structurale de terrains molassiques du piedmont-nord des Pyrenees de Peyrehorade a Carcassonne (D. Sci. thesis): Univ. Bordeaux, Bordeaux, France, no. 346.
———, 1971b, Étude des formations molassiques au sud de Tarbes: Pau-SNPA Cent. Rech. Bull., v. 5, no. 2, p. 165–187.
SOLER, M., AND PUIGDEFABREGAS, C., 1970, Lineas generales de la geologia del alta Aragon occidental: Pireneos, v. 96, p. 5–20.
STEPHAN, W., 1965, Zur faziellen und zyklischen Gliederung der chattischen brackwasser-Molasse in Oberbayern: Geologica Bavarica, no. 55, p. 239–267.
STUDER, B., 1825, Beyträge zu einer Monographie der Molasse: Bern, Switzerland, C. A. Jenni, 427 p.
SUTTON, J., 1965, Some recent advances in our understanding of the controls of metamorphism, *in* PITCHER, W. S., AND FLINN, G. W., Controls of metamorphism: Edinburgh, Scotland, Oliver and Boyd, p. 22–45.

Trümpy, R., 1960, Paleotectonic evolution of the Central and Western Alps: Geol. Soc. Amer. Bull., v. 71, p. 843-908.
Van Eden, J. G., 1970, A reconnaissance of deltaic environment in the middle Eocene of the south-central Pyrenees, Spain: Geol. et Mijnb., v. 49, p. 145-157.
Van der Linden, W. J. M., 1963, Sedimentary structures and facies interpretation of some molasse deposits, Sense-Schwarzwasser area, Canton Bern, Switzerland: Geol. Ultraiectina, D. 12, 42 p.
Vatan, A., 1950, Rythmes de sedimentation en Auqitaine au Cretace et au Tertiere, in Butler, A. J. (ed.), Rhythms in sedimentation: 18th Internat Geol. Cong., Proc. Sec. C, pt. 4, p. 74-82.
Wiedenmeyer, C., 1950, The structural development of areas of Tertiary sedimentation in Switzerland: Bull. Assoc. suisse Géologues et Ingénieurs Pétrol Bull., v. 17, p. 15-35.
Winnock, E., 1971, Géologie succincte du Bassin Aquitaine: Inst. Français Pétrol Pub., Coll. Colloques et Semenaires 22, v. 1, paper 4.1, p. 1-30.

EVOLUTION OF SUCCESSOR BASINS IN THE CANADIAN CORDILLERA

G. H. EISBACHER
Geological Survey of Canada, Vancouver, British Columbia

ABSTRACT

The evolution of the Canadian Cordilleran eugeosyncline can be understood in terms of five rock sequences or assemblages. Three of these, Laberge, Bowser, and Sustut and Georgia Assemblages, fulfill the requirements of epieugeosynclines (or successor basins) as defined by Marshall Kay very closely.

The Laberge Assemblage (Lower and Middle Jurassic) was derived from an island complex having a granitoid core and was probably deposited within a mobile, marginal oceanic basin.

The Bowser Assemblage (Upper Jurassic and Lower Cretaceous) was derived from a mobile orogenic core made up of rocks as old as Precambrian and as young as the Laberge Assemblage. The Bowser sediments were probably deposited within a marginal basin underlain by crust transitional in character between oceanic and continental.

The Sustut and Georgia Assemblages (Upper Cretaceous and Paleogene) are characterized by extensive continental clastics that were deposited in structurally controlled basins underlain by incompletely cratonized continental crust.

Intracrustal spreading of sialic material is invoked to explain the progressive continentalization during the evolution of the successor basins. The transition from predominantly marine deposition (Bowser Assemblage) to predominantly nonmarine deposition (Sustut Assemblage), seems to coincide with a time of peripheral cooling of metamorphic complexes and the emplacement of large discordant plutons. The transition from nonmarine deposition to regional erosion coincides with shallow-level plutonism, explosive volcanism, and uplift of the high-grade metamorphic core zones. It is suggested that the rate of these transitions is directly related to the magnitude of shortening of sialic crust and to the magnitude of the isostatic anomaly resulting therefrom.

INTRODUCTION

The Canadian Cordillera contains one of the best exposed eugeosynclinal domains in the world. Kay (1951) named this domain the Fraser Belt, and his early synthesis established the first orderly sequence of depositional and igneous events for the various eugeosynclinal rock associations. One of the new concepts introduced by Kay that resulted in better understanding of orogenic belts is the concept of "epieugeosyncline" (or "successor basin" of King, 1966). Epieugeosynclines, or successor basins, are "deeply subsiding troughs with limited volcanism associated with rather narrow uplifts, and overlying deformed and intruded eugeosynclines" (Kay, 1951, p. 107). The clastic rocks in the successor basins are "prevalent graywacke suites, but arkoses are relatively common" (p. 87) and contain "plutonic pebbles in many of their coarser sediments" (p. 56).

Most of the successor basins of British Columbia and southern Yukon Territory were discovered years after Kay's classification was first proposed, but they fit his definition nicely. These successor basins contain clastic sedimentary rocks of late Mesozoic and early Cenozoic age overlying unconformably the plutons and volcanogenic rocks of the late Paleozoic and early Mesozoic eugeosyncline.

The concept of "successor basin" as defined above, therefore, gains great significance in the analysis of the eugeosyncline because the depositional record of the successor basins contains the story of how mobile sea floor and volcanic islands emerge from an oceanic crustal setting and are transformed into continental crust. Many of the regional controls of economic metal and fuel resources are probably related to this transitional tectonic history.

TECTONO-STRATIGRAPHIC ASSEMBLAGES

Most of the information that makes it possible to approach the problem of the Cordilleran successor basins in a unified way has been collected during the past twenty years. The first major breakthrough in the understanding of the Mesozoic and Cenozoic tectono-stratigraphic elements of the western Cordillera came with the subregional syntheses of Gabrielse and Wheeler (1961), Souther and Armstrong (1966), Campbell (1966), Jeletzky and Tipper (1968), and Muller and Jeletzky (1970).

In a large and only incompletely mapped region like the western Canadian Cordillera, it is easy to lose the overall tectonic perspective in a maze of stratigraphic names that apply to limited areas. To ease the task of overview, Souther and Armstrong (1966), and Campbell (1966) first introduced the concept of tectono-stratigraphic packets, which they loosely termed "assemblages."

Souther and Armstrong (1966, p. 171) defined assemblages as "units of layered rocks . . . separated from one another by major unconformities . . . containing a variety of lithologies, but reflecting a unique environment of deposition and a unique history of deformation, without being strictly lithostratigraphic map-units." In this respect, the western Cordilleran eugeosyncline can be understood in terms of five assemblages: two eugeosynclinal assemblages and three successor-basin assemblages (fig. 1). The Cache Creek Assemblage (Mississippian to Middle Triassic) is the oldest datable eugeosynclinal unit and is generally found in tectonic contact with the Takla-Hazelton Assemblage (Upper Triassic and Middle Jurassic). The Laberge Assemblage, the oldest successor-basin assemblage, was deposited in a nonvolcanic marine environment astride the volcanic Takla-Hazelton Assemblage during Early Jurassic time. The Laberge Assemblage in turn was succeeded by the Bowser Assemblage of Middle Jurassic and Early Cretaceous age. The final stages of conversion of the Cordilleran eugeosyncline into part of the North American continent are reflected in the deposition of the Sustut and Georgia Assemblages during Late Cretaceous and Paleogene time.

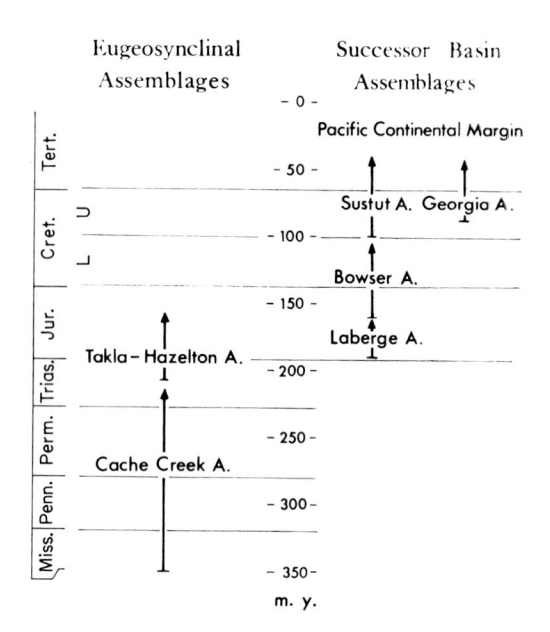

FIG. 1.—Five assemblages of Cordilleran eugeosyncline and its successor basins.

TECTONIC ENVIRONMENT OF SUCCESSOR BASINS

The preservation and distribution of successor basin assemblages within the Canadian Cordillera is the result of orogenic movements within two distinct orogenic belts (fig. 2). Wheeler and Gabrielse (1972) proposed that both the Columbian Orogen in the east and the Pacific Orogen in the west can be understood in terms of crystalline core zones flanked by terrane having tectonic transport directed away from the core zones. It is the Intermontane Belt between the Columbian and the Pacific Orogens that contains most of the preserved and accessible successor-basin assemblages. The successor basins, therefore, represent a depositional record of the interaction between the Columbian Orogen and the Pacific Orogen. The clastic record in the Insular Belt, which has been summarized by Sutherland-Brown (1966, 1968), is not considered here in the context of interaction because of the strong possibility of large-scale lateral displacements during Cenozoic time of crustal segments west of the Coast Crystalline Belt (Monger and others, 1972; Jones and others, 1972). Figure 3 schematically illustrates the interaction between the two orogens, the backdeep of the Columbian Orogen becoming subsequently a foredeep of the Pacific Orogen.

As seen from figure 3, a most important factor in the initiation of successor-basin sedimentation is the contemporaneous uplift and emergence of metamorphic and plutonic complexes along the basin hinges. Erosional unroofing of these uplifted terranes added the nonvolcanic clastic components to the fill of the adjacent successor basins. It is, therefore, the evolution of uplift within the two crystalline belts that should afford one key to the understanding of individual successor-basin assemblages. The following discussion relates the depositional history of the three Cordilleran successor basins to uplift in the adjoining crystalline core zones and demonstrates the effect of the core zones on deformation of the basin fill.

In the light of plate-tectonics hypotheses, the paleogeographic evolution of the eugeosynclinal Cache Creek and Takla-Hazelton terrane has recently been reconsidered by Monger and others (1972).

The upper Paleozoic Cache Creek Assemblage consists of graywacke, chert, argillite, limestone, basaltic lava, and ultramafic bodies, which reflect oceanic and island-arc depositional environments (Monger, 1972). The lower Mesozoic Takla-Hazelton Assemblage contains thick accumulations of andesitic-rhyolitic lavas and associated volcanogenic sediments that were laid down in both marine and nonmarine environments and that are interpreted as island-arc deposits. Syndepositional movement along fundamental fault zones disrupted original basin geometries and resulted in a great jigsaw puzzle of rafted or overthrust crustal blocks. Composition of Cache Creek and Takla-Hazelton sandstones and conglomerates is characterized by paucity of quartz and granitic debris and by

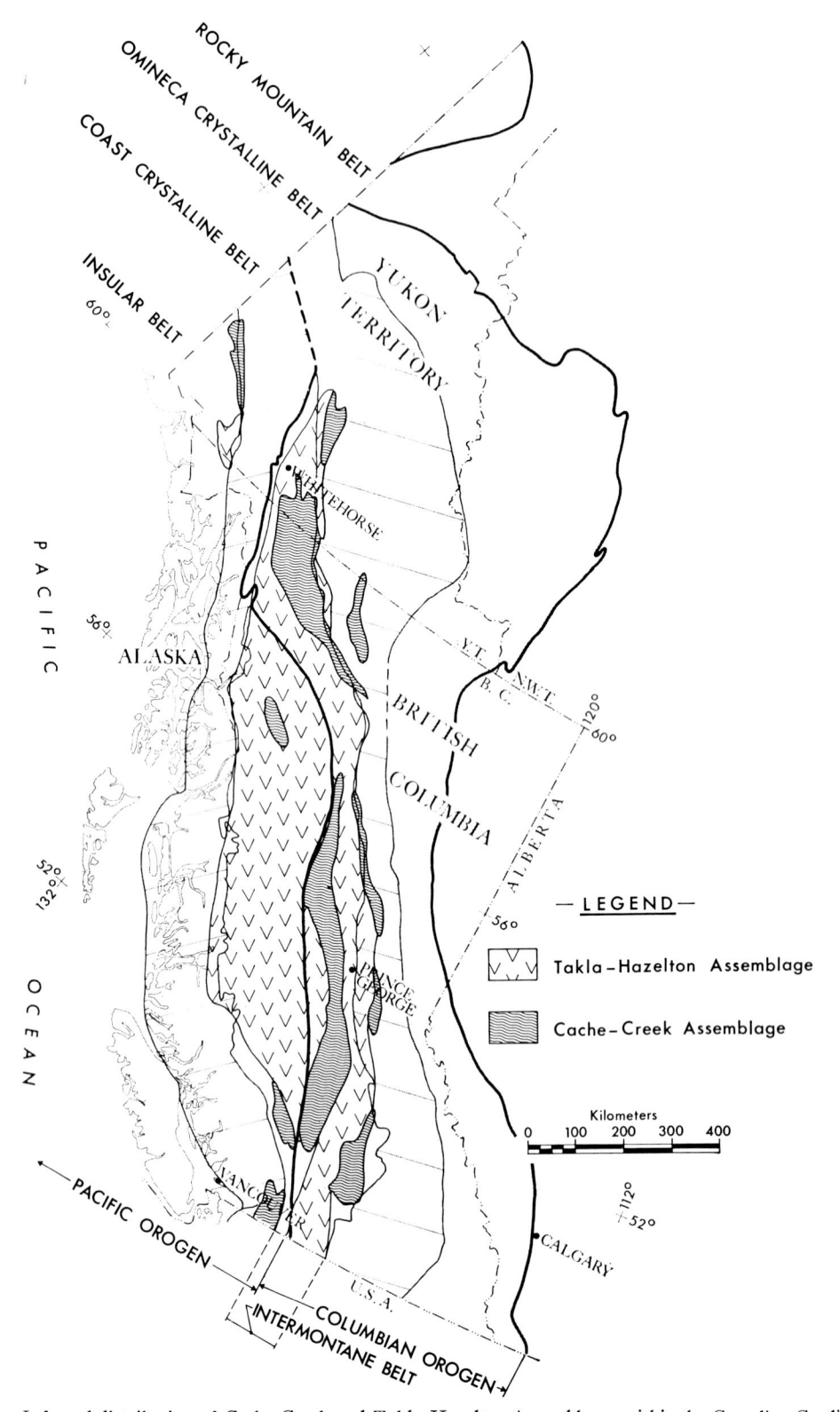

Fig. 2.—Inferred distribution of Cache Creek and Takla-Hazelton Assemblages within the Canadian Cordillera.

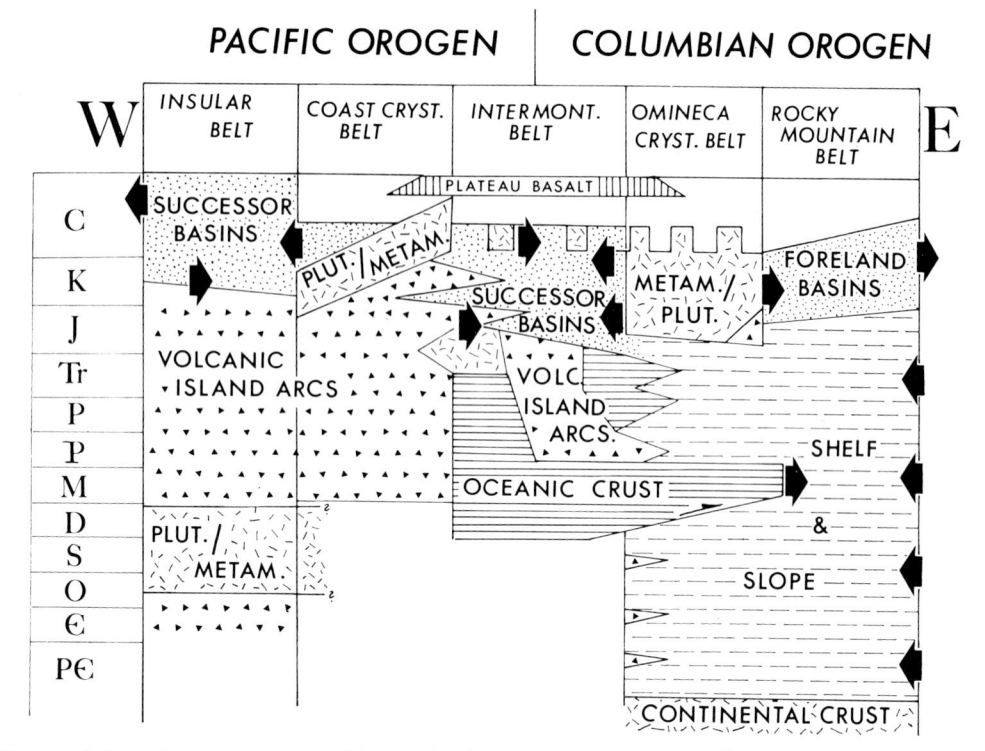

FIG. 3.—Schematic representation of interaction between tectonic belts of Canadian Cordillera. Arrows indicate direction of sedimentary transport (after Monger and others, 1972).

the dominance of volcanic rock fragments. The nonvolcanic Laberge Assemblage was deposited in a subsiding trough alongside a rising island arc and represents the youngest sedimentary sequence that was involved significantly in penetrative deformation of the Columbian Orogen.

LABERGE ASSEMBLAGE

Basin.—The Laberge Assemblage represents synorogenic clastic wedges containing the first major quantities of plutonic debris to be derived from within the eugeosynclinal domain. Rock units of the Laberge Assemblage are dated as Early to Middle Jurassic and signify a marked change in the depositional environment from a predominantly marine volcanic setting to a vigorous nonvolcanic shelf-slope environment. Laberge rocks crop out along the western edge of the Columbian core zone and are well exposed along the northern part of the core. In the more southern parts of the Cordillera, outcrops of the Laberge Assemblage are small and tectonically disrupted (fig. 4). The thickness of the assemblage varies throughout the basin, but in the north-central part of the basin, a representative thickness is about 2,000 meters.

The bulk of the assemblage consists of open-framework subaqueous conglomerates, graywackes, shales, and rare autoclastic limestone blocks. Wheeler (1961, p. 60) has illustrated the interfingering of very coarse, conglomeratic subsea channel deposits with finer grained continental slope deposits made up of graded sandstone and shale sequences, which also display scour, load, slump, and convolute structures (Souther, 1971).

Along most of the Laberge trough, the proximal conglomerate channels occur along the southwest margin, whereas much of the central outcrop belt is made up of repetitive fine-grained subsea-slope and -fan sequences (Souther, 1971; Aitken, 1959). Where the northeastern margin of the basin is exposed, it seems to be characterized also by an increase in the proportion of plant-bearing conglomerate channels, which suggests a continental source area to the north (Bostock and Lees, 1938; Wheeler, 1961; D. J. Tempelman Kluit, personal commun., 1972). The basal contact of the marginal channel facies generally displays a distinct disconformity with the underlying Upper Triassic limestone and volcanoclastics, but this disconformity changes gradually into a conformable contact toward the center of the basin (Aitken, 1959; Souther, 1971).

The channel conglomerates are generally poly-

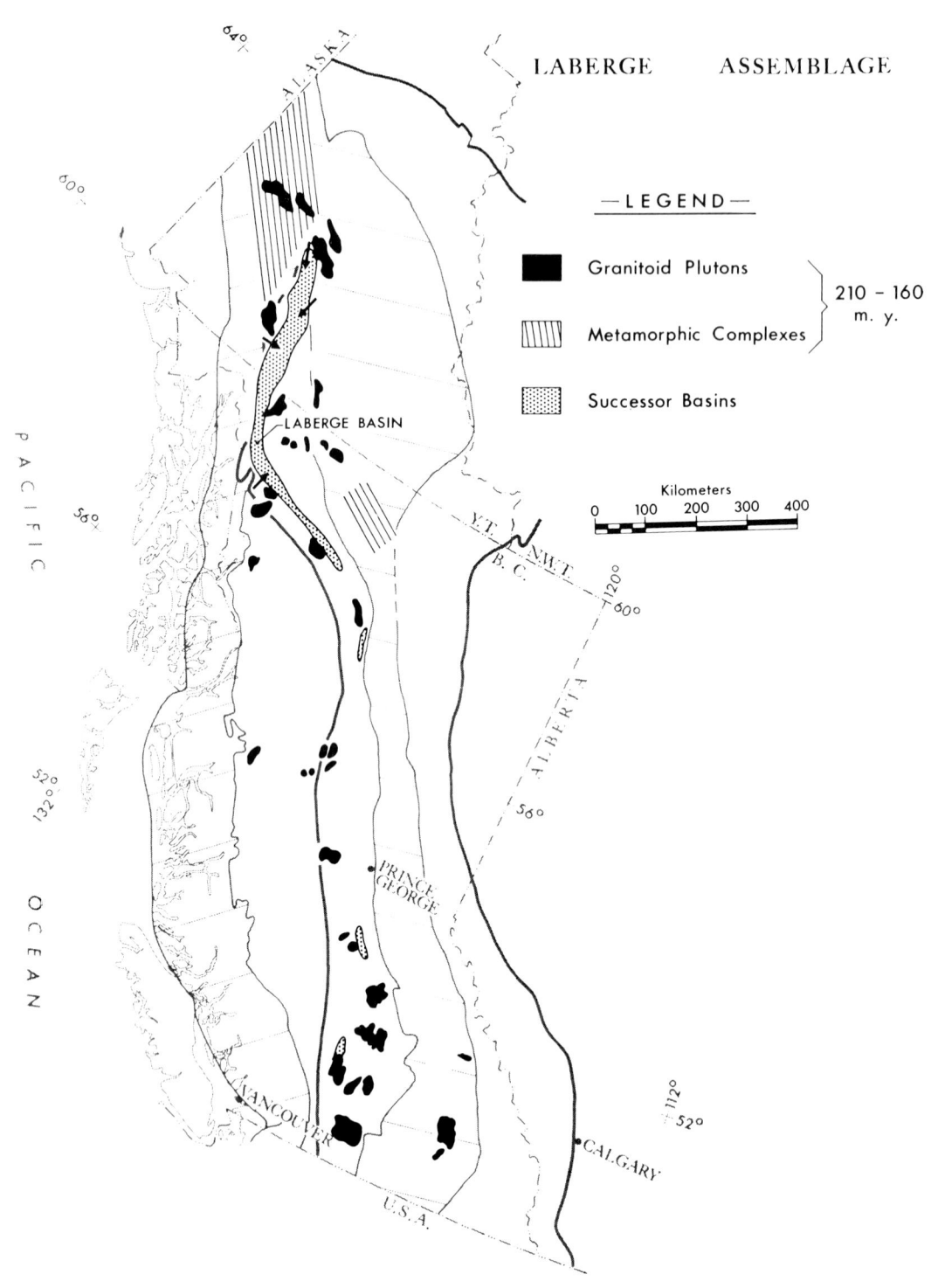

FIG. 4.—Generalized distribution of Laberge Assemblage and inferred contemporaneous uplift of plutonic and metamorphic terrane. Arrows indicate paleocurrent trends.

mictic, containing clasts of porphyritic andesite (30 to 50%), granodiorite-quartz diorite (20 to 50%), limestone, graywacke, argillite, chert, and quartzite. Arenites reflect the composition of the conglomerates and record the first appearances of substantial amounts of detrital quartz and K-feldspar sediments. A modal diagram comparing the composition of pre-Laberge with Laberge sandstones is shown in figure 5.

The interlayering of plutonic-volcanic boulder channels and silts and clays of similar mineralogical composition appears to be typical for the Laberge Assemblage and indicates a rugged orogenic coastline with rivers and submarine canyons feeding diverse clastic debris into the subsiding trough. The fact that plutonic clasts are generally well rounded, whereas most of the sedimentary clasts are angular (Wheeler, 1961), suggests that intensive mixing of fluvially transported clasts and locally eroded debris took place during transport down subsea canyons. Souther and Lambert (1972) also reported granitic and sedimentary clasts suspended in contorted distal siltstones and shales that indicate submarine mudflows and turbidity currents as the means of deposition of at least some of the Laberge sediments. Steep coastlines similar to the ones accompanying Laberge sedimentation apparently existed also during contemporaneous deposition of parts of the volcanic Takla-Hazelton Assemblage. Plutonic clasts occur only sporadically within the Takla-Hazelton Assemblage, however, and the principal direction of transport was to the west (Coates, 1970; Tipper, 1963; Jeletzky, 1972).

The sedimentological characteristics of the highest units in the Laberge Assemblage indicate that quiet depositional environments were established toward the end of Laberge deposition. Souther (1971) reported calcareous shales, argillaceous limestones, and a few beds of resinous coal high in the section from the north-central part of the basin. Wheeler (1961) noted the appearance of well-sorted quartz-chert pebble conglomerate in rocks of possible Middle Jurassic age. This change to possibly shallower marine environments marked the depositional transition to another successor-basin assemblage and preceded the deformational phase of the Laberge Basin.

Uplift.—The oldest widespread K-Ar ages for plutonic rocks within the Cordilleran eugeosyncline are found mainly in discordant granodiorite-quartz diorite bodies that were intruded into shallow crustal levels during Late Triassic time and that were at least partly unroofed by Early Jurassic time (Douglas and others, 1970; Frebold and Tipper, 1969; Campbell and Tipper,

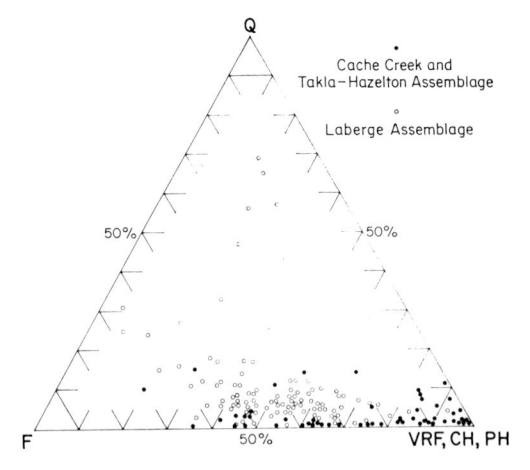

FIG. 5.—Modal composition of pre-Laberge and Laberge sandstones. Q, quartz, quartzite; F, feldspar; VRF, volcanic rock fragments; CH, chert; PH, phyllosilicates. Data from Aitken (1959), Wheeler (1961), Monger (1969), Souther (1961), and H. Gabrielse (personal commun. 1972).

1971). The distribution of plutons of Late Triassic to Early Jurassic age (210–170 my) does not coincide with the outline of either the Coast or Omineca Crystalline Belts (fig. 4). The swell created by the emplacement of these older plutonic complexes probably rose to the west of the future site of the Omineca Crystalline Belt and became part of the Omineca belt only after extensive crustal shortening during post-Early Jurassic time. This swell contained elements of what has been referred to as the Stikine Arch, Skeena Arch, and Pinchi Geanticline (Souther and Armstrong, 1966; Campbell, 1966; Douglas and others, 1970) and probably had the appearance of a narrow chain of islands. The Omineca Crystalline Belt probably existed only along the northernmost part of the swell, where regionally metamorphosed rocks have yielded Triassic-to-Early Jurassic K-Ar ages. In much of British Columbia and the western United States, an emergent Early Jurassic geanticline at the site of the present Omineca belt did not exist; open sea possibly lay between the Laberge depositional area and the prograding Lower Jurassic shales of the passive shelf-slope along the western edge of the North American craton (Frebold, 1957; Stanley and others, 1971). Nevertheless, physical continuity between the Laberge Assemblage and the miogeoclinal shelf rocks has not yet been established.

In terms of current tectonic models, we may speculate that the Laberge Assemblage represents a marginal basin bordered by a rising island arc in the west, a slowly subsiding continental margin to the east, and by an uplifted metamorphic terrane to the north. The width of this basin, however, must have greatly exceeded the present breadth of the assemblage inasmuch as only traces of volcanic tuffs are found among the shales and carbonate sediments of the Upper Triassic and Lower Jurassic shelf rocks (Deere and Bayliss, 1969). Today, only a few tens of kilometers separate volcanic Takla-Hazelton rocks from the miogeoclinal sedimentary rocks. If this distance were the original width of the basin, the volcanic environment should have had a more profound effect on the geologic evolution of the contemporaneous shelf-slope assemblage.

Deformation.—Frebold and Tipper (1970) compiled available information on the distribution of marine Jurassic rocks of the Canadian Cordillera. They emphasized a notable scarcity of fossiliferous rocks of late Bajocian to Bathonian (Middle Jurassic) age and demonstrated the possibility of extensive mid-Jurassic unconformities.

The period of mid-Jurassic nondeposition or erosion was followed by widespread clastic deposition of the Bowser Assemblage in the area southeast of the Laberge trough. Bowser deposition commenced with predominantly fine-grained arenites and shales of Callovian age (Jeletzky and Tipper, 1968; Souther, 1972; Eisbacher, in preparation). Northeast of the Laberge Basin, Precambrian, Paleozoic, and Triassic rocks were penetratively deformed and metamorphosed prior to emplacement of the large discordant plutonic complexes of the Omineca Crystalline Belt in latest Jurassic and Early Cretaceous time. The Laberge Assemblage was involved in this penetrative deformation (H. Gabrielse, personal commun., 1971), the resulting folds being folds overturned to the southwest and the cleavage dipping steeply to the northeast (Wheeler, 1961; Christie, 1957). Nearshore and offshore facies of the Laberge Assemblage were brought into juxtaposition by thrust faults directed to the southwest (Souther, 1971). Along much of the northeastern boundary of the Laberge belt, a steeply dipping fault zone separates the upper Paleozoic Cache Creek Assemblage from the Laberge Assemblage, and structural style and vergence of major folds in Cache Creek terrane are similar to those found in the Laberge Assemblage (Monger, 1969). Chert clasts, derived from the Cache Creek Assemblage, constitute much of the Upper Jurassic conglomerates of the Bowser Assemblage. These relationships suggest that deformation of the Laberge Assemblage and adjacent older rock units immediately preceded or was partly contemporaneous with deposition of the Upper Jurassic units of the Bowser Assemblage. The total shortening of the Laberge successor basin during this period of deformation is unknown.

BOWSER ASSEMBLAGE

Basins.—The Bowser Assemblage represents a series of predominantly marine synorogenic clastic wedges that strongly reflect deformation and initial uplift along the western edge of the Omineca Crystalline Belt (Souther and Armstrong, 1966). The Bowser Assemblage represents the time interval from Middle Jurassic (Callovian) to Early Cretaceous (Albian) and heralded beginning erosion along the Columbian core zone, which was concomitant with regression of the sea to the southwest (Jeletzky, 1971). The outlines of Bowser basins were controlled largely by regional topography inherited from the underlying volcanic terrane and pre-existing deep crustal fracture zones southwest of the former Laberge Basin (Muller, 1967; Eisbacher and Tempelman-Kluit, 1972).

The sedimentary sequences of the Bowser Assemblage were deposited in coastal plain, delta, prodelta, and continental slope environments and are characterized by abrupt changes in facies and thickness and by several intraformational unconformities. The present distribution of Bowser sediments is the result of original basin geometry and later modifications brought about by deformation, erosion, and by intrusion of plutonic complexes. From north to south, the Bowser Assemblage is present in these four sub-basins (fig. 6): the Dezadeash Basin (Kindle, 1953; Muller, 1967); the Tantalus Basin (Bostock and Lees, 1938; Wheeler, 1961); the Bowser Basin, which is the largest and the type area of the Bowser Assemblage (Duffell and Souther, 1964; Souther and Armstrong, 1966); and the Tyaughton Basin (Jeletzky and Tipper, 1968).

Measured or estimated thicknesses of the Bowser Assemblage range from 300 to 6,000 meters, but structurally unbroken sections in excess of 2,000 meters are rare. Precise reconstruction of the basin geometry for parts of the Bowser Assemblage is also hampered by possibly major contemporaneous and postdepositional transcurrent fault movement (Tipper, 1969).

The eastern parts of the basins are characterized by fluvial conglomerates and sandstones (e.g., Tantalus Basin) or by deltaic sequences with a high proportion of conglomeratic, distributary-channel deposits (e.g., northeastern part of Bowser Basin, and Tyaughton Basin), all of which indicate a river-dominated coastal environment. Much of the Upper Jurassic western facies consists of repetitive shale or shale and graywacke sequences that commonly dis-

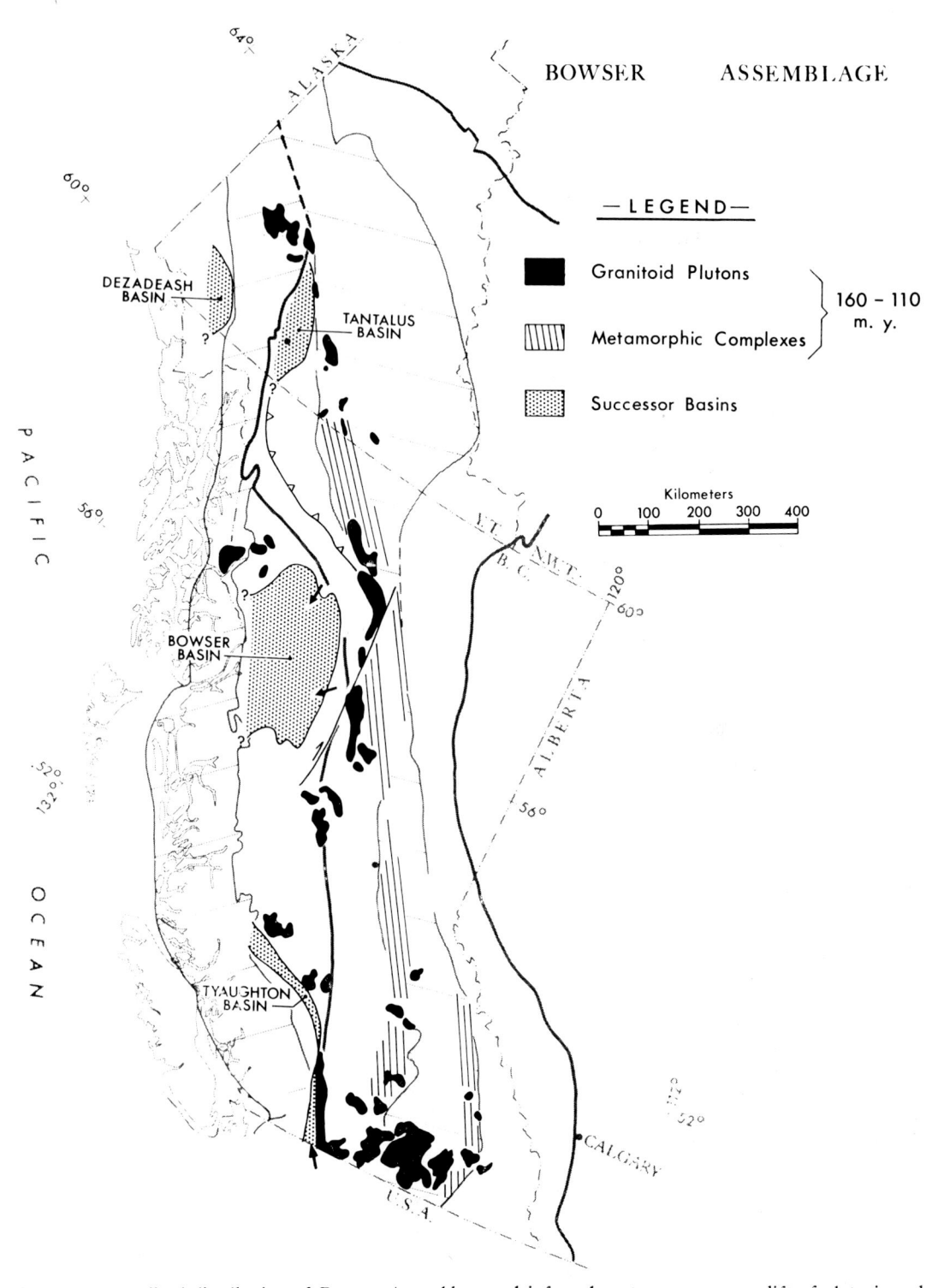

Fig. 6.—Generalized distribution of Bowser Assemblage and inferred contemporaneous uplift of plutonic and metamorphic terrane. Arrows indicate paleocurrent trends.

plays graded bedding and slumping, which suggests continental slope or subsea-fan environments (Dezadeash Basin, southwestern part of Bowser Basin, Tyaughton Basin).

The major source areas lay to the east and northeast and consisted of terranes of varying composition, depending upon the stage of erosional downcutting. In the southwestern part of the Bowser Basin, some of the Upper Jurassic clastics apparently had an intrabasinal or westerly source (Duffell and Souther, 1964; H. W. Tipper, personal commun., 1972), but much of this region is as yet unmapped and only poorly understood. Conglomerate composition varies from basin to basin. Andesite-limestone-granitoid (Dezadeash), chert-quartz (Tantalus), chert (northeast Bowser), volcanic-granitoid (southwest Bowser), and volcanic-granitoid-chert (Tyaughton) are the most common clast associations. Sandstone composition reflects a similar pattern, particularly in the Bowser Basin, where, in addition, a marked change from deposition of cherty and volcanic sands to highly micaceous quartz and feldspar sands occurred during Early Cretaceous sedimentation. This compositional change probably marked the Early Cretaceous emergence and initial unroofing of the Columbian crystalline core and the culminating deposition of thick coal-bearing clastics of Aptian and Albian age along both sides of the Omineca Crystalline Belt (Jeletzky and Tipper, 1968; Tipper, 1972; Norris and others, 1965).

In spite of a widespread and well-documented Albian transgression in the Canadian Cordillera, the highest units of the Bowser Assemblage indicate that by late Early Cretaceous time the successor basins were regionally restricted and had become separated from the synorogenic Columbian foredeep basins to the east (Jeletzky, 1971). Both foredeep and successor basins received progressively more continental deposits (Stott, 1967, 1968; Campbell, 1966; Richards and Dodds, 1973).

Uplift.—The principal Middle-to-Late Jurassic events within the source area of the Bowser Assemblage were deformation of lower Paleozoic Cache Creek, Takla-Hazelton, and Laberge rocks and the probable emplacement of major ultramafic bodies (Irvine, 1968). During this deformation, the Bowser Assemblage accumulated along faults that extended to the mantle and which, at this stage of crustal evolution, must have defined a new continental margin.

Penetrative deformation and metamorphism of sedimentary and volcanic units of the Omineca belt was followed by emplacement of large, elongate granodiorite plutons yielding K-Ar dates ranging from 140 my to 100 my. Radiometric ages from these discordant plutons and ages obtained from low-to-medium grade metamorphic rocks document broad regional uplift and unroofing along the whole length of the Omineca belt during Early Cretaceous time (Douglas and others, 1970).

From the alignment of facies belts and the linear uplifts in the Columbian core zone, the inferred basin-swell model for the Bowser successor basins is one of a structurally thickened and isostatically rising continental margin, which had offshore crustal segments (backdeeps), lagging behind in the uplift of both crystalline terrane and landward basins (foredeeps) along deep crustal fractures.

For southeast Alaska, Berg and others (1972) have proposed a magmatic arc environment for rocks that are time correlative with the Bowser Assemblage. If such a correlation is accepted, and major lateral displacements are ruled out for the area of the present Coast Mountains, a quite reasonable assumption is that strata within a Bowser marginal basin and a magmatic arc in southeast Alaska are part of the same paleogeographic setting.

Deformation.—Deformation of the Bowser Assemblage was mainly passive, was driven by Cretaceous-Tertiary uplift of the Coast Crystalline Belt, and resulted in extensive shallow décollement, recumbent folds, and thrust faults extending less than a few kilometers along strike (Souther and Armstrong, 1966; Eisbacher, 1970; Tipper, 1969). An early northerly structural trend along the northwest side of the Bowser Basin constitutes only a minor deviation from the dominant northwesterly trend, and, as a rule, axial planes, cleavage, and thrust surfaces dip to the southwest (Grove, 1971; Souther and Armstrong, 1966; Tipper, 1969).

In the Bowser and Tyaughton Basins, the strong northwesterly structural trend is probably related to the main linear uplift of the Coast Crystalline Belt, which has yielded K-Ar ages getween 80 my and 40 my (Hutchison, 1970), and is also strongly expressed in the geometry of an Eocene clastic wedge (upper part of Sustut Assemblage) along the eastern edge of the deformed Bowser Assemblage. Although the underlying Cache Creek and Takla-Hazelton Assemblages are known to be involved in the deformation of the Bowser Assemblage (Grove, 1971; Lord, 1958; Eisbacher, 1972a; Tipper, 1972), overall crustal shortening of the Bowser Basin in Late Cretaceous and Paleogene time was probably less than 20 percent. In contrast to the Laberge Assemblage, which was involved to a considerable degree in penetrative deformation of the Columbian core zone, the bulk of the Bowser Assemblage was deformed passively in response to uplift within the Pacific core zone and was affected only by minor penetrative deformation and metamorphism. That folding of the Bowser Assemblage must have occurred before, during, and after deposition of the Sustut Assemblage has been documented (Eisbacher, 1972a).

SUSTUT AND GEORGIA ASSEMBLAGES AND PACIFIC CONTINENTAL MARGIN

The Sustut and Georgia Assemblages (fig. 7) (Upper Cretaceous and Paleogene) constitute late orogenic deposits, their sedimentation, uplift, shallow-level plutonic intrusion, and deformation being intricately interwoven within the confines of the basins. Therefore, distinction between basin, uplift, and deformation in the discussion of these assemblages must be abandoned. Both assemblages display a striking change in depositional pattern between Late Cretaceous and Eocene time. Also, during the Eocene, a number of fault-bounded volcanic basins developed in southern Yukon and British Columbia. These basins were filled almost entirely by Eocene lavas and volcanogenic sediments but contain minor coals (Monger, 1968; Souther, 1970; Church, 1970; Mathews and Rouse, 1963).

Neogene sedimentation and tectonic style along the present Pacific continental margin are distinctly different from the patterns displayed by the Georgia and Sustut Assemblages and seem to define a new successor-basin assemblage. The Neogene sedimentary record of the Pacific continental margin has been discussed recently by Shouldice (1971) and Tiffin, Cameron, and Murray (1972) and, therefore, will not be considered in the context of this paper.

Sustut Assemblage.—Concomitant with the gradual emergence of the Bowser Basin above sea level, much of the Omineca Crystalline Belt of northern British Columbia must have undergone uplift, erosional dissection, and erosional truncation during Early Cretaceous to mid-Cretaceous time. Onto this emerging landscape, the basal nonmarine Sustut Assemblage was deposited unconformably on the northeastern part of the Bowser Basin fill and on the older rock units of the Omineca Crystalline Belt. The undulating pediment beneath the Sustut Assemblage, which in places was deformed during post-Sustut time (Eisbacher, 1972b), is presently being exhumed by erosion along the northeastern side of the main Sustut Basin and on ranges adjacent to the northern Rocky Mountain Trench. The pediment was carved originally into eugeosynclinal volcanic rocks of the Takla-Hazelton Assemblage, into granitoid plutons of mostly Early Cretaceous age, and into metamorphosed miogeoclinal quartzose clastics and carbonates of Proterozoic and Early Paleozoic age. All these rock types are found as clasts in conglomerates at the base of the Sustut Assemblage. Detrital micas in Sustut sandstones have yielded a K-Ar date of 117 my (GSC/ K-Ar age determination 1930), confirming an Early Cretaceous to mid-Cretaceous unroofing of crystalline terrane. Paleocurrents indicate that the lower part of the Sustut Assemblage (Tango Creek Formation) was derived from the east and northeast and that drainage was centripetal, directed towards the center of the Bowser Basin. The present outcrop of Tango Creek Formation probably represents less than half the original basin fill. The Sustut deposits near the Rocky Mountain Trench were probably never connected with deposits of the main Sustut Basin, and Souther and Armstrong (1966) referred to this depositional area as the Sifton Basin (fig. 7).

Pre-Sustut growth of folds in the Bowser Basin seems to have had only a modifying effect on the centripetal drainage pattern during early Sustut deposition. Subsequent northeasterly and easterly directed tectonic transport of Bowser rocks in latest Cretaceous and Paleogene time had a more profound influence on the basin configuration. Folding and thrust faulting involved the Tango Creek Formation, and deposition of the overlying Brothers Peak Formation (Eocene) was limited to a narrow trough along the northeastern margin of the fold belt. Drainage had become predominantly longitudinal, and much of the coarse detritus was derived directly from the growing fold belt. Brothers Peak clastic sedimentation was accompanied by heavy ash falls, which are preserved as extensive tuff sheets sandwiched between coarse alluvial fan conglomerates. These Eocene tuffs are probably the time-equivalent, airborne products of extensive explosive volcanism in northern British Columbia that was described by Aitken (1959) and Souther (1972). A dramatic change in source area for the Brothers Peak Formation is reflected in sandstone composition (Eisbacher, 1971) and in the composition of clasts in conglomeratic units. Initial widening of the basin by sourceward river erosion was followed by northeastward thrusting of the Bowser Assemblage and by a dramatic increase in chert that was derived from Bowser conglomerates. Reestablished drainage to the south initiated a second phase of erosional downcutting into older units made up of andesite, granitoid intrusions, and sedimentary rocks along the northern and northeastern fringe of the Sustut Basin. Locally, as much as 2,500 m of continental clastics were deposited in the Sustut Basin.

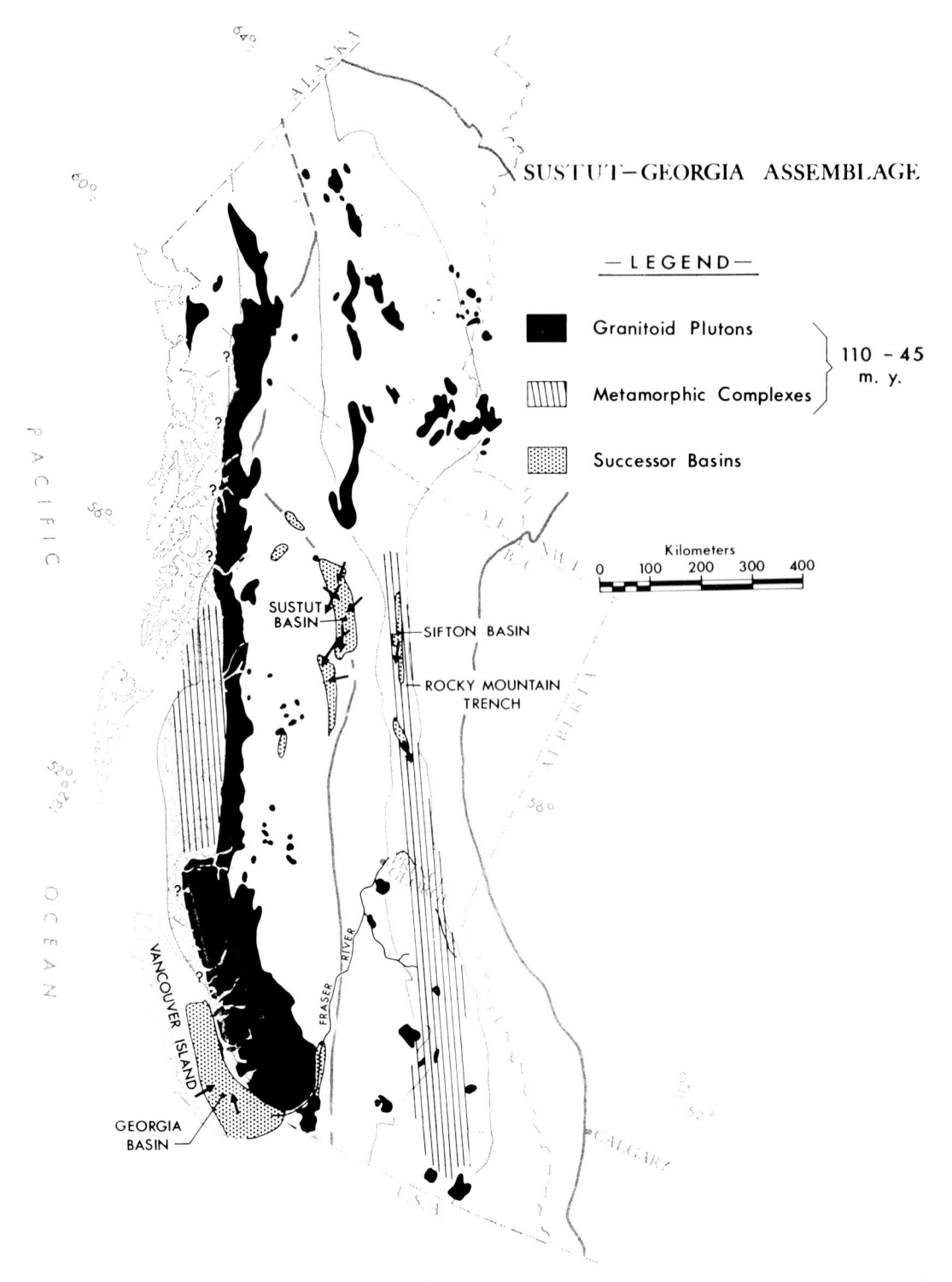

Fig. 7.—Generalized distribution of Sustut and Georgia Assemblages and inferred contemporaneous uplift of plutonic and metamorphic terrane. Arrows indicate paleocurrent trends.

Regionally, these changes in the structural setting and the drainage pattern of the Sustut Basin signify the transition from a Columbian backdeep during deposition of the lower Sustut to a Pacific foredeep during deposition of the upper Sustut. The ages of these two superimposed clastic wedges are somewhat less than, but correspond to, the dominant K-Ar ages in the respective orogenic core zones: Early Cretaceous to mid-Cretaceous K-Ar dates are prevalent for the Omineca Crystalline Belt northeast of the Sustut Basin (H. Gabrielse, personal commun., 1971), and Late Cretaceous and Eocene ages predominate for the linear core zone of the Coast Crystalline Belt (Hutchison, 1970).

Georgia Assemblage.—The continental and marine deposits of the Georgia Assemblage record the gradual establishment of the Georgia Depression between Vancouver Island and the mainland of British Columbia. Geological evolution of the Insular Belt previous to deposition of the Georgia Assemblage was probably not directly related to the mainland of British Columbia. Major transcurrent fault movements in late Mesozoic and Cenozoic time must have profoundly affected most of the present insular and offshore geology of British Columbia (Tipper, 1969; Sutherland-Brown, 1968; Monger and others, 1972). The Georgia Assemblage possibly constitutes the oldest link between the mainland of British Columbia and the Insular Belt.

The basin fill reflects deposition of a series of transgressive sequences during Late Cretaceous (Santonian) and Eocene time (Muller and Jeletzky, 1970; Hopkins, 1969). Deposition commenced along the western margin of the basin with coal-bearing fluvial, lagoonal, barrier-bar, and shallow-marine clastics. The basal units (upper Santonian, lower Campanian) rest unconformably on upper Paleozoic and Mesozoic volcanics and on Middle Jurassic granodiorite intrusions on Vancouver Island, which also constitute the principal source rocks for most of the detrital components (Muller and Jeletzky, 1970). Absence of K-feldspar and a restricted heavy mineral suite rich in mafics (Page, 1972) support the suggestion of westerly source areas for the basal Georgia sediments. Prograding clastic wedges possibly formed in response to syndepositional growth of hinge faults along the western margin of the basin.

During late Campanian and Maestrichtian time, the sea transgressed eastward, tapping new source areas to the southeast and east of the basin (Muller and Jeletzky, 1970; Miller and Misch, 1963). At this stage, paleocurrents flowed predominantly to the northwest into a subsiding delta and prodelta environment (Packard, 1972). The systematic increase in K-feldspar (Page, 1972) suggests a considerable input of acidic plutonic material from the southeast and east.

During Paleogene time, deformation continued and easterly derived mid-Eocene continental deposits were laid down unconformably on the deformed Georgia Assemblage (Miller and Misch, 1963) and onto a peneplane that had been carved into the rising Coast Crystalline Belt (Roddick, 1965; Hopkins, 1969). Eocene deposits of the Georgia Assemblage represent the northernmost outcrops of a fluviatile and deltaic coastal plain complex, which extended at least as far south as southern Oregon (Dott, 1966).

Post-Eocene uplift and tilting of the peneplane was dramatic on both Vancouver Island and along the southern Coast Mountains, and deep continental erosion probably affected all units of the Georgia Assemblage (Tiffin, 1969). About 100 kilometers east of the Georgia basin, Eocene conglomerate and sandstone, correlative with the upper part of the Georgia Assemblage, were intruded by late Oligocene and early Miocene (29 to 26 my) granodiorite plutons that represent the northerly extension of the Cascade igneous complex (Richards and White, 1970). Directional sedimentary structures in Pliocene fluviatile sediments in the Fraser River valley east of the Coast and Cascade Mountains suggest that in Miocene and Pliocene time the Fraser River probably drained to the north (H. W. Tipper, personal commun., 1971), not, as at present, into the Georgia Basin. It is probable, therefore, that during Oligocene and Pliocene time the rising Coast and Cascade Mountains blocked major drainage from the Intermontane Belt to the coast and that only small amounts of sediment entered the Georgia Depression through the reentrant between the southern Coast Mountains and the Cascade Mountains of northern Washington.

The still-active progradation of the Fraser delta into the Strait of Georgia is essentially a postglacial event (Mathews and Shepard, 1962; Tiffin, 1969).

A SPECULATIVE MODEL OF CRUSTAL EVOLUTION FOR THE SUCCESSOR BASINS

The salient point regarding the evolution of successor basins in the Canadian Cordillera is that, at a certain stage in the development of the eugeosyncline, some of the subsiding crustal segments began to receive large quantities of clastic sediments, which succeeded earlier volcanogenic deposits. The sequence of events also suggests a gradual regression of the sea. This

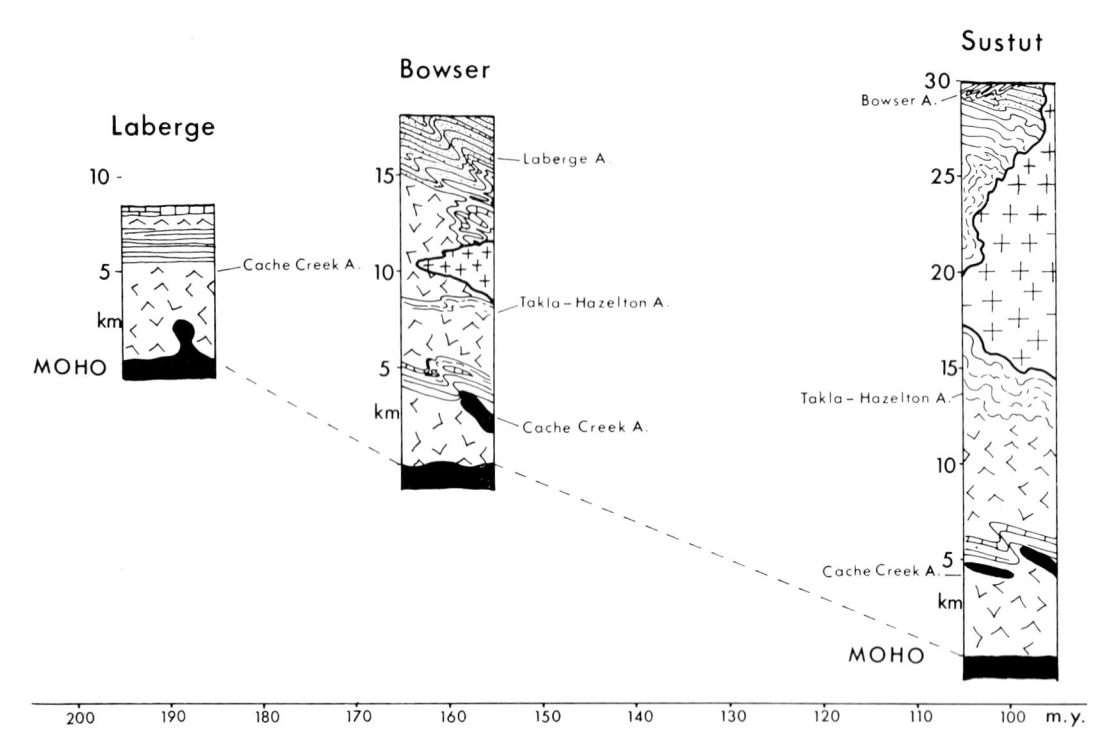

FIG. 8.—Hypothetical scheme for evolution of crust underlying successor-basin assemblages of Canadian Cordillera.

fundamental change in the character of the basins must have been accompanied by changes in the crust underlying the successor basins. It is appropriate, therefore, to speculate on the nature of a representative section through the crust during deposition of each of the three successor-basin assemblages.

The transitions from one crustal regime to the next possibly coincided with times of subdued horizontal stresses in the earth's crust, when vertical movements played a more significant role than did crustal shortening. These transitional periods should be at about 190 my (Hettangian-Sinemurian), 160 my (Callovian), and 110 to 100 my ago (Aptian-Cenomanian). Successor-basin development was terminated about 50 to 40 my ago by broad regional uplift and shallow-level plutonic activity that was accompanied by widespread acidic to intermediate volcanism. A schematic model of the postulated crustal evolution is shown in figure 8.

Laberge crust.—Their close spatial and structural association suggests the possibility that the Laberge Assemblage was deposited at least partly on the Cache Creek Assemblage. Volcanic Cache Creek terrane, therefore, probably foundered prior to clastic deposition of Laberge sediments, which took place contemporaneously with volcanic activity on a rising granodiorite-cored island complex to the southwest. If this scheme is correct, the crust below Laberge sediments can be assumed to represent a rock assemblage made up of a few kilometers of basalt, chert, and graywacke and of some ultramafic bodies whose bulk thickness probably does not exceed 5 to 10 km (Monger and others, 1972). Such a rock association and thickness should designate the crust below the Laberge successor basin as oceanic (Ewing, 1969; Shor and Raitt, 1969), and in terms of recent crustal models the Laberge Basin would represent a marginal oceanic basin (Karig, 1972).

Bowser crust.—Once the Upper Triassic and Lower Jurassic granodiorite swell and the adjacent Laberge Assemblage had become structurally welded to the rising Omineca Crystalline Belt, crustal segments made up of Takla-Hazelton and Cache Creek volcanic deposits continued to subside southwest of the deformed belt, permitting clastic sediment to be ponded in preformed troughs. The crust beneath the Bowser sediments probably consisted of about 10 km of basaltic, andesitic, and dacitic lavas and volcanogenic sediments of the Takla-Hazelton Assemblage, some granitoid complexes, and possibly as much as 5 km of the Cache Creek Assemblage. Bowser sediments, therefore, were deposited on what can be called andesitic crust grading laterally eastward into subcontinental crust. Considering the possibility of a magmatic arc in southeast Alaska (Berg and others, 1972), the Bowser successor basin should be classified, in terms of current nomenclature, as a marginal basin whose crustal thickness is transitional between oceanic and continental (Neprochnov and others, 1970).

Sustut crust.—The transition from marine to nonmarine sedimentation and continued regional uplift during deposition of the Sustut Assemblage suggest further thickening of the crust. This was achieved possibly by mobilization and lateral spreading of sialic material from the tectonically thickened Columbia core zone (Wheeler and Gabrielse, 1972). The 5 to 15 km of sialic crust that had to be added to the Bowser crust in order to effect the emergence of the earth's surface above sea level must essentially comprise the lateral equivalents of Mesozoic plutonic and metamorphic complexes. In spite of plate tectonics, the mechanism of what might be called "lateral continentization," which had to be active over a period of at least 100 my, remains a fundamental geotectonic problem (Beloussov, 1972; Heezen, 1972). Extensive emplacement of small, shallow-level granodiorite plutons of Late Cretaceous and Paleogene age throughout the Intermontane Belt seems to indicate that, by this time, many parts of the former eugeosyncline had developed sialic crustal components.

Discussion.—Like the model inferred here, other ancient eugeosynclinal belts display a sequence beginning with an oceanic stage, followed by a duality of volcanic island arcs and terrigenous flysch marginal basins, and finally an intense piling up of eugeosynclinal material and formation of an intracrustal sialic layer (Peive and others, 1972). Large parts of recent marginal basins are characterized by anomalously high heat flow (Karig, 1971). The motion of such crustal and upper mantle segments having high heat flow toward and underneath adjacent continental slope and shelf deposits could be the cause for the sudden and dramatic increase in heat flow necessary for regional metamorphism of the outer portion of the miogeoclinal wedge. The structure of the Cache Creek and Laberge Assemblages reflects apparent underthrusting of the Laberge crust beneath the miogeoclinal wedge, which preceded large-scale regional metamorphism in the Omineca Crystalline Belt of the Columbia Orogen.

The present crustal thickness of about 30 to 40 kilometers for the southwestern cordillera (Berry and others, 1971), indicates an incomplete stage of craton formation, which is also suggested by high heat flow, possible high electrical conductivity in the lower crust, and extensive Neogene volcanism (Caner, 1970; Souther, 1970). Neogene basaltic volcanism may also represent the beginning of a new geosynclinal cycle brought about by changes in the motion of lithospheric plates on the Pacific Ocean floor (Monger and others, 1972).

THE OROGENIC CYCLE

From the tectono-stratigraphic record of the Cordilleran successor basins, it appears that depositional events within the basins are related in a regular manner to igneous and metamorphic events within the rising crystalline core zone of the Columbian Orogen. This nonrandom interaction has long been implicit in the concept of the orogenic cycle (Bucher, 1933, p. 125–148). The orogenic cycle implies that the evolution of any mountain belt proceeds from a phase of regional subsidence to a phase of regional uplift and that the cycle is accompanied by a regular interaction between igneous and sedimentary events. Seen from the perspective of the Columbian uplift, three distinct sets of phenomena appear to have parallelisms in other mountain belts.

(1) The youngest rocks involved in penetrative deformation of the Columbian core zone represent clastic marine deposits derived mainly from *outside* the core zone and are only slightly older than the clastics derived from the core zone (Laberge Assemblage of Early Jurassic age). The oldest core-derived clastics of Late Jurassic age do not seem to be involved in penetrative deformation or metamorphism of the crystalline core zone. It can be concluded, therefore, that updoming followed crustal shortening, activated a phase of erosion, and may even have inhibted further extensive crustal shortening.

(2) Transition from marine to nonmarine deposition in the successor basins and foredeeps seems to coincide roughly with the emplacement of large discordant granitoid massifs and cooling of low-grade metamorphic rocks of the core zone (140 to 110 my).

(3) Transition from continental deposition to a predominantly erosional regime seems to coincide with emplacement of numerous shallow-level granodiorite and quartz monzonite plutons and the final cooling of high-grade metamorphic rocks (55 to 40 my).

This sequence of events is shown schematically for the north-central part of the Columbian Orogen in figure 9. The depositional sequence of the successor basins (A) is essentially

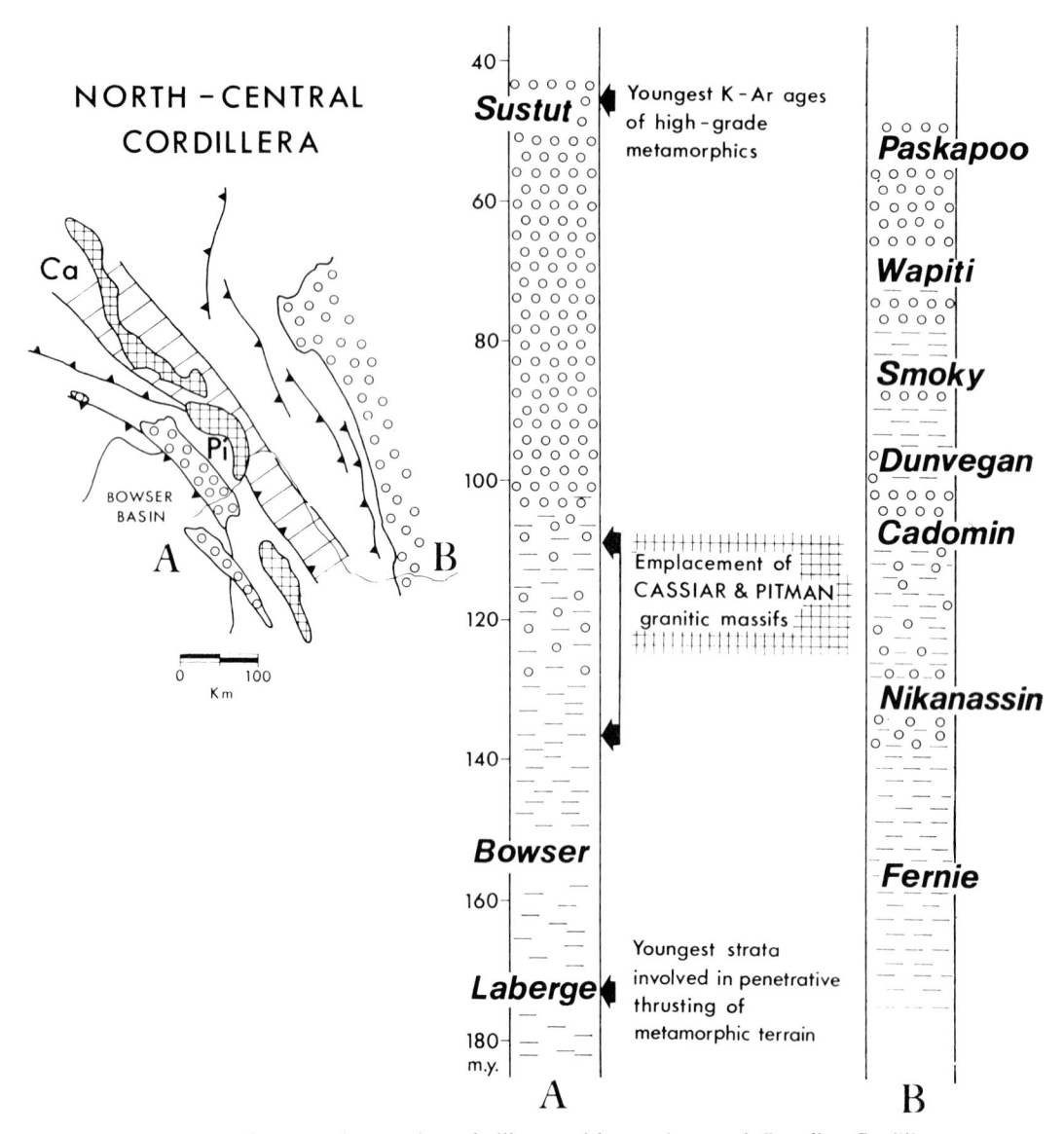

FIG. 9.—Uplift stage of orogenic cycle illustrated by north-central Canadian Cordillera. Diagonal ruling indicates metamorphic terrane.

analogous to the one in the foredeep (B) (see Douglas and others, 1970, for documentation). In the foredeep, however, is a persistent recurrence of marine conditions within the late-orogenic continental succession, which may be due to continued isostatic loading by the advance of sedimentary thrust sheets from the west (Price and Mountjoy, 1970).

The three relationships outlined above may hold for the process of uplift in other orogenic belts as well. Initiation of Devonian continental sedimentation in the foredeeps and successor basins of the Caledonian and Appalachian orogen coincides closely with radiometric ages of metamorphism and plutonism (400 to 350 my) of the core zone (Friend, 1967; Hadley, 1964). The youngest dates from the Appalachian core zone, which range from 250 to 230 my (Hadley, 1964; Lyons and Faul, 1968), also roughly correlate with the transition from continental clastic deposition to an erosional river regime during the Permian Period.

In the Swiss Alps the youngest sedimentary rocks overridden by basement nappes are of early Eocene age (55 my), and the transition from a predominantly marine sequence (Flysch) to shallow-water, marine and nonmarine sediments (Molasse) coincides with the emplacement of the Bergell granite massif at about 30 my ago (Milnes, 1969) and with cooling of low-grade metamorphic rocks at about the same time. Final cooling of the high-grade metamorphic core zone, 15 to 10 my ago, seems to coincide with the transition from the Molasse depositional setting to an erosional regime in the foredeep.

Although time intervals between respective events of the uplift history are different among the Canadian Cordillera, the Appalachians, and the Alps, the general sequence of events is the same.

In the Northern Apennine Orogen, a widespread unconformity associated with the onset of extensive continental deposition at the base of the Pliocene section also seems to coincide with emplacement of large granitoid plutons between 7 and 4 my ago (Sestini, 1970), which were preceded by cooling of low-grade metamorphic rocks between 15 and 10 my ago (Brozolo and Giglia, 1973).

In the Himalayan Orogen widespread freshwater deposition began during late Miocene time and possibly was related to uplift of a large crystalline terrane in the orogenic core zone, which has K-Ar dates of 15 to 10 my (Gansser, 1964).

In the Himalayan chain and the Northern Apennines, uplift has not yet been completed, and continental deposition is active in the foredeeps and successor basins. In these mountain belts metamorphic recrystallization and igneous emplacement probably still are going on at depth.

The geophysical implication of these regularities could be manifold, but isostasy apparently is the most important factor determining the relationship between igneous-metamorphic processes and the depositional record. Large-scale shortening and thickening of sialic crust is followed by high rates of isostatic rebound, by intensive erosion, and the deposition of thick clastic wedges (e.g., Alps, Himalayas); minor shortening and thickening of sialic crust is followed by low rates of isostatic rebound, and lower rates of erosion and foreland deposition (Appalachians, and Canadian Cordillera).

ACKNOWLEDGMENTS

I wish to express my sincere thanks to Dr. H. Gabrielse, Geological Survey of Canada, for his untiring enthusiasm and help in this and related studies. Drs. H. Tipper, J. Monger, J. Souther, and A. Okulitch and others of the Vancouver office of the Geological Survey of Canada have also discussed various aspects of the manuscript and contributed information. Responsibility for possible misinterpretation of any of their data, however, rests entirely with me. Critical comments on the manuscript by Dr. Bates McKee, University of Washington, are gratefully acknowledged.

REFERENCES

AITKEN, J. D., 1959, Atlin map-area, British Columbia: Geol. Survey Canada Mem. 307, 89 p.
BELOUSSOV, V. V, 1972, Basic trends in the evolution of continents: Tectonophysics, v. 13, p. 95–117.
BERG, H. C., JONES, D. L., AND RICHTER, D. H., 1972, Gravina-Nutzotin Belt—Tectonic significance of an upper Mesozoic sedimentary and volcanic sequence in southern and southeastern Alaska: U.S. Geol. Survey Prof. Paper 800-D, p. 1–24.
BERRY, M. T., AND OTHERS, 1971, A review of geophysical studies in the Canadian Cordillera: Canadian Jour. Earth Sci., v. 8, p. 788–801.
BOSTOCK, H. S., AND LESS, E. T., 1938, Laberge map-area, Yukon: Geol. Survey Canada Mem. 217, 32 p.
BROZOLO, F. R. DI, AND GIGLIA, G., 1973, Further data on the Corsica-Sardinia rotation: Nature, v. 241, p. 389–391.
BUCHER, W. H., 1933, The deformation of the earth's crust: Princeton, New Jersey, Princeton Univ. Press, 518 p.
CAMPBELL, R. B., 1966, Tectonics of the south-central Cordillera of British Columbia, in Tectonic history and mineral deposits of the western Cordillera: Canadian Inst. Mining and Metallurgy Special Vol. 8, p. 61–72.
———, AND TIPPER, H. W., 1971, Geology of Bonaparte Lake map-area, British Columbia: Geol. Survey Canada Mem. 363, 100 p.
CANER, B., 1970, Electrical conductivity structure in western Canada and petrological interpretations: Jour. Geomagetism and Geoelectricity, v. 22, p. 113–129.
CHRISTIE, R. L., 1957, Bennett map-area: Geol. Survey Canada Map 19–1957.
CHURCH, B. N., 1970, The geology of the White Lake Basin, in Geology, exploration and mining in British Columbia: British Columbia Dept. Mines and Petroleum Res. p. 396–402.
COATES, J. A., 1970, Stratigraphy and structure of Manning Park area, Cascade Mountains, British Columbia, in Structure of the southern Canadian Cordillera: Geol. Assoc. Canada Special Paper 6, p. 149–154.
DEERE, R. E., AND BAYLISS, P., 1969, Mineralogy of the Lower Jurassic in west-central Alberta: Canadian Petroleum Geology Bull., v. 17, p. 133–153.
DOTT, R. H., JR., 1966, Eocene deltaic sedimentation at Coos Bay, Oregon: Jour. Geology, v. 74, p. 373–420.
DOUGLAS, R. T. W., AND OTHERS, 1970, Geology of western Canada, in Geology and economic minerals of Canada: Geol. Survey Canada, Econ. Geology, v. 1, p. 366–488.

DUFFELL, S., AND SOUTHER, J. G., 1964, Geology of Terrace map-area, British Columbia: Geol. Survey Canada Mem. 329, 117 p.
EISBACHER, G. H., 1970, Tectonic framework of Sustut and Sifton Basins, B.C., in Report of Activities: ibid., Paper 70-1, pt. A, p. 36–37.
———, 1971, A subdivision of the Upper Cretaceous-lower Tertiary Sustut Group, Toodoggone map-area, British Columbia: ibid., Paper 70-68, 16 p.
———, 1972a, Tectonic framework of Sustut and Sifton Basins, B.C., in Report of activities: ibid., Paper 72-1, pt. A, p. 24–26.
———, 1972b, Tectonic overprinting near Ware, northern Rocky Mountain Trench: Canadian Jour. Earth Sci., v. 9, p. 903–913.
———, in preparation, Deltaic sedimentation, northeastern Bowser Basin.
——— AND TEMPELMAN-KLUIT, D. J., 1972, Map of major faults in the Canadian Cordillera and southeast Alaska (map and abs.): Geol. Assoc. Canada, Cordilleran Sec. Mtg. 1972, Abs., p. 13–14, 22–25.
EWING, J., 1969, Seismic model of the Atlantic Ocean, in The earth's crust and upper mantle: Am. Geophys. Union Geophys. Mon. 13, p. 220–225.
FREBOLD, H., 1957, The Jurassic Fernie Group in the Canadian Rocky Mountains and foothills: Geol. Survey Canada Mem. 287, 197 p.
———, AND TIPPER, H. W., 1969, Lower Jurassic rocks and fauna near Ashcroft, British Columbia, and their relations to some granitic plutons: ibid., Paper 69-23, 20 p.
———, AND ———, 1970, Status of the Jurassic in the Canadian Cordillera of British Columbia, Alberta and southern Yukon: Canadian Jour. Earth Sci., v. 7, p. 1–21.
FRIEND, P. F., 1967, Tectonic implications of sedimentation in Spitzbergen and midland Scotland, in International symposium on the Devonian System: Calgary, Alberta, Alberta Soc. Petroleum Geologists, v. 2, p. 1141–1148.
GABRIELSE, H., AND WHEELER, J. O., 1961, Tectonic framework of southern Yukon and northwestern British Columbia: Geol. Survey Canada Paper 60-24, 37 p.
GANSSER, A., 1964, Geology of the Himalayas: London, Interscience Publishers, 289 p.
GROVE, E. W., 1971, Geology and mineral deposits of the Stewart area, British Columbia: British Columbia Dept. Mines and Petroleum Res. Bull. 58, 219 p.
HADLEY, J. B., 1964, Correlation of isotopic ages, crustal heating and sedimentation in the Appalachian region, in LOWRY, W. D. (ed.), Tectonics of the southern Appalachians: Virginia Polytechnic Inst., Dept. Geol. Sci. Mem. 1, p. 33–46.
HEEZEN, B. C., 1972, Inland and marginal seas: Tectono-physics, v. 13, p. 293–308.
HOPKINS, W. S., JR., 1969, Palynology of the Eocene Kitsilano Formation, southwest British Columbia: Canadian Jour. Botany, v. 47, p. 1101–1131.
HUTCHISON, W. W., 1970, Metamorphic framework and plutonic styles in the Prince Rupert region of the central Coast Mountains, British Columbia: Canadian Jour. Earth Sci., v. 7, p. 376–405.
IRVINE, T. N., 1968, Interpretation of GSC K-Ar dates 66–18 and 66–19, in Age determinations and geological studies: Geol. Survey Canada Paper 67-2, pt. A., p. 21–22.
JELETZKY, J. A., 1971, Marine Cretaceous biotic provinces and paleogeography of western and Arctic Canada: illustrated by a detailed study of ammonites: ibid., Paper 70-22, p. 92.
———, 1972, Jurassic and Cretaceous rocks along Hope Princeton Highway and Lookout Road, Manning Park, British Columbia: ibid., Open File Rept. 114, 38 p.
———, AND TIPPER, H. W., 1968, Upper Jurassic and Cretaceous rocks of Taseko Lakes map-area and their bearing on the geological history of southwestern British Columbia: ibid., Paper 67-54, 281 p.
JONES, D. L., IRWIN, W. P., AND OVENSHINE, A. T., 1972, Southeastern Alaska—a displaced continental fragment: U.S. Geol. Survey Prof. Paper 800-B, p. 211–217.
KARIG, D. E., 1971, Origin and development of marginal basins in the western Pacific: Jour. Geophys. Research, v. 76, p. 2542–2561.
KAY, M., 1951, North American geosynclines: Geol. Soc. America Mem. 48, 143 p.
KINDLE, E. D., 1953, Dezadeash map-area, Yukon Territory: Geol. Survey Canada Mem. 268, 68 p.
KING, P. B., 1966, The North American Cordillera, in Tectonic history and mineral deposits of the western Cordillera: Canadian Inst. Mining and Metallurgy, Special Vol. 8, p. 1–25.
LORD, C. S., 1948, McConnell Creek map-area, Cassiar District, British Columbia: Geol. Survey Canada Mem. 251, 72 p.
LYONS, J. B., AND FAUL, H., 1968, Isotope geochronology of the northern Appalachians, in ZEN, E. A., WHITE, W. S., HADLEY, J. B., AND THOMPSON, J. B. (eds.), Studies of Appalachian geology, northern and maritime: New York, Interscience Publishers, p. 305–318.
MATHEWS, W. H., AND SHEPARD, F. P., 1962, Sedimentation of Fraser River delta, British Columbia: Am. Assoc. Petroleum Geologists Bull., v. 46, p. 1416–1438.
———, AND ROUSE, G. E., 1963, Late Tertiary volcanic rocks and plant-bearing deposits in British Columbia: Geol. Soc. America Bull., v. 74, p. 55–60.
MILLER, C. M., AND MISCH, P., 1963, Early Eocene angular unconformity at western front of northern Cascades, Whatcom County, Washington: Am. Assoc. Petroleum Geologists Bull., v. 47, p. 163–174.
MILNES, A. G., 1969, On the orogenic history of the Central Alps: Jour. Geology, v. 77, p. 108–112.
MONGER, J. W. H., 1968, Early Tertiary stratified rocks, Greenwood map-area, British Columbia: Geol. Survery Canada Paper 67-42, 39 p.
———, 1969, Stratigraphy and structure of Upper Paleozoic rocks, northeast Dease Lake map-area, British Columbia: Geol. Survey Canada Paper 68-48, 41 p.
———, 1972, Oceanic crust in the Canadian Cordillera, in The ancient oceanic lithosphere, a symposium: Canadian Dept. Energy, Mines and Res, Earth Physics Br., Pubs. v. 42, no. 3, p. 59–64.

———, Souther, J. G., and Gabrielse, H., 1972, Evolution of the Canadian Cordillera: a plate-tectonic model: Am. Jour. Sci., v. 272, p. 577–602.
Muller, J. E., 1967, Kluane Lake map area, Yukon Territory: Geol. Survey Canada Mem. 340, 137 p.
———, and Jeletzky, J. A., 1970, Geology of the Upper Cretaceous Nanaimo Group, Vancouver Island and Gulf Islands, British Columbia: ibid., Paper 69–25, 77 p.
Neprochnov, Yu. P., Kosminskaya, I. P., and Malovitsky, Ya. P., 1970, Structures of the crust and upper mantle of the Black and Caspian Seas: Tectonophysics, v. 10, p. 517–538.
Norris, D. K., Stevens, R. D., and Wanless, R. K., 1965, K-Ar age of igneous pebbles in the McDougall-Segur Conglomerate, southeastern Canadian Cordillera: Geol. Survey Canada Paper 65-2, 11 p.
Packard, J. A., Jr., 1972, Paleoenvironments of the Cretaceous rocks, Gabriola Island, British Columbia (master's thesis): Corvallis, Oregon State Univ., 101 p.
Page, R. J., 1972, A preliminary petrographic examination of the Gabriola, Decourcy and Comox Formations, Nanaimo Group, Vancouver Island and Gulf Islands, British Columbia (master's thesis): Seattle, Univ. Washington, 28 p.
Peive, A. V., Perfiliev, S. V., and Rushentsev, S. V., 1972, Problems of intracontinental geosynclines: 24th Internat. Geol. Cong., Montreal, sec. 3, p. 486–493.
Price, R. A., and Mountjoy, E. W., 1970, Geologic structure of the Canadian Rocky Mountains between Bow and Athabaska Rivers—a progress report, in Structure of the southern Canadian Cordillera: Geol. Assoc. Canada Special Paper 6, p. 7–25.
Richards, T. A., and Dodds, C. J., 1973, Hazelton (East) map-area, in Report of activity: Geol. Survey Canada Paper 73–1, pt. A, p. 38–42.
———, and White, W. H., 1970, K-Ar ages of plutonic rocks between Hope, British Columbia, and the 49th Parallel: Canadian Jour. Earth Sci., v. 7, p. 1203–1207.
Roddick, J. A., 1965, Vancouver North, Coquitlam and Pitt Lake map-areas, British Columbia: Geol. Survey Canada Mem. 335, 248 p.
Sestini, G., 1970, Development of the Northern Apennines Geosyncline—sedimentation of the late geosynclinal stage: Sed. Geology, v. 4, p. 445–479.
Shor, G. G., Jr., and Raitt, R. W., 1969, Explosion seismic refraction studies of the crust and upper mantle in the Pacific and Indian Oceans, in The earth's crust and upper mantle: Am. Geophys. Union Geophys. Mon. 13, p. 225–230.
Shouldice, D. H., 1971, Geology of the western Canadian continental shelf: Bull. Canadian Petroleum Geology, v. 19, p. 405–436.
Souther, J. G., 1970, Volcanism and its relationship to recent crustal movements in the Canadian Cordillera: Canadian Jour. Earth Sci., v. 7, p. 553–568.
———, 1971, Geology and mineral deposits of Tulsequah map-area, British Columbia: Geol. Survey Canada Mem. 362, 84 p.
———, 1972, Telegraph Creek map-area, British Columbia: ibid., Paper 71–44, 38 p.
———, and Armstrong, J. E., 1966, North-central belt of the Cordillera of British Columbia, in Tectonic history and mineral deposits of the western Cordillera: Canadian Inst. Mining and Metallurgy Special Vol. 8, p. 171–184.
———, and Lambert, M. B., 1972, Volcanic rocks of northern Canadian Cordillera, in Guidebook 24: Internat. Geol. Cong., Field Excursion A12, p. 18.
Stanley, K. O., Jordan, W. H., and Dott, R. H., Jr., 1971, New hypothesis of Early Jurassic paleogeography and sediment dispersal for western United States: Am. Assoc. Petroleum Geologists Bull., v. 55, p. 10–19.
Stott, D. F., 1967, The Cretaceous Smoky Group, Rocky Mountain foothills, Alberta and British Columbia: Geol. Survey Canada Bull. 132; 133 p.
———, 1968, Lower Cretaceous Bullhead and Fort St. John Groups, between Smoky and Peace Rivers, Rocky Mountain foothills, Alberta and British Columbia: ibid., Bull. 152, 279 p.
Sutherland-Brown, A., 1966, Tectonic history of the Insular Belt of British Columbia, in Tectonic history and mineral deposits of the western Canadian Cordillera: Canadian Inst. Mining and Metallurgy Special Vol. 8, p. 83–100.
———, 1968, Geology of the Queen Charlotte Islands, British Columbia: British Columbia Dept. Mines and Petroleum Res. Bull. 54, 226 p.
Tiffin, D. L., 1969, Continuous seismic reflection profiling in the Strait of Georgia, British Columbia (Ph.D. thesis); Vancouver, Univ. British Columbia.
———, Cameron, B. E. B., and Murray, J. W., 1972, Tectonics and depositional history of the continental margin off Vancouver Island, British Columbia: Canadian Jour. Earth Sci., v. 9, p. 280–296.
Tipper, H. W., 1963, Nechako River map-area, British Columbia. Geol. Survey Canada Mem. 324, 59 p.
———, 1969, Mesozoic and Cenozoic geology of the northeast part of Mount Waddington map-area (92N), Coast District, British Columbia: ibid., Paper 68–33, 103 p.
———, 1972, Smithers map-area, British Columbia: in Report of activities: ibid., Paper 72–1, pt. A, p. 39–41.
Wheeler, J. O., 1961, Whitehorse map-area, Yukon Territory: ibid., Mem. 312, 156 p.
———, and Gabrielse, H., 1972, Cordilleran structural province, in Price, R. A., and Douglas, R. J. W. (eds.), Variations in tectonic styles in Canada; Geol. Assoc. Canada Special Paper 11, p. 1–81.

SEDIMENTATION ALONG THE SAN ANDREAS FAULT, CALIFORNIA

JOHN C. CROWELL
University of California, Santa Barbara

ABSTRACT

Active sedimentation along the present San Andreas transform fault occurs primarily at the head of the Gulf of California, including the Salton Trough. Here a vertical thickness of about 6,000 m (20,000 ft) of young sediments, mainly derived from ancestral Colorado rivers, has accumulated on spreading quasi-oceanic floor as the gulf has opened. Northwestward along the trace of the San Andreas, nonmarine sedimentary wedges are more closely related to regional uplift and depression than to the San Andreas itself. Only northwest of San Francisco does the San Andreas mark the shoreline, where presumably shallow marine beds pinch out and change facies at the fault scarp.

In the past, the fault zone has demarked highlands from adjacent depressions, some of which have been deep and narrow. In the Ridge Basin, for example, 12,000 m (40,000 ft) of marine and nonmarine beds accumulated during late Cenozoic time in a moving depression associated with major strike slip. Marked facies changes from shale and sandstone along the basin axis to conglomerate and sedimentary breccias at the margins document fault movements with both strike-slip and dip-slip components occurring concurrently with sedimentation. Other Cenozoic sedimentary units in California probably have similar origins.

Basins and sites of deposition along transform faults are related mainly to bends and irregularities in fault-zone traces; tectonic plates do not slide simply by each other without deforming along their edges. Gapping or pulling apart of the crust results with motion away from bends; the holes formed are as large as giant rhombochasms. With motion of plate margins toward bends, overlapping, thickening, and elevation of the crust on one side may be associated with the sagging and depression of the other plate to make a shallow depositional site.

Where bends of plate margins are gentle, one side may stretch and sag to make a long narrow basin, while the other is squeezed and elevated to form an upland source area. In addition, complex braiding and slicing, and splaying of faults take place in major transform zones, at places on a huge scale, to form crustal blocks that tip and rise and fall. In this regime, several sedimentation sites evolve. Such origins probably can be recognized also for basins in Venezuela, New Zealand, maritime Canada, and the Levant.

INTRODUCTION

Basins and troughs of geosynclinal size occur along some transform faults and may receive great volumes of sediment if continental source areas are nearby. In general, the depositional sites along transform systems are the result of sinking of blocks and wedges where the fault zone is irregular, braided, or crooked. If this broad zone of deformation lies in midoceanic areas, thick accumulations of sediment within the depressions are unlikely because there is no nearby source of voluminous sediment. The depressions then can receive debris only from local high areas, mostly far below sea level, by turbidity currents, debris flows, downslope sliding, volcaniclastic depositional processes, and pelagic settling, for example. On the other hand, if the transform system and its attendant depressions lie near or within a continent, land source areas may shed much detritus to the basins.

In the main, the boundaries between major tectonic plates ignore the boundaries between continents and oceans, and much of the variety in the sedimentary and deformational record is the result of subduction zones, trenches, and island arcs moving toward or away from a continental margin. In addition, transform zones may bring together basement terrane of either continental or oceanic type, and these terranes may carry with them sedimentary masses of nearly any variety. Basins originating along transform systems are not successor basins: such a term implies a sequence within the so-called geotectonic cycle, a concept that now needs modification in the light of plate-tectonic tenets. Large prisms of sedimentary rocks, such as those formed in Atlantic-type miogeoclines or within trenches, may or may not be sliced up by subsequent transform faults.

The style of sedimentation along a major transform fault can be illustrated by the San Andreas system. Although its history is long and complicated and its origins obscure, rocks are reasonably well known for more than 1,200 km along the San Andreas Fault. For at least its late Miocene and younger history we have fair understanding of deformation and sedimentation along the fault, and we can begin to relate these events to plate-tectonic movements. Sedimentary prisms associated by facies and origin to strands of the San Andreas system older than late Miocene include several accumulations, but, with the possible exception of the Gualala Group (Wentworth, 1968) near Point Arena, which is of latest Cretaceous to Eocene age, there is ambiguity.

THE SAN ANDREAS FAULT TODAY

In looking at the active San Andreas Fault today, we can ask: where is sedimentation going on that shows a facies relationship to the fault or to one of its branches? In California northwest of San Francisco the fault more or less demarks the boundary between land and sea (fig. 1). Most of the sediment carried to the sea along this stretch of coast, however, moves down submarine canyons to deep water at the foot of the continental slope, and only a veneer of shallow-water material is deposited near the trace of the San Andreas (see reflection profiles in Curray and Nason, 1967). Locally the fault forms a scarp facing seaward, and presumably beds pinch out and change facies against its base, but published profiles do not extend close enough inshore to record this. At Bolinas, Tomales, and Bodega Bays, however, fault slicing has dropped blocks so that lagoonal and other sediments are accumulating locally (Daetwyler, 1965). In proportion to the areas and volumes of sedimentation well offshore, however, these bay deposits are insignificant. Along extensive parts of this coastline northwest of San Francisco, continental rocks lie west of the San Andreas, as at Point Reyes and Point Arena, and the fault zone does not separate continental basement from oceanic basement (fig. 1). The structural edge of the continent lies in deep water at the base of the continental slope at some distance offshore.

Quite locally at the mouth of San Francisco Bay, marine waters extend across the San Andreas and reach inland. Millions of years hence, if displacement on the fault continues and if the patchy sediments are preserved, offset lithofacies lines and biofacies lines can perhaps be used to measure right slip. But the configuration of the bay and coastline here are obviously controlled far more by regional tectonic processes other than by transection by the San Andreas. Of much greater importance in guiding sedimentation is the sinking responsible for the bay being the gathering locus for drainage from the Great Valley of California.

This lack of effect of the San Andreas transform fault on the present spread of sediments is especially notable, with minor exceptions, southeastward through the Coast and Transverse Ranges and all the way to the Salton Trough. The course of the fault is well inland from the coastline, and at places its surface trace reaches nearly to elevations of 2,000 m. Eroded material is washed away from the vicinity of the fault to temporary sites in the San Joaquin Valley on the east and to the Salinas and Cuyama Valleys on the west. Along the north flank of the San Gabriel Mountains, debris is transported to the Mojave Desert, where it accumulates in alluvial fans, but the heads of these fans do not reach to the San Andreas Fault itself. They reach instead into highlands marginal to the fault zone, which here trends obliquely across the eastern San Gabriel Mountains, into Cajon Pass, and along the southern face of the San Bernardino Mountains. The topography of the east-west trending ranges influences sedimentation in this region far more than does the fault. Along the southern steep scarp of the San Bernardino Mountains, however, the fault separates rugged high ground from thick alluvium accumulating adjacent to the fault. Only here do coarse sediments demark the fault. From this area, the alluvium extends to the south and west along the southern base of the steep San Gabriel Mountains. The fans are thick, and nonmarine sediments lie upon north-sloping basement that is caught in a prismatic depression between the Peninsular Ranges and the Transverse Ranges.

THE SALTON TROUGH AND GULF OF CALIFORNIA

Farther to the southeast lies the Salton Trough, the only region of marked sedimentation along the San Andreas in California today. Splays of the San Andreas system, such as the Banning, San Jacinto, and Imperial Faults, in part form the boundaries of the trough, which opens on the southeast into the Gulf of California (figs. 1 and 2). The Salton Trough is but the northwest tip of the gulf, is complicated structurally, and largely filled by sediments. These sediments now come down steep fans from the bordering high ranges, but at the south end of the Salton Trough, and especially in the immediate past, they come (came) from the Colorado River. This immense drainage system discharges its debris into the gulf, whereas in a similar tectonic environment in the Middle East, the Nile River fails to reach the Red Sea but instead flows somewhat parallel to it to reach the Mediterranean Sea. The Red Sea, in contrast to the Gulf of California, receives very little terrigenous sediment from rivers. Most of its sediment is washed down steep canyons and fans from the immediate desert highlands rimming the depression.

The Gulf of California is almost certainly the result of the splitting of Baja California and western California from the mainland of North America, followed by its movement relatively northwestward (Wegener, 1924, reprinted 1966, p. 198; Carey, 1958, fig. 42; Hamilton, 1961; Larson and others, 1968; Moore and Buffington, 1968; Elders and others, 1972). The process responsible is pictured as sea-floor spreading, in

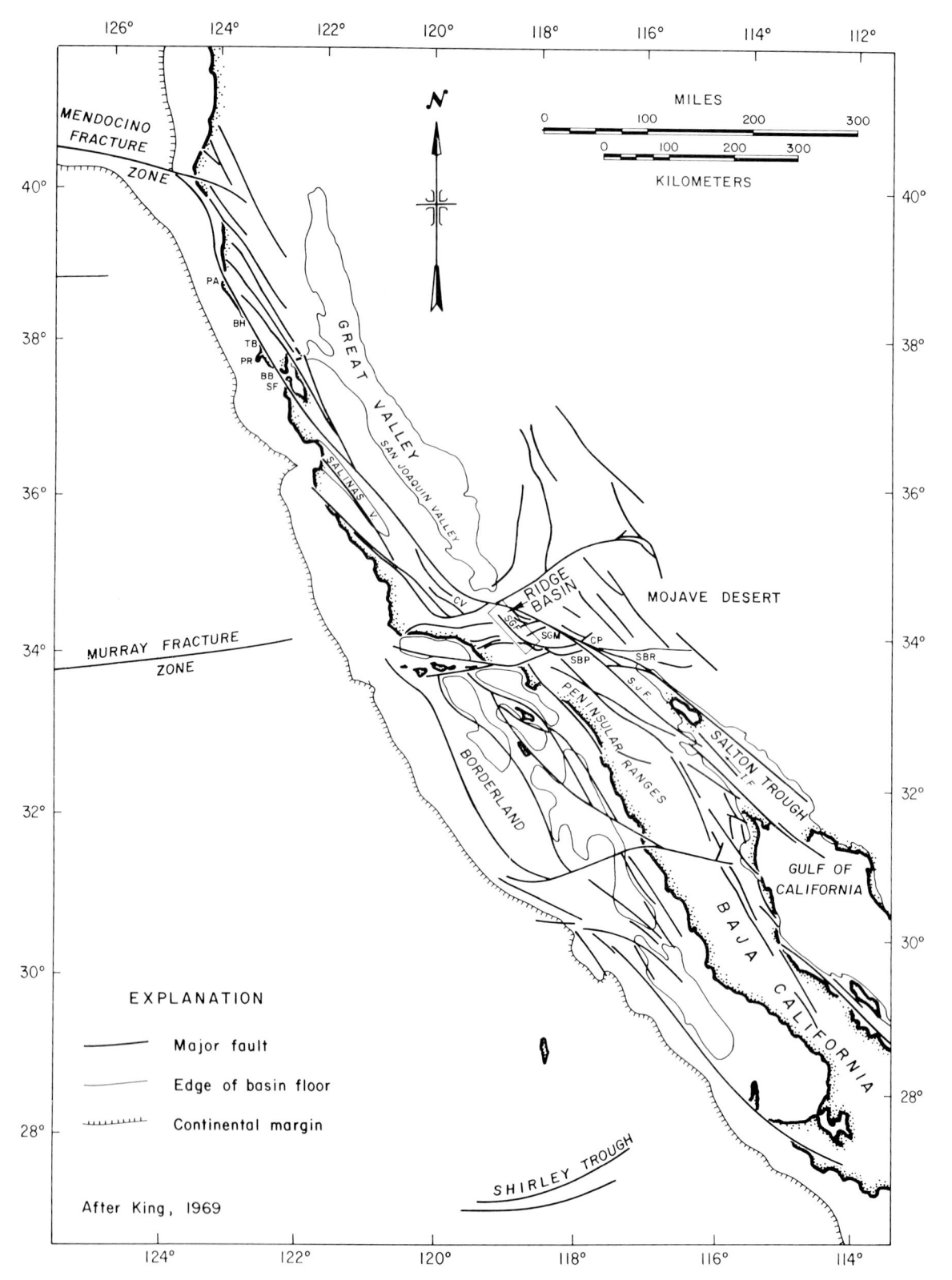

FIG. 1.—Major faults and present sedimentation sites of California and Baja California (traced selectively from King, 1969). Geographic localities from north to south: PA, Point Arena; BH, Bodega Head; TB, Tomales Bay; PR, Point Reyes; BB, Bolinas Bay; SF, San Francisco; CV, Cuyama Valley, SGM, San Gabriel Mountains; CP, Cajon Pass; SBR, San Bernardino Range; SBP, San Bernardino Plain. Faults: SGF, San Gabriel Fault; SJF, San Jacinto Fault; IF, Imperial Fault.

which the floor beneath the gulf is largely simatic and consists either of a pattern of spreading ridges and associated transform faults or a more disorganized arrangement resulting from diapiric intrusions from the upper mantle (fig. 2). Some of the transform faults are confined to the simatic floor and do not cut the continental layer on either side, but the faults may extend beneath the continental margins for unknown distances in the lower crust or upper mantle. Such faults are here termed oceanic transform faults. They contrast with those faults, such as the San Andreas itself, that cut also the continental platform. The latter kind is here called a continental transform fault.

The expected relations between spreading ridges, oceanic and continental transform faults, and other structures are diagrammed in figure 3, which is very much simplified from the actual situation in the Salton Trough. The assumption is made in this diagram that a segmented oceanic rise lies within the Gulf of California and the Salton Trough. On the other hand, extension may be due in part to diapiric intrusion from the upper mantle in a volcanotectonic zone (Karig and Jensky, 1973). Initially the peninsula was split from the continental mainland along a trend nearly parallel to the western coast of North America. The break apparently followed an earlier protogulf (Imperial Formation, latest Miocene) with roughly the same orientation. This long trough was probably floored by attenuated continental material and originated behind a major trench-arc system offshore to the west. It may have begun as either an elongate marginal basin or as a volcanotectonic extensional zone.

According to such models, the margins of the Gulf of California follow previous tectonic trends and are irregular. They are the walls of the pullapart and form the contact between continental basement and quasi-oceanic basement in between (Phillips, 1964; Elders and others, 1972). Faults parallel to the coasts of the gulf, if older than mid-Pliocene, can be expected to have orientations and slips related to the tectonics of the protogulf, and younger faults, largely to gravitational slope failure that in turn is related to extension and to steep differences in elevations between the continental and oceanic basements. These faults would be normal-slip faults. The head of the gulf, in contrast, follows the trace of the boundary transform fault, the San Andreas in this case.

Where the boundary continental transform fault extends between continental rocks on both walls, it is a full fault in the sense that preexisting terrane is cut and displaced and that

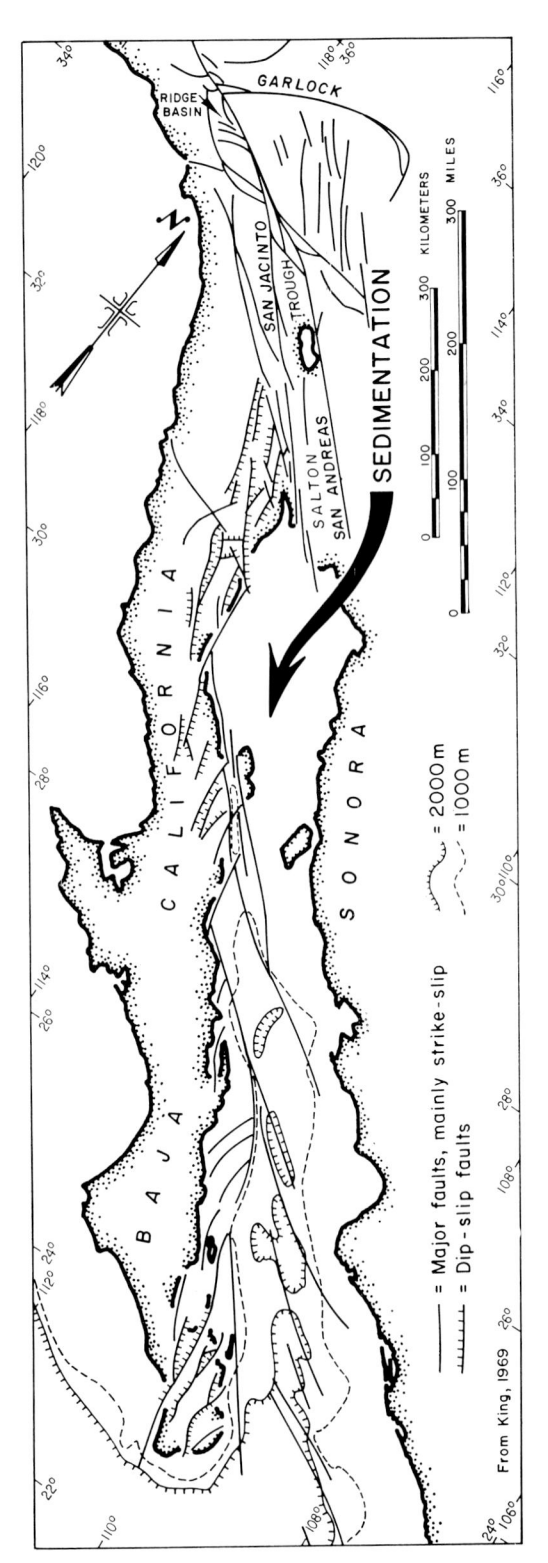

FIG. 2.—Faults and basins in Gulf of California.

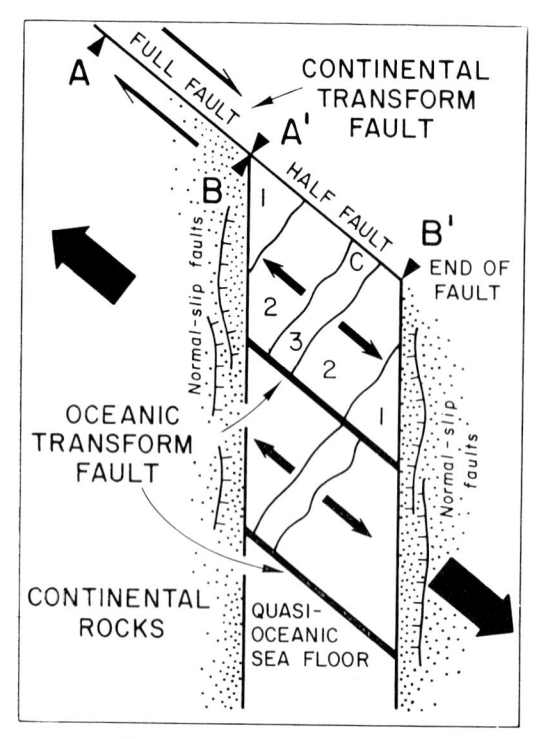

Fig. 3.—Diagram showing pullapart of continental rocks as at head of Gulf of California. 1, 2, 3: order of formation of basin floor. See text for explanation.

correlatable features on the two walls can be sought (fig. 3). At the head of the gulf, in contrast, only one previously existing wall is preserved, and the volcanic or diapiric basement at the floor of the gulf has ages dependent upon the pattern of the spreading mechanism. With only one of two walls of the fault preserved, it seems appropriate to call this a half fault. In theory, where trough floor meets the half fault there should be no fault at all, rather, a complicated zone where volcanic or hypabyssal rocks overlap or intrude the edge of the continental platform. In moving away from the center of active extension and along the contact (along the half fault), the contact should be faulted, and both the age and displacement of the fault or faults should increase systematically. In the diagram (fig. 3), older positions of the edge of the extension center are shown, and, assuming symmetry of growth of the basin floor, there should be a symmetrical and orderly increase in slip and age between positions C and A' on one side of the center and between C and B' on the other. A, B, A', and B' are located on continental rocks older than all the faulting under discussion. C is where the new oceanic ridge meets the half fault and lies on the quasi-oceanic sea floor.

The geometry shown in figure 3 implies that the boundary continental transform fault ends at B'. We therefore have no justification in looking for the San Andreas beyond point B' to the southeast, and if a line of faults can be traced into it with a slightly different trend they may be completely different in type and origin. Northwest of point B', from the field geologist's viewpoint, the fault is a strike-slip fault with systematically decreasing age and slip as far as C (where the active spreading center is encountered), then systematically increasing age and slip as far as B and A'. Northwest of these points, there should be a mismatch in features of the continental basement directly facing across the fault. Details between A and B are correlatable with those between A' and B'. These correlatable geological features of the continental plate should have each component part older than the original movement of the transform fault. In fact, very ancient geological bodies and structures should be displaced the same amount as those originating just before the birth of the transform fault. A graph showing age against displacement for all correlatable features in such a region should show a break in slope for the time of fault origin. This is inherent in the statement that the transform fault was born and then continued to acquire displacement more or less continuously.

This digression into geometric and tectonic considerations was necessary because it was into such an environment that sedimentation took place. As the Salton Trough opened, it was filled largely by fluviatile and lacustrine material from the ancestral Colorado River and less so from eroding local ranges (Muffler and Doe, 1968). Deformation went on hand in hand with widening of the trough, but on a field-mapping scale the imprint of the gross extensional tectonics is often not recognizable. Local folds and faults originated through gravitational sagging into the deepening and widening trough and from squeezing within a branching and braided strike-slip zone. Unconformities and abrupt changes in facies and thickness abound, especially around the margins of the Salton Trough, but mapping and subsurface studies within the trough have not yet advanced far enough to classify the structures observed according to time of origin and growth or to tectonic setting. In general, however, basement rocks rimming the Salton Trough are all older than the development of the protogulf (late Miocene?) and include rocks ranging in age from Precambrian to late Mesozoic and perhaps to Paleogene (Dibblee, 1954; Allen, 1957; Crowell, 1962). Overlying sedimentary strata of Eocene, Oligocene(?), and early and middle Miocene ages probably are displaced on the San Andreas system as much as is the underlying basement and apparently were deposited in a paleogeographic and tectonic environment that preceded the origin of the proto-gulf (Crowell and Susuki, 1959; Crowell, 1962; Ehlig and Ehlert, 1972). The oldest sedimentary unit laid down in a local basin with the same trend as the Gulf of California is the Imperial Formation, which is of latest Miocene or earliest Pliocene age (Woodring, 1932; Allison, 1964). This unit consists of a maximum of 1,000 m of marine claystone and sandstone, which is considered to have been deposited in the northern end of the proto-Gulf of California. Overlying units are clastics including nonmarine fan deposits and coarse conglomerates around the basin margins at the surface. These beds aggregate in outcrop as much as 3,000 m, to which unknown subsurface thicknesses need to be added. Deeper parts of the Salton Trough have not been penetrated by wells, but interpretations based on gravity and seismic-refraction studies suggest infilling of about 6,000 m (20,000 ft) (Biehler and others, 1964, figs. 6, 7; Elders and others, 1972). At the south end of the Salton Sea, this sedimentary section, consisting of Pliocene and Quaternary sands, silts, and clays, is undergoing metamorphism at present into the greenschist facies (Muffler and White, 1969; Elders and others, 1972). The geothermal fields and hot spots apparently lie above active spreading centers that simulate buried midocean ridges or that are centers of diapiric uprisings from the upper mantle.

In summary, sediments related to the opening of the head of the Gulf of California in the Salton Trough probably reached a vertical thickness of 6,000 m (20,000 ft). Most of the material came from the Colorado River and entered both marine and nonmarine environments until prevented by dams and levees. The San Andreas Fault here plays a significant role in bordering this depositional site on the northeast, and the great mass of sediment is the largest that can be considered as fault related along the San Andreas today. The rhombochasmic trough opened as sediments were laid down so that complex overlaps and unconformities abound, especially around the margins, and complex intrusive and extrusive relations must occur at depth, especially in the center near the locus of active sea-floor spreading or stretching. In general just above the basin floor, older units can be expected near the trough margins with progressively younger beds lapping toward the trough center in a manner similar to that found by the Deep Sea Drilling Project in the open ocean. Unfortunately, however, many of these expected complications are deep and out of sight.

PLIOCENE AND PLEISTOCENE DEPOSITION RELATED TO THE SAN ANDREAS SYSTEM

Several basins along the San Andreas system received sediments during parts of the Pliocene and Pleistocene Epochs. Some of these, such as the Salton Trough, continue to accumulate sediments at present within many of the same, earlier interrelationships between sedimentation and deformation; others, however, are no longer sites of active deposition. Among the latter is the basin in which the Mount Eden and San Timoteo Formations were deposited (between the Cities of Riverside and Beaumont) and that is bounded on the southwest by the San Jacinto Fault (Fraser, 1931, p. 511-514; Morton and Gray, 1971). Another is the San Jacinto Trough that is largely in the subsurface and that lies parallel to the San Jacinto Fault for at least 35 km (Woodford and others, 1971). Still another is the basin in which the Mill Creek Formation was laid down, between branches of the San Andreas northeast of Redlands (Gibson, 1971). Along the San Andreas system northwest of Cajon Pass are several other fault-related basinal units of Pliocene age (Anaverde and Crowder Formations, part of the Punchbowl Formation, etc.; Noble, 1954; Dibblee, 1967). Similar Pliocene units still farther northwest are in central and northern California (Dibblee, 1966; Cummings, 1968) and include the Merced Formation near San Francisco (Higgins, 1961).

Ridge Basin.—The Ridge Basin (Neogene), now uplifted and dissected to view, lies at the junction of the San Gabriel and San Andreas Faults and received about 12,000 m (40,000 ft) of both marine and nonmarine beds during the period from late Miocene to early Pleistocene (Crowell, 1950, 1954, 1962). The basin formed as a narrow depression bounded on the southwest by the scarp of the San Gabriel Fault zone and on the northeast by highlands raised by folding and by movements on the Clearwater, Liebre, and other faults (fig. 4). Large quantities of terrigenous debris were washed down alluvial fans to playas, lakes, and marine bays within the trough at different times. Deposition and deformation went on hand in hand so that older beds are deformed more than younger beds. The sediments grade sharply from conglomerate and sandstone on the northeast, to predominantly shale along the axis of the trough, and thence southwestward to coarse sedimentary breccia (Violin Breccia) against the San Gabriel Fault zone. Rocks flooring and surrounding the basin consist of granites,

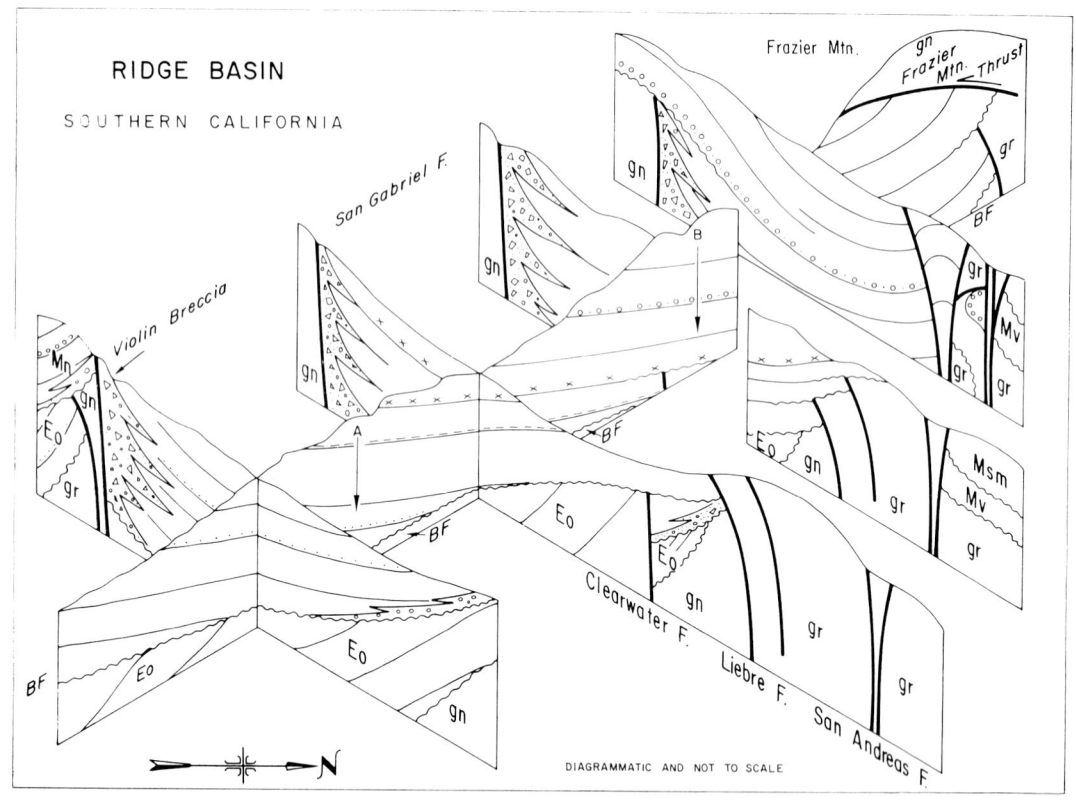

Fig. 4.—Isometric sketch of Ridge Basin, southern California (diagrammatic and not to scale). Strata of Ridge Basin not labelled. Stratigraphic section at A is overlapped at B. Symbols: BF, basin floor; gn, gneiss; gr, granitic rocks; Eo, mainly Eocene; Mv, Miocene volcanic rocks; Msm, Miocene Santa Margarita Formation; Mm, Miocene Modelo Formation.

gneisses, and other metamorphic rocks as well as Paleogene shale, sandstone, and conglomerate.

Stones within the oldest part of Violin Breccia show that they were derived from a limited source area, now offset about 28 km (17 miles) along the San Gabriel Fault. The breccia consists of earthy unsorted debris, including angular gneissic blocks as large as 2 m in size, that accumulated at the base of a continuously rejuvenated fault scarp. Moreover, marine upper Miocene beds on the southwest side of the San Gabriel fault have also been displaced laterally from their source area about the same amount (Crowell, 1952). The basin formed, therefore, within a strike-slip regime but so that its floor was depressed as much as 4,000 m (13,000 ft). Along the axis of the trough and plunging gently northwestward, a 12,000-m (40,000-ft) continuous stratigraphic section can be measured. Downdip along the axis of the trough, these beds overlap against the trough floor so that successively younger beds lie on the unconformity toward the northwest. The depositional hole remained directly across the San Gabriel Fault from the rising source area as the northeastern block carrying the Ridge Basin moved relatively to the southeast in the right-slip regime (fig. 5). In this way and involving a continuously and laterally moving hole, the very thick stratigraphic section was accommodated—a section about three times thicker cumulatively than the vertical thickness to the floor of the basin at any one place. A diagram of these relations, simplified from detailed field studies, is shown in figure 4.

On the north, the youngest beds of the Ridge Basin are truncated by the present San Andreas Fault and were derived from a part of the Mojave Desert region now displaced toward the southeast during the Quaternary Period. The offset source area, however, is not distinctive enough to be pinpointed at present on the basis of Ridge Basin stones. This latest displacement on the present San Andreas, along with associated thrusting and folding, occured after filling of the basin. During the time of sedimentation, however, the San Gabriel Fault was the principal strand of the San Andreas system; there is no evidence that the presently active strand of the San Andreas was active then. It seems likely that the Ridge Basin originated at a bend in the Miocene and Pliocene San Andreas (fig. 5), but at a bend that was considerably less sharp than that in the region today.

Inside such bends the tectonic plate margin was compressed and shortened and elevated, whereas outside the bends it was stretched and depressed. In this moving strike-slip system involving curvature of the major fault boundary between tectonic plates, source areas were apparently elevated and restricted, whereas nearby sediments derived from them were spread out for long distances. The outcrop length of the Violin Breccia along the San Gabriel Fault is about twice as great as the width of the reconstructed source area. This strewing out of coarse debris across a strike-slip fault (having a dip-slip component) from a restricted source area perhaps can be recognized as well for upper Miocene breccias (Santa Margarita) in the Temblor Range of central California (Fletcher, 1967; Huffman, 1972). This strewing process is one that should be entertained for other, similar breccias, such as the San Onofre of the coastal region near Los Angeles (Woodford, 1925).

TECTONIC SETTING OF DEPOSITIONAL SITES ALONG CONTINENTAL TRANSFORM FAULTS

From our brief look at sedimentary basins associated with the San Andreas system we can recognize a few general types. If the continental transform fault is straight, opposing tectonic plates can slide gently by each other without forming uplands or depressions; local sedimentation along the transform fault either does not result or it is uniform. This situation is ideal and unlikely, and actual transform faults display bends and slices. In detail most of them consist of braided zones, and the San Andreas, for example, displays features ranging from anastomosing scarps and sagponds at a scale a few kilometers wide (Sharp, 1954; Vedder and Wallace, 1970) to the system of branching faults in southern California and in the offshore borderland that is nearly 500 km wide. Here we are concerned with basins—some of which are akin to giant sagponds—that are large enough so that the sediments within them may become part of the preserved geologic record. In order to emphasize the ongoing and linked aspect of deformation and sedimentation, participles rather than nouns are employed in the discussion below; processes are accented rather than static results.

Gapping and overriding.—Where a single continental transform fault has a double bend, that is, where the map trace is side stepped but throughgoing, gaps or overrides result with continuing plate motion (fig. 6). Holes or pullaparts originate if the motion of the blocks is away from the doubly bent area. Depending on scale, gap depressions may be deep and defined by two sharp boundaries and become elongated in a direction subparallel to the transform fault. For large gaps, many complicating details are likely, such as faults and folds related in origin to downslope movement rather than to the dynamics of the transform fault itself. The Dead Sea Fault system of the Levant, for example, has long been recognized as displaying such features (Quennell, 1958; Freund, 1965; Freund and others, 1970), and the Dead Sea depression is an excellent example of a pullapart. If such a feature is large enough to be floored by simatic crust, it is defined by Carey (1958, p. 192) as a *rhombochasm*, a term applicable to segments of the Gulf of California. The rhomb-shaped pullaparts, therefore, range in scale from that of such small land forms as sagponds along an active strike-slip fault to that of large expanses of the ocean floor.

If the relative motion of the two tectonic plates along the continental transform fault is toward the double bend, one plate partially oversides the other (fig. 6). With isostatic readjustment, the lower plate bows downward, and the nearby overriding plate rises. The depression on the lower plate forms a site for sediments fining

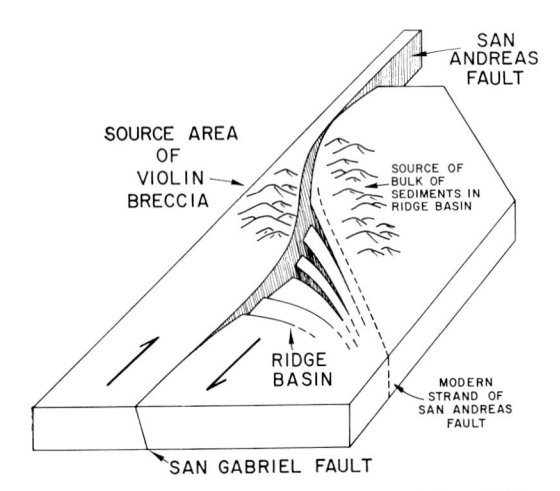

FIG. 5.—Block diagram illustrating origin of Ridge Basin at a curve in San Andreas Fault. During Pliocene Epoch San Gabriel Fault was probably main San Andreas strand.

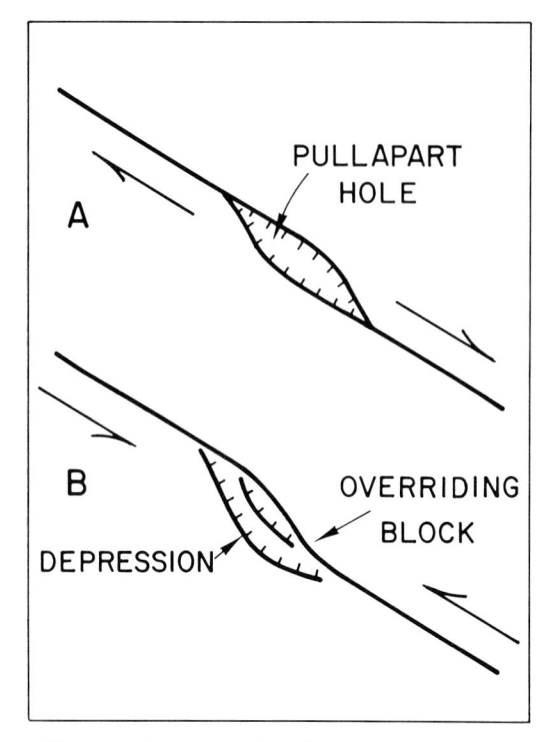

FIG. 6.—Formation of sedimentary sites along a strike-slip fault with double bend. A, Pullapart hole; B, site formed when overriding block depresses overridden block.

outward from the trace of the fault. Such a depression is shallower and has less sharp boundaries than are present at pullapart gaps. The wedge of continental sediments extending westward from the San Andreas face of the San Bernardino Mountains may be an example. In viewing this region broadly, however, compression of the margins of both the Americas and Pacific plates across the broad San Andreas system here (including the San Jacinto Fault) has simultaneously elevated both sides, raising both the San Gabriel and San Bernardino ranges.

Stretching and squeezing.—Curves in continental transform faults may result in stretching the margin of the tectonic plate on one side and in squeezing or compressing the side opposing, such as at the Ridge Basin. Under isostasy, a deep, narrow, and asymmetric basin may result from the stretching, but this process is not thought to extend far enough to bring simatic basement to the basin floor. In a significant, brief paper, Kingma (1958, fig. 2A) presented a diagram, which is similar to figure 5 here, before the introduction of the concept of transform faulting (Wilson, 1965). He offered it to explain the association of deep and narrow basins in New Zealand with faults of the Alpine system. In both California and New Zealand, the stretching and squeezing process is accompanied by much faulting and folding, and, as deposition, deformation, and erosion go on together, geologic details become complex.

The source area for the Violin Breccia, which is across the San Gabriel Fault from the depositional center in the Ridge Basin, not only stands high but is deformed. Squeezing of the margin of the Pacific plate resulted in folding of the basement terrane and culminated in thrusting. The area of uplift and erosion lies inside the curve of the transform fault.

Braiding and splaying.—All major strike-slip faults exhibit slices and wedges in a braided zone along the fault, and here and there faults branch and splay outward. The braided zone may be a fraction of a kilometer in width, or it may be more than a hundred. In fact, the whole of southern California can be likened to a sliced and splayed transform margin between the Americas and Pacific plates and where movement on the branches has been complex throughout the Cenozoic Era. Other regions where similar complex tectonic relations prevail include the Dead Sea area, the faulted Alpine area of New Zealand, the post-Acadian rift areas of Maritime Canada (Belt, 1968), northwestern Scotland (Kennedy, 1946; Dearnley, 1962), and the southern margin of the Caribbean Sea.

In such anastomosing and branching schemes that are dominated by regional strike slip, crustal blocks and wedges rise and fall. Horsts result where curvature of the transform system brings about squeezing in much the same way as described above as overriding (fig. 6). Grabens result where there is gapping. In addition, splay faults lead out from such braided zones into adjoining plates with diminishing vertical separations and are like those east of the Ridge Basin diagrammed in figure 5. Slices within the braided zones, at places long and slender and elsewhere short and stubby, may be tipped up at one end and depressed at the other. We see this at different scales: that of minor landforms within a fault zone (fault-slice ridges or pressure ridges) and associated with adjoining sag-ponds, or that of regional magnitude. At a grand scale, the block between the San Gabriel and San Andreas Faults, including the high San Gabriel Mountains on the southeast and the depressed Ridge Basin at the northwest, is an example. Pressure on this large slice, caught between the Americas and Pacific plates, has raised one end high and caused it to become deeply eroded; the same pressure has tipped down the other end and caused sediments to accumulate there.

These geometric consequences of strike-slip movement have long been recognized in New Zealand. Kingma (1958, fig. 1) showed diagrammatically how continued movement in a strike-slip regime results in horsts and grabens. At some places horsts are inundated by accumulating sediment and, with continuing displacement on the major fault system, are pushed higher. In this way large isolated blocks of older rock, several kilometers long, become surrounded by younger sediments. Kingma referred to these as piercement structures. In several important papers, Lensen (e.g., 1958) has related the shapes of slices and the dips of bounding faults to inferred components of regional tectonic stress, to the principal horizontal stress in particular. His geometric generalizations largely hold for the plate-tectonic model for examples where there is an angle between the fault trace in map view and the regional direction of plate motion.

The ultimate in complexity within a braided transform zone of huge scale probably is exhibited along the southern margin of the Caribbean plate, in the Coast Ranges of Venezuela, and in adjoining areas offshore (Smith, 1962; Stainforth, 1966, 1969; Bell, 1967; Donnelly, 1971; Silver, 1972). As the Caribbean plate moved relatively eastward with respect to the mainland of South America and beginning in the early mid-Tertiary (Malfait and Dinkelman, 1972), perhaps whole mountain ranges were sliced off and lifted and depressed obliquely. Other attenuated blocks tilted downward to form deep basins at their ends. High-standing blocks, including those uplifted by large components of strike slip, collapsed at the surface with the result that low-angle overthrusts and immense slides are most conspicuous structures. This jostling of long, slender crustal blocks caught between megatectonic plates has gone on amidst voluminous sedimentation from the ancestors of such rivers as the Oronoco and Magdalena. Narrow pullapart holes and tipped-down ends of long slices became depositional sites, many of which were continuously deformed as jostling between tectonic segments continued. In addition, high-standing terranes within this braided system shed exotic blocks and thrust plates to nearby low areas across major oblique-slip fault zones and then moved away laterally so that the source of the thrusts and exotic slides was lost. Events recorded in the Coast Range rocks of Venezuela are especially complex because these rocks also hold the earlier stamp of Late Cretaceous and Paleogene sedimentation, deformation, volcanism, plutonism, and metamorphism, probably as the result of southeastward directed plate convergence (Malfait and Dinkelman, 1972).

CONCLUSIONS

Sedimentation sites along transform faults range from small sagponds to huge rhombochasms. To receive and preserve appreciable sediments, however, the basins must be within or near a continent, and maximum sedimentation occurs where major drainage systems pour into the basins. Basins occur where opposing tectonic plates are sliced and fragmented into a braided zone of rising, tipping, and sinking crustal blocks as well as where the transform fault bends only gently. Long, narrow basins, such as the Ridge Basin, result from stretching, thinning, and isostatic sinking of a plate as it moves around a bend. The margins of great tectonic plates are not rigid, but weak and easily bent and broken.

Sediments laid down in sites along ancient transform zones may be recognized on the basis of facies and thickness relations to faults of the transform system, provided that these faults formed scarps at basin margins. With continued movement after deposition, sedimentary prisms are frequently offset from their sources. At the same time, rocks comprising basin floors are offset as well so that mismatches across fault strands ensue. Perhaps the offset counterparts of correlatable feature can be found, but often they become lost, either by erosion or by burial. The pattern of braided and splayed faults and echelon folds also mark such ancient transform zones, but they are not characterized by distinctive volcanism, plutonism, or metamorphism.

ACKNOWLEDGMENTS

Investigations of the geology of southern California have recently been supported by the U.S. National Science Foundation, Grant GA-30901 and the Committee on Research, University of California, Santa Barbara. I am grateful to these organizations and to many students and colleagues for numerous discussions and comments.

REFERENCES

ALLEN, C. R., 1957, The San Andreas Fault zone in San Gorgonio Pass, southern California: Geol. Soc. America Bull., v. 68, p. 315-350.
ALLISON, E. C., 1964, Geology of areas bordering Gulf of California, in VAN ANDEL, T. H., AND SHOR, G. G.,

Jr. (eds.), Marine geology of the Gulf of California—a symposium: Am. Assoc. Petroleum Geologists Mem. 3, p. 3–29.

Bell, J. S., 1967, Geology of the Camatagua area, Estado Aragua, Venezuela (Ph.D. dissertation): Princeton, New Jersey, Princeton Univ., 282 p.

Belt, E. S., 1968, Post-Acadian rifts and related facies, eastern Canada, p. 95–113, in Zen, E-An, and others (eds.), Studies of Appalachian geology: New York, Northern and Maritime, Interscience Publishers, 475 p.

Biehler, Shawn, Kovach, R. L., and Allen, C. R., 1964, Geophysical framework of northern end of Gulf of California structural province, in van Andel, T. H., and Shor, G. G., Jr. (eds.), Marine geology of the Gulf of California—a symposium: Am. Assoc. Petroleum Geologists Mem. 3, p. 126–143.

Carey, S. W., 1958, A tectonic approach to continental drift: Univ. Tasmania Dept. Geology Symposium, p. 177–355.

Crowell, J. C., 1950, Geology of Hungry Valley area, southern California: Am. Assoc. Petroleum Geologists Bull., v. 34, p. 1623–1646.

——, 1952, Probable large lateral displacement on the San Gabriel fault, southern California: ibid., v. 36, p. 2026–2035.

——, 1954, Geology of the Ridge Basin area, Los Angeles and Ventura Counties, California: California Div. Mines Bull. 170, map sheet 7.

——, 1962, Displacement along the San Andreas Fault, California: Geol. Soc. America Special Paper 71, 61 p.

——, and Susuki, T., 1959, Eocene stratigraphy and paleontology, Orocopia Mountains, southeastern California: ibid., Bull, v. 70, p. 581–592.

Cummings, J. C., 1968, The Santa Clara Formation and possible post-Pliocene slip on the San Andreas Fault in central California: Stanford Univ. Pubs. Geol. Sci., v. 11, p. 191–207.

Curray, J. R., and Nason, R. D., 1967, San Andreas Fault north of Point Arena, California: Geol. Soc. America Bull., v. 78, p. 413–418.

Daetwyler, C. C., 1965, Marine geology of Tomales Bay, central California (Ph.D. dissertation): San Diego, Univ. California, 195 p.

Dearnley, Raymond, 1962, An outline of the Lewisian complex of the Outer Hebrides in relation to that of the Scottish mainland: Geol. Soc. London Quart. Jour., v. 118, p. 143–176.

Dibblee, T. W., Jr., 1954, Geology of the Imperial Valley region, California: California Div. Mines Bull. 170, chap. 2, p. 21–28.

——, 1966, Evidence for cumulative offset on the San Andreas Fault in central and northern Calif.: ibid., Bull. 190, p. 375–384.

——, 1967, Areal geology of the western Mojave Desert, California: U.S. Geol. Survey Prof. Paper 522, 153 p.

Donnelly, T. W. (ed.), 1971, Caribbean geophysical, tectonic, and petrologic studies: Geol. Soc. America Mem. 130, 224 p.

Ehlig, P. L., and Ehlert, K. W., 1972, Offset of Miocene Mint Canyon Formation from volcanic source along the San Andreas Fault, California: ibid., Abs. with Programs, v. 4, p. 154.

Elders, W. A., and others, 1972, Crustal spreading in southern California: Science, v. 178, p. 15–24.

Fletcher, G. L. 1967, Post late Miocene displacement along the San Andreas Fault zone, central California, in Gabilan Range and adjacent San Andreas Fault: Am. Assoc. Petroleum Geologists, Pacific Sec., Guidebook, p. 74–80.

Fraser, D. M., 1931, Geology of San Jacinto Quadrangle south of San Gorgonio Pass, California: California Div. Mines and Geology Bull., v. 27, p. 494–540.

Freund, Raphael, 1965, A model of the structural development of Israel and adjacent areas since Upper Cretaceous times: Geol. Mag., v. 102, p. 189–205.

——, and others, 1970, The shear along the Dead Sea Rift: Royal Soc. London, Philos. Trans., ser. A, v. 267, p. 107–130.

Gibson, R. C., 1971, Non-marine turbidites and the San Andreas Fault, San Bernardino Mountains, California: Riverside, Univ. California, Campus Mus. Contr. 1, p. 167–181.

Hamilton, W. B., 1961, Origin of Gulf of California: Geol. Soc. America Bull., v. 72, p. 1307–1318.

Higgins, C. G., 1961, San Andreas Fault north of San Francisco, California: ibid., p. 51–68.

Huffman, O. F., 1972, Lateral displacement of upper Miocene rocks and the Neogene history of offset along the San Andreas Fault in central California: ibid., v. 83, p. 2913–2946.

Karig, D. E., and Jensky, Wallace, 1972, The proto-Gulf of California: Earth and Planetary Sci. Letters: v. 17, p. 169–174.

Kennedy, W. Q., 1946, The Great Glen Fault: Geol. Soc. London Quart. Jour., v. 102, p. 47–76.

King, P. B., 1969, Tectonic map of north America (scale 1:500,000): U.S. Geol. Survey.

Kingma, J. T., 1958, Possible origin of piercement structures, local unconformities and secondary basins in the Eastern Geosyncline, N.Z.: New Zealand Jour. Geology and Geophysics, v. 1, p. 269–274.

Larson, R. L., Menard, H. W., and Smith, S. M., 1968, Gulf of California: a result of ocean-floor spreading and transform faulting: Science, v. 161, p. 781–784.

Lensen, G. J., 1958, A method of horst and graben formation: Jour. Geology, v. 66, p. 579–87.

Malfait, B. T., and Dinkelman, M. G., 1972, Circum-Caribbean tectonic and igneous activity and the evolution of the Caribbean plate: Geol. Soc. America Bull., v. 83, p. 251–272.

Moore, D. G., and Buffington, E. C., 1968, Transform faulting and growth of the Gulf of California since the late Pliocene: Science, v. 161, p. 1238–1241.

Morton, D. M.. and Gray C. H., 1971. Geology of the northern Peninsular Ranges, southern California: Riverside, Univ. California, Campus Mus. Contr. 1, p. 60–93.

MUFFLER, L. J. P., AND DOE, B. R., 1968, Composition and mean age of detritus of the Colorado River delta: Jour. Sed. Petrology, v. 35, p. 384–399.
———, AND WHITE, D. E., 1969, Active metamorphism of upper Cenozoic sediments in the Salton Sea geothermal field and the Salton Trough, southeastern California: Geol. Soc. America Bull., v. 80, p. 157–182.
NOBLE, L. F., 1954, The San Andreas Fault zone from Soledad Pass to Cajon Pass, California: California Div. Mines Bull. 170, chap. 4, p. 37–48.
PHILLIPS, R. P., 1964, Seismic refraction studies in Gulf of California, in VAN ANDEL, T. H., AND SHOR, G. G., JR. (eds.), Marine geology of the Gulf of California—a symposium: Am. Assoc. Petroleum Geologists Mem. 3, p. 90–121.
QUENNELL, A. M., 1958, The structural and geomorphic evolution of the Dead Sea Rift: Geol. Soc. London, Quart. Jour., v. 114, p. 1–24.
SHARP, R. P., 1954, Physiographic features of faulting in southern California: California Div. Mines Bull. 170, chap. 5, p. 21–28.
SILVER, E. A., 1972, Geophysical study of the Venezuelan borderland: Geol. Soc. America Abs. with Programs, v. 4, p. 237.
SMITH, F. D., JR. (compiler), 1962, Mapa geologico-tectonico del Norte de Venezuela (scale 1:1,000,000): I Cong. venezolano Petroleo.
STAINFORTH, R. M., 1966, Gravitational deposits in Venezuela: Assoc. venezolano Geologica, Minería, y Petroleo Bol. Inf., v. 9, p. 277–287.
———, 1969, The concept of seafloor spreading applied to Venezuela: ibid., v. 12, p. 257–274.
VEDDER, J. G., AND WALLACE, R. E., 1970, Recently active breaks along the San Andreas Fault between Cholame Valley and Tejon Pass, California: U.S. Geol. Survey Misc. Inv. Map I-574.
WEGENER, ALFRED, 1924, The origin of continents and oceans: New York, Dover Publications, Inc., 1966 reprint, 246 p.
WENTWORTH, C. M., 1968, Upper Cretaceous and Lower Tertiary strata near Gualala, California, and inferred large right slip on the San Andreas Fault: Stanford Univ. Pubs. Geol. Sci., v. 11, p. 130–143.
WILSON, J. T., 1965, A new class of faults and their bearing on continental drift: Nature, v. 207, p. 343–347.
WOODFORD, A. O., 1925, The San Onofre Breccia, its nature and origin: Univ. California Pub., Bull. Dept. Geol. Sci., v. 15, p. 159–280.
WOODFORD, A. O., AND OTHERS, 1971, Pleiocene-Pleistocene history of the Perris Block, southern California: Geol. Soc. America Bull., v. 82, p. 3421–3448.
WOODRING, W. P., 1932, Distribution and age of the marine Tertiary deposits of the Colorado Desert: Carnegie Inst. Washington Pub. 418, p. 1–25.

PROBLEMS OF PALINSPASTIC RESTORATION

LITHOGENESIS AND GEOTECTONICS: THE SIGNIFICANCE OF COMPOSITIONAL VARIATION IN FLYSCH ARENITES (GRAYWACKES)

KEITH A. W. CROOK
Australian National University, Canberra

ABSTRACT

Flysch arenites (graywackes) vary greatly in their framework mineralogy and volatiles-free chemistry. Three distinct groups are recognized on the basis of abundance: (1) quartz-poor graywackes (<15% quartz, average 58% SiO_2, $K_2O/Na_2O \ll 1$), (2) quartz-intermediate graywackes (15–65% quartz, average 68–74% SiO_2, $K_2O/Na_2O < 1$), and (3) quartz-rich graywackes (> 65% quartz, average 89% SiO_2, $K_2O/Na_2O > 1$).

Mineralogical data from modern deep-sea sands, which are regarded as the contemporary analogues of flysch arenites, together with data from sands of rivers that debouch into the deep sea provide the basis for an actualistic hypothesis to explain the compositional variation of ancient graywackes in terms of their geotectonic settings. This hypothesis is supported by chemical comparisons.

Quartz-poor graywackes are indicative of magmatic island arcs. Their average chemical composition approximates the average composition of these arcs and that of tholeiitic andesite. Quartz-intermediate graywackes are indicators of Andean type continental margins and approximate the upper levels of continental crust in composition. Quartz-rich graywackes indicate Atlantic-type continental margins. Chemically, they are related to the sand fraction of continental platform cover.

INTRODUCTION

This paper proposes an hypothesis to explain the significance of the considerable chemical and mineralogical variation that occurs within a major class of terrigenous arenites, namely those which occur in flysch (or so-called turbidite) sequences. Each of the major compositional groups of flysch arenites (graywackes, see appendix) corresponds to a distinctive geotectonic setting. Graywacke composition provides an important constraint in paleogeotectonic reconstructions.

THEORY

The hypothesis rests on three fundamental premises. The first, which has its ultimate origins in the work of M. S. Shvetsov (1934), is that compositional differences between arenites are not the accidental reflection of "what happens to be around in the source area" but are the result of systematic interaction between several lithogenetic processes that can and should be elucidated.

The second premise is that our understanding of arenite lithogenesis is unlikely to progress greatly as long as we concentrate upon composition as the primary basis of discussion. Comparisons between "all basalts" and "all rhyolites" are no longer meaningful for igneous petrogenesis. Comparisons such as those between "all quartz arenites" and "all lithic arenites" are of equally limited value to lithogenetic studies.

I do not imply by this observation that composition is unimportant. Rather, we must refine our usage of it, just as igneous petrologists have done by recognizing that the category "all basalts" includes members of several suites of igneous rocks, each of which is of more fundamental petrogenetic significance than the category "all basalts." But how do we refine our use of composition?

The question leads to my third premise, which derives from the work of Sir E. B. Bailey (1930). He recognized that arenites occur in two distinct facies, one characterized by graded bedding and the other by medium- to large-scale cross bedding. These correspond to the flysch (or turbidite) and nonflysch (or nonturbidite) facies of contemporary usage. My premise is that we can refine our use of composition, and thereby improve our understanding of first-order lithogenetic processes, by first dividing sedimentary rocks into a number of suites on the basis of their gross sedimentary features (see appendix) and then proceeding to examine the origin of the compositional variations in the arenites within each suite.

The justification for this premise is the hypothesis that the different suites of sedimentary rocks reflect differences in the modes of interaction of first-order lithogenetic processes.

I have chosen to examine arenites of the

flysch (or turbidite) suite for several reasons. These arenites, which are graywackes by definition in Packham's (1954) terminology and mostly graywackes according to other widespread usages (see appendix), show considerable compositional variation and are widespread in space and time. By their occurrence in geosynclines, they are participants in the development of continental crust, in which lithogenetic processes are of fundamental importance.

GRAYWACKE DATA

Graywackes vary widely in framework mineralogy. Contrary to common belief, this variation is not a uniform frequency continuum between the Q-pole (mono- and polycrystalline quartz), F-pole (feldspar), and R-pole (rock fragments, including chert and quartzarenite, and other unstable grains) of the arenite QFR diagram (Crook, 1960a, 1964; Folk and others, 1970).

Figure 1 is a frequency-contoured QFR plot of the framework compositions of 328 arenites from Phanerozoic flysch sequences on five continents, Oceania, and the Caribbean. The data are principally taken from the literature. Representative modal analyses are given in table 1.

In figure 1, frequency discontinuities are evident at about 15 percent Q and 65 percent Q, graywackes at these quartz contents being relatively rarer than at others. The discontinuities divided graywackes into three major groups (fig. 2), at least one of which can be further subdivided.

1. *Quartz-poor graywackes.*—These contain less than 15 percent quartz[1] as framework grains. They are predominantly of basic volcanic and ultramafic provenance. Three subgroups occur: volcanic lithic graywacke (Chappell, 1968), which is the most common; plagioclase graywacke (Okada and Nakao, 1968); and serpentinite graywacke (Zimmerle, 1968).

2. *Quartz-intermediate graywackes.*—These are of mixed provenance and contain 15 to 65 percent quartz as framework grains. Subdivision of the group may be possible after further work. Some are of predominantly granitic and high-grade metamorphic provenance (de Keyser and Lucas, 1968), others are of low-grade metamorphic provenance (McBride, 1962), and yet others have a significant acid volcanic component (Packham, 1967). Some contain a mixture of material of all three provenances (Huckenholz, 1959).

3. *Quartz-rich graywackes.*—These contain

[1] Mono- and polycrystalline quartz, exclusive of chert and quartz arenite.

FIG. 1.—Frequency-contoured plot of the framework compositions of 328 flysch arenites (graywackes) from the Phanerozoic of Australia, North America, South America, Asia, Europe, Oceania, and the Caribbean. Symbols: Q, mono- and polycrystalline quartz exclusive of chert and quartzarenite; F, feldspars; R, rock fragments including chert, quartz arenite, and other unstable framework components. Frequency contours at > 0, 1, 2 and 4 percent per ½ percent of area. Reproduced from K. A. W. Crook (1973), courtesy of *Encyclopaedia Brittanica*.

more than 65 percent and comonly more than 80 percent quartz as framework grains. They appear to be predominantly of sedimentary provenance and have minor granitic and low-grade metamorphic admixtures. Quartz-rich graywackes are common in the Ordovician of the Tasman Geosyncline in eastern Australia (Packham 1969). They also occur in Europe (Logvinenko and others, 1961) and North America (Klein, 1966).

MODERN SEDIMENT DATA

Among modern sediments, the flysch suite is presumably represented by the interstratified sands and muds that occur in the deep sea adjoining continents, microcontinents, and island arcs. Some of these sediments show graded bedding and flysch profiles. The areas in which these sediments occur can be regarded in the light of the new global tectonics as modern geosynclines (see Dewey and Bird, 1970).

Modal analyses of modern deep-sea sands are not abundant. They can be supplemented by data from onland sites located on the principal sediment-transport paths to deep-sea areas, particularly from major river systems.

The author's studies in Papua, New Guinea, now in progress indicate that volcanic lithic

TABLE 1.—MODAL COMPOSITIONS OF SELECTED GRAYWACKES AND DEEP-SEA SANDS (PERCENT)[1]

Component	A	B	C	D	E	F	G	H
Quartz	8.8	0.4	—	29.16	19.57	26.93	48.94	80.2
K-feldspar	—	—	—	2.06	2.34	9.18	4.71	9.4
Plagioclase	67.5	12.7	2.0	2.92	21.05	18.36		1.0
Serpentinite	—	—	29.0	—	—	—	—	—
Granite and schist	—	—	13.0	—	14.22	9.55	2.76	1.0
Volcanic rock fragments	2.5	63.6	13.5	—		11.02		—
Sedimentary rock fragments	—	—	6.0	20.63	5.12	6.12		—
Low-grade metamorphic rock fragments	—	—	12.0	9.96	22.45	[2]		1.0
Muscovite	—	—	—	—	0.75	—	5.65	1.0
Biotite	—	—	—	—	2.14	2.45	1.59	
Chlorite	1.7	—	—	—	7.67	0.37	—	—
Pyroxene and hornblende	—	1.5	—	—	0.07	0.37	—	—
Other heavy minerals	0.4	0.6	1.5	—	—	1.08	—	5.2
Carbonate materials	1.4	0.4	—	—	—	—	—	—
Miscellaneous detritus	—	—	—	—	—	—	0.53	1.0
Matrix	—	20.3	—	35.24	[2]	10.31	15.41	[2]
Cement	17.8	0.5	23.0	—	4.63	3.99	20.12	—
Totals	100.1	100.0	100.0	99.97	100.01	99.73	99.71	99.8[3]

[1] Reproduced from K. A. W. Crook (1973), courtesy of *Encyclopaedia Brittanica*.
[2] Not quoted separately.
[3] Glauconite and mud pellets omitted.
A, Plagioclase graywacke, Carboniferous; New England, Australia (Crook, 1960b)
B, Average of 31 Devonian volcanic lithic quartz-poor graywackes; New England, Australia (Chappell, 1968)
C, Serpentinite graywacke, Tertiary; North Coast Basin, Colombia, (Zimmerle, 1968)
D, Average S-1 (five samples) of lithic quartz-intermediate graywacke, Martinsburg Formation, Ordovician; Appalachians (McBride, 1962)
E, Average of graywacke 7b, lithic quartz-intermediate graywacke, Tanner Graywacke, Devonian; Harz, Germany (Huckenholz, 1959)
F, Average of seven lithic quartz-intermediate graywackes, Axial Facies, upper Paleozoic and Mesozoic; New Zealand Geosyncline (Dickinson, 1971)
G, Average of 17 quartz-rich graywackes, Tavricheskaya Formation, Triassic; Crimea, (Logvinenko and others, 1961)
H, Average of 92 modern deep-sea sands; western province, North Atlantic (Hubert and Neal, 1967)

sands, as might be expected, are characteristic of the volcanic island arcs such as Bougainville, New Britain, and the south flank of the Huon Peninsula.

Plagioclase graywackes have a modern analogue among deep-sea sands, such as the sands adjoining the Marianas arc that were sampled on Deep Sea Drilling Project leg 20 (H. Okada, personal commun.). In ancient sequences these rocks are intimately associated with volcanic lithic graywackes. They clearly indicate a volcanic island arc setting, perhaps one at a different stage of development to those that yield volcanic lithic sands exclusively.

Serpentinite graywackes have no known modern analogues, but sub-Recent examples are known from the British Solomon Islands (Univ. Sydney, 1956), where they are associated with a serpentinite massif that has been interpreted as an obduction zone.

Quartz-intermediate graywackes are represented by the modern deep sea sands of the Astoria Fan off the Oregon coast (Kulm and Fowler, this volume) and by the sediment supply entering the Coral Sea Basin from the Angabunga River in Papua New Guinea (Crook, Taylor, and Belbin, in preparation).

The modern analogues of quartz-rich graywacke occur in the western North Atlantic petrographic province adjoining North America (table 1, column H) (Hubert and Neall, 1967) and probably also in the Tasman Basin off the east coast of Australia. Graded bedded deep-sea sands occur in this basin (Conolly, 1969; Eade and van der Linden, 1970). The sands being supplied by eastern Australian rivers are dominated by quartz as are the sands of east coast beaches.

CHEMICAL DATA AND COMPARISONS

There are marked chemical differences between the three groups of graywackes, the differences in SiO_2 content on a volatiles-free basis being the most obvious (table 2). Of the quartz-poor group, serpentinite graywackes (table 2, column E) are the least siliceous, containing 47 percent SiO_2. Volcanic lithic and plagioclase graywackes (table 2, columns A and D) average 58 or 59 percent SiO_2. Quartz-intermediate graywackes (table 2, G, H, I) range from 68 percent SiO_2, and quartz-rich graywackes (table 2, L) are remarkably siliceous, having 89 percent SiO_2. K_2O/Na_2O ratios vary from about 0.06 in plagioclase graywackes to 1.7 in quartz-rich graywackes. All types show a predominance of ferrous iron.

Comparisons of these data with those of modern deep-sea sands is not yet possible. Suitable data are virtually nonexistent. Although this avenue for comparison is closed, some important insights can be gained by comparing the chemical compositions of graywackes with various geochemical averages, particularly those that represent different parts of the earth's crust (table 2).

A close resemblance between quartz-poor volcanic lithic graywackes and tholeiitic andesite (table 2, A and B) is apparent as is their resemblance to the bulk composition of magmatic island arcs (table 2, C), although the last-mentioned composition has a higher proportion of calcium. Quartz-poor plagioclase graywacke (table 2, D) differs from the others principally in having a much higher ratio of aluminum to total iron. The overall resemblances suggest an island arc source for these graywackes.

The third quartz-poor variety, serpentinite graywacke (table 2, E), is close to mantle ultra-

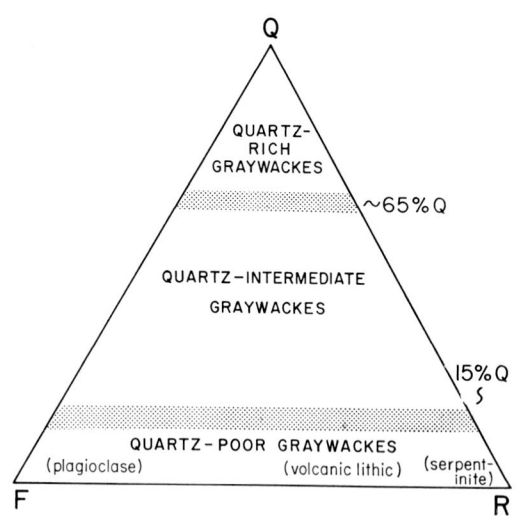

FIG. 2.—Approximate limits of framework composition for the three major types of graywacke. Terms in brackets refer to the subtypes of quartz-poor graywacke.

mafics in composition (table 2, F), reflecting its uncommon source material.

The compositional spectrum of quartz-intermediate graywackes (table 2, G, H, I) lies generally within a range defined by the average composition of Precambrian shields (table 2, J) at the low silica end, and the average composition of the sedimentary platform cover of the

TABLE 2.—CHEMICAL COMPOSITIONS OF REPRESENTATIVE GRAYWACKES AND COMPARATIVE ANALYSES OF OTHER CRUSTAL MATERIALS (PERCENT)

Component	Graywackes and other Materials
	A	B	C	D	E	F	G	H	I	J	K	L	M
SiO_2	57.56	58.61	58.78	59.08	46.47	44.5	68.26	70.1	73.5	66.29	73.05	88.82	86.25
TiO_2	1.17	1.28	0.84	0.52	0.14	0.15	0.66	0.5	0.6	0.51	0.80	0.39	0.76
Al_2O_3	16.53	15.94	15.58	23.75	3.28	2.55	15.75	14.0	14.8	15.44	13.44	5.88	6.17
Fe_2O_3	1.74	3.55	2.63	0.66	5.38	7.3	0.72	1.2	1.5	1.86	3.88	0.39	1.52
FeO	7.69	5.11	5.04	2.65	3.93	1.5	4.67	3.1	2.9	3.53	1.44	1.35	—
Fe Total	(10.28)	(9.22)	(8.22)	(3.60)	(9.74)	(9.0)	(5.90)	(4.6)	(4.7)	(5.78)	(5.48)	(1.89)	(1.52)
MnO	0.17	0.21	0.11	0.37	0.13	0.14	0.06	0.1	0.1	0.07	0.10	0.02	—
MgO	3.90	3.45	4.57	1.19	38.26	41.7	2.82	2.3	1.6	2.31	3.48	0.83	1.17
CaO	5.28	6.27	8.02	4.70	1.74	2.25	1.75	2.5	1.0	3.38		0.39	1.07
Na_2O	5.05	4.29	3.39	6.89	0.25	0.25	3.22	3.7	2.5	3.21	1.00	0.69	1.03
K_2O	0.70	0.44	0.82	0.42	0.12	0.015	1.97	1.8	1.4	3.09	3.21	1.18	1.99
P_2O_5	0.24	0.45	0.22	0.14	0.09	—	0.12	0.1	0.1	0.13	0.17	0.09	0.08
Totals	100.03	99.60	100.00	100.37	99.79	100.36	100.00	99.4	100.0	99.82	100.57	100.03	100.04

A, Average of 10 quartz-poor (volcanic lithic) graywackes; Devonian, New England, Australia (Chappell, 1968)
B, Tholeiitic andesite; Fiji, (Gill, 1970, no. 68-55, recalculated volatiles free)
C, Composition of a "developed island arc" (Jakes and White, 1970)
D, Plagioclase graywacke; Carboniferous, New England, Australia (unpub. analysis, recalculated calcite and water free)
E, Average of five serpentinite graywackes (data from Lockwood, 1971, analyses 1, 4, 5, 6, and 7, recalculated volatiles free)
F, Average upper mantle composition (Wyllie, 1971, table 6-6, analysis 3)
G, Average of three quartz-intermediate Archean graywackes (Henderson, 1972, analyses 1-3, recalculated volatiles free)
H, Average of 21 quartz-intermediate Franciscan graywackes (Bailey and others, 1964, recalculated calcite free)
I, Average of 20 quartz-intermediate graywackes, Culm, Harz Mountains (Huckenholz, 1963, recalculated volatiles free)
J, Average composition of the shields (Ronov and Migdisov, 1971, recalculated volatiles free)
K, Average composition of sedimentary cover of Russian and North American platforms excluding carbonate matter, etc. (Ronov and Migdisov, 1971, recalculated volatiles free)
L, Average of 24 quartz-rich graywackes principally from the Tasman Geosyncline, Australia (unpub. data)
M, Average composition of the Cambrian through Quaternary sands of the North American platform (Ronov and Migdisov, 1971, recalculated calcite free and volatiles free)

shields, excluding the carbonate and evaporite components (table 2, K). Evidently quartz-intermediate graywackes are fair average samples of the silicate materials forming the upper levels of continental crust.

Quartz-rich graywackes (table 2, L) are notably more siliceous than the average sandstone of the Russian Platform (Ronov and Migdisov, 1971) and slightly more so than the average of Phanerozoic sandstones from the North American Platform (table 2, M). These graywackes were evidently derived from the sands of continental platforms.

DISCUSSION

The mineralogical similarities between graywackes and modern deep-sea sands and the results of chemical comparisons together provide a basis for explaining variations in graywacke composition. This basis is developed as follows.

Modern deep-sea sands accumulate adjacent to major topographic breaks that result from various types of fundamental discontinuities in the lithosphere (LD's). The LD's separate major geotectonic elements. Some are boundaries of lithosphere plates. Others are within-plate LD's separating continental from oceanic crust.

Modern deep-sea sand deposits commonly lie on the opposite sides of LD's to their source areas. These sources vary in nature and composition, depending upon the type of adjacent LD. Typically the sources are of regional extent, being continental drainage basins or island arc terranes. Modern deep-sea sands can be expected to resemble closely the average compositions of their source areas. Their composition will therefore indicate the type of adjacent LD. Thus modern deep-sea sand composition is a function of the geotectonic position of the sand deposit.

The compositional variation in ancient flysch arenites (graywackes) can therefore be interpreted on an actualistic basis. It reflects the geotectonic settings in which the flysch sequences accumulated. Specifically, it can be used to infer the nature of adjacent LD's and hence is an important tool in paleogeotectonic reconstructions.

To judge from the locations of their modern deep-sea analogues and from their chemistry, quartz-poor volcanic lithic and plagioclase graywackes are indicators of ancient magmatic island arcs. Some may have accumulated adjacent to emergent midoceanic ridges and sea mounts. This variety remains unrecognized. It is likely to be volumetrically minor unless ocean water depths were very different in the past.

Extensive exposures of mantle material form the source areas for serpentinite graywackes. Their low quartz content suggests ensimatic sources; the sub-Recent analogue of these rocks has such a source. The only mantle material likely to be exposed extensively is material in obducted slabs of lithosphere. Serpentinite graywackes therefore probably indicate the presence of obducted lithosphere.

The modern analogues of quartz-intermediate graywackes adjoin LD's that separate tectonically active continental crust and oceanic crust. Chemically, these graywackes are intimately related to the upper levels of continental crust. They are indicators of Andean-type (i.e., tectonically mobile) continental margins. Chemical and mineralogical variations within this group of graywackes may possibly reflect different varieties of Andean-type continental margins.

The quartz-rich graywackes are a little-studied group. Their modern equivalents adjoin tectonically quiescent continental margins. This, and their chemical relationship to the sand fraction of continental platform cover, strongly suggests that they are indicators of the Atlantic-type continental margins that occur within lithosphere plates on the trailing edges of continents.

Further data on the mineralogy and chemistry of modern deep-sea sands are needed to substantiate the deductions reached above. I intend to pursue this matter in the immediate future and will welcome communications from other workers concerning the composition of modern deep-sea sands.

ACKNOWLEDGMENT

I am indebted to *The Encyclopaedia Brittanica* for permission to draw upon an article (Crook, 1973) that was in press at the time of writing.

REFERENCES

BAILEY, E. B., 1930, New light on sedimentation and tectonics: Geol. Mag., v. 67, p. 77–92.
BAILEY, E. H., IRWIN, W. P., AND JONES, D. L., 1964, Franciscan and related rocks, and their significance in the geology of western California: California Div. Mines and Geology Bull. 183, 178 p.
BOUMA, A. H., 1962, Sedimentology of some flysch deposits: Amsterdam, Elsevier Publishing Co., 168 p.
CHAPPELL, B. W., 1968, Volcanic graywackes from the Upper Devonian Baldwin Formation, Tamworth-Barraba district, New South Wales: Geol. Soc. Australia Jour., v. 15, p. 87–102.
CONOLLY, J. R., 1969, Western Tasman Sea floor: New Zealand Jour. Geology and Geophysics, 12, 310–343.
CROOK, K. A. W., 1960a, Classification of arenites: Am. Jour. Sci., v. 258, p. 419–428.
———, 1960b, Petrology of Parry Group, Upper Devonian-Lower Carboniferous, Tamworth-Nundle district, New South Wales: Jour. Sed. Petrology, v. 30, p. 538–552.

———, 1964, A classification of common sandstones: comment on a paper by Earle F. McBride: *ibid.*, v. 34, p. 696–700.
———, 1973, Graywackes: Encyclopaedia Brittanica, 15th ed., v. 8, p. 295–299.
———, Taylor, G. M., and Belbin, L., in preparation, Petrology of Angabunga River sediments, Papua New Guinea.
de Keyser, F., and Lucas, K. G., 1968, Geology of the Hodgkinson and Laura Basins, North Queensland: Australia Bur. Min. Res. Bull. 84, 254 p.
Dewey, J. F., and Bird, J. M., 1970, Plate tectonics and geosynclines: Tectonophysics, v. 10, p. 625–638.
Dickinson, W. R., 1971, Detrital modes of New Zealand graywackes: Sed. Geology, v. 5, p. 37–56.
Eade, J. V., and Linden, J. M. van der, 1970, Sediments and stratigraphy of deep-sea cores from the Tasman Basin: New Zealand Jour. Geology and Geophysics, v. 13, p. 228–268.
Folk, R. L., Andrews, P. B., and Lewis, D. W., 1970, Detrital sedimentary rock classification and nomenclature for use in New Zealand: *ibid.*, p. 937–968.
Gilbert, C. M., 1954, Sedimentary rocks, *in* Williams, H., Turner, F. J., and Gilbert, C. M., Petrography: San Francisco, W. H. Freeman and Co., 406 p.
Gill, J. B., 1970, Geochemistry of Viti Levu, Fiji, and its evolution as an island arc: Contr. Mineralogy and Petrology, v. 27, p. 179–203.
Henderson, J. B., 1972, Sedimentology of Archean turbidites at Yellowknife, Northwest Territories: Canadian Jour. Earth Sci., v. 9, p. 882–902.
Hubert, J., and Neal, P. F., 1967, Mineral composition and dispersal patterns of deep-sea sands in the western North Atlantic petrologic province: Geol. Soc. America Bull., v. 78, p. 749–772.
Huckenholz, H. G., 1959, Sedimentpetrographische Untersuchungen an Gesteinen der Tanner Grauwacke: Beitr. Mineralogie und Petrologie, v. 6, p. 261–298.
———, 1963, Mineral composition and texture in graywackes from the Harz Mountains (Germany) and in arkoses from the Auvergne (France): Jour. Sed. Petrology, v. 33, p. 914–918.
Jakes, P., and White, A. J. R., 1971, Composition of island arcs and continental growth: Earth and Planetary Sci. Letters, v. 12, p. 224–230.
Klein, G. D., 1966, Dispersal and petrology of sandstones of Stanley-Jackfork boundary, Ouachita fold belt, Arkansas and Oklahoma: Am. Assoc. Petroleum Geologists Bull., v. 50, p. 308–326.
Krumbein, W. C., and Sloss, L. L., 1963, Stratigraphy and sedimentation: San Francisco, W. H. Freeman and Co., 2nd ed., 660 p.
Lockwood, J. P., 1971, Sedimentary and gravity-slide emplacement of serpentinite: Geol. Soc. America Bull., v. 82, p. 919–936.
Logvinenko, N. V., Karpova, V., and Shaposhnikov, D. P., 1961, Sedimentology and genesis of the Tavricheskaya Formation of the Crimea: Kharkov, U.S.S.R., Kharkov Univ., Pub. House, 400 p.
McBride, E. F., 1962, Flysch and associated beds of the Martinsburg Formation (Ordovician), central Appalachians: Jour. Sed. Petrology, v. 32, p. 39–91.
Okada, H., and Nakao, S., 1968, Plagioclase arenite in Lower Cretaceous flysch in the Furano area, Hokkaido: Geol. Soc. Japan Jour., v. 74, p. 451–452.
Packham, G. H., 1954, Sedimentary structures as an important feature in the classification of sandstones: Am. Jour. Sci., v. 252, p. 466–476.
———, 1967, The lower and middle Paleozoic stratigraphy and sedimentary tectonics of the Sofala-Hill End-Euchareena region, N.S.W.: Linnaean Soc. New South Wales Proc., v. 93, p. 111–163.
——— (ed.), 1969, The geology of New South Wales: Geol. Soc. Australia Jour., v. 16, p. 1–654.
Pettijohn, F. J., 1949, Sedimentary rocks: New York, Harper & Bros., 526 p.
———, 1954, Classification of sandstones: Jour. Geology, v. 62, p. 366–374.
———, Potter, P. E., and Siever, R., 1972, Sand and sandstone: New York, Springer-Verlag, 618 p.
Ronov, A. B., and Migdisov, A. A., 1971, Geochemical history of the crystalline basement and the sedimentary cover of the Russian and North American platforms: Sedimentology, v. 16, p. 137–187.
Shvetsov, M. S., 1934, Petrografiya osadochnykh porod: Moscow, Gostoptekhizdat.
Univ. Sydney, Dept. Geology and Geophysics, 1956, Geological reconnaissance of parts of the central islands of the British Solomon Islands Protectorate: Colonial Geology and Mineral Res., v. 6, p. 267–306.
Wyllie, P. J., 1971, The dynamic earth: New York, John Wiley & Sons, Inc., 416 p.
Zimmerle, W., 1968, Serpentine graywackes from the North Coast Basin, Colombia, and their geotectonic significance: Neues Jahrb. Mineralogie, Abh., v. 109, p. 156–182.

APPENDIX

Recognition and Naming of Lithogenetically Significant Classes of Arenites and Other Sedimentary Rocks

Sandy sedimentary rocks have been a peculiarly fertile source of nomenclatural confusion. Even the term "sandstone" carries an implication of mineralogy as well as size, as Pettijohn (1949, p. 226) noted. The term "arenite," suggested by Pettijohn (1949) as an alternative general size term free of any mineralogical connotation, regrettably has been introduced into petrographic nomenclature by Gilbert (1954, p. 289) in a restricted sense to apply to siliciclastic sands containing less than 10 percent matrix. Pettijohn's usage is followed herein.

In his classification of arenites, G. H. Packham (1954), following Bailey (1930), divided these rocks into two great suites, a sandstone suite and a graywacke suite, on the basis of the association of sedimentary structures observed in the arenites and interstratified rocks. The graywacke suite is characterized by the sedimentary structures of the classic flysch profile of A. H. Bouma (1962). The use of "graywacke" in Packham's sense is commonplace in Australia. This usage is, in practice, little different from that of many sedimentologists. Most of the graywacke-suite arenites of Packham have sufficient matrix to fit the definitions of "graywacke" or "wacke" proposed by Gilbert (1954) and Pettijohn (1954), both of which definitions have wide currency.

The matrix content of arenites is unlikely to be a useful parameter on which to base lithogenetically significant classes of arenites, simply because matrix is a polygenetic feature (see Pettijohn and others, 1972, p. 206–211, for a summary). The fact that some flysch arenites and most modern deep-sea sands contain little or no matrix is, therefore, I believe, irrelevant to the end in view, which is an understanding of first-order lithogenetic processes.

The problem of terminology remains, however, Sedimentologists are familiar with the problems surrounding the meaning and usage of the terms "flysch," "turbidite," and "graywacke." I have used the last term because it seems best for describing the rocks in question. Most of them would be called graywackes or wackes by all non-Soviet geologists, particularly when observed in the field. Soviet geologists, many of whom regard graywacke as a lithic arenite, would find a greater terminological problem.

I wonder, in view of this, may not a new terminology be timely? Suites of sediments, each composed of stratigraphically associated sedimentary rocks ranging from lutite to rudite in grade, can be recognized on a purely descriptive basis. The criteria for their recognition are the sedimentary features of the rock sequences observed in the field, *other than their mineralogy and chemistry,* without regard to their inferred tectonic settings, modes of emplacement, or depositional environments. Suites so defined resemble the "observed lithological associations" of Krumbein and Sloss (1963) but differ in not having mineralogy and chemistry as a defining parameter. Probably several of Krumbein and Sloss's associations would be compositionally distinctive subgroupings within these suites.

Terminology might conveniently utilize the names of prominent workers who were responsible for elucidating particular suites. In this vein, the rocks discussed in this paper are members of the baileyitic suite, or simply baileyitic sands, being so-named here for Sir Edward Bailey, whose work already has been mentioned.

As the total number of suites is likely to be small (cf. the suites used in igneous petrology), gains in precision and clarity seem likely to outweigh the disadvantages of introducing new terminology. Reactions to this suggestion will be appreciated.

MIGRATION OF ANCIENT ARC-TRENCH SYSTEMS

HAKUYU OKADA

Kagoshima University, Kagoshima, Japan

ABSTRACT

In Hokkaido, two asynchronous pairs of Jurassic to earliest Tertiary arc-trench systems are recognized. The trench belt of the older pair occupies a part of the Jurassic to Cretaceous Yezo Geosyncline, where the Kamuikotan metamorphic rock belt may represent a west-dipping subduction zone. The trench belt of the younger pair is represented by the latest Cretaceous to earliest Tertiary Small Kuril Geosyncline (new name). In both geosynclines, the trench sediments are characterized by the following fourfold succession, in ascending order: (1) radiolarian chert or siliceous shale and pillow lavas, (2) flysch, (3) shallow-marine muddy sediment, and (4) molasselike sequences. Farther to the south of the Small Kuril Geosyncline, the modern Kuril Trench has been developed. These consecutive developments of arc-trench systems present a good example of oceanward migration patterns of the systems.

INTRODUCTION

In the circum-Pacific orogenic belt, the concept of eugeosynclinal assemblages seems to be defined by the settings of arc-trench systems as modeled by Dickinson (1971) through analytical comparisons between ancient and modern systems. For better understanding of eugeosynclinal evolution, this paper clarifies a pattern of migration of ancient arc-trench systems recognized in Hokkaido, where the Kuril and East Honshu arcs meet each other.

ACKNOWLEDGMENTS

The writer owes much to Professor Tatsuro Matsumoto of Kyushu University, Fukuoka, for his stimulating discussions and encouragement and for his giving in general from his expert knowledge of Cretaceous geology of Hokkaido. Professor Hideo Ishikawa, Kagoshima University, is gratefully acknowledged for his valuable discussions. The writer is heartily grateful to Professor Robert H. Dott, Jr., not only for inviting the writer to attend the Conference on Modern and Ancient Geosynclinal Sedimentation and to contribute to this volume but also for his encouragement with deep interest. The manuscript was critically read by Professors Dott and Keith Crook, to whom the writer is greatly indebted.

PAIRED METAMORPHIC BELTS AND ARC-TRENCH SYSTEMS

The zonal arrangement of paired metamorphic rock belts has been firmly established by Miyashiro (1961) in Honshu, Shikoku, and Kyushu of the Japanese Islands, where metamorphic belts of the high-pressure and low-temperature (high-P/T) type always lie on the oceanic side, whereas the low-P/T-type belts lie on the continental side. This concept of paired metamorphic belts has been regarded as a fundamental principle for the Pacific-type orogenic evolution in terms of plate tectonics (e.g., Miyashiro, 1967; Ernst, 1971; Matsuda and Uyeda, 1971; Mitchell and Reading, 1971).

The paired-belt concept has also been applied to the metamorphic rock belts in Hokkaido (Miyashiro, 1961). Miyashiro pointed out, however, that the disposition of the metamorphic belts in Hokkaido is in sharp contrast to that in the main parts of the Japanese Islands. That is to say, the high-P/T-type Kamuikotan metamorphic rock belt lies on the present continental side of the low-P/T-type Hidaka belt on the oceanic side (fig. 1).

This apparent reversal of the zonal arrangement of the metamorphic rock belts in Hokkaido has been interpreted in various ways: One interpretation is that the Japan Basin plate, underthrusting beneath Hokkaido, produced an east-dipping Benioff zone (Matsuda and Uyeda, 1971; Mitchell and Reading, 1971). Another interpretation is that a clockwise rotation of Hokkaido, in reference to the northeast Honshu arc, brought about the present orientation from a presumed L-shaped, continentward bending of Hokkaido (Matsuda and Uyeda, 1971). The third interpretation is that orogenesis in Hokkaido was a result of collision between two microcontinents, which originally built up the western and the eastern half of Hokkaido, respectively (Horikoshi, 1972). All these arguments are highly speculative, and intensive studies are needed for their verification. Therefore, an alternative that seems geologically more warranted is given in the following: It is argued that these metamorphic belts were not synchronous and paired, but rather represent episodes

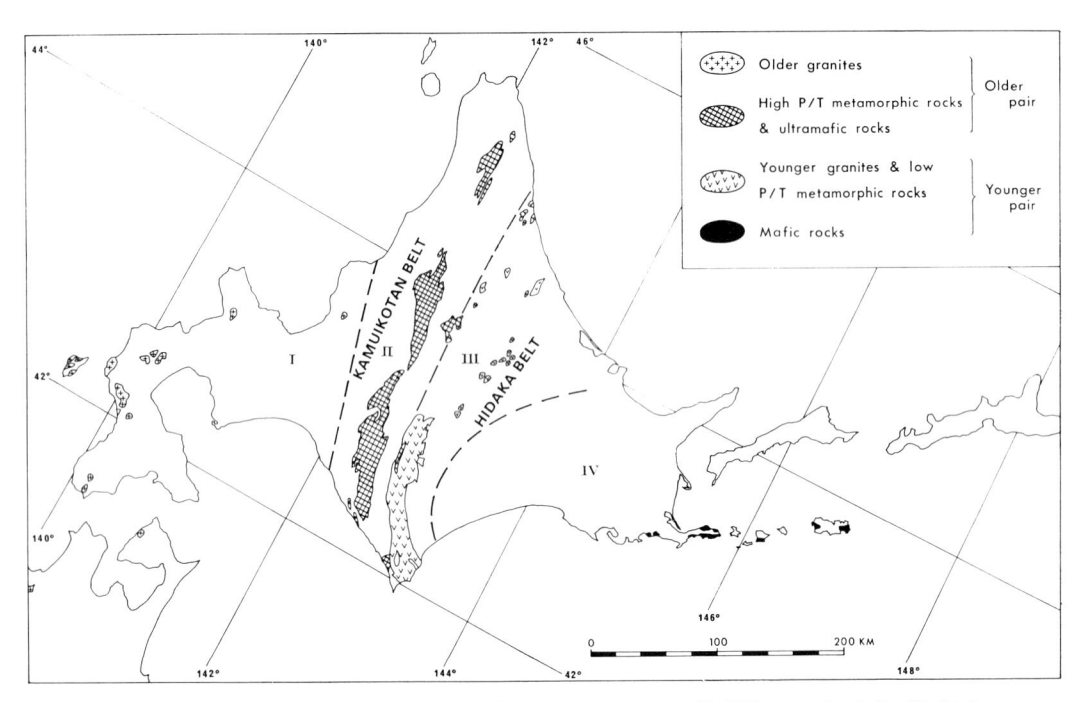

FIG. 1.—Map of Hokkaido showing two pairs of arc-trench systems. I, Older granite belt; II, high-pressure Kamuikotan metamorphic rock belt; III, lower-pressure Hidaka metamorphic rock belt; IV, younger mafic rock belt. North is toward upper right; see figure 2 for locality identifications.

of Pacific-ward migration of arc-trench systems. Radiometric, stratigraphic, and sedimentologic evidence for this hypothesis is presented.

The apparently paired metamorphic rock belts may be separated into independent belts as described below. The Mesozoic plutons, which at present are indicated by scattered exposures in the area west of the Kamuikotan metamorphic rock belt (I of fig. 1), may be considered as a magmatic emplacement due to the thermal activity defined as the magmatic front (Sugimura, 1960; Matsuda and Uyeda, 1971). Thus, the inner granitic belt and the outer Kamuikotan belt can be considered as a pair. The Kamuikotan belt seems to represent the subduction zone in a trench belt. The Hidaka metamorphic rock belt, composed of a plutonic and high-thermal metamorphic complex, constitutes a couple with the pillowed alkaline dolerite belt in the Nemuro area, a southeastern extremity of Hokkaido (figs. 1 and 2). Therefore, the inner Hidaka belt and the outer Nemuro dolerite belt form another pair together.

Although radiometric datings in Hokkaido are still scanty, available data on K-Ar dating prepared by Kawano and Ueda (1967a,b) show that most granitic rocks exposed in the Oshima Peninsula, southwestern Hokkaido, belong to the period of 110 to 120 my (late Early Cretaceous). The granitic rocks of the same age are also found in the Kitakami Massif in the northeastern marginal area of Honshu. K-Ar dating of the garnet amphibolite from the Kamuikotan metamorphic rock belt indicates the value of 109 to 120 my (Bikerman, Minato, and Huahashi, 1971). This value agrees well with that of the granitic rocks in southwestern Hokkaido. Therefore, an asumption of some relationship between these two rock belts is very reasonable.

The radiometric ages of granites and migmatites from the Hidaka metamorphic rock belt, as far as the available data are concerned, fall in the period 30 to 40 my (late Eocene to early Oligocene) (Kawano and Ueda, 1967a). These figures seem to represent the late stage of Hidaka plutonism. Nemuro dolerite, the outer belt coupled with the inner Hidaka belt, is characterized by pillowed sheets, which intruded into the Nemuro Group (Late Cretaceous). The K-Ar ages of the dolerite range from 84 to 88 my (Late Cretaceous; Ueda and Aoki, 1968).

Judging from the age data presented above, the Early Cretaceous acidic plutons in southwestern Hokkaido and the contemporary Kamuikotan metamorphic rock belt represent

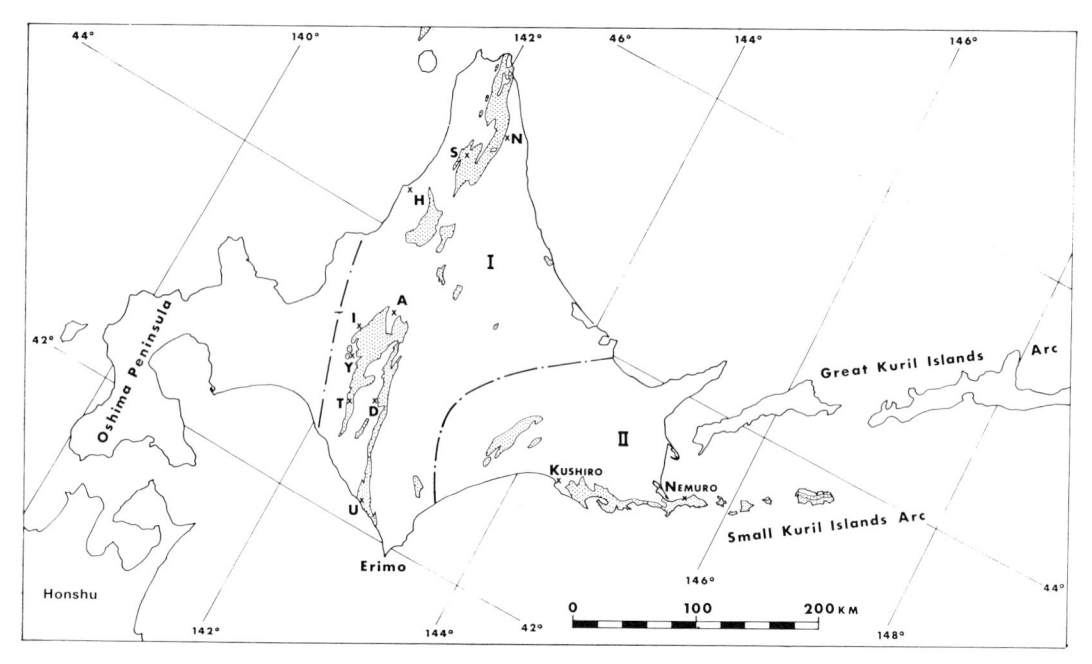

Fig. 2.—Map showing outcrops (dotted) of Cretaceous deposits in Hokkaido and Small Kuril Islands. I, Yezo Geosyncline; II, Small Kuril Geosyncline. Broken lines indicate major geotectonic boundaries. Place names from north to south: N, Nakatombetsu; S, Saku; H, Haboro; A, Ashibetsu; I, Ikushumbetsu; Y, Yubari; T, Tomiuchi; D, Hidaka; U, Urakawa.

the older pair of the arc-trench system, whereas the Late Cretaceous to early Tertiary Hidaka metamorphic rock belt and the Late Cretaceous Nemuro mafic rock belt represent the younger pair (fig. 1).

It should be noted that such ultramafic rocks as amphibolite, peridotite, and serpentinite, all characteristic of the Kamuikotan belt, are also sporadically found even in the Hidaka metamorphic rock belt (fig. 1). Spilitic rocks and mafic pyroclastics of Jurassic and Early Cretaceous ages are also exposed in patches just to the southeast of the Hidaka belt, so that the Kamuikotan rocks and their allies may originally have covered a much wider area than at present.

Another fact is that the Hidaka belt is closest to the Kamuikotan belt at its southernmost extremity (Erimo area; see fig. 2) in the axial zone of Hokkaido and gradually deviates from the Kamuikotan belt northward, turning in the end to the northeast. Such a trend is in harmony with major tectonic structures of Hokkaido, which show a radial feature diverging away from the southern extremity of the axial zone of Hakkaido. Consequently, northeastward the Hidaka belt seems to become almost parallel to the present Kuril arc.

STRATIGRAPHIC SETTINGS OF GEOSYNCLINAL SEDIMENTS IN TWO TRENCH BELTS

Under the tectonic situations mentioned in the foregoing section, the Yezo Geosyncline seems to have developed in the trench belt of the older arc-trench system, and the Small Kuril Geosyncline (new name, defined farther on) in the trench belt of the younger (fig. 2). Stratigraphic characteristics of these geosynclinal sediments are summarized in the following:

The Yezo Geosyncline

The main phase of the Yezo Geosyncline, proposed originally by Matsumoto (1943, p. 183), is represented by a Jurassic-to-Cretaceous sedimentary series. The earliest phase of the geosyncline is concealed by later, severe tectonic disturbances, although the Hidaka Group, composed mainly of metamorphic rocks, is said to represent the early phase. The chief period of the geosynclinal stage is represented by thick marine sediments of the Sorachi Group of Jurassic to Neocomian age, by the Lower Yezo, Middle Yezo, and Upper Yezo Groups of Aptian to early Campanian age, and by the Hakobuchi Group of Campanian to Maastrichtian age (fig. 3).

The Sorachi Group.—The Sorachi Group con-

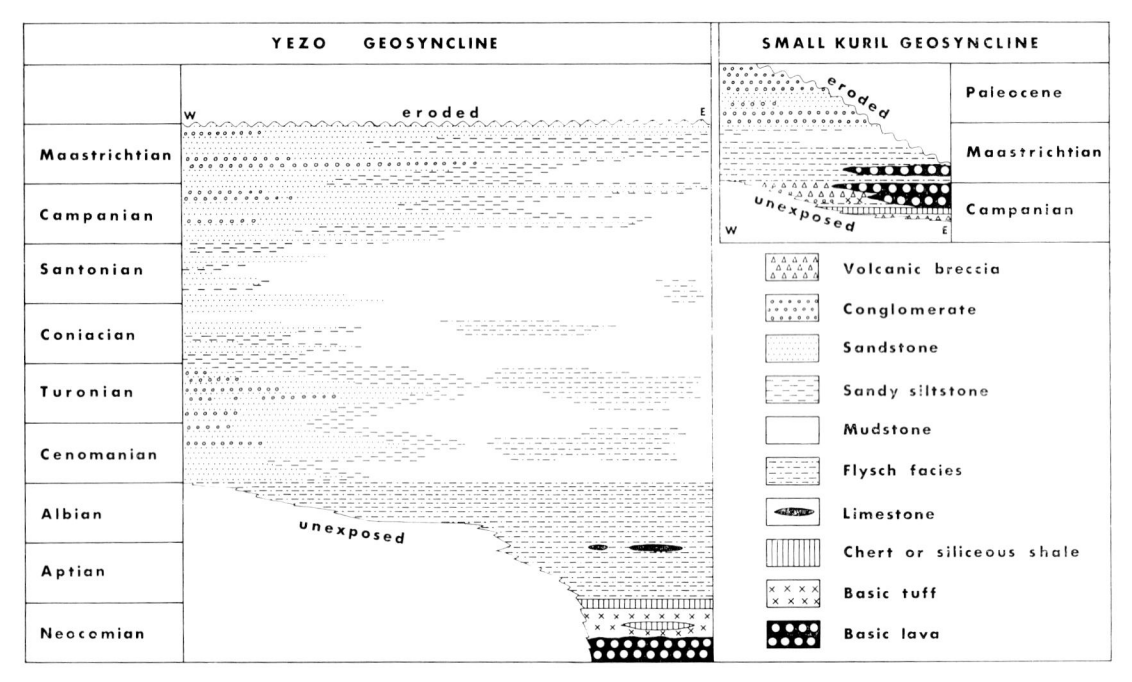

FIG. 3.—Schematic profile showing lithofacies of Cretaceous sequence in standard areas in Yezo Geosyncline (left) (adapted from Matsumoto, 1954, and Matsumoto and Okada, 1971) and lithofacies of the Late Cretaceous to earliest Tertiary deposits in the Small Kuril Geosyncline (right). Not to true scale either horizontally or vertically.

sists mainly of interbeds of basaltic to diabasic flows that commonly display pillow structures, mafic to intermediate pyroclastic rocks, radiolarian chert, and siliceous shale. Volcaniclastic sandstones and micritic limestone lenses are subordinately intercalated. Outcrops of the Sorachi Group are not restricted to the axial zone of Hokkaido, but are scattered over large areas. Chert consists of red, green, and gray varieties that typically contain abundant tests of radiolarians, and some cherty units are tens to hundreds of meters thick. Diabasic rock bodies in some sequences show considerable size, some attaining an apparent thickness of more than 500 m.

Relations between the Sorachi Group and the overlying Lower Yezo Group are apparently conformable in many places but are said to be disconformable in other parts. Further detailed examinations are necessary on the stratigraphical relationship between the two groups in terms of pelagite-turbidite relations as understood from recent ocean researches.

The Lower Yezo Group.—The Lower Yezo Group comprises interbedded sandstone and mudstone representing a flysch facies about 1,400 m thick in the central part of the Yezo Geosyncline, where the basal sequence of the group is represented by a sandy flysch facies (Tomitoi Sandstone). Graded bedding, sole markings such as flute and grooves, and organic trace markings are common features in the sandstone sequence. Channel-fill and other eroded structures are also present at some places.

In the middle and upper sequences of the Lower Yezo Group, shaly flysch predominates. Particularly in the southern part of the Yezo Geosyncline, small masses of *Orbitolina* limestone are met with, in which thick-shelled pachyodont pelecypods, nerinean gastropods, corals, hydrozoas, and calcareous algae are known in addition to *Orbitolina* and other foraminifers. These limestone blocks are associated with pebbly mudstone or slump deposits in many places. These blocks are thought to be of allochthonous origin, owing their present positions to gravitational flow processes. Except for organic remains in limestones, fossils are generally rare throughout the Lower Yezo Group, although some ammonites of Aptian to mid-Albian ages are reported.

A probable marginal facies of the Lower Yezo Group traceable on the Pacific coast of Iwate Prefecture, northeastern extremity of Honshu, where the Miyako Group (Aptian to middle Albian) is present. This sequence is composed chiefly of littoral deposits of very fossiliferous calcareous sandstones, in which abundant remains of *Orbitolina,* corals, pelecypods, gastropods, ammonites, belemnites, and echinoids are found. The thickness of the group is less than 300 m. Therefore, the *Orbitolina* limestone in the Lower Yezo Group, characterized by reef-forming pachyodonts, may have been derived from a very shallow marine environment related to the contemporary Miyako Group.

The Middle Yezo Group.—The Middle Yezo Group rests conformably on the Lower Yezo Group but overlies the lower group unconformably in some restricted areas. The entire thickness of the Middle Yezo Group is 2,000 to 3,000 m.

The lower half of the group (middle to upper Albian) is represented by a sandy to shaly flysch sequence, in which paleocurrents from the north-northwest prevailed. Calcareous nodules, locally contaning ammonites, are frequently found in the shaly sequences. In the northern part (Haboro area; H of fig. 2), wedges of cross-bedded sandstones associated with breccias and conglomerates are intercalated in normal and shaly flysch sequences. Interbedded pyroclastic breccias and pebbly mudstone are also found.

The upper half of the Middle Yezo Group (Cenomanian and Turonian) is variable in lithology. The western marginal facies of the group is represented by the Mikasa Formation, which displays its typical characteristics in the Ikushumbetsu and Yubari areas of central Hokkaido (I and Y of fig. 2). The Mikasa Formation shows lithofacies changes from west to east: In the west, conglomerate and very coarse grained calcareous sandstone predominate. *Trigonia, Ostrea,* and other shallow-marine, thick-shelled bivalves are abundant, and ammonites are found in some parts. Furthermore, thin layers of coaly shale and red beds are intercalated at lower horizons. Cross stratification and scour-and-fill structures are common features. Eastward, the Mikasa Formation tends to increase in finer sediments such as silty fine-grained sandstone and mudstone, which contain abundant ammonites and *Inoceramus.* Farther eastward, the Cenomanian and Turonian strata, equivalent of the Mikasa Formation, are characterized by mudstone. The thickness of the Mikasa Formation and its equivalent ranges from 200 m in the west to 1,600 m in the east.

Further detailed sedimentological information about the Mikasa Formation can be found in Okada (1965).

In the Turonian sequence in the central part of the geosyncline, mudstones are generally interbedded with graded sandstones and pebbly mudstones, resulting in a flysch unit locally called the Saku Formation. This formation is typically present in the northern part of the axial zone of Hokkaido (S of fig. 2).

At some horizons in the Middle Yezo Group, felsic tuff or tuffite layers are intercalated.

The Upper Yezo Group.—This group overlies conformably the Middle Yezo Group, attaining 400 to 1,500 m in thickness. The strata as a whole consist of homogeneous claystone and siltstone, containing a number of calcareous nodules rich in ammonites and *Inoceramus.* This sequence represents a major transgressive phase (Senonian) in the Cretaceous Period that generally is called the Urakawan Transgression (Matsumoto, 1963, p. 112), which represents the maximum inundation phase of Santonian age.

A marginal facies of the Upper Yezo Group is known in the Haboro area (fig. 2), where sandstone beds of varying thicknesses are inserted in Coniacian to Santonian sequences, forming a coarsening-upward sequence (Okada and Matsumoto, 1969). Such a sequence was deposited only during the transition from the Santonian transgressive phase to the Maastrichtian regressive phase. These coarse clastic wedges protrude from north to south. Even in the distal part of the group, where a flysch sequence containing pebbly mudstone also is present, the above-mentioned relationships are still observable. Another feature consists of the very thin layers of acidic to intermediate tuff that are commonly intercalated in the Upper Yezo Group. In the central part of Hokkaido (Ashibetsu area; A of fig. 2), certain volcaniclastic beds become as thick as 250 m (locally called the Tsukimi Member).

The Hakobuchi Group.—The Hakobuchi Group, 450 to 900 m thick, is conformable with the underlying Upper Yezo Group. The transitional sequence between these two groups is characterized by coarsening-upward sedimentation as stated above. The Hakobuchi Group is composed almost wholly of coarse-grained volcanic sandstones having common intercalations of conglomerates. The whole succession is divisible into at least five cyclic units, each of which, beginning with glauconitic beds, is characterized in its lower part by marine sediments containing *Inoceramus* and other pelecypods, brachiopods, lobsters, etc., and in its upper part by on-delta and delta-front deposits bearing thin coal-seams and plant remains. Acidic to intermediate tuff layers are intercalated at some horizons.

Lateral changes of lithology and thickness of the Hakobuchi Group are well shown in the Nakatombetsu area, northernmost part of the axial zone of Hokkaido (N of fig. 2), where the strata tend to become finer in grain size and much thicker eastward (about 3,000 m).

Small Kuril Geosyncline

The new term Small Kuril Geosyncline is applied by the writer to the depositional basin of the latest Cretaceous and earliest Tertiary sequence excellently exposed in the Kushiro and Nemuro areas in the southeastern extremity of Hokkaido (fig. 2). The same sequence is also found on the Small Kuril Islands (see fig. 2; Sasa, 1934; Gnibidenko, 1971). The main part of the geosynclinal stage consists as a whole of the 3,000 m-thick Nemuro Group of Campanian to Paleocene age (fig. 3). The Nemuro Group is subdivided into five conformable units (N_0 to N_4 in ascending order) according to Matsumoto (1963, pp. 112–113).

N_0.—This unit consists mainly of andesitic volcanic breccia and conglomerate, tuff, and basaltic and doleritic flows of alkalic nature and of lesser amounts of mafic volcanic sandstone interbedded with siliceous shale, in which *Inoceramus schmidti* is found. The thickness of this unit is about 500 m.

N_1.—The second unit is characterized by black siliceous shale and some tuff and mafic volcanic sandstone and is intruded by pillowed sheets of alkalic basalt and dolerite. *Inoceramus* is sporadically found in shale. The thickness of this unit is about 700 m.

N_2.—The third unit, 300 to 700 m thick, is composed of thick-bedded mafic volcanic sandstone and siliceous shale. At some horizons alkalic mafic-rock flows exhibiting pillow structures are found. *Inoceramus* and some other marine molluscs are occasionally found.

N_3.—This unit, about 2,000 m thick, is subdivisible into two sequences. The lower sequence is characterized by massive claystone with or without thin sandstone beds, and the upper sequence is composed of normal and sandy flysch. Calcareous nodules containing ammonites and *Inoceramus* are common in the lower sequence. Pebbly mudstone and large-scale contorted structures are characteristic of the flysch sequence.

N_4.—The top unit is represented by conglomerate with intercalations of cross-laminated fine-grained sandstone. Boulders and cobbles of andesite, dolerite, porphyrite, granite, and hornfels are common in the conglomerate. Marine molluscs other than ammonites and *Inoceramus* are found in sandstone. The thickness of this unit is 250 to 300 m.

The geological ages of these units are: N_0, Campanian; N_1 to N_3, Maastrichtian; and N_4, Paleocene.

PETROLOGICAL FEATURES OF THE GEOSYNCLINAL CLASTIC SEDIMENTS

Yezo Geosyncline

Detailed petrological analyses of clastic rocks in the Yezo Geosyncline were performed by Fuiii (1958), Okada and others (1964), Okada (1965), Okada and Nakao (1968), and Matsumoto and Okada (1971). Their data and additional information permits petrological features of clastic rocks of the Yezo Geosyncline to be summarized as follows:

The Sorachi Group (Jurassic to Neocomian).—Sandstones of the Sorachi Group have not been systematically studied. But some samples show that they are characterized by volcaniclastic and basaltic origin. They contain clinopyroxene and small amounts of zircon, garnet, hornblende, and tourmaline. These clastics may have been derived from intrabasinal volcanic swells or uplands.

The Lower Yezo Group (Aptian to middle Albian).—The sandstone of this group contains fragments of such rocks as spilite, diabase, andesite, granite, porphyry, mylonite, radiolarian chert, glass, mudstone, arenite, and wacke. Particularly the Tomitoi Sandstone mentioned above is characterized by abundant granitic derivatives, of which microcline and perthite are diagnostic minerals. Also notably, quartz-schist fragments and alkali amphibole are rarely seen in the Tomitoi Sandstone. The provenance of granitic as well as schistose materials may be traced to terranes probably exposed southwest of the Yezo Geosyncline. The Lower Yezo sequence as a whole is represented by turbidites. In this regard, small masses of *Orbitolina* limestone found in flysch strata may be slumped deposits from nearby shallow-sea environments.

Moreover, either graded or evenly laminated beds of plagioclase arenite are intercalated at some horizons. Volcanic plagioclase of composition An_{21-39} constitutes more than 75 percent of the arenite (Okada and Nakao, 1968).

The Middle Yezo Group (upper Albian to Turonian).—Calcareous lithic arenite is dominant in the Middle Yezo Group, which consists of clasts of older sedimentary rocks, granite, diorite, andesite, porphyries, and small amounts of mafic volcanic rocks. Among them, sedimentary rock clasts, such as chert, arenite, and slate, are more prevalent than others. The Middle Yezo clastics are also characterized by a suite of the heavy minerals zircon, garnet, epidote, and augite.

The Mikasa Formation (Cenomanian to Turonian), representing a main part of the Middle Yezo Group, shows distinct lateral variations in composition and texture of sandstones. Sorting and roundness of sand grains are much better in the western margin of the basin than in the eastern, axial part. Heavy minerals of the Mikasa Formation are characterized by a suite of garnet and zircon in the western part and by epidote and augite in addition to garnet and zircon in the eastern part.

The Saku Formation (Turonian), protruding from the north or northwest to the south, is also characterized by calcareous lithic arenite. Lithic clasts of sandstones comprise andesite, granite, diabase, spilites, porphyries, tuff, radiolarian chert, wacke, and arenite, the last three of which are most important. Fresh microcline, myrmekite, and andesine are also common minerals. Euhedral and rounded zircons, garnet, tourmaline, rutile, augite, hypersthene, apatite, chlorite, etc., are found as heavy minerals. Lateral variation in composition from the proximal to the distal part of the basin is almost negligible.

The Upper Yezo Group (upper Turonian to lower Campanian).—The Upper Yezo Group is composed mainly of mudstone as mentioned in the foregoing chapter. Some interbedded sandstones are more or less glauconitic and are characterized not only by sedimentary rock clasts such as chert, wacke, arenite, marl, siltstone, and hornfels, but also by fragments of granite and volcanic rocks such as rhyolite, porphyry, pumice, and spilite. Thus, sandstones are mainly lithic arenites cemented with sparry calcite.

An uncommonly thick volcaniclastic unit called the Tsukimi Member (Coniacian to Santonian) in the distal, muddy facies is characterized by plagioclase arenite containing abundant biotite and augite that probably were derived from nearby plagiorhyolite bodies. The Tsukimi plagioarenite is regarded mostly as turbidite.

The Hakobuchi Group (Campanian to Maastrichtian).—The sandstone of this group is exclusively of felsic-to-intermediate volcaniclastic arenite. Thus, sandstones consist of plagioclase arenite and volcanolithic arenite, both of which are well sorted and cemented with devitrified glass, chlorite, and (or) calcite. Sand grains consist mainly of corroded β-quartz, oligoclase-to-andesine plagioclases, andesite, porphyry, rhyolite, spherulite, radiolarian chert, sandstones, and so on, of which andesite clasts are dominant. Microcline and myrmekite are rarely found. Heavy minerals in the Hakobuchi sandstones are characterized by an augite and hornblende assemblage, which shows slight lateral variation in composition. Particularly interesting is the fact that the Hakobuchi sandstones in the southern part of Hokkaido (Tomiuchi area; T of fig. 2) are locally intercalated with thin iron beds which are made up of more than 75 percent of ilmenite grains.

The contemporary sandstones in the Hidaka area in the eastern part of the basin (D of fig. 2) are generally similar also to the Hakobuchi sandstones, although the Hidaka sandstone has a little more rhyolite and granitic material, as well as sedimentary fragments, than does the Hakobuchi.

To sum up, Cretaceous sandstones in the Yezo Geosyncline exhibit these vertical compositional changes: those of pre-Aptian age are characterized by mafic to intermediate volcaniclastics, those in the lower sequence (Aptian to Albian) by granitic fragments, those in the middle sequence (late Albian to early Campanian) by sedimentary clasts, and those in the upper sequence by andesitic to rhyolitic volcaniclastics (fig. 4A).

A feature of special note is the presence of many interbeds of plagioclase arenite throughout the whole Cretaceous sequence, not only in flysch facies, as in the Lower and Middle Yezo Groups, but also in nonflysch facies, as in the Upper Yezo and Hakobuchi Groups. They become thick enough in some places to form a stratigraphic unit like the Tsukimi Member.

Small Kuril Geosyncline

Coarse clastics in the Small Kuril Geosyncline of Late Cretaceous to earliest Tertiary age are generally characterized by mafic to intermediate volcaniclastics. Sandstones are subdivisible into two petrological groups (figs. 4B, 5): one group characterizes the lower sequence (N_0, N_1, and N_2) of the Nemuro Group, and the other is characteristic of the upper sequence (N_3 and N_4).

N_0–N_2 sandstones (Campanian to Maastrichtian).—Sandstones of the lower sequence of the Nemuro Group are composed wholly of mafic to intermediate volcaniclastics. They consist of basaltic, diabasic, and andesitic rocks showing intersertal texture and of glass and pumice fragments. Nonopaque heavy minerals include augite, aegirine, hypersthene, amphiboles (including a titaniferous kind), epidote, and apatite, among which pyroxene and amphibole assemblages constitute 98 to 100 percent (fig. 5). Quartz is absent (fig. 4B). The framework constituents are cemented by chlorite, calcite, devitrified glass, and zeolite (chiefly analcite?). Furthermore, plagioarenite beds are also present. The composition of plagioclase grains is andesine to labradorite.

P/T-metamorphic rocks might hardly have been formed in the trench belt, although still another possibility is that a supposed greenschist belt on Kamachatka might extend somewhere into the Small Kuril arc belt (see Dobretsov and others, 1966).

Finally, the youngest arc-trench system, namely the modern Kuril arc and trench, was developed adjacent to the oceanic side of the preceding Small Kuril Geosyncline, probably at the time of the so-called Green-Tuff Orogeny or Mizuho Orogeny strictly defined by Sugimura and others (1963) (late Miocene to Quaternary). That is to say, the Small Kuril Geosyncline belt was eventually uplifted to form the youngest arc belt. Such consecutive migration of ancient arc-trench systems was also observed by Avediko (1970) on Kamchatka.

REFERENCES

AVDEIKO, G. P., 1970, Evolution of geosynclines on Kamchatka: Pacific Geology, v. 3, p. 1–13.
BIKERMAN, MICHAEL, MINATO, MASAO, AND HUNAHASHI, MITSUO, 1971, K-Ar age of the garnet amphibolite of the Mitsuishi district, Hidaka Province, Hokkaido, Japan: Assoc. Geol. Collabor. Japan Jour., v. 25, p. 27–30.
DICKINSON, W. R., 1971, Clastic sedimentary sequences deposited in shelf, slope, and trough settings between magmatic arcs and associated trenches: Pacific Geology, v. 3, p. 15–30.
DOBRETSOV, N. L., AND OTHERS, 1966, The map of metamorphic facies of the USSR: Ministerstva Geologii SSSR, Moskva.
ERNST, W. G., 1971, Metamorphic zonations on presumably subducted lithospheric plates from Japan, California and the Alps: Contr. Mineralogy and Petrology, v. 34, p. 43–59.
FUJII, KOJI, 1958, Petrography of the Cretaceous sandstones of Hokkaido, Japan: Kyushu Univ., Fac. Sci. Mem., ser. D, Geology, v. 6, p. 129–152.
GNIBIDENKO, H. S., 1971, Geology and deep structure of Sakhalin, Kuril Islands and Kamchatka; in ASANO, SHUZO, AND UDINTSEV, G. B. (eds.), Island arc and marginal sea: Tokyo, Tokai Univ. Press, 1st Japan-USSR Symposium Solid Earth Sci. Proc., p. 5–16.
HORIKOSHI, EI, 1972, Orogenic belts and plates in Japanese Islands: Kagaku, v. 42, p. 665–673.
KAWANO, YOSHINORI, AND UEDA, YOSHIO, 1967a, K-Ar dating on the igneous rocks in Japan (VI)—Granitic rocks, summary—: Japanese Assoc. Mineralogy, Petrology and Econ. Geology Jour., v. 57, p. 177–187.
———, AND ———, 1967b, Periods of the igneous activities of the granitic rocks in Japan by K-A dating method: Tectonophysics, v. 4, p. 523–530.
MATSUDA, TOKIHIKO, AND UYEDA, SEIYA, 1971, On the Pacific-type orogeny and its model—extension of the paired belts concept and possible origin of marginal seas: ibid., v. 11, p. 5–27.
MATSUMOTO, TATSURO, 1943, Fundamentals in the Cretaceous stratigraphy of Japan. Parts II and III: Kyushu Univ., Fac. Sci. Mem., ser. D, Geology, v. 2, p. 97–237.
———, 1954, The Cretaceous System in the Japanese Islands: Tokyo, Japan Soc. Promotion Sci. Research, 324 p.
———, 1963, The Cretaceous, in TAKAI, FUYUJI, MATSUMOTO, TATSURO, AND TORIYAMA, RYUZO (eds.), Geology of Japan: Tokyo, Univ. Tokyo Press, p. 99–128.
———, AND OKADA, HAKUYU, 1971, Clastic sediments of the Cretaceous Yezo Geosyncline: Geol. Soc. Japan Mem. 6, p. 61–74.
MITCHELL, A. H., AND READING, H. G., 1971, Evolution of island arcs: Jour. Geology, v. 79, p. 253–284.
MIYASHIRO, AKIHO, 1961, Evolution of metamorphic belts: Jour. Petrology, v. 2, p. 277–311.
———, 1967, Orogeny, regional metamorphism, and magmatism in the Japanese Islands: Dansk Geol. Foren. Meddel., v. 17, p. 390–446.
OKADA, HAKUYU, 1965, Sedimentology of the Cretaceous Mikasa Formation: Kyushu Univ., Fac. Sci. Mem., ser. D, Geology, v. 16, p. 81–111.
———, 1971, Classification of sandstone: Analysis and proposal: Jour. Geology, v. 79, p. 509–525.
———, HARADA, MASATO, AND MATSUMOTO, TATSURO, 1964, Petrographic notes on the Cretaceous sandstones of the Yubari Dome, in MATSUMOTO, TATSURO, AND HARADA, MASATO, Cretaceous stratigraphy of the Yubari Dome, Hokkaido: Kyushu Univ., Fac. Sci. Mem., ser. D, Geology, v. 15, p. 79–115.
———, AND MATSUMOTO, TATSURO, 1969, Cyclic sedimentation in a part of the Cretaceous sequence of the Yezo Geosyncline, Hokkaido: Geol. Soc. Japan Jour., v. 75, p. 311–328.
———, AND NAKAO, SEIZO, 1968, Plagioclase arenite in Lower Cretaceous flysch in the Furano area, Hokkaido: Jour. Geol. Soc. Japan, v. 74, p. 451–452.
SASA, YASUO, 1934, A preliminary note on the geology of the Island of Shikotan, southern Tisima (South Kuril Islands): 5th Pacific Sci. Cong. Proc., v. 2, p. 2479–2482.
SUGIMURA, ARATA, 1960, Zonal arrangement of some geophysical and petrological features in Japan and its environs: Univ. Tokyo, Fac. Sci. Jour., sec. 2, v. 12, p. 133–153.
———, AND OTHERS, 1963, Quantitative distribution of late Cenozoic volcanic materials in Japan: Bull. Volcanology, v. 26, p. 125–140.
SUGISAKI, RYUICHI, 1972, Tectonic aspects of andesite line: Nature, v. 240, p. 109–111.
TSUBOI, CHUJI, 1954, Gravity survey along the lines of precise levels throughout Japan by means of a Worden gravimeter. Part 4, Map of Bouguer anomaly distribution in Japan based on approximately 4500 measurements: Earthquake Research Inst. Bull., Special vol. 4, p. 125–127.
UEDA YOSHIO, AND AOKI, KEN-ICHIRO, 1968, K-Ar dating on the alkaline rocks from Nemuro, Hokkaido: Japanese Assoc. Mineralogy, Petrology and Econ. Geology Jour., v. 59, p. 230–235.
YAGI, KENZO, 1969, Petrology of the alkalic dolerites of the Nemuro Peninsula, Japan: Geol. Soc. America Mem. 115, p. 103–147.

MELANGES AND THEIR DISTINCTION FROM OLISTOSTROMES

K. J. HSÜ
Geologisches Institut, ETH Zürich, Switzerland

ABSTRACT

Mélanges were once considered gravity-slide deposits, but many current workers link their genesis to subduction along convergent plate margins. Franciscan mélanges are the best known examples. They are impressive because of the scale of chaotic deformation. The pervasively sheared mass of ophiolites, cherts, and graywackes in a dominantly pelitic matrix bears witness to an oceanward migration of a late Mesozoic Benioff zone between oceanic and North American plates. Similar mélanges have been reported from other circum-Pacific mountains, but their absence in the central Andean province is noteworthy. There the distribution of igneous rocks suggests a landward migration of the Benioff zone. Mélanges are present in the Swiss Alps, but the Alpine mélanges are dwarfs compared to the Franciscan giants. Andesitic activity, the twin brother of mélange, so to speak, was also inconspicuous during the Alpine orogenesis. Those facts might be construed as evidence that the magnitude of plate consumption in the Alps was modest compared to that along the eastern Pacific margin.

Mélanges are tectonic units bounded by shear surfaces. In contrast, olistostromes are stratigraphic units, generally separated from overlying and underlying formations by depositional contacts. Regional mapping is one of the tools for distinction. In a mélange terrane, one can observe different degrees of severity of fragmentation and mixing, grading from internally intact allochthonous slabs to broken formations, and to pervasively sheared and intimately mixed mélanges. In an olistostrome terrane, one finds gradations from boulder beds to graded turbidites. Mélanges are massive and they developed on a regional scale, although thin mélanges occupying certain tectonic horizons are not uncommon. Olistostromes, as sedimentary units, are commonly limited in size and could be placed in a sedimentary basin within a simple paleogeographic framework. Mélanges are deformed under an overburden, so that the pelitic materials flow and other rocks fracture. Partially broken blocks are commonly bounded by shear fractures, although tectonic transport may ultimately lead to rounding. Olistostrome blocks are commonly sedimentary boulders, already rounded prior to sedimentary transport. Aside from the common pelitic matrix, sandy matrix is present locally in some olistostromes.

Pervasively sheared olistostromes are practically indistinguishable from mélanges. For example, the *argille scagliose* of the Apennines include both mélanges and olistostromes. The olistostrome aggregates there reached proportions comparable to those of the Franciscan. However local specialists have been able to recognize in those units individual olistostrome beds produced by separate events, and those beds are rarely more than 100 m thick. Pervasively sheared olistostromes are practically indistinguishable from mélanges. Where a genetic interpretation of a chaotically deformed deposit cannot be made, I recommend the use of the descriptive term "wildflysch."

INTRODUCTION

Sedimentary formations are commonly characterized by laterally persistent beds. This observation was made by Nicolas Steno in the 17th Century and was formulated as one of the three fundamental laws of sedimentary sequences. The law states (Gilluly and others, 1968, p. 92):

"A water-laid stratum, at the time it is formed, must continue laterally in all directions until it thins out as a result of non-deposition, or until it abuts against the edge of the original basin of deposition."

In some deformed terranes, one is impressed by the fact that few rock bodies can be traced to any large lateral extent. Rocks of different types lie jumbled in close juxtaposition. A classical example is the wildflysch in the Swiss Alps. The wildflysch consists of blocks of flysch sandstones and other rocks in a shaly matrix. Kaufmann (1886) assigned to descriptive prefix "wild" to those rock bodies, because of the wild, or undisciplined nature of the bedding in the wildflysch in contrast to the so-called "quiet" (*ruhig*) stratification of the type Flysch, which is characterized by a regular alternation of very evenly bedded graywackes and shales.

The origin of the wildflysch has been a subject of unending controversy during the last hundred years. Even today, my colleagues and I might stand on the same outcrop and not agree on its genesis. Several reviews have been published (e.g., Cadisch, 1953, p. 182–186; Badoux, 1967). The dispute has been focused on the mode of emplacement of the blocks in the wildflysch. One school considered the blocks to be sedimentary, being mixed in a muddy matrix during submarine slumping, sliding, or during some other sedimentary events (Schardt, 1898; Lugeon, 1916). Another school pointed out that the wildflysch also included tectonic blocks that were picked up and modified during some tectonic event, such as folding, thrusting, or pervasive shearing (e.g., Adrian, 1915; Häfner, 1924). As a matter of fact, what has been called wildflysch includes both sedimentary olistostromes and tectonic mélanges. Furthermore, pervasively sheared olistostromes might appear identical to a mélange, and the two might be indistinguishable in the field. If a wildflysch deposit is to be considered as a sedimentary formation, the law of original continuity applies. On the other hand, if a body of wildflysch is a tectonic unit, the principles of mélanges must be invoked (see Hsü, 1968). The two alternative interpretations for a given wildflysch unit could thus lead to completely different geological conclusions. Thus, it is important to attempt to distinguish them. This fact did not escape Marshall Kay (1970) when he pondered if his Newfoundland wildflysch should be considered as a slump breccia (e.g., Patrick, 1956) or a tectonic mélange (e.g., Dewey, 1969). On the occasion of the symposium honoring Kay, I was invited by the convener to outline some characteristic features of mélanges, to discuss their significance, and to present some criteria for differentiating mélanges from olistostromes.

NATURE OF MÉLANGES

Mélanges are bodies of deformed rocks characterized by the inclusion of tectonically mixed fragments or blocks, which may range up to several miles long, in a pervasively sheared matrix (Greenly, 1919; Hsü, 1968).[1] Each mélange includes both exotic and native blocks and a matrix. Native blocks are disrupted brittle layers, which were once interbedded with the ductilely deformed matrix. Exotic blocks are tectonic inclusions detached from some rock-stratigraphic units foreign to the main body of the mélange. A body of pervasively sheared strata that contains no exotic elements may be called a broken formation, because such a body, regardless of its broken state, functions as a rock-stratigraphic unit.

To make a mélange involves two processes: fragmentation and mixing of broken fragments in a ductilely deformed matrix. A broken formation resulting from fragmentation, but no mixing of exotic elements, may be considered as a preparatory stage in the genesis of a mélange.

The broken formations intercalated within the Franciscan mélanges of the Santa Lucia Range, California, provide some of the best illustrations of progressive fragmentation. Those formations

[1] Editors' note: Some workers urge use of "mélange" as a nongenetic term for any chaotic deposit; they would invoke genetic modifiers, for example, "tectonic" or "sedimentary" mélange. In such usage, "mélange" is synonymous with Hsü's definition of "wildflysch." (See also papers in this volume by Blake and Jones, Wood, and Kay.)

are interbedded graywackes and shales. Graywackes appear to be the more brittle and tend to fracture, whereas the shales have been deformed by flowage. Fragmentation commonly occurred along shear fractures, which formed under compression subparallel to the bedding planes. Thus, they may be recognized as miniature thrust faults.

In contrast, fragmentation resulting from extension parallel to the bedding surface resulted in shear fractures inclined at a high angle (60° or more) to the bedding plane (fig. 1). In the field, one might recognize examples representing progressive stages of separation: starting from shear joints, to miniature-grabens, to lozenge-shaped boudins (name given by Rast, 1956), and to completely separated, isolated, and rotated fragments. Typical boudins are bounded by extensional fractures, normal to the bedding. Yet the boudins in the Franciscan mélanges are almost exclusively bounded by extensional shear fractures. I did observe, however, typical boudins in the *argille scagliose* of the Apennines. The occurrence of extensional fracture is indicative of brittle, or very brittle behavior, which in turn attests to the very low effective confining pressure during the time of fragmentation.

Broken formations commonly show evidence of having been subjected to alternate tension and compression parallel to bedding surfaces. Lenticular masses bounded by compressional shears are often internally traversed by extensional shear joints (fig. 2).

Fragmentation *per se* is not sufficient proof of brittle deformation, as brittleness is a relative measure of deformation behavior. A material is regarded as very brittle, brittle, moderately brittle, moderately ductile, or ductile, if the total strain before the fracturing is less than 1, 1 to 5, 2 to 8, 5 to 10, and more than 10 percent respectively (Handin, 1966, p. 226). Not only brittle rocks fracture, for rocks undergoing ductile deformation may eventually fracture after the total strain has exceeded a certain undefined limit. The occurrence of phacoids and boudins is a measure of the relative ductility between the fragmented layers and the host matrix.

The fragments of a mélange are autoclasts. Autoclasts bounded by compressional shear surfaces have commonly been modified to produce phacoids (Greenly, 1919; Cadisch, 1953, p. 185). Autoclasts produced by stretching were originally boudins. Many, if not most of the autoclasts no longer betray their tectonic heritage. Ductile stretching of phacoids produced tails. Rotation of lozenges led to tectonic rounding. Ultimately, the autoclasts may assume such ir-

FIG. 1.—Lozenge-shaped boudin in a broken formation, Franciscan in age.

FIG. 2.—Boudins traversed by extensional shear fractures and bounded by compressional shears, indicative of alternate extension and compression, Franciscan in age.

FIG. 3.—Partially rounded exotics in mélange, Franciscan in age.

regular shapes of rounded outlines that they may become indistinguishable from sedimentary boulders in an olistostrome (fig. 3).

The matrix of the mélange is characteristically a material that could undergo very large permanent deformation without fracturing. The shales were apparently more ductile than graywackes or limestones under the conditions of mélange genesis. The matrix of mélanges is most commonly a pelitic rock subjected to different degrees of regional metamorphism. The matrix of the Franciscan mélanges in the Santa Lucia Range is a shale or a slightly metamorphosed phyllite (Hsü, 1969); The Franciscan mélange matrix in the northern Coast Ranges has been locally altered to lawsonite schist or to metamorphic rocks of higher grade. Locally it contains a high proportion of sheared serpentinite. Rarely some other rock may form the mélange matrix. The famous Colored Mélange of central Iran (e.g., Gansser, 1955) has a serpentinite matrix (Davoudzadeh, 1969). In metamorphic terranes, tectonic fragments of calc-silicate rocks or of dolomite are found in marble matrix. In rare instances, the matrix of a mélange is gypsum (Leine and Egeler, 1962).

TECTONIC SIGNIFICANCE OF MÉLANGES

The fragmented nature of a mélange testifies to the fact that such a body is incapable of transmitting stress. This consideration played a critical role in the interpretation of the *argille scagliose* of the Apennines. Obviously, those scaly shales are too weak to have been moved as a sheet by a push on one side (Page, 1963, p. 669). One school of thought considers the *argille scagliose* as an olistostrome, or a "gigantic submarine gravity slide" (Maxwell, 1959b, p. 2711; written communication, 1972). Yet, the *alberesi* blocks in some of the *argille scagliose* are "mechanically derived," as Page (1963, p. 664) demonstrated. My own field observations in the Apennines convinced me that a large part of this allochthonous complex is a tectonic mélange, although olistostrome beds are present as allochthonous slabs.

If a mélange has been displaced as a body of jumbled blocks, an interpretation of tectonic gravity sliding seems to be the only possible mechanism. Page (1963) reached this conclusion for the *argille scagliose*. I once also invoked this hypothesis to account for the genesis of the Franciscan mélanges (e.g., Hsü, 1967a; 1968; Hsü and Ohrbom, 1969). On the other hand, if we regard a mélange as a composite schuppen zone at the toe of a thrust plate, we no longer have to appeal to gravity as the prime driving force. Instead, allochthonous materials could have been transported for long distances as an integral part of a lithospheric plate. The plate was broken as it was descending along a subduction zone. We might picture a freight train carrying piles of lumber whistling its way into a tunnel with a low ceiling. The part of the lumber piled too high would be sheared off. The broken and splintered planks would accumulate as a mélange in front of the tunnel. Such a mélange would be formed *in situ,* although the lumber would have come a long way from its original supply depot.

In a paper analyzing thrusting mechanics, I have demonstrated that the toe effect of thrusting under compression is to produce a mélange having densely spaced shear planes if the thrust plate is displaced along a zone of flowage (Hsü, 1970). This mechanical analysis is in agreement with the postulate of plate-tectonics theory that interprets mélanges as products of pervasive shearing along consuming plate margins (e.g., Hamilton, 1969; Ernst, 1970; Dewey and Bird, 1970; Dickinson, 1970; Hsü, 1971a).

The Franciscan mélanges of the California Coast Ranges were very probably produced as a late Mesozoic subduction zone migrated oceanwards (e.g., Hamilton, 1969; Hsü, 1971a). The presence of a mélange could thus indicate the prior existence of a convergent plate margin. Its absence from some circum-Pacific mountains, such as the central Andes, is noteworthy. A probable explanation lies in the fact that the subduction zone there may have migrated landward (e.g., James, 1971); older mélanges went down with the descending plate into the mantle. I have also pointed out that the Alpine Wildflysch is volumetrically insignificant compared to the Franciscan mélanges. In the Alps the cover thrusts constitute coherent nappes, and the deformed sediments between thrust planes have largely retained their stratal continuity. The subordinate role of mélange deformation might be considered as evidence that plate subduction was far less important as a mechanism in the Alps than in California (Hsü, 1972).

In contrast to the largely underthrusting movement of the oceanic plates under the North American plate, the plate movement between Africa and Europe, which was responsible for the genesis of the Alps and the Mediterranean, had a large lateral component (e.g. Smith, 1971; Hsü, 1971b; Pitman and Talwani, 1972).

OLISTOSTROMES

The name "olistostrome" was introduced by Flores in 1955 during the discussion of a paper presented by E. Beneo on the oil geology of Sicily. His (Flores') definition is quoted as follows (p. 122):

"By olistostromes, we define those sedimentary deposits occurring within normal geological sequences that are sufficiently continuous to be mappable, and that are characterized by lithologically or petrographically heterogenous material, more or less intimately admixed, that were accumulated as a semifluid body. They show no true bedding, except for possible large inclusions of previously bedded material. In any olistostrome we distinguish a 'binder' or 'matrix' represented by prevalently pelitic, heterogenous material containing dispersed bodies of harder rocks. The latter may range in size from pebble to boulder and up to several cubic km."

The name "olistolith" was applied to the masses included as individual elements within the binder. These masses had been previously referred to as "exotics" or "erratics."

There is no doubt that Flores intended the term to designate sedimentary deposits. In 1959, Flores referred to olistostromes as chaotic accumulations, "vertically delimited by underlying and overlying series of normal marine sediments, . . . that contain fossils in situ of identifiable age and environment" (p. 261). He added, "the evidence collected indicates that the olistostrome emplacement must have occurred in an *aqueous medium,* as shown by the associated turbidity current and mudflow depositional phenomena observed" (p. 261).

The Tertiary strata of Sicily include both normally deposited sedimentary sequences and chaotic rocks. Beneo (1951) considered the whole of Sicily, with the exceptions of two small areas, as being one overthrust mass, resulting from gravity sliding. We see that Flores and later Marchetti agreed to the gravity-sliding origin of the chaotic deposits. However, they emphasized that the chaotic rocks constitute olistostromes, which are formations, or "regular and mappable units of the sedimentary series" (Marchetti, 1957, p. 220). Instead of one large mélange envisioned by Beneo, the other authors found in Sicily a normal sedimentary series including a few chaotic intercalations.

From the sketches and descriptions provided by Flores and Marchetti, we might attempt to delineate the difference between a conceptual olistostrome and a conceptual mélange (table 1).

Olistostromes do exist in the Apennines, as I was shown during the 1964 American Geological Institute field trip to Italy. Those deposits were illustrated by some superb photographs by Abbate and others (1970). Identical or similar deposits have been called pebbly mudstones, tilloids, fluxoturbidites, or simply slides or slump beds. Badoux, (1967, p. 403) suggested that the olistostromes contain exotic blocks, whereas the submarine slides have a matrix that encloses blocks of the same age. On the other hand, the Italian school prefers to call both olistostromes

TABLE 1.—CONCEPTUAL DIFFERENCE BETWEEN MELANGES AND OLISTOSTROMES

Item	Mélange	Olistostrome
Is underlain by	a formation not pervasively sheared	a formation slightly older than the olistostrome
Is overlain by	a formation not pervasively sheared	a formation slightly younger than the olistostrome
At time of chaotic emplacement	tectonic emplacement is under overburden of hundreds or thousands of meters	Sedimentary emplacement of upper surface of olistostrome was then subaerial or submarine
Site of chaotic emplacement	is a major shear zone, in some cases the consuming plate margin	is a major submarine topographic depression as a suitable receptacle for a large olistostrome
Duration of chaotic emplacement of one unit	is duration of shearing movement, which may have been short or long	is geologically very short, such as for a sedimentary layer

(e.g., Elter and Raggi, 1965).

Abbate and others (1970) modified the definition of Flores. However, they agreed on the essential principle that olistostromes are sedimentary bodies. Those authors "usually regarded as olistostromes chaotic units up to 100–200 m thick' (p. 524), and very large chaotic structures were excluded. The term was applied to designate several different kinds of coarse sedimentary deposits. Typically, they have subrounded cobbles and pebbles embedded in a marly or shaly matrix (fig. 4). These are similar to pebbly mudstones (Crowell, 1957). Others have isolated or scattered angular fragments in an argillite. These are comparable to tilloids (cf. Lindsay, 1966) or boulder-bearing shales (cf. Cline 1970). Some are simply calcareous breccias containing angular clasts and little binding material. Still others consist of subangular or subrounded cobbles and boulders cemented by muddy sands (fig. 5).

Görler and Reutter (1968, figs. 1 and 2) proposed to restrict the term to those coarse deposits having an argillaceous matrix and suggested viscous flow as the mechanism of transport to distinguish olistostrome from other coarse sedimentary deposits. The authors also recognized that olistostromes of the Apennines are commonly composite deposits, each consisting of many individual olistostrome beds. Thus, a unit might be more than 1 km thick, but each olistostrome bed only a few meters thick. The attempt to characterize olistostromes by their postulated transport mechanism is not satisfactory, especially in view of the fact that Dott (1963) had emphasized the plastic behavior of olistostrome transport (by subqueous sliding) in contrast to the viscous behavior of turbidity flows.

Despite the somewhat varied usages and interpretations, "olistostrome" is defined as a sedimentary deposit. Such chaotic masses move downslope under gravity, are emplaced as a sedimentary stratum of finite dimensions, and are intercalated between other, more normal sedimentary units. The different types of olistostromes illustrated by Abbate and others and by Görler and Reutter (1968) are indisputably sedimentary deposits. Their distinction from a mélange should constitute no problem. Yet the *argille scagliose* includes not only olistostromes; also present in this chaotic complex are large allochthonous slabs as much as hundreds of cubic kilometers in volume (see Maxwell, 1959b; Hsü 1967b). Those large allochthonous sheets were commonly not considered olistoliths, and "the movement of the sheets . . . is generally attributed to a tectonic process" (Abbate and others, 1970, p. 526). Görler and Reutter (1968) also attempted to discriminate a large olistolith from a gliding thrust sheet on the basis of the thickness of the slab, those thicker than 1 km being called thrust sheets.

Is a tectonic process to be distinguished from a sedimentary process solely on the basis of a geometric scale? Is there a fundamental genetic difference? If so, what are the criteria for distinction? Those are the questions I explore in the following sections.

NEED FOR DISTINCTION

It may not seem particularly necessary to make a great distinction between olistostromes and mélanges inasmuch as both have been referred to as the product of gravitysliding. Abbate and others (1970, p. 524) commented on this problem:

"In the common geological practice, tectonics and

FIG. 4.—Olistostrome in *argille scagliose* resembling pebbly mudstone (after Abbate and others, 1970).

FIG. 5.—Olistostrome in *argille scagliose* having sandy matrix.

sedimentation are considered two classes of distinct phenomena. When slides are involved, however, no sharp boundary can be drawn between the two: a small olistostrome is unquestionably a sedimentary feature and the emplacement of a large, even chaotic nappe is commonly considered a tectonic process. Yet there exists a continuous transition between these extreme cases."

They chose to distinguish the two primarily on the basis of size. However, this criterion is arbitrary. Some geologists have made no distinctions at all and referred to tectonic mélanges as olistostromes (e.g., Rigo di Rhigi and Cortesini, 1964). For paleogeographical reconstructions, mélanges have to be distinguished from olistostromes. Consider the zone of carbonate blocks under the Vermont high Taconic sequence as an example. If those blocks are olistoliths, we could interpret the high taconic material as a normally superposed sequence above the autochton, a hypothesis favored by one school of geologists (e.g., McFayden, 1956). If, however, those blocks are exotics in a mélange, the high Taconic sequence must be allochthonous (Zen and Ratcliff, 1966).

The wildflysch problem provides another illustration. Wildflysch rocks were originally thought to be sedimentary breccias or olistostromes (e.g., Kaufmann, 1886). The discovery of the exotic nature of some blocks argued against a normal stratigraphic succession (Schardt, 1898; Beck, 1908). Micropaleontological research proved that the wildflysch of central Switzerland is thrust under older flysch formations (e.g., Boussac, 1912; Leupold, 1942). Meanwhile, the origin of the exotics continued to be a favorite subject of dispute (see Trümpy, 1960, p. 888–890). One school of thought traced their origin to an advancing nappe front (Schardt, 1898). The opposition invoked cordillera in the Ultrahelvetic realm itself (e.g., Lugeon, 1916; Tercier, 1928; Leupold, 1942; Hsü, 1960). The question has been particularly puzzling concerning the source of a pelagic limestone, known locally as the Leimernkalk, that occurs as phacoids in a wildflysch deposit. The phacoids range from Turonian to Paleocene in age and are in a matrix that yielded late Eocene (Priabonian) fossils. The classical approach assumed the existence of a late Eocene wildflysch trench fringed by a coastal range of Leimernkalk terranes. Because the Leimern phacoids resemble the *couches rouges* of the Klippen Nappe, some Alpine geologists postulated that the front of the Klippen had advanced to the southern margin of the wildflysch trench to supply the phacoids (e.g., Gigon, 1952). Another school adopted the concept of an Ultrahelvetic cordillera and considered the resemblance of the *Leimern* and *couches rouges* coincidental. Their key argument was that the Klippen could not have come so far north during the late Eocene to dump exotic blocks into the wildflysch trench (e.g., Herb, 1966). My own field observations led me to the belief that the Leimern phacoids have been tectonically mixed with the Helvetic *Globigerina* Mergel to produce the wildflysch. If the wildflysch is a mélange, not an olistostrome, then we need not postulate the existence of an Eocene wildflysch trench in central Switzerland as the depositional site for the Leimern phacoids. Instead, this mélange was formed from tectonic mixing of rocks at the base of the Klippen Nappe with the Priabonian *Globigerina* Mergel at the top of the Helvetic sequence. The time of the mixing, or the arrival of the Klippen into the Helvetic realm, could be dated as post-Eocene, thus avoiding the necessity of assuming an overhasty itinerary for the Klippen travel.

It is not my intention to settle the controversial wildflysch problem in this article. I have only mentioned the various alternative interpretations to emphasize the need to distinguish a mélange from an olistostrome and to counter the arguments of some of my colleagues that such an effort is merely a semantic exercise.

CRITERIA FOR DISTINCTION

General statment.—Typical mélanges and typical olistostromes are so different that a glance at figures 1 through 5 would suffice to make a distinction. Yet, large and prevasively sheared olistostromes are all but indistinguishable from mélanges. I cannot offer any infallible rules but shall discuss several different lines of approach to the problem.

Shape of clasts.—Broken formations and mélanges can be distinguished from olistostromes by the shape of fragments. Broken strata exhibit progressive fragmentation through boudinage. Some mélange autoclasts are also bounded by fracture surfaces, especially by compressional and extensional shears. They may be modified by pervasive shearing and converted to phacoids having delicate tails trailing into the matrix. The occurrence of very rounded autoclasts in a mélange would suggest derivation from disintegration of an olistostrome (fig. 6).

Nature of clasts.—An overthrust, like a bulldozer, may pick up exotic blocks from an underlying tectonic element. The Taconic wildflysch blocks, for example, have been derived from the underlying autochthon (Zen and Ratcliff, 1966). The Alpine wildflysch may include blocks from both the underlying and the overlying tectonic

Fig. 6.—A pervasively sheared olistostrome in Franciscan rocks. Subangular fragment in middle is glaucophane schist.

elements (see Cadisch, 1953). In contrast, olistoliths are derived from an advancing thrust or gravity-slide mass, but not from a buried autochthon.

Blocks in a narrow mélange zone are commonly recognizable as having been detached from adjacent tectonic elements. Blocks that cannot be identified as such, like those of the Habkern Granite in the Alpine Wildflysch, are most probably olistoliths introduced by gravity deposition. This particular interpretation can be reinforced by the fact that the same granite furnished considerable debris to turbidite sandstones associated with the wildflysch (see Hsü, 1960).

Blocks in a broad mélange zone, such as the Franciscan, may also trace their origins to so-called "phantom-stratigraphical units," which had been broken and mixed to yield the mélange (see Hsü and Ohrbom, 1969). Exotic blocks completely foreign to other components in the mélange may have been derived from a disintegrated olistostrome. For example, only a few scattered blocks of glaucophane schist have been found in the Franciscan mélange of the Santa Lucia Range, California (fig. 6). They might have been introduced as blocks in Cretaceous olistostromes, which have since been broken up, leaving the originally sedimentary clasts of high-pressure metamorphics in an unmetamorphosed mélange matrix.

Blocks yielding fossils younger than the matrix fauna are in all probability mélange exotics. Not recognizing this simple rule has led to confusion, giving rise to the misnomer "Einsiedler Flysch." In Sihltal, near Einsiedeln, Switzerland, lenses of nummulitic limestones are apparently intercalated in a shaly formation, and the Einsiedler complex was mapped by Kaufmann (1886) as part of his Wildflysch. The nummulitids are typically early Tertiary image. Yet, Late Cretaceous faunas, including an ammonite specimen, were eventually found in the shales (Rollier, 1912; Heim, 1923). Making the implicit assumption that the blocks were sedimentary and that they could not carry a fauna younger than the matrix, Rollier (1923) and Heim (1923) questioned the stratigraphical value of the nummulitic faunas and spoke of Late Cretaceous *Nummulites*. Eventually, the geology was worked out and the Einsiedler Flysch was recognized as a tectonic mixture in a schuppen zone, where Tertiary limestone lenses were shoved into an underlying Upper Cretaceous shale.

Fabric of matrix.—The matrix of mélanges is pervasively sheared. The ductilely deformed materials have flowed into the irregular spaces between blocks that are arranged chaotically or in a subparallel fashion. In places the pelitic ma-

Fig. 7.—Olistostrome in *argille scagliose,* a serpentinite breccia containing ophiolite blocks in matrix of ophiolite debris (after Abbate and others, 1970).

terials were intruded into the cracks or fractures of disintegrating blocks. The orientation of shear surfaces tends to be parallel to edges of the blocks. Mélanges containing irregular blocks show random orientation of local shear surfaces (see Hsü and Ohrbom, 1969, fig. 4). Those containing phacoids commonly acquired a preferred orientation parallel to regional tectonic trend.

The matrix of olistostromes is not necessarily pervasively sheared (e.g. fig. 4). The shaly matrix may acquire a fissility under compaction, or it may acquire a cleavage during tectonic deformation (Abbate and others, 1970).

Nature of matrix.—Both mélanges and olistostromes typically have a pelitic matrix. However, some olistostromes have a sandy matrix. Abbate and others (1970) referred to serpentinite breccias as olistostromes (fig. 7). I have found similar breccias in the Franciscan rocks (Hsü, 1969). The breccia beds, each a few tens of feet thick, are composed of blocks of ophiolite in a sandy matrix of similar composition. Those olistostromes are interbedded with coarse-grained sandstones, and they are easily distinguished from serpentinite-bearing mélanges, which have pelitic or antigoritic matrix.

Nature of contacts with adjacent units.—Mélanges are tectonic units. Their contacts with adjacent units are almost invariably shear surfaces. In certain regions like the Pacific Coast Ranges, the base of a mélange is nowhere to be seen, and the thickness of the mélange piles may reach several miles. The upper contact is, as a rule, definable but may seem locally anomalous to geologists schooled in overthrust tectonics. One of the cardinal rules in mapping overthrust terranes states that the highest tectonic element in an allochthonous pile has been transported the farthest (see Bailey, 1935). It was thus most puzzling that the highest tectonic element in western California is the Great Valley sequence, which is autochthonous with respect to the North American continent. A parallel situation exists in the Northern Apennines. The Mount Antola unit is the highest element of the Ligurian Nappe, which is the highest nappe of the Northern Apennines. Yet the Mount Antola is autochthonous with respect to the Po Valley block (Abbate and Sagri, 1970, p. 319). This paradox is now resolved if we assume the underthrusting of mélanges, for the highest tectonic element in an underthrust pile is autochthonous or paratochthonous with respect to the upper or the overthrust plate.

Another paradox is the observation that the upper contact of a mélange with an upper autochthonous element seems to be a shear surface at one place and a depositional contact at another (Hsü, 1968, p. 1070). The long controversy over the Franciscan-Knoxville problem could be traced to this observation. In the northern Coast Ranges, slabs of the Great Valley sequence were detached and mixed with the Franciscan rocks. The slabs are separated from the mélange by shear surfaces, which have been collectively described as an overthrust (e.g., Irwin, 1964; Bailey and others, 1964; Blake and others, 1967). However, on the west side of the Great Valley the contact between the Great Valley sequence and its underlying ophiolite, which has been traditionally mapped as Franciscan, is depositional (Hsü, 1969; Bailey and others, 1970; Page, 1972). For this reason, some current workers propose to include the ophiolite slab in the Great Valley sequence (e.g., Blake and Jones, this volume).

Olistostromes are stratigraphic units intercalated in a normal sedimentary sequence. The contact with overlying and underlying units should be depositional and conformable. Flores (1959) noted a discordant lower contact of the Sicilan *argille scagliose,* apparently overlapping rocks ranging from Aquitanian to Tortonian in age. He rationalized the observation by assuming uncommon erosive power for the postulated submarine slide.

Regional synthesis.—A distinction between a mélange and an olistostrome is not always possible on the basis of observable criteria. A final judgment depends upon consideration of all available geological data. The current consensus interpreting the Franciscan as a mélange is based not only upon field observations, but also upon regional synthesis. Such an interpretation clarified a long-standing paradox of Coast Range geology. The olistostrome hypothesis took root in the Apennines because it accounted for many facets of the regional geology there. Nevertheless, the hypothesis also proved inadequate to solve some puzzles. Would a mélange hypothesis lead to a better understanding there?

SUMMARY

Mélanges are among the most important bodies of rocks in the mountain ranges of the world. Yet I have found no reference to mélanges, in any of the current textbooks on structural geology. The concept of tectonic mixing has been much neglected by our profession, considering that the term was coined by Greenly in 1919. Prior to 1965, only a few European geologists resorted to the use of this expression (e.g., Bailey and McCallien, 1950; 1963; Gansser, 1955). The term became popular after parts of the Franciscan were identified (e.g., Hsü, 1967a) and were referred to as such (e.g., Ernst, 1965; Hsü, 1968; Hamilton, 1969; Page, 1970; Swe and Dickinson, 1970). The idea of mélange genesis through underthrusting found a cozy niche in the scheme of plate-tectonic theory (e.g., Crowell, 1968; Bailey and Blake, 1969; Page, 1969; Hamilton, 1969; Ernst, 1970; Dewey and Bird, 1970; Dickinson, 1970; Dott, 1971; Hsü, 1971a). Today, one rarely picks up a geological journal without finding frequent recurrence of "mélange." Meanwhile, the concept of olistostrome survives to describe the deposition of coarse, chaotic sedimentary units. Both mélanges and olistostromes are common in regions of underthrusting tectonics and so are often indistinguishable in zones of pervasive shearing. This article defines their distinction. Where distinction is impossible, some nongenetic, collective term such as "wildflysch" is suggested.

ADDENDUM

During an international excursion attended by more than a hundred colleagues in early June 1973 to examine the ophiolitic mélanges in Tienshan and Caucasus, it became apparent that an unsheared olistostrome could be distinguished easily from a mélange. On the other hand, it is not always clear in the field if a mélange is a sheared olistostrome or if the fragmentation and mixing have been entirely tectonic. The use of the term wildflysch as I suggested in the text of my paper has been objected to by many colleagues because the alpine Wildflysch, in contrast to most mélanges, generally does not include ophiolite blocks. I now tend to agree to a suggestion made by the editors in the footnote of my text that the term mélange *sensu lato* be used as a descriptive term to designate all pervasively sheared chaotic deposits with exotic blocks. Sedimentary mélanges are sheared olistostromes, which include blocks fragmented and mixed by sedimentary processes that are now embedded in a pervasively sheared matrix.

REFERENCES

ABBATE, E., AND SAGRI, M., 1970, Development of the Northern Apennines Geosyncline—the eugeosynclinal sequences: Sed. Geology, v. 4, p. 251-340.
———, BORTOLOTTI, V., AND PASSERINI, P., 1970, Development of the Northern Apennines Geosyncline—olistostromes and olistoliths: *ibid.,* v. 4, p. 521-557.

ADRIAN, H., 1915, Geologische Untersuchung der beiden Seiten des Kandertals im Berner Oberland: Eclog. Geol. Helvetiae, v. 13, p. 238–351.
BADOUX, H., 1967, De quelques phénomènes sédimentaires et gravifiques liés aux orogenèse: *ibid.*, v. 60, p. 399–406..
BAILEY, E B., 1935, Tectonic essays, mainly alpine: Oxford, England, Oxford University, 200 p.
──────, 1963, Liguria nappe: Northern Apennines: Royal Soc. Edingburgh, Trans., v. 65, p. 315–333.
──────, AND MCCALLIEN, W. J., 1950, The Ankara mélange and the Anatolian thrust: Nature, v. 166, p. 938–940.
BAILEY, E. H., AND BLAKE, M. C., 1969, Tectonic development of western California during the late Mesozoic: Geotektonika, v. 3, p. 17–30, v. 4, p. 24–34.
──────, ──────, AND JONES, D. L., 1970, On-land Mesozoic oceanic crust in California Coast Ranges: U.S. Geol. Survey Prof. Paper 700-C, p. 70–81.
──────, IRWIN, W. P., AND ──────, 1964, Franciscan and related rocks and their significance in the geology of western California: California Div. Mines Geol. Bull. 183, 177 p.
BECK, P., 1908, Vorläufige Mitteilung über Klippen and exotische Blöcke östlich des Thunersees: Naturf. Gesell. Bern, Mitt., fur 1908.
BENEO, E., 1951, Sull' identià tettonica esistente fra la Sicilia e il Rif: Boll. Service Geol. d'Italia, v. 72, Note 1, Roma.
BLAKE, M. C., IRWIN, W. P., AND COLEMAN, R. G., 1967, Upside-down metamorphic zonation, blueschist facies, along a regional thrust in California and Oregon: U.S. Geol. Survey Prof. Paper 575-C, p. 1–9.
BOUSSAC, J., 1912, Études stratigraphiques sur le Nummulitique alpin: Carte Géol. France, Mém., 662 p.
CADISCH. J., 1953, Geologie der Schweizer Alpen: Basel, Switzerland, Wepf et Cie., 480 p.
CLINE, L. M., 1970, Sedimentary features of late Paleozoic flysch, Ouachita Mountains, Oklahoma, *in* LAJOIE, J. (ed.), Flysch sedimentology in North America, Geol. Assoc. Canada, Special Paper 7, p. 85–102.
CROWELL, J. C., 1957, Origin of pebbly mudstones: Geol. Soc. America Bull., v. 67, p. 993–1010.
──────, 1968, Movement histories of faults in the transverse ranges and speculations on the tectonic history of California, *in* DICKINSON, W. R., AND GRANTZ, A. (eds.), Proceedings of conference on geologic problems of San Andreas Fault System, 11: Stanford, California, Stanford Univ. Pub. Geol. Sci., 374 p.
DAVOUDZADEH, M., Geologie und Petrographie des Gebietes nördlich von Näin, Zentral-Iran: Geol. Inst. ETH, Mitt., new ser., v. 98, 91 p.
DEWEY, J. F., 1969, Evolution of the Appalchian/Caledonian Orogen: Nature, v. 222, p. 124–129.
──────, AND BIRD, J.,1970, Mountain belts and the new global tectonics: Jour. Geophys. Research, v. 75, p. 2625–2647.
DICKINSON, W. R., 1970, Second Penrose Conference: The new global tectonics: Geotimes, v. 15, no. 4, p. 18–22.
DOTT, R. H., 1963, Dynamics of subaqueous gravity depositional processes: Am. Assoc. Petroleum Geologists Bull., v. 47, p. 104–128.
──────, 1971, Geology of the southwestern Oregon coast west of the 124th Meridian: Oregon Geol. and Min. Industries Bull. 69, 63 p.
ELTER, P., AND RAGGI, G., 1965, Contributo alla cónoscenza dell'Appennino ligure: 1. Osservazioni sul problema degli olistostromi: Soc. Geol. Italia, Boll., v. 84, p. 303–322.
ERNST, W. G., 1965, Mineral paragenesis in Franciscan metamorphic rocks, Panoche Pass, California: Geol. Soc. America Bull., v. 76, p. 879–914.
──────, 1970, Tectonic contact between the Franciscan melange and the Great Valley sequence—crustal expression of a late Mesozoic Benioff zone: Jour. Geophys. Research, v. 75, p. 887–901.
FLORES, G., 1955, Discussion, *in* BENEO, E., Les resultats des etudes pour la recherche petrolifere en Sicilie (Italie): 4th World Petroleum Cong., Rome, Proc. sect. 1, p. 121–122.
──────, 1959, Evidence of slump phenomena (olistostromes) in areas of hydrocarbons exploration in Sicily: 5th World Petroleum Cong., New York, Proc. sect. 1, p. 259–275.
GANSSER, A., 1955, New aspects of the geology in central Iran: 4th World Petroleum Cong., Rome, Proc. sect. 1, p. 279–300.
GIGON, W., 1952, Geologie des Habkerntales und des Quellgebiets der Grossen Emme: Naturf. Gesell. Basel, Verh., v. 63, p. 49–136.
GILLULY, J., WATERS, A. C., AND WOODFORD, A. O., 1968, Principles of geology, 2nd ed.: San Francisco, Freeman and Co., 534 p.
GÖRLER, K., AND REUTTER, K. J., 1968, Entstehung und Merkmale der Olistostrome: Geol. Rundschau, v. 57, p. 484–514.
GREENLY, E., 1919, The geology of Angelsey: Great Britain Geol. Survey Mem., 980 p.
HÄFNER, W., 1924, Geologie des südöstlichen Rätikon (zwischen Klosters und St. Antönien): Geol. Karte Schweiz, Beitr., new ser., v. 54, p. 1–33.
HAMILTON, W., 1969, Mesozoic California and the underflow of Pacific mantle: Geol. Soc. America Bull., v. 80, p. 2409–2430.
HARDIN, J., 1966, Strength and ductility, *in* CLARK, S. P., JR. (ed.), Handbook of physical constants: Geol. Soc. America Mem. 97, p. 223–289.
HEIM, A., 1923, Der Alpenrand zwischen Appenzell und Rheintal (Fähnern-Gruppe) und das Problem der Kreide-Nummuliten: Geol. Karte Schweiz, Beitr., new ser., v. 53, p. 1–51.
HERB, R., 1966, Geologie von Amden mit besonerer Berücksichtigung der Flyschbildung: *ibid.*, 144, p. 1–130.
HSÜ, K. J., 1960, Paleocurrent structures and paleogeography of the Ultrahelvetic Flysch basins: Geol. Soc. America Bull., v. 71, p. 577–610.
──────, 1967a, Mesozoic geology of California Coast Ranges—a new working hypothesis, *in* SCHAER, J. P. (ed.), Étages tectoniques: Neuchâtel, Switzerland, a la Baconnière, p. 279–296.

———, 1967b, Origin of large overturned slabs of Apennines, Italy: Am. Assoc. Petroleum Geologists Bull., v. 51, p. 65–72.
———, 1968, Principles of mélange and their bearing on the Franciscan-Knoxville paradox: Geol. Soc. America Bull., v. 79, p. 1063–1074.
———, 1969, Preliminary report and geologic guide to Franciscan mélanges of the Morro Bay-San Simeon area, California: California Div. Mines and Geology Special Pub. 35, 46 p.
———, 1970, Cohesive strength, toe effect, and the mechanics of imbricated thrusts: 2nd Internat. Rock Mechanics Conf., Proc. Paper 3–36, 4 p.
———, 1971a, Franciscan mélanges as a model for eugeosynclinal sedimentation and underthrusting tectonics: Jour. Geophys. Research, v. 76, p. 1162–1170.
———, 1971b, Origin of the Alps and western Mediterranean: Nature, v. 233, p. 44–48.
———, 1972, Alpine flysch in a Mediterranean setting: 24th Internat. Geol. Cong., Montreal, Proc. sect. 6, p. 67–74.
———, AND OHRBOM, R., 1969, Mélanges of San Francisco Peninsula: geologic reinterpretation of type Franciscan: Am. Assoc. Petroleum Geologists Bull., v. 53, p. 1348–1367.
IRWIN, W. P., 1964, Late Mesozoic orogenies in the ultramafic belts of northwestern California and southwestern Oregon: U.S. Geol. Survey Prof. Paper 501C, p. 1–9.
JAMES, D. E., 1971, Plate tectonic model for the evolution of the central Andes: Geol. Soc. America Bull., v. 82, p. 3325–3346.
KAUFMAN, F. J., 1886, Emmen- und Schlierengegend nebst Kantone Schwyz und Zug und Bürgenstocks bei Stanz: Geol. Karte Schweiz, Beitr., new ser., v. 24.
KAY, M., 1970, Flysch and bouldery mudstone in northeastern Newfoundland, in LAFOIE, J. (ed.), Flysch sedimentology in North America: Geol. Assoc. Canada, Special Paper 7, p. 155–164.
LEINE, L., AND EGELER, C. G., 1962, Preliminary note on the origin of the so-called "konglomeratische Mergel" and associated "Rauhwackes," in the region of Menas de Seròn, Sierra de los Filabres (SE Spain): Geologie en Mijnb., v. 41, p. 305–314.
LEUPOLD, W., 1942, Neue Beobachtungen zur Gliederung der Flyschbildung der Alpen zwischen Reuss und Rhein: Eclog. Geol. Helvetiae, v. 35, p. 247–291.
LINDSAY, J. F., 1966, Carboniferous subaqueous mass-movement in the Manning-Macleay Basin, Kempsey, New South Wales: Jour. Sed. Petrology, v. 36, p. 719–732.
LUGEON, M., 1916, Sur l'origin des blocs exotiques du Flysch préalpin: Eclog. Geol. Helvetiae, v. 14, p. 217–221.
McFADYEN, J. A., JR., 1956, The geology of the Bennington area, Vermont: Vermont Geol. Survey Bull. 7, 72 p.
MARCHETTI, M. P., 1957, The occurrence of slide and flowage materials (olistostromes) in the Tertiary series of Italy: 20th Internat. Geol. Cong., Mexico City, 1956, sec. 5, v. 1, p. 209–225.
MAXWELL, J. C., 1959, Turbidite, tectonics and gravity transport, northern Apennine Mountains, Italy: Am. Assoc. Petroleum Geologists Bull., v. 43, p. 2701–2719.
PAGE, B. M., 1963, Gravity tectonics near Passo della Cisa, northern Apennines, Italy: Geol. Soc. America Bull., v. 74, p. 655–672.
———, 1969, Relation between ocean floor spreading and structure of the Santa Lucia Range, California: ibid., Abs. with Programs, pt. 3 (Cordilleran Sec.), p. 51–52.
———, 1970, Sur-Nacimiento Fault zone of California: continental margin tectonics: ibid., Bull., v. 81, p. 667–690.
———, 1972, Oceanic crust and mantle fragment in subduction complex near San Luis Obispo, California: ibid., v. 83, p. 957–971.
PATRICK, T. O. H., 1956, Comfort Cove, Newfoundland (geologic map with marginal notes): Geol. Survey Canada Paper 55-31.
PITMAN, W. C., AND TALWANI, M., 1972, Sea-floor spreading in the North Atlantic: Geol. Soc. America Bull., v. 83, p. 619–646.
RAST, N., 1956, The origin of significance of boudinage: Geol. Mag., v. 93, p. 401–408.
RIGO DI RHIGI, M., AND CORTESINI, A., 1964, Gravity tectonics in foothills structure belt of southeast Turkey: Am. Assoc. Petroleum Geologists Bull., v. 48, p. 1911–1937.
ROLLIER, L., 1912, Ueber obercretazischen Pyritmergel (Wang und Seewener-Mergel) der Schwyzeralpen. Eclos. Geol. Helvetiae, v. 22, p. 178–180.
———, 1923, Supracrétacique et Nummulitique dans les Alpes suisses orientales: Karte Geol. Schweiz, new ser. 53, p. 53–85.
SCHARDT, H., 1898, Les régions exotiques du versant Nord des Alpes Suisse (Préalpes du Chablais et du Stockhorn et les Klippes): Soc. vaudoise Sci. nat. Bull., v. 34, p. 113–219.
SMITH, A. G., 1971, Alpine deformation and the oceanic areas of the Tethys, Mediterranean and Atlantic: Geol. Soc. America Bull., v. 82, p. 2039–2070.
SUPPE, J., 1969, Times of metamorphism in the Franciscan terrain of the northern Coast Ranges, California: ibid., Bull., v. 80, p. 134–142.
SWE, W., AND DICKINSON, W. R., 1970, Sedimentation and thrusting of late Mesozoic rocks in the Coast Ranges, California: ibid., v. 81, p. 165–187.
TERCIER, J., 1928, Géologie de la Berra: Karte Geol. Schweiz, new ser. 60, 111 p.
TRÜMPY, R., 1960, Paleotectonic evolution of the central and western Alps: Geol. Soc. America Bull., v. 71, p. 843–908.
ZEN, E. A., AND RATCLIFF, N. M., 1966, A possible breccia in southwestern Massachusetts and adjoining areas and its bearing on the existence of the Taconic allochton: U.S. Geol. Survey Prof. Paper 550-D, p. 39–46.

OPHIOLITES, MELANGES, BLUESCHISTS, AND IGNIMBRITES: EARLY CALEDONIAN SUBDUCTION IN WALES?

DENNIS S. WOOD
University of Illinois, Urbana

ABSTRACT

North Wales appears to be unique among pre-Mesozoic terranes in possessing a temporal and spatial association of all phenomena that are presently considered to be indicative of plate boundaries and zones of crustal subduction.

More than 30,000 feet of late Precambrian sedimentary and volcanic rocks belonging to the Monian System are preserved in Anglesey and Caernarvonshire. These rocks accumulated in a subsiding basin, which probably was one of several marking the onset of the Caledonian cycle. The lower part of this eugeosynclinal sequence consists of flysch sediments, whereas the upper part contains limestones, arenites, cherts, and basic pillowed lavas, and is characterized by the presence of mélange over a region of several hundred square miles. The mélange, which is the original example named by Edward Greenly in 1919, attains a thickness of at least 3,000 feet and is a deformed sedimentary slide rather than a tectonic breccia. The term mélange should be descriptive only. The lower part of the Monian contains a suite of ultramafic and mafic intrusions, which are serpentinized and carbonated to varying degrees. Late Precambrian deformation was accompanied by metamorphism of variable intensity over short distances, so that unmetamorphosed sedimentary and volcanic rocks pass into sillimanite-bearing migmatites within a distance of a few hundred yards. Blueschist metamorphism proceeded in relation to major zones of contemporaneous shearing (slides) in eastern Anglesey where the Precambrian now passes beneath the lower Paleozoic rocks to the east. Between the upper Precambrian of Anglesey and the Lower Cambrian Series of the Welsh mainland is a thick sequence of ignimbrites, which pass conformably upwards into the Lower Cambrian. Both the Monian System and the earliest Cambrian (Arvonian) ignimbrites are intruded by potash granites, which, like the Monian metamorphic rocks, yield isotopic ages within the range of 580 to 610 my. These ages, for a region famed for its thick classical Cambrian sequence, demonstrate that metamorphism and intrusive activity occurred at a high crustal level, were probably shortlived, were virtually contemporaneous with ignimbrite activity, and were immediately followed by the development of the lower Paleozoic Welsh Basin to the southeast. The Arvonian ignimbrites are interpreted as recycled upper crustal material derived by melting from the Monian rocks during subduction near the edge of the incipient Welsh Basin. A continuation of the same process could well explain the great thickness of basic to acidic Ordovician volcanic rocks of the Snowdonia region of North Wales, which probably constituted a Middle Ordovician island arc system. It could also explain the lower Paleozoic gold and copper mineralization of North Wales. The Anglesey region was a moderately to strongly positive area throughout the lower Paleozoic, constituting an important cordilleran zone of internal sediment supply within the Caledonian mobile belt. Structures in the Monian basement controlled the development of large fault-bounded rotational blocks, which helped to localize the positions of several major unconformities.

The regional relationships invite analogy with the Mesozoic sequences of California. The Monian would be equivalent to the Franciscan Sequence, and the lower Paleozoic would be equivalent to the Great Valley Sequence. The greater time separation between oldest Monian and the youngest rocks of the Welsh Basin than between oldest Franciscan and youngest Great Valley rocks may be accounted for by invoking in North Wales a smaller amount of plate consumption and a slower rate of subduction.

INTRODUCTION

Since 1831, when Adam Sedgwick in company of the youthful Charles Darwin commenced his geological investigations, North Wales has been a classical region. The remarkable variety of its geology is above all proportion to its size, but for many well described regions some interpretations that now may be offered in the light of new developments in geological thought have been either delayed or overlooked. The associations within a late Precambrian eugeosynclinal sequence that includes pillowed olivine-basalt lavas, cherts, jaspers, mélanges, serpentinites, gabbros, and blueschist metamorphic rocks provide adequate scope for speculation. When the responsible processes are considered in a space-time context together with regional structure, subsequent effusive ignimbrite activity, and with the overall aspect of the lower Paleozoic Welsh Basin, a unified view of this region unfolds that may prove to be an ideal small-scale model for one of the more important crustal processes.

Edward Greenly (1919) provide most of our modern knowledge of the Precambrian of Anglesey. Marshall Kay is one of few geologists to have been shown this region by Greenly. It is, therefore, appropriate that a symposium to honor Professor Kay and to recognize his contribution to geology should refer to this fascinating region.

THE PRECAMBRIAN (MONIAN) ROCKS OF NORTH WALES

Precambrian rocks occur in four main regions of the island of Anglesey and in one area in the southwestern promontory of Caernarvonshire, the Lleyn Peninsula. The area of outcrop is approximately 180 square miles in Anglesey and 35 square miles in Caernarvonshire. The stratigraphic thickness is at least 35,000 ft. Equivalent rocks are present in Ireland on Carnsore Head, County Wexford.

Of the earliest investigators, Henslow (1822) referred to Anglesey as containing "the earliest stratified rocks," Sedgwick (1838, 1843) regarded the "metamorphic slates" of Anglesey and southwest Caernarvonshire as coeval and older than the Paleozoic systems; and Sharpe (1846) recorded the presence of jaspers and serpentines. The term "Monian System" was introduced by Blake (1887, 1888a), who described the importance of lavas and large-scale breccias and discovered glaucophane in Anglesey (Blake, 1888b). Gabbros and "spheroidal basalts" were recorded by Raisin (1893), the basalts being recognized as spilitic pillowed lavas by Dewey and Flett (1911). The most detailed descriptions of the Monian of Caernarvonshire are by Matley (1913, 1928) and Greenly, whose 25 years of research and mapping of the whole of Anglesey were published in 1919.

Stratigraphy.—The lower part of the pertinent sequence is best preserved on Holy Island at the western extremity of Anglesey. It consists in upward succession of turbidite graywackes interbedded with pelites (South Stack Series); thick orthoquartzites (Holyhead Quartzite); turbidite graywackes, pelites and thin quartzites (Rhoscolyn Beds); and a thick finely laminated flysch group of pelites and semipelites (New Harbour Group).

The uper part of the sequence, which is best seen in northern and southeastern Anglesey and in southwest Caernarvonshire, consists of basic to acid tuffs and volcanic conglomerates (Skerries Group) and is overlain by an extremely varied assemblage not less than 12,000 ft thick (Gwna Group). The Gwna contains oolitic limestones, dolomites, orthoquartzites, graywackes, pelites, cherts, jaspers, basic pillow lavas, tuffs, and hyaloclastites. Much of the Gwna Group is a mélange on scales ranging from microscopic to megascopic. Above the Gwna Group at the top of the succession is the Fydlyn Group of unlayered acidic volcanic rocks and tuffs. The stratigraphic terms are, with one exception, those assigned by Greenly, but the stated order of succession is almost the reverse of his. Greenly's interpretation of the Anglesey structure invoked enormous recumbent folds with westerly vergence. This interpretation, together with his discovery in the Holyhead Quartzite of jasper fragments, which he presumed to be derived from the Gwna Group, led him to conceive of the succession in the reverse order from that given here. Shackleton (1953, 1954) used sedimentary structures and applied the innovative concept of facing (Shackleton, 1957) to demonstrate the true sequence.

Provenance of sediments.—The graywackes of the lower Monian of Holy Island are in graded beds as much as 20 ft thick. They are best seen at South Stack (204823)[1] and at Rhoscolyn (258753–264750). The great thickness of many units and the observable consistency over hundreds of yards, imply that the sediments accumulated in relatively deep water. Channeling and grooving show that they were emplaced by currents that flowed along a northeast-southwest trough; the sense of flow is not known. Associated pelites slumped down slopes which inclined to the northwest. Hence, axial transport and lateral supply from the southeast were involved. This graywacke facies is uncommon, being interbedded with extremely pure quartzites. In 5 miles, from Holyhead to Rhoscolyn, a single quartzite, at least 800 ft thick, splits into three thinner units, one of which is 120 ft thick and devoid of internal stratification. Wood (1960) found gigantic flute casts as much as 17 ft long and 3 ft deep at the base, indicating flow from the northeast (see fig. 1). The upper few feet of this quartzite is cross bedded and indicates that reworking currents also flowed from the northeast. The quartzites were interpreted as relatively deep water sand-flow deposits.

The mélange.—The mélange of the Gwna Group is perhaps the most extraordinary and spectacular aspect of the North Wales Precambrian. It is best seen on Wylfa Head (353946–358945) and at Cemaes Bay (369938) in northern Anglesey and on Bardsey Island (120220) some 2 miles off the southwest coast of Caernarvonshire. Matley (1913) gave the first clear description of the mélange from Bardsey Island. Matley and, subsequently, Greenly (1919) both regarded the mélange as a tectonic breccia as did Sir Edward Bailey (personal communication, 1961). Matley used the terms "crush-breccia" and "crush-conglomerate" before Greenly introduced the term "autoclastic mélange." Greenly intended no genetic implication for "mélange"; otherwise, he would not have needed to prefix it. For the term to carry

---

[1] Good field examples are referred to in the text by six-figure National Grid References of the Kilometer grid of the Ordnance Survey of Great Britain. All are in the 100-kilometer square SH and are located on the 1-inch-to-1-mile sheets 106 (Anglesey), 107 (Snowdon), and 115 (Pwllheli).

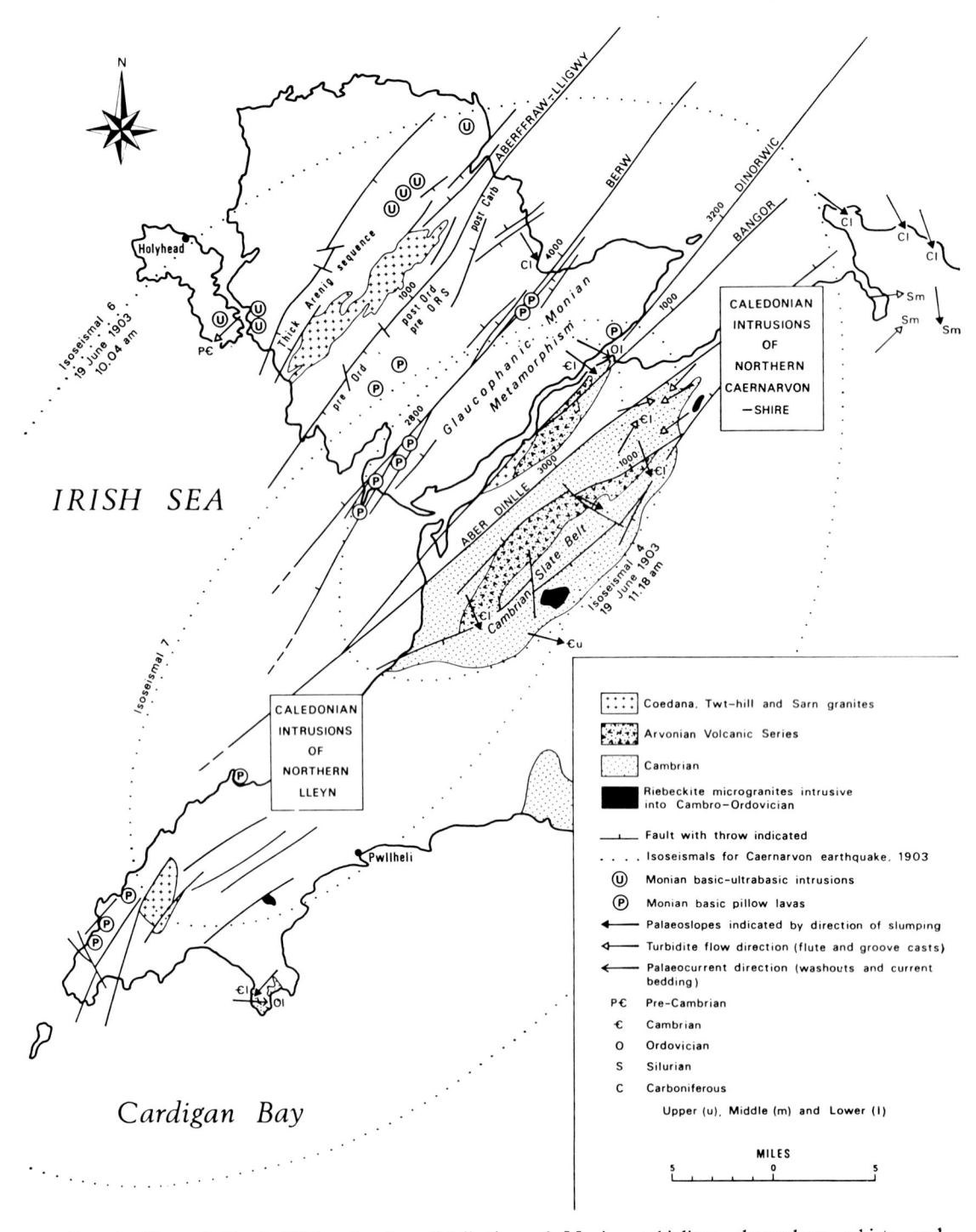

FIG. 1.—Map of North Wales showing distribution of Monian ophiolites, glaucophane schists, and Arvonian ignimbrites together with major acidic intrusions and Caledonian faults of the region and sedimentary transport directions.

the connotation of either a purely tectonic or purely sedimentary origin would be very unfortunate. The Gwna mélange is appreciably deformed in many places, but Shackleton (1956, 1969) and the writer believe it to be ultimately of sedimentary origin, probably the result of submarine sliding. The mélange has regional extent, for, as Greenly (1919, p. 66) noted, "the Gwna Group as a whole is usually in the condition of an Autoclastic General Mélange in which all the members of the group are involved."

In most places the mélange shows a convincing ghost stratigraphy, the individual blocks of one lithology being separated by distances ten to twenty times greater than the block size. It may be argued that tectonic boudinage could account for such a relationship. For such to be the case, the ductility contrasts between the materials and the mean ductility of all the materials involved would need to be impossibly high in view of the very low greenschist level of metamorphism. Lithologies having contrast as great as those in the mélange occur elsewhere in Anglesey without such associated disruption. Furthermore, the ooliths of the limestones lack penetrative deformation in most examples; nor does the matrix of the mélange show ductile flow. Many separated blocks within an individual ghosted unit are very angular. More cogently, much of the mélange is still relatively flat lying. For boudinage to account for the structure would require that the whole of the mélange outcrop had been subjected to a vertical principal compressive force acting perpendicularly to the bedding and under an enormous confining pressure. The undoubted principal deformation features of the Monian, both above and below the mélange, are steep structures, not easily reconcilable with intervening flat structures. The penetrative deformation in the mélange is everywhere less than that characteristic of the deeper levels of the Monian stratigraphy. In many places the only tectonic structure is a rather crude cleavage, which cuts across the intervals containing separated blocks of the same lithology, as at Porth Cadwaladr (362664) in southern Anglesey. Sedimentary contacts with undisturbed sediments are present at the top and bottom of the mélange. The undisturbed beds consist of strongly variable, well-layered materials, which would have responded to strong deformation with distinct ductility contrast and would have resulted in disharmonic relationships. Well-exposed lower contacts occur at Porth Cadwaladr, Anglesey (362664) and well-exposed upper contacts at Braich y Pwll, Caernarvonshire (136258).

Though effects of tectonic deformation can be evaluated and removed, the mélange still remains as a highly disturbed stratigraphic unit of regional extent and nontectonic origin. The ghost stratigraphy and lack of materials exotic to the Gwna Group preclude the mélange from being a tillite. The great areal extent and paucity of volcanic materials preclude it from being a volcanic mudflow. The remaining possibility is that the mélange is an olistostrome. Much complication may pertain because the materials were in varying stages of lithification at the time of their initial movement, arenitic members being more completely lithified than argillites but less lithified than carbonate members. Near Porth Gwylan (214367), the sense of rotation of highly angular blocks of arenite and limestone and the deformation of enclosing layered matrix suggest that relative movement of the stratigraphically higher material was toward the west.

The ophiolites.—The term "ophiolite" may be used in the original strict sense (Steinmann, 1905) to include commonly serpentinized mafic and ultramafic rocks ranging from spilite and basalt to gabbro and periodotite. In view of Steinmann's emphasis upon association of ophiolite with radiolarian cherts, the Steinmann Trinity of serpentinite, spilite, and radiolarian chert came into prominence (Hess, 1955; Bailey and McCallien, 1960), and both "ophiolite" and "Steinmann Trinity" have been used synonymously. Dewey and Bird (1970) have urged that the term "ophiolite" be "restricted to a full sequence of ultramafic rocks, gabbro dike complexes, pillow lava and chert, a sequence that, if fully developed, almost certainly represents oceanic mantle and crust." Any of these usages would apply to the Precambrian of North Wales. The distribution of mafic-ultramafic intrusives and pillow lavas is shown in figure 1.

Mafic and ultramafic intrusives, best developed in the southern part of Holy Island (267772) and the adjacent part of the main island of Anglesey, consists of a variety of largely serpentinized ultramafic rocks and altered gabbros emplaced prior to the regional late Precambrian deformation and metamorphism. The serpentinites were not emplaced cold, so to speak, because they preserve good igneous textures and have caused appreciable contact effects in the enclosing sediments. The ultramafics may be small layered bodies. The majority of lithic types were dunite and harzburgite, but pyroxenite, lherzolite, and minor amounts of wehrlite were also present (Wood, 1960). The dunites, formerly consisting of 90 percent olivine and 10 percent orthopyroxene and accessories, and harzburgites have been replaced by serpentine minerals. Olivine and orthopyroxene did not survive in the lherzolites but clinopyroxene did. The serpentinization of ferromagnesian components of the peridotites and pyroxenites ranges from 10 percent for dunite and harzburgite to about 85 percent for lherzolite and to 65 to 60 percent for pyroxenites. The serpentine minerals are dominantly lizardite and antigorite. Carbonate metasomatism of the serpentinites has formed rocks that, except for surviving magnetite and chromite-picotite, are wholly dolomite. In some cases the nature of both the preexisting serpentine minerals and the earlier ferromagnesian minerals can be recognized. The carbonation ranges from the mere presence of isolated dendritic dolomite rhombs to complete replacement. Dunite and harzburgite serpentinites may be found in all conditions of carbonation from zero to 100 percent (Wood, 1960). The associated gabbros are equally strongly altered, pyroxene being replaced by tremolite and the remaining feldspar being oligoclase. The original feldspar altered to epidote-group minerals, talc, prehnite, scapolite, and calcic garnet. Remaining problems with regard to interpretation of structural form, mode of emplacement, and mineralogical and textural changes should be clarified by the current mapping and petrographic work of A. J. Maltman.

The Monian pillow lavas are exceptionally well preserved and have their finest expressions in the Newbrough Forest and Llandwyn Island area of Anglesey (391635) and on the headland west of Porth Dinllaen (276420) and at Dinas Bach (158293) in the Lleyn Peninsula of Caernarvonshire. The lavas, at least 2,000 ft thick, are spilitic albite basalts containing either original olivine or its alteration products. The basalts are generally vesicular and variolitic, much fresh augite remains, and the albite seems to be original. Except that alumina (17.5 percent) is enriched at the expense of total iron (9.5 percent), the lavas have the chemical character of typical oceanic basalts. The interpillow interstices and the central vacuoles of many pillows are occupied by jasper or carbonate. The pillowed sequences are interbedded with tuffs, basic hyaloclastites, dolomite, dolomite breccias, and massive ferruginous cherts. The pillow lavas have been replaced by carbonate so extensively that some are reduced virtually to ellipsoidal dolomites. Greenly (1919, p. 84) recognized these as metasomatic pseudomorphs after spilitic agglomerate. Every stage in the replacement of complete pilows can be recognized in the Newbrough Forest (395641), commencing with the interpillow spaces and vacuoles being filled with dolomite and ending with a thin relict film of epidote and hematite marking the outermost skin of the former pilows. Some axinite and prehnite are present as minor metasomatic products.

Structure and Metamorphism.—The overall structure of the Monian of Anglesey (figure 2) consists of a series of large asymmetric folds with southeasterly vergence. The basic structure is simple, but many small scale details are complex, particularly in the lower part of the sequence as in the cliff sections of the Rhoscolyn Anticline (260763–265749). Deformation was polyphase. An early phase, accompanied by regional metamorphism, produced folds that have northeasterly trending axes and cross folds that have northwesterly trending axes. A postmetamorphic phase was responsible for most of the small-scale structural complexity. This later phase may have been post-Ordovician. As seen from figure 2, the Ordovician rocks are appreciably deformed in parts of Anglesey.

The Monian metamorphism is extraordinary in several respects. The lower stratigraphic units of western Anglesey are nowhere above chlorite grade, the higher grades being restricted to the younger formations of central and eastern Anglesey, where the grade varies greatly. The higher grade rocks are acidic and mafic gneisses and migmatites, generally proximal to the Coedana Granite of central Anglesey (fig. 2). In places the granite is intrusive, but elsewhere, as at Llandrygarn (383796) and Gwyndy (395794), acidic gneisses seem to pass gradually into the granite. Garnet and sillimanite gneisses occur at Llechcynfarwy (381811). The variable intensity of metamorphism and the telescoped metamorphic zonation were explained by Shackleton (1953, 1969) as the probable result of rapid heat loss, which is suggestive of magmatically introduced heat. The sudden changes in metamorphic character may result from the rocks being held at elevated temperatures for a period insufficient for complete thermal equilibration.

The glaucophane schists, restricted to that part of Anglesey east of the Berw Fault (fig. 1), are best examined at the Anglesey Monument (538716) and in the Mynydd Llwydiarth area (550790). Glaucophane schists are interleaved with hornblende schists and garnetiferous semipelitic schists. Glaucophane is restricted to the steep limb of the largest fold structure in Anglesey, the Aethwy Anticline (fig. 2). In the steep limb of this fold, the stratigraphic units must be standing nearly vertically for a distance of several miles. The small scale tectonic effects are complicated by folds produced by two phases of coaxial folding. The glaucophane schists are in enormous lenses bounded by tectonic slides

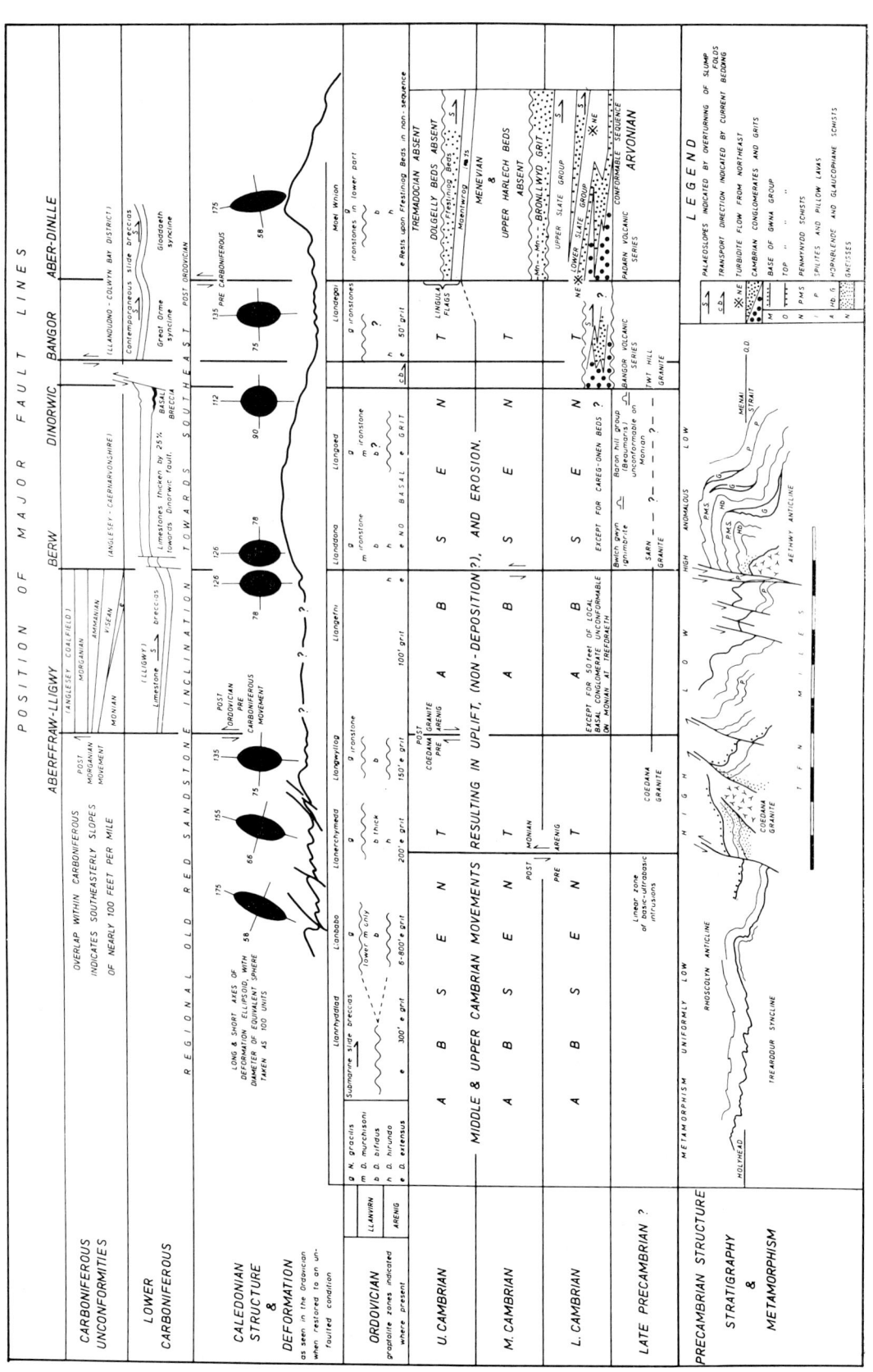

FIG. 2.—Precambrian and Caledonian structure of northwest Wales and diagramatic representation of lower Paleozoic stratigraphic changes across major basement-controlled fault lines.

(fold faults), which are indicative of high deformation rates and, presumably, of high tectonic overpressures. Glaucophane formed during the later part of a polyphase metamorphism and always replaced earlier blue-green amphiboles. The mineralogy has been carefully examined by T. S. Nataraj (1967), who discovered lawsonite in these rocks. Pumpellyite has not been found. Whole-rock analyses by Nataraj demonstrate almost complete similarity of bulk composition between glaucophane schists and associated nonglaucophanic amphibole schists. They differ from the unmetamorphosed spilitic pillows lavas of the Gwna Group, from which they were presumably derived, only by a slight decrease in calcium relative to iron.

LATEST PRECAMBRIAN-EARLIEST CAMBRIAN (ARVONIAN) IGNIMBRITES

Acidic volcanics (Arvonian Volcanic Series) are found in two main areas of Caernarvonshire: (1) the Padarn Ridge, the core of a major periclinal fold, of which the southeastern limb is the Lower Cambrian Slate Belt of North Wales, and (2) the district between Caernarvon and Bangor (see fig. 1). The base is not seen in either area. Similar pre-Ordovician rocks occur in Anglesey near Beaumaris (597755), where they rest unconformably upon Monian greenschists, and at Bwlch Gwyn (483731), where they rest unconformably upon Monian amphibolitic gneisses. Greenly (1919, p. 348) referred to these as the Baron Hill Group and Bwlch Gwyn Felsite respectively.

The Arvonian consists of moderately to strongly welded ignimbrites, welded breccias, agglomerates, and tuffs. The ignimbrites of the Llanberis district, which are best observed along the northern shores of Lake Padarn (562622-568618), are visibly over more than 3,000 ft thick and consist of several flows that include abundant devitrified glass shards. They are cut occasionally by tuffisite breccia pipes. Pyroclastic members of the Arvonian are abundant in the Bangor district (579714) and also across the Menai Strait at Beaumaris. The felsite at Bwlch Gwyn is also an ignimbrite, containing preserved vitroclastic textures.

The Arvonian passes upward into the Lower Cambrian, conformably in the Llanberis district and with slight disconformity in the Bangor district (Wood, 1969). The lowest conglomerates of these localities defined the unconformable type base of Sedgwick's Cambrian System. In view of the absence of lowermost Cambrian fossils in the Cambrian Slate Belt of Caernarvonsire, the Arvonian possibly could be included within the Cambrian. A conglomerate, remarkably similar to the lowest Cambrian conglomerates of Caernarvonshire, occurs in southern Anglesey at Trefdraeth (404706), where it rests unconformably upon Monian metasediments (Greenly, 1919, p. 399; Wood, 1969, p. 57–59). Therefore, both Arvonian rocks and Cambrian rest unconformably upon the Monian, the Arvonian being of local important only and being overstepped completely by the Lower Cambrian between the Llanberis district of Caernarvonshire and central Anglesey.

TIMING OF METAMORPHISM, INTRUSION, AND IGNIMBRITES

If the Monian System is a sequence that accumulated rapidly, was metamorphosed during a short period, and that deformed at relatively high strain rates, the metamorphism probably followed shortly after deposition. The time relations of metamorphism, granitic intrusion, and the Arvonian ignimbrites may be established by reference to the field relationships of three pre-Ordovician granites and to radiometric ages.

The granites in question are the Coedana Granite of central Anglesey, the Sarn Granite of southwest Caernarvonshire and the Twt Hill Granite at Caernarvon (see fig. 1). All are high-level potash granites that are petrographically and geochemically similar and that are overlain unconformably by the lowest Ordovician rocks. The strongly linear Coedana Granite in places passes outwards into Monian migmaties, which in turn grade into schists; in other places the granite appears to have acted intrusively with respect to the Monian. When the relationship is intrusive, the granite possesses a contact aureole of post-Monian hornfels, which includes thin units of subgraywacke. Like other problematic rocks in Anglesey, these may well be of earliest Paleozoic age; they are certainly no younger than Cambrian. The Sarn Granite is also intrusive into the Monian, but in places it is a foliated granite that merges into Monian migmatitic gneisses and schists (217293). Both are, therefore, essentially metamorphic granites, so to speak. In contrast, the Twt Hill Granite evidently intruded the Arvonian, with which it is closely associated in time, because pebbles of the granite occur in the lowest Cambrian conglomerates at Llanberis (565613). It is interpreted as a subvolcanic granite. All three intrusions may have been essentially contemporaneous with the final stage of a short-lived metamorphic event and could, at the same time, have been both metamorphic and subvolcanic granites. Present erosion levels are such that the Coedana and Sarn Granites show intrusive relations and metamorphic affinities with the Monian, whereas the Twt Hill Granite is said to be sub-Cambrian and yet intruded rapidly accumulated ignimbrite volcanics that pass conformably into the Cambrian. It may be that metamorphically derived granite magma broke through its own metamorphic and nonmetamorphic cover, gave rise to effusive ignimbrite activity, and finally behaved intrusively with respect to the base of its own volcanic cover.

Verification of these possibilities with reference to the available radiometric ages remains. Some attempts to date certain of the pre-Ordovician events of North Wales have been thwarted by inappropriate isotopic compositions. Others have produced largely meaningless ages as a result of the complete or partial overprinting effects of Middle Ordovician (490 my), so-called main Caledonian (420 − 410 ± 10 my), or younger events (Fitch, Miller, and Meneisy, 1963; Fitch and Miller, 1964; Fitch, Miller, and Brown, 1964). A potassium-argon apparent age, not obviously too low, is a muscovite age of 611 ± 16 my[2] obtained from the Gwna Group of the Monian (Fitch, Miller, and Brown, 1964). This date appeared, at the time of computation, to be disturbingly close to the beginning of the Cambrian. Ages subsequently determined by a variety of methods for the Coedana Granite have given remarkably consistent results (Moorbath and Shackleton, 1966). These include potassium-argon muscovite ages of 579 ± 15 and 580 ± 11 my, a rubidium-strontium muscovite age of 576 ± 13 my, and a rubidium-strontium whole-rock age of 581 ± 14 my. In addition, Moorbath and Shackleton recorded a muscovite whole-rock age of 580 ± 18 my from the hornfels surrounding the Coedana Granite, which is in close accord with three subsequent muscovite potassium-argon ages of 596 ± 15, 597 ± 15, and 598 ± 10 my reported by Fitch and others, (1969). Other ages include a muscovite age of 593 ± 15 for the Coedana Granite itself and one of 580 ± 30 my for the Arvonian ignimbrites of Bwlch Gwyn in Anglesey (Fitch and others, 1969, p. 34–35).

The uniformity of these eleven ages produced from different laboratories and falling within the range of 580 to 610 my cannot be fortuitous. Neither can they be the result of uniform partial overprinting. Moorbath and Shackleton (1966, p. 115) regarded 610 my as the maximum possible age of intrusion for the Coedana Granite. If uplift and cooling to below the radiogenic diffusion threshold occurred between 580 and 600 my, there is only a short interval remaining to accommodate the eruption of the Arvonian volcanics. This is the situation to be expected in view of the upward passage of Arvonian into Lower Cambrian and the intrusion of the Twt Hill Granite into the Arvonian. There appears to be a very close age similarity between the Monian metamorphism and migmatization, granite intrusion, and acid effusive activity, all these events having ages slightly greater than the presently accepted age of 570 my (Cowie, 1964) or possibly 550 my (Fairbairn and others, 1967) for the beginning of the Cambrian Period.

DEVELOPMENT OF THE LOWER PALEOZOIC WELSH BASIN

The Anglesey region formed the northwest margin of the Welsh Paleozoic basin, otherwise known as the Welsh Geosyncline. The basin received a Cambro-Silurian sedimentary and volcanic sequence, estimated by O. T. Jones (1956) to have been 50,000 feet thick. This is probably an overestimate, failing to consider tectonic thickening of an original thickness of no less than 30,000 feet. The Anglesey area was an important source of sediment for the Welsh Basin.

Monian basement structures controlled the development of a series of major faults along which important movements occurred at intervals during the Paleozoic Era. The positions of these faults together with the magnitude of some of their individual throws are shown in figure 1. The relationship of the faults to the Precambrian structures together with the Paleozoic stratigraphic variations across fault lines are shown in figure 2. Rotational block faulting either precluded Cambrian deposition in the Anglesey region or enabled uplift to result in rapid postdepositional erosion. Most striking are the Cambrian changes across the Aber-Dinlle and Dinorwic Faults in northwest Caernarvonshire. These are such that in Snowdonia the Arenig Series (lowest Ordovician) rests unconformably upon Ffestiniog Beds (Upper Cambrian); between the Aber-Dinlle and Dinorwic Faults the Arenig Series rests unconformably upon Lower Cambrian; and, to the northwest of the Dinorwic Fault in northeast Anglesey, the Arenig rests unconformably upon Monian glaucophane schists. The linear zone where Monian glaucophane schists pass southeastward beneath the Arvonian and Paleozoic of Caernarvonshire is one of attenuation of the whole of the Cambrian System, and it marked the precise margin of the initiating Welsh Basin. Slump structures in the Cambrian of Caernar-

---

[2] All ages quoted in this paper are either directly as given by the respective authors or recalculated according to the following decay constants:
K/Ar data: $\lambda = 4.72 \times 10^{-10} \text{yr}^{-1}$; $e = 0.584 \times 10^{-10} \text{yr}^{-1}$.
Rb/Sr data: $\lambda = 1.47 \times 10^{-11} \text{yr}^{-1}$.

vonshire show that paleoslopes were inclined to the southeast. Flute casts in the Lower Cambrian graywackes of the Bethesda district demonstrate that axial turbidite flow was mainly from the northeast (fig. 1).

There was pre-Ordovician slip on the major faults, but the faults behaved as hinge lines during the Ordovician. This is reflected by smaller stratigraphic variations than those of the Cambrian. Basal, Arenig sandstones vary in thickness, and oolitic ironstones change stratigraphic positions across fault lines. Unconformities and nonsequences within the Ordovician, as recognized by absence of certain graptolite zones, change horizon in a similar manner (fig. 2).

There is little evidence of fault movements during the Silurian Period, although there was at least 1,000 feet of post-Ordovician and pre-Middle Devonian movement on the Aberffraw-Lligwy Fault of Anglesey.

The fault system, which is a dominant feature of North Wales, was thus initiated in Early Cambrian time. It was controlled by the structure of the Precambrian basement and strongly influenced Paleozoic sedimentation. It helped the Anglesey region to become an internal Caledonian cordilleran supply area so to speak, for early Paleozoic sedimentation in the manner proposed by Ksiakiewicz (1960) for the late Mesozoic and Tertiary Carpathian Geosyncline. Important movements along existing faults continued into the late Paleozoic. The latest recorded effects of slip on the Dinorwic Fault are shown in figure 1 by the isoseismals for the 1903 Caernarvon earthquake (Davison, 1924). A smaller earthquake in 1906 resulted from slip along the Aber-Dinlle Fault Davison, 1908).

INTERPRETATIONS AND ANALOGIES

In view of the association of a full ophiolite suite with mélanges together with a steeply inclined zone of blueschists seen to pass below the almost contemporaneous edge of an initiating sedimentary basin, an analogy between North Wales on a small scale and western California on a larger scale can be drawn.

The North Wales relationships may be interpreted as the result of late Precambrian (Monian) deposition in a trench above a southeasterly descending subduction zone, which generated its own largely subaerial volcanic arc and gave rise to the Arvonian ignimbrites. The scale is comparable with that envisaged by Marshall Kay (1972) in his interpretation of the Dunnage Mélange of northeastern Newfoundland as the result of early Paleozoic subduction. That this involved the collision of two major crustal plates is not implied, but at least one of the plates was a small oceanic one (e.g., Dewey and Bird, 1970, p. 2641, fig. 12 A-D). It may be that limited subduction, perhaps along a series of contemporaneous *en echelon* zones of restricted linear extent, could have been a common occurrence during pre-Mesozoic time.

A limited and slow continuation of subduction could have accompanied the development of the Welsh Paleozoic basin during the Ordovician. The mid-Ordovician volcanicity of Snowdonia, the Caledonian alkali-rich intrusions, which are located at either end of the Padarn Pericline (fig. 1), and the riebeckite microgranite intrusions of Caernarvonshire could also be genetically related. The associated Late Ordovician copper mineralization of North Wales together with the mineralization of the Dolgelly gold belt could be ascribed to the same fundamental subduction.

It is equally possible that the Monian was partly consumed beneath continental crust, depressing the continental crustal margin to initiate the Welsh Basin. In either event, the older Monian rocks are comparable with the Franciscan Sequence of the California Coast Ranges, and the younger Welsh Basin rocks are comparable with the Great Valley Sequence. The model is similar to that proposed by Ernst (1970, fig. 3) for the Franciscan and explains the near synchroneity of glaucophane metamorphism and ignimbrite extrusion on the edge of the Welsh Basin and deposition of lowest Paleozoic sediments within the basin. The time separation between the oldest Monian sediments and youngest lower Paleozoic rocks of the Welsh Basin is greater than that between oldest Franciscan and youngest Great Valley rocks and may be accounted for by supposing a smaller amount of plate consumption and a slower rate of subduction for North Wales. Also, later sedimentation within the Welsh Basin may not have been directly related to events that delineated the basin's northwestern margin and provided its early sediment source.

The timing of late Precambrian metamorphism together with extrusive and intrusive activity in Wales, dated within the range of 610 to 580 my, finds parallels in the Appalachian region of North America. Certain stratigraphic similarities of basal Cambrian relationships between Wales and the Appalachian region have been noted previously (Wood, 1969). Events of the same age as those discussed here for North Wales occurred throughout the length of the Appalachian region. The rocks, relationships, and timing of events, which most closely simulate those of North Wales, are to be found in the Avalon Peninsula of eastern Newfoundland. As has been pointed out (e.g., Rodgers, 1972, p. 514), the volcanics and sediments of the Avalon Peninsula belong to a belt "whose northeastern extension is to be sought in southwestern Wales and Shropshire and perhaps also in southeastern Ireland and Anglesey."

The Harbour Main Group of the Avalon Peninsula consists of varied sediments and largely subaerial volcanics, which include ignimbrites. These are intruded by the Holyrood Granite, which in turn is unconformably overlain by the Lower Cambrian. The Holyrood Granite has yielded a rubidium-strontium isochron age of 574 ± 11 my (McCartney and others, 1966), which is in reasonable agreement with a recalculated feldspar age of 565 ± 42 my obtained by Fairbairn (1965). Somewhat similar ages have been obtained farther south, where Hills and Dasch (1969) reported an age of 610 my for the crystallization of the Stony Creek Granite of southeastern Connecticut and where Fairbairn and others, (1967) recorded an isochron age of 559 ± 28 my for the Dedham Granodiorite of southeastern Massachusetts. The latter intrusion is overlain by the Lower Cambrian. In Virginia and North Carolina, Glover and others, (1971) have obtained lead ages of 620 my on zircons from the late Precambrian volcanics of the Virgilina Synclinorium and a minimum age of 570 my for the Roxboro Granodiorite, which is intrusive into the synclinorium.

There can be no doubt that the late Precambrian events of North Wales are part of a widespread major episode involving thick sedimentary accumulation, volcanicity, metamorphism, deformation, and plutonism. It is becoming clear for most parts of the Appalachian-Caledonian zone where relationships have not been obscured by Ordovician (Taconic), late Silurian (main Caledonian), or Acadian events, that there is a widely recognizable earlier event dated from 610 to 570 my. It is termed Avalonian in North America, Monian in Britain, and Cadomian in Brittany. The Avalonian-Monian event marked the initiation of the Appalachian-Caledonian mobile belt, and as such it deserves to stand as one of the four important activity phases of the Eocambrian and early Paleozoic. In Wales, the earliest phase may have been characterized by small-plate tectonics, so to speak.

REFERENCES

Bailey, E. B., and McCallien, J. W., 1960, Some aspects of the Steinmann Trinity: mainly chemical: Geol. Soc. London Quart. Jour., v. 111, p. 365–395.
Blake, J. F., 1887, Introduction to the Monian System of rocks: British Assoc. Rept. (1886), p. 669.
———, 1888a, The Monian System of rocks: Geol. Soc. London Quart. Jour., v. 44, p. 463–547.
———, 1888b, The occurrence of a glaucophane-bearing rock in Anglesey: Geol. Mag., ser. 3, v. 5, p. 125–127.
Cowie, J. W., 1964, The Cambrian Period: Geol. Soc. London Quart. Jour., v. 120s, p. 225–228.
Davison, C., 1908, On some minor British earthquakes of the years 1904–1907: Geol. Mag., v. 45, p. 296–309.
———, 1924, A history of British earthquakes: Cambridge, England, Cambridge Univ. Press, 416 p.
Dewey, H., and Flett, J. S., 1911, Some British pillow-lavas and the rocks associated with them: Geol. Mag., v. 8, p. 202–209, 241–248.
Dewey, J. F., and Bird, J. M., 1970, Mountain belts and the new global tectonics: Jour. Geophys. Research, v. 75, p. 2625–2647.
Ernst, W. G., 1970, Tectonic contact between the Franciscan Mélange and the Great Valley Sequence—Crustal expression of a late Mesozoic Benioff Zone: *ibid.*, v. 75, p. 886–901.
Fairbairn, H. W., 1965, Personal communication in McCartney and others, 1966.
———, and others, 1967, Rb-Sr age of granitic rocks of southeastern Massachusetts and the age of the Lower Cambrian at Hoppin Hill: Earth and Planetary Sci. Letters, v. 2, p. 321–328.
Fitch, F. J., and Miller, J. A., 1964, Age of the paroxysmal Variscan Orogeny in England: Geol. Soc. London Quart. Jour., v. 120s, p. 159–173.
———, ———, and Brown, P. E., 1964, Age of the Caledonian Orogeny and metamorphism in Britain: Nature, v. 203, p. 275–278.
———, ———, and Meneisy, M. Y., 1963, Geochronological investigations on rocks from North Wales: *ibid.*, v. 199, p. 449–451.
———, and others, 1969, Isotopic age determinations on rocks from Wales and the Welsh borders, *in* Wood, A. (ed.), The Precambrian and lower Palaeozoic rocks of Wales: Cardiff, Univ. Wales Press, p. 23–45.
Glover, L., Sinha, A. K., and Higgins, M. W., 1971, Virgilina phase (Precambrian and Early Cambrian?) of the Avalonian Orogeny in the central Piedmont of Virginia and North Carolina (abs.): Geol. Soc. America Abs. with Programd, v. 3, p. 581–582.
Greenly, E., 1919, The geology of Anglesey: Geol. Survey Great Britain Mem., 2 v.
Henslow, J. S., 1822, Geological description of Anglesea: Cambridge Philos. Soc. Trans., v. 1, p. 359–452.
Hess, H. H., 1955, Serpentines, orogeny and epeirogeny, *in* Poldervaart, A. (ed.), The crust of the earth: Geol. Soc. America Special Paper 62, p. 391–408.
Hills, F. A., and Dasch, E. J., 1969, Rb-Sr evidence for metamorphic remobilization of the Stony Creek Granite, southeastern Connecticut (abs.): Geol. Soc. America Special Paper 121, p. 136–137.
Jones, O. T., 1956, The geological evolution of Wales and the adjacent regions: Geol. Soc. London Quart. Jour., v. 111, p. 323–350.

KAY, M., 1972, Dunnage Mélange and lower Paleozoic deformation in northeastern Newfoundland: 24th Internat. Geol. Cong., Sec. 3, p. 122–133.

KSIAKIEWICZ, M., 1960, Pre-orogenic sedimentation in the Carpathian Geosyncline: Geol. Rundschau, v. 50, p. 8–31.

MCCARTNEY, W. D., AND OTHERS, 1966, Rb-Sr age and geological setting of the Holyrood Granite, southeast Newfoundland: Canadian Jour. Earth Sci., v. 3, p. 947–957.

MATLEY, C. A., 1913, The geology of Bardsey Island: Geol. Soc. London Quart. Jour., v. 29, p. 514–533.

———, 1928, The Precambrian complex and associated rocks of southwestern Lleyn: *ibid.*, v. 84, p. 440–504.

MOORBATH, S., AND SHACKLETON, R. M., 1966, Isotopic ages from the Precambrian Mona Complex of Anglesey, North Wales, Great Britain: Earth and Planetary Sci. Letters, v. 1, p. 113–117.

NATARAJ, T. S., 1967, Glaucophanic metamorphism in Anglesey (thesis); Leeds, England, Univ. Leeds, 103 p.

RAISIN, C. A., 1893, Variolite of the Lleyn and associated volcanic rocks: Geol. Soc. London Quart. Jour., v. 44, p. 145–165.

RODGERS, J., 1972, Latest Precambrian (post-Grenville) rocks of the Appalachian region: Am. Jour. Sci., v. 272, p. 507–520.

SEDGWICK, A., 1838, Synopsis of the English stratified rocks inferior to the Old Red Sandstone: Geol. Soc. London Proc., v. 2, p. 675–685.

———, 1843, An outline of the geological structure of North Wales: *ibid.*, v. 4, p. 212–224.

SHACKLETON, R. M., 1953, The structural evolution of North Wales: Liverpool and Manchester Geol. Jour., v. 1, p. 261–297.

———, 1954, The structure and succession of Anglesey and the Lleyn Peninsula: British Assoc. Adv. Sci., v. 11 (41), p. 106–108.

———, 1956, Notes on the structure and relations of the Precambrian and Ordovician rocks of southwestern Lleyn (Caernarvonshire): Liverpool and Manchester Geol. Jour., v. 1, p. 400–409.

———, 1957, Downward-facing structures of the Highland Border: Geol. Soc. London Quart. Jour., v. 113, p. 361–392.

———, 1969, The Precambrian of North Wales, in WOOD, A. (ed.), The Precambrian and lower Palaeozoic rocks of Wales: Cardiff, Univ. Wales Press, p. 1–22.

SHARPE, D., 1846, Contributions to the geology of North Wales: Geol. Soc. London Quart. Jour., v. 2, p. 283–316.

STEINMANN, G., 1905, Gelogische Beobachtungen in den Alpen: Naturf. Gesell. Freiburg, Ber., v. 16, p. 18–67.

WOOD, D. S., 1960, The geology and structure of the Rhoscolyn district, Holy Island, Anglesey (thesis): Liverpool, England, Univ. Liverpool, 65 p.

———, 1969, The base and correlation of the Cambrian rocks of North Wales, in WOOD, A. (ed.), The Precambrian and lower Palaeozoic rocks of Wales: Cardiff, Univ. Wales Press, p. 47–66.

ORIGIN OF FRANCISCAN MELANGES IN NORTHERN CALIFORNIA

M. C. BLAKE, JR., AND DAVID L. JONES
U. S. Geological Survey, Menlo Park, California

ABSTRACT

In northern California, chaotic Franciscan mélange occurs beneath the overlying ophiolite and Great Valley Sequence. Identical mélanges occur to the west, separating well-bedded, coherent Franciscan units that differ markedly in age. Detailed studies in several places indicate that these mélanges mark the boundaries of imbricate thrust sheets, and they appear to occur at several discrete structural horizons.

The mélange comprises blocks of graywacke, greenstone, chert, serpentinite, and isolated so-called knockers of high-grade blueschist and eclogite set in a matrix of sheared and quartz-veined mudstone and minor sandstone. Except for the blocks of high-grade schist, these rocks are similar to, but more deformed than, the orderly sedimentary, volcanic, and other rocks that occur immediately above the Coast Range thrust at the base of the Great Valley Sequence. Unlike the other Franciscan units, the mélanges contain relatively abundant fossils, mainly *Buchia*, radiolarians, and dinoflagellates. Significantly, all of these fossils are of Tithonian to Valanginian age.

We suggest, on the basis of similarity of lithology and fossil content, that the matrix of the mélanges represents a distal, or seaward, portion of the basal sediments of the Great Valley Sequence and that the abundant greenstone, chert, and serpentinite found as tectonic blocks within the mélanges were derived from the underlying oceanic crust and upper mantle.

Formation of the mélanges must be related to multiple subduction of separate plates, the mélange being generated repeatedly from the ultramafic-mafic-chert and *Buchia*-bearing shale and minor graywacke sequence that constitutes the oldest rocks of the Coast Ranges. This process of subduction probably began in the Early Cretaceous and continued into the Tertiary, as Eocene fossils have been found recently in deformed Franciscan (coastal belt) rocks structurally below the mélanges.

The tectonic blocks of high-grade blueschist and eclogite were formed during an earlier period of subduction, then embedded in serpentinite and carried westward by flow in the upper mantle. During subsequent subduction, the serpentinite and embedded blocks of schist were tectonically mixed with the overlying volcanic rocks, chert, graywacke, and fossiliferous shale.

INTRODUCTION AND GEOLOGIC BACKGROUND

A popular concept of the Franciscan assemblage is a mélange comprising blocks of ultramafic rocks, gabbro, diabase, volcanic rocks, radiolarian chert, limestone, graywacke, and metamorphic rock. According to plate tectonic models proposed by several geologists, late Mesozoic oceanic crust was generated at a midocean ridge and was subsequently covered by radiolarian oozes formed on abyssal plains, and limestone formed on rises and seamounts as it was swept toward the continent by ocean-floor spreading. As this oceanic plate encountered a postulated trench marking the western margin of the North American plate, it was further covered by clastic sediments. "Mixing of these rocks under the trench and along the Benioff zone, produced the mélanges" (Hsü, 1971, p. 1168). This hypothesis may be correct for some parts of the Franciscan but does not appear to fit known geologic data for the Coast Ranges north of San Francisco. Published mapping of the Coast Ranges by Brown (1964a), Rich (1971), Swe and Dickinson (1970), Fox and others (1973), Berkland (1969, 1972), Sims and others (1973), Blake and others (1971 and in preparation), and Suppe (1972) and unpublished reports by geologists of the California Department of Water Resources indicate that the Franciscan rocks comprise a number of subparallel, north-northwest trending lithologic belts. These include (1) relatively well-bedded graywacke and mudstone with interbedded chert and volcanic rocks; (2) metagraywacke and metachert and metagreenstone, characterized by a faint to pronounced metamorphic fabric and development of new metamorphic minerals such as lawsonite, jadeite, and glaucophane; and (3) massive arkosic sandstone and mudstone alternating with thin-bedded flyschlike sandstone and mudstone. The contacts between these distinctive lithologic units are shear zones defined by mélanges made up of blocks and slabs of graywacke, greenstone, chert, serpentinite, and their metamorphosed equivalents set in a matrix of dark, sheared, and quartz-veined mudstone and minor sandstone. Structurally overlying all these lithologic belts is a discontinuous sheet of ultramafic rocks, gabbro, diabase, pillow lavas, and chert that in turn is overlain by the sedimentary rocks of the Great Valley Sequence of Late Jurassic and Cretaceous age. The major outcrop area of these upper plate rocks is along the west side of the Great Valley; they also occur as scattered tectonic outliers (klippen) west of the valley. The Coast Range thrust marks the base of this upper plate.

DESCRIPTION OF LITHOLOGIC BELTS

Franciscan assemblage.—The structurally highest Franciscan unit in northern California is informally named the Yolla Bolly belt after the Yolla Bolly Mountains in the northern part of the area mapped (fig. 1). Rocks of this belt are predominantly graywacke and mudstone but include significant interbedded radiolarian chert and volcanic rocks.

The medium- to coarse-grained graywacke occurs in beds that range in thickness from centimeters to tens of meters but most commonly are about 30 cm to 1 m thick. The darker mudstone is generally subordinate, forming interbeds only a few centimeters thick. Other rocks present are mappable beds of pebble conglomerate and pebbly to bouldery mudstone. The mudstone locally is several hundred meters thick. These rocks contain these abundant sedimentary structures that suggest deposition by turbidity currents: graded beds, sole markings, and small-scale current-ripple laminations.

Radiolarian chert in a number of localities is clearly interbedded with coarse-grained graywacke and mudstone without associated volcanic rocks. In the Yolla Bolly area (fig. 1), three chert beds, each more than 50 m thick, have been mapped along strike for more than 15 km and extend northwest and southeast far beyond the mapped area (Blake, 1965).

Volcanic rocks in the Yolla Bolly belt range in composition from quartz keratophyre to basalt; tuff is particularly abundant. Pillow structures are extremely rare and may be restricted to bouldery mudstone intervals. Only in a very few places are volcanic rocks and radiolarian cherts found together; this suggests that in this area there is no genetic relation between them.

Fossils are extremely rare in this belt. Several collections of *Buchia* of Late Jurassic and Early Cretaceous age are known from the pebbly and bouldery mudstone beds; one collection of Lower Cretaceous *Buchia* is known from coarse-grained, foliated, and metamorphosed metagraywacke containing blueschist minerals.

Exotic rocks such as high-grade glaucophane schist, amphibolite, and eclogite are unknown from this belt, and serpentinite is extremely rare.

The Yolla Bolly belt is continuous along the west side of the Great Valley north of Stonyford, where it is overlain by the ophiolite and sedimentary rocks of the Great Valley Sequence (fig. 1). It extends north at least as far as the Pickett Peak Quadrangle (Irwin and others, in press) and possibly into southwestern Oregon (Blake and others, 1967), where it is structurally overlain by the older rocks of the Klamath Mountains province. Detailed study in the Yolla Bolly area (Blake, 1965, and unpublished data, 1973) indicates that the belt includes a number of mappable lithologic subunits upon which a regional metamorphic fabric is superposed. This fabric ranges from faint in the south and west to pronounced in the east and north, increasing in intensity toward the thrust fault that marks the upper boundary of the belt. The part of this belt texturally highest in grade has been named the South Fork Mountain Schist (Blake and others, 1967).

The lower contact of the Yolla Bolly belt has been mapped in detail near its southern end by Brown (1964a), who clearly showed that the metamorphic rocks (Brown's phyllonite unit) structurally overlie, above a low-angle fault contact, a mélange, termed a friction carpet by Brown. Later work by geologists of the California Department of Water Resources (unpublished mapping by Michael Dwyer, James Vantine, and others) extended Brown's contact to the north and west, where the friction carpet has been given the informal structural term, Skunk Rock mélange.

Several metagraywacke units lie west of the Skunk Rock mélange. Some of these are probably klippen of the Yolla Bolly belt, but at least one differs both lithologically and paleontologically, informally called the Hull Mountain belt (fig. 1) for the prominent mountain that it underlies. To the west is another metagraywacke belt, informally referred to as the Mount Sanhedrin belt, which consists of a number of other graywacke-rich Franciscan units, each bounded by mélange. These units, including the bounding mélanges, are largely unnamed, and very little is known concerning their age or metamorphic state except that they generally lack a metamorphic fabric and contain the mineral assemblage quartz-albite-pumpellyite-chlorite-white mica-celadonite ± aragonite, but not lawsonite. One of these belts, along the Eel River near English Ridge (fig. 1), yielded palynomorphs of late Early Cretaceous (Albian) age (J. O. Berkland, written commun., 1972).

The westernmost mapped unit is the coastal belt Franciscan, long known to contain fossils of Late Cretaceous age (Bailey and others, 1964). Rocks of the coastal belt differ from other Franciscan units in that the sandstones are notably more arkosic and contain very little lithic volcanic or chert detritus. The southernmost known occurrence of this unit is in the San Francisco Bay area. Fossils found in the coastal belt are of Late Cretaceous (Turonian and Campanian), Paleocene, and Eocene ages (fig. 1). Metamorphic grade appears to be lower than in the other Franciscan rocks, laumontite being abundant and prehnite-pumpellyite being locally present. Because of the widespread later faulting related to the San Andreas system, all rocks in the western part of the map area are fragmented, and an adequate stratigraphic or structural sequence in these youngest known Franciscan rocks has not yet been established. Near Cloverdale (fig. 1, west-central), however, the coastal belt structurally underlies a chaotic mélange unit and is isoclinally folded and sheared, suggesting that it was involved in deformation of the same kind of, but less intense than, deformation seen along the Coast Range thrust to the east (Blake and others, 1971).

Great Valley Sequence.—Rocks of the Great Valley Sequence differ from those of the Franciscan assemblage in many ways (see Bailey and others, 1964, and Bailey and Blake, 1969). The most pronounced differences are the great continuity of individual sandstone and shale beds, the broad open style of deformation, the striking lack of small-scale crumpling, folding, or faulting, and the lower sand/shale ratio. The base of the Great Valley Sequence consists of ophiolite and minor amounts of chert, but in many places these basal rocks have been faulted out and the sedimentary rocks of the Great Valley Sequence are in fault contact with sheared serpentinite or metamorphosed Franciscan rocks. Overlying the ophiolite is 16,000 m or more of mudstone, sandstone, and conglomerate ranging in age from Late Jurassic (mid-Tithonian) to latest Cretaceous. Sedimentary rocks low in the sequence are dominantly dark mudstone and some thin-bedded sandstone, including basaltic sandstone and tuffs (Brown, 1964a), pebbly mudstone, and conglomerate that locally aggregate 5,000 m in thickness.

Graded bedding and sole markings are common in these strata, suggesting deposition by turbidity currents. Current-direction indicators, such as flute casts, groove casts, and small-scale crossbedding, are abundant along the west edge of the northern half of the Great Valley. They show remarkably uniform, nearly north to south transport through the entire sequence except for a local reversal in Turonian time (Ojakangas, 1968). Their abundance and consistency suggest that the currents were longitudinal and that through much of the time of deposition this area was the central part of a northward trending trough in which the Great Valley Sequence accumulated.

In addition to significant differences in the sand/shale ratio between the coeval older Franciscan rocks (Yolla Bolly belt) and the *Buchia* beds of the Great Valley Sequence (sandstone/mudstone ratio 3:1 in Yolla Bolly belt as compared with 1:3 in Great Valley Sequence), there are important compositional differences. The petrology of the sandstone of the Great Valley Sequence in northern California has been detailed by Ojakangas (1968) and Dickinson and Rich (1972), and similar studies have been made in the southern Coast Ranges (Gilbert and Dickinson, 1970). These studies indicate that sediments of the Great Valley Sequence fall into several distinctive petrologic units that closely approximate mapped stratigraphic units. The oldest petrofacies (Tithonian and Neocomian) consists of quartz-poor and feldspatholithic sandstones containing very little potassium feldspar and mica. These rocks are considered by most workers to have been derived largely from the Upper Jurassic volcanic and associated plutonic rocks of the western Sierra Nevada, but our own studies suggest that much of the volcanic detritus near the base was derived from the underlying ophiolite sequence.

That ophiolite was locally present at the surface during the filling of the late Mesozoic basins is proven by the extensive development of what may be called sedimentary serpentine interbedded with Lower Cretaceous strata of the Great Valley Sequence. These unusual sedimentary beds, composed almost entirely of serpentine detritus, locally contain fossils. Although they have been known for many years, their significance seems to have been largely overlooked. They are nowhere adequately described but are mentioned by Taliaferro (1943, p. 206–207), Bailey and others (1964, p. 164), Lawton (1956), Brice (1953, p. 25), Averitt (1945, p. 73), and Moiseyev (1966).

Undoubted Sierran detritus appears near the Jurassic-Cretaceous boundary in the form of conglomerates containing granitic pebbles and cobbles dated as Late Jurassic (Irwin, 1966). Composition of Cretaceous sandstones younger than Neocomian shows a good correlation with the dated intrusive episodes in the Sierra Nevada. There seems to be little doubt that the batholiths and related calcalkaline volcanoes were the source of most of the Cretaceous sediments (Dickinson and Rich, 1972).

West of the Great Valley, a number of iso-

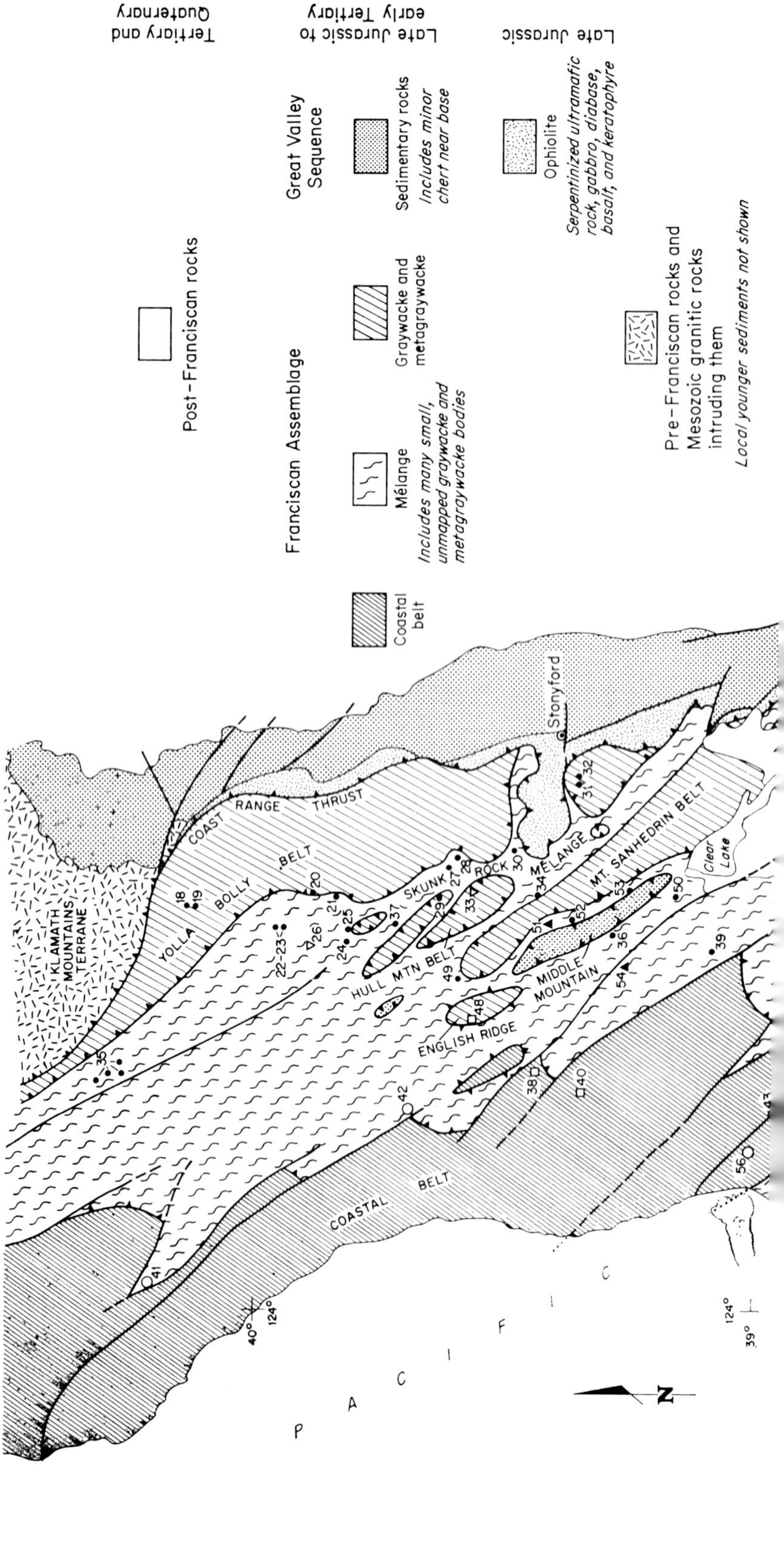
Tertiary and Quaternary
Post-Franciscan rocks
Late Jurassic to early Tertiary
Franciscan Assemblage
Great Valley Sequence
Coastal belt
Mélange
Includes many small, unmapped graywacke and metagraywacke bodies
Graywacke and metagraywacke
Sedimentary rocks
Includes minor chert near base
Late Jurassic
Ophiolite
Serpentinized ultramafic rock, gabbro, diabase, basalt, and keratophyre
Pre-Franciscan rocks and Mesozoic granitic rocks intruding them
Local younger sediments not shown
KLAMATH MOUNTAINS TERRANE
YOLLA BOLLY BELT
COAST RANGE THRUST
Stonyford
SKUNK ROCK MÉLANGE
MT. SANHEDRIN BELT
Clear Lake
HULL MTN BELT
MIDDLE MOUNTAIN
ENGLISH RIDGE
COASTAL BELT
P A C I F I C
40°
124°
39°
N

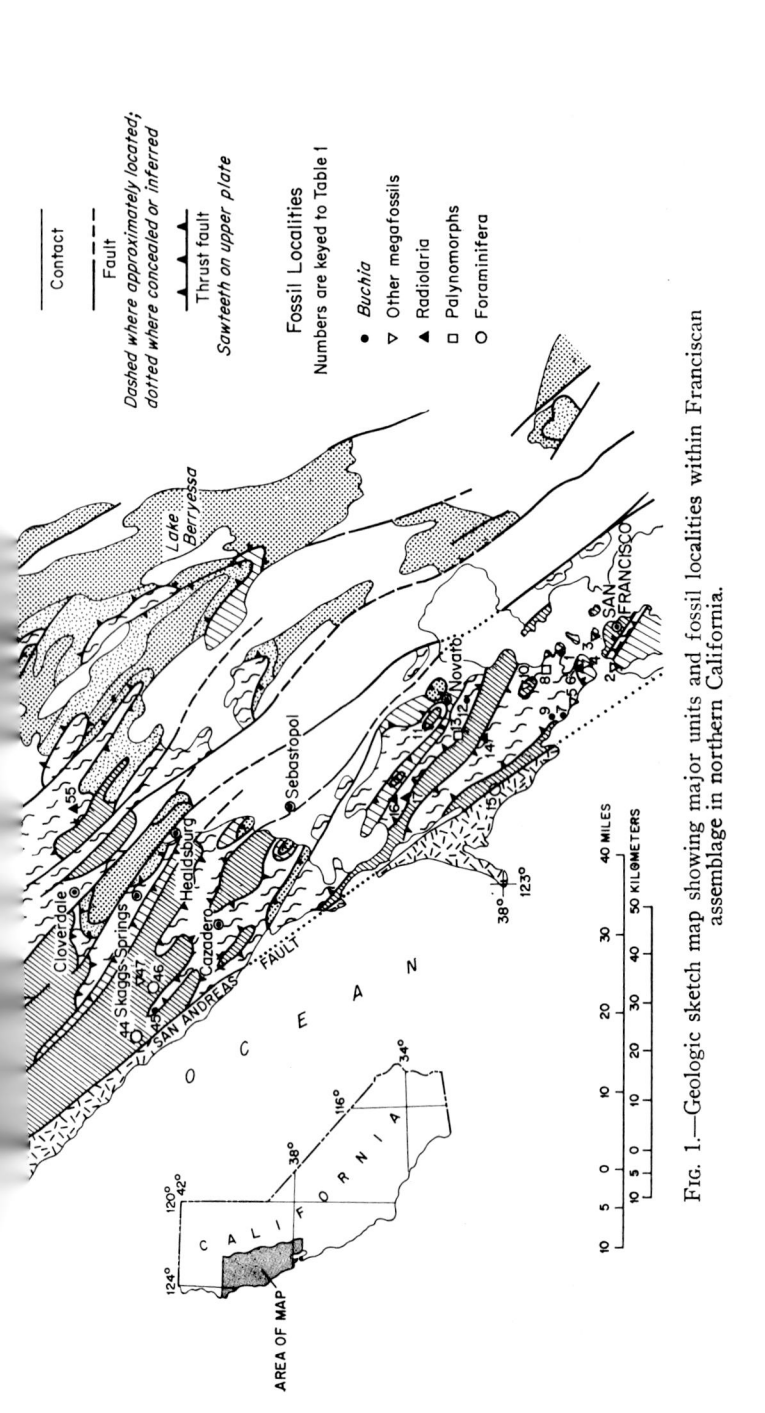

Fig. 1.—Geologic sketch map showing major units and fossil localities within Franciscan assemblage in northern California.

tophyre and albite granite than has been reported from the present active ridge crests;

(c) generally lacks abundant dikes comparable to the sheeted dike complexes described elsewhere and attributed to formation at a spreading ridge crest; and

(d) lacks the thick pelagic cover that would be expected if it formed in midocean far to the west of the continental mass (Scholl and Marlow, this volume).

A more reasonable environment for generation of new crust is an interarc basin or marginal sea near the present Great Valley, although an inferred remnant arc to the west has not been identified. One possible solution to this dilemma has already been suggested by Karig (1972, p. 1065). He pointed out that the siliceous and intermediate igneous rocks, which locally make up much of the Great Valley ophiolite, may be related to arc volcanism rather than representing midocean ridge material. Chemical analyses of the California ophiolite (M. C. Blake, Jr., and E. H. Bailey, unpublished data, 1972) indicate that many of these rocks are close to andesite, dacite, and rhyolite in composition except for their extremely high soda-to-potash ratios, which may be due to later metasomatism.

Because of the lack of reliable radiogenic data and the great area covered by younger sediments in the Great Valley itself, the exact sequence of events and even the polarity of the inferred subduction zones are still highly speculative. An east-dipping zone is favored because this seems to fit best the overall tectonic style of the Pacific coast region. For example, possible fossil subduction zones in the Klamath Mountains all appear to have dipped to the east (Irwin, 1966, fig. 6, p. 32), and K_2O contents in Mesozoic granitic plutons appear to increase eastward at some places, extending into Nevada (Dickinson, 1970; Hotz, 1971; Bateman and Dodge, 1970; Evernden and Kistler, 1970).

According to the most recent sedimentary and tectonic models (Bailey and Blake, 1969; Hamilton, 1969; Ernst, 1970; Dickinson, 1971; and many others) and beginning in mid-Tithonian time, vast quantities of sediment were derived from the uplifted ancestral Sierran and Klamath mountains, were carried into the Great Valley basin by west-flowing rivers, then redeposited by longitudinal turbidity currents. At the same time, somewhat coarser grained sediments were periodically carried across the Great Valley basin in submarine canyons and dumped into the Franciscan trench. During periods of relative quiescence, thick lenses of radiolarian chert accumulated in the trench. Concurrent with sedimentation but probably taking place during several distinct episodes or pulses, the trench material was subducted to form high-pressure lawsonite and jadeite-bearing metagraywackes as well as forming the mélanges.

Recognition that the mélanges of the northern Coast Ranges formed from a restricted stratigraphic succession that belongs to the lowest part of the upper Mesozoic sedimentary sequence of the Coast Ranges imposes definite constraints on interpretation of the mode of origin and significance of these remarkable tectonostratigraphic units as well as on the previously described model. We interpret their significant features as follows:

(1) The matrix of the mélanges is thought to represent a distal, or seaward, portion of the so-called Knoxville Formation on the basis of similarity of lithology and fossil content with the Knoxville at the base of the Great Valley Sequence.

(2) The abundant greenstone, chert, and serpentinite found as tectonic blocks within the mélanges probably were derived from the immediately underlying oceanic crust and upper mantle.

(3) This oceanic crust and serpentinite probably was not formed at an oceanic ridge lying far to the west, but may have formed in an interarc basin or a marginal sea.

(4) Some or all the siliceous to intermediate igneous rocks, which intrude and overlie mafic portions of the ophiolite, may represent the products of island arc volcanism west of the marginal basin.

(5) The tectonic blocks, or knockers as they have been called, of high-grade blueschist and eclogite that occur in the mélanges were formed during an earlier period of subduction that deformed the Galice and Mariposa Formations 150 to 160 my ago, that is, at about the time the ophiolite was formed (Lanphere, 1971).

(6) Because the fossiliferous (*Buchia*-bearing) mélanges in places are in contact with much younger rocks (Late Cretaceous and younger), yet do not contain abundant blocks of these younger rocks, it seems unlikely that they could have formed through sedimentary sliding, for such a process would lead to an intimate mixing of rocks of widely differing ages throughout the mélange. Only along the sheared contacts between mélange and Franciscan graywacke or metagraywacke units is much mixing of rock types observable.

Formation of the mélanges must involve multiple subducted plates, the mélange being generated repeatedly from the ultramafic, mafic, chert, and *Buchia*-bearing shale sequence that constitutes the oldest rocks of the Coast Ranges. The mélanges could have formed in two ways: (1) by tearing up and shearing out the subducted plate composed of these rocks or (2) by abrading the base of a previously subducted ophiolite and shale sequence by passage of another younger plate beneath the overlying older rocks. According to the paleontologic evidence, this process of subduction probably began in late Valanginian or Hauterivian time. It continued through the Eocene, as shown by the recent discovery of Eocene fossils in deformed Franciscan rocks (coastal belt) lying structurally below a typical mélange.

In order to fit these petrologic, paleontologic, tectonic, and radiogenic data into a reasonable geologic history, we have prepared a series of diagrams of hypothetical models based on the plate tectonic model and in particular on the recent studies by Karig (1972) in the southwest Pacific. The first diagram (fig. 2a) shows conditions during Tithonian to Valanginian time, which is considered to be about the same time as the Yosemite intrusive epoch (148 to 132 my, Evernden and Kistler, 1970). We suggest in our highly generalized cartoon that the basal part of the Great Valley Sequence was deposited in an interarc basin above an east-dipping subduction zone. The entire ophiolite at the base of the Great Valley Sequence probably formed prior to the deposition of the Knoxville Formation, but this is not clear. The older Franciscan sediments (Yolla Bolly belt and probably the Tithonian to Valanginian metagraywacke and chert of the Diablo Range) are seen as formed in an arc-trench gap rather than in a trench environment. This concept is partly based on radiogenic ages on blueschists, indicating that the type III metabasalts and metacherts of Cazadero (Lee and others, 1964; Suppe and Armstrong, 1972) and other areas were being subducted at about the time that the Knoxville and older Franciscan beds were being deposited. The diagram (fig. 2a) shows a possible remnant volcanoplutonic arc that is inferred to have rifted away from the Mariposa-Amador arc at the site of the present Sierra Nevada during the initial formation of the ophiolite-floored Great Valley basin. This arc would have been in the right position to provide the coarse-grained volcanoplutonic detritus to the westernmost Great Valley Sequence, seen in klippen today north of San Francisco, and also to the older Franciscan graywackes. The problem here is that the inferred arc is now completely gone and if once present must have been subsequently eroded away or subducted. [A similar remnant arc seems to be required to explain the geologic data recently presented by Ross, Wentworth, and McKee (1973). Their data indicate that the Late Cretaceous Gualala Formation of Weaver (1943) was derived from a granitic source area on the west and oceanic crust to the east. Prior to Tertiary offset on the San Andreas Fault, the inferred Gualala basin was believed to lie near the extreme southern margin of the Great Valley, west of the Sierra Nevada Batholith.)]

During the Early Cretaceous (fig. 2b) a pronounced change occurred in the tectonic regime as the subduction zone flattened and began to consume the older Franciscan rocks deposited in the former arc-trench gap. This event is inferred from radiogenic dating of Sierran plutonic rocks (the Huntington Lake epoch of Evernden and Kistler, 1970) and from the dating of blueschists in the Yolla Bolly and other areas (Suppe and Armstrong, 1972). The change in dip of the subduction zone to near horizontal is inferred from the present geometry of the Coast Range thrust (Bailey and others, 1970), from the wide east-west extent of calcalkaline plutonic rocks of this age, and the apparent lower P-T conditions inferred in the mineral assemblages of the 110 my blueschists as compared with the older ones. Detailed studies of the Great Valley Sequence in northern Sacramento Valley (Jones, Bailey, and Imlay, 1969; Jones and Irwin, 1971) indicate that the North American plate was being deformed by tear faults related to absolute westward movement of the Klamath Mountains relative to the Great Valley Sequence at the time the older Franciscan sediments were being underthrust along the oceanic plate. The timing of these events coincides closely with a pronounced acceleration in plate motion inferred from magnetic studies in both the Pacific and Atlantic (Larson and Chase, 1972; Larson and Pitman, 1972). It has recently been proposed by Hyndman (1972) that, if the absolute motion of *both* plates is convergent, there would be a pronounced flattening of the 45° or greater dip in most presently active Benioff zones.

It is at this time that the development of the mélanges is believed to have commenced. The oldest *in situ* Franciscan unit is the Yolla Bolly belt, which, as described, directly underlies both the older rocks of the Klamath Mountains province and, to the south, the Great Valley Sequence, including the basal ophiolite. Following subduction of the Yolla Bolly belt, the rapid westward motion of the overriding North American plate apparently resulted in periodic underthrusting of part of the Great Valley Sequence and underlying ophiolite. This material represents the tectonic mélange that marks the

(a)

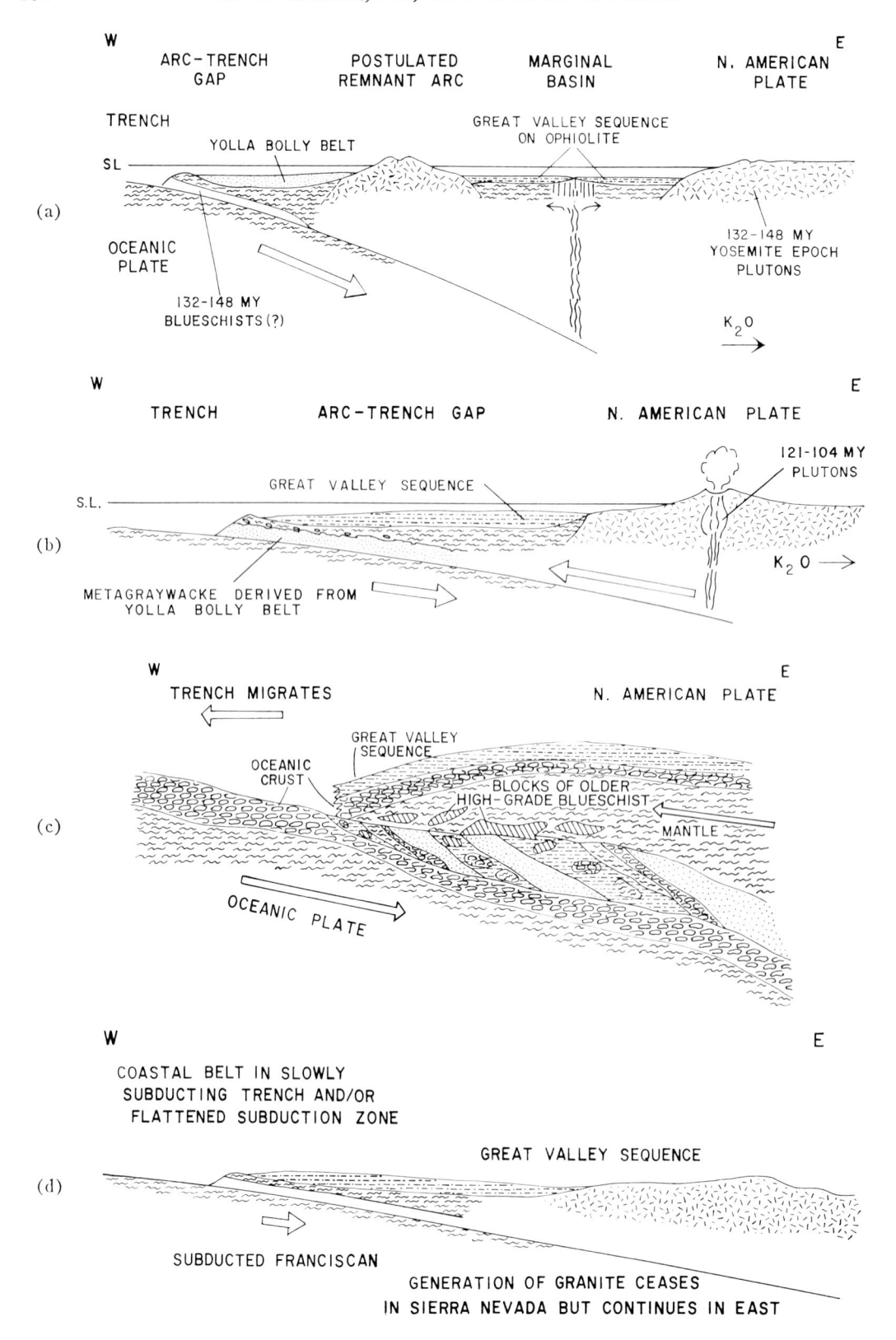

W — ARC–TRENCH GAP — POSTULATED REMNANT ARC — MARGINAL BASIN — N. AMERICAN PLATE — E

TRENCH
YOLLA BOLLY BELT
GREAT VALLEY SEQUENCE ON OPHIOLITE
SL
OCEANIC PLATE
132–148 MY BLUESCHISTS (?)
132–148 MY YOSEMITE EPOCH PLUTONS
$K_2O \rightarrow$

(b)

W — TRENCH — ARC–TRENCH GAP — N. AMERICAN PLATE — E

GREAT VALLEY SEQUENCE
121–104 MY PLUTONS
S.L.
METAGRAYWACKE DERIVED FROM YOLLA BOLLY BELT
$K_2O \rightarrow$

(c)

W — TRENCH MIGRATES — N. AMERICAN PLATE — E

OCEANIC CRUST
GREAT VALLEY SEQUENCE
BLOCKS OF OLDER HIGH-GRADE BLUESCHIST
MANTLE
OCEANIC PLATE

(d)

W — E

COASTAL BELT IN SLOWLY SUBDUCTING TRENCH AND/OR FLATTENED SUBDUCTION ZONE
GREAT VALLEY SEQUENCE
SUBDUCTED FRANCISCAN
GENERATION OF GRANITE CEASES IN SIERRA NEVADA BUT CONTINUES IN EAST

boundaries between successive Franciscan subducted plates. The previously subducted high-grade metamorphic rocks, such as eclogite, glaucophane-epidote gneiss, and amphibolite, were moved up and westward as part of the upper plate (embedded in the serpentinized basal ophiolite) and were resubducted to form the so-called polyphase knockers (Coleman and Lanphere, 1971). Our proposed model for these structures is shown in more detail in figure 2c. Eventually, the subduction slowed or nearly ceased, possibly as a result of a decrease in plate motions about 80 my ago as inferred from the lack of calcalkaline volcanoplutonic activity as well as from the lack of evidence for blueschist formation younger than that age (fig. 2d). That subduction continued through the Eocene is shown by the presence of deformed Eocene Franciscan rocks (coastal belt). No granitic plutons of this age are known from the Sierra Nevada Batholith, although Laramide and younger events farther east are known to be of this same age. Apparently, then, some kind of change in tectonic activity occurred about 80 my ago, when subduction slowed or when some other related activity took place to cause igneous activity to be shifted far to the east and high-pressure metamorphism apparently to be replaced by a shallower phenomenon. This change possibly was brought about by an even greater flattening of the subduction zone in such a manner that the underthrust rocks in California did not attain the depths necessary for generation of either blueschists or calcalkaline magmas.

ACKNOWLEDGMENTS

We are greatly indebted to numerous individuals for providing geologic and paleontologic data. For their generous premission to use unpublished data, we thank James Berkland, Appalachian State University; Salem Rice, California Division of Mines and Geology; and M. J. Dwyer and James Vantine, California Department of Water Resources. For identifying the palynomorphs and radiolarians, we thank respectively W. R. Evitt, Stanford University, and E. A. Pessagno, Jr., University of Texas. We acknowledge much valuable criticism and review from our colleagues of the U.S. Geological Survey and from the participants at the "Conference on Modern and Ancient Geosynclinal Sedimentation," Madison, Wisconsin, November 10-11, 1972.

Publication has been authorized by the Director, U.S. Geological Survey.

REFERENCES

AVERITT, PAUL, 1945, Quicksilver deposits of the Knoxville district, Napa, Yolo, and Lake Counties, California: California Jour. Mines and Geology, v. 41, no. 2, p. 65–89.
BAILEY, E. H., AND BLAKE, M. C., JR., 1969, Tektonicheskoe razvitiye zapadnoy kalifornii v pozdnem mezozoe (pts. 1 and 2) (Tectonic development of western California during the late Mesozoic): Geotektonika, no. 3, p. 17–30; no. 4, p. 24–34 (in Russian).
——, ——, AND JONES, D. L., 1970, On-land Mesozoic oceanic crust in California Coast Ranges: U.S. Geological Survey Prof. Paper 700-C, p. 70–81.
——, IRWIN, W. P., AND ——, 1964, Franciscan and related rocks and their significance in the geology of western California: California Div. Mines and Geology Bull. 183, 177 p.
BATEMAN, P. C., AND DODGE, F. C. W., 1970, Variations of major chemical constituents across the central Sierra Nevada Batholith: Geol. Soc. America Bull., v. 81, p. 409–420.
BERKLAND, J. O., 1964, Notes on the geology of the Alder Creek area near Point Arena, California: Calif. Div. Mines and Geology Min. Inf. Service, v. 17, p. 139–141.
——, 1969, Geology of the Novato Quadrangle, Marin County, California (M.S. thesis): San Jose, California, San Jose State Coll., 146 p.
——, 1972, Paleogene "frozen" subduction zone in the Coast Ranges of northern California, in Tectonics: 24th Internat. Geol. Cong., Montreal, 1972, Repts. Sec., no. 3, p. 99–105.
BLAKE, M. C., JR., 1965, Structure and petrology of low-grade metamorphic rocks, blueschist facies, Yolla Bolly area, northern California (Ph.D. thesis): Stanford California, Stanford Univ., 91 p.
——, IRWIN, W. P., AND COLEMAN, R. G., 1967, Upside-down metamorphic zonation, blueschist facies, along a regional thrust in California and Oregon, in Geological Survey Research, 1967: U.S. Geol. Survey Prof. Paper 575-C, p. 1–9.

FIG. 2.—Hypothetical cross sections in northern California showing progressive changes in plate tectonic regime and development of Franciscan mélanges. a, Tithonian to Valanginian: formation of hypothetical marginal ocean basin at present site of the Great Valley; b, Early Cretaceous; acceleration of westward movement of North American plate leads to overriding of trench and flattening of subduction zone. c, Schematic diagram showing formation of imbricate mélange and metagraywacke units. d, Late Cretaceous to post-Eocene; note very much flattened subduction zone.

———, AND OTHERS, 1971, Preliminary geologic map of western Sonoma county and northernmost Marin County, California: *ibid.*, open-file report.

———, AND ———, 1973, (in preparation) Preliminary geologic map of Marin County, California: *ibid.*, map, scale 1:62,500.

BRICE, J. C., 1953, Geology of the Lower Lake Quadrangle, California: California Div. Mines Bull. 166, 72 p.

BROWN, R. D., JR., 1964a, Geological map of the Stonyford Quadrangle, Glenn, Colusa, and Lake Counties, California: U.S. Geol. Survey Min. Inv. Field Studies Map MF-279, scale 1:48,000.

———, 1964b, Thrust-fault relations in the northern Coast Ranges, California: *ibid.*, Prof. Paper 475-D, p. 7–13.

COLEMAN, R. G., AND LANPHERE, M. A., 1971, Distribution and age of high-grade blueschist, associated eclogites, and amphibolites from Oregon and California: Geol. Soc. America Bull., v. 82, p. 2397–2412.

DICKINSON, W. R., 1970, Relations of andesites, granites, and derivative sandstones to arc-trench tectonics: Rev. Geophysics and Space Physics, v. 8, p. 813–860.

———, 1971, Clastic sedimentary sequences deposited in shelf, slope, and trough settings between magmatic arcs and associated trenches: Pacific Geology, v. 3, p. 15–30.

———, AND RICH, E. I., 1972, Petrologic intervals and petrofacies in the Great Valley Sequence, Sacramento Valley, California: Geol. Soc. America Bull., v. 83, p. 3007–3024.

ERNST, W. G., 1970, Tectonic contact between the Franciscan mélange and the Great Valley Sequence, crustal expression of a late Mesozoic Benioff zone: Jour. Geophys. Research, v. 75, p. 886–902.

EVERNDEN, J. F., AND KISTLER, R. W., 1970, Chronology of emplacement of Mesozoic batholithic complexes in California and western Nevada: U.S. Geol. Survey Prof. Paper 623, 42 p.

FOX, K. F., AND OTHERS, 1973, Preliminary geologic map of eastern Sonoma, Napa, and Solano Counties, California: *ibid.*, Map MF-483, scale 1:62,500.

GEALEY, W. K., 1951, Geology of the Healdsburg Quadrangle, California: California Div. Mines Bull. 161, p. 7–50.

GHENT, E. D., 1963, Fossil evidence for maximum age of metamorphism in part of the Franciscan Formation, northern Coast Ranges, California: *ibid.*, Special Rept. 82, p. 41.

GILBERT, W. G., AND DICKINSON, W. R., 1970, Stratigraphic variations in sandstone petrology, late Mesozoic Great Valley Sequence, southern Santa Lucia Range, California: Geol. Soc. America Bull., v. 81, p. 949–954.

HAMILTON, WARREN, 1969, Mesozoic California and the underflow of Pacific mantle: *ibid.*, v. 80, p. 2409–2429.

HOTZ, P. E., 1971, Plutonic rocks of the Klamath Mountains, California and Oregon: U.S. Geol. Survey Prof. Paper 684B, p. 1–20.

HSÜ, K. J., 1971, Franciscan mélanges as a model for eugeosynclinal sedimentation and underthrusting tectonics: Jour. Geophys. Research, v. 76, p. 1162–1170.

HYNDMAN, R. D., 1972, Plate motions relative to the deep mantle and the development of subduction zones: Nature, v. 238, p. 263–265.

IRWIN, W. P., 1957, Franciscan Group in Coast Ranges and its equivalents in Sacramento Valley, California: Am. Assoc. Petroleum Geologists Bull., v. 41, p. 2284–2297.

———, 1964, Late Mesozoic orogenies in the ultramafic belts of northwestern California and southwestern Oregon, *in* Geological Survey Research 1964: U.S. Geol. Survey Prof. Paper 501-C, p. 1–9.

———, 1966, Geology of the Klamath Mountains province: California Div. Mines and Geology Bull. 190, p. 19–38.

———, AND OTHERS, in press, Geologic Map of the Pickett Peak Quadrangle, Trinity County, California: U.S. Geol. Survey Map GQ-1111.

JONES, D. L., BAILEY, E. H., AND IMLAY, R. W., 1969, Structural and stratigraphic significance of the *Buchia* zones in the Colyear Spring-Paskenta area, California: *ibid.*, Prof. Paper 647-A, p. 1–24.

———, AND BLAKE, M. C., JR., in preparation, Fossil localities in Franciscan Assemblage, northern California.

———, AND IRWIN, W. P., 1971, Structural implications of an offset early Cretaceous shoreline in northern California: Geol. Soc. America Bull., v. 82, p. 815–822.

KARIG, D. E., 1972, Remnant arcs: *ibid.*, v. 83, p. 1057–1068.

LANPHERE, M. A., 1971, Age of the Mesozoic oceanic crust in the California Coast Ranges: *ibid.*, v. 82, p. 3209–3212.

LARSON, R. L., AND CHASE, C. G., 1972, Late Mesozoic evolution of the western Pacific Ocean: *ibid.*, v. 83, p. 3627–3644.

———, AND PITMAN, W. C., III, 1972, World-wide correlation of Mesozoic magnetic anomalies, and its implications: *ibid.*, p. 3645–3662.

LAWTON, J. E., 1956, Geology of the north half of the Morgan Valley Quadrangle and the south half of the Wilbur Springs Quadrangle, California (Ph.D. thesis): Stanford, California, Stanford Univ.

LEE, D. E., AND OTHERS, 1964, Isotopic ages of glaucophane schists from the area of Cazadero, California: U.S. Geol. Survey Prof. Paper 475-D, p. 105–107.

MOISEYEV, A. N., 1966, The geology and geochemistry of the Wilbur Springs quicksilver district, Colusa and Lake Counties, California (Ph.D. thesis): Stanford, California, Stanford Univ., 214 p.

O'DAY, MICHAEL, AND KRAMER, J. C., 1972, Geologic guide to the northern Coast Ranges—Lake, Sonoma, and Mendocino Counties, California: Sacramento, California, Sacramento Geol. Soc. Ann. Field Trip 1972, Guidebook, p. 51–56.

OJAKANGAS, R. W., 1968, Cretaceous sedimentation, Sacramento Valley, California: Geol. Soc. America Bull., v. 79, p. 976–1008.

RICH, E. I., 1971, Geologic map of the Wilbur Springs Quadrangle, Colusa and Lake Counties, California: U.S. Geol. Survey Misc. Geol. Inv. Map I-538, scale 1:48,000.

Ross, D. C., Wentworth, C. M., and McKee, E. H., 1973, Cretaceous mafic conglomerate near Gualala offset 350 miles by San Andreas Fault from oceanic crustal source near Eagle Rest Peak, California: U.S. Geol. Survey Jour. Research, v. 1, p. 45–52.

Sims, J. D., and others, 1973, Preliminary geologic map of Solano County and parts of Napa, Contra Costa, Marin and Yolo Counties, California: *ibid.*, Map MF-484, scale 1:62,500.

Soliman, S. M., 1965, Geology of the east half of the Mount Hamilton Quadrangle, California: California Div. Mines and Geology Bull. 185, 32 p.

Suppe, John, in press, Geology of the Leech Lake Mountain-Ball Mountain region, California: a cross section of the northeastern Franciscan belt: California Univ. Publs. Geol. Sci.

——, and Armstrong, R. L., 1972, Potassium-argon dating of Franciscan metamorphic rocks: Am. Jour. Sci., v. 272, p. 217–233.

Swe, Win, and Dickinson, W. R., 1970, Sedimentation and thrusting of late Mesozoic rocks in the Coast Ranges near Clear Lake, California: Geol. Soc. America Bull., v. 81, p. 165–188.

Taliaferro, N. L., 1943, Franciscan-Knoxville problem: Am. Assoc. Petroleum Geologists Bull., v. 27, p. 109–219.

Weaver, C. E., 1943, Point Arena-Fort Ross region: California Div. Mines Bull. 118, pt. 3, p. 628–632.

OPHIOLITE GENERATION AND EMPLACEMENT: A KEY TO ALPINE EVOLUTION

J. F. DEWEY
State University of New York at Albany, New York

ABSTRACT

Fully developed ophiolite sequences are widespread in the Alpine System from the Betics to the Oman. Ophiolite occurrences vary from giant nappes, through slabs and blocks in mélanges, to small clasts in flysch and wildflysch. Contact relations invariably involve tectonic juxtaposition of ultramafics against the country rocks. Where ophiolite sheets occur as high level nappes obducted onto shelf and (or) exogeosynclinal terranes (e.g., Oman), full, fresh ophiolite sequences are well preserved. Basal contact relationships of such sheets may involve a complex combination of mylonitic ultramafics, ophiolitic wildflysch, transported blueschists, garnet amphibolites, greenschists, and rodingites. Such rocks may represent a variety of tectonic-metamorphic environments from subduction zones to wasting mélanges developed and telescoped into juxtaposition during progressive obduction. High-level ophiolite nappes involve an obduction problem that is poorly understood. Such sheets may originate from Mediterranean Ridge-like welts, or from ophiolitic basements of arc-trench gaps that are driven from the advancing jaws of continental blocks, or by the development of subduction zones that are adjacent to passive continental margins and that have a continent-facing polarity. All such nappes appear to involve some gravity sliding, at least during the final stages of obduction. Some ophiolite assemblages (e.g., those of the Platta Nappe) are intensely shredded, deformed, and metamorphosed in narrow flat-lying zones between major crystalline nappe sequences. Such assemblages, whose emplacement postdates blueschist metamorphism, were probably driven from oceanic realms by the collision of continental blocks and may root, as do obducted high-level sheets, at considerable distances from their present outcrops. A third type of ophiolite occurrence is that of disjunct slabs in steeply dipping mélange blueschist zones (e.g., Zagros Crush Zone); these appear to represent collisional sutures that did not involve extensive nappe development.

Two particular problems relate to Alpine ophiolites. First, the time interval between origin and emplacement is frequently very short. For example, in the Zagros Ocean of Asia Minor, ophiolite generation and emplacement appear to have occurred during one stage of the Late Cretaceous, suggesting the development of a convergent plate boundary very close to, and shortly after, an accreting plate boundary. Second, there is an almost complete absence of pre-Late Triassic ophiolites that could represent portions of the pre-Alpine Tethys. Alpine ophiolites are dominantly Mesozoic and represent sea-floor spreading along accreting plate boundaries within the Tethys in spite of the gross convergence of Africa and Europe. The likeliest candidate for a remnant of the early Tethys is the zone defined by the Black Sea, Colkhide Depression, and south Caspian area, on both sides of which active volcanic arcs faced an oceanic realm for most of the Mesozoic and Tertiary. Alpine ophiolites mainly represent smaller, younger oceans, some of which (e.g., Pontide Ophiolite Belt) may have been rear-arc, or interarc, marginal basins.

The distribution and age of origin and emplacement of Alpine ophiolites are consistent with the relative motion of Africa and Europe from 178 my onward as deduced from Atlantic-plate accretion history. Most appear to represent short lengths of accreting margins that joined transform faults and subduction zones within a complex small-plate mosaic. Theoretically, dike orientation in sheeted diabase complexes would lie normal to the divergent motion vector across an accreting margin and thus, in favorable circumstances (e.g., Cyprus), correlate with the vector deduced from the Africa-Europe motion model. However, in most examples, the complex deformations and rotations developed during ophiolite emplacement prohibit such a correlation.

EUGEOSYNCLINAL BASEMENT AND A COLLAGE CONCEPT OF OROGENIC BELTS

JAMES HELWIG
Case Western Reserve University, Cleveland, Ohio

ABSTRACT

The perception that orogens may be divided into eugeosynclines having abundant volcanic rocks and miogeosynclines lacking volcanics triggered the hypothesis that eugeosynclines are perhaps ancient island arcs. Subsequent geophysical and geological studies supported a concept of eugeosynclines as at least partly comprising ancient oceanic terranes as well as arcs. It has become increasingly evident that to understand the role of orogenesis in crustal evolution the extent of oceanic crust in the basement of eugeosynclinal belts must be determined.

A consistent feature of eugeosynclines is their composite nature as manifest in elongate tectono-stratigraphic units or tectonic elements. The stratigraphic, tectonic, and plutonic and metamorphic evolution of each element is distinct and yet is partly related to adjacent elements. Major tectonic elements are separated by long-lived faults. The lithological sequence of each element is correlated with its basement type and the nature and history of its boundaries. Ancient eugeosynclinal tectonic elements may be elucidated by comparison with modern tectonic elements clearly related to plate motions. The basement of such tectonic elements is highly varied, and thus all eugeosynclinal zones are not ensimatic.

Tectonic processes play a key role in determining preservation and mode of occurrence of oceanic lithosphere in orogens. To survive orogenesis, oceanic crust must have a thick low-density cap or be tectonically intercalated with thick lower density materials. Preservation of oceanic crust and mantle within orogens therefore requires some mechanism of crustal thickening, commonly by sedimentation (to form buried basement of sedimentary furrows), magmatism (to form basement of oceanic arcs), or tectonic imbrication (to form ophiolite sequences and mélange belts). In the absence of such mechanisms, gravity and subduction can efficiently remove dense oceanic lithosphere from the crust. The crust of continental rifts, rhombochasms, sphenochasms, marginal basins, oceanic arcs, and remnant basins is susceptible to the crust-thickening mechanisms mentioned above and is, therefore, more abundantly preserved in orogenic belts than is normal ridge-generated oceanic crust.

The diagnosis of ensimatic tectonic elements within ancient orogens is difficult, but the composition of igneous rocks and other data can be related to basement composition. The composition of detritus can also indicate the nature and time of linkage of source blocks.

The sialic vs. simatic nature and extent of the initial basement of eugeosynclinal zones are highly varied and are dependent upon the evolution of the individual orogen in terms of geometry and nature of starting conditions, rifts, arcs, marginal and remnant basins, subduction zones, and strain history. The addition of oceanic and mantle materials to the continents by orogen accretion is complex due to the interaction and evolution of many processes. Single processes may be described but not single theories or finite models of orogenesis. Each orogen is a unique time-space collage of mappable elements, all generated, assembled, and rearranged by tectonic processes.

INTRODUCTION

The most interesting scientific questions invariably seem to deal with that which is just beyond our powers of observation. Such is true with the deep structure of orogenic belts, in particular the question of the nature and origin of the basement of eugeosynclinal blets. This question is fundamental to our understanding of orogenesis and crustal growth.

Eugeosynclinal belts have been identified as the sites of continental accretion, either because they rarely have any presumed basement exposed, or because the oldest exposed rocks are ophiolites or greenstones. If eugeosynclinal belts develop on preexisting sialic crust (ensialic), continental growth is dominantly vertical; this would be true if orogens are essentially tectonized continental shelf-slope-rise complexes. If eugeosynclinal belts are ensimatic, forming beyond continental margins on oceanic crust, then the continents have expanded laterally at the expense of ocean basins by accretion of eugeosynclines (Stille, 1941; Wells, 1949; Wilson, 1949; Kay, 1951; Drake and others, 1959; Dietz, 1963). According to plate tectonic concepts, both types of continental growth occur, although the problem remains to distinguish the tectonic elements, igneous rocks, and sediments of orogens that represent oceanic crust or upper mantle contributions (Dewey, 1969).

Stille (1941), Kay (1944), Wilson (1949), and Hess (1962) fully realized that a most important component of the eugeosyncline is the island arc. Few geologists realized the significance of this comparison or the need for rigorous study of island arcs. The great strides of marine geology, seismology, and volcanic geochemistry over the past decade have led to the new global tectonics (Isacks, Oliver, and Sykes, 1968) whereby island arcs and eugeosynclines are related to the motions of large crustal plates (Dewey and Bird, 1970).

There have been many recent attempts to integrate orogenic geosynclinal concepts with plate tectonics. Most of these attempts have been strongly inferential on the basis of limited new observations of Mesozoic and Cenozoic mountain belts rather than being strongly deductive on the basis of new critical observations of pre-Mesozoic orogeny. In fact, a greater reality seems to be attached to what may be called plate models of orogenic belts than to the belts themselves (Helwig, 1973). Plate tectonics is not yet a complete orogenic theory (Smith, 1971). The great contributions of legions of geologists who have worked in mountain belts are in danger of being cast aside (Trümpy, 1971), and there is a real need to rediscover old ideas and concepts as shown by White and others' 1970 reintroduction of the term subduction and by Hoffman's (this volume) use of "aulacogen." Perhaps this is rather like the biologists' need to rediscover Mendel sixty years ago. A major contribution of this book, and of the conference on which it is based, should be its emphasis on the significance of past and future research in mountain belts as a means of contributing to tectonic theory.

Two unifying concepts of orogenic theory are tectonic elements and tectonic processes. The ultimate driving forces of plate motions and mountain building (McKenzie, 1972) are presently unknown. This paper proposes that the initial nature and distribution of basement types are correlated with the initial and subsequent differentiation of tectonic elements and with the sequence of tectonic processes affecting the elements during the evolution of mountain belts. Thus, we may conclude that knowledge of initial basement, especially "zones of geosynclinal systems which can be demonstrated to have been initiated directly on an oceanic crust" (Khain, 1972, p. 211), is of fundamental importance in describing and understanding orogenic belts. A conception of orogenic belts as collages of basement-controlled tectonic elements may be considered as an extremely useful approach toward description and understanding of orogens. It must be understood, nevertheless, that the ultimate origins of tectonic elements, and of the sequence of tectonic processes whereby they are assembled, is a problem beyond present skills and knowledge.

ACKNOWLEDGMENTS

This paper is contribution no. 94 of the Department of Geology, Case Western Reserve University. I am grateful to my colleagues and students, especially Frank Stehli, Alan Nairn, and Steve Franks for providing a stimulating environment for thinking. I owe a great debt to Marshall Kay for inspiring and encouraging this work, and also to Robert Dott and the conferees for insights into geosynclines relative to plate tectonics. Ian Dalziel's comments contributed substantially to the final revision of this paper. The interpretations are mine.

NATURE OF TECTONIC ELEMENTS
IN OROGENIC BELTS

Even an untutored but careful observer may note that mountain chains contain parallel elongate zones of contrasting topography and lithology. These I call tectonic elements. They are an intrinsic feature of mobile belts. Tectonic elements have many names, for example, belts, units, zones, isopic zones (Aubouin, 1965), and assemblages (Monger, Souther, and Gabrielse, 1972). (See King, 1969, p. 1–30 for discussion.) Tectonic elements here may be defined as mostly elongate regions of relatively homogeneous deformational, igneous, sedimentary, and metamorphic history as contrasted to adjacent elements within a mountain belt. The essence of a tectonic element is continuity in space and time, but a given tectonic element may be either a composite or a distinct part of several superposed elements representing a temporal progression and (or) a spatial migration. Similarly, tectonic elements may be recognized on different scales. Thus, the term "eugeosyncline," or "eugeosynclinal belt," is admitted to include many smaller tectonic elements (Kay, 1951).

This concept of continuity, in structure, stratigraphy, metamorphism, and in plutonism, is evident in every published map and paper concerned with mountain belts, and it demonstrates the need for simultaneous solving of many problems when studying orogenic belts. Whatever the scheme of subdivision, contrasts of tectonic elements are real. For example, structural units coincide with stratigraphic units, and element boundaries often prove to be major faults (Borukayev, 1970), all of which demonstrates fundamental contrast between adjacent elements in space and time.

If tectonic elements are universal features of mountain belts, one immediately reasons that, if similar elements are found in different mountain belts and if their spatial and temporal relations are similar, it is possible to formulate a general descriptive theory of mountain building, a theory of the orogenic cycle. Such reasoning underlay all orogenic theories until the time of plate tectonics (Coney, 1970). Each theory was an attempt to provide a general explanation of mountain belts on the basis of sedimentation (Hall, 1859), crustal nature (Haug, 1900), deformation (Kober, 1928), or magmatism (Stille, 1941), but strong emphasis was given to the determinative role of geosynclines (e.g., as elegantly proposed by Aubouin, 1965).

With the advent of plate tectonics, the major objective of orogenic synthesis has been to establish the relation between tectonic elements that were so beautifully analyzed by orogenic geologists and the Mesozoic-Cenozoic tectonic elements that were inferred to be clearly related to plate models (Mitchell and Reading, 1969; Bird and Dewey, 1970).

TECTONIC ELEMENTS: PLATE MODELS VS. REAL MOUNTAIN BELTS

A summary comparison.—Tectonic elements have been defined in both plate models and orogenic syntheses. Each element, therefore, has overlapping meanings and commonly has two terms applied to it. Some elements can be clearly identified only in the modern, others in the ancient. That all important species of tectonic elements have been recognized is uncertain, and that more than two terms exist for the same type of element exposed or preserved in contrasting geological contexts or at different structural levels is equally unclear.

Three observational classes of tectonic elements exist:

(a) The first class of tectonic elements comprises those elements that are recognized clearly in both active tectonic environments and ancient orogenic belts, but it includes nonorogenic elements. This class includes tectonic elements formed at consuming plate margins, accreting plate margins, midplate positions, and at postorogenic successor basins (table 1, A).

(b) The second class of tectonic elements encompasses those elements that probably are identified correctly in both active and inactive orogenic belts, but these may evoke dispute. This class includes a variety of tectonic elements: trenches, marginal basins, midocean ridges, miogeoclinal and eugeoclinal ridges, and paired metamorphic belts (table 1, B).

(c) The third class of tectonic elements comprises those elements which are attributed to plate models or are found in mountain belts but not both. These elements are significant, but often elusive, and further work should be done in order that we may better recognize them. Tectonic elements of this kind in plate models include the sediment prism of the ensimatic continental rise, the ocean floor itself, the outer sedimentary arcs of island arcs, and the East Indies-type of isolated small ocean basin. Tectonic elements that are difficult to identify in plate models but that are obvious in orogenic belts include, for example, deep-water ensialic foredeep furrows, nappes, and gneiss-dome belts. In addition, tectonic elements of questionable definition or smaller scale exist that are not considered here. Many tectonic elements in orogenic belts may be unique to the extent that they possess a polyphase history.

Resolving the differences.—The complexity of

TABLE 1.—TECTONIC ELEMENTS INCORPORATED INTO OROGENIC BELTS

Tectonic element	Young example	Other possible or probable terms
A. Oceanic arc*	Tonga arc	Island arc, eugeosyncline
Cordilleran arc	Andes arc	Same
Detached arc	Japan arc	Same
Pericratonic foredeep	Andean foredeep	Exogeosyncline
Successor basin	Great Valley, California	Epieugeosyncline
Miogeocline	Atlantic coastal plain	Paraliageosyncline, miogeosyncline
B. Marginal basin*	Philippine Sea	Eugeosyncline, small ocean basin
Arc-trench gap*	?	Eugeosyncline, high P/T belt
Trench*	Tonga Trench	Eugeosyncline, high P/T belt
Rift basin*	East African Rift, Red Sea	Taphrogeosyncline
Aulacogen	Benue Trough	Taphrogeosyncline
Miogeoclinal ridge	Outer basement ridge, Atlantic shelf	Barrier en creux
Eugeanticlinal ridges* sedimentary inactive arc	Andaman arc, New Hebrides	Tectonic and eugeanticlinal ridges
Microcontinent	Corsica	Block
Midocean ridge*	Iceland	
Oceanic island*	Hawaii	
Sphenochasm*	Bay of Biscay	Small ocean basin
Rhombochasm*	Cayman Trough	Eugeosyncline, small ocean basin
Remnant sea*	Black Sea	Small ocean basin

Asterisk indicates elements with oceanic basement or partly oceanic basement.

TABLE 2.—FEATURES OF BOUNDARIES OF TECTONIC ELEMENTS

Essential feature	Major (plate) boundaries			Examples, boundaries of other tectonic elements		
	Accreting	Consuming	Transform	Marginal basin	Ocean arc	Pericratonic foredeep
Subduction		+			+	
Normal faulting	+			+	+	
Strike-slip fault		+	+			
Obduction/thrusting		+		+	+	+
Truncation of tectonic trends	+	+	+			
High P/T metamorphism		+				
Ophiolite	+	+		+	+	
Bounds plutonic arc		+		+	+	
Bounds shelf deposits	+					+
Bounds pelagites		+		+		
Bounds submarine fans	+	+		+		+
Bounds flysch		+		+		+
Bounds molasse	+	+	+	+		+

assembled and crushed tectonic elements comprising each mountain belt has led to a complex terminology for classifying tectonic elements. Many terms overlap or have multiple meanings, and some definitions have changed or lost meaning with time and usage (Coney, 1970). Plate-model terminology for tectonic elements that is based upon use of specific named active tectonic elements as analogues appears to offer a relatively objective descriptive and genetic framework for classification and is recommended. However, there are several reasons why an eclectic terminology retaining old purely descriptive terms (table 1) should be maintained. (See Khain, 1972, and Zonenshain, 1972, for other views.)

The first reason is ignorance. Much is yet to be learned of the presently active mountain belts before they can be characterized and causally related to plate motions and (or) other tectonic processes (Gilluly, 1972).

The second reason is time, as related to orogeny. In order to understand tectonic elements, progressive stages of development at different structural levels of exposure must be known. Hence the study of deeply eroded ancient mountain belts must be integrated with study of young tectonics.

The third reason is that tectonic elements may be polygenetic (table 1). There may be several generative settings for similar classified tectonic elements and their sediments. For example, flysch nappes may be elements of foredeep troughs that originated as marginal basin fill, or continental rise fill, or perhaps by some other means. Thus, generic classification of a tectonic element may omit, clarify, or obscure its origin.

Resolving differences of terminology depends on new knowledge, critically gathered. New syntheses of orogenesis hopefully should be biased neither toward so-called "plate models" as has been true recently (Helwig, 1973), nor toward the locally conceived frameworks of orogenic specialists (e.g., Aubouin, 1965). In this context, it would be useful to consider a collage concept of orogenesis whereby orogenic evolution consists of the bringing together of diverse tectonic elements without regard to ruling theories, but by employing identified young analogs and the principles of plate tectonics. This concept is considered further at the end of this paper.

Nature of boundaries of tectonic elements.—The largest scale tectonic elements of the earth's crust are the rigid plates as conceived in the new global tectonics (Le Pichon, 1968; Isacks, Oliver, and Sykes, 1968). Seismology has shown that the boundaries of these plates include fault systems of three types: rifts, oceanic underthrusts (i.e., subduction zones), and strike-slip or transform faults. Hybrid rift-transform and underthrust-transform boundaries are also possible (Harland, 1971).

Tectonic elements of orogenic belts have contrasting rock suites and structure and are always topographically distinguished from adjacent elements (table 2). These features immediately suggest that the boundaries between tectonic elements are major faults or abrupt flexures. This proposition may be verified by examining any zonal subdivision or historical interpretation of a mountain belt.

The boundaries of the largest scale tectonic elements of orogenic belts are therefore the likely sites of defunct plate boundaries. The boundary between the eugeosynclinal and miogeosynclinal belts of orogens is generally thrust faulted, but it may represent the edge of a sediment-filled marginal basin having oceanic crust (e.g., Churkin, this volume) or a continental edge broken at an accreting plate margin (Dietz, 1963; Rodgers, 1970; Dewey and Bird, 1970). The boundaries of cordilleran ridges upheld by sialic microcontinental blocks would be similarly faulted as has been shown for the Sardinian block in the Mediterranean. Serpentine belts, attendant mélanges, and median tectonic lines have been attributed to ancient trenches and subduction zones (Dickinson, 1971). Transform faults may be attached to convergent plate boundaries as well as to divergent ones. In both situations, the abrupt strike termination of orogenic tectonic elements may be explained by such faults (Bird and Dewey, 1970, p. 1049–1050); Hoffmann, this volume).

The topographic differentiation of tectonic elements in their formative stages, for example, the remnant arcs, marginal basins, and active arcs of the western Pacific (Karig, 1971), has generally been explained by block faulting (Mitchell and Reading, 1971). The igneous and sedimentary contributions to these tectonic elements both control and reflect the topography. The boundaries of the elements are thus zones of facies change, for example, from volcanic to pelagic sediments or from submarine fan turbidites to bathyal or abyssal plain finer turbidites and pelagites. Russian geologists have emphasized the persistent, so-called deep faults and block structure of geosynclines (Peyve, 1945).

The relative positions of tectonic elements may change with time. In ancient orogenic belts, the establishment of sedimentary facies and provenance linkages between adjacent tectonic elements in effect establishes that they do indeed behave as independent blocks at times and are not only fault bounded but also differentially translated or transported (Monger, Souther, and Gabrielse, 1972). Linkages between blocks may be difficult to establish unless transitional facies are preserved within or adjacent to fault zones (e.g., Kay and Crawford, 1964; Eisbacher, this volume; Crowell, this volume).

The persistent, consistent differentiation of tectonic elements during contrasting phases of orogenesis probably reflects basement contrasts.

The basement of tectonic elements.—By "basement" I mean crustal rocks formed prior to (in the example of inherited basement) or during the initial stages of (generative basement) orogenesis. Implicit in this definition is the assumption that orogenesis is episodic, probably involving the shifting of crustal consumption from one plate boundary to another (Mitchell and Reading, 1969).

An attractive interpretation is that contrasting tectonic elements are underlain by contrasting basement rocks because tectonic differentiation implies deep-seated causes. Also, there are data justifying this view.

Detailed geophysical surveys have established that the crust of stable continental margins changes over a distance of several tens of kilometers from thick typically continental sialic crust to thin simatic oceanic crust in the vicinity of the continental rise (Drake and others, 1959). Even stronger and more abrupt crustal changes take place across Pacific-type continental margins. Oceanic arcs show a twofold to fourfold thickened oceanic crust separating typical oceanic crust on the trench side from volcanic and sedimentary-thickened transitional oceanic crust of the marginal basin behind the arc (Karig, 1971; Vogt, Schneider, and Johnson, 1969). Comparable geophysical contrasts are preserved in young orogenic belts that have been thoroughly studied (Thomspon and Talwani, 1964).

It is not possible, however, to compare unambiguously geophysical data for *ancient* orogens, that have undergone complex polyphase history and intense crushing followed by stress relaxation and isostatic adjustment. More indirect evidence may be used to evaluate basement (see farther on). To understand continental growth and fragmentation, our principal problem is to recognize tectonic elements within orogens and to evaluate their basement. The fate of tectonic elements depends upon the interaction of basement, cover, and tectonic processes.

NATURE OF TECTONIC PROCESSES

Principles of orogenic tectonics.—Orogenic belts are belts of crustal shortening. The principal tectonic consequences of this shortening are folding, crustal thickening, and transport of orogenic zones toward stable forelands. Seismology, paleomagnetism, and structural studies have confirmed that processes other than shortening are also significant: longitudinal displacements (strike-slip faulting), lateral displacements involving the removal of crustal material from below (subduction), large-scale rotations, and lateral displacements involving extension. These displacements are not mutually exclusive (Packham and Falvey, 1971; Harland, 1971).

Displacements.—The potential magnitude of crustal displacements between elements within orogens is fantastic considering that all pre-Mesozoic oceanic crust has apparently been consumed or incorporated in orogenic belts and circum-Pacific subduction zones. Hence, crustal blocks may be transported from one continental block to another, and island arcs and microcontinents may be swept in, so to speak, to an orogenic belt (Wilson, 1967; Dewey and Bird, 1970). Large displacements may also occur across transcurrent faults. Large rotations and crustal extension may accompany lateral displacements as indicated for the Mediterranean region (Smith, 1971; Hsü, 1972). Crustal extension may occur within successor basins (Eisbacher, this volume), possibly proceeding so far as to form marginal seas with a new ocean crust (Packham and Falvey, 1971). The separation of closely related tectonic elements and the juxtaposition of tectonic elements formed hundreds or even thousands of kilometers apart is possible, therefore, and consequently necessitates extremely abrupt changes in basement, sedimentary cover, magmatism, metamorphism, fossils, paleomagnetism, and paleoclimatology from one tectonic element to another (Wilson, 1966; Monger, Souther, and Gabrielse, 1972; Helwig, 1972; Ernst, 1973). The arrival of a displaced or newly generated block is marked by sedimentary-erosional linkage with its neighbor (Monger, Souther, and Gabrielse, 1972). If two tectonic elements exhibit sharply disjunctive histories, their boundary is probably a transcurrent fault, deep normal fault or suture belt, or some superposed combination thereof.

The more traditional types of displacement in orogens may also be considerable. The Alpine nappes are tectonic elements showing considerable absolute and relative displacement and may be reconstructed in time and space by establishing facies-paleogeographic linkages (Trümpy, 1960). Displacement could be especially great if thrust sequences are transported across suture belts (Oxburgh, 1972) or transcurrent faults. Thrust transport may preserve tectonic elements in such manner that they commonly are bottomless, lacking part or all of their basement, but they could also be topless. In either event, thrust sequences could preserve tectonic elements that might otherwise have been lost to view by subduction.

Plate-tectonic orogenic models (Dewey and Bird, 1970) have emphasized the ordered, organized spatial-chronological development of orogenic belts, as have nearly all orogenic theories. However, recent studies (e.g., Monger, Souther, and Gabrielse, 1972; Gastil and others, 1972) have emphasized the disordering potential of tectonic processes. Regretfully, the reductionist approach toward orogenic theory, including plate tectonics, by emphasis on ordering processes has left little room for recognition that tectonic elements in mountain belts may be highly disordered in anything other than highly complex palinspastic-chronological frameworks. Evidently, an important consequence of displacements is that tectonic elements become disordered with respect to simple plate tectonic models (fig. 1).

Crustal thickening and crustal loss.—Crustal thickening is one essential property of mountain belts (Eisbacher, this volume). In orogenic belts crustal thickening is produced by mechanical, sedimentary, or magmatic means. Compressive stress across an orogen mechanically thickens its crust by ductile flow, folding, and thrust and reverse faulting: uplift above sea level is accompanied by physical depression of the Moho. Oceanic underthrusting could accrete seaward and (or) underplate an orogen with any type of tectonic element, including ocean crust and sediments and continental or microcontinental blocks. Sedimentary thickening occurs in geosynclines, particularly in the flanking clastic wedges of oceanic and cordilleran arcs, but also in furrows and successor basins between eugeoanticlinal ridges and in pericratonic foredeep troughs. The arcs themselves and their pyroclastic and erosional products may more appropriately be considered as juvenile magmatic contributions to eugeosynclinal belt thickening (Dickinson, this volume).

In any place that geosynclinal thicknesses of sediments accumulate, crustal thickening is occurring, and thus miogeoclines, marginal basins, and even rift basins, transcurrent pullapart basins (Crowell, this volume), and Bahama-type platforms are all tectonic elements thickened by sedimentation. (Note that these regions commonly are attributed to oceanization by some schools of geology.) Any of them may be incorporated into orogenic belts. The pulling apart of continental crust seems involved in all of these tectonic elements. Extension of this crust does not involve actual loss, but only thinning and subsidence due to coupling of sial to new oceanic crust (Bott 1971). There is no real oceanization, but rather a stretching of continental crust. Eventually, subsided semioceanic areas will accumulate great thickness of sediment (Hutchinson and Engels, 1972). Hence, real crustal thickening by volcanism and chemical sedimentationn occurs; these rocks may be incorporated into orogens during subsequent compression (Trümpy, 1971). Similarly, the pulling apart of marginal seas forms a deep sediment trap, the potential site of threefold thickening of oceanic crust by sedimentation (Packham and Falvey, 1971).

So-called "crustal loss" in compressed orogenic belts likewise mostly constitutes crustal thickening or redistribution. To understand this misleading phraseology, consider that there are two types of presumed loss, one kind from below and one from above. Plate consumption, if it involves driving out of ocean crust below orogens, yields melts to produce overall chemical thickening of orogenic crust. If it involves other types of tectonic elements, they are of lower density than the mantle and thus contribute to thickening by seaward accretion, underthrusting, and underplating, or perhaps by overthrusting and obduction. The second type of presumed loss, from above, involves erosion and outward gravity gliding from the orogenic root. These processes serve merely to redistribute the thickened crustal mass of the orogen, although admittedly a considerable loss of erosion products could occur to build distant miogeoclines, such as the Mississippi delta or Bengal cone, if foredeeps are not efficient in trapping sediment. The principal effect of what has been called crustal loss may well be to produce the high-pressure, low-temperature metamorphic belt above the suture zone of crustal consumption at a given time.

From the discussion above and from knowledge of the history of mountain belts, we may conclude that the development of the continental crust is an essentially irreversible process (Mouratov, 1972). Such irreversibility is attributed to the inherent buoyancy and volatile geochemistry (following many authors) of continental crust. These factors also place an upper limit on maximum thickness of continental crust (Fyfe, 1973). Oceanization (Van Bemmelen, 1969) is viewed as more of a semantic than geologic problem (Trümpy, 1971, p. 312). The thinning of continental crust in one region is inevitably accompanied by its thickening or redistribution elsewhere.

Effects of orogenic processes: summary.—The principal effect of orogenic processes is to generate what is called the granitic layer of the earth's crust, such generation being an irreversible process (Peyve and others, 1972). The genesis of tectonic elements involved is initially in some orderly spatial-chronological sequence; but continuing orogenesis may modify, disperse, reassemble, and strain tectonic elements to yield a consolidated orogenic belt that often is extremely difficult to decipher. Crustal shortening,

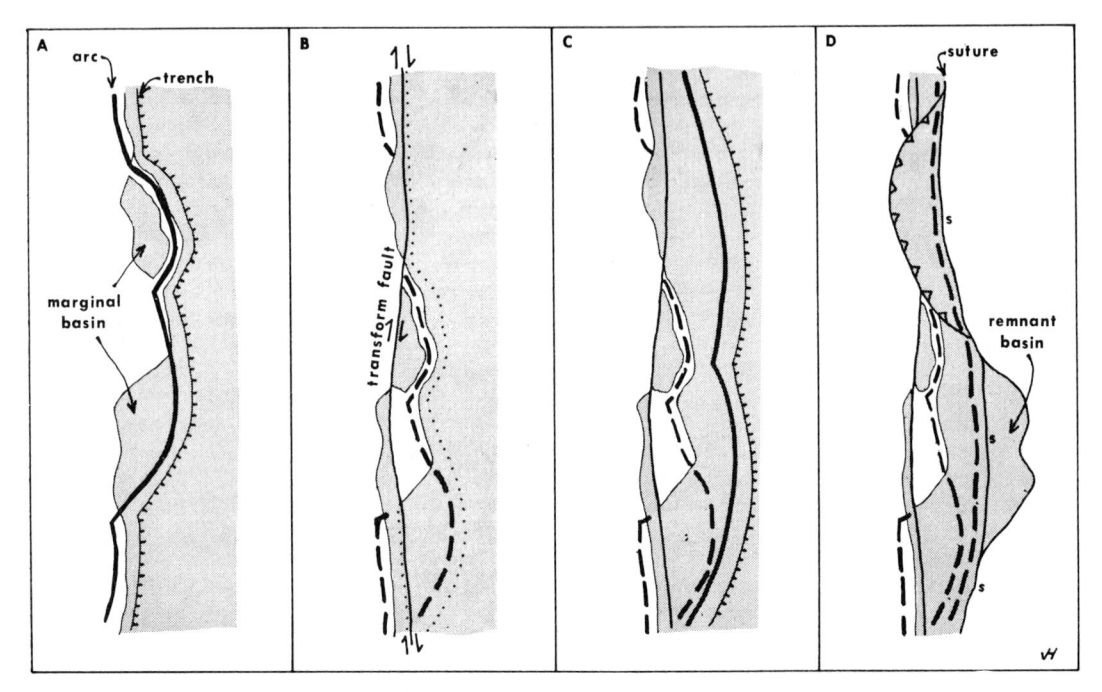

FIG. 1.—A simple hypothetical example of an orogenic collage (shaded areas are underlain by oceanic crust). A, Development of subduction zone (trench) and orderly arrangement of related arcs and marginal basins. B, Lateral displacement; initial configuration of arcs, basins, and sediment-source linkages becomes disordered; transform fault uses preexisting line of weakness (subduction zone dotted line) to produce hybrid structure in some regions; note also omission and repetition of inactive arc (dashed line) due to faulting. C, Renewed crustal convergence; a new arc and trench are formed. D, Continental collision; obduction of tectonic elements occurs at continental salient, and remnant basin forms in continental recess.

thickening, and suturing are perhaps the most important effects of tectonic stresses, particularly as they lead to termination of old generative configurations and to beginning of new ones (e.g., new arcs, new subduction zones, plate flips, and conversion to transform boundaries). Such changes inevitably lead to hybrid structures. In particular, the boundaries of tectonic elements are anisotropies likely to be reactivated under tension, compression, and shear. Consequently, it is profitable to think of major structural boundaries as superposed transcurrent faults, subduction zones, and normal faults.

Any change in plate-margin movement or process is potentially a disordering process with regard to an existing configuration of tectonic elements. Oceanic crust and sediments seem particularly susceptible to complete disordering to the point of mélange (Peyve, 1969; Hsü, this volume).

At convergent plate margins, gravity can act to remove efficiently the dense oceanic lithosphere by subduction. Oceanic remnants also could be preserved by favorable geometry by cover of thick sediments, by cover of oceanic arcs or volcanoes, or by tectonic imbrication (see farther on). Conversely, gravity forces spread thickened rising orogens by erosion and outward thrusting, thereby producing displacement and strain of assembled tectonic elements and, ultimately, more uniform crustal thickness in response to isostasy.

Orogenic processes are thus simultaneously ordering and disordering processes, and the longest lived orogenic belts are rationally seen as the most disordered; that is, as a *collage*, or collection of diverse crustal elements complexly assembled (fig. 1). By fully investigating tectonic processes and tectonic elements, we have the means to understand orogenic collages. The most important features of orogens are tectonic elements, their boundaries, and their mutual relations through time as established by linkages and disjunctions of igneous, metamorphic, sedimentary, and structural events. Paleontology, geochronology, and paleomagnetism are important tools of stratigraphic-structural studies needed to establish linkages. A knowledge of the nature and age of the basement rocks in each tectonic element is most critical perhaps to establish the locale of origin of the element and to allow evaluation of the evolution of the earth's crust in terms of the kinematics of plates, history of individual tectonic elements, and of mechanics of continental growth.

KEY ROLE OF BASEMENT

It can be argued that basement contrast, in terms of gross composition of crust, has little influence upon inception of orogenic belts because convergent plate boundaries are not all located at continental margins (Dott, this volume). (Presumably, inception of an orogenic belt requires plate convergence by definition.) This is conceivably true in terms of the initial spatial position of arcs which contribute new sialic material to the crust; however, the work of Karig (1971), Matsuda and Uyeda (1971), and Packham and Falvey (1971) shows that the western Pacific arcs have migrated away from the Australian and Asian land masses, as originally suggested by A. Wegener, demonstrating that the oceanic positions of these arcs are secondary.

Furthermore, if continental edges originate by rifting, they must be the loci of oldest ocean crust that is also the coolest, most dense, and topographically lowest (Sclater and others, 1971). Hence, continental edges are the sites of minimum energy input required for onset of subduction. Conversely, initiation of subduction within continents is highly unlikely, and, in fact, continental suturing halts convergence.

Whether basement contrasts exert a role in the siting of extensional and transcurrent boundaries within orogens is debatable. The San Andreas Fault appears to ignore, so to speak, the boundary of Franciscan and Sierran basement (Gastil and others, 1972). On the other hand, the Appalachian-Caledonian belt certainly determined the site of opening of the North Atlantic Ocean. In Eurasian mountain chains, the repeated activation of deep faults is noteworthy (Khain, 1972).

Orogenic belts are incorporated into the continents by being crushed between or against continental basement cratonic blocks. In this respect, the fate of all tectonic elements is predestined. At every stage of development between the generative and terminal stages of orogenic evolution, the type of crust, including both cover and basement rocks composing each tectonic element, plays a key role in determining its history.

Ensimatic versus ensialic tectonic elements.—The stable regions of the earth's crust are of two types (Dietz, this volume): oceanic and continental. They are each remarkably homogeneous from a geophysical point of view, and each has distinctive thickness and composition that is reflected in mean elevation and sedimentary accumulation.

Orogenic belts are the unstable regions of the earth's crust that are characterized by geophysical, compositional, and topographic diversity. For active orogens, the delineation of ensimatic and ensialic tectonic elements is possible by direct geological and geophysical study (table 1). These belts are the sites of modification of the two end-member crustal types by means of tectonic processes. By actualistic principles and geologic evidence, it is clear that ensimatic tectonic elements do occur in ancient orogens. They must behave much differently than ensialic elements because their thickness, composition, and density, and thus their mechanical and thermal properties as well, are greatly different.

The general effect of orogenic processes through geologic time has been to thicken and to add modified or totally new crust to the continents. The added strips constitute a diverse assemblage of deformed tectonic elements. To survive orogenic processes, tectonic elements must generally have a thick low-density cap. Any sizable tectonic elements composed of essentially unaltered oceanic crust cannot be preserved in orogenic belts because gravity and subduction can efficiently remove dense oceanic lithosphere from consuming plate margins. Any sizeable tectonic elements composed of modified or sialic crust, that is, crust having mean density not exceeding about 2.8 to 2.9 gm/cc, becomes incorporated into orogens because their compositional and density contrast assures that they cannot be made to sink into the underlying asthenosphere. It is also unreasonable to consider that such elements could be oceanized because the silica and alkali content could readily melt and rise again to maintain a high crustal level. Nevertheless, tectonic elements of modified or sialic crust possibly could subside deeply if thinned by extension or could be partially subducted by coupling to descending lithosphere plates, which would result in continental underplating and in great crustal thickening as demonstrated in the Himalaya and the Andes (Plafker, 1972).

In summary, oceanic crust must be considerably modified to survive orogenic processes as ensimatic tectonic elements. The principal means of modification are igneous and sedimentary and mechanical thickening. Therefore, the area of ensimatic tectonic elements within an orogenic belt, which represents the best estimate of continental growth and also contains the limited record of the earlier ocean basins, tends to be deeply subsided and (or) difficult to recognize.

Ensimatic tectonic elements: definition.—Ensimatic tectonic elements are those crustal domains in an orogenic belt which are underlain by oceanic crust generated during orogenesis or inherited essentially unmodified from an earlier period of formation. Table 3 lists types of occurrence of ocean crust; all types could be preserved in orogenic belts. Preservation of ocean crust is possible either in outcrop as ophiolite or mélange suites or at depth under cover of stratigraphically or structurally higher rock sequences.

The definition of ensimatic tectonic elements entails a semantic problem related to the inherently allochthonous character of an orogenic collage. The joining of sialic blocks along a suture zone involves squeezing out of materials. Ensimatic tectonic elements may be subducted or obducted (overthrust), or ocean crust and its cover may be tectonically separated. Ensimatic tectonic elements may be similarly allochthonous. Ultimately, a mountain chain thereby may become entirely ensialic but still incorporate many ensimatic tectonic elements or sedimentary sequences initially ensimatic (for example, in accord with Gansser's interpretation of the Himalayan belt, 1966, and Glennie and others interpretation of the Oman Mountains, 1973). It is difficult to conceive of the intense deformation and crustal shortening involved in the nappe structure of the Alpine-Himalayan belt unless squeezing out of oceanic domains occurred (Trümpy, 1971). In addition, if oceanic crust can be thrust over continental rocks, it seems even more probable that continental blocks could be thrust over oceanic crust; the degree of allochthony of sialic basement blocks within orogens (Peyve, 1969) is problematical.

Therefore, it appears necessary to qualify the ensimatic nature of a tectonic element as being allochthonous where tectonically transported over ocean crust and as being detached if inferred to be detached from an original oceanic basement. A tectonic element may thus be: (1) stratigraphically ensimatic (autochthonous on ocean crust), (2) an ensimatic detachment (or paleogeographically ensimatic, Lemoine, 1972), or (3) structurally ensimatic (allochthonous). Condition 1 or 2 may also apply where condition 3 applies.

Ensimatic tectonic elements: recognition.— The identification of ophiolite sequences as oceanic crust seems reasonably well established on geophysical, geochemical, and petrologic grounds (Dewey and Bird, 1971; Pearce and Cann, 1971). However, information on the composition and sequence of modern ocean crust is inadequate to allow distinction of ophiolites generated at midocean ridges from those formed in marginal basins or in other places. Certain alpine-type peridotites and serpentinites may be confused with ophiolites (Chidester and Cody, 1972), and some ophiolites could speculatively represent the lower part of continental crust and adjacent upper mantle. Thus, the origin of some ultramafic sequences called "ophiolite" is disputable.

TABLE 3.—OCEANIC CRUST: ORIGIN AND OCCURRENCE IN OROGENS

Descriptive name	A. Origin and distinctive features (see table 1 for examples)
Rift	Crustal extension, incipient ocean crust formation; block faulting; prisms of clastic, carbonate and evaporite sediments; sialic detritus; peralkaline basalts
Sphenochasm	Crustal extension with rotation of sialic blocks, directly coupled to zone of compression; wedge-shaped ocean basin; submarine fans and turbidites; sialic detritus
Rhombochasm	Crustal extension limited by coupled transform faults; block faulting; submarine fans and turbidites; sialic detritus
Normal ocean basin	Prolonged development of first three above; pelagic oozes, cherts
Marginal basin	Crustal extension behind arc; block faulting; turbidites and tuffs; both pyroclastic and sialic detritus; alkali olivine basalts
Remnant basin	Remnant of normal ocean crust preserved between translated sialic blocks; passes into deformed rocks along strike; turbidites and restricted basin facies are postorogenic; sialic detritus
Ocean arc	Ocean crust preserved beneath arc volcanics
Oceanic island	Ocean crust preserved beneath Hawaiian-type volcanoes

Descriptive name	B. Occurrence in orogens	
	Nature of preservation	Example
Ophiolite sheet	Allochthonous, obducted (Coleman, 1971)	New Caledonia
Ophiolitic mélange (thalassogeosyncline)	High P/T metamorphism and accretion in subduction zones (Hamilton, 1969; Bogdanov, 1969)	Franciscan
Ophiolitic suture zone	Intense deformation in narrow belt	Indus suture
Ophiolitic cordillera	Compression without subduction or obduction (Gansser, 1973)	Western Cordillera of Colombian Andes?
Hidden beneath eugeosynclinal tectonic elements	Buried beneath thick sedimentary, volcanic, or structural cover	Central Newfoundland Appalachians
Subducted paleogeographic realms	Inferred to originally underlie allochthonous nappes, etc., but not preserved	Alpine belt

Review of the ophiolite problem is beyond the scope of this paper. The writer concurs with Coleman (1971) and Dewey and Bird (1971) that the origin of true ophiolite sequences is ocean crust and upper mantle, but more thorough studies of such rocks and their modern analogs undeniably are needed.

The problem at hand is to evaluate the basement of tectonic elements where basement is not exposed or where perhaps only the uppermost part of an ophiolite sequence is exposed. That eugeosynclinal furrows (Aubouin, 1965) are the most extensive ensimatic fraction of orogenic belts seems likely. This inference follows from consideration that most oceanic crust of the geologic past has vanished without a trace (Smith, 1971), and the sutures marking its disappearance (subduction zones) are subsident by nature. The effect of relative buoyancy buries these oceanic sutures in sediments that are derived from rebounding sialic terranes (Ernst, 1973).

Some plausible approaches to evaluation of the basement of eugeosynclinal tectonic elements are designated (a) through (f) in the discussions that follow. An effort is made in these discussions to evaluate both the presence and the origin of the oceanic crust, which can be preserved in a variety of ways (table 3).

(a) Stratigraphic continuity: If a stratigraphic unit can be continuously traced across the strike of an orogenic belt, the presence of younger ocean crust or subduction zones within the belt is ruled out (Cady, 1972). The presence of older ocean crust is not ruled out, even if identical basement rocks are exposed on opposite flanks of the belt or within the belt, because such basement rock relations could be attributed to rifting, development of marginal seas, or transcurrent faulting. Stratigraphic continuity of older rocks across orogenic belts is inevitably interrupted by synclinoria or by successor basins of younger clastic fill. This observation allows relatively free interpretation of the nature of basement and position of ancient subduction zones beneath such synclinoria.

Where allochthonous relations are dominant, vertical stratigraphic continuity is critical. In the Alpine nappes, the sedimentary and ophiolite sequences apparently show both allochthonous and autochthonous relations with each other and with pre-Triassic sialic basement (Lemoine, 1972). Hence, we may conclude that the nappes are stratigraphically both ensimatic and ensialic, and some are ensimatic detachments but structurally ensialic.

(b) Initial sedimentary environments: The sedimentary environments of rifts (table 3) may be distinctive. The East African Rift, the Afar Triangle of Ethiopia, and the adjacent Red Sea may be considered collectively as an example of rifted sialic crust. The distinctive sediments here are fluviatile clastics, shallow-water carbonates (including reefs) and as much as 3 km of evaporites. In the Afar Triangle, the complex mosaic of Precambrian crystalline and young volcanic blocks indicates crustal thinning but widespread sialic crust (Hutchinson and Engels, 1972). In contrast, the fully oceanic central basin of the Red Sea Rift displays distinctive thermal saline brines and heavy metal-enriched pelagic sediments (Degens and Ross, 1969). Sialic rifts may develop into normal ocean basins or become arrested to form grabens or aulacogens (table 3).

Ocean basins such as the post-Hercynian and post-Alpine basins of the Mediterranean are probably sphenochasms (Carey, 1958), which form between rotating sialic blocks (Smith, 1971). Development of deep-water clastic and carbonate facies follows probable initial rift-facies deposition (Hsü, 1972). The succession of sedimentary environments in rhombochasms (table 3) is likely similar.

Typical deep-sea sediments are turbidites, cherts, and pelagic oozes. Such sediments should be encountered in sequences overlying normal oceanic crust, including crust beneath oceanic volcanoes, or crust preserved in reentrants (table 3). Proof of deposition at true oceanic depths, however, is difficult (Hallam, 1967). The similarity of sediments overlying or tectonically emplaced with ophiolites to those overlying the basaltic layer of the ocean, as determined by deep-sea drilling, corroborates their ocean-crust origin (Glennie and others, 1973). The occurrence of flysch-type facies is possible in various tectonic settings (Reading, 1972), including all varieties of ocean crust given in table 3, but is not diagnostic of oceanic tectonic elements.

The sedimentary environments of oceanic arcs and interarc basins (table 3) accumulate fluviatile to deep-water volcanogenic sediments, which are not diagnostic of oceanic crust as they also may be found in cordilleran arcs or indetached arcs. If the pyroclastic and associated sedimentary rocks are largely submarine, low oceanic topography is suggested, although true oceanic basement is not proven. Walker and Croasdale (1972) have presented criteria for diagnosing submarine mafic pyroclastic activity.

(c) Composition of detritus: Kay (1951) first emphasized the significance of plutonic pebble conglomerates for eugeosynclinal development. The composition of detritus reflects source area. The sources of detrital quartz and K-feldspar within eugeosynclinal furrows may be cratons, microcontinental sialic blocks, oceanic arcs, cordilleran arcs, or deformed sedimentary ridges (producing recycled detritus). The acid plutonic detritus of oceanic arcs is characteristically tonalitic, rich in silica but low in potash (Helwig and Sarpi, 1969; Mitchell and Reading, 1971), and generally much less important than associated quartz-poor detritus of intermediate to mafic volcanic provenance. On the other extreme, microcontinental fragments predictably yield more abundant and more potassic detritus as well as metamorphic rock fragments and minerals (de Booy, 1966). Cordilleran arcs yield detritus compositionally intermediate between ocean arcs and microcontinents, depending on the degree of unroofing of the basement.

Crook (this volume) and Dickinson (1970) have summarized the genetic significance of these compositional contrasts. The detritus provides substantial evidence of the nature and proximity of uplifted tectonic elements. The basement of the trough receiving the detritus may be evaluated in terms of its paleotectonic relationship to the source area (fig. 2).

Ophiolites also can be emergent, as on Cyprus, and shed mafic and ultramafic detritus. This is well known in Cuba and New Caledonia and was more recently described from Newfoundland (Church, 1969) and the western cordillera of North America (Monger, Souther, and Gabrielse, 1972; Churkin, this volume). If the source block shows continuous subsequent linkage with the depositional block, the source block contains ensimatic tectonic elements, although consideration must be given to the possibility that the source block is inevitably displaced and also may be allochthonous (obducted) (see fig. 2 d, f).

(d) Igneous rock geochemistry: Compositional gradients are observed in igneous rocks where detailed sampling profiles have been made across island arcs (Kuno, 1959; Hatherton and Dickinson, 1967), major young plutonic belts (Moore, 1959), and batholiths (Bateman and Dodge, 1970). The most notable gradient is an increase in potassium and potassium-type elements (Rb, Ba, Sr) at constant silica level from the volcanic front of the arc toward its foreland. The gradient occurs in rocks of basaltic to andesitic composition, both plutonic and volcanic. This geochemical gradient, when first observed in Pacific arcs, was correlated with

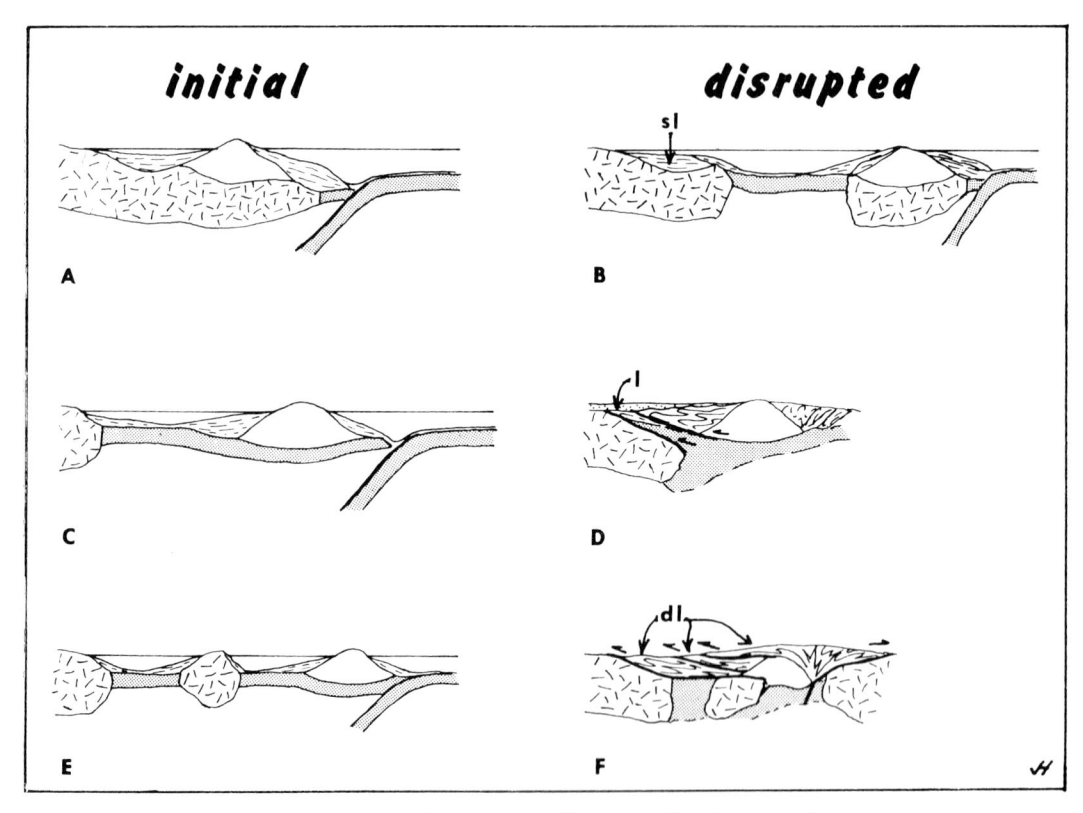

FIG. 2.—Some possible relations of sediment-source linkages. (Symbols: shaded, oceanic crust; random dashes, continental crust; dashed, sediments; open, arcs and sea water). Diagrams on left show initial configurations; on right, disrupted situations. A, Cordilleran arc and flanking ensialic pericratonic and arc-trench geosynclines. B, Same as A after opening of marginal basin; separated linkage (sl) exists between sediments of pericratonic geosyncline and source in detached arc. C, Oceanic arc and flanking ensialic sediment wedges. D, Same as C after closing of marginal basin; linkage (L) of cratonic margin sediments from oceanic arc and microcontinent source areas. E, Small ocean basins receiving sediments from oceanic arc and microcontinent source areas. F, Same as E after suturing by continental collision; disjunctive linkages (dl) of thrust sheets indicates disordering and covering of originally linked source areas.

depth of melting in the mantle as related to the inclined seismic zones or arcs (Kuno, 1966; Hatherton and Dickinson, 1967). Such geochemical gradients, however, now appear to occur in time as well as space (Jakes and White, 1972) and thus could reflect crustal thickness and composition (Condie and Potts, 1970). If this interpretation is correct, we then have a means of evaluating eugeosynclinal basement.

Data on abundances of major elements, trace elements, and Sr isotopes justify the view that basement types can be inferred chemically (table 4). The compositional data for volcanic rocks allow distinction of three important basement types: generative ocean crust (abyssal tholeiites), ocean arc crust (distinctive arc tholeiites and calc-alkaline suite), and continental crust of cordilleran arcs (calc-alkaline and silicic rocks). [Detached arcs having continental crust as basement (Japan type), and ensialic volcanic provinces of the Basin and Range-type do not appear to be geochemically distinctive at present.] Undoubtedly this approach will prove applicable to plutonic rocks as well, but at present sufficient data are not widely available.

The underlying cause of these compositional distinctions involves the origin of magmas, a problem not without opposing views (Boettcher, 1973). Controversy centers about the relative contributions of oceanic crust, upper mantle, oceanic sediments, and continental crust to the various magmas. From this viewpoint it appears that the three types of magmatic evolution summarized in table 4 involve three different sources that are nevertheless interrelated. The abyssal tholeiites are the direct products of partial melting of the upper mantle at midocean ridges. The arc tholeiites form by partial melting and dehydration of abyssal tholeiites at the top of the descending lithosphere slab below island arcs (Fitton, 1971; Wyllie, 1973). The differences of arc tholeiites from abyssal tholeiites are thus explained, whereas the similarities may be due to contamination by upper mantle material during ascent of the magma (White, Jakes, and Christie, 1971). The upper stratigraphic levels of ocean arcs contain calc-alkaline and shoshonitic rocks because with time, basal arc tholeiites apparently are remelted coincident with upper mantle rocks becoming depleted of low-melting fractions. The Cordilleran-type calc-alkaline series may be contaminated by remelting of lower crustal rocks, which are at least as differentiated as is a mature oceanic arc. Even upper crustal rocks may be remelted in such Andean-type arcs, producing young rhyolitic ignimbrites containing Sr^{87}/Sr^{86} in excess of 0.710 (Pushkar, McBirney and Kudo, 1972). Such ignimbrites are only known to occur in areas of pre-Mesozoic crystalline sialic basement rocks such as in the western United States, Central America, and the Andes. In summary, formation and thickening of continental crust apparently involves magma generation by partial melting of mantle, followed by increasing partial remelting or cannabilization of basaltic sequences and, eventually, through superposition, by remelting of previously consolidated typically continental lower crust of mostly gabbroic, but partly granitic composition (Glikson, 1972).

The data and interpretations above indicate that igneous rock geochemistry may provide the best evidence for direct comparison of modern and ancient tectonic elements located above subduction zones. The spatial variation of igneous rocks can be related in part to depth to a subduction zone (Dickinson, 1970). The stratigraphic succession and composition of igneous rocks can be related to basement type (table 4). The study of trace elements may circumvent compositional modifications produced by metamorphism (Pearce and Cann, 1971).

As yet, few systematic regional studies of igneous rocks in pre-Mesozoic orogenic belts have been done that are sufficiently detailed to allow definition of subduction zones or of basement types. (However, see the many studies of Archaean greenstones, which suggest ocean-arc origin, or the study by Fitton and Hughes, 1970, suggesting a southeast-dipping subduction zone beneath the early Paleozoic Welsh geosyncline-marginal basin). Assuming metamorphism plays a negligible role in compositional zoning of crust, data presently available show that arc tholeiites are major components of Archaean greenstones and that ensimatic tectonic elements probably were widespread in ancient eugeosynclines.

(e) Geophysics: Gravity and seismic data show that active consuming plate margins contain every transition between continental and oceanic crust. In ancient crushed orogenic belts, interpreting such geophysical data in terms of continental and oceanic crust is nearly impossible because of the inherently altered character and complexity of tectonic elements; longitudinal, rather than across-strike geophysical sur-

TABLE 4.—OROGENIC VOLCANIC GEOCHEMISTRY RELATED TO BASEMENT TYPE[1]

Parameter	Abyssal tholeiite	Oceanic arc		Cordilleran Arc (Andean)
Existing crust, that is, inherited basement	None: generative setting of ophiolite	Ocean crust (ensimatic) plus arc tholeiites		Continental crust (ensialic)
SiO_2 (%)	<50	50–66		56–75 (including silicic ignimbrites)
		Arc tholeiites	Calc-alkaline	Calc-alkaline
K_2O (%)	<0.3	0.3–0.5	1.0–2.7	1.1–3.4
K_2O/Na_2O	<0.1	<0.5	<0.8	0.6–1.1
Al_2O_3 (%)	12–18	14–18	15–19	15.2–18.2
Fe enrichment	High	Moderate	Low	None
Rare earth elements	Chondritic	Chondritic	Light rare earth elements enriched	—
K/Rb	1000	1000	400	230
Rb, Ba, Sr (ppm)	2, 30, 120	5, 100, 200	30, 270, 385	80, 680, 700
Th, U, Zr (ppm)	—	0.5, 0.3, 70	2.2, 0.7, 110	—, —, 210
Sr^{87}/Sr^{86}	0.701–0.702	0.703–0.706	0.703–0.706	0.704–0.707

[1] Data generalized from: Ewart and Bryan, 1972; Jakes and White, 1972; Kay, Hubbard, and Gast, 1970; Peterman and Hedge, 1971; Pichler and Zeil, 1972; Pushkar, 1968; and White, Jakes, and Christie, 1971.

veys within tectonic elements, however, possibly can delineate to some extent the crustal contrasts between adjacent elements. That steep gravity gradients are characteristic of contrasts between oceanic and continental tectonic elements (Case and others, 1971) is noteworthy.

The most useful application of geophysics has been to the study of marginal basins (Packham and Falvey, 1971), reentrant small ocean basins, sphenochasms, and rhombochasms inasmuch as these features may be altered only by having great sediment cover (Fedynsky and others, 1972). Studies of the Black Sea and the Gulf of Mexico show that negative gravity anomaly, weak magnetic relief, and deep thin gabbroic crust are characteristic of such remnant ensimatic basins so long as they have not undergone significant deformation or magmatism. Remnant ocean basins may have an arc on their plate, which could make them similar to marginal basins. However, marginal basins characteristically have positive gravity anomalies (Packham and Falvey, 1971).

(f) Diatreme sampling: Late orogenic or postorogenic igneous rocks, particularly in diatremes, may bring xenoliths of basement and even mantle rocks to the surface. There has been little effort to analyze such samples, but that would seem to be well worth the effort inasmuch as the nature and age of both crust and mantle beneath orogens conceivably could be determined. Two examples of this approach are: (1) presence of gneiss, amphibolite, and granodiorite xenoliths in volcanic rocks of the Aeolian Islands show the existence of sialic basement (Honnorez and Keller, 1968) and (2) presence of ultramafic inclusions in intrusions cutting the Dunnage Mélange of Newfoundland (Kay, 1972) suggest the presence of an ophiolitic basement.

CONCLUSIONS

The addition of oceanic and mantle materials to the continent by accretion of orogens, an irreversible process, is complex due to the interaction and evolution of tectonic elements affected by complex tectonic processes. The portions of eugeosynclinal belts that are ensimatic may be delineated by several approaches, none of which is adequately demonstrated as being definitive for both young and ancient examples. These approaches attempt to relate the rock compositions of tectonic elements to their basement types. Taken together, available data of this type suggest that ophiolitic elements and a considerable portion of the arcs and furrows of eugeosynclines are probably ensimatic. The actual processes that generate tectonic elements are not understood, but clearly the mutual interrelations of elements during orogenesis strongly reflect basement type. Thus, basement exerts a key role in orogenesis.

The sialic vs. simatic nature and extent of the inherited and generative basement of eugeosynclinal belts is highly varied and dependent upon the evolution of the individual orogen in terms of the starting configuration of rifts, trenches, transform faults, arcs and marginal basins, and reentrants and in terms of their subsequent termination as new configurations evolve. Tectonic processes that continue in a stable configuration generally lead to an orderly array of tectonic elements. But the establishment of new configurations by displacements and the ultimate suturing and crushing of major orogenic zones may produce a disordering of tectonic elements. Thus, single tectonic processes or tectonic elements may be described but not single theories, cycles, or finite models of orogenesis (Coney, 1970). Each orogen is a unique time-space collage of tectonic elements.

The objection may be raised that a collage "model" (a misnomer) of orogenesis is scientific anarchy, telling us that each mountain belt has its own absolutely unique history and that no general theory of orogenesis is possible. This objection cannot be sustained in a historical science like geology for two reasons. First, the collage concept is not a model in conflict with either plate models or orogenic theories but is a word picture that attempts to convey the idea that various models and theories can be integrated and reconciled to yield insight into the astonishing complexities, similarities, and differences of mountain belts. The so-called odd rocks of mountain belts (Fischer, this volume) are not really odd; they simply show that every type of crustal rock (and some mantle rocks) can be exposed in eugeosynclines. Any orogenic paradigm apparently cannot involve limitations on the kind of rock that occurs in orogens. Second, just as the theory of evolution provides a simple explanation for the diversity of life, it also explains why no species of plant or animal ever arose twice. By analogy, a collage conception of orogenesis offers a simple explanation for the diversity of orogens, and it also explains why no two crustal configurations are ever exactly the same—a truism to which we would all admit (Trümpy, 1971, p. 294).

The fact that there may be a limited number of tectonic elements (table 1), particularly in the beginning stages of accreting and consuming plate margins (Dewey and Bird, 1970), does suggest that the initial stages of development of orogenic belts can be represented by a limited number of models. For example, the conversion of a midplate continental margin to consuming plate margin produces the transition from miogeocline to eugeosyncline-exogeosyncline. However, as orogenic evolution proceeds, tectonic elements are linked or separated by displacements governed by boundary geometry, crustal conposition, and plate motions. Elements are modified by segmentation, changes of boundary faulting, and superposition of tectonic processes, and are affected by one or more episodes of coupling and decoupling and compression and extension. All these processes would have to be repeated in precisely the same order without addition or subtraction in order to produce two mountain belts essentially alike (fig. 1). Orogenesis is noncommutative.

The evolution of each orogen is thereby different. Plate tectonic models are process models, that allow understanding of each step in evolution of an orogen, just as all models in geology represent methodological, rather than substantive uniformitarianism (Gould, 1965). Plate models should not be construed to portray adequately anything other than a generalized single stage of development of part of a specific orogen. Thus, the understanding of each tectonic element may be gained most readily by analogy with specific modern tectonic elements (table 1). Therefore, understanding an entire orogen requires superposition of many plate models, and it is doubtful that much is learned by considering that an entire orogen fits a single plate model. The assembling, including disordering, of tectonic elements can be understood by returning the pieces of the collage to "tectonically reasonable positions, revealing major igneous, metamorphic and deformational belts" (Gastil and others, 1972) for specific times in the past. The task of understanding orogenic belts thus assumes the broadest conceivable palinspastic problems, which are not easily solved (Monger, Souther, and Gabrielse, 1972; Schenk, 1972). In making palinspastic reconstructions, geologic fits of all kinds may be especially strengthened if geologic *gradients* can be reconstructed (Whitten, 1969), as Ross (1973) has attempted for the compositional gradients in the granitic rocks of the Sierra Nevada and Salinian block in California.

The remaining fundamental question now appears to be directed to what controls the sequence of processes in plates and orogenic belts, not what are the models of orogens (Khain, 1972). Do plate boundaries and orogenic collages follow any particular rules, so to speak, of migration and assemblage respectively that we may observe in the crustal records? Have these process rules changed through geologic time? How are the similarities of orogens (Dott, 1964) explained? What about processes and rules of orogenesis that are overlooked by present emphasis on plate boundaries and plate models (Gilluly, 1972; Ramberg, 1972)? It seems probable to me that some quantifiable rules are followed in which energy is dissipated (Griggs, 1972) with maximum efficiency and minimum work, following a pathway toward continentalization.

REFERENCES

AUBOUIN, F., 1965, Geosynclines: developments in geotectonics: Amsterdam, Elsevier Publishing Co., v. 1, 335 p.
BATEMAN, P. C., AND DODGE, F. C. W., 1970, Variations of major chemical constituents across the central Sierra Nevada Batholith: Geol. Soc. America Bull., v. 81, p. 409–420.
BEMMELEN, R. S. VAN, 1969, Origin of the western Mediterranean Sea: K. Nederlandsch Geol. Mijnb. Genoot. Verh., v. 26, p. 13–52.
BIRD, J. M., AND DEWEY, J. F., 1970, Lithosphere plate-continental margin tectonics and the evolution of the Appalachian Orogen: Geol. Soc. America Bull., v. 81, p. 1031–1059.
BOETTCHER, A. L., 1973, Volcanism and orogenic belts-the origin of andesites: Tectonophysics, v. 17, p. 223–240.
BOGDANOV, N. A., 1969, Thalassogeosynclines of the Circumpacific ring: Geotectonics, 1969, p. 141–147.
BOOY, T. DE, 1967, Nue Daten fur die Annahme einer sialischen Kruste unter den fruhgeosynklinalen Sedimenten der Tethys: Geol. Rundschau, v. 56, p. 94–102.
BORUKAYEV, C. B., 1970, Palinspastic representations: Geotectonics, 1970, p. 343–346.
BOTT, M. H. P., 1971, Evolution of young continental margins and formation of shelf basins: Tectonophysics, v. 11, p. 319–327.
CADY, W. M., 1972, Are the Ordovician northern Appalachians and the Mesozoic Cordilleran system homologous?: Jour. Geophys. Research, v. 77, p. 3806–3815.
CAREY, S. W., 1958, The tectonic approach to continental drift, *in* Continental drift—a symposium: Hobart, Tasmania, Univ. Hobart, p. 177–355.
CASE, J. E., AND OTHERS, 1971, Tectonic investigations in western Colombia and eastern Panama: Geol. Soc. America Bull., v. 82, p. 2685–2712.
CHIDESTER, A. H., AND CADY, W. M., 1972, Origin and emplacement of Alpine-type ultramafic rocks: Nature Phys. Sci., v. 240, no. 98, p. 27–31.
CHURCH, W. R., 1969, Metamorphic rocks of Burlington Peninsula and adjoining areas of Newfoundland,

and their bearing on continental drift in North Atlantic, *in* KAY, M. (ed.), North Atlantic—Geology and continental drift: Am. Assoc. Petroleum Geologists Mem. 12, p. 212–235.

COLEMAN, R. G., 1971, Plate tectonic emplacement of upper mantle peridotites along continental edges: Jour. Geophys. Research, v. 76, p. 1212–1222.

CONDIE, K. C., AND POTTS, M. J., 1969, Calc-alkaline volcanism and the thickness of the early Precambrian crust in North America: Canadian Jour. Earth Sci., v. 6, p. 1179–1184.

CONEY, P. J., 1970, The geotectonic cycle and the new global tectonics: Geol. Soc. America Bull., v. 81, p. 739–748.

DEGENS, E. T., AND ROSS, D. A. (eds.), 1969, Hot brines and recent heavy metal deposits in the Red Sea: New York, Springer Verlag, 600 p.

DEWEY, J. F., 1969, Continental margins: a model for conversion of Atlantic type to Andean type: Earth and Planetary Sci. Letters, v. 6, p. 189–197.

———, AND BIRD, J. M., 1970, Mountain belts and the new global tectonics: Jour. Geophys. Research, v. 75, p. 2625–2647.

———, AND ———, 1971, Origin and emplacement of the ophiolite suite: Appalachian ophiolites in Newfoundland: *ibid.*, v. 76, p. 3179–3206.

DICKINSON, W. R., 1970, Relations of andesites, granites, and derivative sandstones to arc-trench tectonics: Rev. Geophysics and Space Physics, v. 8, p. 813–860.

———, 1971, Plate tectonic models of geosynclines: Earth and Planetary Sci. Letters, v. 10, p. 165–174.

DIETZ, R., 1963, Collapsing continental rises: an actualistic concept of geosynclines and mountain building: Jour. Geology, v. 71, p. 314–333.

DOTT, R. H., JR., 1964, Superimposed rhythmic stratigraphic patterns in mobile belts: Kansas Geol. Survey Bull. 169, p. 69–85.

DRAKE, C. L., EWING, M., AND SUTTON, G. H., 1959, Continental margins and geosynclines: the east coast of North America north of Cape Hatteras, *in* Physics and chemistry of the earth: London, Pergamon Press, v. 3, p. 110–198.

ERNST, W. G., 1973, Interpretative synthesis of metamorphism in the Alps: Geol. Soc. America Bull., v. 84, p. 2053–2078.

EWART, A., AND BRYAN, W. B., 1972, Petrography and geochemistry of the igneous rocks from Eua, Tongan Islands: *ibid.*, v. 83, p. 3281–3298.

FEDYNSKY, V. V., AND OTHERS, 1972, The earth's crust of the inland seas and continental depressions of the west Tethys region: 24th Intnat. Geol. Cong., Sec. 3, p. 51–57.

FITTON, J. G., 1971, The generation of magmas in island arcs: Earth and Planetary Sci. Letters, v. 11, p. 63–67.

———, AND HUGHES, D. J., 1970, Volcanism and plate tectonics in the British Ordovician: *ibid.*, v. 8, p. 223–228.

FYFE, W. S., 1973, The generation of batholiths: Tectonophysics, v. 17, p. 273–283.

GANSSER, A., 1966, The Indian Ocean and the Himalaya: Eclogae Geol. Helvetiae, v. 59, p. 831–848.

———, 1973, Facts and theories on the Andes: Geol. Soc. London Quart. Jour., v. 129, p. 93–131.

GASTIL, G., PHILLIPS, R. P., AND RODRIGUEZ-TORRES, R., 1972, The reconstruction of Mesozoic California: 24th Intnat. Geol. Cong., Sec. 3, p. 217–229.

GILLULY, J., 1972, Tectonics involved in the evolution of mountain ranges, *in* ROBERTSON, E. C. (ed.), The nature of the solid earth: New York, McGraw-Hill Book Co., p. 414–447.

GLENNIE, K. W., AND OTHERS, 1973, Late Cretaceous nappes in Oman Mountains and their geologic evolution: Am. Assoc. Petroleum Geologists Bull., v. 57, p. 5–27.

GLIKSON, A. Y., 1972, Early Precambrian evidence of a primitive ocean crust and island nuclei of sodic granite: Geol. Soc. America Bull., v. 83, p. 3323–3344.

GOULD, S. J., 1965, Is uniformitarianism necessary?: Am. Jour. Sci., v. 263, p. 223–228.

GRIGGS, D. T., 1972, The sinking lithosphere and the focal mechanism of deep earthquakes, *in* ROBERTSON, E. C. (ed.), The nature of the solid earth: New York, McGraw-Hill Book Co., p. 361–384.

HALL, JAMES, 1859, Description and figures of the organic remains of the lower Helderberg Group and the Oriskany Sandstone: New York Geol. Survey, National History of New York, pt. 6, Paleontology, v. 3, 532 p.

HALLAM, A. (ed.), 1967, Depth indicators in marine sedimentary environments: Marine Geology, v. 5, no. 5, 6, 240 p.

HAMILTON, W., 1969, Mesozoic California and the underflow of Pacific mantle: Geol. Soc. America Bull., v. 80, p. 2409–2430.

HARLAND, W. B., 1971, Tectonic transpression in Caledonian Spitsbergen: Geol. Mag., v. 108, p. 27–41.

HATHERTON, T., AND DICKINSON, W. R., 1967, Andesitic volcanism and seismicity around the Pacific: Science, v. 157, p. 801–803.

HAUG, E., 1900, Les géosynclinaux et les aires continentales: Soc. géol. France Bull., v. 28, p. 617–711.

HELWIG, J., 1972, Stratigraphy, sedimentation, paleogeography, and paleoclimates of Carboniferous ("Gondwana") and Permian of Bolivia: Am. Assoc. Petroleum Geologists Bull., v. 56, p. 1008–1033.

———, 1973, Plate tectonic model for the evolution of the central Andes: discussion: Geol. Soc. America Bull., v. 84, p. 1493–1496.

———, AND SARPI, E., 1969, Plutonic-pebble conglomerates, New World Island, Newfoundland, and history of eugeosynclines, *in* KAY, M. (ed.), North Atlantic—Geology and continental drift: Am. Assoc. Petroleum Geologists Mem. 12, p. 443–466.

HESS, H. H., 1962, History of ocean basins, *in* ENGLE, A. E. J., JAMES, H. L., AND LEONARD, B. L. (eds.), Petrologic studies: a volume in honor of A. F. Buddington: Boulder, Colorado, Geol. Soc. America, p. 599–620.

Honnorez, J., and Keller, J., 1968, Xenolithen in vulkanischen Gesteinen der Aölischen Inseln (Sizilien): Geol. Rundschau, v. 57, p. 719–736.
Hsü, K. J., 1972, Alpine flysch in a Mediterranean setting: 24th Intnat. Geol. Cong., Sec. 6, p. 67–74.
Hutchinson, R. W., and Engels, G. G., 1972, Tectonic evolution in the southern Red Sea and its possible significance to older rifted continental margins: Geol. Soc. America Bull., v. 83, p. 2987–3002.
Isacks, B. I., Oliver, J., and Sykes, L. R., 1968, Seismology and the new global tectonics: Jour. Geophys. Research, v. 73, p. 5855–5899.
Jakes, P., and White, A. J. R., 1972, Major and trace element abundances in volcanic rocks of orogenic areas: Geol. Soc. America Bull., v. 83, p. 29–40.
Karig, D. E., 1971, Origin and development of marginal basins in the western Pacific: Jour. Geophys. Research, v. 76, p. 2542–2561.
Kay, M., 1944, Geosynclines in continental development: Science, v. 99, p. 461–462.
——, 1951, North American geosynclines: Geol. Soc. America Mem. 48, 143 p.
——, 1972, Dunnage Mélange and lower Paleozoic deformation in northeastern Newfoundland: 24th Intnat. Geol. Cong., Sec. 3, p. 122–133.
——, and Crawford, J. P., 1964, Paleozoic facies from the miogeosynclinal to the eugeosynclinal belt in thrust slices, central Nevada: Geol. Soc. America Bull., v. 75, p. 425–454.
Kay, R., Hubbard, N. J., and Gast, P. W., 1970, Chemical characteristics and origin of oceanic ridge volcanic rocks: Jour. Geophys. Research, v. 75, p. 1585–1613.
Khain, V. Ye., 1972, Present status of theoretical geotectonics and related problems: Geotectonics, 1972, p. 199–214.
King, P. B., 1969, The tectonics of North America—a discussion to accompany the tectonic map of North America, scale 1:5,000,000: U.S. Geol. Survey Prof. Paper 628, 94 p.
Kober, L., 1928, Der Bau der Erde: Berlin, Borntraeger, 500 p.
Kuno, H., 1966, Lateral variation of basaltic magma across continental margins and island arcs, in Poole, W. H. (ed.), Continental margins and island arcs: Geol. Survey Canada Paper 66–15, p. 317–336.
Lemoine, M., 1972, Eugeosynclinal domains of the Alps and the problem of past oceanic areas: 24th Intnat. Geol. Cong., Sec. 3, p. 476–485.
Le Pichon, X., 1968, Sea-floor spreading and continental drift: Jour. Geophys. Research, v. 73, p. 3661–3697.
McKenzie, D. P., 1972, Plate tectonics, in Robertson, E. C. (ed.), The nature of the solid earth: New York, McGraw-Hill Book Co., p. 323–360.
Matsuda, T., and Uyeda, S., 1971, On the Pacific-type orogeny and its model-extension of the paired belts concept and possible origin of marginal seas: Tectonophysics, v. 11, p. 5–27.
Mitchell, A. H., and Reading, H. G., 1969, Continental margins, geosynclines, and ocean floor spreading: Jour. Geology, v. 77, p. 629–646.
——, 1971, Evolution of island arcs: ibid., v. 79, p. 253–284.
Monger, J. W. H., Souther, J. G., and Gabrielse, H., 1972, Evolution of the Canadian Cordillera: A plate-tectonic model: Am. Jour. Sci., v. 272, p. 577–602.
Moore, J. G., 1959, The quartz diorite boundary line in the western United States: Jour. Geology, v. 67, p. 198–210.
Mouratov, M., 1972, Main structural features of the crust on continents, their interrelations and age: 24th Intnat. Geol. Cong., Sec. 3, p. 71–78.
Oxburgh, E. R., 1972, Flake tectonics and continental collision: Nature, v. 239, p. 202–204.
Packham, G. H., and Falvey, D. A., 1971, An hypothesis for the formation of marginal seas in the western Pacific: Tectonophysics, v. 11, p. 79–109.
Pearce, J. A., and Cann, J. R., 1971, Ophiolite origin investigated by discriminant analysis using Ti, Zr and Y: Earth and Planetary Sci. Letters, v. 12, p. 339–349.
Peterman, Z. E., and Hedge, C. E., 1971, Related strontium isotopic and chemical variations in oceanic basalts: Geol. Soc. America Bull., v. 82, p. 493–500.
Peyve, A. V., 1967, Faults and tectonic movements: Geotectonics, v. 5, p. 268–275.
——, 1969, Oceanic crust of the geologic past: ibid., 1969, p. 210–224.
——, Perfiliev, A. S., and Ruzhentsev, S. V., 1972, Problems of intracontinental geosynclines: 24th Intnat. Geol. Cong., Sec. 3, p. 486–493.
Pichler, H., and Zeil, W., 1972, The Cenozoic rhyolite-andesite association of the Chilean Andes: Bull. Volcanol., v. 35, p. 424–452.
Plafker, G., 1972, Alaskan earthquake of 1964 and Chilean earthquake of 1960: implications for arc tectonics: Jour. Geophys. Research, v. 77, p. 901–925.
Pushkar, P., 1968, Strontium isotope ratios in volcanic rocks of three island arc areas: ibid., v. 73, p. 2701–2714.
——, McBirney, A. R., and Kudo, A. M., 1972, The isotopic composition of strontium in Central American ignimbrites: Bull. Volcanol., v. 35, p. 265–294.
Ramberg, H., 1972, Theoretical models of density stratification and diapirism in the earth: Jour. Geophys. Research, v. 77, p. 877–889.
Reading, H. G., 1972, Global tectonics and the genesis of flysch successions: 24th Intnat. Geol. Cong., Sec. 6, p. 59–65.
Rodgers, J. 1968, The eastern edge of the North American continent during the Cambrian and Early Ordovician, in Zen, E-An, White, W. S., and Hadley, J. B. (eds.), Studies of Appalachian geology: northern and maritime: New York, Interscience Publishers, p. 141–150.
Ross, D. C., 1973, Are the granitic rocks of the Salinian block trondhjemitic?: U.S. Geol. Survey Jour. Research, v. 1, p. 251–254.

SCHENK, P. E., 1972, Eastern Canada as successive remnants of northwestern Africa: 25th Internat. Geol. Cong., Sec. 6, p. 14–23.

SCLATER, J. G., ANDERSON, R. N., AND BELL, M. L., 1971, Elevation of ridges and evolution of the central Pacific: Jour. Geophys. Research, v. 76, p. 7888–7915.

SMITH, A. G., 1971, Alpine deformation and the oceanic areas of the Tethys, Mediterranean and Atlantic: Geol. Soc. America Bull., v. 82, p. 2039–2070.

STILLE, H., 1941, Einführung in den Bau Amerikas: Berlin, Borntraeger, 717 p.

THOMPSON, G. A., AND TALWANI, M., 1964, Geology of the crust and mantle, Western United States: Science, v. 146, p. 1539–1549.

TRÜMPY, R., 1960, Paleotectonic evolution of the central and Western Alps: Geol. Soc. America Bull., v. 71, p. 843–908.

———, 1971, Stratigraphy in mountain belts: Geol. Soc. London Quart. Jour., v. 126, p. 293–318.

VOGT, P. R., SCHNEIDER, E. D., AND JOHNSON, G. L., 1969, The crust and upper mantle beneath the sea, *in* HART, P. J. (ed.), The earth's crust and upper mantle: Am. Geophys. Union Geophys. Mon. 13, p. 556–617.

WALKER, G. P. C., AND CROASDALE, R., 1972, Characteristics of some basic pyroclastics: Bull. Volcanol., v. 35, p. 303–317.

WELLS, F. G., 1949, Ensimatic and ensialic geosynclines: Geol. Soc. America Bull., v. 60, p. 1927.

WHITE, A. J. R., JAKES, P., AND CHRISTIE, D. M., 1971, Composition of greenstones and the hypothesis of sea-floor spreading in the Archaean: Geol. Soc. Australia Special Pub. 3, p. 47–56.

WHITE, D. A., AND OTHERS, 1970, Subduction: Geol. Soc. America Bull., v. 81, p. 3431–3432.

WHITTEN, E. H. T., 1969, Continental-drift models and correlation of geologic features across North Atlantic Ocean, *in* KAY, M. (ed.), North Atlantic—Geology and continental drift: Am. Assoc. Petroleum Geologists Mem. 12, p. 919–930.

WILSON, J. T., 1949, The origin of continents and Precambrian history: Royal Soc. Canada Trans., v. 43, p. 157–184.

———, 1966, Did the Atlantic close and then reopen?: Nature, v. 211, p. 676–681.

———, 1967, Theories of building of continents, *in* GASKELL, T. F. (ed.), The earth's mantle: London, Academic Press, p. 445–473.

WYLLIE, P. J., 1973, Experimental petrology and global tectonics—a preview: Tectonophysics, v. 17, p. 189–209.

ZONENSHAIN, L. P., 1972, Similarities in the evolution of geosynclines of different types: 24th Intnat. Geol. Cong., Sec. 3, p. 494–502.

REFLECTIONS

GEOSYNCLINES, FLYSCH, AND MELANGES

MARSHALL KAY
New York, New York

INTRODUCTION

My first field work was at Baraboo, Wisconsin, sixty miles north of Madison. And my first stratigraphic research, if one may call it that, was in correlation of sections of Black Earth, St. Lawrence, and Mendota dolomites (Cambrian) in the vicinity of Madison at a time when E. O. Ulrich maintained that they belonged in different systems. In the preceding summer I had studied fresh-water plankton in northern Iowa under the instruction of Gilbert M. Smith, then Professor of Botany at Wisconsin. And my doctorate thesis pertained to shales and limestones in southwestern Wisconsin and neighboring states. So I am delighted that we could hold the symposium in the fine setting of the University of Wisconsin at Madison.

I am particularly grateful to R. H. Dott, Jr., for his having arranged the symposium. It was a pleasure to visit again with so many friends and former students and to learn of the continuing progress of study of geosynclines and their sediments. The preceding papers are such excellent summaries that a concluding synthesis would be somewhat redundant. I plan to comment briefly on some of the events that led to the formulation of the description and classification of geosynclinal belts and, then, on misgivings on some models that have been erected.

VOLCANIC BELTS

The view of paleotectonics that prevailed in North America until thirty years ago was one of the so-called "borderlands" of ancient metamorphic and igneous rocks separated by geosynclines from the interior shield. It may be of interest to trace the development of concepts (Kay, 1967). Volcanic rocks were something exceptional that were difficult to fit into the model. Their sites commonly were entered on maps as embayments into the borderlands.

My interest in the source of volcanism came with the discovery of widespread altered volcanic ash beds in Ordovician rocks in eastern North America and, specifically, of ash interbeds in the Decorah Formation of northeastern Iowa. At first there was "Nelson's volcano," a notion, which I accepted naively (1931), that there had been but a single great ash fall. But when two clays were seen in the Lemont Quarry near State College, Pennsylvania, the assumption had to be revised (1935). Soon there were too many ash beds to comprehend (1943; 1956)! In 1934, I made a paleogeographic map of New England having the orthodox borderland Appalachia (1935, p. 241–243). Then Billings (1934) determined that the presumed Precambrian of New Hampshire shown on the geologic map of 1933 had Devonian fossils. My map of 1935 (1937, p. 289) had a trough, the Magog Trough, in place of the borderland. The idea that the geosynclinal Magog belt was volcanic became clearer in 1936 when I was shown the pillow lavas in the Devonian along the Gale River in New Hampshire.

In my graduate course in stratigraphy the volcanic rocks that I discussed were seen to be concentrated in belts along the two margins of the American continent, each volcanic geosynclinal belt contrasting with nonvolcanic ones toward the interior. The lava and radiolarian cherts at Nizhni-Serghinsk in the eastern Urals that I saw on an International Geological Congress trip in 1937 (Nalivkin, 1937, p. 31) further confirmed the contrast between volcanic and nonvolcanic belts. With publication of the Russian Congress proceedings, I learned that Hess (1939) had observed serpentine belts in positions similar to those in which these volcanic rocks had been reported. Thus, the specific observations in many localities on several continents led to generalization.

Meanwhile, Stille's article in the AAPG Bulletin (1936a) used "orthogeosynclinal." I had come to know Stille from his visit to Columbia in the early thirties. I assumed that "orthogeosyncline" was intended for what I called volcanic geosynclinal belts (1941). When I inquired, he referred me to his original reference (1936b) pertaining to the American Cordillera. I wrote that I was using "orthogeosynclinal" for both volcanic and nonvolcanic belts and that the rocks in his eastern Cordilleran *miomagmatisch* zone were more like those of the craton than those of the *pliomagmatisch* zone. He wrote in May 1941 that he was introducing "miogeosynclinalen" and "eugeosynclinalen" for the two zones in a new book (Stille, 1941; Kay, 1942,

p. 1642). Very few, if any, of you will have seen Stille's book. Knopf (1960) reported that only four were sent to North America—I have one of these, and our rare book library at Columbia University has one I bought in Holland. Their rarity results from the destruction of the printed books by fire before they could be distributed. Stille used the terms very rarely, for his interest was directed more toward the chronology of orogenies than the paleogeography of geosynclinal belts.

The type section for the pliomagmatic and miomagmatic belts was the western American Cordillera. The term miogeosynclinal zone was applied to a belt having tectonic swells and swales suffering appreciably greater relative subsidence than those on the craton (Kay, 1966); the 300-km-wide type miogeosynclinal belt of Utah-Nevada in an early stage was broadly synclinal (Bentley, 1958), but this was not known by those who defined the term. "Miogeocline" (Dietz and Holden, 1966) seems to refer to marginal platforms rather than to broad belts with differential subsidence. Eugeosynclinal belts were stated to have volcanic troughs, island arcs, and tectonic lands (Kay, 1944, 1951). It has become popular to assume that there is always a couple, but the section across western New England (1951, p. 26) was not intended as a model and is not to scale. The section shows lower Paleozoic carbonate rocks in the miogeosynclinal belt grading into offshore argillites, fulfilling the concept of carbonate banks so well described by Rodgers (1968). The tectonic separation of the miogeosynclinal and eugeosynclinal belts came later. Moreover, my Ordovician paleogeographic map (1951) shows the Atlantic and Pacific Oceans that we now know had not come into existence; the volcanic belt on the east relates to the earlier Protacadic Ocean. Those who reproduce this map should emphasize the anachronism.

PLATE-TECTONIC ANALOGUES

In these days when plate tectonics are in the forefront, there is clarification of the nature of volcanic belts. Twenty years ago, the restriction of volcanic rocks had been recognized. Such eugeosynclinal belts frequently bordered oceans and vaguely related to ocean basins. But ocean basins were thought to be permanent, so volcanic belts seemed to be derived from them. They were stated to contain volcanic troughs, island arcs, and tectonic lands, but the analogies were obscure. Now one can see that some of the so-called "ophiolite" suites within continents were taken to lie within the troughs, whether as the basement or not. Volcanic arc assemblages were accepted as geosynclinal and now may be compared to the intraoceanic arcs. Thus, the arcs can be on oceanic or continental crust, depending on whether oceanic crust is subducting oceanic crust, as in the Marianas, or is subducting continental crust, as beneath the Andes or Alaska. The first suggestion of subduction based on deep-focus earthquakes and on consequent volcanism that came to my attention was that of Gunn (1947) on the Andean side of the Andean Trench. Though the continuity of the Cascade volcanic belt into the Aleutian arc was recognized (1951, p. 53–57), there was hesitation to include such continental margins as the Cascade belt in the eugeosynclinal belts. Now we can surmise that arcs can be either on oceanic or continental crust.

Tectonic lands seemed to be raised by orogeny within the volcanic belts to produce plutonic pebble conglomerates. Now one may attribute some of these plutonic rocks to hypabyssal intrusions, diatremes; others might relate to what have been called microcontinents. I am very dubious about some of these presumed microcontinents in ancient reconstructions, for we do not know the restoration or magnitude of wrench faults that can repeat ensialic belts, as is true for the concept of Salinia in California.

FLYSCH

This brings me to comment on certain models for the sedimentation in eugeosynclinal belts. For instance, I read "the separation between eugeosynclinals and miogeosynclines is based on the early appearance of flysch and the allochthonous position of the former" (Abbate and others, 1970); such is not the case. The term flysch commonly is applied to strata having such characters as repeated graded beds, sole marks, and poor sorting. Such rocks were deposited in water deep enough to permit turbid flow, and such depths depend on subsidence such as geosynclines require. The graywacke also usually requires unstable sources where erosion yields varied constituents, and it requires transportation that allows only limited sorting.

In some belts, such graywackes succeed argillite and chert, which in turn overlie mafic pillow lavas. Some refer to this flysch sequence as the orogenic cycle, but I think the mixing of parameters is unfortunate. Frequency of association need not imply genetic relationship. The lavas frequently are called ophiolites, which at the present time are likely to be considered as possible ocean floor. Such sequences may well be common—I am studying one in the Campbellton block in central Newfoundland now (Kay, 1972). *But* flysch does not uniquely characterize volcanic geosynclinal belts; it characterizes depositional sites having transportation with limited sorting from source lands that yield varied mineral and rock fragments into appropriate water depths. There are other areas that have plenty of pillow lava, argillite, and chert, but *no* flysch. Lower Paleozoic sequences in Nevada in Stille's type-eugeosynclinal belt have, as the dominant arenaceous rocks, as pure orthoquartzite as can be found, and in hundred of meters of thickness. There were no known tectonic or volcanic lands to supply such sediment within the eugeosynclinal belt, but well-sorted sand on the cratonal margin could flow into the deeps in the volcanic geosynclinal belts (Kay, 1966; Churkin and Kay, 1967; Ketner, 1966), as Churkin commented in his paper in the symposium. If early Ordovician seas could have buried some of the sand-producing Precambrian and Cambrian source rocks, there would not have been St. Peter sand to have carried into these Cordilleran troughs! Other flysch deposits were laid in quite different tectonic belts, such as in the exogeosynclinal or foredeep troughs on the margins of the early Paleozoic craton of eastern United States (McBride, 1962; McIver, 1970) and of the Ouachitas in the south (Walthall, 1967).

SOURCE, SORTING, AND SITE

The sediments within a geosynclinal belt are dependent on factors at the depositional site but also on factors quite separate from those at the site. Crook has given us a summary of the range in composition of graywackes in his paper in this volume. Eugeosynclinal belts are distinguished by their volcanic rocks. Those lacking flysch may have had sufficient depths but did not have suitable source volcanic and tectonic lands; but they *are* volcanic rock bearing. In the present analogies, we learn that abyssal plains may have sediment that slumped from ocean shores. The deeps may contain anything that was free to slump from the beaches and deltas—the material has nothing to do with the site of deposition, but with the source. The sediment is the product of *source, sorting,* and *site!*

MELANGES

Mélanges and olistostromes have been discussed by Hsü and by Wood (this volume). Wood has shown how the mélange in Anglesey, north Wales, which was studied by Greenly (1919), may contain slide masses, olistostromes. The phrase tectonic mélange of Greenly suggests that he used the term in the normal English sense for a class of mixtures of which the one at hand in Anglesey seemed to be tectonic. The lower Paleozoic Dunnage Melange of Newfoundland (Horne, 1969; Kay, 1970, 1972), with which I am most familiar, has bouldery mudslide masses or olistostromes within layered, bedded sediments that are interpreted as laid in a trench above a subduction zone in the Protacadic Ocean (Dewey, 1969). The whole was deformed by shearing so as to produce tectonic breccias in some of its parts. The mélange contains stocks of dacite porphyry with apophyses that were deformed before they had fully solidified. The intrusions are such as are attributed conventionally to magma formed at depth, rising from the descending subduction zone through overlying continental or oceanic crust to form a magmatic arc. Perhaps magma, given the option, would rise through the mud of a trench rather than through the directly overlying solid crust! Some magmas evidently did choose such a course.

"One of the greatest merits of plate tectonics is not to give us now the solution to all of our problems, but to induce new thoughts, and above all new research in mountain belts" (Lemoine, 1972). We have been learning the results of some of these studies.

REFERENCES

ABBATE, ERNESTO, BORTOLLETI, V., PASSERINI, P., AND SAGRI, M., 1970, The geosynclinal concept and the northern Apennines; Sed. Geology, v. 4, p. 625–636.
BENTLEY, C. B., 1958, Upper Cambrian stratigraphy of western Utah: Brigham Young Univ. Geol. Studies, v. 5, no. 6, 70 p.
BILLINGS, M. P., 1934, Paleozoic age of the rocks of central New Hampshire: Science, v. 79, p. 55–56.
CHURKIN, MICHAEL, JR., AND KAY, MARSHALL, 1967, Graptolite-bearing Ordovician siliceous and volcanic rocks, northern Independence Range, Nevada: Geol. Soc. America Bull., v. 78, p. 651–668.
CLINE, L. M., 1960, Late Paleozoic rocks of the Ouachita Mountains: Oklahoma Geol. Survey Bull. 85, 113 p.
DIETZ, R. S., AND HOLDEN, J. C., 1966, Miogeoclines (miogeosynclines) in space and time: Jour. Geology, v. 74, p. 566–583.
DEWEY, J. F., 1969, Evolution of the Appalachian-Caledonian Orogen: Nature, v. 222, p. 124–129.
GREENLY, E., 1919, The geology of Anglesey: Great Britain Geol. Survey Mem., 980 p.
GUNN, ROSS, 1947, Quantitative aspects of juxtaposed ocean deeps, mountain chains and volcanic ranges: Geophysics, v. 12, p. 238–255.
HESS, H. H., 1939, Island arcs, gravity anomalies and serpentine intrusions, a contribution to the ophiolite problem: Internat. Geol. Cong., 17th Sess. Rept., v. 2, p. 263–282.

HORNE, G. S., 1969, Early Ordovician chaotic deposits in the Central Volcanic Belt of northeastern Newfoundland: Geol. Soc. America Bull., v. 80, p. 2451–2464.
KAY, MARSHALL, 1931, Stratigraphy of the Hounsfield Metabentonite: Jour. Geology, v. 39, p. 361–376.
———, 1935, Ordovician altered volcanic materials and related clays: Geol. Soc. America Bull., v. 46, p. 225–244.
———, 1937, Stratigraphy of the Trenton Group: *ibid.,* v. 48, p. 233–302.
———, 1941, Classification of the Artinskian Series in Russia: Am. Assoc. Petroleum Geologists Bull., v. 25, p. 1396–1404.
———, 1942, Development of the northern Allegheny synclinorium and adjoining regions: Geol. Soc. America Bull., v. 53, p. 1601–1658.
———, 1943, Middle Ordovician of central Pennsylvania: Jour. Geology, v. 52, p. 1–23, 97–116.
———, 1944, Geosynclines in continental development: Science, v. 99, p. 461–62.
———, 1951, North American geosynclines, Geol. Soc. America Mem. 48, 143 p.
———, 1956, Ordovician limestones in the western anticlines of the Appalachians in West Virginia and Virginia northeast of the New River: *ibid.,* Bull., v. 67, p. 55–106.
———, 1966, Comparison of the lower Paleozoic volcanic and non-volcanic geosynclinal belts in Nevada and Newfoundland: Bull. Canadian Petroleum Geology, v. 14, p. 579–599.
———, 1967, On geosynclinal nomenclature: Geol. Mag., v. 104, p. 311–316.
———, 1970, Flysch and bouldery mudstone in northeast Newfoundland: Geol. Assoc. Canada Special Paper 7, p. 155–164.
———, 1972, Dunnage Melange and lower Paleozoic deformation in northeastern Newfoundland: Internat. Geol. Cong., 24th Sess., Sec. 3, p. 122–133.
KETNER, K. B., 1966, Comparison with Ordovician eugeosynclinal and miogeosynclinal quartzites of the Cordilleran Geosyncline: U.S. Geol. Survey Prof. Paper 550C, p. 54–60.
KNOPF, ADOLPH, 1960, Analysis of some recent geosynclinal theory: Am. Jour. Sci., v. 258A, p. 126–136.
LEMOINE, M., 1972, Eugeosynclinal domains of the Alps and the problem of past oceanic areas: Internat. Geol. Cong., 24th Sess., Sec. 3, p. 476–485.
McBRIDE, E. F., 1962, Flysch and associated beds of the Martinsburg Formation (Ordovician), central Appalachians: Jour. Sed. Petrology, v. 32, p. 39–91.
McIVER, N. L., 1970, Appalachian tubidites, *in* FISHER, G. W., AND OTHERS (eds.), Studies of Appalachian geology: central and southern: New York, Interscience, p. 69–81.
NALIVKIN, B. V., 1937, The Ufa amphitheatre, *in* Guidebook, Permian excursion, northern part: Internat. Geol. Cong., 17th Sess., p. 5–33.
RODGERS, JOHN, 1968, The eastern edge of the North American continent during the Cambrian and Early Ordovician, *in* ZEN, E., AND OTHERS (eds.), Studies of Appalachian geology: northern and maritime: New York, Interscience, p. 141–149.
STILLE, HANS, 1936a, The present tectonic state of the earth: Am. Assoc. Petroleum Geologists Bull., v. 20, p. 849–880.
———, 1936b, Wege und Ergebnisse der geologisch-tektonischen Forschung: 25 Jahre Kaiser Wilhelm Gesell., Förh. Wiss., v. 2, p. 84–85.
———, 1941, Einführung in den Bau Amerikas: Berlin, Borntraeger, p. 717.
WALTHALL, B. H., 1967, Stratigraphy and structure, part of the Athens Plateau, southern Ouachitas, Arkansas: Bull. Am. Assoc. Petroleum Geologists Bull., v. 51, p. 504–528.